ORGANIC FUNCTIONAL GROUP PREPARATIONS

Second Edition

VOLUME II

This is Volume 12-II of
ORGANIC CHEMISTRY
A series of monographs
Editor: HARRY H. WASSERMAN

A complete list of the books in this series can be obtained from the publisher on request.

ORGANIC FUNCTIONAL GROUP PREPARATIONS

Second Edition

Stanley R. Sandler
Pennwalt Corporation
King of Prussia, Pennsylvania

Wolf Karo
Polysciences, Inc.
Warrington, Pennsylvania

II

1986

ACADEMIC PRESS, INC.

Harcourt Brace Jovanovich, Publishers
Orlando   San Diego   New York   Austin
Boston   London   Sydney   Tokyo   Toronto

This book is a guide to provide general information concerning its subject matter; it is not a procedural manual. Synthesis of chemicals is a rapidly changing field. The reader should consult current procedural manuals for state-of-the-art instructions and applicable government safety regulations. The Publisher and the authors do not accept responsibility for any misuse of this book, including its use as a procedural manual or as a source of specific instructions.

ACADEMIC PRESS, INC.
Orlando, Florida 32887

United Kingdom Edition published by
ACADEMIC PRESS INC. (LONDON) LTD.
24–28 Oval Road, London NW1 7DX

Library of Congress Cataloging in Publication Data
(Revised for vol. 2)

Sandler, Stanley R., Date
    Organic functional group preparations.

    (Organic chemistry ; v. 12)
    Includes bibliographical references and indexes.
    1. Chemistry, Organic—Synthesis.   I. Karo, Wolf,
Date   . II. Title. III. Series. IV. Series:
Organic chemistry (New York, N.Y.) ; v. 12.
QD262.S23   1983   547'.2   83-2555
ISBN 0–12–618602–2 (v. 2 : alk. paper)

PRINTED IN THE UNITED STATES OF AMERICA

86 87 88 89   9 8 7 6 5 4 3 2 1

CONTENTS

v

Contents

Chapter 15/AZOXY COMPOUNDS

Chapter 16/C-NITROSO COMPOUNDS

Chapter 17/N-NITROSO COMPOUNDS

PREFACE

The purpose of this series is to provide the organic chemist with a convenient and updated source of useful preparative procedures. In this volume, we cover synthetic methods for the generation of 17 functional groups.

For this second edition, the literature has been reviewed since 1970–1971 to include new information, new and expanded tables of data, additional preparations, and recent references from both the journal and patent literature. A unique feature of this work is that the U.S. and foreign patent literature is frequently cited. It is these latter references which usually are of great value to the industrial chemist.

Information about safety precautions is given where available. We particularly urge the laboratory worker to check the toxicity and other potential hazards of all reagents, intermediates, and final products. The reader is urged to check known toxicity data and also to become familiar with the *NIOSH Registry of Toxic Effects of Chemical Substances* 1981–1982 Edition (edited by R. L. Tatken and R. J. Lewis, Sr., published by the U.S. Department of Health and Human Services, Public Health Service, Centers for Disease Control, National Institute for Occupational Safety and Health, Cincinnati, Ohio 40226, June 1983). We caution those anticipating scale-up work first to undertake extensive tests to ensure that this can be done safely. Many of the preparations have not been checked either by us or by an independent laboratory for hazards and toxicity and are given here only for information purposes. We do not warrant the preparations against any safety or toxic hazards and assume no liability with respect to the use of the preparations or of the products.

We wish to express our gratitude to our wives and our families for their patience, understanding, and encouragement during the preparation of this manuscript. We also express our appreciation to the production staff of Academic Press for their help in all phases of publication of this volume.

We again were fortunate to have Ms. Emma Moesta to type the chapters of the second edition and to provide us with a highly professional manuscript. Her sincere efforts are much appreciated.

PHILADELPHIA   STANLEY R. SANDLER
APRIL 1986   WOLF KARO

FROM THE PREFACE TO
THE FIRST EDITION

Volume II describes 17 additional functional groups and presents a critical review of their available methods of synthesis with preparative examples of each. Attention is especially paid to presenting specific laboratory directions for the many name reactions used in describing the synthesis of these functional groups.

The unique features of this work, as in Volume I, are that each chapter deals with the preparation of a given functional group by various reaction types (condensation, elimination, oxidation, reduction) and a variety of starting materials. In many cases the available data are summarized in tables to make them more useful to the reader. The literature has been checked up to 1970 and each chapter abounds with references.

The procedures for inclusion in this text, as in Volume I, had to meet the following requirements:

1. The laboratory operations should be safe and free from the danger of explosion.

2. The procedures should afford the highest yield possible of compounds of reliable structure.

3. The procedures should be relatively uncomplicated.

4. The procedures should be generally useful for a wide range of organic structures.

In several cases the preparations have been repeated in our laboratories, and supplementary information is given. As a general rule it is recommended that the purity of the products should be checked by gas chromatography especially if the procedure was originally described prior to 1960.

Some of the functional groups (ynamines, enamines, allenes, etc.) described in this volume have only of recent date been the subject of intensive investigation. For example, this volume contains the first preparative review of ynamines with synthesis procedures from the recent literature. In some chapters where reliable procedures for some compound types still do not exist, it is hoped that the reader will be stimulated to undertake research in these problem areas and report upon them.

STANLEY R. SANDLER
WOLF KARO

CONTENTS OF VOLUMES I AND III

CHAPTER 1 / **ALLENES**

1. INTRODUCTION

Reports on the synthesis of allenes by various methods have been increasing in the past few years. The synthesis of allenes involves some methods different from those described for the synthesis of olefins in Volume I, Chapter 2.

Allenes have the 1,2-diene structure (I), where R^1, R^2, R^3, R^4 = H, alkyl, aryl, halogen, heterocyclic, ether, etc. Since the terminal methylene groups lie

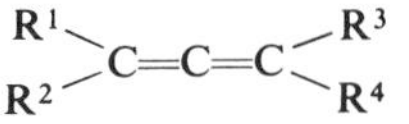

$$R^1\diagdown_{R^2}C{=}C{=}C\diagup^{R^3}_{\diagdown R^4}$$

(I)

in mutually perpendicular planes, optical isomers are possible [1a].

Allene is the generally accepted class name for all such compounds. However, the systematic name of 1,2-diene is also used as described for (II) and (III).

$CH_2{=}C{=}CH_2$ $CH_3{-}CH{=}C{=}CH{-}CH_3$

(II) Allene (Propadiene)   (III) 1,3-Dimethylallene (2,3-Pentadiene)

The first synthesis of allene was reported in 1887–1888 [1b–d]. Recently allenes have been screened for industrial applications in polymers [2,a b], antioxidants [3a, b], drugs [3a], dyes [3b], and fibers [4]. Allenes have also been found to occur in the structure of compounds derived from natural organisms [5].

Allenes have been reviewed recently by Taylor [6], and earlier reviews [7a–f] are also worth consulting.

Allenes are generally prepared by elimination of halogens, hydrogen halides, or water from adjacent carbon atoms, the dehalogenation of *gem*-dihalocyclopropanes, rearrangement of acetylenes, and the 1,4-addition to vinylacetylenes, as described in Scheme I.

Richey [8], and then Pittman and Olah [9a], observed the NMR spectra of acidic solutions of tertiary ethynyl carbinols, and their data support the idea that allenyl carbonium ions contribute significantly to the following ion structure:

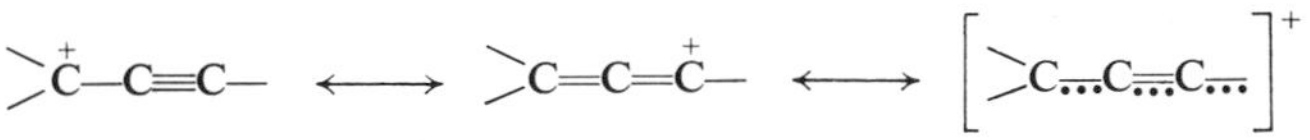

$$>\overset{+}{C}{-}C{\equiv}C{-} \longleftrightarrow >C{=}C{=}\overset{+}{C}{-} \longleftrightarrow \left[>C_{\cdots}C_{\overline{\overline{\cdots}}}C_{\cdots}\right]^+ \qquad (1)$$

1

1. Allenes

SCHEME 1

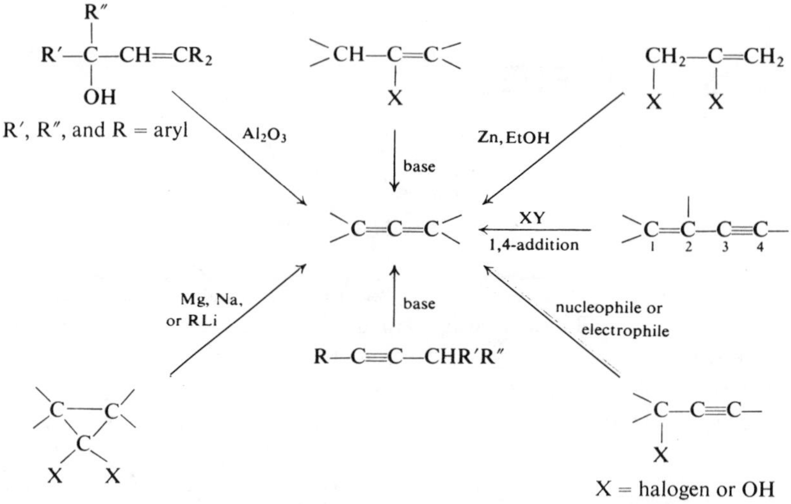

R′—C(R″)—CH=CR₂ with OH ... (chemical scheme)

$$\underset{\substack{\overset{\displaystyle R''}{|}\\ \overset{\displaystyle |}{OH}}}{R'-C-CH=CR_2}$$

$$\underset{\substack{|\\ X}}{>CH-C\equiv C<}$$

$$\underset{\substack{|\ \ |\\ X\ \ X}}{CH_2-C=CH_2}$$

R′, R″, and R = aryl

Al_2O_3

base

Zn, EtOH

$$>C=C=C<$$

$\xrightarrow[\text{1,4-addition}]{XY}$

$$\underset{\substack{|\\ 1\ \ \ 2\ \ \ 3\ \ \ 4}}{>C=C-C\equiv C-}$$

Mg, Na, or RLi

base

nucleophile or electrophile

$$R-C\equiv C-CHR'R''$$

$$\underset{\substack{X\ \ \ X}}{>C-C<\ \text{with}\ C}$$

$$\underset{\substack{|\\ X}}{>C-C\equiv C-}$$

X = halogen or OH

Secondary ethynyl alcohols gave poor spectra because of the production of large amounts of by-products.

The structure of the propargyl-allenyl anion is still uncertain but by analogy to the carbonium ion above it may be that shown in Eq. (2). The

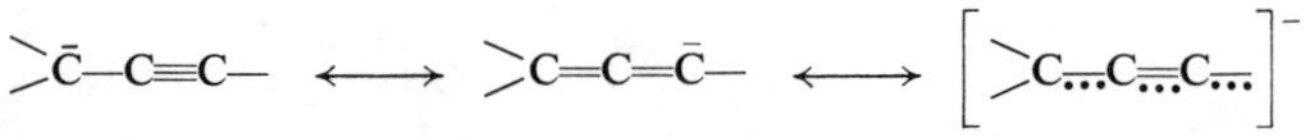

$$>\bar{C}-C\equiv C- \longleftrightarrow >C=C=\bar{C}- \longleftrightarrow \left[>C\overset{...}{}C\overset{...}{\equiv}C\overset{...}{}\right]^{-} \qquad (2)$$

photoelectron spectra (photodetachment spectra) of this ion have been studied and it has been shown to be a relatively labile species [9b].

2. ELIMINATION REACTIONS

The elimination reactions involving dehalogenation, dehydrohalogenation, and dehydration are often laborious compared to the more recent techniques involving dehalogenation of *gem*-dihalocyclopropanes [10a, b]. However, the availability of the starting materials is the deciding factor.

A. Dehalogenation of *gem*-Dihalocyclopropanes

Doering and LaFlamme [10b] were the first to report that sodium and magnesium metal are capable of converting substituted *gem*-dibromocyclo-propanes to allenes in varying yield. However, it was found that sodium reacts best in the form of a high surface dispersion on alumina. At a later date, Moore and Ward [11a] and then Skattebøl [12a] reported that methyllithium or *n*-butyllithium reacts with *gem*-dibromocyclopropanes to give allenes in high yield. The related dichloro compounds were found to be inert to methyllithium but reacted slowly with *n*-butyllithium. Several examples of the preparation of allenes from *gem*-dibromocyclopropanes are shown in Table I.

The *gem*-dibromocyclopropanes are treated with an etheral solution of either methyllithium or *n*-butyllithium at 0° to −80°C. Methyllithium is preferable to *n*-butyllithium because occasionally difficulties are encountered in completely separating the *n*-butyl bromide from the allene product.

TABLE I

PREPARATION OF ALLENES FROM *gem*-DIBROMOCYCLOPROPANES

Starting olefin	Product allene	Dehalogenating reagent	Yield (%)	Ref.
trans-2-Butene	2,3-Pentadiene	Na–alumina	44	10
1-Pentene	1,2-Hexadiene	Na–alumina	64	10
2-Methyl-2-butene	2-Methyl-2,3-pentadiene	Mg	34	10
		CH_3Li	69	12a
Isobutylene	3-Methyl-1,2-butadiene	CH_3Li	92	12a
1-Ethoxy-2-methyl 2-propene	1-Ethoxy-3-methyl-1,2-butadiene	CH_3Li	71	12a
1-Pentene	1,2-Hexadiene	Mg	45	10
2-Hexene	2,3-Heptadiene	CH_3Li	88	11
		C_4H_9Li	83	11
1-Octene	1,2-Nonadiene	CH_3Li	81	11
		C_4H_9Li	49	11
1-Decene	1,2-Undecadiene	CH_3Li	68	11
		C_4H_9Li	43	11
cis-Cyclooctene	1,2-Cyclononadiene	CH_3Li	81	11
cis-Cyclononene	1,2-Cyclodecadiene	C_4H_9Li	78	11
cis-Cyclodecene	1,2-Cycloundecadiene	C_4H_9Li	89	11
Styrene	1-Phenylpropadiene	CH_3Li	82	12a
1,1-Diphenylethylene	1,1-Diphenylpropadiene	CH_3Li	43	12a
Cyclooocta-1,5-diene	1,2,6-Cyclononatriene	CH_3Li	80	12a
1-Dodecene	Decylallene	Mg	45	12b
Bicyclobutylidene	1,3-Bis(trimethylene) propradiene	CH_3Li	83	12c
4,4,9,9-Tetramethoxy cyclododeca-1,6-diene	5,5,11,11-Tetramethoxy-1,2,7,8-cyclododeca tetraenes	CH_3Li	—	12d
tert-Butylethylene	*tert*-Butylallene	CH_3Li	56	12e
Cyclotetradecene	Cyclopentadeca-1,2-diene	*n*-BuLi	—	12f

The C_8–C_{11} cyclic allenes [11a, b] which have been synthesized by this route are obtained only with difficulty by other methods [13] as mixtures with the corresponding acetylenic isomer (Eqs. 3).

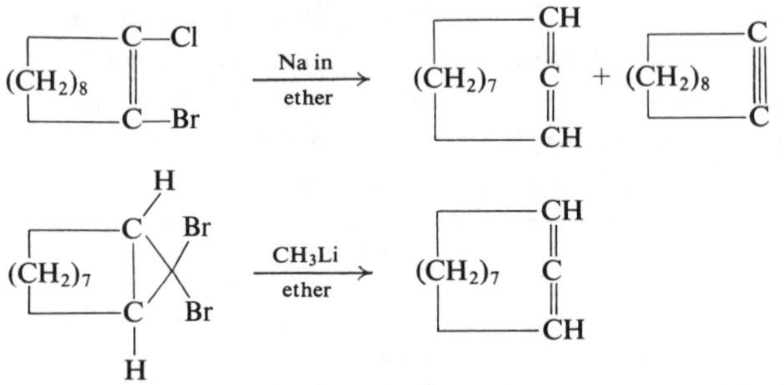

$$
\begin{array}{ccc}
\text{(CH}_2\text{)}_8 \overset{\text{C—Cl}}{\underset{\text{C—Br}}{\Vert}} & \xrightarrow[\text{ether}]{\text{Na in}} & \text{(CH}_2\text{)}_7 \overset{\text{CH}}{\underset{\text{CH}}{\Vert}} \text{C} \quad + \quad \text{(CH}_2\text{)}_8 \overset{\text{C}}{\underset{\text{C}}{\Vert}}
\end{array}
$$

$$
\begin{array}{ccc}
\text{(CH}_2\text{)}_7 \overset{\overset{H}{\diagup}}{\underset{\underset{H}{\diagdown}}{\overset{\text{C}\diagdown \text{Br}}{\underset{\text{C}\diagup \text{Br}}{\times}}}} & \xrightarrow[\text{ether}]{\text{CH}_3\text{Li}} & \text{(CH}_2\text{)}_7 \overset{\text{CH}}{\underset{\text{CH}}{\Vert}} \text{C}
\end{array}
$$

(3)

Smaller cyclic olefins react with dibromocarbene to give the *gem*-dibromobicyclic systems, but these have not been reported to give allenes on reaction with methyl- or *n*-butyllithium. One possible reason may be the severe ring strain of cyclic allenes with less than seven carbon atoms. For example, Moore and Ward [14] found that 7,7-dibromobicyclo[4.1.0]heptane reacts with methyllithium to give bicyclic carbene intermediates which can be trapped with olefins (Eq. 4).

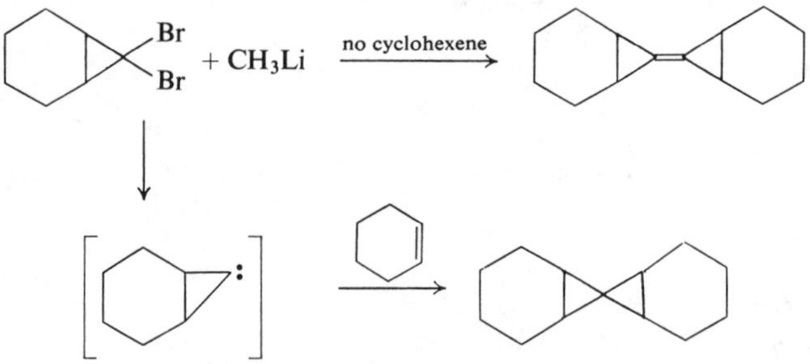

(4)

Skattebøl [12], on the other hand, could not detect such carbene intermediates by their addition product (spiropentanes) to olefins using *gem*-dibromocyclopropanes derived from noncyclic olefins.

In a few cases, allenes are not always the sole product. For example, 1,1-dibromotetramethylcyclopropane does not give tetramethylallene but a mixture of products which is mainly 1-methyl-1-isopropenylcyclopropane [12] (Eq. 5).

The preparation and dehalogenation of *gem*-dibromocyclopropanes to give allenes can be carried out in one step by the reaction an excess of olefins with

TABLE II
ONE-STEP OLEFIN-TO-ALLENE CONVERSION USING CARBON TETRABROMIDE–
OLEFIN–LITHIUM ALKYL

Olefin	R–Li	Temp. (°C)	Allene	Yield (%)
cis-Cyclooctene	CH$_3$	−65	1,2-Cyclononadiene	74
1,5-Cyclooctadiene	CH$_3$	−65	1,2,6-Cyclononatriene	64–71
	C$_4$H$_9$	−65	1,2,6-Cyclononatriene	37

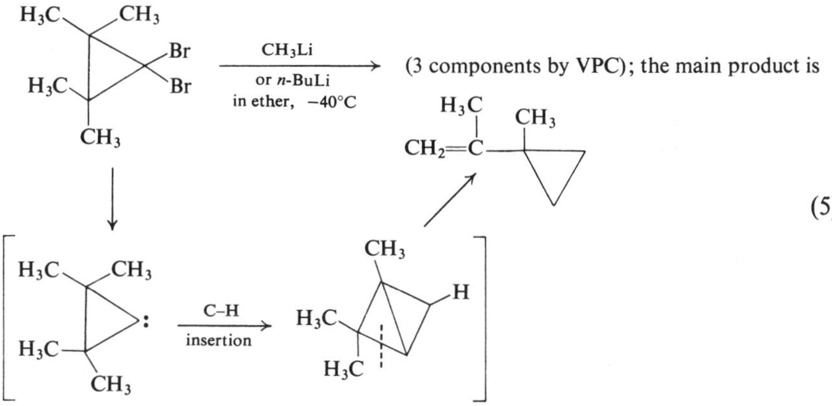

(5)

carbon tetrabromide and 2 equivalents of methyllithium in ether at −65°C [15].
Using n-butyllithium in hexane gives reduced yields, as in the two-step process
[11, 12] (Eq. 6). (Table II lists a few cyclic allenes prepared by this one-step
process.)

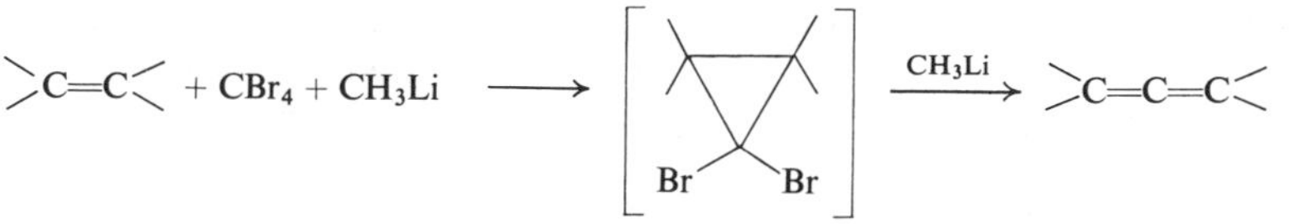

(6)

(±)-Bicyclo[10.8.1]-heneicosa-1(21),12(21)-diene is a doubly bridged allene synthesized by reacting the dichlorocyclopropane shown in Eq. (6a) with
n-butyllithium [12g].

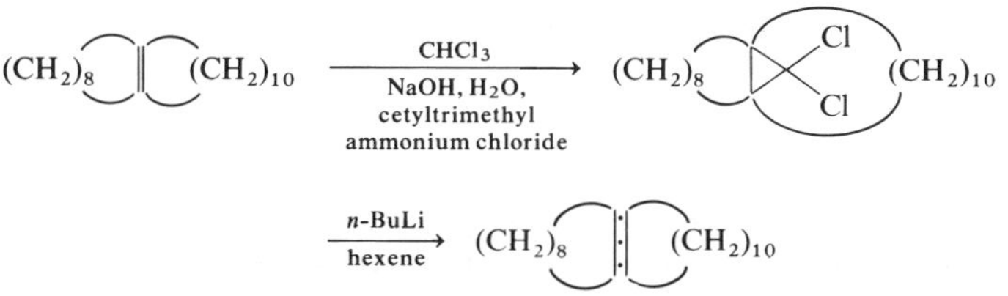

(6a)

2-1. Preparation of 3-Methyl-1,2-butadiene [12]

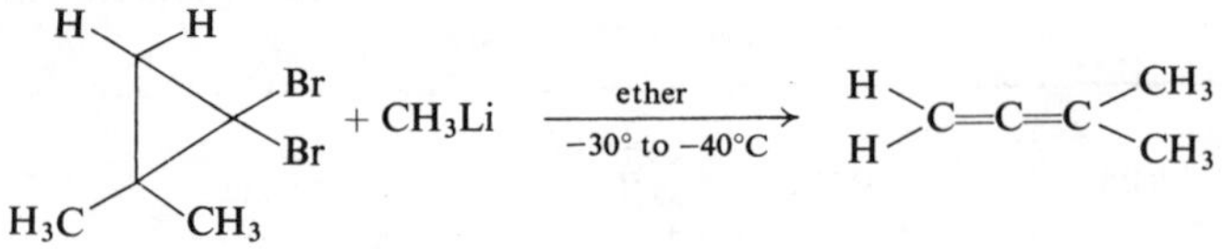

$$\text{(7)}$$

To a flask cooled with Dry Ice–acetone (−30° to −40°C) and containing 22.8 gm (0.10 mole) of 1,1-dibromo-2,2-dimethylcyclopropane and 25 ml of dry ether is added 2.2 gm (0.10 mole) of methyllithium in 80 ml of ether over a $\frac{1}{2}$ hr period. The reaction mixture is stirred for $\frac{1}{2}$ hr, hydrolyzed with water, the ether separated, washed with water, dried, and distilled to afford 6.2 gm (92%), b.p. 40°C, n_D^{24} 1.4152.

1-Phenyl-1,2-propadiene, b.p. 64°–65°C (11 mm), n_D^{24} 1.5809, is obtained in a similar manner in 82% yield by the reaction of 1,1-dibromo-2-phenyl-cyclopropane with methyllithium at −60°C.

2-2. Preparation of 1-Ethoxy-3-methyl-1,2-butadiene [16]

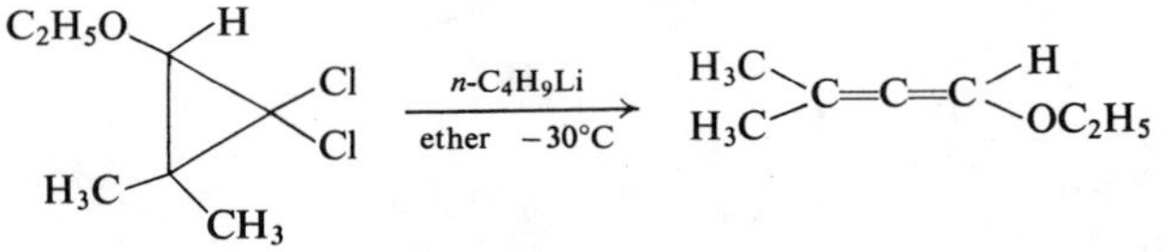

$$\text{(8)}$$

To a flask cooled to −30° to −40°C and containing 20.0 gm (0.11 mole) of 1,1-dichloro-2-ethoxy-3,3-dimethylcyclopropane in 30 ml of dry ether is added dropwise 100 ml of a 1.3 M ethereal n-butyllithium solution (0.13 mole) over a 1 hr period. The reaction mixture is then stirred for an additional $\frac{1}{2}$ hr while it is warmed to room temperature. Water is added, the ether is separated, washed with water, dried over sodium carbonate, and then fractionally distilled to afford 7.3 gm (59%) of impure allene, b.p. 55°–57°C (65 mm), n_D^{25} 1.4344. Redistillation gives a pure product, b.p. 51°C (60 mm), n_D^{20} 1.4400.

2-3. One-Step Olefin-to-Allene Conversion of cis-Cyclooctene to 1,2-Cyclo-nonadiene [16]

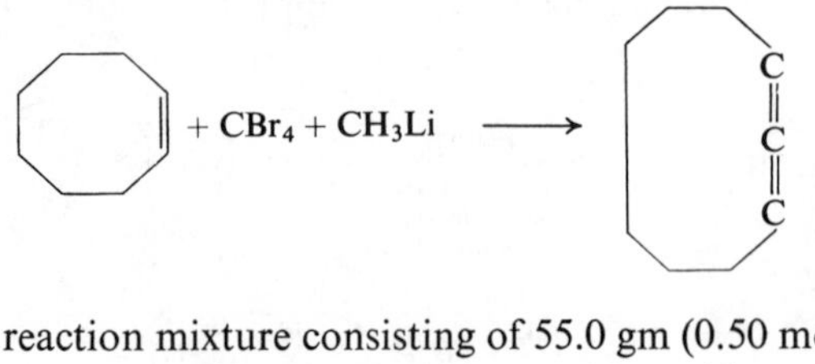

$$\text{(9)}$$

To a stirred reaction mixture consisting of 55.0 gm (0.50 mole) of cis-cyclo-octene and 41.5 gm (0.125 mole) of carbon tetrabromide cooled to −65°C is added 74.0 ml (0.125 mole) of methyllithium in ether over a 45 min period. After stirring at $-65°$ to $-68°C$ for an additional $\frac{1}{2}$ hr, another 80.5 ml (0.135 mole) of methyllithium is added over a $\frac{1}{2}$-hour period. The reaction mixture is stirred for another $\frac{1}{2}$ hr at $-68°C$, warmed to $0°C$, water added, the ether separated, washed with water until neutral, dried over sodium sulfate, and then fractionally distilled to afford 11.3 gm (74 %), b.p. $62°-63°C$ (16 mm), n_D^{20} 1.5060.

B. Dehalogenation of Vicinal Dihalo Compounds

The use of zinc in ethanol to dehalogenate haloallyl halides to allenes was reported earlier by Gustavson and Demjanoff [17a, b]. This method is still a useful one and has been applied when the starting dihalides are available. Recently chromous sulfate has also been shown to be an effective reagent at room temperature [17c] (Eq. 10).

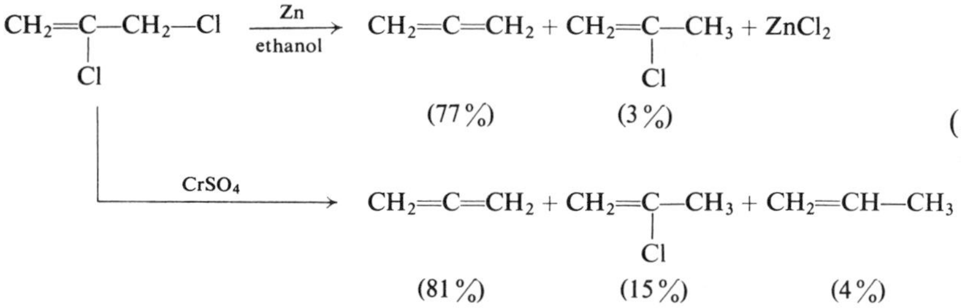

$$CH_2=\underset{\underset{Cl}{|}}{C}-CH_2-Cl \xrightarrow[\text{ethanol}]{Zn} CH_2=C=CH_2 + CH_2=\underset{\underset{Cl}{|}}{C}-CH_3 + ZnCl_2$$

$$(77\%) \qquad (3\%) \qquad (10)$$

$$\xrightarrow{CrSO_4} CH_2=C=CH_2 + CH_2=\underset{\underset{Cl}{|}}{C}-CH_3 + CH_2=CH-CH_3$$

$$(81\%) \qquad (15\%) \qquad (4\%)$$

The main drawback to the dehalogenation method is that the haloallyl halides are usually difficult to obtain. Some of the available methods of synthesis of haloallyl halides are the following.

(1) Thermal rearrangement of *gem*-dihalocyclopropanes [17d]:

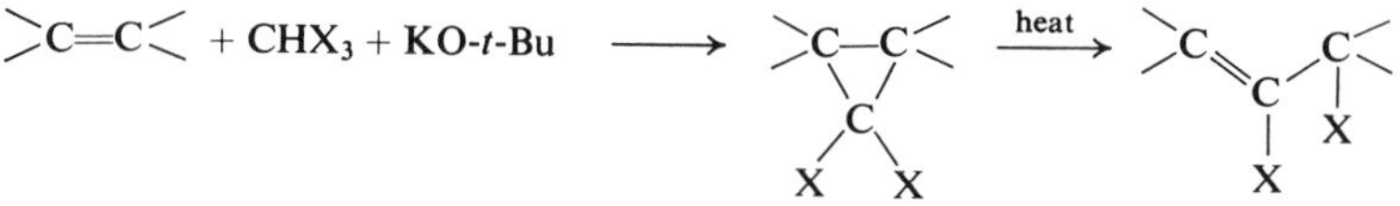

$$\underset{}{>}C=C\underset{}{<} + CHX_3 + KO\text{-}t\text{-}Bu \longrightarrow \underset{X \quad X}{>C-C<} \xrightarrow{heat} \underset{X}{>C}=\underset{X}{C<} \qquad (11)$$

(2) Reaction of allyl alcohols [7a]:

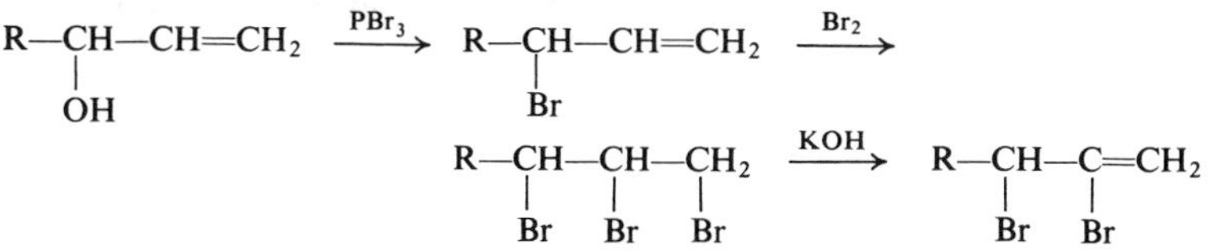

$$R-\underset{\underset{OH}{|}}{CH}-CH=CH_2 \xrightarrow{PBr_3} R-\underset{\underset{Br}{|}}{CH}-CH=CH_2 \xrightarrow{Br_2}$$

$$R-\underset{\underset{Br}{|}}{CH}-\underset{\underset{Br}{|}}{CH}-\underset{\underset{Br}{|}}{CH_2} \xrightarrow{KOH} R-\underset{\underset{Br}{|}}{CH}-\underset{\underset{Br}{|}}{C}=CH_2 \qquad (12)$$

(3) Reaction of vinylacetylene with HBr [18]:

$$CH_2{=}\underset{\displaystyle \overset{\displaystyle R}{|}}{C}{-}C{\equiv}CH + HBr \xrightarrow{-10°C} CH_3{-}\underset{\displaystyle \overset{\displaystyle R}{|}}{C}{=}\underset{\displaystyle \underset{\displaystyle Br}{|}}{C}{-}\underset{\displaystyle \underset{\displaystyle Br}{|}}{CH_2} \qquad (13)$$

(4) Free radical reaction of alkenes with tetrahalomethanes [19]:

$$CH_2{=}CH_2 + CF_2Br_2 \xrightarrow{peroxide} CF_2{-}\underset{\displaystyle \underset{\displaystyle Br}{|}}{CH_2}{-}\underset{\displaystyle \underset{\displaystyle Br}{|}}{CH_2} \xrightarrow{KOH}$$

$$\underset{\displaystyle \underset{\displaystyle Br}{|}}{CF_2}{-}CH{=}CH_2 \xrightarrow[hv]{Br_2} \underset{\displaystyle \underset{\displaystyle Br}{|}}{CF_2}{-}\underset{\displaystyle \underset{\displaystyle Br}{|}}{CH}{-}\underset{\displaystyle \underset{\displaystyle Br}{|}}{CH_2} \xrightarrow{KOH} \underset{\displaystyle \underset{\displaystyle Br}{|}}{CF_2}{-}\underset{\displaystyle \underset{\displaystyle Br}{|}}{C}{=}CH_2 \qquad (14)$$

(5) Free radical addition of trifluoroiodomethane with propargyl halides [20]:

$$R{-}\underset{\displaystyle \underset{\displaystyle X}{|}}{CH}{-}C{\equiv}CH + CF_3I \xrightarrow{hv} R{-}\underset{\displaystyle \underset{\displaystyle X}{|}}{CH}{-}\underset{\displaystyle \underset{\displaystyle I}{|}}{C}{=}CH{-}CF_3 \qquad (15)$$

(6) Free radical addition of trifluoroiodomethane with propargyl alcohols [20]:

$$R{-}\underset{\displaystyle \underset{\displaystyle OH}{|}}{CH}{-}C{\equiv}CH + CF_3I \xrightarrow{hv}$$

$$\underset{\displaystyle \underset{\displaystyle OH}{|}}{RCH}{-}\underset{\displaystyle \underset{\displaystyle I}{|}}{C}{=}CH{-}CF_3 \xrightarrow{SOCl_2} R{-}\underset{\displaystyle \underset{\displaystyle Cl}{|}}{CH}{-}\underset{\displaystyle \underset{\displaystyle I}{|}}{C}{=}CH{-}CF_3 \qquad (16)$$

2-4. *Preparation of 1,1-Difluoroallene* [19a]

$$CF_2Br{-}CBr{=}CH_2 + Zn \rightarrow CF_2{=}C{=}CH_2 + ZnBr_2 \qquad (17)$$

To a flask equipped with stirrer, condenser, and dropping funnel is added 40.0 gm (0.61 gm-atom) of zinc dust in 40 ml of absolute ethanol. After the condenser outlet is connected to a series of Dry Ice–acetone traps cooled to −80° to −70°C, the mixture is heated to gentle reflux while 20.3 gm (0.0852 mole) of 1,2-dibromo-1,1-difluoropropene in 30 ml of 95% ethanol is added dropwise over a 2 hr period. The reaction mixture is heated for an additional hour, and then the gaseous products which are condensed in the Dry Ice traps are purified by several vaporization distillations to afford 3.6 gm (56%) b.p.

−21° to −20°C (thermocouple immersion), ir 4.95 μ $\left(\rangle C{=}C{=}C\langle \right)$ and no absorption at 4.6 μ (—C≡C—). The addition of bromine at −80°C yields the starting material. At room temperature under pressure in the absence of oxygen the product allene slowly polymerizes to a water-white viscous liquid.

Another synthesis of 1,1-difluoroallene (or 1,1-difluoropropadiene) has recently been reported using KOH followed by BuLi as shown in Eq. (18).

$$CF_3CH{=}CH_2 + Br_2 \longrightarrow CF_3{-}CHBr{-}CH_2Br \xrightarrow{\text{KOH}}$$

$$CF_3CBr{=}CH_2 \xrightarrow[\text{2. } -LiF]{\text{1. BuLi}} CF_2{=}C{=}CH_2 \tag{18}$$

Allene has been prepared recently [17b] in a similar manner by adding 260 gm (2.34 moles) of 2,3-dichloropropene dropwise over a 2–3 hr period to a refluxing mixture of 400 ml of 95% ethanol, 80 ml of water, and 300 gm (4.6 gm-atoms) of zinc dust. The allene is trapped and purified in a manner similar to 1,1-difluoro-allene.

Sodium metal in ether has also been used to dehalogenate cyclic 1,2-dihalo-alkenes to cyclic allenes with some cyclic alkyne as a by-product [13].

2-5. *Preparation of 1,2-Cyclodecadiene* [13]

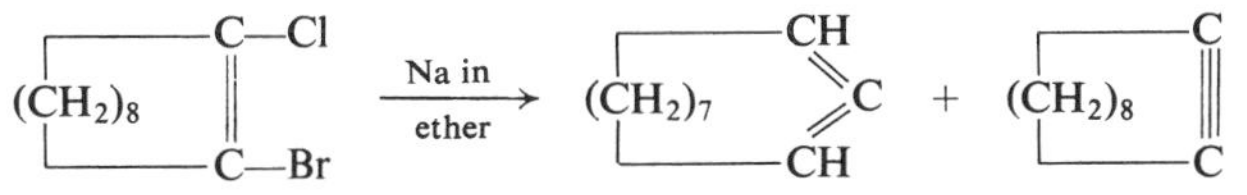

$$\tag{19}$$

To a glass vial is added a solution of 3.75 gm (0.015 mole) of 1-bromo-2-chlorocyclodecene in 35 ml of absolute ether. Then 1.0 gm (0.045 gm-atom) of fine sodium wire, cut into 1 cm lengths, is added. The tube is sealed after flushing with nitrogen, and after 10 days at room temperature, with occasional shaking, the tube is broken open, filtered, washed once with a saturated salt solution, twice with hydrochloric acid, once again with salt solution, dried, and then fractionally distilled to afford 1.0 gm (49%), b.p. 72°–75°C (13 mm), n_D^{20} 1.5024, d_4^{20} 0.897.

C. Dehydrohalogenation

The base-promoted dehydrohalogenation of vinyl or allylic halides affords poor yields of allenes because of the competing side reaction leading to acetylenes [21, 22]. The presence of tertiary hydrogens α to the halogen also helps to give conjugated dienes (Eqs. 20 and 21).

$$\begin{array}{c}>\!CH{-}CX{=}CH{-} \xrightarrow{\text{base}} \;>\!CH{-}C{\equiv}CH + \;>\!C{=}C{=}C\!< \end{array} \tag{20}$$

$$\begin{array}{c}>\!CH{-}CX{-}CH{=}CH{-} \xrightarrow{\text{base}} \;>\!C{=}C{-}CH{=}C\!< + \;>\!CH{-}C{=}C{=}C\!< \end{array} \tag{21}$$

In addition, an allene can rearrange to an acetylene or conjugated diene if it contains labile hydrogens (Eqs. 22, 23). This method is discussed in further detail in Section 3A.

$$>C=C=CH- \xrightarrow{\text{base}} >CH-C\equiv C- \tag{22}$$

$$>C=C=C-\overset{\frown}{C}H \xrightarrow{\text{base}} >C=C-C=C< \tag{23}$$

Rearrangements are less likely when the allene to be formed is highly substituted with either phenoxy [23a], aryl, or halogen groups [23b, 24]. For example,

$$(C_6H_5)_2C=CH_2 + (C_6H_5)_2CHX \xrightarrow[\text{in } C_6H_5]{\text{reflux}} (C_6H_5)_2C=CH-CH(C_6H_5)_2 \xrightarrow[\text{ether}]{\text{Br}_2}$$

$$\underset{\underset{Br}{|}}{(C_6H_5)_2-C=C-CH(C_6H_5)_2} \longrightarrow (C_6H_5)_2C=C=C(C_6H_5)_2 \tag{24}$$

$$X = Br \text{ or } Cl \text{ [23]}$$

$$(C_6H_5)_2CH-CCl_3 \xrightarrow{(C_6H_5)_2CHK} (C_6H_5)_2C=CCl_2 \xrightarrow[-78°C]{(C_6H_5)_2CHK, \text{ liq. } NH_3}$$

$$\left[\underset{\underset{Cl}{|}}{(C_6H_5)_2C=C-C-CH(C_6H_5)_2} \right] \xrightarrow{-HCl} (C_6H_5)_2C=C=C(C_6H_5)_2 \tag{25}$$

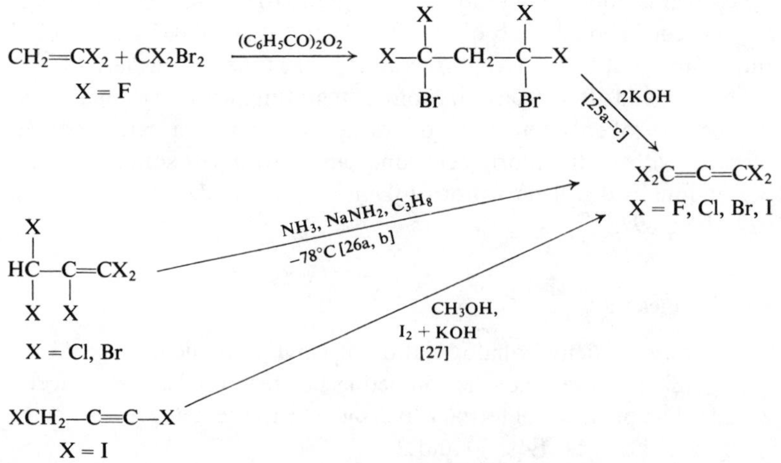

$$CH_2=CX_2 + CX_2Br_2 \xrightarrow{(C_6H_5CO)_2O_2} \underset{\underset{Br}{|}}{X-\overset{\overset{X}{|}}{C}-CH_2-\overset{\overset{X}{|}}{C}-X}$$

$$X = F$$

$$\xrightarrow[\text{[25a-c]}]{2KOH}$$

$$X_2C=C=CX_2 \tag{26}$$

$$X = F, Cl, Br, I$$

$$\underset{\underset{X \quad X}{|\quad|}}{\overset{\overset{X}{|}}{HC}-C=CX_2} \xrightarrow[-78°C \text{ [26a, b]}]{NH_3, NaNH_2, C_3H_8}$$

$$X = Cl, Br$$

$$\xrightarrow[\text{[27]}]{CH_3OH, I_2 + KOH}$$

$$XCH_2-C\equiv C-X$$

$$X = I$$

Dehydrohalogenation of 9–10 carbon atom rings leads mainly to cyclic allenes with some cycloalkyne by-product [28, 29]. The cycloalkyne of only 9–10 carbon atom rings is more highly strained than that of the allenes. Larger rings give higher amounts of acetylenes. Rings containing less than nine carbon atoms do not yield allene or acetylene but dimers or polymers because of the high degree of strain [28].

2-6. Preparation of Phenoxypropadiene [23a]

$$\underset{H}{\overset{C_6H_5O}{>}}C = C \overset{CH_3}{\underset{Br}{<}} + KOH \longrightarrow \underset{H}{\overset{C_6H_5O}{>}}C = C = C \overset{H}{\underset{H}{<}} \qquad (27)$$

To a 125-ml Claisen flask is added 40.0 gm (0.71 mole) of powdered potassium hydroxide. The flask is heated to 50°C in an oil bath while the distillation receiver is cooled with Dry Ice. Then 18.0 gm (0.085 mole) of 2-bromo-1-phenoxy-1-propene is added dropwise to the hot alkali while heating at 17 mm. A vigorous reaction occurs at 90°C, and the product distills over rapidly. The crude allene is dried over sodium sulfate and then fractionally distilled to afford 5.5 gm (50%), b.p. 45.5–46.2°C (2 mm), n_D^{25} 1.5490, d_4^{25} 1.0127.

2-7. Preparation of Tetraphenylallene [23b]

$$\underset{C_6H_5}{\overset{C_6H_5}{>}}CH - \underset{\underset{Br}{|}}{C} = C \overset{C_6H_5}{\underset{C_6H_5}{<}} + KOH \longrightarrow \underset{C_6H_5}{\overset{C_6H_5}{>}}C = C = C \overset{C_6H_5}{\underset{C_6H_5}{<}} \qquad (28)$$

A flask containing 1.0 gm (0.00235 mole) of 2-bromo-1,1,3,3-tetraphenyl-propene, 2.0 gm (0.0357 mole) of potassium hydroxide, and 30 ml of ethanol is heated to reflux for 1 hr. After this time, the reaction mixture is cooled, diluted with water, the solid filtered and then recrystallized from ethanol:acetone (2:1) to afford 0.81 gm (100%), m.p. 165°C.

Using a similar technique, tetrafluoroallene is prepared in 34% yield from 3-bromo-1,1,3,3-tetrafluoropropene and potassium hydroxide at 110°C [25a].

2-8. General Procedure for Preparation of Tetraarylallenes by the Reaction of Potassium Diphenylmethide and 1,1,1-Trichlorodiphenylethanes and 1,1-Dichloro-2,2-diphenylethenes [24]

Preparation of tetraphenylallene. (See Eq. 25.) To 200 ml of stirred liquid ammonia containing 7.28 gm (0.0255 mole) of 1,1,1-trichloro-2,2-diphenyl-ethane is added a solution of 0.0255 mole of potassium diphenylmethide in 300 ml of liquid ammonia. The mixture turns an orange-red color and then discharges immediately. After 1 hr the mixture is worked up to give 4.46 gm (75%) of 1,1-dichloro-2,2-diphenylethene.

To a stirred solution of 0.02 mole of an orange-red solution of potassium diphenylmethide in 150 ml of liquid ammonia and 50 ml of dry ethyl ether, prepared from 0.02 mole of KNH_2 and 3.36 gm (0.02 mole) of diphenyl-methane, is added 2.5 gm (0.01 mole) of 1,1-dichloro-2,2-diphenylethene. The reaction mixture darkens and then a precipitate forms. The reaction mixture is stirred until all the ammonia evaporates to leave a solid. The solid is recrystal-lized from chloroform–methanol to afford 3.1 gm (90%) of tetraphenyl-allene, m.p. 164°C.

D. Dehydration

Dehydration of alcohols, diols, or glycols does not give allenes in appreci-able yield because conjugated dienes [30] are the main product. Even allyl alcohol cannot be suitably dehydrated to allene in reasonable yields. When elimination to the conjugated diene is prevented, as for 2,2-dimethyl-4-hexene-3-ol, the allene is still obtained in only 19% yield [31].

2-9. Preparation of 2,2-Dimethyl-3,4-hexadiene [31]

$$
\underset{\substack{|\\ \text{OH} \quad \text{CH}_3}}{\text{CH}_3-\text{CH}=\text{CH}-\overset{\displaystyle\overset{\text{CH}_3}{|}}{\underset{}{\text{C}}}-\text{CH}_3} \xrightarrow[\text{H}_2\text{O}]{\text{oxalic acid}} \text{CH}_3-\text{CH}=\text{C}=\text{CH}-\overset{\displaystyle\overset{\text{CH}_3}{|}}{\underset{\underset{\text{CH}_3}{|}}{\text{C}}}-\text{CH}_3 \qquad (29)
$$

To a single-necked, round-bottomed flask equipped with a distillation column are added 2,2-dimethyl-4-hexen-3-ol and one-half its weight of oxalic acid. The mixture is heated so that it slowly distills through the long column. The water layer is separated and the organic layer is dried and distilled to afford a 19% yield of 2,2-dimethyl-3,4-hexadiene, b.p. 107.4°–108.0°C, n_D^{25} 1.4425, d_4^{25} 0.7375.

Only polyarylallenes can be made in fair yield by heating the corresponding alcohols with either acetic anhydride or potassium bisulfate since the hydroxyl group is relatively labile [32]. The competing side reaction is the formation of indenes and dimers [33–36] (Eq. 30).

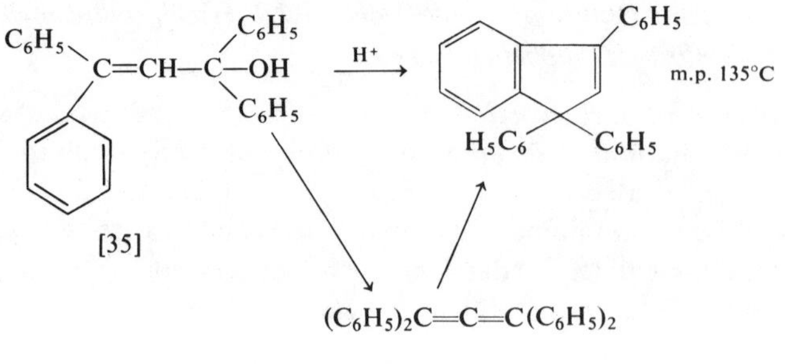

$$(C_6H_5)_2C=C=C(C_6H_5)_2$$

Polyphosphoric ethyl ester has been used to convert 1,1-diphenyl-3-hydroxy-3,3-biphenylene to 1,1-diphenyl-3,3-biphenylene allene in 21% yield [37] (Eq. 31).

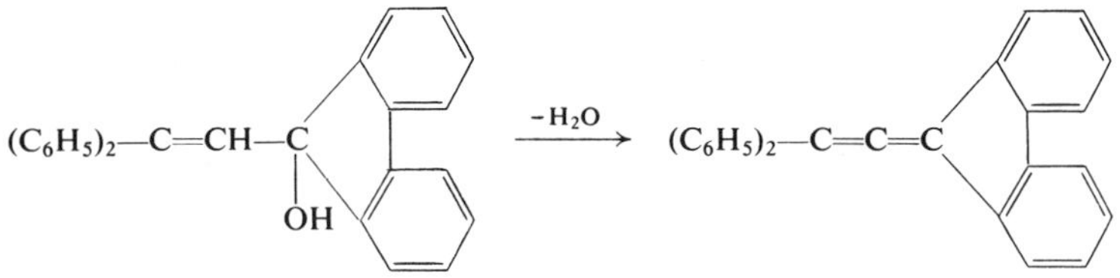

$$(C_6H_5)_2-C=CH-\underset{\underset{OH}{|}}{C}\begin{matrix} \end{matrix} \quad \xrightarrow{-H_2O} \quad (C_6H_5)_2-C=C=C\begin{matrix} \end{matrix} \quad (31)$$

3. REARRANGEMENT REACTIONS

The various types of acetylene–allene rearrangements [1a, 38–41] have been described by Jacobs [38] to involve the following processes:

$$-C\equiv C-\underset{|}{\overset{|}{C}}-X \quad \rightleftharpoons \quad Y-\overset{|}{C}=C=\overset{|}{C}- \quad (32)$$

(A) X = Y = H   prototropic
(B) X = Y = anion   anionotropic
(C) X ≠ Y   displacement
(D) Intramolecular rearrangements

The prototropic rearrangement (Eq. 33) was first discovered by Favorskii in 1886 [42], but the very general nature of this type of rearrangement has only been recognized recently.

$$(CH_3)_2CH-C\equiv CH \quad \xrightarrow[150°C, 6\ hr]{KOH,\ C_2H_5OH} \quad (CH_3)_2C=C=CH_2 \quad [41] \quad (33)$$

The rearrangement occurs more readily when activating groups (aryl, carboxyl, etc.) are attached to the triple bond. Jacobs [38] reports that a reaction involving adsorption of an acetylenic compound on an active basic surface has led to the practical synthesis of arylallenes, allenyl ethers, allenyl halides, and other substituted allenes.

An example of the anionotropic rearrangement is the reaction of propargyl halides with cuprous salts (Eq. 34). The mechanism and the role of the cuprous salt have not been completely elucidated.

$$CH\equiv C-CH_2Br \quad \xrightarrow[70°C]{Cu_2Br_2} \quad Br-CH=C=CH_2 \quad (34)$$

The displacement rearrangements are very similar and comprise those in which the incoming and outgoing groups are different anions:

$$X-C-C\equiv C-C + Y^- \longrightarrow X^- + \underset{}{>}C=C=CY- \qquad (35)$$

For example,

$$(CH_3)_2C-C\equiv CH \quad \xrightarrow[\text{2. } H_2O]{\text{1. } C_2H_5MgBr, \text{ ether}} \quad (CH_3)_2C=C=CHC_2H_5 \qquad (36)$$
$$\underset{Br}{|}$$

A typical intramolecular rearrangement involves 1,4-addition across the conjugated system of alkenynes to give allenes (Eq. 37).

$$\underset{}{>}C=C-C\equiv C- \quad \xrightarrow{X-Y} \quad \underset{X}{>}C-C=C=C-Y \qquad (37)$$

A. Prototropic Rearrangement

Jacobs [42] in 1951 showed that the rearrangement of 1-pentyne, 1,2-pentadiene, and 2-pentyne by means of 3.7 N ethanolic potassium hydroxide in a sealed tube at 175°C for about 3 hr gave the same equilibrium mixture.

$$CH_3CH_2CH_2C\equiv CH \rightleftharpoons CH_3CH_2CH=C=CH_2 \rightleftharpoons CH_3CH_2C\equiv C-CH_3$$

$$\quad (1.3\%) \qquad\qquad\qquad (3.5\%) \qquad\qquad\qquad (95.2\%) \quad (38)$$

The rearrangement cannot involve the addition and elimination of the alcohol because the rearrangement also takes place with powdered potassium hydroxide in the absence of alcohol. In addition, 2-ethoxy-1-pentene was shown to remain largely unchanged under the rearrangement conditions, and it affords no alkynes or allenes. A probable mechanism involves the removal of a proton by the base to give the anion.

Cyclic alkynes (C_9, C_{10}, or C_{11}) also rearrange in the presence of bases to an equilibrium mixture containing cyclic allene [43]. The bases used were NaNH$_2$–liq. NH$_3$ at −33.4°C, KOH–C$_2$H$_5$OH at 131°–134°C (sealed tubes), KO-t-Bu–t-BuOH at 79.4°–120.0°C (sealed tubes) [44]. A solution of sodamide in liquid ammonia gave the most rapid ($\frac{1}{2}$ to 3 hr) allene–acetylene interconversions of all systems examined.

For acetylenes with adjacent acidic hydrogens the formation of substituted allenes is favored by prototropic rearrangement (Eq. 39).

$$R-C\equiv C-CH_2-R' \quad \xrightarrow{\text{base}} \quad R-CH=C=CHR'$$

R′ = aryl [45–47], CH=CH$_2$ [44, 48, 49], C≡CH [50], CN [51],   (39)
COOH [52], COOR [52], SR [53].

R = H, alkyl, aryl, or carboxyl

The use of $DMSO/CH_3OH/KOH$ in the ratio of 96/3/1 has recently been reported for the batch and continuous interconversion of propyne to propadiene (allene) [44b]. Propyne has also been reported to be converted to allene in 21–22% by passing propyne over sodium or potassium aluminite ($NaAlO_2$ or $KAlO_2$) at 293–344°C [44c]. Other references to related processes are worth consulting [44d–f].

Recently, functionally substituted allenes have been reported to be prepared from methylacetylene via propargylic lithium alanate or lithium borate intermediates [44g, h] as described in Eq. (40).

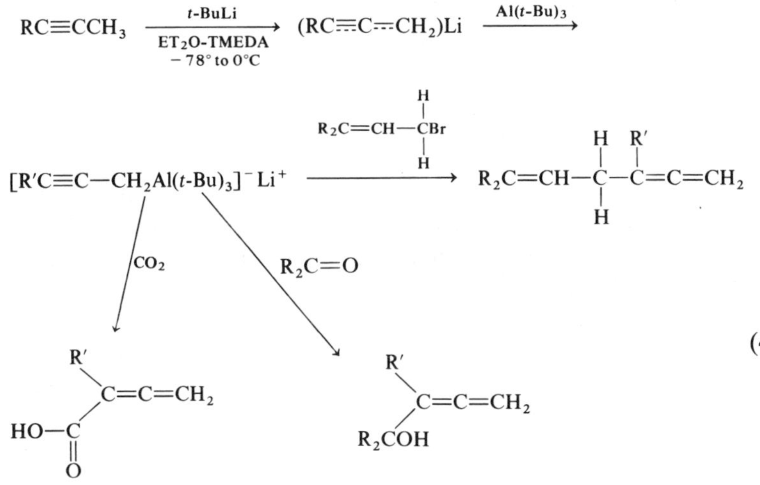

$$RC{\equiv}CCH_3 \xrightarrow[\substack{ET_2O\text{-}TMEDA \\ -78° \text{ to } 0°C}]{t\text{-BuLi}} (RC{\equiv}C{\cdots}CH_2)Li \xrightarrow{Al(t\text{-Bu})_3}$$

$$R_2C{=}CH{-}\underset{H}{\overset{H}{C}}Br$$

$$[R'C{\equiv}C{-}CH_2Al(t\text{-Bu})_3]^- Li^+ \xrightarrow{\hspace{2cm}} R_2C{=}CH{-}\underset{H}{\overset{H}{C}}{-}\overset{R'}{C}{=}C{=}CH_2$$

with branches to CO_2 and $R_2C{=}O$ giving:

$$\underset{\underset{O}{\|}}{HO{-}C}\overset{R'}{\diagdown}C{=}C{=}CH_2 \qquad \underset{R_2COH}{\diagup}\overset{R'}{\diagdown}C{=}C{=}CH_2 \tag{40}$$

3-1. Preparation of 1-(p-Biphenyl)-3-phenylallene [46]

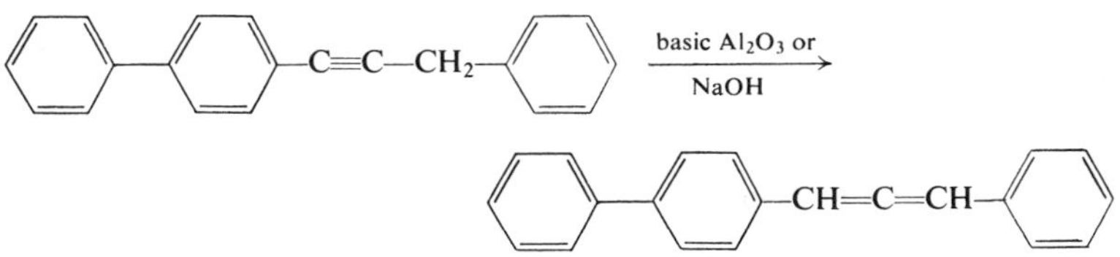

$$\xrightarrow[NaOH]{basic\ Al_2O_3\ or} \tag{41}$$

Method A. To an alumina (Harshaw Chemical Co.; 80 mesh alumina, slightly basic type used for chromatography, is dried at 160°–170°C for 5 days before use) column (1.7 × 11 cm) is added a solution of 0.77 gm of 1-(p-biphenyl)-3-phenyl-1-propyne in n-pentane. After 45 min the product is eluted with n-pentane and concentrated at room temperature using a nitrogen atmosphere (with a water aspirator) to afford 0.53 gm (69%), m.p. 84.8°–86.4°C. (See Method B for the preparation of a higher melting product.)

Method B. To a pressure bottle is added a solution of 0.100 gm of 1-(*p*-biphenyl)-3-phenyl-1-propyne in 50 ml of petroleum ether (b.p. 20°–40°C) with 1.0 gm of sodium hydroxide pellets. The solution is kept for 11 hr at room temperature. After this time the solution is filtered and concentrated as in Method A to afford 0.058 gm (58 %) of 1-(*p*-biphenyl)-3-phenylallene, m.p. 91°–94°C. On recrystallization from petroleum ether (b.p. 20°–40°C), the product melted at 94.6°–95.5°C, ir 1944 cm^{-1} $\left(\text{>C=C=C<} \right)$.

3-2. Preparation of 2,3-Butadienoic Acid [54]

$$CH{\equiv}C{-}CH_2{-}COOH + K_2CO_3 \xrightarrow[40°C]{3\ hr}$$

$$CH_2{=}C{=}CH{-}COO^-K^+ \xrightarrow{H^+} CH_2{=}C{=}CH{-}COOH \quad (42)$$
$$(92\%)$$

A flask containing 5.0 gm (0.06 mole) of 3-butynoic acid and 200 ml of an 18 % aqueous potassium carbonate solution is heated at 40°C for 3 hr. After acidification the product is extracted with ether, dried, and concentrated to afford 4.6 gm (92 %) of 2,3-butadienoic acid, m.p. 65°–66°C (recrystallized from petroleum ether).

α,β-Acetylenic acid ester or amide derivatives which do not possesss a γ-hydrogen atom undergo basic cleavage in good yield to the terminal acetylene [55]. However, if the derivatives above possess two γ-hydrogens, then there is a rapid rearrangement to the α,β-allenic acids [56] (Eqs. 42, 43).

$$R{-}C{\equiv}C{-}\overset{\overset{\textstyle O}{\|}}{C}OR \longrightarrow RC{\equiv}CH \quad (43)$$

$$RCH_2C{\equiv}C{-}\overset{\overset{\textstyle O}{\|}}{C}OR \longrightarrow RCH{=}C{=}CH{-}COOH \quad (44)$$

B. Anionotropic Rearrangements

Primary propargyl halides can be rearranged in the presence of cuprous salts to allenes in fair to good yields (Eq. 45)

$$HC{\equiv}C{-}CH_2{-}X \xrightarrow{aq.\ HX,\ CuX} XCH{=}C{=}CH_2 \quad (45)$$
$$X = Br\ or\ Cl$$

Heating 3-bromopropyne at 105°–110°C for 24 hr in the absence of cuprous bromide gave only 2 % isomerization.

3-3. Preparation of Chloropropadiene [57]

$$CH{\equiv}C{-}CH_2Cl \xrightarrow[\substack{NH_4Cl,\\HCl}]{CuCl} CH_2{=}C{=}CHCl + CH{\equiv}C{-}CH_2Cl + polymer \qquad (46)$$
$$\qquad\qquad\qquad\qquad\qquad\quad (14.5\%)\qquad\quad (49.5\%)$$

To a flask equipped with a stirrer and condenser is added 37.2 gm (0.5 mole) of 3-chloropropyne, 5.0 gm (0.051 mole) of cuprous chloride, 4.0 gm (0.075 mole) of ammonium chloride, 4.0 ml of concentrated hydrochloric acid, and 12.0 ml of water. The mixture is stirred for 5 days and then the organic layer is separated (weight 31.0 gm), dried with potassium carbonate, and fractionated to afford 5.5 gm (14.5%) of product, b.p. 44°C, n_D^{20} 1.4617, d_4^{20} 1.4344, and higher-boiling material, probably polymeric in nature. The yield of chloropropadiene is 29% based on the reacted 3-chloropropyne.

3-4. Preparation of Bromopropadiene [57]

$$CH{\equiv}C{-}CH_2Br \xrightarrow[\substack{72.8{-}73.5°C,\\24\ hr}]{CuBr} CH_2{=}C{=}CHBr + CH{\equiv}C{-}CH_2Br \qquad (46a)$$
$$\qquad\qquad\qquad\qquad\qquad (71\%)\qquad\qquad (13\%)$$

To a distillation flask is added 29.0 gm (0.244 mole) of 3-bromopropyne and 2.5 gm (0.0174 mole) of dry cuprous bromide. The flask is attached to a concentric-tube column (25–30 theoretical plates), and the temperature of the flask is controlled so that the takeoff temperature at the head remains at 72.8°–73.5°C. In 24 hr, 24.4 gm (84%) of bromopropadiene of 75–85% purity is obtained. The remaining 3-bromopropyne (propargyl bromide) is removed by washing the product with a 40% aqueous solution of diethylamine. Three to four moles of diethylamine is used for each mole of propargyl bromide in the product as calculated from VPC or refractive index data. After swirling the mixture (acidified with 15% hydrochloric acid) for $\frac{1}{2}$ hr, the organic layer is separated, washed with water, dried over potassium carbonate, and distilled quickly under reduced pressure into a Dry Ice-cooled receiver to afford pure bromopropadiene, b.p. 72.8°C (760 mm), n_D^{20} 1.5212, d_4^{20} 1.5508.

Method **3-4** is an improvement over the method of making bromopropadiene by shaking 3-bromopropyne for 6 days with cuprous bromide, concentrated hydrochloric acid, ammonium bromide, and copper bronze to afford a 28% yield of bromopropadiene.

Secondary and tertiary chlorides can be rearranged to allenes with the aid of concentrated hydrochloric acid in the presence of cuprous chloride and ammonium chloride [57].

When 1-chloro-3-methyl-1,2-butadiene is allowed to stand for several days with the catalyst above, it is converted to 4-chloro-2-methyl-1,3-butadiene [58] (Eq. 47).

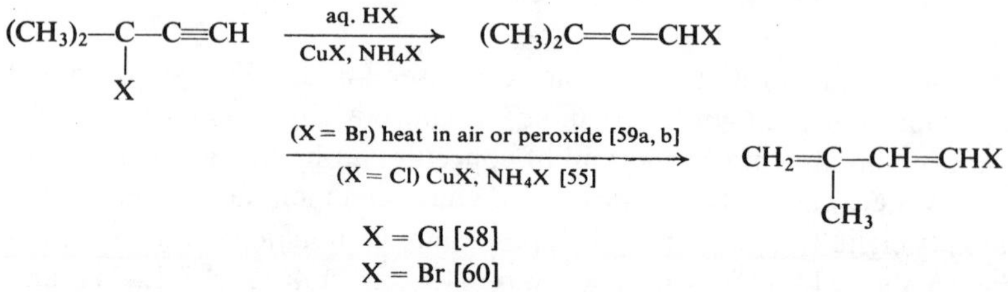

$$(CH_3)_2-\underset{\underset{X}{|}}{C}-C\equiv CH \xrightarrow[\text{CuX, NH}_4\text{X}]{\text{aq. HX}} (CH_3)_2C=C=CHX$$

$$\xrightarrow[\text{(X = Cl) CuX, NH}_4\text{X [55]}]{\text{(X = Br) heat in air or peroxide [59a, b]}} \underset{\underset{CH_3}{|}}{CH_2=C}-CH=CHX \quad (47)$$

$$X = Cl \text{ [58]}$$
$$X = Br \text{ [60]}$$

1-Bromo-3-methyl-1,2-butadiene can be made either by the anionotropic rearrangement of 3-bromo-3-methyl-1-butyne using cuprous bromide and 48% hydrobromic acid or by letting 2-methyl-3-butyne-2-ol react directly with a mixture of 45% hydrobromic acid, ammonium bromide, and cuprous chloride.

3-5. *Preparation of 1-Chloro-1,2-hexadiene* [55]

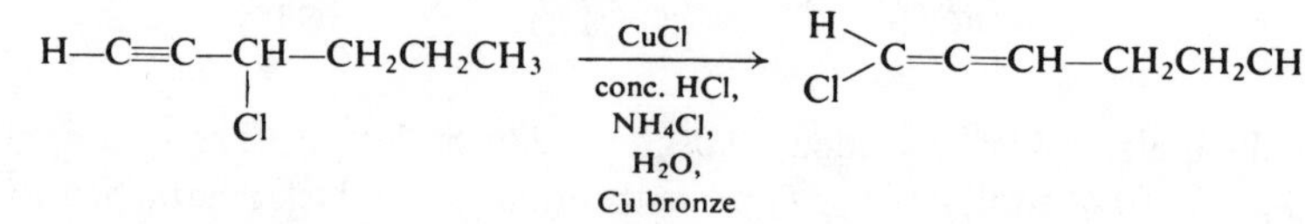

$$H-C\equiv C-\underset{\underset{Cl}{|}}{CH}-CH_2CH_2CH_3 \xrightarrow[\substack{\text{conc. HCl,}\\ \text{NH}_4\text{Cl,}\\ \text{H}_2\text{O,}\\ \text{Cu bronze}}]{\text{CuCl}} \underset{Cl}{\overset{H}{>}}C=C=CH-CH_2CH_2CH_3 \quad (48)$$

To a large bottle are added 116.5 gm (1.0 mole) of 3-chloro-1-hexyne, 20.0 gm (0.20 mole) of cuprous chloride, 16.0 gm (0.30 mole) of ammonium chloride, 10 ml of concentrated hydrochloric acid, 50 ml of water, and 0.6 gm of copper bronze. The bottle is sealed and shaken at room temperature for 14 days. The organic layer is separated, dried over potassium carbonate, and fractionally distilled through a glass-helix-packed column to afford 61.0 gm (52%) of nearly pure 1-chloro-1,2-hexadiene, b.p. 54°–57°C (50 mm), n_D^{25} 1.4567–1.4680, 9.3 gm (8%) of 3-chloro-1-hexyne, and some polymeric material. Shorter periods of reaction give lower yields. The product is purified as described for bromopropadiene.

C. Displacement Rearrangements Involving Propargyl Halides

The propargyl rearrangement in which the incoming anion and outgoing anion are different is called a displacement (Eq. 49). Some examples are summarized in Scheme 2.

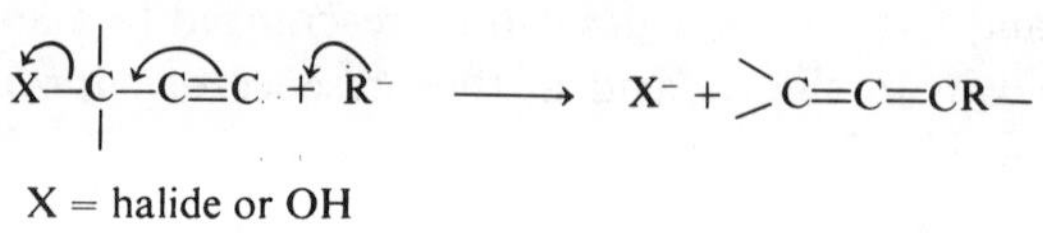

$$X-\underset{\underset{|}{C}}{C}-C\equiv C + R^- \longrightarrow X^- + >C=C=CR- \quad (49)$$

$$X = \text{halide or OH}$$

SCHEME 2

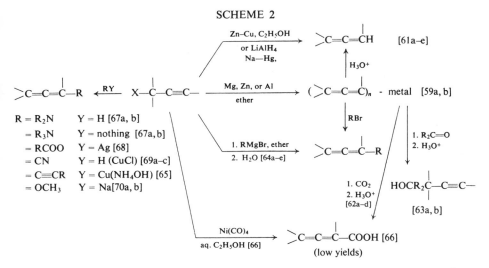

The zinc–copper couple technique [61a–e] is generally used to give good yields of terminal allenes and also of allenes with neighboring double or triple bonds. Ethanol or butanol is used as the solvent, and deuterium oxide is used if deuterioallenes are desired (see Table III).

The use of lithium aluminum hydride gives slightly lower yields and probably involves a displacement reaction by hydride ion. The zinc–copper couple technique probably involves formation of an organozinc intermediate. Sodium, magnesium, and aluminum metal may be used to replace the zinc–copper couple [59a, b]. These organometal intermediates react with aldehydes and ketones to give mainly acetylenic alcohols [63a, b]. However, carbonation of the organometal intermediates gives acetylenic and allenic carboxylic acids [62a–d].

Propargyl bromides can react with alkyl Grignard reagents [64a–e] to give good yields of alkyl-substituted allenes only if the final hydrolysis is effected with the minimum amount of pure water rather than with acid [64a–e].

Serratosa [64c] reported that the data in the literature suggest that two reaction paths are operative in the reaction of propargyl bromide with alkyl-magnesium bromide. These reactions are dependent on temperature, solvent, and the structure of the Grignard reagent (Eqs. 50, 51).

$$CH{\equiv}C{-}CH_2Br + RMgBr \xrightarrow[\text{ether}]{\text{reflux}} CH{\equiv}C{-}CH_2R + CH_2{=}C{=}CHR \qquad (50)$$

$$\downarrow \text{0°C, ether}$$

$$BrMg{-}C{\equiv}C{-}CH_2Br \longrightarrow [:C{=}C{=}CH_2] \xrightarrow{RMgBr}$$

$$\overset{R}{\underset{MgBr}{\diagdown}}C{=}C{=}CH_2 \xrightarrow{H_2O} R{-}CH{=}C{=}CH_2 \qquad (51)$$

Using tetrahydrofuran (THF) in place of ether, alkyl Grignard reagents give exclusively the alkyne product regardless of the temperature. In ether at 0°C the alkyl Grignard reagents give mainly the allene product. Under reflux conditions these Grignards give a mixture of alkyne and allene. However, aryl Grignard reagents react in ether at 0°C with propargyl bromide to give a mixture of allene and alkyne products.

Propargyl bromides can also react with other acetylenes in the presence of 1–2% of cuprous chloride (based on stoichiometry), a small quantity of hydroxylamine hydrochloride, and an amine [65] (Eq. 52). Table IV gives several examples.

$$
\underset{\text{(IV)}}{R-C\equiv CH} + \underset{\text{(V)}}{R'-\underset{X}{\overset{R''}{\underset{|}{\overset{|}{C}}}}-C\equiv C-R''} \xrightarrow{\ Cu^+\ }
$$

$$
\underset{\text{(VI)}}{R-C\equiv C-\overset{R''}{\underset{|}{C}}=C=C\overset{R'}{\underset{R''}{\diagup}}} + \underset{\text{(VII)}}{R-C\equiv C-\overset{R'}{\underset{R''}{\overset{|}{\underset{|}{C}}}}-C\equiv C-R''} \quad (52)
$$

As shown in Scheme 2, propargyl bromides can also react with various nucleophiles (RY) [65, 67–69] to give variously substituted allenes.

Jacobs and Hoff [70a] recently reported that, when sodium methoxide in methanol is the displacing reagent for a series of mixtures of substituted propargyl and allenyl chlorides, the position of attack depends on the side of the groups attached to the propargyl moiety (Eq. 53).

$$
\underset{\text{(IX)}}{(CH_3)_3C\underset{OH}{\overset{R}{\underset{|}{\overset{|}{C}}}}C\equiv CH} \longrightarrow \underset{\text{(X)}}{(CH_3)_3C\underset{Cl}{\overset{R}{\underset{|}{\overset{|}{C}}}}C\equiv CH} + \underset{\text{(XI)}}{(CH_3)_3C\overset{R}{\underset{|}{\overset{|}{C}}}=C=CHCl} \longrightarrow
$$

$$
\left[\underset{\text{(VIII)}}{(CH_3)_3C\overset{R}{\underset{+}{\overset{|}{C}}}C\equiv C{:} \longleftrightarrow (CH_3)_3C\overset{R}{\underset{|}{\overset{|}{C}}}=C=C{:}^{-}} \right] \xrightarrow{\text{styrene}} \underset{\underset{C_6H_5}{|}}{\overset{H_2C}{\underset{HC}{\diagup}}}{\overset{}{\diagdown}} C=C=\overset{R}{\underset{|}{\overset{|}{C}}}-C(CH_3)_3
$$

prepared using   [70b]
KO-t-Bu + styrene

$$
\underset{\text{(XII)}}{(CH_3)_3C\overset{CHR'}{\overset{\|}{C}}-C\equiv CH} + \underset{\text{(XIII)}}{(CH_3)_3C\underset{OCH_3}{\overset{R}{\underset{|}{\overset{|}{C}}}}C\equiv CH} + \underset{\text{(XIV)}}{(CH_3)_3C\overset{R}{\underset{|}{\overset{|}{C}}}=C=CHOCH_3} \quad (53)
$$

TABLE III

REACTION OF ZINC–COPPER COUPLE WITH PROPARGYL HALIDES TO GIVE ALLENES

Propargyl halide (gm)	Zn–Cu couple (gm)	Solvent (ml)	Temp. (°C)	Time (hr)	Product	Yield (%)	B.p., °C (mm Hg)	n_D (°C)	Ref.
3-Chloropenta-1,4-diyne (6.0)	5.0	n-Butanol (40)		5	Penta-1,2-diene-4-yne	61	35–37 (250)	1.4790 (20)	61b
trans-5-Chloropenta-3-one-1-yne (3.1)	3.0	n-Butanol (5)		2.5	Penta-1,2,4-triene	71	46.5–47.5	1.4710 (19)	61b
3-Bromopropyne (6.0)	7.0	Ethanol (35)		3	Propyne (⅓) and allene (⅓)	82	—	—	61c
3-Chloro-1-hexyne (56.0)	70.	Ethanol (150)		4	1,2-Hexadiene	71	74.6–74.8 (740)	1.4252 (25)	61a

TABLE IV

REACTIONS OF VARIOUS ALKYNES WITH PROPARGYL HALIDES [65]

Alkyne (IV)	Propargyl halide (II)			X	Amine	Product (VI) or (VII)	Proportion	Yield (%)	B.p., °C (mm Hg)
	R'	R''	R'''						
H	H	H	H	Cl	NH_4OH	VII	95	60	70 (2)
CH_2OH	H	H	H	Cl	NH_4OH	VII	95	75	60 (2)
CH_3—CH—(OH)—	CH_3	H	H	Cl	NH_4OH	VII	100	50	80 (2)
	H	H	H	Cl	NH_4OH	VII	95	80	65 (2)
	CH_3	H	H	Cl	NH_4OH	VII	100	60	80 (2)
$(CH_3)_2C$—(OH)—	CH_3	CH_3	H	Cl	$(CH_3)_3C$—NH_2	VII	100	60	80 (1)
CH_2=$C(CH_3)$—	H	H	H	Cl	NH_4OH	VII	100	—	—
C_6H_5	H	H	H	Cl	NH_4OH	VII	100	—	—

As seen in Table V, elimination to form enyne XII accompanies the displacement in such a way that it increases in importance as the temperature is raised, and decreases in importance when R increases in size.

TABLE V[e]

REACTIONS OF MIXTURES OF $(CH_3)_3CCRClC\equiv CH$ (X) AND
$(CH_3)_3CCR=C=CHCl$ (XI) WITH 100% EXCESS OF CH_3ONa IN CH_3OH [70a]

Starting material		Conditions		Products, %[a] (VPC data)			
R	X:XI	Temp. (°C)	Time (hr)	X and XI	XII	XIII	XIV
CH_3	67:33	Room temp.	288	12[b]	40	48	
		65	15		Mainly		
C_2H_5	60:40[c]	Room temp.	288	28[b]	30	42	
		65	15		Mainly		
$CH(CH_3)_2$	2:98	Room temp.	288	Mainly	Minor	[d]	[d]
		65	15	~5	40	11	44
$C(CH_3)_3$	2:98	Room temp.	288	Mainly		Trace	Trace
		65	15	Trace		10	90

[a] The numbers represent proportions of products, and errors may be as great as ±10% of the figures given. A single distillation of the reaction mixture usually gave a distillate that accounted for over 90% of starting materials, but with R = $C(CH_3)_3$ the yield was about 80%.

[b] This appeared to be entirely allenyl chloride (XI).

[c] A second experiment with X:XI = 78:22 under the same conditions gave the same product distribution but recovered (XI) was somewhat less.

[d] The XIII:XIV ratio could not be determined as precisely as in the next experiment, but was approximately 20:80, in agreement with the latter.

[e] Reprinted from T. L. Jacobs and S. Hoff, *J. Org. Chem.* **33**, 2987 (1968). Copyright 1968 by The American Chemical Society. Reprinted by permission of the copyright owner.

The halide mixtures X and XI were produced by treatment of the propargyl alcohol IX with hydrochloric acid, calcium chloride, and copper bronze.

3-6. *Preparation of 1,2-Hexadiene (Zn–Cu couple)* [61a]

$$C_3H_7-CH-C\equiv CH \xrightarrow[C_2H_5OH]{Zn-Cu} C_3H_7-CH=C=CH_2 \qquad (54)$$
$$|$$
$$Cl$$

(*a*) *Preparation of Zn–Cu couple.* Zinc dust, 62 gm (0.95 gm-atoms), is placed in a Büchner funnel and then washed four times with 50 ml portions of 3% hydrochloric acid solution, twice with 50 ml portions of distilled water, twice with 100 ml portions of 2% copper sulfate solution, twice with distilled water, once with 100 ml of 95% ethanol, and once with 100 ml of absolute ethanol (do not dry; use directly below).

(*b*) *Preparation of 1,2-hexadiene.* The zinc–copper couple above, along with 150 ml of absolute ethanol, is added to a 500 ml round-bottomed flask equipped with a ground-glass stirrer, dropping funnel, Vigreaux column, and distillation head. Then 56.0 gm (0.48 mole) of 3-chloro-1-hexyne is added dropwise while stirring and occasionally cooling. After the addition, the mixture is heated to reflux for 4 hr and then distilled at 63°–66°C to collect the allene–alcohol azeotrope. The distillate is washed three times with water and the resulting organic layer dried and fractionally distilled to afford 28.0 gm (71 %), b.p. 73.8°–74.6°C (744 mm). Redistillation through a Todd column affords a product with the following properties: b.p. 74.6°–74.8°C (740 mm), n_D^{25} 1.4252, d_4^{25} 0.7102.

3-7. *Preparation of 1,2-Hexadiene (LiAlH₄ Technique)* [61c]

$$C_3H_7-CH-C\equiv CH + LiAlH_4 \xrightarrow{\text{ether}} C_3H_7-CH=C=CH_2 + C_4H_9-C\equiv CH \quad (55)$$
$$\underset{Cl}{|}$$

To a flask containing 22.0 gm (0.58 mole) of lithium aluminum hydride in 500 ml of ether is added dropwise 58.5 gm (0.5 mole) of 3-chloro-1-hexyne. The reaction is not exothermic. After the addition the mixture is refluxed for 3 hr, allowed to stand overnight, and then hydrolyzed with water. The ether layer is separated, dried, and fractionally distilled to afford 2.87 gm (7.0 %) of 1-hexyne, b.p. 70°–72°C, n_D^{25} 1.3990; 20.1 gm (49 %) of 1,2-hexadiene, b.p. 75°–76°C, n_D^{25} 1.4260–1.4267; and 10.0 gm (17.1 %) of 3-chloro-1-hexyne, b.p. 63°–64°C (100 mm), n_D^{25} 1.4375, d_4^{25} 0.9240.

3-8. *Preparation of 1,1-Diphenyl-4,4-dimethyl-1,2-pentadiene* [59b]

$$CH_3-\underset{\underset{CH_3}{|}}{\overset{\overset{CH_3}{|}}{C}}-C\equiv C-\underset{\underset{C_6H_5}{|}}{\overset{\overset{C_6H_5}{|}}{C}}-Br \xrightarrow[\text{Na(Hg)}]{40\%}$$

$$(CH_3)_3C-C\equiv C-\underset{\underset{C_6H_5}{|}}{\overset{\overset{C_6H_5}{|}}{C}}-Na \xrightarrow{H_2O} CH_3-\underset{\underset{CH_3}{|}}{\overset{\overset{CH_3}{|}}{C}}-CH=C=C(C_6H_5)_2 \quad (56)$$

To a flask containing 20.0 gm (0.061 mole) of 1-bromo-1,1-diphenyl-4,4-dimethyl-2-pentyne, 500 ml of dry ether, and 50 ml of 40 % sodium amalgam, the air is displaced by nitrogen, the flask is stoppered, and then placed in a mechanical shaker for 48 hr. The flask is cooled, and previously boiled water is slowly added to discharge the red color. Approximately 200 ml of water is added, the ether layer separated, washed with water, dried, concentrated, and distilled to afford 5.2 gm (34%) of product, b.p. 115°–118°C (0.55 mm) n_D^{20} 1.5690, d_4^{20} 0.9661. The distillation residue is crystallized from petroleum ether, b.p. 25°–65°C, to afford 4.2 gm (20.0%) of the allene dimer, m.p. 178.8°–179.8°C.

The allene monomer dimerizes after standing for a few weeks at room temperature.

In a similar manner 2,2,6,6-tetramethyl-3-phenyl-3,4-heptadiene is prepared in 82% yield [59a] (Eq. 57).

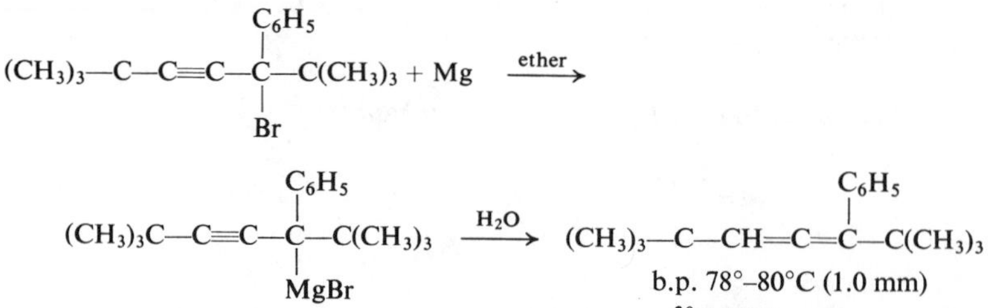

$$
(CH_3)_3-C-C{\equiv}C-\overset{\overset{\displaystyle C_6H_5}{|}}{\underset{\underset{\displaystyle Br}{|}}{C}}-C(CH_3)_3 + Mg \xrightarrow{\text{ether}}
$$

$$
(CH_3)_3C-C{\equiv}C-\overset{\overset{\displaystyle C_6H_5}{|}}{\underset{\underset{\displaystyle MgBr}{|}}{C}}-C(CH_3)_3 \xrightarrow{H_2O} (CH_3)_3-C-CH{=}C{=}\overset{\overset{\displaystyle C_6H_5}{|}}{C}-C(CH_3)_3 \quad (57)
$$

$$
\begin{array}{c}
\text{b.p. } 78°\text{–}80°C \text{ (1.0 mm)} \\
n_D^{20} \text{ 1.5039} \\
d_4^{20} \text{ 0.8808}
\end{array}
$$

3-9. *Preparation of Cyanoallene* [69b]

$$
CH{\equiv}C-CH_2Cl + NaCN \xrightarrow[\text{HCl}]{\text{CuCl}} CH_2{=}C{=}CHCN + CH{\equiv}C-CH_2CN \quad (58)
$$

CAUTION: Because HCN may be evolved, this reaction should be carried out in a well-ventilated hood. Safety precautions in handling NaCN should be strictly observed.

To a flask equipped with two dropping funnels and containing $2\frac{1}{2}$ liters of saturated sodium chloride solution, 50.0 gm (0.51 mole) of cuprous chloride, 2.0 gm of copper powder, and 50 ml of concentrated hydrochloric acid warmed to 75°C is added a 30% sodium cyanide solution until the pH approaches approximately 3–4. At this time 150.0 gm (2.0 mole) of propargyl chloride is added dropwise over a 4-hr period. At the same time, more of the aqueous 30% sodium cyanide is added to keep the pH constant at 3–4. The reaction product is later steam-distilled from the catalyst solution, separated from the water, dried, and fractionally distilled to afford 96.0 gm (73%), b.p. 60°–67°C (95 mm), n_D^{20} 1.44–1.45. This product is contaminated with propargyl cyanide and is refractionated to afford pure cyanoallene, b.p. 50°–51.5°C (50 mm), n_D^{20} 1.4612, λ_{max} 46,500 cm^{-1}, ϵ_{max} 14,200 (methanol).

TABLE VI

DISPLACEMENT REARRANGEMENTS INVOLVING PROPARGYL HALIDES

Propargyl halide	Reactant	Product	Ref.

$(CH_3)_2C-C\equiv CR'$
     |
    Cl
	RMgBr	$(CH_3)_2C=C=CR'R$	a

	(FeCl$_3$ catalyst) where R = C$_2$H$_5$, n-C$_4$H$_9$, i-C$_4$H$_9$, t-C$_4$H$_9$, and CH$_3$		

$R_2C-C\equiv CH$
  |
  X
	Pyridine or CH$_3$NR$_2$
	‿‿‿‿‿‿‿‿‿‿‿‿‿
	Am
		$R_2C=C=C^+AmX^-$	b

$R_2C-C\equiv CR$
  |
  Cl
	$(CH_3)_2CuLi$	$R_2C=C=CRCH_3$	c

$(CH_3)_2C-C\equiv CH$
     |
    Cl
	Cu_2Cl_2/NH_4Cl	$(CH_3)_2C=C=CHCl$	d

[a] D. J. Pasto, S-K. Chou, A. Waterhouse, R. H. Shults, and G. F. Hennion, *J. Org. Chem.* **43**, 1385 (1978).

[b] G. F. Hennion, U.S. Pat. 3,481,982 (1969).

[c] D. F. Pasto, S-K. Chou, E. Fritzen, R. H. Shults, A. Waterhouse and G. F. Hennion, *J. Org. Chem.*, **43**, 1389 (1978).

[d] W. R. Benn, *J. Org. Chem.* **33**, 3113 (1968).

Some additional examples of the displacement rearrangement involving propargyl halides are shown in Table VI.

D. Displacement Rearrangements Involving Propargyl Alcohols

Propargyl alcohols may be converted to allenes by several methods, for example, (a) through the intermediate formation of propargyl halides which are not isolated but react directly with cuprous salts and hydrogen halide [60, 72–73] or cyanide [71]; (b) typical alcohol reactions with thionyl chloride [74a–d] phosphorus halides [75–77], and miscellaneous reagents (see Scheme 3).

The propargyl alcohols react with trivalent phosphorus halides to give allenic phosphorus esters as described in Scheme 3 and Table VII. In the case of aryl-substituted alkynols or highly hindered *t*-propargyl alcohols which contain no free acetylenic —H, thionyl halides or phosphorus trihalides yield bromo- or chloroallenes [74d]. Thionyl chloride also reacts in a similar fashion

SCHEME 3

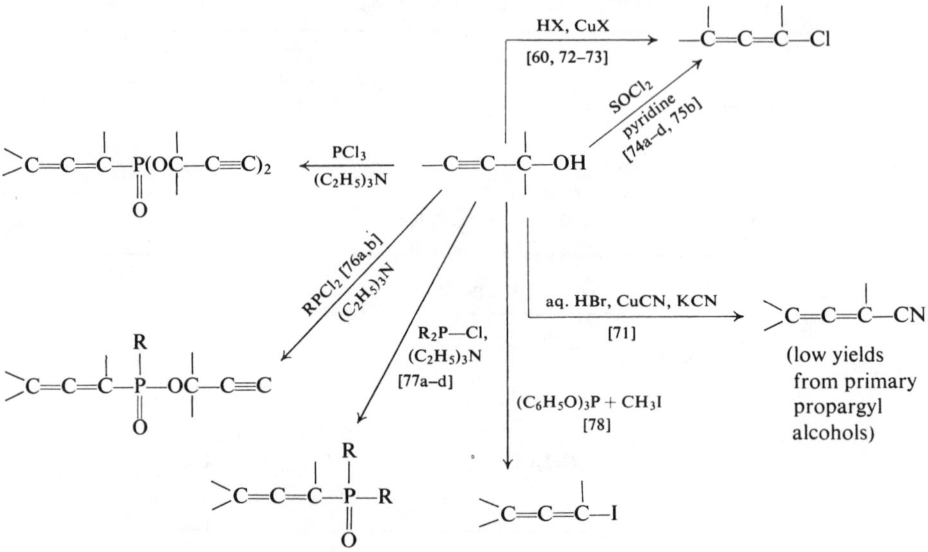

with a wide variety of unhindered secondary alcohols (structure XV) to give a mixture of the chloroallene and chloroalkyne [74a–d].

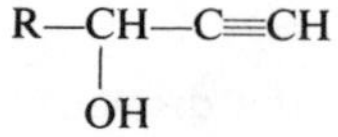

$$R—CH—C{\equiv}CH$$
$$\overset{|}{OH}$$

(XV)

Propargyl alcohols also react with the ether of an enol of carbonyls and lactones to give allenic carbonyl compounds [79a–d] (Eq. 59).

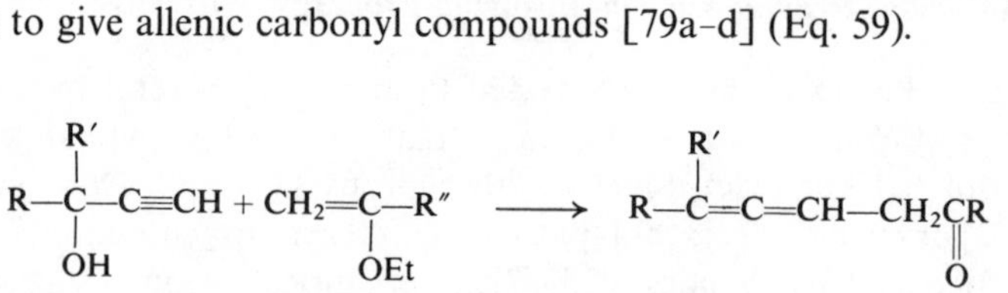

Iodoallenes are prepared from propargyl alcohols by means of phosphonium iodide produced *in situ* from triphenyl phosphite and methyl iodide. In addition, small amounts of the isomeric iodoacetylenes are formed in this process.

The latter phosphonates can be hydrolyzed under alkaline conditions to afford the mono allenic phosphoric esters [76a] (Eq. 60).

(60)

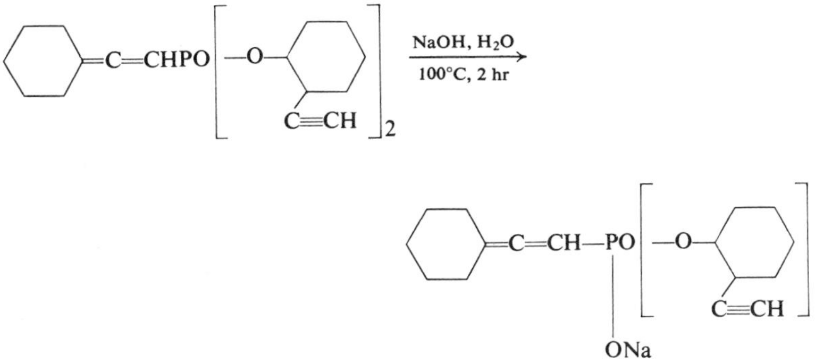

Secondary and tertiary propargyl alcohols are directly converted to haloallenes on reaction with concentrated aqueous hydrogen halides in the presence of the corresponding cuprous halide [60, 72–73]. (See Table VIII.) Better yields are obtained with hydrogen bromide. Hydrogen chloride yields chloroallene, propargyl chloride, and the chloro-1,3-diene isomers (Eq. 61).

$$(CH_3)_2C(OH){-}C{\equiv}CH \xrightarrow{\text{aq. HCl, CuCl}}$$

$$(CH_3)_2{-}C{=}C{=}CHCl + (CH_3)_2C(Cl){-}C{\equiv}CH + CH_2{=}C(CH_3){-}CH{=}CH{-}Cl \quad (61)$$

The chloroallenes are prepared in a similar manner using hydrochloric acid, ammonium chloride, cuprous chloride, and copper powder.

3-10. Preparation of 1-Bromo-3-methyl-1,2-butadiene [60]

$$(CH_3)_2{-}C(OH){-}C{\equiv}CH \xrightarrow{\text{HBr}} \left[(CH_3)_2{-}C(Br){-}C{\equiv}CH\right] \xrightarrow[\text{CuBr}]{\text{HBr, NH}_4\text{Br}}$$

$$(CH_3)_2C{=}C{=}CHBr \quad + \text{trace of } CH_2{=}C(CH_3){-}CH{=}CHBr \quad (62)$$

To a large bottle are added 470.0 gm (5.6 moles) of 2-methyl-3-butyne-2-ol, 1000 ml of 48% technical grade hydrobromic acid, 200.0 gm (2.04 moles) of ammonium bromide, and 70.0 gm (0.71 mole) of cuprous chloride. The bottle

TABLE VII

REACTION OF TRIVALENT PHOSPHORUS HALIDES WITH ALKYNOL AT −20°C IN THE PRESENCE OF BASE TO GIVE ALLENIC ALKYNYL PHOSPHONATES [76a]

Phosphorus halide (moles)	Alkynol RO—H (moles)	R_3N (moles)	Time (hr)	Product	Yield (%)	B.p., °C (mm Hg)
PCl_3 0.1	$HC{\equiv}C—CH_2OH$ 0.3	$(C_2H_5)_3N$ 0.3	24	$CH_2{=}C{=}CH—PO(OR)_2$	92	115–116 (0.25)
0.1	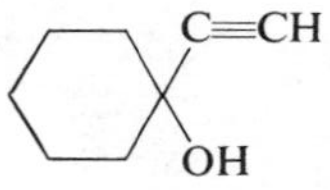 C≡CH / OH (cyclohexanol) 0.6	Pyridine 0.9	18	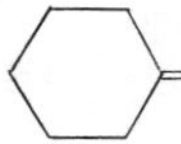 =C=CHPO(OR)₂ (cyclohexylidene)	49	m.p. 78–79
0.1	$HC{\equiv}C—C(CH_3)_2OH$ 0.9	$(C_2H_5)_3N$ 1.0	4	$(CH_3)_2C{=}C{=}C{=}CHPO(OR)_2$	90	—

					No.	Product	Yield (%)	bp (°C) (mm)
C₆H₅PCl₂ 0.1	HC≡C—CH₂OH 0.2		0.2		24	(C₆H₅)(CH₂=C—CH)PO(OR)	23	128–132 (0.3)
0.2	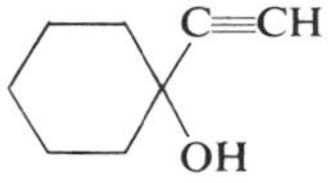 0.4	Pyridine 0.5			29	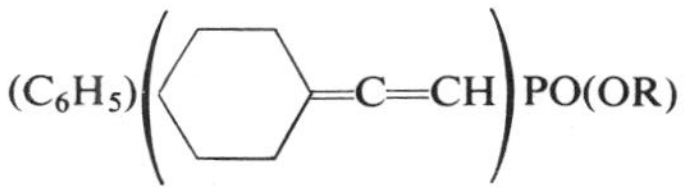	34	—
0.17	HC≡C—C(CH₃)₂OH 0.35		0.4		20	(C₆H₅)[(CCH₃)₂—C≡C—CH)]PO(OR)	30	—

(C₂H₅O)₂PCl 0.1	HC≡C—CH—C₃H₇
	       |
	      OH 0.1
		(C₂H₅)₃N 0.15			4	C₃H₇CH—C=CHPO(OC₂H₅)₂	90	—

0.1	C₆H₅—CH—C≡CH
	      |
	      OH 0.1
		0.15			18	C₆H₅—CH=C=CH—PO(OC₂H₅)₂	90	—

TABLE VIII

Preparation of Allenes by the Reaction of Acetylenic Alcohols with Hydrobromic Acid–Cuprous Bromide [73a]

Alcohol (mole)	HBr (ml)	CuBr gm	NH₄Br gm	Cu gm	Temp./time (°C/hr)	Product (yield %)	B.p., °C (mm Hg)
3-Methylbut-1-yn-3-ol (0.8)	192 (48%)	40	32	2	30/1	1-Bromo-3-methylbuta-1,2-diene (77%)	53–54 (60)
3-Methylpent-1-yn-3-ol (0.15)	36 (48%)	7.5	6	0.3	40/¾	1-Bromo-3-methylpenta-1,2-diene (73%)	51.0–52.5 (24)
3-Ethylpent-1-yn-3-ol (0.5)	124 (45%)	28.9	20.0	2.8	40/1½	1-Bromo-3-ethyl-penta-1,2-diene (65%)	74 (30)
3,4,4-Trimethylpent-1-yn-3-ol (0.5)	180 (48%)	85	42	6.0	40/2	1-Bromo-3,4,4-trimethylpenta-1,3-diene (78%)	45–47 (5)
3-Isopropyl-4-methylpent-1-yn-3-ol (0.5)	144 (48%)	68	36	6.0	40/3	1-Bromo-3-isopropyl-4-methyl-penta-1,2-diene (82%)	49–50 (6)
3-Isobutyl-5-methylhex-1-yn-3-ol	72 (45%)	20	12	2.0	40/20	1-Bromo-3-isobutyl-5-methyl-hexa-1,2-diene (52%)	66 (1.7)
3-t-Butyl-4,4-dimethylpent-1-yn-3-ol (0.1)	36 (48%)	17	9.0	1.0	25/2 wk	1-Bromo-3-t-butyl-4,4-dimethyl-penta-1,2-diene (88%)	80–81 (6.0)
3-Methyloct-1-yn-3-ol (0.25)	62 (45%)	12.5	10.0	1.3	40/2½	1-Bromo-3-methylocta-1,2-diene (55%)	68 (1)
1-Ethynylcyclohexanol (0.5)	40 (48%)	7.0	7.0	1.5	25/6	Cyclohexylidene vinyl bromide (63%)	39–41 (0.3)
But-1-yn-3-ol (0.5)	180 (48%)	72	45	5.0	25/4	1-Bromobuta-1,2-diene (41%)	62.5–63.0 (168)
Pent-1-yn-3-ol (0.25)	96 (60%)	36	24.5	4.0	25/9	1-Bromo-penta-1,2-diene (56%)	62 (66)
Hex-1-yn-3-ol (0.5)	180 (60%)	67	45	5.0	25/10	1-Bromo-hexa-1,2-diene (67%)	51.5–52.5 (22)
4-Methylpent-1-yn-3-ol (0.2)	72 (60%)	27	18	2	25/23	1-Bromo-4-methylpenta-1,2-diene (31%)	60–62 (35)

is sealed, shaken at room temperature for $4\frac{1}{2}$ hr, opened, and the organic layer is separated. The organic layer is washed twice with sodium bicarbonate solution, once with a saturated sodium bisulfite solution, dried over calcium chloride, and fractionally distilled through a glass-helix-packed column to afford 500 gm (61 %) of almost pure product (ir 1956 cm^{-1} allene), b.p. 34°C (18 mm), n_D^{25} 1.5163. The ir showed that the product contained a trace of 1-bromo-3-methyl-1,3-butadiene (1580 and 1620 cm^{-1}).

3-11. Preparation of 3-Chloro-2,2,6,6-tetramethyl-5-phenyl-3,4-heptadiene [75b]

$$\underset{\substack{|\\H_3C\ \ OH}}{\overset{\substack{H_3C\ \ C_6H_5\\|\ \ \ \ \ |}}{CH_3-C-C}}-C\equiv C-\underset{\substack{|\\CH_3}}{\overset{\substack{CH_3\\|}}{C}}-CH_3 + SOCl_2 \longrightarrow \underset{\substack{|\\CH_3}}{\overset{\substack{H_3C\ \ C_6H_5\\|\ \ \ \ \ |}}{CH_3-C-C}}=C=\underset{\substack{|\ \ |\\Cl\ CH_3}}{\overset{\substack{CH_3\\|}}{C}}-CH_3 \quad (63)$$

To 5.0 gm (0.02 mole) of 2,2,6,6-tetramethyl-3-phenyl-4-heptyn-3-ol is added 2.46 gm (0.02 mole) of thionyl chloride. The solution is warmed to reflux, gas is evolved, the solution turns yellow at first and then magenta in color after standing for 2 hr. The reaction mixture is distilled to afford 4.5 gm (83 %) of product, b.p. 105°–108°C (2–3 mm), n_D^{25} 1.5215, ir absorption at 1961 cm^{-1}. The ir spectra of the crude material and that of the distilled material are identical.

The reaction of phosphorus tribromide with the alcohol above has been reported by Marvel to give the analogous bromide. The use of phosphorus trichloride did not yield the chloride [80].

3-12. Preparation of 1-Chloro-3-(α-naphthyl)-1,3-diphenylpropadiene [74d]

$$C_6H_5-C\equiv C-\underset{\substack{|\\C_6H_5}}{\overset{\substack{\text{(naphthyl)}\\|}}{C}}-OH + PCl_3 \longrightarrow C_6H_5-C=C=\underset{\substack{|\ \ \ \ |\\Cl\ \ C_6H_5}}{\overset{\substack{\text{(naphthyl)}\\|}}{C}} \quad (64)$$

To a flask containing 12.4 gm (0.037 mole) of 1-(α-naphthyl)-1,3-diphenyl-2-propyn-1-ol in 50 ml of dry ether is added 2.0 gm (0.015 mole) of phosphorus trichloride in 10 ml of ether. The solution is stirred at room temperature for 2 hr, filtered, the solid washed with dry ether, and dried to afford 6.5 gm (50%), m.p. 124°–125°C, ir 1935 cm^{-1} absorption $\left(\text{>C=C=C<} \right)$ and none at 2240

cm^{-1} (—C=C—). The chloroallene appears to be stable on standing at room temperature for at least 2 days. Further stability data are not reported. An additional 2.0 gm (15%) of crude product is obtained on concentration of the mother liquor, m.p. 118°–120°C.

The same product is obtained when thionyl chloride and pyridine are used in place of phosphorus trichloride [74d].

3-13. Preparation of Allene Dipropargyl Phosphonate [76a, b]

$$PCl_3 + 3CH\equiv C-CH_2OH \xrightarrow[20°C, 24\ hr]{(C_2H_5)_3N} CH_2=C=CH-\overset{O}{\overset{\uparrow}{P}}(OCH_2-C\equiv CH)_2 \quad (65)$$

A solution of 13.8 gm (0.10 mole) of phosphorus trichloride in 30 ml of ethyl ether is added dropwise over a 3–4 hr period at −20°C to a stirred solution of 16.8 gm (0.3 mole) of propargyl alcohol, 30.0 gm (0.3 mole) of triethylamine, and 100 ml of ether. After the addition, the mixture is stirred for an additional hour at 0°–5°C, warmed slowly to room temperature overnight, water is added, the ether separated, and the resulting water layer extracted with two 100 ml portions of ether. The combined ether layer is washed with two 100 ml portions of dilute hydrochloric acid, two 100 ml portions of 10% aqueous sodium carbonate, dried, and concentrated under reduced pressure; distillation of the residue affords 18.0 gm (92%), b.p. 115°–115°C (0.25 mm).

In the case of the reaction of propargyl alcohol with R_2P—Cl, the intermediate propargyl phosphate can be isolated. The latter are rearranged by slowly removing by distilling the solvent and heating carefully to 90°–100°C to effect the rearrangement.

CAUTION: The rearrangement reaction has been reported to be highly exothermic [77a]. If left uncontrolled, the heat of the reaction may cause the reaction to occur with explosive force and with ignition of the products. Some examples of this rearrangement are shown in Table IX and Preparation 3-14.

3-14. Preparation of Diethyl Propadienylphosphonate [77a]

$$CH\equiv C-CH_2OH + (C_2H_5O)_2PCl \xrightarrow[ether]{(C_2H_2)_3N} (CH\equiv C-CH_2O)P(OC_2H_5)_2$$

$$\xrightarrow{\Delta} CH_2=C=CH-\overset{O}{\overset{\|}{P}}(OC_2H_5)_2 \quad (66)$$

To a solution of 156.6 gm (1.0 mole) of diethyl phosphorochloridate in 2 liters of ether was slowly added a solution of 56.1 gm (1.0 mole) of propargyl

TABLE IX

PREPARATION OF ALLENE PHOSPHONATES BY THERMAL REARRANGEMENT OF PROPARGYL PHOSPHITES [77a]

Phosphorus halide	Alkynol	R_3N	Solvent	Time (hr.)	Temp. (°C)	Product	B.p., °C (mm Hg)	n_D^{25}
$(C_2H_5O)_2PCl$	$CH{\equiv}C{-}CH_2OH$	$(C_2H_5)_3N$	Ether	4	0–5 90–100	$CH_2{=}C{=}CHP(OC_2H_5)_2$ (=O)	90 (0.4)	1.4544
$(C_2H_5O)_2PCl$	$CH_3C{\equiv}C{-}CH_2OH$					$CH_3CH{=}C{=}CHP(OC_2H_5)_2$ (=O)	73 (0.1)	1.4587
$(C_2H_5O)_2PCl$	$CH_3{-}CH{-}C{\equiv}CH$ | OH		Ether	4	90–100	$CH_2{=}C{=}C{-}P(OC_2H_5)_2$ with CH_3, (=O)	105 (1)	1.4514
PCl_3	$CH{\equiv}C{-}CH_2OH$	$(C_2H_5)_3N$	Ether	4	90–100	$CH_2{=}C{=}CH{-}P(OCH_2CCH)_2$ (=O)	117 (0.34)	1.4857
$(CH_3)_2PCl$	$CH{\equiv}C{-}CH_2OH$	$(C_2H_5)_3N$	Ether	4	90–100	$CH_2{=}C{=}CHP(CH_3)_2$ (=O)	(m.p. 58–60°)	
$(CH_3S)_2PCl$	$CH{\equiv}C{-}CH_2OH$	$(C_2H_5)_3N$	Ether	4	90–100	$CH_2{=}C{=}CHP(SCH_3)_2$ (=O)	100–105 (0.3)	1.5980
$[(CH_3)_2N]_2PCl$	$CH{\equiv}C{-}CH_2OH$	$(C_2H_5)_3N$	Ether	4	90–100	$CH_2{=}C{=}CHP[N(CH_3)_2]_2$ (=O)	85–88 (0.17)	1.5046

alcohol (2-propyn-1-ol) and 101.2 gm (1.0 mole) of triethylamine dissolved in 1 liter of ether while stirring vigorously at 0°–5°C. The addition required about 4 hr and the precipitated triethylamine hydrochloride was filtered off, the filtrate washed with ether, and the yellow filtrate added dropwise to a 1-liter steam bath-heated three-necked flask equipped with a stirrer and distillation head. In this way the ether was slowly removed and the diethyl-propargyl phosphite $(C_2H_5O)_2P(OCH_2—C\equiv CH)$ was heated in small portions to about 90°–100°C. Afterward the product was analyzed and the ir showed strong absorption at 5.0–5.2 and 7.9–8.1 microns, indicating the presence of the 1,2-propadienyl (allenyl) and phosphonyl (P=O) groups, respectively. The ir showed the absence of absorption at 3.2 and 4.7 microns characteristic of terminal acetylenic groups. The product was obtained as a colorless liquid with a boiling point of 90°C. (0.4 mm Hg), n_D^{25} 1.4544.

Sequential treatment of an ethynyl alcohol to give the acetate followed by reaction with n-BuLi and R_3B gives $R_2B(R)C=C=CR_2$ (allenic borane). This can be protonated to either an allene ($RCH=C=CR_2$) or an acetylene ($RC\equiv CCHR_2$) [76c].

Some additional examples of the preparation of allenes from various propargyl alcohols and their derivatives are shown in Table X.

E. Rearrangements to Allenes Involving 1,4-Addition to Vinylacetylenes

The 1,4-addition to the conjugated system of 1,3-alkenynes yields allenes in many reactions. When the addition occurs 1,2- or 3,4- acetylenes or dienes are produced. Substituents at C_1, C_2, and C_4 greatly influence the course of the addition reaction [81a, b].

$$\overset{\curvearrowright}{\underset{1\quad 2\quad 3\quad 4}{>C=C-C\equiv C-}} \longleftrightarrow \overset{+}{\underset{1\quad 2\quad 3\quad 4}{>C-C=C=C-}} + \overset{+}{Z}-\overset{-}{X} \longrightarrow \underset{1\quad 2\quad 3\quad 4}{X>C-C=C-C-Z}$$

(67)

$\overset{+}{Z}-\overset{-}{X}$	=	$\overline{\overset{+}{H}-\overset{-}{H}}$	(H_2 [82a, b] or $LiAlH_4$ [83a–c])
	=	Li—R	[84a–c]
	=	H—X	(X = Cl, Br) [85a–d]
		Cl—CH₂—OR	[86a–d]
		Br—Br	(or I—Cl) [87a–c]
		H—NR₂	[88a–e] or RNH₂ [89]
		R₃C—Cl	(free radical) [90a, b]

The reaction of R–Li in Eq. (67) allows several reagents (CO_2, O_2, RCH=O, $R_2C=O$, R—CN, $CH_2\overset{\diagdown\diagup}{\underset{O}{—}}CH_2$, C_6H_5COOH, C_6H_5—NCO, etc.) to react with

TABLE X

PREPARATION OF ALLENES FROM PROPARGYL ALCOHOLS AND DERIVATIVES

Propargyl alcohol or derivative	Coreactant	Product	Ref.
$CH_3C{\equiv}C$, 1-hydroxycyclohexyl (CH₃C≡C—C(OH)(cyclohexane))	$(C_2H_5O)_2CHN(CH_3)_2$	CH_3—C=C=(cyclohexylidene) with $CON(CH_3)_2$	a
$\underset{HC{\equiv}C}{\overset{R}{}}$C H, $OC(=O)NHR$	R'_2CuLi	$\underset{H}{\overset{R'}{}}C{=}C{=}C\underset{H}{\overset{R}{}}$	b
$ArC{\equiv}C{-}CH_2Ar$	Brucine or alumina	$ArCH{=}C{=}CHAr$	c
$\underset{R}{\overset{AcO}{}}C\underset{R}{\overset{C{\equiv}CH}{}}$	$(CH_3)_2CuLi$	CH_3—C=C=C(R)(R) with H ($\underset{R}{\overset{CH_3}{}}C{=}C{=}C\underset{R}{\overset{H}{}}$)	d
$HC{\equiv}C{-}CH_2OH$	$X{-}P$ (catechol cyclic phosphate, $X{-}P(=O)(O)(O{-}C(=O))$benzo)	$CH_2{=}C{=}CH{-}\overset{O}{\underset{OH}{P}}{-}O{-}$(2-carboxycyclohexyl, COOH) or $CH_2{=}C{=}CH{-}\overset{O}{P}{-}(OH)_2$	e
$R{-}\overset{OH}{\underset{|}{CH}}{-}C{\equiv}CCH_2N^+(CH_3)_3I^-$	$LiAlH_4$	$R{-}\overset{OH}{\underset{|}{CH}}{-}CH{=}C{=}CH_2$	f
$CH{\equiv}C{-}CH_2OH$	$CH_3CH_2CH(OC_2H_5)_3$ CH_3CH_2COOH	$CH_2{=}C{=}C{=}\overset{CH_3}{\underset{|}{CH}}{-}\overset{O}{\underset{||}{C}}OC_2H_5$	g

[a] K. A. Parker, J. J. Petraitis, R. W. Kosley, Jr., and S. L. Buchwald, *J. Org. Chem.* **47**, 389 (1982).
[b] W. H. Pirkle and C. W. Boeder, *J. Org. Chem.* **43**, 1950 (1978).
[c] T. L. Jacobs and D. Danker, *J. Org. Chem.* **22**, 1424 (1957).
[d] P. Rona and P. Crabb'e, *J. Amer. Chem. Soc.* **90**, 4733 (1968).
[e] J. Kollonitsch and G. Gal, U.S. Pat. 3,769,364 (1973).
[f] E. E. Galantay and D. A. Habeck, U.S. Pat. 3,798,215 (1974).
[g] J. B. Hall, D. E. Hruza, M. H. Vock, J. Vinals, and E. J. Shuster, U.S. Pat. 4,041,069 (1977).

SCHEME 4

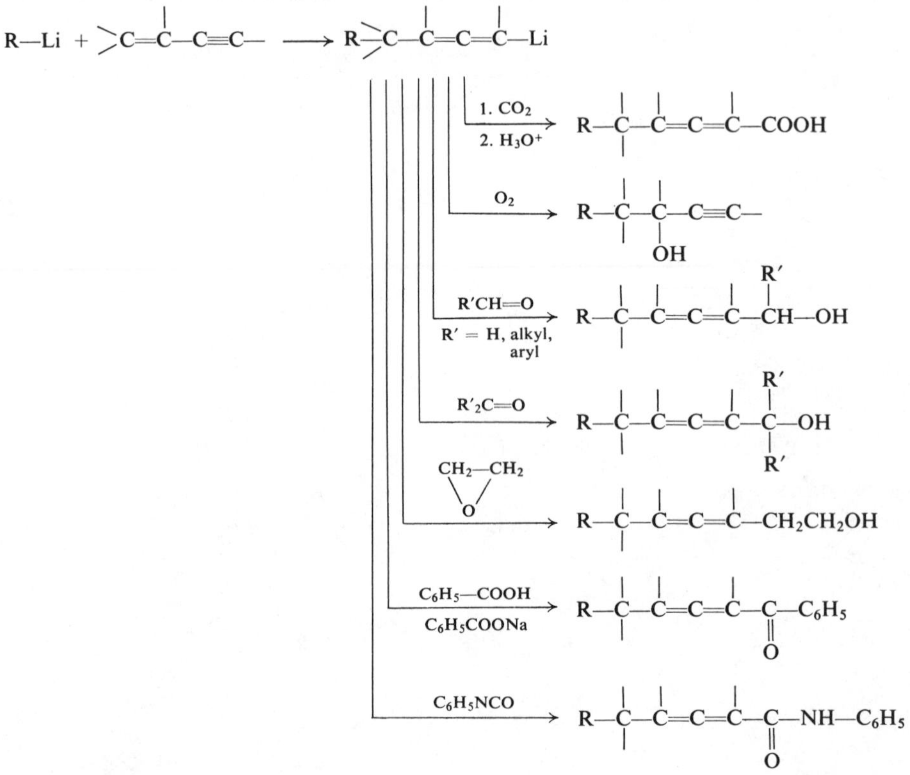

$$R-Li + \underset{/}{>}C\overset{|}{-}\overset{|}{C}-C\equiv C- \longrightarrow R\underset{/}{>}C\overset{|}{-}\overset{|}{C}=C=\overset{|}{C}-Li$$

$$\xrightarrow[\text{2. } H_3O^+]{\text{1. } CO_2} \quad R-\overset{|}{\underset{|}{C}}-\overset{|}{C}=C=\overset{|}{C}-COOH$$

$$\xrightarrow{O_2} \quad R-\overset{|}{\underset{OH}{C}}-\overset{|}{C}-C\equiv C-$$

$$\xrightarrow[\substack{R' = H,\, alkyl,\\ aryl}]{R'CH=O} \quad R-\overset{|}{\underset{|}{C}}-\overset{|}{C}=C=\overset{|}{C}-\overset{R'}{\underset{}{C}}H-OH$$

$$\xrightarrow{R'_2C=O} \quad R-\overset{|}{\underset{|}{C}}-\overset{|}{C}=C=\overset{|}{C}-\overset{R'}{\underset{R'}{C}}-OH$$

$$\xrightarrow[O]{CH_2-CH_2} \quad R-\overset{|}{\underset{|}{C}}-\overset{|}{C}=C=\overset{|}{C}-CH_2CH_2OH$$

$$\xrightarrow[C_6H_5COONa]{C_6H_5-COOH} \quad R-\overset{|}{\underset{|}{C}}-\overset{|}{C}=C=\overset{|}{C}-\underset{\overset{||}{O}}{C}-C_6H_5$$

$$\xrightarrow{C_6H_5NCO} \quad R-\overset{|}{\underset{|}{C}}-\overset{|}{C}=C=\overset{|}{C}-\underset{\overset{||}{O}}{C}-NH-C_6H_5$$

the intermediate lithium alkyl to give other allenic derivatives as shown in Scheme 4 [91a–e]. Substituted vinylacetylenes with alkyl groups at C_1, C_2, and C_3 tend to give more allenic products and less acetylenic products (or 1,3-dienes). Unsubstituted vinylacetylenes usually tend to give a mixture of products.

The reactions in Scheme 4 are typical of those for alkyl lithium or Grignard reagents. Those shown illustrate the possible utility of this reaction to form 1,4-substituted allenes.

3-15. Preparation of 1-Chloro-5-ethoxy-2,3-pentadiene [86c]

$$ClCH_2OC_2H_5 + CH_2=CH-C\equiv CH \xrightarrow{BCl_3}$$
$$C_2H_5-OCH_2CH=C=CHCH_2Cl + C_2H_5OCH_2CH=\underset{\underset{Cl}{|}}{C}-CH=CH_2 \quad (68)$$

To a stainless steel pressure reactor are added 283.5 gm (3.0 moles) of ethyl chloromethyl ether (CAUTION: May be extremely carcinogenic.) and

3.0 gm (0.0257 mole) of boron trichloride. The mixture is cooled to $-10°C$, 170 gm (3.27 moles) of vinylacetylene is added, and the resulting mixture is stirred for 9–10 hr at 5°–10°C while boron trichloride is periodically added. The reactor is vented and the product is distilled to afford 36 gm (8.2%) of 3-chloro-5-ethoxy-1,3-pentadiene, b.p. 56°–57°C (10 mm), n_D^{20} 1.4772, and 71.0 gm (16.2%) of 1-chloro-5-ethoxy-2,3-pentadiene, b.p. 70°–71°C, n_D^{20} 1.4792.

3-16. *Preparation of 4-Chloro-1,2-butadiene* [86d]

$$CH_2=CH-C\equiv CH + HCl \xrightarrow{\ AlCl_3\ }$$

$$ClCH_2-CH=C=CH_2 + CH_2=\underset{\underset{Cl}{|}}{C}-CH=CH_2 \qquad (69)$$

$$(77.3\%) \qquad\qquad (3.4\%)$$

A mixture of 700 ml of concentrated hydrochloric acid and 25 gm (0.189 mole) of aluminum chloride is stirred while 120 gm (2.31 moles) of vinylacetylene is passed in during 4 hr at room temperature. The upper organic layer is separated, washed with water, dried, and distilled to afford 158 gm (77.3%) of 4-chloro-1,2-butadiene, b.p. 88°–90°C, n_D^{20} 1.4760; 7.0 gm (3.4%) of 2-chloro-1,3-butadiene (chloroprene), b.p. 59.4°C, n_D^{20} 1.4538; and 10 gm of a tarry residue.

Substitution of the $AlCl_3$ by $FeCl_3$, BCl_3, $CoCl_2$ or $ZnCl_2$ gives similar results. The highest yield is obtained with $AlCl_3$ and $ZnCl_2$.

Carothers and Berchet [85a] reported earlier that 4-bromo-1,2-butadiene was prepared in 54% yield by adding 1.0 mole of vinylacetylene to 3.0 moles of liquid hydrogen bromide ($-50°C$) over a 5–7 hr period. In addition, a 36% yield of 2,4-dibromo-2-butene was obtained. Carothers [85c] also reported that the reaction of vinylacetylene with hydrogen chloride gave no 4-chloro-1,2-butadiene when cuprous chloride was the catalyst. Only chloroprene was obtained. Separate experiments showed that cuprous chloride–hydrogen chloride catalyzes the isomerization of 4-chloro-1,2-butadiene to chloroprene. Reaction of vinylacetylene with hydrogen bromide and cuprous bromide gives only bromoprene [85b].

The reaction of vinylacetylene with amines gives 2,3-butadienylamines as described in Table XI and Preparation 3-17 [88a].

3-17. *Preparation of N,N-Dimethyl-2,3-butadienylamine* [88a, b]

$$CH_2=CH-C\equiv CH + (CH_3)_2NH \rightarrow (CH_3)_2N-CH_2CH=C=CH_2 \qquad (70)$$

A 1 liter, stainless steel pressure reactor containing 540 gm (3.0 moles) of 25% aqueous dimethylamine is flushed with nitrogen, cooled in Dry Ice, and

TABLE XI

PREPARATION OF 2,3-BUTADIENYLAMINES [88a]

Amine	Vinylacetylene[a]	Pressure (psig)	Temp. (°C)	Time (hr)	Yield (%)	Structure	B.p., °C (mm Hg)	n_D^{25}
$(CH_3)_2NH$	$CH_2=CH-C\equiv CH$	200	100	10	56	$(CH_3)_2NCH_2-CH=C=CH_2$	58–60 (155) 106.5–107 (760)	1.4477
CH_3NH_2	$CH_2=CH-C\equiv CH$	200	100	10	13	$CH_3NHCH_2CH=C=CH_2$	59–59.6 (141) 105–106 (760)	1.4468
$(C_2H_5)_2NH$	$CH_2=CH-C\equiv CH$	200	100	15	20	$(C_2H_5)_2NCH_2CH=C=CH_2$	91–92.5 (147)	—
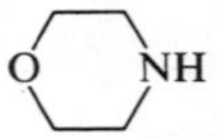 (morpholine) NH	$CH_2=CH-C\equiv CH$	200	100	10	34	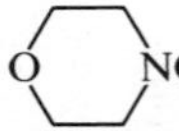 (morpholine) NCH_2CH=C=CH_2	70.5–71.5 (9)	1.4917

[a] CAUTION: Monovinylacetylene polymerizes readily at elevated temperatures and temperature ranges of 90–125°C are preferred.

evacuated. Then 156 gm (3.0 moles) of vinylacetylene at 200 psi is added. The reactor is sealed and heated at 100°C for 10 hr. The reactor is then cooled, vented, opened, the product saturated with potassium carbonate and then extracted with ether. The ether is dried over potassium hydroxide first, then over magnesium sulfate, and is later distilled to afford 164 gm (56%), b.p. 58°–60°C (155 mm), n_D^{25} 1.4468.

The use of acetylenic alcohols in place of the vinylacetylene affords allenic amino alcohols on condensation with amines [88e].

4. CONDENSATION REACTIONS

The condensation of metalloallenes with various functional compounds to give new allenes is a rather unexplored area:

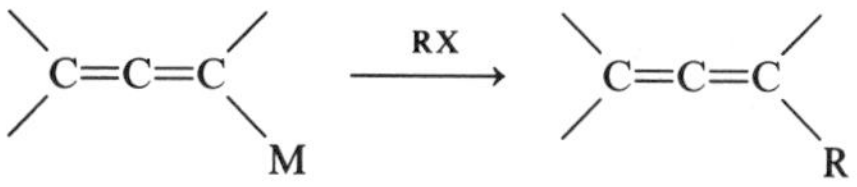

$$\begin{array}{c}\diagdown\\[-2pt]\diagup\end{array}C{=}C{=}C\begin{array}{c}\diagup\\[-2pt]\diagdown\end{array}M \xrightarrow{\text{RX}} \begin{array}{c}\diagdown\\[-2pt]\diagup\end{array}C{=}C{=}C\begin{array}{c}\diagup\\[-2pt]\diagdown\end{array}R \qquad (71)$$

R = aryl, vinyl, etc.
X = halogen

Preparations of metalloallenes such as allenyllithium [88f, g], allenylcopper [88h, i], and allenylsilver [88j] have recently been reported. The utility of the above reaction is illustrated by the examples shown in Table XII.

TABLE XII

PREPARATION OF SUBSTITUTED ALLENES BY REACTION WITH METALLOALLENES [88k]

Allenyl metal	RX	Catalyst $PdP(C_6H_5)_3$	Solvent	Reaction conditions Temp. (°C)	Time (min)	Yield (%)
$(CH_3)_2C{=}C{=}CHLi$	C_6H_5I	5.0	THF	25	60	$<3^a$
t-BuCH$=$C$=$CHMgCl	(Z) BuCH$=$CHI	4.0	THF	35	120	$>98^b$
t-BuCH$=$C$=$CHCu	p-NO$_2$—C$_6$H$_4$I	2.0	THF	45	2	$>98^c$
t-BuCH$=$C$=$CHZnCl	(E)-BuCH$=$CHI	4.0	THF	45	120	$>98^d$

[a] B.p. 93°–95°C (20 mm Hg), n_D^{20} 1.5623.
[b] B.p. 84°–87° (20 mm Hg), n_D^{20} 1.4754.
[c] B.p. 110°C (0.001 mm Hg).
[d] B.p. 65°–67°C (1.0 mm Hg), n_D^{20} 1.4789.

Substituted allenes can be prepared by substitution of the mesylate group. For example [88],

$$
\begin{array}{c}
\diagdown \\
\diagup
\end{array}
C{=}C{=}CHCH_2OSO_2CH_3 \quad \xrightarrow[CH_2Cl_2]{R_2NH} \quad
\begin{array}{c}
\diagdown \\
\diagup
\end{array}
C{=}C{=}CHCH_2NR_2 \qquad (72)
$$

5. MISCELLANEOUS METHODS

(1) Reduction of propargylic chlorides with tri-*n*-butyltin hydride to a mixture of the corresponding acetylene and isomeric allene [92].

(2) Thermal propargyl rearrangement–Claisen rearrangement of propargyl vinyl ethers and Cope-type rearrangements of propargyl malonates [93].

(3) Pyrolysis of methylenecyclobutanes to a mixture of allene, 1,3- and 1,4-dienes [94–97].

(4) Pyrolysis of diketene [98, 99].

(5) Pyrolysis of pyrazolenines [100].

(6) Pyrolysis of esters [101, 102].

(7) Pyrolysis of carboxylic acid salts [103].

(8) Pyrolysis of bis(hydroxytrimethylammonium)alkanes [104].

(9) Wittig synthesis of allenes [105–110].

(10) Pyrolysis of hydrocarbons [111, 112].

(11) Radiolysis of gaseous isobutylene by ^{60}Co γ-rays [113].

(12) Propylene cracking [114, 115].

(13) Ultraviolet irradiation of methane and ethylene to give allene [116].

(14) Addition of diethyllithium and dibutyllithium amides to vinylacetylene and vinylalkylacetylenes [117].

(15) Treatment of 3-allyl-3-(2-propynyl)-2,4-pentanedione with methylhydrazine to give condensation and propargyl–allene rearrangement [118].

(16) Reaction of 1-bromo-2-butyne with hexylmagnesium bromide to give a mixture containing equal amounts of 1-methyl-1-hexylallene and 2-decyne [119].

(17) Formation of tetrakis(trimethylsilyl)allene by an unusual reaction from hexachlorobenzene and derivatives [120].

(18) Decomposition of cyclopropyldiazonium ions [121].

(19) Allene–carbene reactions with olefins [122, 123].

(20) Pyrolytic isomerization of propyne to allene [124, 125].

(21) Acetylene–allene isomerization of 1,4-nonadiyne [126].

(22) Equilibration of cyclic allenes and acetylenes [127].

(23) Pyrolysis of allyl chloride to allene [128].

(24) Preparation of tetraalkylallenes by the pyrolysis of 2,2,4,4-tetraalkyl-3-hydroxy-3-butenoic acid β-lactones at 150°–550°C [129].

(25) Reactions of atomic carbon with ethylene to produce allene and methylacetylene [130].

(26) Reactions of triatomic carbon with olefins [131].

(27) Synthesis of tetramethylallene via the cycloelimination reaction of 4-keto-3,3,5,5-tetramethylpyrazolinehydrazone using nickel peroxide at room temperature [132].

(28) Reaction of phosphinalkenes with Schiff bases [133].

(29) Synthesis of allenic phosphonium salts and ylids [134].

(30) The hydroboration of substituted propargyl halides—a convenient synthesis of terminal allenes [135].

(31) Displacement rearrangements involving propargyl derivatives: allenes from α-acetoxyalkynes [136].

(32) Preparation of allenic acetals from unsaturated carbenes [137].

(33) Preparation of allenes by the basic reaction of N-nitroso cyclopropylcarbamates [138].

(34) Preparation of optically active phenylallene-3-d [139].

(35) Preparation of 15-deoxy-16-hydroxy-16-allenyl prostane derivatives of the E and F series [140].

(36) Preparation of allenes by the reaction of Grignard reagents with propargyl ethers in the presence of cuprous bromide as catalyst [141].

$$RC\equiv C{-}CH{-}OR'' + R'''MgX \xrightarrow[\text{THF}]{\text{CuBr}} \begin{array}{c} R' \\ \diagdown \\ \diagup \\ R''' \end{array} C{=}C{=}CHR' \qquad (73)$$

(37) Thermal conversion of diketene to allene [142].

(38) Preparation of allenes by the condensation of a ketal and a 1,1-disubstituted propargyl alcohol [143].

(39) Preparation of tetralkylallenes by the pyrolysis of 2,2,4,4-tetralkyl-3-hydroxy-3-butenoic acid β-lactones [144].

(40) Preparation of dimethyl allene-1,3-dicarboxylate [145].

(41) Preparation of allene and methylacetylene from propylene [146].

(42) Preparation of allene by cracking propylene and isobutylene in the presence of HBr [147].

(43) Conversion of aldehydes and ketones to terminal allenes in allylic chlorides by reaction with α-silylvinyl carbanions followed by subsequent transformation [148].

(44) Titanium vinylidene route to substituted allenes [149].

(45) Propargylic titanium reagents used for the preparation of allenes by condensation with aldehydes [150].

(46) Synthesis of cyclic allenic esters [151].

(47) Silver perchlorate catalyzed rearrangement of proparyl esters to allenes [152].

(48) Allenes by the alkaline treatment of 3-nitroso-4,5,5-trialkyl-2-oxazolidones [153].

(49) Polylithium compounds converted to substituted allenes [154].

(50) Preparation of 9-deoxy-9-methylene derivatives of (DL)-16-phenoxy and 16-phenoxy substituted prostatriene compounds [155].

(51) Preparation of allene from acetone [156].

REFERENCES

1a. A. Rank, A. F. Drake, and S. F. Mason, *J. Amer. Chem. Soc.* **101**, 2284 (1979); M. Nakazaki, K. Yamamoto, M. Maeda, P. Sato, and T. Tsutsui, *J. Org. Chem.* **47**, 1435 (1982).

1b. A. E. Favorsky, *J. Russ. Phys. Chem. Soc.* **19**, 414 (1887).

1c. G. G. Gustävson, *J. Russ. Phys. Chem. Soc.* **20**, 615 (1888).

1d. L. M. Norton and A. A. Noyes, *J. Amer. Chem. Soc.* **10**, 430 (1888).

2a. I. G. Farbenindustrie Akt.-Ges., Fr. Pat. 850,492 (1942).

2b. O. Nicodemus, H. Lange, and O. Horn, U.S. Pat. 2,125,685 (1938).

3a. J. M. Quinn and A. K. Schneider, U.S. Pat. 3,003,994 (1962).

3b. W. H. Carothers and G. J. Berchet, U.S. Pat. 2,136,177 (1939).

4. F. D. Marsh, U.S. Pat. 2,628,221 (1953).

5. W. D. Celmer and I. Solomons, *J. Amer. Chem. Soc.* **74**, 1870 (1952).

6. D. R. Taylor, *Chem. Rev.* **67**, 317 (1967).

7a. M. Bouis, *Ann. Chim. Phys.* (10), **9**, 402 (1928).

7b. A. A. Petrov and A. V. Federova, *Russ. Chem. Rev.* **33**, 1 (1964).

7c. J. D. Bu'Lock, *Progr. Org. Chem.* **6**, 123 (1964).

7d. H. Fischer, "The Chemistry of Alkenes" (S. Patai, ed.), p. 1025. Wiley (Interscience), New York 1964.

7e. K. Griesbaum, *Angew. Chem. Int. Ed.* **5**, 933 (1966).

7f. S. Patai, ed., "The Chemistry of Ketenes, Allenes and Related Compounds," Wiley-Interscience, New York, 1980.

8. H. G. Richey, J. C. Phillips, and L. E. Rennick, *J. Amer. Chem. Soc.* **87**, 1381 (1965).

9a. C. U. Pittman, Jr., and G. A. Olah, *J. Amer. Chem. Soc.* **87**, 5632 (1965).

9b. J. M. Oakes and G. B. Ellison, *J. Amer. Chem. Soc.* **105**, 2969 (1983).

10a. S. R. Sandler, *J. Org. Chem.* **32**, 3876 (1967).

10b. W. von E. Doering and P. M. LaFlamme, *Tetrahedron* **2**, 75 (1958).

11a. W. R. Moore and H. R. Ward, *J. Org. Chem.* **27**, 4179 (1962).

11b. E. T. Marquis and P. D. Gardner, *Tetrahedron Lett.* 2793 (1966).

11c. L. Skattebol and S. Solomon, *Org. Synth., Collect. Vol.* **5**, 306 (1973).

12a. L. Skattebol, *Acta Chem. Scand.* **17**, 1683 (1963); J. Leland, J. Boucher, and K. Anderson, *J. Polym. Sci. Chem. Ed.* **15**, 2785 (1977).

12b. T. J. Logan, U. S. Pat. 3,096,384 (1963).

12c. L. K. Bee, J. Beeby, J. W. Everett, and P. T. Garratt, *J. Org. Chem.* **40**, 2212 (1975).

12d. P. J. Garratt, K. C. Nicolaou, and F. Sondheimei, *J. Am. Chem. Soc.* **95**, 4582 (1978).

12e. K. C. L. Lje and R. S. Macomber, *J. Org. Chem.* **39**, 3600 (1974).

12f. P. A. Verbrugge and W. Brunmayer-Schilt, U.S. Pat. 4,025,562 (1977).

12g. M. Nagazaki, K. Yamamato, M. Maedra, O. Sato, and T. Tsutsui, *J. Org. Chem.* **47**, 1435 (1982).

13. A. T. Blomquist, R. E. Bruge, Jr., and A. C. Sucsy, *J. Amer. Chem. Soc.* **74**, 3636 (1952).

14. W. R. Moore and H. R. Ward, *J. Org. Chem.* **25**, 2073 (1960).

15. K. G. Untch, D. J. Martin and N. T. Castellucci, *J. Org. Chem.* **30**, 3572 (1965).

16. L. Skattebøl, *J. Org. Chem.* **31**, 1554 (1966).

17a. G. G. Gustävson and N. Demjanoff, *T. Prakt. Chem.* [2]**38**, 202 (1888).

17b. H. N. Cripps and E. F. Kiefer, *Org. Syn.* **42**, 12 (1962).

17c. W. C. Kray, Jr. and C. E. Castro, *J. Amer. Chem. Soc.* **86**, 4606 (1964).

17d. S. R. Sandler, *Chem. Ind.* 1881 (1967).

18. M. Bertrand, Y. Pasternak, J. C. Traynard, J. LeGras, and A. Guillemonat, *Ann. Fac. Sci. Univ. Marseille* **35**, 85 (1964).

19a. A. T. Blomquist and D. T. Longone, *J. Amer. Chem. Soc.* **79**, 4981 (1957).

19b. W. R. Dolbier, Jr., C. R. Burkhilder, and C. A. Piedrahita, *J. Fluorine Chem.* **20**, 637 (1984).

20. G. V. Asolkar, Doctoral Thesis, Univ. of Cambridge, 1955.

21. K. V. Auwers, *Chem. Ber.* **51**, 1116 (1918).

22. J. A. Tebboth and M. K. Johnson, Brit. Pat. 785,727 (1957).

23a. L. F. Hatch and H. D. Weiss, *J. Amer. Chem. Soc.* **77**, 1798 (1955).

23b. W. Tadros, A. D. Sakla, and A. A. A. Helmy, *J. Chem. Soc.* 2687 (1961).

24. F. B. Kirby, W. G. Kofron, and C. R. Hauser, *J. Org. Chem.* **28**, 2176 (1963).

25a. R. E. Banks, R. N. Haszeldine, and D. R. Taylor, *J. Chem. Soc.* 978 (1965).

25b. R. E. Banks, W. R. Deem, R. N. Haszeldine, and D. R. Taylor, *J. Chem. Soc.* C 2051 (1966).

25c. T. L. Jacobs and R. J. Bauer, *J. Amer. Chem. Soc.* **81**, 606 (1959).

26a. A. Roedig, F. Bischoff, B. Heinrich, and G. Markl, *Ann. Chem.* **670**, 8 (1963).

26b. A. Roedig, N. Detzer and H. J. Friedrich, *Angew. Chem.* **76**, 379 (1964).

27. F. Kai and S. Seki, *Chem. Pharm. Bull.* **13**, 1374 (1965).

28. W. J. Ball and S. R. Landor, *J. Chem. Soc.* 2298 (1962).

29. W. R. Moore and R. C. Bertelson, *J. Org. Chem.* **27**, 4182 (1962).

30. J. M. Dumoulin, *C. R. Acad. Sci. Paris* **182**, 974 (1926).

31. S. P. Mulliken, R. L. Wakeman, and H. T. Gerry, *J. Amer. Chem. Soc.* **57**, 1605 (1935)

32. D. Vörlander and C. Siebert, *Chem. Ber.* **39**, 1024 (1900).

33. J. English and F. V. Brutcher, *J. Amer. Chem. Soc.* **74**, 4279 (1952).

34. R. Wizinger and G. Renckhoff, *Helv. Chim. Acta* **24**, 369E (1941).

35. K. Ziegler, *Ann. Chim.* **434**, 34 (1923).

36. K. Ziegler, H. Grabbe, and F. Ulrich, *Chem. Ber.* **57B**, 1283 (1924).

37. H. Fischer and H. Fischer, *Chem. Ber.* **97**, 2975 (1964).

38. T. L. Jacobs, Abstracts of Papers, National Meeting Amer. Chem. Soc., 138th, New York, Sept. 1960, p. 65-P.

39. A. J. Favorskii, *Bull. Soc. Chim. Fr.* **45**, 247 (1886).

40. A. J. Favorskii, *Chem. Zentralbl.* **18**, 1539 (1887).

41. A. J. Favorskii, *J. Prakt. Chem.* **37**, 382 (1888).

42. T. L. Jacobs, R. A. Kawie, and R. G. Cooper, *J. Amer. Chem. Soc.* **73**, 1273 (1951).

43. W. R. Moore and H. R. Ward, *J. Amer. Chem. Soc.* **85**, 86 (1963).

44a. M. Bertrand, *C. R. Acad. Sci. Paris* **247**, 824 (1958).

44b. R. T. Dickerson, U.S. Pat. 3,579,600 (1971).

44c. M. C. Day, U.S. Pat. 3,052,740 (1962).

44d. E. S. Rothman and S. Serota, U.S. Pat. 3,745,195 (1973).

44e. E. S. Rothman and S. Serota, U. S. Pat. 3,832,415 (1974).

44f. E. L. Kay, D. T. Roberts, Jr., L. E. Callhan, and L. B. Wakefield, U.S. Pat. 3,751,511 (1973).

44g. N. R. Pearson, G. Hahn, and G. Zweifel, *J. Org. Chem.* **47**, 3364 (1982).

44h. G. Zweifel, S. J. Backland, and T. Leung, *J. Amer. Chem. Soc.* **100**, 5561 (1978).

45. T. L. Jacobs and S. Singer, *J. Org. Chem.* **17**, 475 (1952).

46. T. L. Jacobs and D. Dankner, *J. Org. Chem.* **22**, 1424 (1957).

47. T. L. Jacobs, D. Dankner, and S. Singer, *Tetrahedron* **20**, 2177 (1964).

48. K. L. Mikolajezak, M. O. Bagby, R. B. Bates, and I. A. Wolff, *J. Org. Chem.* **30**, 2983 (1965).

49. W. Oroshnik, A. D. Mebane, and G. Karmas, *J. Amer. Chem. Soc.* **75**, 1050 (1953).
50. W. J. Gensler and J. Casella, Jr., *J. Amer. Chem. Soc.* **80**, 1376 (1958).
51. L. I. Smith and J. S. Swenson, *J. Amer. Chem. Soc.* **79**, 2962 (1957).
52. G. Eglinton, E. R. H. Jones, G. H. Mansfield, and M. C. Whiting, *J. Chem. Soc.* 3197 (1954).
53. G. Pourcelot, *C. R. Acad. Sci. Paris* **260**, 2847 (1965).
54. G. Eglinton, E. R. H. Jones, G. H. Mansfield, and M. C. Whiting, *J. Chem. Soc.* 3199 (1954).
55. T. L. Jacobs, E. G. Teach, and D. Weiss, *J. Amer. Chem. Soc.* **77**, 6254 (1955).
56. J. C. Craig and M. Moyle, *J. Chem. Soc.* 130 (1963).
57. T. L. Jacobs and W. F. Brill, *J. Amer. Chem. Soc.* **75**, 1314 (1953).
58. T. A. Favorskaya, *J. Gen. Chem. USSR* **9**, 386 (1939).
59a. J. H. Ford, C. S. Marvel, and C. D. Thompson, *J. Amer. Chem. Soc.* **57**, 2619 (1935).
59b. J. G. Stampfli and C. S. Marvel, *J. Amer. Chem. Soc.* **53**, 4057 (1931).
60. T. L. Jacobs and W. L. Petty, *J. Org. Chem.* **28**, 1360 (1963).
61a. G. F. Hennion and J. J. Sheehan, *J. Amer. Chem. Soc.* **71**, 1964 (1949).
61b. E. R. H. Jones, H. H. Lee, and M. C. Whiting, *J. Chem. Soc.* 341 (1960).
61c. T. L. Jacobs, E. G. Teach, and D. Weiss, *J. Amer. Chem. Soc.* **77**, 6254 (1955).
61d. T. L. Jacobs and R. D. Wilcox, *J. Amer. Chem. Soc.* **86**, 2240 (1964).
61e. A. D. Petrov and E. P. Kaplan, *J. Gen. Chem. USSR* **25**, 1269 (1955).
62a. J. H. Wotiz, *J. Amer. Chem. Soc.* **72**, 1639 (1950).
62b. J. H. Wotiz, C. A. Hollingsworth and R. E. Dessy, *J. Amer. Chem. Soc.* **78**, 1221 (1956).
62c. J. H. Wotiz, J. S. Matthews and J. A. Lieb, *J. Amer. Chem. Soc.* **73**, 5503 (1951).
62d. J. H. Wotiz and R. J. Pulchak, *J. Amer. Chem. Soc.* **73**, 1971 (1951).
63a. C. Prevost, M. Gaudemar, L. Miginiac, F. Gaudemar, and M. Andrac, *Bull. Soc. Chim. Fr.* 679 (1959).
63b. M. Gaudemar, *Ann. Chim. (Paris)* **1**, 161 (1956).
64a. A. I. Zakharova, *Zh. Obshch. Khim.* **19**, 83, 1297 (1949).
64b. A. I. Zakharova and R. A. Sapozhnikova, *Zh. Obsch. Khim.* **22**, 1804 (1952).
64c. F. Serratosa, *Tetrahedron Lett.* 895 (1964).
64d. Y. Pasternak, *C. R. Acad. Sci. Paris* **255**, 3429 (1962).
64e. J. Gore and M. L. Roumestant, *Tetrahedron Lett.* 891 (1970).
65. A. Sevin, W. Chodkiewicz, and P. Cadiot, *Tetrahedron Lett.* 1953 (1965).
66. G. P. Chiusoli, *Angew. Chem.* **72**, 74 (1960).
67a. S. A. Vartanyan and S. O. Badanyan, *Izv. Akad. Nauk. Arm. SSR Khim Nauk* **12**, 37 (1959).
67b. G. F. Hennion and C. V. DiGiovanna, *J. Org. Chem.* **31**, 1977 (1966).
68. A. C. Day and M. C. Whiting, *J. Chem. Soc. C* 464 (1966).
69a. P. M. Greaves and S. R. Landor, *Chem. Commun.* 322 (1966).
69b. P. Kuritz, H. Gold, and H. Disselnkötter, *Ann. Chem.* **624**, 1 (1959).
69c. L. I. Smith and J. S. Swenson, *J. Amer. Chem. Soc.* **79**, 2962 (1957).
70a. T. L. Jacobs and S. Hoff, *J. Org. Chem.* **33**, 2986 (1968).
70b. V. J. Shiner, Jr. and J. S. Humphrey, Jr., *J Amer. Chem. Soc.* **89**, 622 (1967).
71. P. M. Greaves, S. R. Landor, and D. R. J. Laws, *Chem. Commun.* 321 (1965).
72a. T. A. Favorskaya, *Zh. Obshch. Khim.* **9**, 386 (1939).
72b. T. A. Favorskaya and I. A. Favorskaya, *Zh. Obshch. Khim.* **10**, 451 (1950).
72c. T. A. Favorskaya and A. I. Zakharova, *Zh. Obshch. Khim.* **10**, 446 (1940).
72d. G. F. Hennion, J. J. Sheehan and D. E. Maloney, *J. Amer. Chem. Soc.* **72**, 3542 (1950).
72e. T. D. Nagibina, *Zh. Obshch. Khim.* **10**, 427 (1940).
72f. T. A. Favorskaya, *J. Gen. Chem. USSR* **9**, 386 (1939).
73a. S. R. Landor, A. N. Patel, P. F. Whiter, and P. M. Greaves, *J. Chem. Soc. C* 1223 (1966).

73b. D. K. Black, S. R. Landor, A. N. Patel, and P. F. Whiter, *Tetrahedron Lett.* 483 (1963).
73c. T. L. Jacobs and W. L. Petty, *J. Org. Chem.* **28**, 1360 (1963).
74a. T. L. Jacobs, W. L. Petty, and E. G. Teach, *J. Amer. Chem. Soc.* **82**, 4094 (1960).
74b. Y. R. Bhatia, P. D. Landor, and S. R. Landor, *J. Chem. Soc.* **24**, (1959).
74c. C. W. Shoppee, J. C. Craig, and R. E. Lack, *J. Chem. Soc.* 2291 (1961).
74d. T. L. Jacobs, C. Hall, D. A. Babbe, and P. Prempree, *J. Org. Chem.* **32**, 2283 (1967).
75a. H. Doupeux and P. Martinet, *C. R. H. Acad. Sci. Ser. C* **262**, 588 (1966).
75b. T. L. Jacobs and D. M. Fenton, *J. Org. Chem.* **30**, 1808 (1965).
75c. P. Martinet and H. Doupeux, *C. R. Acad. Sci. Paris* **261**, 2498 (1965).
75d. C. W. Shoppee, J. C. Craig, and R. E. Lack, *J. Chem. Soc.* 1311 (1961).
75e. H. Tani and F. Toda, *Bull. Chem. Soc. Japan* **37**, 470 (1964).
76a. E. Cherbuliez, S. Jaccard, R. Prince, and J. Rabinowitz, *Helv. Chim. Acta* **48**, 632 (1965).
76b. A. B. Boisselle and N. A. Meinhardt, *J. Org. Chem.* **27**, 1828 (1962).
76c. M. M. Midland, *J. Org. Chem.* **42**, 2650 (1977).
77a. V. Mark, U.S. Pat. 3,197,497 (1965).
77b. A. N. Pudovik, *Zh. Obshch. Khim.* **20**, 92 (1950).
77c. A. N. Pudovik and I. M. Aledzheva, *J. Gen. Chem. USSR* **33**, 700 (1963).
77d. A. N. Pudovik, I. M. Aledzheva, and L. N. Yakovenko, *J. Gen. Chem. USSR* **35**, 1214 (1965).
78. C. S. L. Baker, P. D. Landor, S. R. Landor, and A. N. Patel, *J. Chem. Soc.* 4348 (1965).
79a. B. Thompson, Brit. Pat. 1,012,475 (1966).
79b. B. Thompson, U. S. Pat. 3,236,869 (1966).
79c. F. Hoffmann-La Roche & Co., Akt-Ges., French Pat. 660,099 (1965).
79d. S. Julia, M. Julia, and P. Graffin, *Bull. Soc. Chim. Fr.* 3218 (1964).
80. J. H. Ford, C. D. Thompson, and C. S. Marvel, *J. Amer. Chem. Soc.* **57**, 2619 (1935).
81a. A. A. Petrov, *Russ. Chem. Rev.* **29**, 489 (1960).
81b. S. A. Vartanyan, *Russ. Chem. Rev.* **31**, 529 (1962).
82a. F. W. Breuer, U.S. Pat. 2,366,311 (1945).
82b. A. F. Thompson and E. N. Shaw, *J. Amer. Chem. Soc.* **64**, 363 (1942).
83a. E. B. Bates, E. R. H. Jones, and M. C. Whiting, *J. Chem. Soc.* 1854 (1954).
83b. K. R. Bharucha and B. C. L. Weedon, *J. Chem. Soc.* 1584 (1953).
83c. R. J. D. Evans, S. R. Landor, and J. P. Regan, *Chem. Commun.* 397 (1965).
84a. A. A. Petrov and V. A. Kormer, *J. Gen. Chem. USSR* **30**, 231 (1960).
84b. A. A. Petrov and V. A. Kormer, *J. Gen. Chem. USSR* **30**, 3846 (1960).
84c. A. A. Petrov and V. A. Kormer, *Izv. Vyssh. Ucheb. Zaved. Khim. Khim. Tekhnol.* **3**, 112 (1960).
85a. W. H. Carothers and G. J. Berchet, *J. Amer. Chem. Soc.* **55**, 2807 (1933).
85b. W. H. Carothers, A. M. Collins and J. E. Kirby, *J. Amer. Chem. Soc.* **55**, 786 (1933).
85c. W. H. Carothers, G. J. Berchet, and A. M. Collins, *J. Amer. Chem. Soc.* **54**, 4066 (1932).
85d. J. C. Traynard, *Bull. Soc. Chem. Fr.* 19 (1962).
86a. S. A. Vartanyan and F. V. Dangyan, *Izv. Akad. Nauk Arm. SSR Khim. Nauki* **18**, 269 (1965).
86b. S. A. Vartanyan and A. O. Tosunyan, *Izv. Akad. Nauk Arm. SSR Khim. Nauki* **16**, 499 (1963).
86c. S. I. Sadykhzade, A. Z. Shikhmamedbekova, J. D. Yul'chevstaya, S. K. Salakhava, and A. S. Rzaeva, *Azerb. Khim. Kh.* 37 (1963).
86d. S. A. Vartanyan and S. O. Badanyan, *Izv. Akad. Nauk Arm. SSR Khim. Nauki* **15**, 231 (1962).
87a. A. A. Petrov, G. I. Semenov, and N. P. Sopov, *J. Gen. Chem. USSR* **27**, 1009 (1957).
87b. A. A. Petrov, *Russ. Chem. Rev.* **29**, 489 (1960).

87c. A. A. Petrov and V. I. Porfir'eva, *Zh. Obshch. Khim.* **23**, 1867 (1953).

88a. V. A. Engelhardt, U.S. Pat. 2,647,147 (1953).

88b. V. A. Engelhardt, *J. Amer. Chem. Soc.* **78**, 107 (1956).

88c. S. A. Vartanyan and S. O. Badanyan, *Izv. Akad. Nauk Arm. SSR Khim. Nauki* **13**, 141 (1960).

88d. A. A. Petrov, T. V. Stadnichuk, and V. A. Kormer, *J. Gen. Chem. USSR* **34**, 3324 (1964).

88e. S. A. Vartanyan and S. O. Badanyan, *Izv. Akad. Nauk Arm. SSR Khim. Nauki* **12**, 37 (1959).

88f. G. Linstrumelle and D. Michelot, *J. Chem. Soc., Chem. Commun.* p. 561 (1975).

88g. G. Balme, A. Doutheau, J. Goré, and M. Malacria, *Syntheni* p. 508 (1979).

88h. D. Michelot and G. Linstrumelle, *Tetrahedron Lett.* p. 275 (1976).

88i. K. Ruitenberg, J. Meijer, R. J. Bullee, and P. Vermeer, *J. Organometal. Chem.* **217**, 267 (1981).

88j. J. Meijer, K. Ruitenberg, H. Westmijze, and P. Vermeer, *Synthesis* p. 551 (1981).

88k. K. Ruitenberg, H. Kleijn, J. Meijer, E. A. Oostveen, and P. Vermeer, *J. Organometal. Chem.* **224**, 399 (1982).

88l. C. Sahlberg, S. B. Ross, I. Fugervall, A.-L. Ask, and A. Claesson, *J. Med. Chem.* **26**, 1036 (1983).

89. S. A. Vartanyan and S. O. Badanyan, *Angew. Chem.* **75**, 1034 (1963).

90a. A. A. Petrov, M. D. Stadnichuk, and Y. I. Kheruze, *Dokl. Akad. Nauk SSSR* **139**, 1124 (1961).

90b. A. F. Thompson and D. M. Surgenor, *J. Amer. Chem. Soc.* **65**, 486 (1943).

91a. L. N. Cherkasov and K. V. Bal'yan, *Zh. Org. Khim.* **1**, 1811 (1965).

91b. L. N. Cherkasov, V. A. Kormer, and K. V. Bal'yan, *J. Gen. Chem. USSR* **35**, 618 (1965)

91c. O. V. Perepelkin, L. N. Cherkasov, V. A. Kormer, and K. V. Bal'yan, *J. Gen. Chem. USSR* **35**, 571 (1965).

91d. O. V. Perepelkin, V. A. Kormer, and K. V. Bal'yan, *J. Gen. Chem. USSR* **35**, 963 (1965).

91e. O. V. Perepalkin, V. A. Kormer, K. V. Bal'yan, and A. A. Petrov, *Zh. Org. Khim.* **1**, 1705 (1965).

92. R. M. Fantazier and M. L. Poutsma, *J. Amer. Chem. Soc.* **90**, 5490 (1968).

93. D. K. Black and S. R. Landor, *J. Chem. Soc.* 6784 (1965).

94. J. J. Drysdale, H. B. Stevenson, and W. H. Sharkey, *J. Amer. Chem. Soc.* **81**, 4908 (1959).

95. J. L. Anderson, U.S. Pat. 2,733,278 (1956).

96. D. D. Coffman, P. L. Barrick, R. D. Cramer, and M. S. Raasch, *J. Amer. Chem. Soc.* **71**, 490 (1949).

97. W. H. Knoth and D. D. Coffman, *J. Amer. Chem. Soc.* **81**, 3873 (1960).

98. J. C. Martin, U.S. Pat. 3,131,234 (1964).

99. R. T. Conley and T. F. Rutledge, U.S. Pat. 2,818,456 (1957).

100. A. C. Day and M. C. Whiting, *Chem. Commun.* 292 (1965).

101. R. J. P. Allan, J. McGee, and P. D. Ritchie, *J. Chem. Soc.* 4700 (1957).

102. W. Kimel, and N. W. Sox, U.S. Pat. 2,661,368 (1953).

103. D. Vorländer and C. Siebert, *Chem. Ber.* **39**, 1024 (1906).

104. C. D. Hurd and L. R. Drake, *J. Amer. Chem. Soc.* **61**, 1943 (1939).

105. Y. A. Cheburkov, Y. E. Aronov, and I. L. Knunyants, *Izv. Akad. Nauk SSSR Otd. Khim. Nauk* 582 (1966).

106. G. Wittig and A. Haag, *Chem. Ber.* **96**, 1535 (1963).

107. W. S. Wadsworth, Jr., and W. D. Emmons, *J. Amer. Chem. Soc.* **83**, 1733 (1961).

108. H. J. Bestmann and H. Hartung, *Chem. Ber.* **99**, 1198 (1966).

109. H. J. Bestmann, and F. Seng, *Tetrahedron* **21**, 1373 (1965).

110. H. Gilman and R. A. Tomasi, *J. Org. Chem.* **27**, 3647 (1962).
111. S. B. Zdonik, E. J. Green, and L. P. Hallee, *Oil. Gas. J.* **65**, 133, 137 (1967).
112. J. Happel and C. J. Marsel, U.S. Pat. 3,198,848 (1965).
113. G. J. Collin and J. A. Herman, *Can. J. Chem.* **45**, 3097 (1967).
114. J. Happel and C. J. Marsel, U.S. Pat. 3,270,076 (1966).
115. J. Happel and C. J. Marsel, U.S. Pat. 3,198,848 (1965).
116. W. R. Grace & Co., Brit. Pat. 1,042,240 (1966).
117. A. A. Petrov and V. A. Kormer, *Dokl. Akad. Nauk SSSR* **126**, 1278 (1959).
118. D. T. Manning, H. A. Coleman, and R. A. Langdale-Smith, *J. Org. Chem.* **33**, 4413 (1968).
119. L. Brandsma and J. F. Arens, *Rec. Trav. Chim. Pays-Bas* **86**, 734 (1967).
120. K. Shiina and H. Gilman, *J. Amer. Chem. Soc.* **88**, 5307 (1966).
121. W. Kirmse and H. Schutte, *J. Amer. Chem. Soc.* **89**, 1284 (1967).
122. R. F. Bleiholder and H. Shechter, *J. Amer. Chem. Soc.* **86**, 5032 (1964).
123. H. D. Hartzler, *J. Org. Chem.* **29**, 1311 (1964).
124. J. F. Cordes and H. Günzler, *Chem. Ber.* **92**, 1055 (1959).
125. J. F. Cordes and H. Günzler, Ger. Pat. 1,093,350 (1960).
126. W. J. Gensler and J. Casella, Jr., *J. Amer. Chem. Soc.* **80**, 1376 (1963).
127. W. R. Moore and H. R. Ward, *J. Amer. Chem. Soc.* **85**, 86 (1963).
128. A. M. Goodall and K. E. Howlett, *J. Chem. Soc.* 2596 (1954).
129. J. C. Martin, U.S. Pat. 3,131,234 (1964).
130. M. Marshall, C. Mackay, and R. Wolfgang, *J. Amer. Chem. Soc.* **86**, 4741 (1964).
131. P. S. Skell, L. D. Westcott, Jr., J. P. Goldstein, and R. R. Engel, *J. Amer. Chem. Soc.* **87**, 2829 (1965).
132. R. Kalish and W. H. Pirkle, *J. Am. Chem. Soc.* **89**, 2781 (1967).
133. H. J. Bestmann and F. Seng, *Angew. Chem.* **75**, 475 (1963).
134. K. W. Ratts and R. D. Partos, *J. Amer. Chem. Soc.* **91**, 6112 (1969).
135. G. Zweifel, A. Horng, and J. T. Snow, *J. Amer. Chem. Soc.* **92**, 1427 (1970).
136. P. Rona and P. Crabbé, *J. Amer. Chem. Soc.* **90**, 4773 (1969).
137. M. S. Newman and C. D. Beard, *J. Org. Chem.* **35**, 2412 (1970).
138. J. M. Walbrick, J. W. Wilson, Jr., and W. M. Jones, *J. Amer. Chem. Soc.* **90**, 2895 (1968).
139. A. Viola, G. F., Dudding and R. J. Proverb, *J. Amer. Chem. Soc.* **99**, 7390 (1977).
140. M. B. Floyd, Jr., U.S. Pat. 4,233,453 (1980).
141. J. L. Moreau and M. Gaudemar, *J. Organometal. Chem.* **108**, 159 (1976).
142. R. T. Conley, U.S. Pat. 2,818,456 (1957).
143. R. Marbet and G. Saucy, U.S. Pat. 3,029,287 (1962).
144. J. C. Martin, U.S. Pat. 3,131,234 (1964).
145. T. A. Bryson and T. Dolak, *Org. Synth.* **54**, 155 (1974).
146. S. Kunichika, Y. Sukukibara, and M. Taniuchi, U.S. Pat. 3,454,667 (1969).
147. J. Happel and C. J. Marsel, U.S. Pat. 3,315,004 (1967); U.S. Pat. 3,198,848 (1965); U.S. Pat. 3,270,076 (1966).
148. T. H. Chan, W. Mychajlowskij, B. S. Org, and D. N. Harpp, *J. Org. Chem.* **43**, 1526 (1978).
149. S. L. Buchwald and R. H. Grabbs, *J. Amer. Chem. Soc.* **105**, 5490 (1983).
150. M. Ishiguro, N. Ikeda, and H. Yamamoto, *J. Org. Chem.* **47**, 2225 (1982).
151. A. Silveira, Jr., M. Angelastro, R. Israel, F. Totino, and P. Williamsen, *J. Org. Chem.* **45**, 3522 (1980).
152. D. G. Oelberg and M. D. Schiavelli, *J. Org. Chem.* **42**, 1804 (1977).
153. M. S. Newman and V. Lee, *J. Org. Chem.* **38**, 2435 (1973).
154. W. Priester, R. West and T. L. Chwang, *J. Amer. Chem. Soc.* **98**, 8413 (1976).
155. D. L. Wren, U.S. Pat. 4,418,206 (1983).
156. J. L. Greene, U.S. Pat. 4,301,319 (1981).

1. INTRODUCTION

Carboxylic ortho esters can be regarded as ester–acetal derivatives of the hydrates of carboxylic acids (ortho acids).

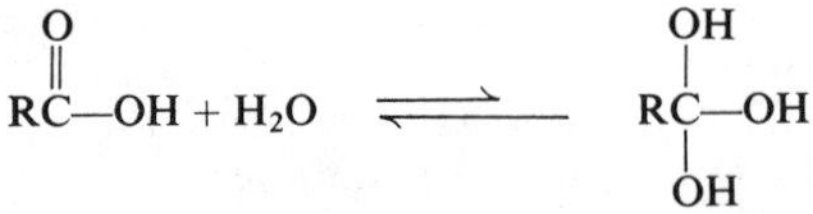

$$
\underset{\substack{\| \\ RC-OH}}{O} + H_2O \quad \rightleftharpoons \quad \underset{\substack{| \\ OH}}{\overset{OH}{RC-OH}} \tag{1}
$$

The free ortho acids have not been isolated because of their thermodynamic instability, and attempts aimed at their detection have not been reported [1]. However, the ortho esters are stable derivatives which have the general structure I. Related to the ortho esters are derivatives in which the R′, R″,

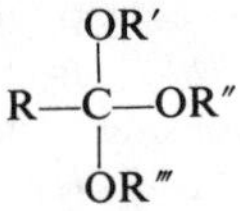

$$
\underset{\substack{| \\ OR''}}{\overset{OR'}{R-C-OR''}}
$$

R, R′, R″, R‴ = alkyl or aryl groups

(I)

R‴ = —OR, —OH, —COR. The thio ortho esters have sulfur in place of oxygen in the structures above. The orthocarbonates $ROC(OR)_3$ and $C(OR)_4$ can also be considered ortho esters.

Other ortho-acid derivatives which will not be considered in detail in this chapter are the thio ortho esters, amide acetals, ester aminals, and the ortho amides which have the following structures, respectively, $RC(SR)_3$, $RC(OR)_2$-NR_2, $RC(OR)(NR_2)_2$, and $RC-(NR_2)_3$.

The chemistry of ortho acid esters was reviewed earlier by Post [2a] and Cordes [2b], and more recently by DeWolfe [3] in a monograph which also reviewed the other ortho acid derivatives described above.

The most important synthetic methods for preparing ortho esters are shown in Eqs. (2)–(7).

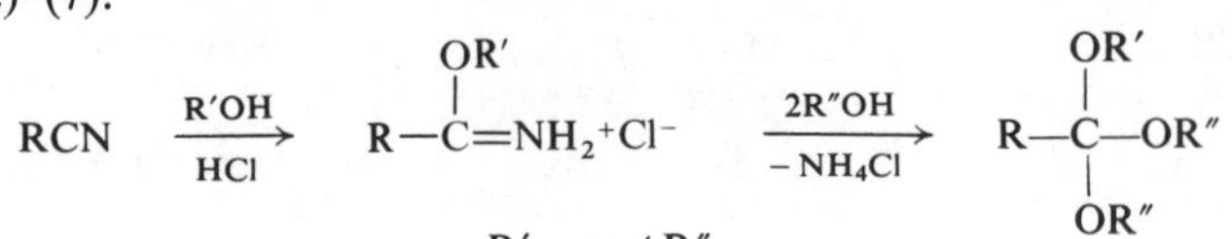

$$
RCN \xrightarrow[\text{HCl}]{\text{R'OH}} \underset{\substack{| \\ R-C=NH_2{}^+Cl^-}}{\overset{OR'}{}} \xrightarrow[-NH_4Cl]{2R''OH} \underset{\substack{| \\ OR''}}{\overset{OR'}{R-C-OR''}} \tag{2}
$$

R′ = or ≠ R″

$$\underset{\substack{\parallel \\ O}}{\text{HCOC}_2\text{H}_5} + \text{RSH} \xrightarrow{\text{HCl}} \text{HC(SR)}_3 \xrightarrow{\text{3ROH + ZnCl}_2} \text{HC(OR)}_3 + 3\text{RSH} \quad (3)$$

$$\text{HCOOH} + 3\text{RSH} \nearrow$$

$$\text{RC(OR)}_3 + 3\text{R}'\text{OH} \underset{\text{H}^+}{\rightleftharpoons} \text{RC(OR}')_3 + 3\text{ROH} \quad (4)$$

$$\text{R}-\text{CX}_3 + 3\text{NaOR}' \longrightarrow \text{R}-\text{C(OR}')_3 + 3\text{NaCl} \quad (5)$$

$$\text{R} = \text{H, aryl, halogen, } -\text{NO}_2, -\text{SCl}$$

$$\text{RO}-\overset{.}{\text{CH}}-\text{X}_2 + 2\text{NaOR}' \longrightarrow \text{ROCH(OR}')_2 + 2\text{NaX} \quad (6)$$

$$(\text{RO})_2\text{CHX} + \text{NaOR}' \longrightarrow (\text{RO})_2\text{CHOR}' + \text{NaX} \quad (7)$$

Ortho esters are either colorless liquids or solids, depending on their molecular weight and structure. They are slightly soluble or very slightly soluble in neutral to basic water. They are soluble in many organic solvents and decompose under acidic conditions as shown in Eqs. (8) and (9). The ortho ester

$$\text{RC(OR}')_3 + \text{HCl} \longrightarrow \text{RCOOR}' + \text{R}'\text{OH} + \text{R}'\text{Cl} \quad (8)$$

$$\text{RC(OR}')_3 + \text{H}_2\text{O} \xrightarrow{\text{H}^+} \text{RCOOR}' + 2\text{R}'\text{OH} \quad (9)$$

functional group has characteristic absorption in the infrared spectra at 1100 cm^{-1} for the C—O stretching band, and the NMR spectrum shows no unusual effects. The ortho-ester group does not give any characteristic ultraviolet absorption.

Some areas where ortho esters have found use involve the elucidation of the structure proof of polyhydroxy alkaloids [4–8], the protection of the esterified hydroxyl groups during chemical transformations of the other parts of the molecule [9], the area of pharmacological screening [10], the preparation of novel polymers [11a], and the stabilization of organic isocyanates [11b].

2. ALCOHOLYSIS REACTIONS

Alcoholysis of nitriles, of ortho and thio ortho esters (transesterification), and of halides is the most common method of preparing the ortho ester functional groups (see Eqs. 2–7). A less practical method is the addition of alcohols to ketene acetals. The latter method is used only when the other methods are not found applicable to the synthesis of specially substituted mixed ortho esters.

A. Alcoholysis of Nitriles and Imidates

A useful method for the preparation of ortho esters was first reported by Pinner [12], who prepared simple and mixed ortho esters by the reaction of alkyl formimidate salts with an excess of alcohols at room temperature; for example, see Eq. (10). With ethyl alcohol, the reaction was exothermic and

$$HC(=\overset{+}{N}H_2\overset{-}{C}l)OC_2H_5 + 2C_2H_5OH \rightarrow NH_4Cl + HC(OC_2H_5)_3 \quad [12] \qquad (10)$$

complete in a few hours. However, with amyl alcohol the reaction was not perceptively exothermic and required 24–48 hr for completion to give the mixed ortho ester ethyl diamyl orthoformate [12].

Later investigators alcoholyzed imidate salts of other monobasic acids to obtain ortho esters of acetic [13, 14], propionic [15], butyric, valeric, caproic, isocaproic, benzoic [16], and phenylacetic acids [17]. For the latter alcoholysis reactions, the reaction time varies from a few days for the production of methyl orthopropionate to six weeks for ethyl orthobenzoate. McElvain reported that the reaction time is drastically cut by carrying out the reaction in boiling ether [18] or petroleum ether [19]. These conditions provide a reaction temperature below the decomposition point of the imidate salt to the amide.

The synthesis of ortho esters from nitriles is usually a two-step process involving first the formation of the imino ester hydrochloride and subsequent reaction with an alcohol. Several examples are described in Tables I and II. Even a glycol such as ethylene glycol can be used to obtain heterocyclic ortho esters, as shown in Eq. (11).

Several investigators have reported a one-step synthesis wherein the imino ester is not isolated. Two examples of this method are described in Table I.

McElvain and Nelson [18] reported that several factors can optimize the yield of imino ester hydrochlorides from the nitrile (see Table III and Eq. 12)

$$RCN + C_2H_5OH + HCl + ether \xrightarrow{\ 0°\ } RC(=\overset{+}{N}H_2\overset{-}{C}l)OC_2H_5 \qquad (12)$$

To an ice-cooled solution of the nitrile, in absolute alcohol (see Table III) is added dry hydrogen chloride until 1.1 moles has been taken up. The resulting solution is allowed to stand at 0°C for the times shown in Table III, column 2.

TABLE I

Preparation of Ortho Esters by Alcoholysis of Nitriles

RCN	ROH	Temp. (°C)	Time (days)	$RC(=\overset{+}{N}H_2\overset{-}{Cl})OR$ (yield %)	R'OH	Solvent	Temp. (°C)	Time (days)	$RC(OR)(OR')_2$ (yield %)	B.p., °C (mm Hg)	n_D	Ref.
H	CH₃	5	1/8	One-step preparation	CH₃	CH₃OH	21	3	50	102	1.3770 (25°)	20
CH₃	C₂H₅	One-step preparation			—	CHCl₃	35–40	4	60–80	70–80 (60)	—	21
C₆H₅OCH₂ ‖O	C₂H₅	5–10	2	82	C₂H₅	C₂H₅OH	25	28	30	99–100 (1.5)	—	16
C₂H₅OC—CH₂	C₂H₅	5–10	16	93	C₂H₅	C₂H₅OH	80	1	82	120–121 (18)	1.4220 (25)	22
CN—CH₂	CH₃	5–10	1	87	CH₃	CH₃OH	65	1	65	98–102 (13)	1.4215 (25)	22
Cl—CH₂	C₂H₅	5–10	3	79	C₂H₅	C₂H₅OH	80	1/24	69	78–80 (10)	1.4988 (20°)	23
C₂H₅	CH₃	5–10	2	88	CH₃	CH₃OH	25	6	69	126–128	—	16
CH₃—CH(C₆H₅)	CH₃	5–10	2	91	CH₃	CH₃OH	25	3	60	70–71.5 (0.5)	1.4928 (25°)	24
n-C₃H₇	CH₃	5–10	3	67	CH₃	CH₃OH	25	28	13	145–157	—	16
i-C₃H₇	CH₃	5–10	2	99	CH₃	CH₃OH–ether	36	3/4	43	135–136	1.4003 (25°)	25
i-C₃H₇	CH₃	5–10	14	99	CH₃	CH₃OH–pet. ether	35	2	70	134–136	—	19
n-C₄H₉	CH₃	5–10	2	63	CH₃	CH₃OH	25	28	12	167–170	1.4090 (24°)	16
n-C₄H₉	CH₃	5–10	2	79	CH₃	CH₃OH–ether	35	3/4	79	164–166	1.4090 (24°)	26
n-C₅H₁₁	CH₃	5–10	4	75	CH₃	CH₃OH	25	5	40	187–190	—	16
i-C₅H₁₁	CH₃	5–10	3	71	CH₃	CH₃OH	25	35	9	178–181	—	16
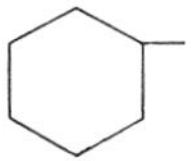 (cyclohexyl)	CH₃	5–10	14	100	CH₃	CH₃OH–pet. ether	65	6	58	82–88	1.4432 (25°)	27
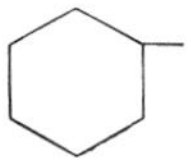 (1-methylcyclohexyl)	CH₃	5–10	14	100	CH₂—CH₂ with OH OH	Pet. ether	35	2	58	81–82.5 (7)	1.4545 (25°)	27

TABLE II

Some Additional Preparations of Ortho Esters by the Alcoholysis of Nitriles

RCN	ROH	Solvent	Acid	Temp. (°C)	Time (hr)	RC(OR')$_3$ (yield %)	B.p., °C (mm Hg) or m.p., °C	n_D^t (°C)	Ref.
HCN	CH$_3$OH	Cl—C$_6$H$_5$	HCl	−5 to −10	3.5	76.6 R = H	98–100 (760)	—	[a]
HCN	CH$_3$OH	CH$_3$OH	HCl	45 1 to 25	12 100	R' = CH$_3$ 50.0	—	—	[b]
HCN	CH$_3$OH	(C$_2$H$_5$)$_2$C$_6$H$_4$	HCN	−10 to 0 15	23 6	R = H, R' = CH$_3$ 77	—	—	[c]
HCN	CH$_3$OH	ClC$_2$H$_5$	HCN	−20 to −15	3	R = H, R' = CH$_3$ 84.9	100	—	[d]
CH$_3$CN	C$_2$H$_5$OH	—	HCN	−5 to 0	2	R = H, R' = CH$_3$ 72.8	(760)	—	[e]
CH$_3$CN	CH$_3$OH (or ROH, R = C$_{2-8}$)	o-Xylene or isooctane	HCl	—	—	R = CH$_3$, R' = C$_2$H$_5$ 90.0	—	—	[f]
RCN R = Me, Et, Me$_2$CH, cyclohexyl, etc.	R'OH R' = Me, Et	R'OH	HCl (pH 3.0, NH$_3$)	5 20 30	1 15 15	R = CH$_3$, R' = CH$_3$ 89	—	—	[g]

[a] J. W. Copenhaver, U.S. Pat. 2,527,494 (1950).

[b] J. G. Erickson, U.S. Pat. 2,567,927 (1951).

[c] G. Kesslin, A. C. Flisik, and R. W. Handy, U.S. Pat. 3,121,751 (1964); U.S. Pat. 3,258,496 (1966).

[d] K. Sennewald, A. Ovorodulk, and H. Neumaier, U.S. Pat. 3,641,161 (1972).

[e] Y. Omura, S. Aihara, F. Morii, Y. Tamai, T. Hosogai, Y. Fujita, F. Wada, T. Nishida, and K. Itoi, Japan Kokai 76,108,011 (1976); Chem. Abstr. 86, 120800h (1977); Y. Omura, F. Yamamoto, K. Kikuchi, T. Kawaguchi, and K. Itoi, Japan Kokai 77,125,108 (1977); Chem. Abstr. 88, 61978p (1978).

[f] H. G. Schmidt, G. Davin, and W. Vogt, Ger. Offen 2,645,477 (1978); Chem. Abstr. 88, 190116e (1978); also U.S. Pat. 4,182,910 (1980).

[g] Nippon Synthetic Chem. Ind. Co. Ltd., Japan Kokai Tokkyo Koho. 80 87,734 (1980); Chem. Abstr. 94, 83603p (1981).

TABLE III

PREPARATION OF IMINO ESTER HYDROCHLORIDES, $RC(OEt)=NH \cdot HCl$

R =	Reaction time at 0°C	Alcohol–nitrile to ether ratio[a]	Hydrochloride (yield %)
CH_3	2 hr	$1:0.5^b$	85–95
C_2H_5	6 hr	1:4	85–95
$n\text{-}C_3H_7$	4 days	1:4	65–70
$i\text{-}C_3H_7$	4 days	1:4	70–90
$n\text{-}C_4H_9$	5 days	1:4	70–80
$i\text{-}C_4H_9$	6 days	1:6	35–40
$ClCH_2$	—	$1:8^b$	80–90

[a] This is the ratio *by volume* of the alcohol–nitrile mixture to ether.

[b] In these runs the ether was added to the alcohol solution of the nitrile before the addition of the hydrogen chloride.

[c] Reprinted from S. M. McElvain and J. W. Nelson, *J. Amer. Chem. Soc.* **64**, 1825 (1942). Copyright 1942 by the American Chemical Society. Reprinted by permission of the copyright owner.

After this time, ether is added in the amounts shown in column 3 for the purpose of preventing the formation of a hard cake of the salt. In the case of very reactive nitriles such as acetonitrile and chloroacetonitrile it is advisable to have the ether present before the hydrogen chloride is added in order to prevent solidification of the reaction mixture. After allowing the reaction mixture to stand for 15–20 hr in a refrigerator, it is cooled to −30°C to hasten crystallization. The product salt is filtered, washed with cold (−40°C) ether, and dried under reduced pressure. The salt at this point should not give an acid reaction toward moistened Congo red test paper.

The ortho esters are prepared as described in Table IV starting with 0.2 mole of the imino ester hydrochloride and 3.0 moles of absolute alcohol (Eq. 13). It will be:

$$RC(\overset{+}{=}NH_2\overset{-}{Cl})OC_2H_5 + C_2H_5OH \xrightarrow{\text{ether}} RC(OC_2H_5)_3 + NH_4Cl \qquad (13)$$

noticed from Table IV that where R = secondary alkyl, the yields of ortho ester are less than those obtained from the straight-chain alkyl. When two α- or β-substituents are present in group R, the yields of ortho ester are only 20–30% [18, 28]. Steric effects are also important in the synthesis of imino esters from aromatic systems [29–31].

The chemistry of imidates and their methods of synthesis have been reviewed [32]. Imidates are mainly available via the Pinner synthesis, via iminochlorides, from amides, from aldehydes and ketones, and via unsaturated systems [32].

The most serious competing reaction involved in the conversion of nitriles to imino ester hydrochloride is the decomposition of the latter to an amide

TABLE IV

Alcoholysis of Imino Ester Hydrochlorides, $RC(OEt)=NH \cdot HCl$

R	Reaction time (hr)	Alcohol–ether ratio	Reaction temp. (°C)	Yield (%)		Properties of $RC(OC_2H_5)_3$			
				NH₄Cl	RC(OEt)₃	B.p. °C	mm	d_4^{25}	n_D^{25}
CH_3	6	1:1	46	100	75–78	144–146	740	—	—
C_2H_5	9	1:2	42	95	75–78	70–72	32	—	—
$n\text{-}C_3H_7$	18	1:3	41	96	60–63	58–59	7	0.875	1.4028
$i\text{-}C_3H_7$	24	1:5	39	54	27–30	50–51	7	0.871	1.4002
$n\text{-}C_4H_9$	12	1:3	42	88	59–61	49–50	3	0.873	1.4086
$i\text{-}C_4H_9{}^a$	28	1:5	39	56	21–23	57–59	7	0.869	1.4056
$ClCH_2$	6	1:0	40	89	70–73	68–70	10	—	—

[a] This ester was prepared by Robert L. Clarke. In addition to the ortho ester a 14% yield of isovaleramide and a 21% yield of ethyl isovalerate were isolated.

[b] Reprinted from S. M. McElvain and J. W. Nelson, *J. Amer. Chem. Soc.* **64**, 1825 (1942). Copyright 1942 by the American Chemical Society. Reprinted by permission of the copyright owner.

and an alkyl chloride [28]. This decomposition can be avoided or minimized by using low reaction temperatures (preferably below about 60°C). The decomposition is favored in highly polar solvents [28].

When R has two or more α-substituents, the amide is the major reaction product [23, 33]. However, if R carries only one substituent such as an alkyl, ethoxy, or halogen, then amide formation is only a minor reaction [23] (Eq. 14).

$$RC(\overset{+}{=}NH_2\overset{-}{Cl})OR \longrightarrow RCONH_2 + RCl \tag{14}$$

2-1. *Preparation of Methyl Orthoformate (One-Step Synthesis)* [20]

$$HCN + CH_3OH + HCl \longrightarrow$$

$$[HC(\overset{+}{=}NH_2\overset{-}{Cl})OCH_3] \xrightarrow[-NH_4Cl]{CH_3OH} HC(OCH_3)_3 \tag{15}$$

CAUTION: Extreme care must be exercised in handling HCN. A well-ventilated hood must be used.

Into a flask containing 410 gm (12.8 moles) of methanol cooled to 0°–5°C is bubbled in anhydrous hydrogen chloride over a $1\frac{1}{2}$ hr period until 55.9 gm (1.53 moles) is dissolved. To this cold solution is rapidly (1 min) added an ice-cold solution of 202 gm (7.50 moles) of hydrogen cyanide in 600 gm (18.8 moles) of methanol. The temperature rises to 9°C and then to 25°C over a $1\frac{1}{2}$ hr period. When the reaction mixture reaches 25°C, it is occasionally cooled to keep the temperature from rising. After $3\frac{1}{2}$ hr at 25°C, ammonium chloride starts to precipitate. After 90 hr at 21°–25°C the reaction mixture is filtered, distilled, and then fractionated through a 4 ft column packed with $\frac{1}{8}$ stainless steel helices to afford 81.0 gm (50.2%), b.p. 102°C, n_D^{25} 1.3770, and 11.0 gm of high-boiling materials of unknown structure.

2-2. Preparation of Ethyl Orthoacetate (One-Step Synthesis) [21]

$$CH_3CN + 3C_2H_5OH \xrightarrow[CHCl_3]{HCl} CH_3C(OC_2H_5)_3 + NH_4Cl \qquad (16)$$

Into a water-cooled mixture of 1025 gm (25.0 moles) of acetonitrile, 1150 gm (25.0 moles) of absolute ethanol and 900 ml of chloroform is slowly bubbled 913 gm (25.0 moles) of anhydrous hydrogen chloride. The temperature rises to 35°–40°C and then is recooled to 20°C. After 48 hr at 20°–25°C, 5 liters of absolute ethanol is added and then the mixture is left for 2 days. The precipitated ammonium chloride is filtered, washed with ethanol, and the filtrate and washings added to 20 liters of 5% sodium hydroxide solution. The product is extracted with chloroform, concentrated and fractionated under reduced pressure to afford 2430–3250 gm (60–80%), b.p. 70°–80°C (60 mm).

2-3. Preparation of Methyl Orthovalerate (Two-Step Synthesis) [26]

$$C_4H_9CN + CH_3OH + HCl \longrightarrow$$

$$C_4H_9C(=\overset{+}{N}H_2\overset{-}{Cl})OCH_3 \xrightarrow{CH_3OH} C_4H_9C(OCH_3)_3 + NH_4Cl \qquad (17)$$

Into a flask containing an ice-cold solution of 166.0 gm (2.0 moles) of valeronitrile and 90 ml (2.2 moles) of anhydrous methanol in 750 ml of anhydrous diisopropyl ether is bubbled dry hydrogen chloride until 78 gm (2.2 moles) is absorbed. The flask is stoppered, placed in the refrigerator for 48 hr, filtered of the white crystals, and the mother liquor concentrated to afford a combined yield of 240 gm (79%) of methyl imidovalerate hydrochloride (after washing with fresh diisopropyl ether, and drying under reduced pressure).

To a solution, at room temperature of 151 gm (1.0 mole) of methyl imidovalerate in 400 ml of anhydrous methanol is added 1300 ml of anhydrous ether and the contents refluxed for 18 hr. The reaction mixture is cooled, filtered of the ammonium chloride (48 gm, 90%), concentrated, and the residue fractionally distilled to afford 129 gm (79%), b.p. 164°–166°C; n_D^{24} 1.4090, d_4^{27} 0.9413. The distillation residue yields 10 gm (10%) of valeramide, m.p. 101°–105°C (recrystallized from ethyl acetate–petroleum ether, b.p. 60°–68°C).

B. Transesterification Reactions of Ortho Esters

Alcohols react with ortho esters by exchange of their alkoxy groups to form new ortho esters. The equilibrium is shifted to the right by removal of the volatile alcohol (or mercaptan) to give high yields. The reaction is usually, but not necessarily, catalyzed by the Friedel-Crafts type of catalyst (Eq. 18). In some cases, only mixed ortho esters are produced, depending on the nature of R′ and

$$RC(XR')_3 + 3R''OH \underset{}{\overset{\text{catalyst}}{\rightleftharpoons}} RC(OR'')_3 + 3R'XH \tag{18}$$

$$X = O \text{ or } S$$
$$R = H, \text{ alkyl, or aryl}$$

R″, reaction time, catalyst, catalyst concentration, and efficiency of removal of the volatile alcohol or mercaptan.

The catalysts generally used are sulfuric, hydrochloric, and p-toluenesulfonic acids, and the reaction probably involves dialkoxycarbonium ion intermediates $[\overset{+}{R}C(XR')_2]$.

Since the reaction is subject to severe steric limitations, the nature of the R′ and R″ is very important. The reaction proceeds readily with primary alcohols, less readily with secondary, and hardly at all with tertiary alcohols.

The transesterification of ethyl orthoformate was earlier reported by Hunter [34], Post [35], and Mkhitaryan [36] and was developed later into a general synthesis by Alexander and Busch [37] (see Table V).

TABLE V[a]

PREPARATION OF TRIALKYL ORTHOFORMATES BY THE UNCATALYZED
TRANSESTERIFICATION OF ETHYL ORTHOFORMATE [37]

Starting alcohol	Orthoformate (yield %)	B.p., °C (mm Hg)	n_D^{20}
1-Propanol	95.3	96–97 (20)	1.4078
2-Methyl-1-propanol	86.0	118–120 (22)	1.4122
2-Propanol	0	—	—
2-Methyl-2-propanol	0	—	—
1-Butanol	86.3	141 (25)	1.4184
3-Methyl-1-butanol	88.2	81–83 (0.3)	1.4251
2-Butanol	87.5	115 (23)	1.4141
1-Pentanol	74.0	101–103 (0.3)	1.4290
1-Hexanol	77.4	127–128 (0.35)	1.4344
1-Octanol	[b]		
Benzyl	[c]	m.p. 8°C	1.5645
Allyl	[d]		

[a] Data taken from Alexander and Busch [37, Table I].

[b] Decomposed on distillation.

[c] Decomposed on distillation, but the product can be purified by crystallization in cold storage (4°–10°C).

[d] Gives a mixture of products difficult to separate by distillation.

2-4. Preparation of 2-Butyl Orthoformate [37]

$$HC(OC_2H_5)_3 + 3CH_3-\underset{\underset{OH}{|}}{CH}-CH_2CH_3 \longrightarrow HC\left(O\underset{\diagdown C_2H_5}{\overset{\diagup CH_3}{CH}}\right)_3 + 3C_2H_5OH \qquad (19)$$

To a 500-ml flask equipped with 14-inch glass-packed distillation column and distillation head are added 74.1 gm (0.5 mole) of ethyl orthoformate and 128.2 gm (2.0 moles) of 2-butanol. The flask is heated for 24 hr or until the ethanol (1.5 moles) is removed. The resulting mixture is distilled under reduced pressure to afford 101.5 gm (87.5%), b.p. 115°C (23 mm), n_D^{20} 1.4141.

Using (+)-(S)-2-butanol having $[\alpha]_D^{25}$ +8.402° (optical purity 61.64%) and letting it react with methyl orthoformate affords a 67% yield of 2-butyl ortho-formate having $[\alpha]_D^{25}$ +28.40° (61.58% optical purity) [38].

2-5. Preparation of Ethyl Orthoformate from Ethyl Orthothioformate [39]

$$HC(SC_2H_5)_3 + 3C_2H_5OH \xrightarrow{ZnCl_2} HC(OC_2H_5)_3 + 3C_2H_5SH \qquad (20)$$

To a 500-ml flask equipped as in Preparation 2-4 are added 98.0 gm (0.5 mole) of ethyl orthothioformate, 92 gm (2.0 moles) of absolute ethanol, and 2.0 gm of fused zinc chloride. The reaction mixture is heated until 83.7 gm (90%) of ethanethiol is removed by distillation at 35°–37°C. The residue is fractionally distilled to afford 49 gm (66%), b.p. 144°–146°C, n_D^{24} 1.3971.

2-Propanol does not react with ethyl orthoformate but reacts with methyl orthoformate to give 75% ortho ester when catalyzed by a small amount of sulfuric acid [40]. Tertiary butyl alcohol under the catalyzed conditions above led to extensive dehydration. Without a catalyst, no reaction occurred in this case after a 48 hr heating period. The best procedure is to use acid catalysts (H_2SO_4) during the removal of either methanol or ethanol and then to neutralize with alkali before distilling. The use of an acid catalyst (H_2SO_4) gave shorter reaction times for the transesterification of ethyl orthoformate with yields equal to or higher than that reported in Table V for primary and secondary alcohols.

2-6. Preparation of n-Propyl Orthoformate [40]

$$HC(OC_2H_5)_3 + 3CH_3CH_2CH_2OH \xrightarrow{H_2SO_4} HC(OC_3H_7)_3 + 3C_2H_5OH \qquad (21)$$

To a 500 ml flask equipped as in Preparation 2-3 are added 54.2 gm (0.37 mole) of ethyl orthoformate, 88.2 gm (1.47 moles) of n-propyl alcohol, and two drops of concentrated sulfuric acid. The mixture is heated, the ethanol (65 mol) is distilled over in approximately 3 hr, and the residue is distilled under reduced pressure to afford 66.0 gm (95%), b.p. 106°–108°C (140 mm), n_D^{25} 1.4058.

Partial alcoholysis of triethyl orthoformate with tertiary butyl alcohol has been reported recently to yield ethyl di-t-butyl and diethyl-t-butyl orthoformate [41].

Alcohols with electronegative substituents in the α-position (phenol [42] and hydroxyacetonitrile [43]) undergo only partial alcoholysis with ethyl orthoformate. The use of ethyl orthoacetate also allows only partial transesterification to take place [42, 44].

2-Chloroethanol with an electronegative substituent in the β-position gives complete transesterification of ethyl orthoformate [44].

The transesterification reaction does not involve fission of the C—O bond of the alcohol, and therefore optically active secondary alcohols yield optically active orthoformates [34, 38, 45–48].

The use of 1,2- and 1,3-diols in the transesterification reaction provides a means of providing monocyclic ortho esters [49–54]. Cycloalkanediols, catechol [54c], or acyclic triols result in bicyclic [55] and polycyclic ortho esters [56, 57] (Eqs. 22, 23).

$$HOCH_2CH_2OH + HC(OC_2H_5)_3 \longrightarrow \quad \underset{\substack{O \quad O \\ H \quad OC_2H_5}}{\boxed{}} \quad + \quad 2C_2H_5OH \qquad (22)$$

$$\underset{—OH}{\overset{—OH}{\bigcirc\hspace{-1.2em}\Big]}} + HC(OC_2H_5)_3 \longrightarrow \underset{\substack{—O \quad H \\ —O \quad OC_2H_5}}{\bigcirc\hspace{-1.2em}\Big]} + 2C_2H_5OH \qquad (23)$$

The reaction of trimethylolethane and trimethylolpropane with triethyl orthoformate in the presence of BF_3 etherate leads to bicyclic orthoformate by a transesterification reaction [58b].

$$CH_3—C(CH_2OH)_3 + (C_2H_5O)_3CH \xrightarrow[\text{etherate}]{BF_3} \underset{\substack{CH_2—O \\ CH_2—O}}{CH_3—C{\overset{\diagup}{\diagdown}}CH_2—O{\overset{\diagup}{\diagdown}}CH} + 3C_2H_5OH$$

$$(24)$$

2-7. *Preparation of 2-Methoxy-1,3-dioxolane* [58a]

$$\underset{\substack{H_2C—OH \\ H_2C—OH}}{|} + \underset{\substack{CH_3O \\ CH_3O}}{\diagdown}CH—OCH_3 \longrightarrow$$

$$\underset{\substack{H_2C—O \\ H_2C—O}}{\diagdown}CH—OCH_3 + 2CH_3OH \qquad (25)$$

To a 500 ml flask equipped as previously described in Preparation 2-3 are added 106.5 gm (1.0 mole) of methyl orthoformate, 65 gm (1.05 moles) of ethylene glycol, and 1 drop of concentrated sulfuric acid. The mixture is heated, the methanol (81 ml, 100%) is removed by distillation, and the residue is distilled to afford 73 gm (70%), b.p. 129.5°C, n_D^{25} 1.4070.

Bicyclic ortho esters are also obtained by starting with bicyclic diols and ortho esters. For example, the reaction of cis-exo-2,3-norbornanediol with triethyl orthoformate and a catalytic amount of benzoic acid gives the bicyclic orthoformate ester in 75% yield as a distilled product, b.p. 47°-52°C (0.025-0.050 mm Hg) [58c].

$$\text{(26)}$$

The synthesis and hydrolysis of other bicyclic ortho esters has also been recently reported [58d].

Steroids [59–69] and a number of carbohydrates [70–73] also react with acyclic ortho esters to give cyclic ortho esters by transesterification.

Alkyl formates or formic acid and its esters can be converted to trialkyl orthothioformates [74–77] which in turn can be converted to trialkyl orthoformates in good yields [78, 79]. It has been reported that acid chlorides of higher carboxylic acids can also be converted to trialkyl orthothioformates [80], but thus far no reports appear in the literature on attempts to convert them to trialkyl ortho esters.

Since tetraalkyl orthocarbonates are related to the trialkyl orthoformates, attempts have been made to apply to them the transesterification reaction conditions discussed above. Results to date are discouraging. The reaction is difficult to drive to completion, and thus only mixtures of orthocarbonates are produced [81].

C. Alcoholysis of Trihalomethyl Compounds and α-Halo Ethers

The reaction of trihalomethyl compounds lacking an α-hydrogen atom with alkoxides or phenoxides affords trialkyl and triaryl orthoformates, trialkyl orthobenzoates, and tetraalkyl orthocarbonates (Eq. 27). Most of these re-

$$\text{R}'\text{CCl}_3 + 3\text{NaOR} \longrightarrow [\text{R}'\text{CCl}_2\text{OR}] \longrightarrow [\text{R}'\text{CCl(OR)}_2] \longrightarrow \text{R}'\text{C(OR)}_3 \quad \text{(27)}$$

R′ = H [35, 82–88], alkyl [89–92], aryl [25, 93–97], halogen [98–100]

R = alkyl [85, 101a], chloroalkyl [101b], aryl [102–105]

actions are thought to occur via carbene intermediates and are a field of active investigation.

When $R' = NO_2$ [108–110a] or SCl [109], the tetraalkyl orthocarbonates are obtained. When $R' =$ halogen, mainly orthoformates are formed except for the recent case describing the reaction of polyfluoroalkanols with carbon tetrachloride catalyzed by ferric chloride to give polyfluoroalkyl orthocarbonates [110b, c] (Eq. 28).

$$4CHF_2(CF_2)_nCH_2OH + CCl_4 \xrightarrow{\text{FeCl}_3} C[OCH_2(CF_2)_nCHF_2]_4 + 4HCl \quad (28)$$

The yields of orthoformates by the reactions above are usually poor; better yields are obtained at elevated temperatures or under special conditions [111].

α-Dihaloalkyl ethers [112] and monohaloalkyl ethers [113, 114] also readily react with alcoholic alkoxides to give ortho esters. (See Table VI.) The former are probably intermediate in the reaction of the trihalomethyl compounds with alkoxides as shown in Eq. (27). These reactions are useful in preparing mixed ortho esters which can be obtained in most cases only by this method (Eqs. 29, 30).

$$R'CX_2OR + 2NaOR'' \longrightarrow R'C(OR)(OR'')_2 \quad (29)$$

$$R'CX(OR)_2 + NaOR'' \longrightarrow R'C(OR)_2(OR'') \quad (30)$$

Alkyl alkoxydichloroacetates undergo a similar reaction with alkoxide to give hemiorthooxalates [115, 116]. (See Table VI.)

$$ROCX_2COOR' + R''OH \longrightarrow ROC(OR'')_2COOR' \quad (31)$$

2-8. Preparation of Methyl Orthoformate [117]

$$3NaOCH_3 + CHCl_3 \xrightarrow{\text{solvent}} HC(OCH_3)_3 + 3NaCl \quad (32)$$

A mixture consisting of 100 gm (1.85 moles) of powdered sodium methoxide suspended in 120 gm of benzene (or crude methyl orthoformate) is heated to 50°C and then 74 gm (0.62 mole) of chloroform is added dropwise over a 1 hr period, the reaction temperature being kept between 60°–80°C. The precipitated sodium chloride is removed by filtration and the filtrate distilled to afford 552 gm (84%), b.p. 103°–105°C, n_D^{25} 1.3770.

Ethyl orthoformate is prepared in 45% yield by adding sodium metal portionwise to a mixture of excess absolute alcohol and chloroform [118]. (See Kaufmann and Dreger [118a] for earlier references to this reaction.) A recent reference describes this preparation from chloroform with 3 moles of EtONa in excess EtOH [118b].

TABLE VI

REACTION OF TRIHALOMETHYL COMPOUNDS AND HALOGENATED ETHERS WITH ALKOXIDES OR CARBOXYLATES TO GIVE ORTHO ESTERS OR CARBOXYLATE DERIVATIVES

$RCCl_3$	$RCCl_2OR$	$RCCl(OR)_2$	RCOONa	RONa	Product	Yield (%)	B.p., °C (mm Hg)	n_D (°C)	Ref.
$HCClF_2$	—	—	—	CH_3ONa	$(CH_3O)_3CH$	—	98–99	1.377 (25)	a
$HCCl_3$	—	—	—	C_2H_5ONa	$(C_2H_5O)_3CH$	45	140–146	1.391 (25)	b
$HCCl_3$	—	—	—	CH_3ONa	$(CH_3O)_3CH$	84	103–105	—	c–e
$HCCl_3$	—	—	—	$o\text{-}CH_3C_6H_4ONa$	$(o\text{-}CH_3C_6H_4O)_3CH$	8.1	m.p. 96°C	—	f
$C_6H_5CCl_3$	—	—	—	C_2H_5ONa	$C_6H_5C(OC_2H_5)_3$	22	108–112 (13)	1.4930 (25)	g
	$CHCl_2OCH_3$	—	—	CH_3ONa	$(CH_3O)_3CH$	43	100.5–101	1.3787 (20)	h
	$CHCl_2OCH_3$	—	—	C_2H_5ONa	$(C_2H_5O)_2(CH_3O)CH$	56	133–134	1.3868 (20)	h
	$CHCl_2OCH_3$	—	—	$n\text{-}C_3H_7ONa$	$(C_3H_7O)_2(CH_3O)CH$	58	61.5–62 (11)	1.4010 (20)	h
	$CHCl_2OCH_3$	—	—	$cycl\text{-}C_6H_{11}ONa$	$(C_6H_{11}O)_2(CH_3O)CH$	36	107–109 (0.4)	1.4671 (20)	h
	$CHCl_2OCH_3$	—	—	C_6H_5ONa	$(C_6H_5O)_2(CH_3O)CH$	41.5	114 (0.05)	1.5517 (20)	h
	$CHCl_2OCH_3$	—	CH_3COONa	—	$(CH_3COO)_2CHOCH_3$	61	85 (7)	1.4052 (20)	h
	$CHCl_2OCH_3$	—	C_2H_5COONa	—	$(C_2H_5COO)_2CHOCH_3$	61	100–101.5 (12)	1.4136 (20)	h
	$CHCl_2O\text{-}n\text{-}C_4H_9$	—	CH_3COONa	—	$CH_3COO)_2CHO\text{-}n\text{-}C_4H_9$	53	111–112 (12)	1.4153 (20)	h
	[benzo-1,3-dioxole, 2,2-dichloro (catechol with O–CCl₂–O ring)]	—	—	C_2H_5ONa	[benzo-1,3-dioxole, 2,2-bis(ethoxy), O–C(OC₂H₅)₂–O ring]	60	123 (15)	1.4943 (20)	i
NO_2CCl_3	—	—	—	C_2H_5ONa	$(C_2H_5O)_4C$	46–49	158–161	1.3905 (25)	j
$Cl_3C\text{—}CCl_3$	—	—	—	CH_3ONa	$(CH_3O)_4C$	48	113.5	1.3858 (20)	k
CCl_4	—	—	—	{ $H(CF_2)_6CH_2OH$; $FeCl_3$ }	$[H(CF_2)_6CH_2O]_4C$	35	170 (0.008)	—	l
		$(C_6H_5O)_2CHCl$	—	{ Et_3N ; CH_3OH }	$(C_6H_5O)_2CH\text{—}OCH_3$	86	145 (3)	—	m
		$(C_6H_5O)_2CHCl$	—	{ Et_3N ; C_2H_5OH }	$(C_6H_5O)_2CHOC_2H_5$	84	150 (3)	—	m
		$(C_6H_5O)_2CHCl$	—	{ Pyridine ; C_6H_5OH }	$(C_6H_5O)_3CH$	96	m.p. 77°C	—	m

a J. Hine and J. J. Porter, J. Amer. Chem. Soc. 79, 5493 (1957).
b W. E. Kaufmann and E. E. Dreger, Org. Syn. Coll. Vol. I, 258 (1932).
c C. Lenz, K. Hass, and H. Epler, Ger. Pat. 1,217,943 (1966).
d Feldmuehle Papier and Zellstoffwerke A.G., Belg. Pat. 613,988 (1962).
e T. A. Weidlich and W. Schulz, Ger. Pat. 919,465 (1954).
f J. E. Driver, J. Amer. Chem. Soc. 46, 2090 (1924).
g S. M. McElvain, H. I. Anthes, and S. H. Shapiro, J. Amer. Chem. Soc. 64, 2525 (1942).
h H. Gross and A. Rieche, Chem. Ber. 94, 538 (1961).
i H. Gross, J. Rusche, and H. Bornowski, Ann. Chem. 675, 146 (1964).
j J. D. Roberts and R. E. McMahon, Org. Syn. Coll. Vol. 4, 457 (1963).
k H. Tieckelmann and H. W. Post, J. Org. Chem. 13, 265 (1948).
l M. E. Hill, D. T. Carty, D. Tegg, J. C. Butler, and A. F. Strong, J. Org. Chem. 30, 411 (1965).
m H. Scheibler and M. Depner, J. Prakt. Chem. 7, 60 (1958); Chem. Ber. 68B, 2151 (1935).

2-9. Preparation of Ethyl Orthobenzoate [96]

$$C_6H_5CCl_3 + 3C_2H_5OH + 3Na \longrightarrow C_6H_5C(OC_2H_5)_3 + 3NaCl \qquad (33)$$

To a freshly prepared solution of 3 moles of sodium ethoxide in 1200 ml of absolute ethanol is added dropwise benzotrichloride until 195 gm (1.0 mole) has been added. The reaction mixture is stirred for 5–6 hr without heating and then refluxed for 11 hr. The salt is filtered off, washed with ether, and the combined ether washings and filtrate are concentrated. Distillation of the residue affords 50 gm (22%), b.p. 108°–112°C (13 mm). The yield is poor, and further work is required to ascertain the optimum conditions for this reaction. Whether other products are produced in the reaction was not reported in this earlier investigation.

2-10. Preparation of 2,2-Diethoxy-1,3-benzodioxolane [119a]

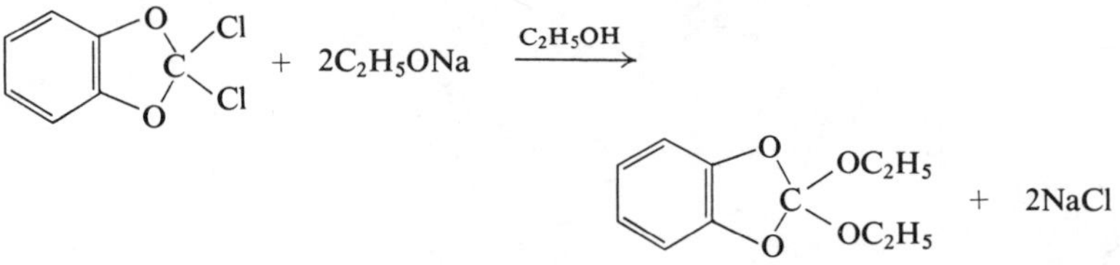

To a flask containing 45 ml (0.77 mole) of absolute ethanol is added portionwise 2.5 gm (0.11 gm-atom) of sodium metal. While stirring vigorously and cooling, 9.55 gm (0.05 mole) of 2,2-dichlorobenzo-1,3-dioxolane in 25 ml of ether is added dropwise. After 18 hr the salt is filtered off, the filtrate diluted with ether, washed with cold water, dried over potassium carbonate, and fractionally distilled to afford 6.4 gm (60%), b.p. 123°C (15 mm), n_D^{20} 1.4943.

The reaction of trichloroacetonitrile and 4 moles of alkali metal or alkaline earth metal salt of alcohols containing at least one α-hydrogen is reported to give orthocarbonic acid esters [119b].

3. CONDENSATION REACTIONS

A. Condensation of Alcohols with Substituted Olefins and Acetylenes to Give Ortho Esters

a. ADDITION OF ALCOHOLS TO KETENE ACETALS

Alcohols add to certain substituted olefins (ketene acetals such as 1,1-diaryloxy-1-alkenes) to form ortho esters [120a–c]. The reaction has been reported to be acid-catalyzed [121a, b] but it may be carried out under basic conditions [120a–c, 122a, b, 123a, b] without any other catalysts.

$$RR'C{=}C(OR'')_2 + R''OH \longrightarrow RR'CHC(OR'')_2OR''' \qquad (35)$$

The addition of alcohols to ketene acetals allows the synthesis of mixed ortho esters [96, 120a–c, 121a, b, 124, 125a, b]. α-Haloaldehydes may be converted to ortho esters by the following process: (a) acetal formation, (b) dehydrohalogenation, and (c) reaction with alcohols via addition reaction (33). In general, the method above, using ketene acetals, is not practical since ketene acetals are not readily available and are difficult to prepare. However, the method is useful because it allows the synthesis of mixed ortho esters and other ortho esters more difficult to synthesize [122–127]. Recently a simple one-step synthesis of ketene acetals and ortho esters has been reported (see p. 65).

3-1. Preparation of Divinyl β-Nitroethyl Orthoacetate [126]

$$H_2C=C(OCH=CH_2)_2 + O_2NCH_2CH_2OH \xrightarrow{H^+}$$
$$CH_3C(OCH=CH_2)_2OCH_2CH_2NO_2 \quad (36)$$

To a 50 ml flask immersed in an ice bath are added 6.0 gm (0.053 mole) of ketene divinyl acetal and 2 drops of a solution of hydrochloric acid in nitroethanol (0.1 gm concentrated HCl/ml nitroethanol). Nitroethanol, 4.8 gm (0.053 mole), is added dropwise with stirring, and after 30 min of reaction the mixture is distilled under reduced pressure to afford 5.6 gm (52%), b.p. 80°–85°C (2 mm), n_D^{25} 1.4492.

The orthoacetate above polymerizes by heating alone or in the presence of benzoyl peroxide to give a dark, viscous liquid. A solid polymer, of unknown structure, may be isolated by dissolving the latter in acetone and precipitating in cold water.

Divinylethyl orthoacetate is prepared in 80% yield in a similar manner by reacting ketene divinyl acetal and ethanol under acid catalysis at 0°–5°C. (See Table VII [124].)

b. Addition of Alcohols to Alkoxyacetylenes

Ethoxyacetylene reacts with refluxing ethanolic sodium ethoxide to give a 44% yield of triethyl orthoacetate [128a]. In addition, aldoximes and ketoximes react with ethoxyacetylene to give orthoacetic acid derivatives shown in Eqs. (37) and (38) [129]. More work is required to establish the generality of this reaction.

$$HC\equiv C-OC_2H_5 + 2C_2H_5OH \xrightarrow{C_2H_5O^-} CH_3C(OC_2H_5)_3 \quad (37)$$

$$\xrightarrow[\substack{RCH=N-OH \\ or \\ RR'C=N-OH}]{} CH_3C(OC_2H_5)(ON=CRR')_2 \quad (38)$$
$$R = or \neq R'$$

Recently it was reported that the addition of phenol to ethoxyethyne gives orthoacetates when catalyzed by Hg(OAc)$_2$ or ZnCl$_2$ [128b].

TABLE VII

REACTIONS OF KETENE ACETALS WITH ALCOHOLS

Ketene acetal	Alcohol	Product	Yield (%)	B.p., °C (mm Hg)	n_D (°C)	Ref.
$CH_2=C(OC_6H_5)_2$	C_6H_5OH	$CH_3C(OC_6H_5)_3$	78	148–153 (0.5) m.p. 61°–62°C	—	[a]
$CH_2=C(OCH=CH_2)_2$	C_2H_5OH	$CH_3C(OCH=CH_2)_2(OC_2H_5)$	80	144–145.4	1.4221 (25)	[b]
	$O_2N{-}CH_2CH_2OH$	$CH_3C(OCH=CH_2)_2(OCH_2CH_2NO_2)$	52	80–85 (2.0)	1.4492 (25)	[c]
$CH_2=C(OCH{-}CH_2OCH_3)_2$ $\;\;\;\;\;\;\;\;\;\;\;\|$ $\;\;\;\;\;\;\;\;\;CH_3$	$CH_3{-}CHCH_2OCH_3$ $\;\;\;\;\;\;\|$ $\;\;\;\;\;OH$	$CH_3C\!\left(OCH{-}CH_2OCH_3\atop CH_3\right)_3$	74	78–80 (0.25)	—	[d]
$Br{-}CH=C(OC_2H_5)_2$	C_2H_5OH	$\begin{cases} CH_3C(OC_2H_5)_3 \\ BrCH_2C(OC_2H_5)_3 \end{cases}$	$\begin{cases} 52 \\ 42 \end{cases}$	68–70 (50) 78–80 (10)	— —	[e] —

[a] S. M. McElvain and B. F. Pinzon, J. Amer. Chem. Soc. **67**, 650 (1945).
[b] S. M. McElvain and A. N. Bolstad, J. Amer. Chem. Soc. **73**, 1988 (1951).
[c] H. Feuer and W. H. Gardner, J. Amer. Chem. Soc. **76**, 1375 (1954).
[d] W. C. Kuryla and D. G. Leis, J. Org. Chem. **29**, 2773 (1964).
[e] S. M. McElvain and P. M. Walters, J. Amer. Chem. Soc. **64**, 1963 (1942).

c. Reaction of 1,1-Disubstituted Olefins with Alcohols

Several 1,1-disubstituted olefins with electrophilic groups react with alcohols to undergo addition reactions, and with alkoxide ions to undergo substitution reactions to afford ortho esters [120a–c, 125a, b, 130–137] (Eq. 39).

$$RRC{=}CX_2 + ROH + 2RONa \longrightarrow RRCH{-}C(OR)_3 + 2NaX \qquad (39)$$

$$X = Cl, I, SO_2C_2H_5, CN$$
$$R = H, alkyl, aryl, fluorine, or X$$

These reactions may first form the ketene acetals which react with alcohols to give the ortho esters. The generality of these reactions is still unknown; thus further research on these methods is required.

Kuryla and Leis [125a] recently reported that ortho esters are readily produced by the slow addition of vinylidene chloride to a sodium β-alkoxy-alcoholate, dissolved or suspended in a solvent. The reaction is exothermic and produces either the ketene acetal or the ortho ester derivative while sodium chloride precipitates (Eq. 40).

$$2ROCH_2CH_2{-}\overset{-}{O}\overset{+}{N}a + CH_2{=}CCl_2 \xrightarrow[100°\text{--}130°C]{NaCl} (ROCH_2CH_2O)_2C{=}CH_2 + NaCl$$

$$\xrightarrow{ROCH_2CH_2OH} (ROCH_2CH_2O)_3C{-}CH_3 \quad (40)$$

Alcohols used as both reactant and solvent (see Eq. 40).

Tetrahydrofurfuryl alcohol	Only ortho ester
2-Ethoxyethanol	isolated
Diethylene glycol monomethyl ether	
Methoxypropanol-2	Only ketene acetal isolated
CH$_3$OH	
CH$_3$CH$_2$OH	
CH$_3$CH$_2$CH$_2$CH$_2$OH	No reaction
CH$_3$CH$_2$CH$-$CH$_3$	
OH	

Earlier, Ruh [120a] reported a similar reaction for 1,2-dichloro-1,2-difluoro-ethylene on reaction with sodium ethoxide to give ethyl chlorofluoroortho-acetate (Eq. 41).

$$FClC{=}CClF + NaOC_2H_5 + C_2H_5OH \rightarrow (C_2H_5O)_3C{-}CHFCl + C_2H_5OCF{=}CFCl$$
$$\tag{41}$$

B. Halogenation of the Acyl Substituents of Ortho Esters

In the presence of pyridine, direct bromination of ortho esters having one or more α-hydrogen atoms affords α-bromo or dibromo ortho esters. (See Table VIII.) The bromo esters can be converted to the iodo esters by heating the

TABLE VIII

REACTION OF BROMINE WITH ORTHO ESTERS

Ortho ester (moles)	Br_2 (moles)	Pyridine (moles)	Product	Yield (%)	B.p., °C (mm Hg)	n_D (°C)	Ref.
$CH_3C(OC_2H_5)_3$ (0.25)	0.25	0.25	$BrCH_2C(OC_2H_5)_3$	74	77–79 (9)	1.4393 (25)	[a]
$CH_3C(OC_2H_5)_3$ (1.0)	2.0	2.0	$Br_2CHC(OC_2H_5)_3$	53	102–104 (8)	1.4691 (25)	[b]
$C_3H_7CH_2C(OCH_3)_3$ (1.0)	1.0	1.1	$C_3H_7CHBrC(OCH_3)_3$	79	92–93 (15)	1.4507 (25)	[c]

[a] F. Beyerstedt and S. M. McElvain, J. Amer. Chem. Soc. **59**, 1273 (1937).
[b] S. M. McElvain and P. M. Walters, J. Amer. Chem. Soc. **64**, 1963 (1942).
[c] S. M. McElvain, R. E. Kent, and C. L. Stevens, J. Amer. Chem. Soc. **68**, 1922 (1946).

former with an alcoholic solution of sodium iodide. These halogenated ortho esters react with magnesium metal to give a nondistillable mixture. From ethyl orthobromoacetate and magnesium there were obtained 29 % ethyl bromide, trace amounts of ethyl alcohol, ethyl acetate, and mainly a nondistillable residue (Eq. 42).

$$RR'CH(OR'')_3 + Br_2 \xrightarrow{\text{pyridine}} RR'CBrC(OR'')_3 \longrightarrow$$

$$(\text{If } R' = H) \downarrow 2Br_2 \qquad \qquad \downarrow NaI \qquad \qquad \Big| Mg \qquad (42)$$

$$R—CBr_2(OR'')_3 \qquad RR'CIC(OR'')_3 \qquad \overset{\downarrow}{} \text{nondistillable residue}$$

$$\underset{Mg}{\underline{}} \nearrow \qquad \text{(polymer ?)}$$

3-2. Preparation of Tri(2-ethoxyethyl) Orthoacetate [125a]

$$2C_2H_5OCH_2CH_2ONa + CH_2=CCl_2 \xrightarrow{C_2H_5OCH_2CH_2OH}$$

$$(C_2H_5OCH_2CH_2O)_3CCH_3 + 2NaCl \quad (43)$$

To a flask containing 1000 gm (11.1 moles) of 2-ethoxyethanol under a nitrogen atmosphere is added portionwise 100 gm (4.35 gm-atoms) of sodium metal. The resulting sodium alcoholate solution is warmed to 100°C, and 250 gm (2.58 moles) of vinylidene chloride is slowly added with rapid stirring. The reaction is exothermic and the temperature rises to 160°C while sodium chloride is being precipitated. The reaction mixture is filtered and the salt is washed with ether and dried to afford 230 gm (91 %). The filtrate is concentrated and the residue distilled under reduced pressure to afford 362 gm (56.6%), b.p. 98°–100°C (0.5 mm).

3-3. Preparation of Ethyl Orthobromoacetate [138]

$$CH_3C(OC_2H_5)_3 + Br_2 + C_5H_5N \longrightarrow Br—CH_2C(OC_2H_5)_3 + C_5H_5\overset{+}{N}H\overset{-}{B}r \quad (44)$$

To a stirred mixture of 20.8 gm (0.25 mole) of pyridine and 40.5 gm (0.25 mole) of ethyl orthoacetate is added dropwise 40 gm (0.25 mole) of bromine over a period of $\frac{1}{2}$ hr while the temperature is kept at 10°C. The pale yellow brominated ester is filtered, the pyridine hydrobromide washed with ether, the ether washings combined with the bromo ester and then distilled to afford 44 gm (74%), b.p. 77°–79°C (9 mm), n_D^{25} 1.4393, d_{25}^{25} 1.2639. A small amount of ethyl orthodibromoacetate, b.p. 102°–104°C (8 mm), n_D^{25} 1.469, d_{25}^{25} 1.5272, is also isolated.

Using 2 moles of bromine and 2 moles of pyridine per mole of ethyl orthoacetate, a 53% yield of ethyl orthodibromoacetate can be isolated [123a].

4. ELIMINATION REACTIONS

Elimination Reactions Involving the Alkoxy Substituents of Ortho Esters

Alkoxy substituents containing reactive groups can also undergo elimination
reactions. For example, 2-chloroethyl groups can be converted to 2-vinyl
groups [124, 139]. The 2-chloroethyl group can also react with a variety of
groups such as trimethylamine to give quaternary salts [140]. (See Eqs. 45,
46.)

$$RR'CHC(OCH_2CH_2Cl)_3 \longrightarrow RR'CHC(OCH=CH_2)_3 \qquad (45)$$

$$\xrightarrow{3(CH_3)_3N} RR'CH[OCH_2CH_2\overset{+}{N}(CH_3)_3]_3\ 3Cl^- \qquad (46)$$

Stettler and Reske reported that potassium t-butoxide is an effective base
for the dehydrohalogenation reaction [139b].

5. REDUCTION REACTIONS

A. Reduction Reactions Involving the Acyl Substituents of Ortho Esters

The ortho ester function is quite stable under neutral and basic conditions
and is resistant to catalytic reduction [141]. However, acyl substituents with
acid chloride [142] are reduced to aldehyde, and unsaturated centers [143–145]
are saturated. The acyl side chain can also undergo the usual addition and
substitution reactions without affecting the ortho ester function [145–147].
(See Eqs. 42, 47–49, and Table VIII.)

$$Cl-COCH_2-C(OR)_3 \xrightarrow{LiAlH(O-t-Bu)_3} OCH-CH_2C(OR)_3 \qquad (47)$$

$$RCH=CH-CH_2C(OR)_3 \xrightarrow[H_2]{Raney\ Ni} RCH_2CH_2CH_2C(OR)_3 \qquad (48)$$

$$R'CH_2C(OR'')_3 + RCCl_2CHO \longrightarrow \underset{\underset{Cl\ \ OH}{|\quad\ |}}{\overset{\overset{Cl}{|}}{RC}-CH-CHR'C(OR'')_3} \qquad (49)$$

5-1. Preparation of 1-Methoxy-2,10-dioxabicyclo[4.4.0]decane [144]

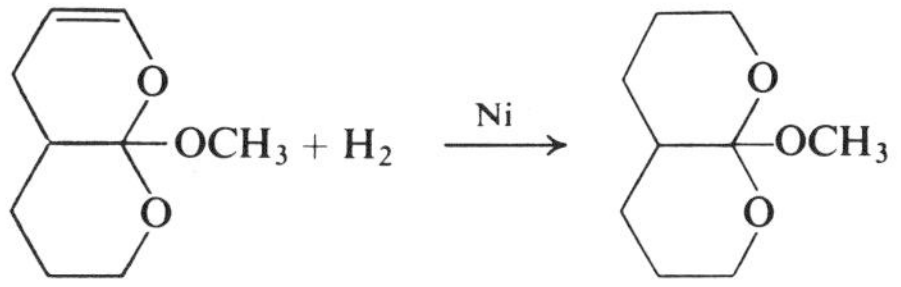

$$\text{(50)}$$

To a glass pressure bottle are added 275 gm (1.62 moles) of 1-methoxy-2,10-dioxabicyclo[4.4.0]-3-decene and 2 tablespoons of W-7 Raney nickel [148] which has been washed free of ethanol with anhydrous ether. Over a 24 hr period a hydrogen atmosphere is maintained at 1–3 atm. The catalyst is removed by centrifugation, washed with ether, and the combined organic layer distilled to afford 268 gm (96.3%), b.p. 38°C (0.1 mm), n_D^{25} 1.4653, d_4^{25} 1.087.

B. Reduction Reactions Involving the Alkoxy Substituents of Ortho Esters

2-Cyanoethyl groups can be conveniently reduced to 2-aminoethyl groups using Raney nickel–hydrogen to give 2-aminoethyl-substituted ortho esters [149] (Eq. 51).

$$RC(OC_2H_5)_2OCH_2CH_2CN \xrightarrow[\text{Ni}]{H_2} RC(OC_2H_5)_2OCH_2CH_2NH_2 \qquad \text{(51)}$$

6. MISCELLANEOUS METHODS

(1) Reaction of alcohols with dioxolenium salts [150].

(2) Reaction of alcohols with dioxenium salts [151–153].

(3) Reaction of alcohols with O-alkyllactonium salts [154, 155].

(4) Reaction of formate esters with epoxides to give orthoformates [156, 157].

(5) Preparation of spirocyclic ortho esters by the addition of epoxides to γ- and δ-lactones using BF₃ or SnCl₄ as catalyst [158, 159].

(6) Preparation of 2-alkoxy-1,3-dioxolanes and 2-alkoxy-1,3-dioxanes from acyloxyarene sulfonates [160–163].

(7) Preparation of polycyclic ortho esters and hydrogen ortho esters by acylation of diols and triols [164–170].

(8) Reaction of dialkoxycarbenes with alcohols to give carboxylic ortho esters [171–174].

(9) Transesterification of ortho esters using an acid (sulfonated polystyrene) ion-exchange resin [175].

(10) Addition of ketene acetals to α,β-unsaturated carbonyl compounds to give ortho esters [18, 144].

(11) Reaction of 2,2-dichloro-3,3-dimethylcyclopropanone dimethyl ketal with trimethyl orthoisobutyrate to give trimethyl α-chloro-β-methylorthocrotonate [176].

(12) Reaction of 2,2-dichlorocyclopropanone ketals with potassium *t*-butoxide to form dialkyl *t*-butyl orthopropiolates [176].

(13) Reaction of *N,N*-dimethyl-*N*-allyl-2,2,3,3-tetrachlorobutylammonium bromide with ethanolic sodium ethoxide to give trialkyl orthopropiolate [177].

(14) Reaction of diazomethyl ketone with ketene acetals to form 2,2-dialkoxy-1,2-dihydropyrans [178].

(15) Reaction of ketene acetals with nitrones to give 5,5-dialkoxyisoxazolidines [179].

(16) Addition of ketene dimethylacetals to 2,2-dimethylcyclopropane to give 1,1-dimethyl-5,5-dimethoxy-4-oxaspiro[2.3]hexene [180].

(17) Preparation of α-iminoorthothiocarboxylates [181].

(18) Preparation of ortho esters by the reaction of diphenylacetylene, chloroform, and ethylene oxide at elevated temperatures in the presence of tetramethylammonium bromide [182].

(19) Preparation of 1,1-dimethoxy-4-oxoisochroman by the reaction of methyl and phenyl *o*-diazoacetylbenzoates with methanolic HCl [183].

(20) Preparation of polycyclic ortho esters by the reaction of 1,2-hydroxybenzene thionocarbonate with trimethyl phosphite [184, 185].

(21) Preparation of tetraalkyl orthocarbonates by the reaction of phenyl cyanate with alcohols in the presence of acids [186].

(22) Preparation of polycyclic orthocarbonate by reaction of 2,2-dihydroxybiphenyl with thiophosgene [187].

(23) Preparation of tetramethyl monoorthooxalate by the oxidation of tetramethoxyethylene with oxygen [188].

(24) Preparation of tetramethyl thionomonoorthooxalate by reaction of tetramethoxyethylene with sulfur in chloroform solution [188].

(25) Preparation of 2,7,8-trioxabicyclo[3.2.1]octanes by the oxidation of 2,7-dioxabicyclo[2.2.1]heptanes with *m*-chloroperbenzoic acid [189].

(26) Preparation of hexamethyl orthomuconate by the electrochemical methoxylation of 1,2-dimethoxybenzene [190].

(27) Preparation of trimethyl ortholevulinate by the reaction of 2,5-dimethoxy-2,5-dihydro-2-methylfuran with methanolic HCl [191].

(28) Preparation of trimethyl perfluoroorthocarboxylates by the methylation of the sodium methoxide addition products of methyl perfluoroalkanoates [192].

(29) Thermal decomposition of 1,2,3,4-tetrachloro-7,7-dimethoxy-5-phenylbicyclo-[2.2.1]hepta-2,5-diene in the presence of methanol yields trimethyl orthoformate [193].

(30) Preparation of peroxy ortho esters by the reaction of trialkyl ortho-formates, orthoacetates, orthobenzoates, ketene acetals, and 2-phenyl-2-methoxy-1,3-dioxolanes with hydroperoxides or oxygen [194–197].

(31) Preparation of acyclic acyloxy ortho esters by the reaction of acid anhydrides with trialkyl orthoformates and trialkyl orthoacetates [198–200].

(32) Preparation of tetraethyl monoortho-2-carbethoxysuccinate by the reaction of diethyl maleate with triethyl orthoformate [201].

(33) Preparation of 1-p-nitrophenyl-4-p-nitrobenzoxy-2,6,7-trioxabicyclo-[2.2.2]octane by the reaction of 2-phenyl-5,5-bis(p-nitrobenzoxymethyl)-1,3-dioxane with p-nitrophenylhydrazine and acetic acid [202].

(34) Preparation of alkoxydiacyloxymethanes by reaction of carboxylic acids with alkyl dichloromethyl ethers in the presence of tertiary amines [203].

(35) Use of acidic ion-exchange resins for the transesterification of ortho esters [175].

(36) Formation of mixed ortho esters by a disproportionation reaction between two ortho esters [35].

(37) Disproportionation reactions of ortho esters [203].

(38) Reaction of dihalo- and trihaloacetic acid and α,α-dihalopropionic acid with trimethylolethane (2-hydroxymethyl-2-methyl-1,3-propanediol) to give bicyclic ortho esters [204].

(39) Reaction of 1,2-diiodoacetylene with ethanolic sodium ethoxide to form triethyl iodoorthoacetate in low yield [205].

(40) Tschitschibabin synthesis involving the reaction of a Grignard reagent with ethyl orthocarbonate [18, 206].

(41) Preparation of triethyl orthoacetate by the reaction of ethyl 1,1-diazido-ethyl ether with sodium ethoxide [207].

(42) Reaction of 1,2-diaminoethane and 1,3-diaminopropane with 1,2-diethoxy-1,1,2,2-tetrahaloethanes in ethanol solution to give heterocyclic ortho esters [208].

(43) Reaction of 2-chloromethylene-1,3-dioxolane to give ortho esters [209].

(44) Preparation of 2-(2-chloroethyl)-1,3-dioxolane by the low-temperature photochlorination of 1,3-dioxolane in the presence of ethylene oxide [210].

(45) Insertion reactions of methylene with methyl orthoformate or 2-methoxy-1,3-dioxolane in the gas phase and in solution [211].

(46) Preparation of spiro ethercarbonates [212, 213].

(47) Preparation of orthocarbonates from thallous alkoxides and carbon disulfide [214].

(48) Preparation of orthocarboxylic acid ester by reacting trichloromethyl isocyanide dichloride with hydroxy compounds [215].

(49) A novel method of preparing orthocarboxylates from dithiocarboxy-lates and dialkoxydibutylstannanes [216].

(50) Preparation of ortho esters by the electrochemical methoxylation of acetals [217].

(51) Reaction of methyl perfluoroacetate with sodium methylate followed by reaction with methyl sulfate to give $CF_3C(OCH_3)_3$ [218].

(52) Preparation of m-iodoorthothiobenzoate by the reaction of m-iodobenzoyl chloride with ethanethio and $AlCl_3$. Treatment of the ortho thioester with silver nitrate in methanol gives trimethyl m-iodoorthobenzoate [219].

(53) Preparation and separations involving chiral ortho esters [220].

(54) Synthesis of aryl ortho esters from benzanilide acetals [221].

(55) Meerwein ortho ester synthesis [222, 223].

REFERENCES

1. W. Colles, *J. Chem. Soc.* **89**, 1246 (1906).
2a. H. W. Post, "Chemistry of the Aliphatic Ortho Esters," pp. 11–44. Reinhold, New York, 1943.
2b. E. H. Cordes, *In* "Chemistry of Carboxylic Esters" (S. Patai, ed.), pp. 623–667. Wiley (Interscience), New York, 1969.
3. R. H. DeWolfe, "Carboxylic Ortho Acid Derivatives." Academic Press, New York, 1970.
4. D. H. R. Barton, C. J. W. Brooks, and J. S. Fawcett, *J. Chem. Soc.* 2137 (1954).
5. S. M. Kupchan, S. P. Erickson, and Y. T. S. Liang, *J. Amer. Chem. Soc.* **88**, 347 (1966).
6. A. Stoll and E. Seebeck, *Helv. Chim. Acta* **37**, 824 (1954).
7. H. Auterhoff and H. Möhrle, *Arch. Pharm.* **291**, 288 (1958).
8. Z. Valenta and K. Wiesner, *Experientia* **18**, 111 (1962).
9. P. A. Diassi and J. Fried, U.S. Pat. 3,073,817 (1963).
10. J. P. Dusza and S. Bernstein, U.S. Pat. 3,069,419 (1962).
11a. S. Inoue and T. Kataoka, Toyo Rayon Co., Ltd., Japan. Pat. 3708 (1965).
11b. D. H. Chadwick, G. K. Rockstroh, and E. L. Powers, U.S. Pat. 3,535,359 (1970).
12. A. Pinner, *Chem. Ber.* **16**, 352 (1883).
13. H. Reitter and E. Hess, *Chem. Ber.* **40**, 3020 (1907).
14 P. P. T. Sah, *J. Amer. Chem. Soc.* **50**, 516 (1928).
15. F. Sigmund and S. Herschdörfer, *Monatsh. Chem.* **58**, 280 (1931).
16. L. G. S. Brooker and F. L. White, *J. Amer. Chem. Soc.* **57**, 2480 (1935).
17. P. P. T. Sah, S. Y. Ma, and C. H. Kao, *J. Chem. Soc.* 305 (1931).
18. S. M. McElvain and J. W. Nelson, *J. Amer. Chem. Soc.* **64**, 1825 (1942).
19. S. M. McElvain and C. L. Aldridge, *J. Amer. Chem. Soc.* **75**, 3987 (1953).
20. J. G. Erickson, *J. Org. Chem.* **20**, 1573 (1955).
21. VEB Filmfabrik Agfa Wolfen, Belg. Pat. 617,666 (1962).
22. S. M. McElvain and J. P. Schroeder, *J. Amer. Chem. Soc.* **71**, 40 (1949).
23. S. M. McElvain and B. F. Pinzon, *J. Amer. Chem. Soc.* **67**, 690 (1945).
24. S. M. McElvain and C. L. Stevens, *J. Amer. Chem. Soc.* **69**, 2663 (1947).
25. S. M. McElvain and J. T. Venerable, *J. Amer. Chem. Soc.* **72**, 1661 (1950).
26. S. M. McElvain, R. E. Kent, and C. L. Stevens, *J. Amer. Chem. Soc.* **68**, 1922 (1946).
27. S. M. McElvain and R. E. Starn, Jr., *J. Amer. Chem. Soc.* **77**, 4574 (1955).
28. S. M. McElvain and B. E. Tate, *J. Amer. Chem. Soc.* **73**, 2233 (1951).
29. G. D. Lander and F. T. Jewson, *J. Chem. Soc.* **83**, 766 (1903).
30. A. Pinner, *Chem. Ber.* **23**, 2917 (1890).

31. F. E. King, K. G. Latham, and M. W. Partridge, *J. Chem. Soc.* 4268 (1952).
32. R. Roger and D. G. Neilson, *Chem. Rev.* **61**, 179 (1961).
33. P. E. Peterson and C. Niemann, *J. Amer. Chem. Soc.* **79**, 1389 (1957).
34. H. Hunter, *J. Chem. Soc.* **125**, 1389 (1924).
35. H. W. Post and E. R. Erickson, *J. Amer. Chem. Soc.* **55**, 3851 (1933).
36. V. G. Mkhitaryan, *J. Gen. Chem. USSR* **8**, 1361 (1938).
37. E. R. Alexander and H. M. Busch, *J. Amer. Chem. Soc.* **74**, 554 (1952).
38. F. Piacenti, M. Bianchi, and P. Pino, *J. Org. Chem.* **33**, 3653 (1968).
39. W. E. Mochel, C. L. Agre, and W. E. Hanford, *J. Amer. Chem. Soc.* **70**, 2268 (1948).
40. R. M. Roberts, T. D. Higgins, Jr., and P. R. Noyes, *J. Amer. Chem. Soc.* **72**, 3801 (1955).
41. R. P. Narain and R. C. Mehrotra, *Proc. Nat. Acad. Sci. India* **A33**, 45 (1963).
42. B. Smith, *Acta Chem. Scand.* **10**, 1006 (1956).
43. D. J. Loder and W. F. Gresham, U.S. Pat. 2,409,699 (1946).
44. J. Hebky, *Collec. Czech. Chem. Commun.* **13**, 442 (1948).
45. V. G. Mkhitaryan, *Zh. Obshch. Khim.* **8**, 1361 (1938).
46. F. Piacenti, *Gazz. Chim. Ital.* **92**, 225 (1962).
47. R. Rossi, P. Pino, F. Piacenti, L. Lardicci, and G. Del Bino, *J. Org. Chem.* **32**, 842 (1967).
48. R. Rossi, P. Pino, F. Piacenti, L. Lardicci, and G. Del Bino, *Gazz. Chim. Ital.* **97**, 1194 (1967).
49. R. J. Crawford and R. Raap, *Proc. Chem. Soc.* 370 (1963).
50. M. J. Astle, J. A. Zaslowski, and P. G. Lafyatis, *Ind. Eng. Chem.* **46**, 787 (1954).
51. H. Baganz and L. Domaschke, *Chem. Ber.* **91**, 650 (1958).
52. V. G. Mkhitaryan, *Zh. Obshch. Khim.* **10**, 667 (1940).
53. L. E. Tenenbaum and J. V. Scudi, U.S. Pat. 2,636,884 (1953).
54a. G. Crank and F. W. Eastwood, *Austr. J. Chem.* **17**, 1392 (1964).
54b. E. L. Eliel and F. W. Nader, *J. Amer. Chem. Soc.* **92**, 584 (1970).
54c. F. Vellaccio, J. M. Phelan, R. L. Trottier, and T. W. Napier, *J. Org. Chem.* **46**, 3087 (1981).
55. G. Crank and F. W. Eastwood, *Aust. J. Chem.* **17**, 1385 (1964).
56. H. Stetter and K. H. Steinacker, *Chem. Ber.* **86**, 790 (1953).
57. O. Vogel, B. C. Anderson, and D. M. Simons, *J. Org. Chem.* **34**, 204 (1969).
58a. H. Baganz and L. Domaschke, *Chem. Ber.* **91**, 65 (1958).
58b. G. Kesslin and R. W. Handy, U.S. Pat. 3,415,846 (1968).
58c. R. R. Sauers and P. A. Odorisio, *J. Org. Chem.* **44**, 2980 (1979).
58d. R. A. McClelland, K. S. Godge, and J. Bahanek, *J. Org. Chem.* **46**, 886 (1981).
59. P. A. Diassi and J. Fried, U.S. Pat. 3,073,817 (1963).
60. J. P. Dusza and S. Bernstein, U.S. Pat. 3,069,419 (1962).
61. R. Gardi, R. Vitali, and A. Ercoli, *Gazz. Chim. Ital.* **93**, 413 (1963).
62. R. Gardi, R. Vitali, and A. Ercoli, *Gazz. Chim. Ital.* **93**, 431 (1963).
63. R. Gardi, R. Vitali, and A. Ercoli, *Tetrahedron* **21**, 179 (1965).
64. Roussel-UCLAF, Belg. Pat. 615,766 (1962).
65. J. P. Dusza and S. Bernstein, *J. Org. Chem.* **27**, 4677 (1962).
66. A. Ercoli and R. Gardi, U.S. Pat. 3,139,425 (1964).
67. R. Joly and C. Warnant, U.S. Pat. 3,017,409 (1962).
68. K. Morita, H. Nawa, and T. Miki, Jap. Pat. 5827 (1965).
69. Roussel-UCLAF, Fr. Pat. M1407 (1962).
70. H. G. Bott, W. N. Haworth, and E. L. Hirst, *J. Chem. Soc.* 1395 (1930).
71. G. Crank and F. W. Eastwood, *Aust. J. Chem.* **17**, 1392 (1964).
72. F. Eckstein and F. Cramer, *Chem. Ber.* **98**, 995 (1965).
73. A. Holy and K. H. Scheit, *Chem. Ber.* **99**, 3778 (1966).
74. A. Frohling and J. F. Arens, *Rec. Trav. Chim.* **81**, 1009 (1962).

75. B. Holmberg, *Chem. Ber.* **40**, 1740 (1907).
76. J. Houben and K. M. L. Schultze, *Chem. Ber.* **44**, 3235 (1911).
77. J. D. Kendall and J. R. Majer, *J. Chem. Soc.* 687 (1948).
78. W. E. Hanford and W. E. Mochel, U.S. Pat. 2,229,651 (1941).
79. W. E. Mochel, C. L. Agre and W. E. Hanford, *J. Amer. Chem. Soc.* **70**, 2268 (1948).
80. L. C. Rinzema, J. Stoffelsma, and J. F. Arens, *Rec. Trav. Chim.* **78**, 354 (1959).
81. B. Smith and S. Delin, *Sv. Kem. Tidskr.* **65**, 10 (1953).
82. H. Kopp, *Jahresber. Chem.* 391 (1860).
83. A. W. Williamson and G. Kay, *Ann. Chem.* **92**, 346 (1854).
84. A. W. Williamson and G. Kay, *Proc. Roy. Soc. London* **7**, 135 (1854).
85. A. Deutsch, *Chem. Ber.* **12**, 115 (1879).
86a. A. Lenz, K. Hass, and H. Epler, Ger. Pat. 1,214,943 (1966).
86b. A. Lenz, O. Ackermann, and O. Bleh, U.S. Pat. 3,901,946 (1975).
87. P. P. T. Sah and T. S. Mah, *J. Amer. Chem. Soc.* **54**, 2964 (1932).
88. F. Beilstein and E. Wiegand, *Chem. Ber.* **18**, 482 (1885).
89. P. Fritsch, *Ann. Chem.* **297**, 315 (1897).
90. A. Roedig and E. Degener, *Chem. Ber.* **86**, 1469 (1953).
91. T. C. Daniels and R. E. Lyons, *J. Amer. Chem. Soc.* **58**, 2646 (1936).
92. S. Cohen, E. Thom, and A. Bendich, *J. Org. Chem.* **27**, 3545 (1962).
93. H. Kwart and M. B. Price, *J. Amer. Chem. Soc.* **83**, 5123 (1960).
94. R. A. McDonald and R. A. Krueger, *J. Org. Chem.* **31**, 488 (1966).
95. H. Limpricht, *Ann. Chem.* **135**, 87 (1865).
96. S. M. McElvain, H. I. Anthes and S. Shapiro, *J. Amer. Chem. Soc.* **64**, 2525 (1942).
97. S. J. Lapporte, *J. Org. Chem.* **27**, 3098 (1962).
98. J. U. Nef, *Ann. Chem.* **308**, 329 (1899).
99. C. S. Cleaver, U.S. Pat. 2,853,531 (1958).
100. W. G. Kofron, F. G. Kirby, and C. R. Hauser, *J. Org. Chem.* **28**, 873 (1963).
101a. M. Arnhold, *Ann. Chem.* **240**, 192 (1887).
101b. J. J. Porter, U.S. Pat. 3,407,236 (1968).
102. K. Auwers, *Chem. Ber.* **18**, 2655 (1885).
103. H. Baines and J. Driver, *J. Chem. Soc.* **125**, 907 (1924).
104. F. Tiemann, *Chem. Ber.* **15**, 2685 (1882).
105. A. Weddige, *J. Prakt. Chem.* [2]**26**, 444 (1882).
106. W. Kirmse, "Carbene Chemistry," Chapter 9. Academic Press, New York, 1964.
107. S. R. Sandler, *J. Org. Chem.* **32**, 3876 (1967).
108. H. Bassett, *Ann. Chem.* **132**, 54 (1864).
109. H. Tieckelmann and H. W. Post, *J. Org. Chem.* **13**, 265 (1948).
110a. B. Rose, *Ann. Chem.* **205**, 249 (1880).
110b. M. E. Hill, D. T. Carty, D. Tegg, J. C. Butler, and A. F. Strong, *J. Org. Chem.* **30**, 411 (1965).
110c. M. E. Hill, U.S. Pat. 3,426,078 (1969).
111. W. Kaufman and E. E. Dreger, *Org. Syn. Coll. Vol.* **1**, 258 (1941).
112. H. Baganz and K. E. Kruger, *Chem. Ber.* **91**, 807 (1959).
113. H. Scheibler and M. Depner, *Chem. Ber.* **68**, 2151 (1935).
114. H. Scheibler and M. Depner, *J. Prakt. Chem.* [4]**7**, 60 (1958).
115. R. Anschutz, *Ann. Chem.* **254**, 31 (1889).
116. R. G. Jones, *J. Amer. Chem. Soc.* **73**, 5168 (1951).
117. A. Lenz, K. Hass, and H. Epler, Ger. Pat. 1,217,943 (1966).
118a. W. E. Kaufmann and E. E. Dreger, *Org. Syn. Coll. Vol.* **1**, 258 (1932).
118b. W. Grabowicz and Z. Cybulska, Pol. Pat. 125,872 (1984); *Chem. Abstr.* **102**, 45490a (1985).

119a. H. Gross, J. Ruscheand, and H. Bornowski, *Ann. Chem.* **675**, 146 (1964).

119b. P. Speh and W. Kantlehner, U.S. Pat. 3,876,708 (1975).

120a. R. P. Ruh, U.S. Pat. 2,737,530 (1956).

120b. F. Beyerstedt and S. M. McElvain, *J. Amer. Chem. Soc.* **58**, 529 (1936).

120c. S. M. McElvain, S. B. Mirviss, and C. L. Stevens, *J. Amer. Chem. Soc.* **73**, 3807 (1951).

121a. R. M. Roberts, J. Corse, R. Boschan, D. Seymour, and S. Winstein, *J. Amer. Chem. Soc.* **80**, 1247 (1958).

121b. B. G. Yasnitskii, S. A. Sarkisyants, and E. G. Ivanyuk, *Zh. Obshch. Khim.* **34**, 1940 (1964).

122a. F. Beyerstedt and S. M. McElvain, *J. Amer. Chem. Soc.* **59**, 2266 (1937).

122b. S. M. McElvain and L. R. Morris, *J. Amer. Chem. Soc.* **74**, 2657 (1952).

123a. S. M. McElvain and P. M. Walters, *J. Amer. Chem. Soc.* **64**, 1963 (1942).

123b. U. Faas and H. Hilgert, *Chem. Ber.* **87**, 1343 (1954).

124. S. M. McElvain and A. N. Bolstad, *J. Amer. Chem. Soc.* **73**, 1988 (1951).

125a. W. C. Kuryla and D. G. Leis, *J. Org. Chem.* **29**, 2773 (1964).

125b. S. M. McElvain and B. F. Pinzon, *J. Amer. Chem. Soc.* **67**, 650 (1945).

126. H. Feuer and W. H. Gardner, *J. Amer. Chem. Soc.* **76**, 1375 (1954).

127. S. M. McElvain and M. J. Curry, *J. Amer. Chem. Soc.* **70**, 3781 (1948).

128a. T. R. Rix and J. F. Arens, *Kon. Ned. Akad. Wetensch. Proc.* **B56**, 364 (1953).

128b. R. J. Broekema, *Rec. Trav. Chim. Pays-Bas* **94** (9–10), 209 (1975); *Chem. Abstr.* **84**, 30605c (1976).

129. H. D. A. Tigchelaar-Lutjebaer, H. Bootsma, and J. F. Arens, *Rec. Trav. Chim.* **79**, 888 (1960).

130. R. Meier and F. Bohler, *Chem. Ber.* **90**, 2342 (1957).

131. J. D. Park, W. M. Sweeney and J. R. Lacher, *J. Org. Chem.* **21**, 1035 (1956).

132. T. G. Miller and J. W. Thenassi, *J. Org. Chem.* **25**, 2009 (1960).

133. A. Roedig, K. Grahe, and W. Mayer, *Chem. Ber.* **100**, 2946 (1967).

134. J. F. Harris, *J. Org. Chem.* **32**, 2063 (1967).

135. L. I. Zakharkin, *Izv. Akad. Nauk SSSR Otd. Khim. Nauki* 1064 (1957).

136. O. W. Webster, M. Brown, and R. E. Benson, *J. Org. Chem.* **30**, 3223 (1965).

137. W. R. Hertler and R. E. Benson, *J. Amer. Chem. Soc.* **84**, 3474 (1962).

138. F. Beyerstedt and S. M. McElvain, *J. Amer. Chem. Soc.* **59**, 1273 (1937).

139a. S. G. Matsoyan, G. M. Pogosyan, and M. A. Eliazyan, *Vysokomol. Soedin.* **5**, 777 (1963).

139b. H. Stetter and E. Reska, *Chem. Ber.* **103**, 639 (1970).

140. J. Hebky, *Collect. Czech. Chem. Commun.* **13**, 442 (1948).

141. T. Kariyone and Y. Kimura, *Yakugaku Zasshi* **500**, 746 (1923).

142. F. Bohlmann and W. Sacrow, *Chem. Ber.* **97**, 1839 (1964).

143. F. Lacasa, J. Pascual, and L. V. del Arco, *An. Real Soc. Espan. Fis. Quim.* **B52**, 549 (1956).

144. S. M. McElvain and G. R. McKay, *J. Amer. Chem. Soc.* **77**, 560 (1955).

145. F. Bohlmann and W. Sucrow, *Chem. Ber.* **97**, 1846 (1964).

146. H. Stetter and K. H. Steinacker, *Chem. Ber.* **86**, 790 (1953).

147. J. M. Osbond, P. G. Philpott, and J. C. Wickens, *J. Chem. Soc.* 2779 (1961).

148. H. Adkins and H. R. Billica, *J. Amer. Chem. Soc.* **70**, 695 (1948).

149. D. J. Loder and W. F. Gresham, U.S. Pat. 2,409,699 (1946).

150. H. Meerwein, K. Bodenbenner, P. Borner, F. Kunert, and K. Wunderlich, *Ann. Chem.* **632**, 38 (1960).

151. R. Gardi, R. Vitali, and A. Ercoli, *Tetrahedron* **21**, 179 (1965).

152. G. Schneider and L. K. Lang, *Chem. Commun.* 13 (1967).

153. G. Schneider, *Tetrahedron Lett.* 5921 (1966).

154. H. Meerwein, P. Borner, O. Fuchs, H. J. Sasse, H. Schrodt, and J. Spille, *Chem. Ber.* **89**, 2060 (1956).
155. H. Meerwein, *in* "Methoden der Organischen Chemie" (E. Mueller, Ed.), Vol. 6, Part 3, p. 361. Thieme, Stuttgart, 1965.
156. K. Bodenbenner, Dissertation, Univ. of Marburg (1953).
157. H. Meerwein, *Angew. Chem.* **67**, 374 (1955).
158. K. Bodenbenner, *Ann. Chem.* **623**, 183 (1959).
159. K. Bodenbenner and O. Beyer, Ger. Pat. 1,084,733 (1960).
160. S. Winstein and R. E. Buckles, *J. Amer. Chem. Soc.* **65**, 613 (1943).
161. O. J. Kovacs, G. Schneider, and K. Lang, *Proc. Chem. Soc.* 374 (1963).
162. G. Schneider and K. J. Kovacs, *Chem. Commun.* 202 (1965).
163. E. Pascu, *Advan. Carbohyd. Chem.* **1**, 77 (1945).
164. T. Holm, U.S. Pat. 2,611,787 (1952).
165. H. Hibbert and M. E. Grieg, *Can. J. Rev.* **4**, 254 (1931).
166. H. Meerwein and G. Hinz, *Ann. Chem.* **484**, 1 (1930).
167. P. Bladon and G. C. Forrest, *Chem. Commun.* 481 (1966).
168. R. B. Woodward and J. Z. Gougoutas, *J. Amer. Chem. Soc.* **86**, 5030 (1964).
169. H. A. Weidlich and W. Schulz, Ger. Pat. 919,465 (1954).
170. A. Schonberg and A. Mustafa, *J. Chem. Soc.* 997 (1947).
171. D. M. Lemal, E. P. Gosselink, and S. D. McGregor, *J. Amer. Chem. Soc.* **88**, 582 (1966).
172. R. W. Hoffmann and H. Hauser, *Tetrahedron Lett.* 197 (1964).
173. R. J. Crawford and R. Raap, *Proc. Chem. Soc.* 370 (1963).
174. R. W. Hoffmann and J. Schneider, *Tetrahedron Lett.* 4347 (1967).
175. Dynamit Nobel A.-G. Neth. Pat. 6,609,612 (1967).
176. S. M. McElvain and P. L. Weyna, *J. Amer. Chem. Soc.* **81**, 2579 (1959).
177. A. T. Babayan, G. T. Martirosyan, and R. B. Minasyan, *Dokl. Akad. Nauk Arm. SSR* **39**, 99 (1964).
178. R. Scarpati, M. Cioffi, G. Scherillo, and R. A. Nicolaus, *Gazz. Chim. Ital.* **96**, 1164 (1966).
179. R. Scarpati, D. Sica, and C. Santacroce, *Gazz. Chim. Ital.* **96**, 375 (1966).
180. N. J. Turro and J. R. Williams, *Tetrahedron Lett.* 321 (1969).
181. K. R. Henery-Logan and T. L. Fridinger, *J. Amer. Chem. Soc.* **89**, 5724 (1967).
182. F. Nrdel, J. Buddrus, J. Windhoff, W. Bodrowski, D. Klamann, and K. Ulm, *Ann. Chem.* **710**, 77 (1967).
183. P. M. Duggleby and G. Holt, *J. Chem. Soc.* 3579 (1962).
184. E. J. Corey, R. A. F. Winter, *Chem. Commun.* 208 (1965).
185. R. Hull and R. Farrand, *Chem. Commun.* 164 (1965).
186. D. Martin, *Chem. Ber.* **98**, 3286 (1965).
187. C. M. S. Yoder and J. J. Zuckerman, *J. Heterocycl. Chem.* **4**, 166 (1967).
188. R. W. Hoffmann and J. Schneider, *Chem. Ber.* **100**, 3698 (1967).
189. Y. Gaoni, *J. Chem. Soc. C* 2925, 2934 (1968).
190. B. Belleau and N. L. Weinberg, *J. Amer. Chem. Soc.* **85**, 2525 (1963).
191. E. C. Sherman and A. P. Dunlop, *J. Org. Chem.* **25**, 1309 (1960).
192. T. Holm, U.S. Pat. 2,611,787 (1952).
193. R. W. Hoffmann and H. Hauser, *Tetrahedron* **21**, 891 (1965).
194. A. Rieche, E. Schmitz, and E. Beyer, *Chem. Ber.* **91**, 1942 (1958).
195. E. Schmitz, A. Rieche, and E. Beyer, *Chem. Ber.* **94**, 2921 (1961).
196. H. E. Seyfarth and A. Hesse, *Chem. Ber.* **100**, 2491 (1967).
197. N. A. Milas and R. J. Klein, *J. Org. Chem.* **33**, 848 (1968).
198. H. W. Post and E. R. Erickson, *J. Org. Chem.* **2**, 260 (1937).

199. J. W. Scheeren and W. Stevens, *Rec. Trav. Chim.* **85**, 793 (1966).
200. A. R. Mattocks, *J. Chem. Soc.* 1918 (1964).
201. A. Nagasaki, R. Oda, and S. Nukina, *J. Chem. Soc. Jap.* **57**, 169 (1954).
202. E. D. Bergmann, E. Biograchov, and S. Pinchas, *J. Amer. Chem. Soc.* **73**, 2352 (1951).
203. H. Gross and A. Rieche, *Chem. Ber.* **94**, 538 (1961).
204. R. A. Barnes, G. Doyle, and J. A. Hoffman, *J. Org. Chem.* **27**, 90 (1962).
205. J. U. Nef, *Ann. Chem.* **298**, 350 (1897).
206. A. E. Tshitschibabin, *Chem. Ber.* **38**, 561 (1905).
207. Y. A. Sinnema and J. F. Arens, *Rec. Trav. Chim.* **74**, 901 (1955).
208. H. Baganz, L. Domaschke, J. Fock, and S. Rabe, *Chem. Ber.* **95**, 1832 (1962).
209. H. Griss, J. Freiberg, and B. Costisella, *Chem. Ber.* **101**, 1250 (1968).
210. J. Jonas, T. P. Forrest, M. Kratochvil, and H. Gross, *J. Org. Chem.* **33**, 2126 (1968).
211. W. Kirmse and M. Buschoff, *Chem. Ber.* **102**, 1087 (1969).
212. S. Sakai, Y. Kiyohara, K. Itoh, and Y. Ishii, *J. Org. Chem.* **35**, 2347 (1970).
213. S. Sakai, Y. Kobayashi, and Y. Ishii, *J. Org. Chem.* **36**, 1176 (1971).
214. S. Sakai, Y. Kuroda, and Y. Ishii, *J. Org. Chem.* **37**, 4198 (1972).
215. K. Findeisen and K. Wagner, U.S. Pat. 3,857,897 (12-31-74).
216. S. Sakai, T. Fujinami, K. Kunio, and K. Matanaga, *Chem. Lett.* (8), 891 (1976); *Chem. Abstr.* **85**, 159598c (1976).
217. J. W. Scheeren, H. J. M. Goossens, and A. W. H. Top, *Synthesis* (4), 283 (1978).
218. T. Holm, U.S. Pat. 2,611,787 (1952).
219. R. Breslow and P. S. Pandy, *J. Org. Chem.* **45**, 740 (1980).
220. G. Saucy, R. Borer, and D. P. Trullinger, *J. Org. Chem.* **42**, 3206 (1977).
221. R. A. McClelland, G. Patel, and P. W. K. Lam, *J. Org. Chem.* **46**, 1011 (1981).
222. H. Meerwein, P. Borner, O. Fuchs, H. J. Sasse, H. Schrodt, and J. Spille, *Chem. Ber.* **89**, 2060 (1956).
223. B. Mir-Mohamad-Sadeghy and B. Rickborn, *J. Org. Chem.* **49**, 1477 (1984).

1. INTRODUCTION

The first report on sulfites appeared in 1846, and several synthetic procedures were later published in 1858–1859 [1]. The literature contains numerous reports on sulfites from 1909 to the present, yet sulfites have not even been mentioned briefly in some well-known texts on sulfur chemistry. However, the first comprehensive review appeared in 1963 in *Chemical Reviews* [1]. Sulfites are characterized by the structure (I) where R = aryl or alkyl groups. The

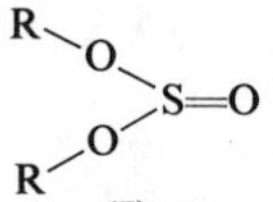

$$
\begin{array}{c}
R \diagdown O \diagdown \\
 S{=}O \\
R \diagup O \diagup
\end{array}
$$

(I)

sulfites may have two similar groups, or two different groups. Recently the use of sulfites as insecticides [2], pesticides [2], plant-growth regulators [3], and plasticizers [4] and their ability to form polyesters when dicarboxylic acids react with ethylene sulfite [5] have revived great interest in this functional group. For example, 2-(*p-t*-butylphenoxy) 1-methylethyl-2-chloroethyl sulfite has been reported for insect control on currants [6].

Chemical Abstracts refers to sulfites as sulfurous acid esters. Simple esters (benzyl, phenyl, ethyl, etc.) are listed as sulfurous acid esters under the names of the corresponding hydroxy compound. All mixed esters are indexed separately under the heading "sulfurous acid esters."

The best methods of preparing sulfites involve the reactions outlined in Eq. (1).

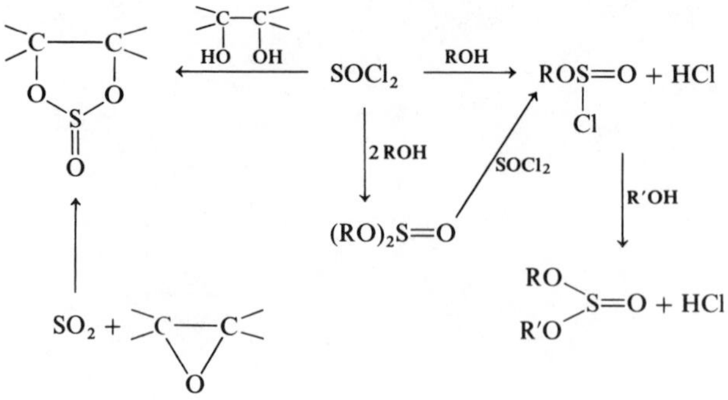

$$
\text{SOCl}_2 \xrightarrow{\text{ROH}} \text{ROS}{=}\text{O} + \text{HCl} \tag{1}
$$

Recently, a brief review of organic sulfites and their reactions has appeared [6a].

2. CONDENSATION REACTIONS

A. Reaction of Thionyl Chloride with Alcohols

Thionyl chloride is known to react with alcohols in the absence of hydrogen chloride acceptors to give reactions (2)–(7). Optimization of reactions (2), (3),

$$SOCl_2 + ROH \longrightarrow ROSOCl + HCl \qquad (2)$$

$$ROSOCl + ROH \longrightarrow (RO)_2SO + HCl \qquad (3)$$

$$SOCl_2 + 2ROH \longrightarrow (RO)_2SO + 2HCl \qquad (4)$$

$$(RO)_2SO + SOCl_2 \longrightarrow 2ROSOCl \qquad (5)$$

$$ROSOCl \longrightarrow RCl + SO_2 \qquad (6)$$

$$SOCl_2 + ROH \longrightarrow olefin + 2HCl + SO_2 \qquad (7)$$

and (4) is of primary concern in this section.

The need for a hydrogen chloride acceptor was shown by Bissinger and Kung [7], who found that primary and secondary sulfites decompose to the halide if hydrogen chloride is not removed (Eq. 8). The effect of hydrogen

$$R_2SO_3 + HCl \longrightarrow RCl + ROH + SO_2 \qquad (8)$$

chloride can be minimized by the following techniques: (a) passage of an inert gas through the mixture [7, 8a, b], (b) use of reduced pressure to remove hydrogen chloride [9], (c) use of a solvent in which hydrogen chloride is insoluble (CH_2Cl_2, $CHCl_3$, CCl_4, C_6H_5Cl, and $o\text{-}C_6H_4Cl_2$) [10], or (d) use of a tertiary amine (triethylamine, pyridine, or quinoline) as a hydrogen chloride scavenger [11].

TABLE I

CHANGE IN THE PRODUCT RATIOS ON USING VARIOUS AMOUNTS
OF THIONYL CHLORIDE TO REACT WITH 1.0 MOLE OF 2-OCTANOL [13]

Thionyl chloride (moles)	Alkyl chloride (%)	Alkyl chlorosulfinate (%)	Sulfite (%)
0.5	—	—	84
1.0	Trace	29	43
2.0[a]	0	50	34

[a] 2-Octanol was added to thionyl chloride in this experiment.

TABLE II

Preparation of Cyclic Sulfites

1,2-Diol	(moles)	SOCl$_2$ (moles)	R$_3$N (moles)	Solvent (ml)	Yield (%)	B.p., °C (mm Hg) or m.p., °C	n_D (°C)
			Pyridine	CH$_2$Cl$_2$			
Cycloheptane [14]							
cis-	0.1	0.1	0.24	250	41	90 (0.5)	1.4860 (24)
trans-	0.1	0.1	0.24	250	38	92 (0.5)	1.4865 (24)
cis-Indane [14]	0.1	0.1	0.24	250	72	70	—
1,4-Anhydroerythritol [14]	0.1	0.1	0.24	250	55	106–108	—
Ethane [15]	1.35	1.35	—	—	90	86–88 (38)	—
1,1-Dimethylol-3-cyclopentene [16]	0.274	0.823	0.55	Ether 150	59	117–119 (8) m.p. 49–50 (ether)	—
2,3-Diphenyl-2,3-butanediol [17]	0.06	0.15	0.12	C$_6$H$_6$ 120	62	163 (1.3)	1.5752 (24)
3,3-Diphenyl-2-methyl-2,3-propanediol [17]	0.06	0.15	0.12	120	57	83–84	—

Compound				Ether			
Hydrobenzoin [18]							
meso-	0.094	0.093	0.19	540 / C$_6$H$_6$ / 420 Dioxane	83	126–128	—
dl-	0.023	0.024	0.051		83	84–86	—
Cyclohexane [18]							
cis-1,2-	0.086	0.087	0.18	100	76	90 (2.0) m.p. 6–8.0	1.4832 (20)
trans-1,2-	0.17	0.17	0.34	200 / CS$_2$ / 230	76	94–96 (2.0) m.p. −15	1.4847 (20)
Catechol [19]	1.0	1.0	2.0	ClCH$_2$CH$_2$Cl / 300	26	137–138 (105 mm)	—
2,2-Dinitro-1,3-propanediol [19a]	0.5	0.5	12 drops	—	81	m.p. 37–38	—
Trichloropropylene glycol [19b]	1.0	1.0	—	—	100	68 (8)	1.5500 (25)
3,4-Dimethyl-3,4-hexanediol [19c]	0.021	0.12	0.24	Ether / 23	43	67–68 (0.6)	—

Gerrard [11] had shown earlier that the slow addition of 0.5 mole of thionyl chloride to a mixture of pyridine (1.0 mole) and hydroxy compounds (n-butyl, n-amyl, or ethyl lactate—0.1 mole) gives pyridine hydrochloride and good yields of the sulfite (see Eq. 4). Primary and secondary alcohols with an aromatic nucleus in the α-position give chlorides in the absence of catalysts [12]. For example, diphenylmethanol gives the chloride even at −78°C. Sulfites derived from tertiary alcohols are not known. The further addition of thionyl chloride converted the sulfite to the chlorosulfinate (see Eq. 5 and Table I). On heating, the chlorosulfinate is catalytically decomposed by pyridine hydrochloride to the corresponding alkyl chloride and sulfur dioxide (see Eq. 6). Secondary chlorosulfites give olefins even under the mildest conditions [11]. The use of excess pyridine reduces the yield of sulfite by the method described in Eq. (4).

Secondary alkyl sulfites are produced if the formed hydrogen chloride is removed by carrying out the reaction at reduced pressure (aspirator) [9].

Examples of the preparation of cyclic sulfites from 1,2-diols are shown in Table II. Other examples involving the preparation of cyclic sulfites are found in a recent patent [20].

2-1. Preparation of Dibenzyl Sulfite [12]

$$2C_6H_5CH_2OH + SOCl_2 \xrightarrow{\text{pyridine}} (C_6H_5CH_2O)_2SO \qquad (9)$$

To a stirred solution of 5.4 gm (0.05 mole) of benzyl alcohol in 4.0 gm (0.05 mole) of pyridine and 30 ml of ether at −78°C is added dropwise a solution of 3.0 gm (0.025 mole) of thionyl chloride in 15 ml of ether over a 20–30 min period. After 1 hr the mixture is filtered, concentrated under reduced pressure, and distilled to afford 11.4 gm (87%), b.p. 152°C (0.4 mm), n_D^{25} 1.5590.

Table III lists the alcohol used, the yields of sulfites obtained, and their physical properties in a similar procedure.

TABLE III[a]

PREPARATION OF SULFITES USING PROCEDURE 2-1

Alcohol	Sulfite (yield %)	B.p., °C (mm Hg)	n_D (°C)
2-Phenylethanol	86	162–165 (0.5)	1.5510 (15)
3-Phenylpropanol	85	185–190 (0.3)	1.5423 (18)
1-Phenyl-2-propanol	86	158–160 (0.1)	1.5351 (21.5)

[a] Data taken from Gerrard and Shepherd [12].

2-2. The Preparation of the Cyclic Sulfite of Phenylethane-1,2-diol [20]

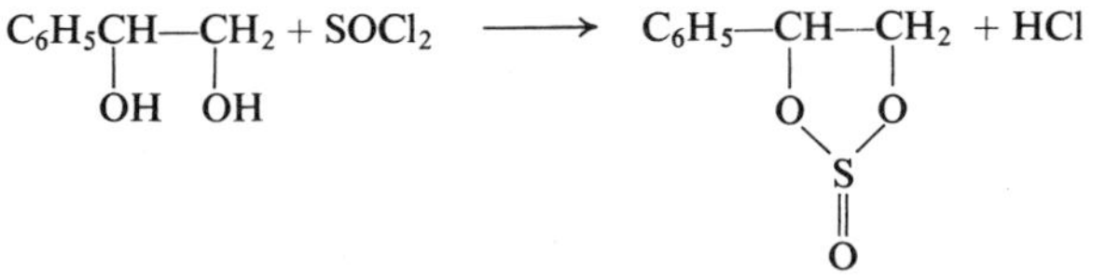

$$C_6H_5CH{-}CH_2 + SOCl_2 \longrightarrow C_6H_5{-}CH{-}CH_2 + HCl \qquad (10)$$

To a flask containing 6.9 gm (0.05 mole) of phenylethane-1,2-diol in 25 ml of chloroform is added dropwise at room temperature 5.95 gm (0.05 mole) of thionyl chloride. After the initial reaction in which hydrogen chloride is evolved, the solution is refluxed for 1 hr. The solution is cooled, washed with water, then with 2 % sodium bicarbonate solution, dried, and distilled to afford 5.7 gm (62 %) b.p. 62°–64°C (0.15 mm), $n_D^{24.6}$ 1.5421.

Some other examples of cyclic sulfites prepared by this method are shown in Table IV.

TABLE IV

PREPARATION OF CYCLIC SULFITES USING 1,2- OR 1,3-DIOLS
ACCORDING TO PROCEDURE 2-2

Diol	Physical properties of isolated cyclic sulfite[a]	
	B.p., °C (mm Hg)	n_D (°C)
2,2-Diethylpropane-1,3-	66 (0.5)	1.4594 (22.5)
1,1-Diphenylethane-1,2-	m.p. 67.5–68.5	—
2-Methyl-2-n-propylpropane-1,3-	110 (18)	1.4530 (26.5)
2-n-butyl-2-ethylpropane-1,3-	80 (0.4)	1.4611 (22.5)
2-Methyl-2-phenylpropane-1,3-	78–80 (0.07)	1.5412 (23.5)
2-Ethyl-2-phenylpropane-1,3-	108–109 (0.4)	1.5384 (23)
2-n-Amyl-2-methylpropane-1,3-	60 (0.0)	1.4563 (23)
2-sec-Amyl-2-methylpropane-1,3-	124.5 (9)	1.4626 (23.5)
2-Allyl-2-ethylpropane-1,3-	120–122 (15)	1.4712 (22.5)
2,2-Diallylpropane-1,3-	81.5–84.5 (0.35)	1.4842 (22.5)
2-Benzyl-2-phenylpropane-1,3-	m.p. 146–147 (cis isomer)	—
	m.p. 75.5–76.5 (trans isomer)	—
1-Benzyl-3-methylbutane 2,3-	128.5–129.5 (0.55)	1.5242 (23.8)
1-Allyl-1,2-dimethylethane-1,2-	98 (14)	1.4584 (21)
1-Ethyl-1-phenylethane-1,2-	78 (0.1)	1.5280 (21.8)

[a] Data taken from Wiggins [20].

2-3. Preparation of the Cyclic Sulfite of Catechol [19]

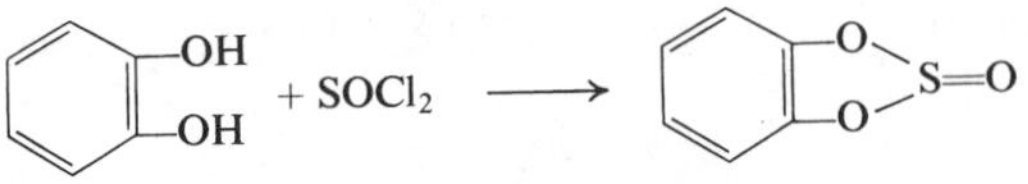

$$\text{(catechol)} + SOCl_2 \longrightarrow \text{(cyclic sulfite)} \qquad (11)$$

To a cooled (10°C) stirred solution of 30 gm (0.27 mole) of catechol in 150 ml of dry carbon disulfide and 42.4 gm (0.54 mole) of pyridine is added dropwise 33.7 gm (0.28 mole) of thionyl chloride in 80 ml of dry carbon disulfide. After the addition the reaction mixture is allowed to stand at 10°C for $\frac{1}{2}$ hr, refluxed for $\frac{1}{2}$ hr, separated from the warm syrupy layer of pyridine hydrochloride, concentrated, and the residue distilled to afford 41 gm (97%), b.p. 137°–138°C (105 mm).

2-4. Preparation of the Cyclic Sulfite of trans-Cyclohexane-1,2-diol [14]

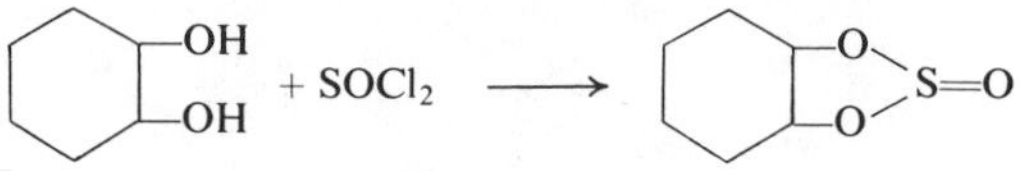

$$\text{(trans-cyclohexane-1,2-diol)} + SOCl_2 \longrightarrow \text{(cyclic sulfite)} \qquad (12)$$

To a flask containing 11.6 gm (0.10 mole) of *trans*-cyclohexane-1,2-diol and 18.6 gm (0.24 mole) of pyridine in 200 ml of methylene chloride is added dropwise, over a 2 hr period at 0°C, 23 gm (0.19 mole) of thionyl chloride dissolved in 50 ml of methylene chloride. The pyridine hydrochloride is filtered and the methylene chloride solution is successively washed with 0.01 N hydrochloric acid, aqueous sodium bicarbonate, and water. After drying over $MgSO_4$, distillation affords 11.7 gm (72%), b.p. 72°–75°C (0.1 mm), n_D^{21} 1.4245. Earlier, Price and Berti [21] reported the same compound with b.p. 94°–96°C (2 mm), n_D^{20} 1.4817. The cis compound has b.p. 90°C (2 mm), n_D^{20} 1.4832, m.p. 6°–8°C.

The preparation of polymeric sulfites involves the reaction of most diols such as diethylene glycol, 1,4-butanediol, 1,1-decanediol, dipropylene glycol, and triethylene glycol [14a]. The reaction of 2,2,4,4-tetraethylcyclobutane-1,3-diol with thionyl chloride has also been reported to give a polysulfite [14b]. Surprisingly, 2,2,3,3-tetrachloro-1,4-butanediol is reported to react with thionyl chloride to give the seven-membered cyclic sulfite with a melting point of 59°–61°C [14c].

B. Reaction of Sulfinyl Chlorides with Alcohols

Unsymmetrically substituted sulfites are prepared by the reaction of chlorosulfinates with alcohols as shown in Eq. (3). The chlorosulfinates may be prepared by the methods described in Eqs. (13)–(16) [22–23].

$$ROH + SOCl_2 \longrightarrow ROSOCl + HCl \tag{13}$$

$$RONa + SOCl_2 \longrightarrow ROSOCl + NaCl \tag{14}$$

$$(RO)_2SO + PCl_5 \longrightarrow ROSOCl + POCl_3 + RCl \tag{15}$$

$$(RO)_2SO + SOCl_2 \longrightarrow 2ROSOCl \tag{16}$$

An equivalent amount of base is necessary to effect the reaction in Eq. (3). Solvents in which the hydrochloride is insoluble helps to give improved yields. Otherwise the product may disproportionate into two symmetrically substituted compounds [8a] (Eq. 17).

$$2(RO)(R'O)SO \longrightarrow (RO)_2SO + (R'O)_2SO \tag{17}$$

Some examples illustrating the preparation and yields of unsymmetrically substituted sulfites are shown in Table V.

2-5. *Preparation of trans-2-p-Tolylsulfonylcyclohexyl Sulfite* [26]

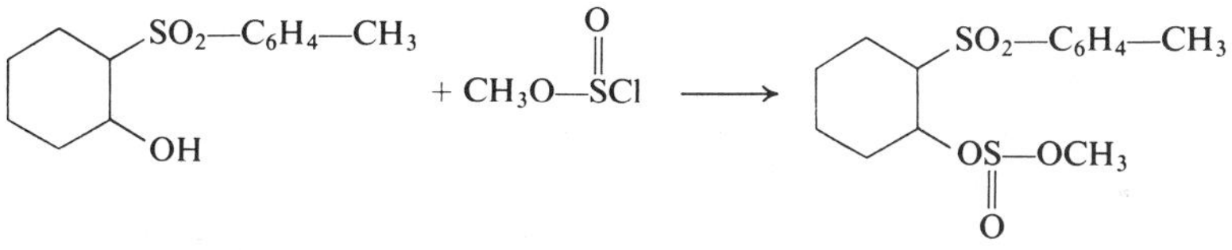

$$\tag{18}$$

(*a*) *Preparation of methyl chlorosulfinate* [24]. To a 500 ml three-necked flask equipped with a mechanical stirrer, condenser, drying tube, and dropping funnel is added 260 gm (2.2 mole) of thionyl chloride. Then 64 gm (2.0 mole) of methanol is added dropwise over a 45 min period. The reaction mixture is left for 2 days at room temperature and then distilled under reduced pressure to yield 199 gm (88%), b.p. 35°–36°C (65 mm).

(*b*) *Reaction of methyl chlorosulfinate with trans-2-p-tolylsulfonylcyclohexanol* [26]. To 3.8 gm (0.015 mole) of *trans-2-p*-tolylsulfonylcyclohexanol and 1.4 gm (0.0177 mole) of dry pyridine dissolved in 20 ml of ether at 0°C is added dropwise 2.0 gm (0.0175 mole) of methyl chlorosulfinate in 10 ml of ether. The mixture is stirred for 1 hr at 0°C, filtered, the ether filtrate is washed in turn with water, 5% hydrochloric acid, and then 5% sodium hydroxide. The dry ether extract is concentrated to give 4.1 gm (82%), m.p. 82°–84°C (recrystallized from hexane).

TABLE V

Preparation of Unsymmetrically Substituted Sulfites Using Chlorosulfinates

R—OH	R'OSOCl, R'=	React. temp. (°C)	(RO)(R'O)SO (yield %)	B.p., °C (mm Hg) or m.p., °C	n_D^{20}	Ref.
1-Methyl-2-phenylethanol	CH_3	0	75[a]	115–120 (2)	1.5068	24
3-Phenylpropanol	CH_3	10	58[b]	132–134 (2.5)	1.5702	24
2-Phenylcyclohexanol						
cis-	CH_3	−5	80	148 (1.0)	1.5313	24
trans-	CH_3	−5	75[c]	145 (0.9) m.p. 48–50 (from pet. ether)	—	24
β-Menthol	CH_3	0	80	105–108 (1.0)	1.5640	24
Chlolesterol	CH_3	5	85	m.p. 115–117 (from ethanol)	—	24
2-Propanol	CH_3	0	48	48–52 (17)	1.4122	25
Cyclopentanol	CH_3	0	44	97–100 (15)	1.4600	25
Cyclohexanol	CH_3	0	63	105–110 (17)	1.4630	25
2-p-Tolylsulfonylcyclohexanol						
trans-	CH_3	0	82	m.p. 82–84	—	26
cis-	CH_3	5	85	m.p. 106–108	—	26
Methanol	C_2H_5	0	—	140–143 (760)	—	22
Methanol	C_4H_9	0	—	86–88 (14)	—	22
Ethanol	C_4H_9	0	—	94–96 (14)	—	22
Propanol	C_4H_9	0	—	102–104 (15)	—	22
n-Butanol	$Cl—CH_2CH_2$	0	—	149–152 (40)	—	22
Allyl	C_4H_9	0	—	92–93 (18)	—	22
	$Cl—CH_2CH_2$	0	—	103–105 (22)	—	22
n-Heptanol	C_3H_7	0	—	148–150 (18)	—	22

				b.p. °C (mm)	n_D	Ref.
Lauryl	Cl—CH$_2$CH$_2$	40	52	168–170 (1)	1.4589	22a
2-Methoxycyclohexanol	CH$_3$OCH$_2$CH$_2$	0	82	115–118 (0.7)	—	22b
Propargyl	o-Phenylphenoxy ethoxyethyl	15	78	—	1.5620 (at 25°)	22c
n-Decanol	C$_2$H$_5$	0	—	188–190 (30)	—	22
Methanol	C$_6$H$_5$CH$_2$	0	—	137–138 (18)	—	22
Benzyl	C$_2$H$_5$	0	—	160–161 (30)	—	22
2-Propanol	n-C$_4$H$_9$	0	—	99–101 (21)	—	22
Cyclohexanol	n-C$_3$H$_7$	0	—	175–178 (60)	—	22
t-Butanol	n-C$_3$H$_7$	0	—	99–100 (30)	—	22
Methanol	C$_6$H$_5$	0	—	67–68 (0.04)	1.5400	22, 27
Ethanol	C$_6$H$_5$	0	—	142–144 (25)	—	22
n-Butanol	C$_6$H$_5$	0	—	170–173 (20)	—	22
Ethanol	o-CH$_3$—C$_6$H$_4$	0	—	145–147 (20)	—	22
n-Butanol	o-CH$_3$—C$_6$H$_4$	0	—	170–173 (13)	—	22
Methanol	m-CH$_3$—C$_6$H$_4$	0	—	137–140 (25)	—	22
Ethanol	m-CH$_3$—C$_6$H$_4$	0	—	147–150 (25)	—	22
n-Butanol	m-CH$_3$—C$_6$H$_4$	0	—	178–180 (20)	—	22
Methanol	p-CH$_3$—C$_6$H$_4$	0	—	134–136 (20)	—	22
Ethanol	p-CH$_3$—C$_6$H$_4$	0	—	146–149 (20)	—	22
n-Propanol	p-CH$_3$—C$_6$H$_4$	0	—	175–177 (20)	—	22
Ethanol	p-Cl—C$_6$H$_4$	0	—	155–158 (25)	—	22
C$_6$H$_5$	p-Cl—C$_6$H$_4$	0	—	195–198 (16)	—	22

[a] Some di(1-methyl-2-phenylethyl) sulfite was formed in the distillation, b.p. 175° (0.5 mm).

[b] Approximately 25% of di(3-phenylpropyl) sulfite was formed in the distillation, b.p. 221–223°C (2.5 mm), n_D 1.5428 (20°C).

[c] A solid residue remained in the distillation flask, m.p. 130°–131°C (from methanol), probably di(trans-2-phenylcyclohexyl) sulfite but not adequately characterized.

C. Reaction of Epoxides with Sulfur Dioxide

The reaction of alkylene oxides or epoxides with sulfur dioxide to give cyclic sulfites is effected by carrying out the reaction at about 150°C for 4 hr at 2000 atm of SO_2 [28]. Pyridine is used in small amounts as a polymerization inhibitor. In addition, it has been reported that free radical-producing catalysts give improved yields and allow the reaction to be carried out at lower temperatures [28] (Eq. 19).

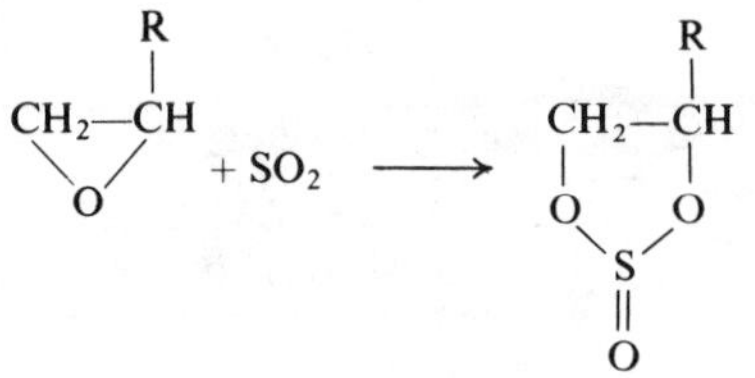

$$
\begin{array}{ccc}
& R & \\
& | & \\
CH_2\!-\!\overset{|}{CH} & & \\
\diagdown\!\diagup + SO_2 & \longrightarrow & \\
O & &
\end{array}
\qquad (19)
$$

R = CH$_3$, b.p. 68°–78°C (5 mm)
R = C$_6$H$_5$, b.p. 125°C (15 mm)

Other examples illustrating the utility of this method are shown in Table VI.

TABLE VI

REACTION OF EPOXIDES WITH SULFUR DIOXIDE TO GIVE CYCLIC SULFITES

Epoxide	SO_2	Temp. (°C)	Time (hr)	Yield (gm)	B.p., °C (mm Hg)	n_D^t	Ref.
Ethylene oxide							
25 gm	2000 atm	150	4	50	—	—	28
Propylene oxide[a]	2000 atm	150	4	—	68–70 (5)	—	28
Styrene oxide[a]	2000 atm	150	4	—	125 (15)	—	28
Ethylene oxide[b]							
200 gm	291 gm	5–10	120	81	53 (12)	1.4470 (20)	29
0.9 mole	0.9 mole	220	1	—	173 (760)	—	30
500 gm[c]	793 gm	106	8	500	—	—	31
Propylene oxide							
2720 gm[d]	2260 gm	105	12	1870	61 (10)	1.437 (22)	31
Ethylene oxide–							
SO$_2$ adduct[e]	—	80	2	17.5	—	—	32
36 gm							

[a] Pyridine (1.0 gm) added as a polymerization inhibitor.
[b] S(CH$_2$CH$_2$OH)$_2$ (6.0 gm) added as a catalyst.
[c] Activated carbon with silver oxide or Ni–W sulfide catalyst added.
[d] Anion-exchange resin (40–100 gm) added; see U.S. Pat. 2,614,099 (1952).
[e] Absolute ethanol (62 gm) and 1 gm of FeCl$_3$ or TiCl$_3$ added.

2-6. Preparation of Ethylene Sulfite [28]

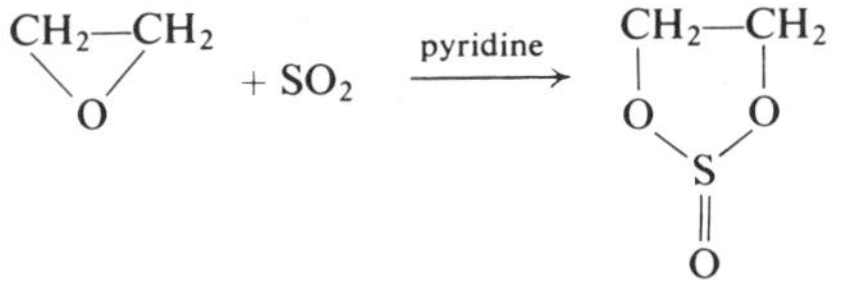

$$\begin{array}{c} CH_2\!-\!CH_2 \\ \diagdown\!\!\diagup \\ O \end{array} + SO_2 \xrightarrow{\text{pyridine}} \begin{array}{c} CH_2\!-\!CH_2 \\ | \quad\quad | \\ O \quad\quad O \\ \diagdown_S\diagup \\ \| \\ O \end{array} \qquad (20)$$

To a high-pressure vessel of 100 ml capacity are added 25 gm (0.57 mole) of ethylene oxide, 1.0 gm (0.0127 mole) of pyridine, and sulfur dioxide of an amount sufficient to give 2000 atm at 150°C. The reaction is heated for 4 hr, cooled, and vented. The product is distilled to afford 50 gm (82%), b.p. 53°C (12 mm), n_D^{20} 1.4470.

NOTE: If the pyridine catalyst is omitted, substantial amounts of dioxane are obtained.

Another reference describes this same synthesis in the absence of catalyst using 25 atm in a pressure reactor (pipe coil type) heated at 140°C to give a 92% yield [28a].

Ethylene sulfite polymerized at 0°–5°C in the presence of pyridine to give a viscous material in almost quantitative yield [28b].

Alkylene oxide–sulfur dioxide copolymers have recently been reported to be useful for surface-active agents [28c].

D. Reaction of Alcohols with Halogen–Pyridine–Sulfur Dioxide

Primary and secondary alcohols have been reported to be converted to dialkyl sulfites by reaction with iodine (0.5 mole equiv.) in pyridine (1.5 mole equiv.) in liquid sulfur dioxide (15 mole equiv.) at room temperature in 80–95% yield (see Table VII) [28d].

$$2ROH + I_2 + \text{pyridine} + SO_2 \longrightarrow (RO)_2SO$$

2-7. Preparation of Dinorbornan-2-endo-yl Sulfite [28d]

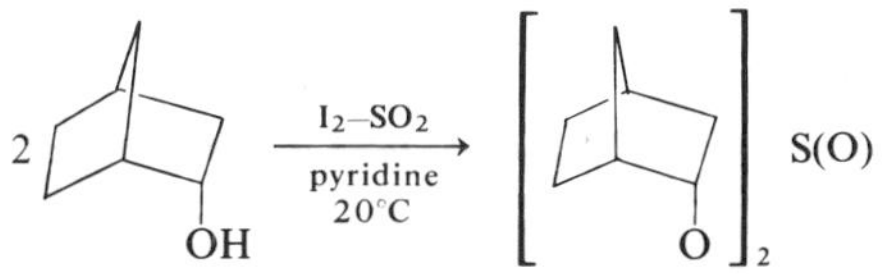

To a high-pressure vessel containing 30 ml of sulfur dioxide and cooled to −70°C is added 2.67 gm (0.015 mole) of iodine and 3.84 gm (0.045 mole) of pyridine. Then 3.36 gm (0.03 mole) of norbornan-2-endo-ol is added and the temperature kept at 20°C for 12 hr. The reaction is cooled in an acetone–Dry Ice bath and then the contents poured into aqueous sodium hydroxide. The sulfite product is extracted with ether and the extract then washed with hydrochloric acid and saturated brine. The extract is first dried over sodium sulfate and then concentrated. The product is isolated by distillation to give 3.85 gm (95%), b.p. 160°–162°C at 2 mm Hg.

Substitution of bromine for iodine gives lower yields. In addition, tertiary alcohols do not give sulfites but give mainly olefins.

TABLE VII

CONVERSION OF ALCOHOLS TO DIALKYL SULFITES BY USING REACTION WITH
IODINE–PYRIDINE–SULFUR DIOXIDE

ROH	Reaction time (hr)	Yield (%)	B.p. (°C)	mm Hg
R =				
C_2H_5	1	84	156–158	760
C_4H_9	1	92	118–120	20
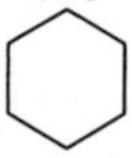 ⬡	12	95	159–160	10
Norbornane-2-endo-yl[a]	12	95	160–162	2

[a] See Preparation 2-7.

E. Alcoholysis of Sulfites

The alcoholysis of sulfites such as dimethyl sulfite offers a convenient method for the preparation of high-boiling dialkyl sulfites [33]. Earlier, Voos and Blanke [8a] reported that dimethyl sulfite is converted to diethyl sulfite in 44% yield. The reaction was shown to be acid-catalyzed and failed when barium carbonate was present. However, a patent refers to the use of lithium hydride in the transalcoholysis of 2,2-(4,4'-dihydroxyphenyl)propane with diphenyl sulfite or di-o-cresyl sulfite [32]. Recently Mehrotra and Mathur [34] reported that the alcoholysis reaction proceeds in the absence of catalysts. Their results are summarized in Eqs. (21)–(23) and Table VIII. Tertiary butanol did not

$$(C_2H_5O)_2SO + 2ROH \longrightarrow (RO)_2SO + 2C_2H_5OH \qquad (21)$$
$$R = CH_3, \, n\text{-}C_3H_7, \, n\text{-}C_4H_9, \, i\text{-}C_4H_9, \text{ and } i\text{-}C_5H_{11}$$

$$(C_2H_5O)_2SO + ROH \longrightarrow (RO)(C_2H_5O)SO + C_2H_5OH \qquad (22)$$
$$R = i\text{-}C_3H_7, \, s\text{-}C_4H_9$$

$$(CH_3O)_2SO + i\text{-}C_3H_7OH \longrightarrow (CH_3O)(i\text{-}C_3H_7O)SO + CH_3OH \qquad (23)$$

TABLE VIII

TRANSALCOHOLYSIS OF SULFITES[a]

Dialkyl sulfite (moles)	Alcohol	Moles	Product	Yield (%)	B.p., °C (mm Hg)	n_D^{20}
Ethyl						
0.106	CH_3	Excess	$(CH_3O)_2SO$	54	121.5 (760)	1.4560
0.0326	$n\text{-}C_3H_7$	0.20	$(n\text{-}C_3H_7O)_2SO$	52	65 (9)	1.4016
0.0407	$n\text{-}C_4H_9$	0.122	$(n\text{-}C_4H_9O)_2SO$	75	76 (10)	1.4458
0.0313	$i\text{-}C_4H_9$	0.149	$(i\text{-}C_4H_9O)_2SO$	79	65 (7)	1.4320
0.0234	$i\text{-}C_5H_{11}$	0.148	$(i\text{-}C_5H_{11}O)_2SO$	80	86 (10)	1.4126
0.030	$i\text{-}C_3H_7$	0.25	$(i\text{-}C_3H_7O)(C_2H_5O)SO$	77	55 (7)	1.4560
0.045	$s\text{-}C_4H_9$	0.175	$(s\text{-}C_4H_9O)(C_2H_5O)SO$	76	65 (10)	1.4430
CH_3						
0.0833	$i\text{-}C_3H_7$	0.615	$(i\text{-}C_3H_7O)(CH_3O)SO$	67	48 (10)	1.4210

[a] Data taken from Mehrotra and Mathur [34].

react with ethyl sulfite even after prolonged periods of refluxing and careful fractionation.

Another recent reference reports the uncatalyzed reaction of dimethyl sulfate with diols to give the cyclic sulfite shown in Eq. (24) [34a].

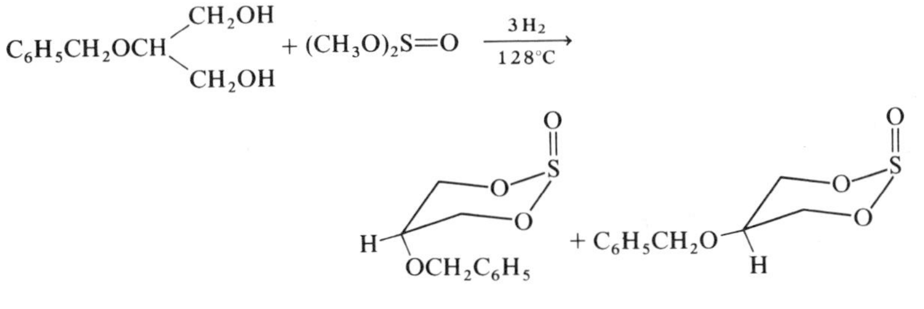

$$C_6H_5CH_2OCH \begin{matrix} CH_2OH \\ \\ CH_2OH \end{matrix} + (CH_3O)_2S{=}O \xrightarrow[128°C]{3H_2} \qquad (24)$$

2-8. Alcoholysis of Ethyl Sulfite with n-Propanol to Give Di-n-propyl Sulfite [34]

$$(CH_3CH_2O)_2S{=}O + 2CH_3CH_2CH_2OH \rightarrow (CH_3CH_2CH_2O)_2SO + 2CH_3CH_2OH \qquad (25)$$

To 4.50 gm (0.0326 mole) of ethyl sulfite are added 12.0 gm (0.20 mole) of n-propanol and 100 ml of benzene. The reaction mixture is refluxed for 8 hr, and during the course of the reaction the ethanol formed is removed as a benzene–ethanol azeotrope boiling at 68°C. The remaining liquid is further distilled under reduced pressure to afford 3.0 gm (55%), b.p. 65°C (9 mm), n_D^{20} 1.4616. Other alcoholysis examples, in which the same conditions are used, are described in Table VIII.

2-9. Preparation of Erythryl Sulfite [33]

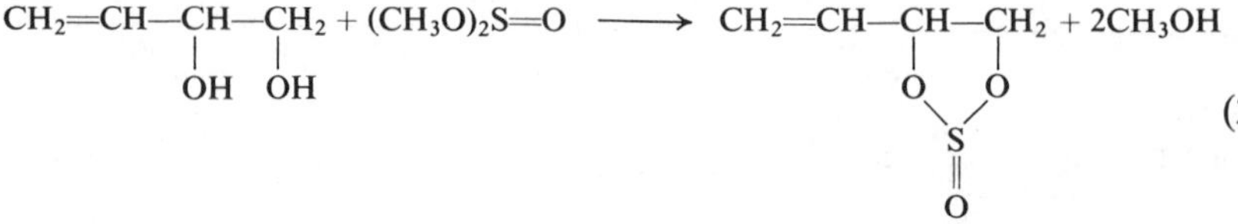

$$CH_2{=}CH{-}CH{-}CH_2 + (CH_3O)_2S{=}O \longrightarrow CH_2{=}CH{-}CH{-}CH_2 + 2CH_3OH$$

with OH, OH groups on the left compound, and the right compound bearing

$$\underset{\underset{O}{\parallel}}{\overset{O\diagdown\diagup O}{S}}$$

(26)

A mixture of 22.0 gm (0.20 mole) of dimethyl sulfite and 17.6 gm (0.20 mole) of erythrol (3-butene-1,2-diol) is heated to 120°C, whereupon a reaction commences which, within 20 min, yields 13.6 gm of methanol, b.p. 65°–66°C. The residue is distilled under reduced pressure and affords 18.2 gm (86%) of erythryl sulfite, b.p. 56°–68°C (3 mm), n_D^{20} 1.4588.

3. MISCELLANEOUS METHODS

(1) Oxidation of sulfur monoxide diethyl acetal with ozone [35].

(2) Reaction of disodium alkoxides with thionyl chloride [36].

(3) Reaction of diazo compounds with sulfur dioxide [37].

(4) Preparation of diethyl sulfate by the reaction of ethylene with $SO_2 +$ H_2O using an Ag_2SO_4 catalyst. Other catalytic methods are also described [38].

(5) Preparation of sulfites of β-halo alcohols by reaction of thionyl chloride with a 1,2-epoxides [39]. (See Procedure 3-1.)

(6) The reaction of sulfur monochloride with 1,2-epoxide to give β-chlorinated symmetrical sulfites [40].

(7) Reaction of an ether of glycidol with thionyl chloride to prepare bis(chloroalkyl) sulfites [41].

3-1. Preparation of Bis(2-chloroethyl) Sulfite [39]

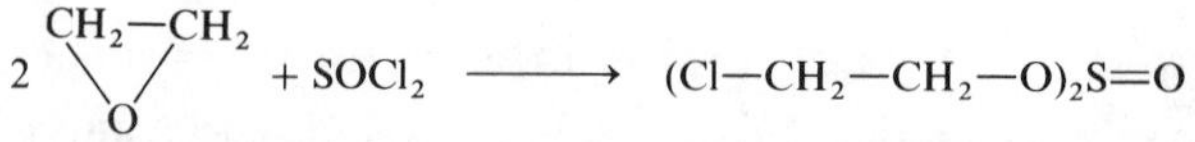

$$2 \underset{O}{\overset{CH_2{-}CH_2}{\diagup\diagdown}} + SOCl_2 \longrightarrow (Cl{-}CH_2{-}CH_2{-}O)_2S{=}O \qquad (27)$$

To an ice-cooled flask equipped with a Dry Ice condenser, mechanical stirrer, and dropping funnel and containing 135.6 gm (1.15 mole) of thionyl chloride is added over a 2.5 hr period 100 gm (2.30 moles) of ethylene oxide while the temperature is kept between 10°C and 15°C. The mixture is stirred for 1 hr at room temperature and then fractionally distilled to afford 198 gm (83%), b.p. 129°–133°C (10 mm), n_D^{20} 1.4814.

Propylene oxide under similar conditions yields bis(2-chloropropyl) sulfite, b.p. 122°–124°C (6 mm), n_D^{20} 1.4705, and 1,2-epoxy-3-butene affords bis(2-chloro-3-butenyl) sulfite, b.p. 140°–145°C (6 mm), n_D^{20} 1.4945. The use of epichlorohydrin (2-chloropropylene oxide) affords bis(2,3-dichloropropyl) sulfite, b.p. 150°–157°C (1 mm), n_D^{20} 1.5070.

REFERENCES

1. H. F. Van Woerden, *Chem. Rev.* **63**, 557 (1963).
2. H. French, H. Goebel, H. Staudermann, and W. Finkenbrink, U.S. Pat. 2,799,685 (1958).
3. S. B. Richter, U.S. Pat. 2,901,338 (1960).
4. A. Pechukas, U.S. Pat. 2,576,138 (1952).
5. Chemstrand Corp., Brit. Pat. 769,700 (1957).
6. W. W. Cone, *J. Econ. Entomol.* **60**, 436 (1967).
6a. K. K. Andersen, *in* "Comprehensive Organic Chemistry, the Synthesis and Reactions of Organic Compounds" (D. N. Jones, ed.), Vol. 3, pp. 367–369, 372, Pergamon, New York, 1979.
7. W. E. Bissinger and F. Kung, *J. Amer. Chem. Soc.* **69**, 2958 (1947).
8a. W. Voss and E. Blanke, *Ann. Chem.* **485**, 258 (1931).
8b. C. Barkenbus and J. J. Owen, *J. Amer. Chem. Soc.* **56**, 1204 (1934).
9. L. P. Kyrides, *J. Amer. Chem. Soc.* **66**, 1006 (1944).
10. W. E. Bissinger and F. E. Kung, *J. Amer. Chem. Soc.* **70**, 2664 (1948).
11. W. Gerrard, *J. Chem. Soc.* 218 (1940).
12. W. Gerrard and B. D. Shepherd, *J. Chem. Soc.* 2069 (1955).
13. W. Gerrard, *J. Chem. Soc.* 85 (1944).
14. J. S. Brimacombe, A. B. Foster, E. B. Hancock, W. G. Overend, and M. Stacey, *J. Chem. Soc.* 201 (1960).
14a. W. J. Myles and J. H. Prichard, U.S. Pat. 2,497,135 (1950).
14b. R. H. Garst and J. P. Henry, U.S. Pat. 3,554,986 (1971).
14c. M. E. Chiddix and R. W. Wynn, U.S. Pat. 3,169,130 (1965).
15. W. W. Carlson and L. H. Cretcher, *J. Amer. Chem. Soc.* **69**, 1952 (1949).
16. E. J. Grubbs and D. J. Lee, *J. Org. Chem.* **29**, 3105 (1964).
17. S. Hauptmann and K. Dietrich, *J. Prakt. Chem.* **19**, 174 (1963).
18. C. C. Price and G. Berti, *J. Amer. Chem. Soc.* **76**, 1211 (1954).
19. A. Green, *J. Chem. Soc.* 500 (1927).
19a. E. E. Hamel, U.S. Pat. 3,492,311 (1970).
19b. H. C. Vogt and P. Davis, U.S. Pat. 3,394,147 (1968).
19c. W. Reeve and S. K. Davidsen, *J. Org. Chem.* **44**, 3430 (1979).
20. L. F. Wiggins, C. C. Beard, and J. W. James, Brit. Pat. 944,406 (1963).
21. C. C. Price and G. Berti, *J. Amer. Chem. Soc.* **76**, 1207, 1211 (1954).
22. P. Carre and D. Libermann, *Bull. Soc. Chem. Fr.* **53** (5), 1050 (1933).
22a. W. D. Harris, M. D. Tate, and J. W. Zukel, U.S. Pat. 2,529,493 (1950).
22b. R. A. Covey, A. E. Smith, and W. L. Hubbard, U.S. Pat. 3,463,859 (1969).
22c. R. A. Covey and R. E. Grahame, Jr., U.S. Pat. 4,003,940 (1977).
23. P. Carre and P. Mauclere, *C. R. Acad. Sci. Paris* **192**, 1738 (1931).
24. G. Berti, *J. Amer. Chem. Soc.* **76**, 1213 (1954).
25. A. B. Foster, E. B. Hancock, W. G. Overend, and J. C. Robb, *J. Chem. Soc.* 2589 (1956).
26. F. G. Bordwell and P. S. Landis, *J. Amer. Chem. Soc.* **80**, 6379 (1958).

94   3. Sulfites

27. H. F. Herbrandsom, R. T. Dickerson, Jr., and J. Weinstein, *J. Amer. Chem. Soc.* **78**, 2576 (1956).
28. A. J. Shipman, Brit. Pat. 898,630 (1962).
28a. W. Dietrich and H. Höfermann, U.S. Pat. 2,833,785 (1958).
28b. M. J. Viard, U.S. Pat. 2,798,877 (1957).
28c. J. H. McCain, Jr., U.S. Pat. 4,304,732 (1981).
28d. S. Hasegawa, M. Nojima, and N. Tokura, *J. Chem. Soc., Perkin Trans.* **1**(1), 108 (1976).
29. H. Distler and G. Dittus, Ger. Pat. 1,217,970 (1966).
30. Farbwerke Hoechst A.-G., Brit. Pat. 753,872 (1956).
31. W. A. Rogers, Jr., J. E. Woekst, and R. M. Smith, U.S. Pat. 3,022,315 (1962).
32. K. Stuerzer, Ger. Pat. 1,212,072 (1966).
33. W. E. Bissinger, R. H. Fredenburg, R. G. Kadesch, F. Kung, J. H. Langston, H. C. Stevens, and F. S. Train, *J. Amer. Chem. Soc.* **69**, 2955 (1947).
34. R. C. Mehrotra and S. N. Mathur, *J. Indian Chem. Soc.* **44**, 651 (1967).
34a. J. S. Baran, D. D. Langford, and I. Laos, *J. Org. Chem.* **42**, 2260 (1977).
35. A. Meuwsen and H. Gebhardt, *Chem. Ber.* **69**, 937 (1936).
36. M. Allan, A. F. Janzen, and C. J. Willis, *Chem. Commun.* 55 (1968).
37. G. Hesse and S. Majmudar, *Chem. Ber.* **93**, 1129 (1960).
38. A. S. Ramage, U.S. Pat. 2,472,618 (1949).
39. A. Pehukas, U.S. Pat. 2,576,138 (1951).
40. M. S. Malinowski, *Zh. Obshch. Khim.* **9**, 835 (1939).
41. A. Pechukas, U.S. Pat. 2,684,380 (1954).

CHAPTER 4 / ENAMINES

1. INTRODUCTION

Wittig and Blumenthal [1] in 1927 introduced the term "enamine" to designate the nitrogen analog of the term "enol" (structures I and II). The enamine

$$\overset{\diagup}{\underset{\diagdown}{C}}=\overset{|}{\underset{}{C}}-N\overset{\diagup}{\diagdown} \qquad \overset{\diagup}{\underset{\diagdown}{C}}=\overset{|}{\underset{}{C}}-OH$$

enamine   enol

(I)   (II)

structure had been known in the early literature (pyrrole, indole, etc.), but it was not until 1954 [2, 3] that the chemical potential of this group was emphasized.

Enamines are capable of resonance, and electrophilic attack may occur at either the nitrogen atom or the β-carbon atom [2] (Eq. 1). In most cases

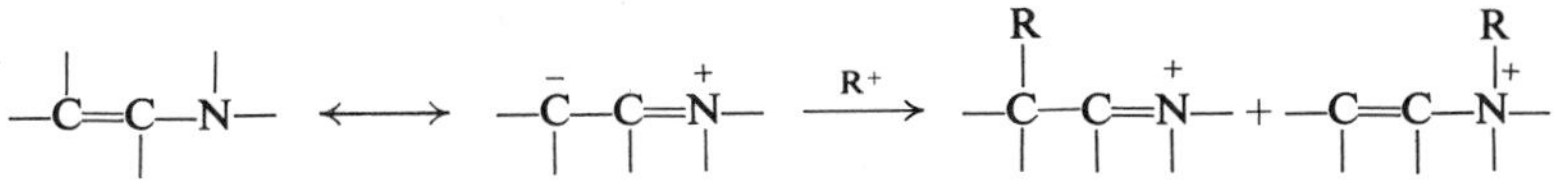

$$-\overset{|}{\underset{|}{C}}=\overset{|}{\underset{}{C}}-\overset{|}{N}- \longleftrightarrow -\overset{-}{\underset{|}{C}}-\overset{}{\underset{|}{C}}=\overset{+}{\underset{|}{N}}- \overset{R^+}{\longrightarrow} -\overset{R}{\underset{|}{\underset{}{C}}}-\overset{}{\underset{|}{C}}=\overset{+}{\underset{|}{N}}- + -\overset{}{\underset{}{C}}=\overset{}{\underset{}{C}}-\overset{R}{\underset{|}{\overset{|+}{N}}}-$$

(A)   (B)

(1)

electrophilic attack occurs on the β-carbon atom to yield the quaternary salt (A) which is decomposed by water to produce 2-alkyl-substituted ketones as illustrated in Eq. (2).

Owing to the low reactivity of substituted enamines, reaction (2) is useful mainly for the preparation of monoalkylated ketones.

In some instances an intermediate N-alkylation product may be formed first; it is then converted to a C-alkylation product.

Tertiary enamines are the most stable ones and are the main subject of this chapter. Secondary and primary enamines tend to revert to the tautomeric imino form [4] (Eq. 3).

95

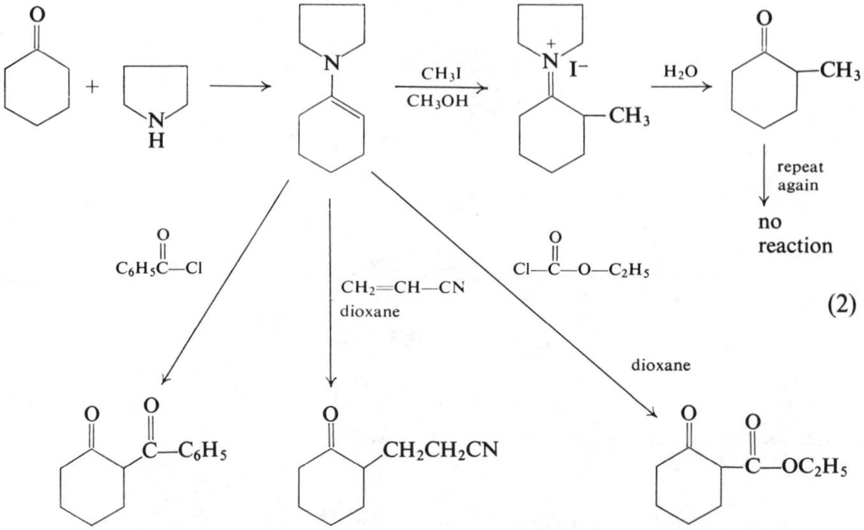

$$\text{(reaction scheme, Eq. 2)}$$

Cyclohexanone + pyrrolidine (N–H) $\longrightarrow$ enamine $\xrightarrow[\text{CH}_3\text{OH}]{\text{CH}_3\text{I}}$ iminium salt ($\overset{+}{\text{N}}$ I$^-$, –CH$_3$) $\xrightarrow{\text{H}_2\text{O}}$ 2-methylcyclohexanone $\xrightarrow{\text{repeat again}}$ no reaction

(2)

Branch reactions of the enamine:

$\text{C}_6\text{H}_5\overset{\text{O}}{\overset{\|}{\text{C}}}\text{–Cl}$ $\longrightarrow$ cyclohexanone with $\overset{\text{O}}{\overset{\|}{\text{C}}}\text{–C}_6\text{H}_5$

$\text{CH}_2\text{=CH–CN}$, dioxane $\longrightarrow$ cyclohexanone with $\text{CH}_2\text{CH}_2\text{CN}$

$\text{Cl–}\overset{\text{O}}{\overset{\|}{\text{C}}}\text{–O–C}_2\text{H}_5$, dioxane $\longrightarrow$ cyclohexanone with $\overset{\text{O}}{\overset{\|}{\text{C}}}\text{–OC}_2\text{H}_5$

Enamines with a tertiary nitrogen atom may be more basic than tertiary amines or enamines with a primary or secondary nitrogen atom [5–9]. The presence of α-alkyl substituents increases basicity, whereas β-alkyl substituents decrease basicity [10]. Stamhuis [11a] and co-workers found that the morpholine, piperidine, and pyrrolidine enamines of isobutyraldehyde in aqueous solutions are 200–1000 times weaker bases than the starting secondary amines and 30–200 times less basic than the corresponding saturated tertiary amines [11a]. For further discussion of enamine basicity see Stollenberger and Martin [11b].

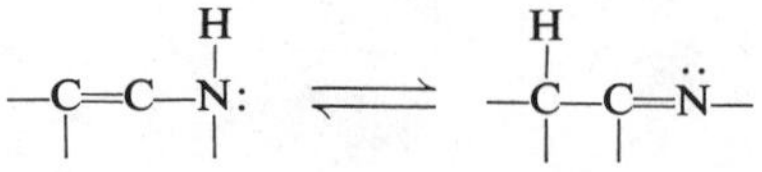

$$-\overset{\text{H}}{\underset{|}{\text{C}}}=\text{C}-\overset{\text{H}}{\underset{|}{\text{N}}}: \;\rightleftharpoons\; -\overset{\text{H}}{\underset{|}{\text{C}}}-\text{C}=\overset{..}{\text{N}}- \qquad (3)$$

Enamines are useful intermediates in organic synthesis, and they undergo some of the following reactions: alkylation [12a, b], arylation [13], oxidation [14], reduction [15], halogenation [16], cyanoethylation [12a, b, 17] (reaction with electrophilic olefins is quite general) [12a, b], reaction with ketenes [18], isocyanate [19], acylation [20, 31], Michael addition reactions [22], and cycloaddition reactions [22a]. Enamines also find use as a protecting group in peptide synthesis [23]. Since the hydrolysis of enamines gives ketones in good yield, the alkylation of enamines is useful for the preparation of alkyl-substituted ketones [12a, b].

In addition to the condensation of secondary amines with aldehydes or ketones (Eq. 2), the other important methods [24] of synthesizing enamines are briefly outlined in Eq. (4).

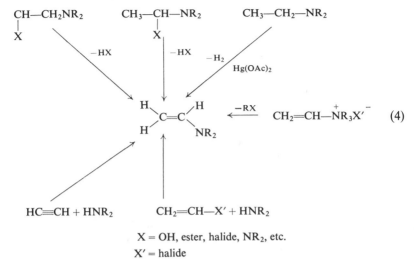

$$X = OH, \text{ ester, halide, } NR_2, \text{ etc.}$$
$$X' = \text{halide}$$

Another useful method involves the base-catalyzed isomerization of tertiary allylamines to *cis*- and *trans*-propenylamines (Eq. 5) and this is also described in Section 6.

$$CH_2=CH-CH_2NR_2 \xrightarrow[DMSO]{K-O\text{-}t\text{-}Bu} CH_3-CH=CH-NR_2 \qquad (5)$$

Enamides and thioenamides are prepared by similar methods and involve the condensation reaction of the appropriate amides with aldehydes [24a] to give

$$X = O \text{ or } S$$

The chemistry of enamines has been described in several reviews [25–29] and in a monograph [30].

2. CONDENSATION REACTIONS

A. Condensation of Secondary Amines with Aldehydes and Ketones

The most widely used method for preparing enamines involves the condensation of aldehydes and ketones with secondary amines (Eq. 6). In 1936,

Mannich and Davidsen [31] were the first to describe the reaction of secondary amines with aldehydes in the presence of anhydrous potassium carbonate at approximately 0°C to give enamines (Eqs. 6, 7).

$$R-CH_2CH{=}O + 2HNR'_2 \xrightarrow{-H_2O} RCH_2CH(NR'_2)_2 \xrightarrow{\text{heat}}$$

$$RCH{=}CH-NR'_2 + R_2'NH \qquad (6)$$

$$\underset{\displaystyle \underset{O}{\|}}{R-C-CH_2R'} + HNR''_2 \longrightarrow \underset{\displaystyle \underset{NR''_2}{|}}{R-C{=}CHR'} + H_2O \qquad (7)$$

Mannich and Davidsen [31] also reported that most of the aldehydes were first converted to a diamino intermediate which on distillation afforded the enamine. Less successful was the attempt to extend the reaction to ketones. Only with the use of calcium oxide at elevated temperatures were the enamines directly obtained, but in poor yields. Herr and Heyl [32, 33] found that better yields of enamines from ketones can be obtained by the azeotropic removal of water in the presence or absence [32, 33, 33a] of catalysts [2a, b, 34, 35]

$$R_2NH + R'-\underset{\displaystyle \underset{\|}{R'}}{CH_2C}{=}O \xrightarrow[\substack{\text{solvent or} \\ \text{drying agent}}]{\text{catalyst}} \underset{R}{\overset{R}{>}}N-\underset{\displaystyle \underset{R}{\overset{H}{|}}}{C}{=}C\underset{R'}{\overset{R'}{<}} + H_2O \qquad (8)$$

where the catalyst is K_2CO_3 [20], CaO [31], BaO [36], $p\text{-CH}_3\text{-C}_6\text{H}_4\text{SO}_3\text{H}$ [34, 35], Dowex 50 [37], acetic acid [38], or montmorillonite catalyst K10 [39]; and the solvent for azeotropic removal of water is C_6H_6 [32, 33 (less preferred because of toxicity)], $CH_3C_6H_5$ [12a, b] or $(CH_3)_2C_6H_4$ [12a, b]. Solvents for the nonazeotropic removal of water are methanol [40], acetone [41], pyridine [40], dimethylformamide [40], or tetrahydrofuran [41]. The drying agent consists of molecular sieves [42], calcium oxide [31], calcium chloride [43], or calcium carbide [40] (see Table I).

White and Weingarten [44] recently reported a novel method for converting carbonyl compounds to enamines using titanium tetrachloride and the free amine (Eq. 9). In this reaction the titanium tetrachloride acts as an effective

$$2RCH_2\overset{\displaystyle \overset{O}{\|}}{C}R' + 6NHR''_2 + TiCl_4 \longrightarrow 2RCH{=}C[NR_2'']R' + 4R_2''NH_2Cl + TiO_2$$
$$(9)$$

drying agent and as a Lewis acid catalyst to polarize the carbonyl bond. The yields of enamines ranged from 55 to 94% when dimethylamine, morpholine, or pyrrolidine reacted with ketones. Under the reaction conditions, 2,2,4-trimethyl-3-pentanone required several days for any visible reaction and 2,6-di-t-butylcyclohexanone was inert. (See Table II).

Several other Lewis acid metal halides are active ($AlCl_3$, $SnCl_4$, $FeCl_3$, $AsCl_3$, $SbCl_3$), but none appeared to be more effective than titanium tetrachloride.

For the general condensation reaction of secondary amines with ketones to yield enamines, pyrrolidine, piperidine, or morpholine is generally used. The rate of enamine formation depends on the basicity of the secondary amine and the steric environment of the carbonyl group [12a, b, 29]. Pyrrolidine, which is more basic, usually reacts faster than morpholine. The investigation of piperazine, a disecondary amine, has only been reported recently by Benzing [45, 46] and Sandler [41]. Surprisingly, the reaction of excess n-butyraldehyde with piperazine in tetrahydrofuran at −20°C to 0°C gave mainly N-1-butenyl-piperazine [41] (see Eq. 13).

It has been reported that some aldehydes react with secondary amines to give first a diamino derivative (aminal) which then decomposes on distillation to give the enamine and the starting secondary amine. Some aromatic aldehydes have been reported to give isolatable carbinolamines first [47].

The synthesis of enamines by the reaction of some ketones or aldehydes with a secondary amine can lead to saturated by-products when the enamine is heated in a nitrogen atmosphere with a catalytic amount of p-toluenesulfonic acid [48–50] (Eq. 10).

$$(50\%) \qquad (50\%) \tag{10}$$

Whether it is true that saturated by-product amines are formed in the reactions of other aldehydes and ketones needs further investigation.

Cyclic ketones react faster than the aliphatic ketones in the order: cyclopentanone > cyclohexanone > higher-membered cyclic ketones. α-Substituted ketones give the less substituted enamines, which in turn could be alkylated to put a substituent on this position of the ketone [51] (Eq. 11).

$$R_2CH-\underset{\underset{O}{\|}}{C}-CH_2R + HNR'_2 \longrightarrow R_2CH-\underset{\underset{NR_2'}{|}}{C}=CHR \xrightarrow[\text{2. } H_2O]{\substack{1.\ R''I, \\ CH_3OH}} R_2CH-\underset{\underset{O}{\|}}{C}-\underset{\underset{R''}{|}}{C}HR \tag{11}$$

Recently [52] it was reported that enamines derived from 2-substituted ketones exist as a mixture of the more and less substituted double-bond isomers. The isomer ratio is dependent on the various steric and electronic factors of the overlap of the lone pair of electrons on the nitrogen and the double bond of the

TABLE I

Preparation of Enamines by the Reaction of Secondary Amines with Aldehydes and Ketones

RCH=O (moles)	R'$_2$C=O (moles)	R''$_2$NH (moles)	R''$_2$N—C=C<	Catalyst (gm)	Solvent (ml)	Time (hr)	Temp. (°C)	B.p., °C (mm)	n_D^t	Yield (%)	Ref.
CH₃—CH—CH=O CH₃ (4.0)	—	(CH₃)₂NH (4.4)	(CH₃)₂C=CH—N(CH₃)₂	K₂CO₃ 150	Xylene 500	4.0	100	87–88	1.4221 (20°C)	55	53
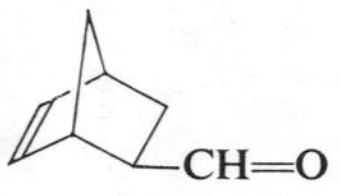 [norbornene]—CH=O (0.5)	—	 HN[morpholine] (0.5)	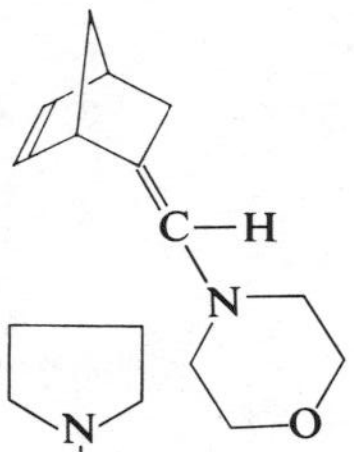 [norbornylidene-morpholine] C—H	—	Toluene 150	3.0	120	132–135 (14)	1.5346 (23°C)	81	54
—	Cyclopenta-none (1.0)	[pyrrolidine] N—H (1.05)	[pyrrolidine]N—[(CH₂)ₙ ring]	PTSA 1.0	C₆H₆ 375	2.5	80	85–86 (11)	1.5147 (20°C)	75	55
—	Cyclohexa-none (1.0)	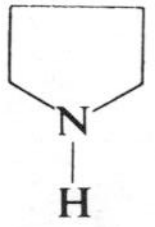 [pyrrolidine] N—H (1.05)	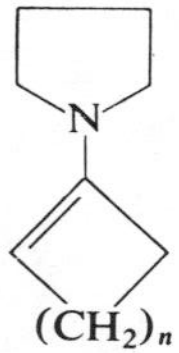 [pyrrolidine]N—[(CH₂)ₙ ring]	1.0	375	3.0	80	92–93 (5)	1.5223	93	55

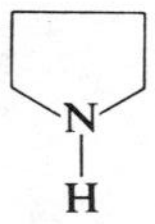

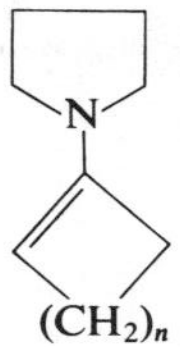

100

Ketone	Amine	Enamine	PTSA	C_6H_6			bp (mm)	n_D^{20}	Yield	Ref.
2-Methylcyclo-hexanone (1.0)	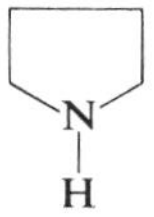 N—H (1.05)	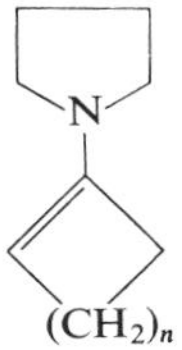 (CH$_2$)$_n$	1.0	375	40	80	91–92 (5)	1.5145	53	55
2-Phenylcyclo-hexanone (1.0)	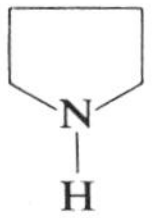 N—H (1.05)	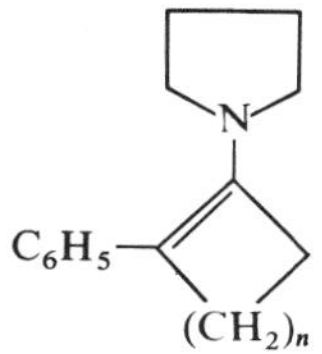 C$_6$H$_5$ (CH$_2$)$_n$	1.0	375	20	80	125–126 (0.005)	1.5755	73	55
Cycloheptanone (1.0)	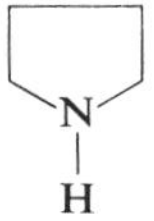 N—H (1.05)	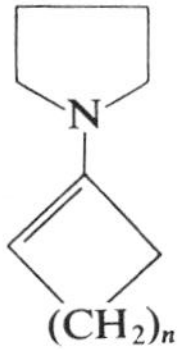 (CH$_2$)$_n$	1.0	375	20	80	100–102 (5)	1.5195	42	55
Cyclooctanone (1.0)	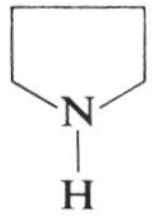 N—H (i.05)	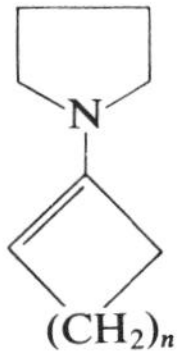 (CH$_2$)$_n$	1.0	375	20	80	113–114 (5)	1.5255	60	55

(continued)

101

TABLE I (continued)

RCH=O (moles)	R'₂C=O (moles)	R''₂NH (moles)	R''₂N—C=C<	Catalyst (gm)	Solvent (ml)	Time (hr)	Temp. (°C)	B.p., °C (mm)	n_D^t	Yield (%)	Ref.
				PTSA							
—	Cyclononanone (1.0)	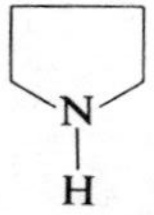 pyrrolidine N—H (1.05)	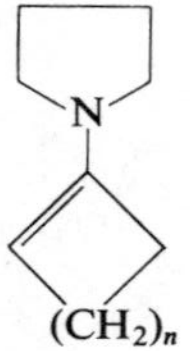	1.0	375	18	80	133–134 (5)	1.5270	72	55
—	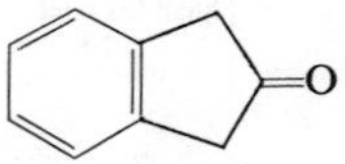 indanone (0.1)	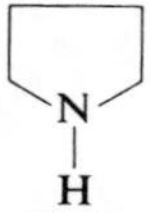 pyrrolidine N—H (0.2)	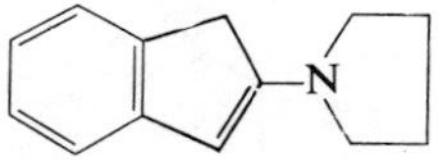	—	100		80	m.p. 120°–121°C	—	92	56
—	Cyclopentanone (450 ml)	 morpholine HN (350 ml)	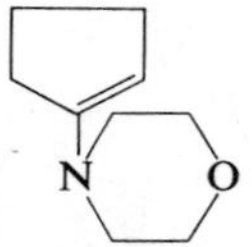	—	450	5	80	104–107 (12)	—	73	57

102

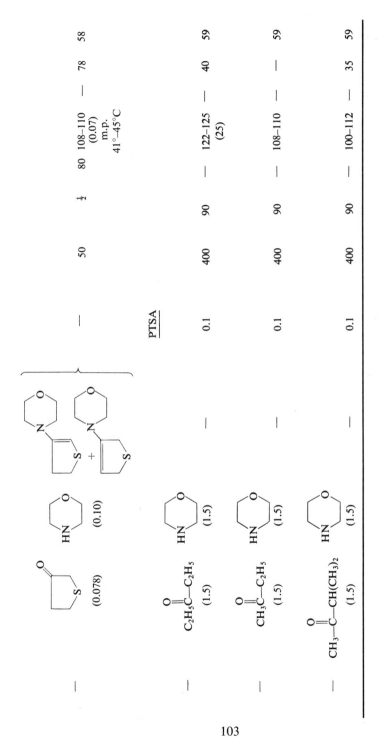

	PTSA								
—		50	$\frac{1}{2}$	80	108–110 (0.07) m.p. 41°–45°C	—	78	58
—	0.1	400	90	—	122–125 (25)	—	40	59
—	0.1	400	90	—	108–110	—	—	59
—	0.1	400	90	—	100–112	—	35	59

103

TABLE II

PREPARATION OF ENAMINES BY THE REACTION OF PROPIOPHENONES WITH CYCLIC AMINES OR DIMETHYLAMINE

$R_2O{=}O$	R_2NH	Catalyst	Solvent	Time	Temp. (°C)	B.p., °C (mm)	M.p., °C	n_D^t (°C)	Yield (%)	Product $R{-}C_6H_5\ C{=}CH{-}CH_2 / R_2'N$	Ref.
Conditions: 1.0 equiv.	1.5–2.0 equiv.										
$C_6H_5{-}\underset{\underset{O}{\|}}{C}{-}CH_2CH_3$	morpholine H	p-Toluene-sulfonic acid	Toluene	10 days	141	160–160.5 (14)	—	1.5523 (23.2)	53	R = H R_2' 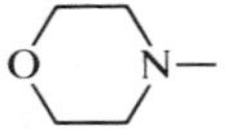 = O	a
$p\text{-}CH_3{-}C_6H_4\underset{\underset{O}{\|}}{C}{-}CH_2CH_3$	morpholine H	p-Toluene-sulfonic acid	Toluene	8 days	111	—	54–57°	—	42	R = H R_2' 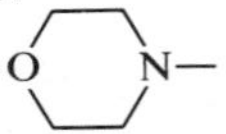 = O	a
$p\text{-}Cl{-}C_6H_4\underset{\underset{O}{\|}}{C}{-}CH_2CH_3$	morpholine H	p-Toluene-sulfonic acid	Toluene	7 days	111	—	43.2–44.2	—	54	R = Cl R_2' 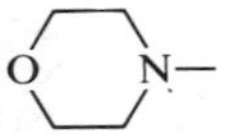 = O	a
$p\text{-}NO_2{-}C_6H_4\underset{\underset{O}{\|}}{C}{-}CH_2CH_3$	morpholine H	p-Toluene-sulfonic acid	Toluene	5 days	111	—	—	—	—	R – NO_2 R_2' 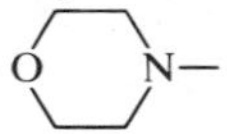 = O	a

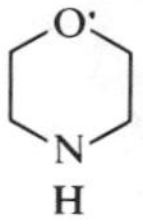

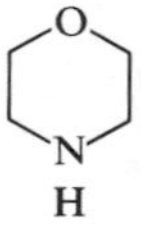

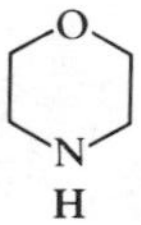

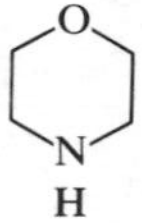

104

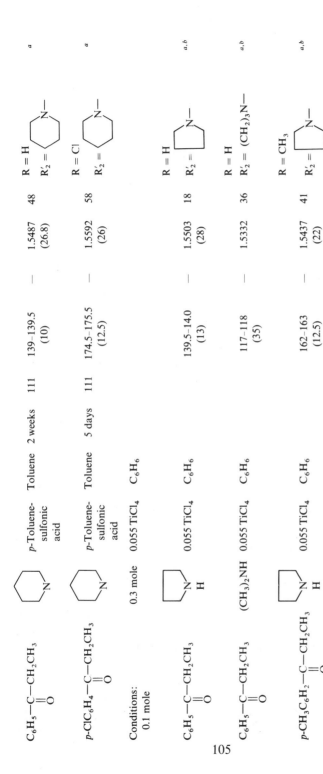

Ketone	Amine	Catalyst	Solvent	Time	Temp (°C)	b.p. °C (mm)		n_D (°C)	Yield (%)	Product	Ref.
$C_6H_5-C(=O)-CH_2CH_3$	(piperidine)	p-Toluene-sulfonic acid	Toluene	2 weeks	111	139–139.5 (10)	—	1.5487 (26.8)	48	R = H, R′₂ = (piperidine)	a
$p\text{-}ClC_6H_4-C(=O)-CH_2CH_3$	(piperidine)	p-Toluene-sulfonic acid	Toluene	5 days	111	174.5–175.5 (12.5)	—	1.5592 (26)	58	R = Cl, R′₂ = (piperidine)	a
Conditions: 0.1 mole	0.3 mole	0.055 TiCl₄	C₆H₆								
$C_6H_5-C(=O)-CH_2CH_3$	(pyrrolidine, N–H)	0.055 TiCl₄	C₆H₆			139.5–14.0 (13)	—	1.5503 (28)	18	R = H, R′₂ = (pyrrolidine)	a,b
$C_6H_5-C(=O)-CH_2CH_3$	$(CH_3)_2NH$	0.055 TiCl₄	C₆H₆			117–118 (35)	—	1.5332	36	R = H, R′₂ = $(CH_2)_3N-$	a,b
$p\text{-}CH_3C_6H_2-C(=O)-CH_2CH_3$	(pyrrolidine, N–H)	0.055 TiCl₄	C₆H₆			162–163 (12.5)	—	1.5437 (22)	41	R = CH₃, R′₂ = (pyrrolidine)	a,b

[a] P. Y. Sollenberger and R. S. Martin, *J. Amer. Chem. Soc.* **92**, 4261 (1970).

[b] W. A. White and H. Weingarten. *J. Org. Chem.* **32**, 213 (1967).

enamine. On the other hand, alkylation of ketones via their anions affords substitution on the most substituted carbon position [51].

Tables I and II give many examples of the preparation of enamines by the reaction of secondary amines with aldehydes and ketones.

2-1. Preparation of Dimethylamino-1-cyclohexene [60]

$$
\text{(cyclohexanone)} + (CH_3)_2NH \xrightarrow{\text{ether}} \text{(dimethylamino-1-cyclohexene)} + H_2O \qquad (12)
$$

To a flask equipped with a mechanical stirrer and Dry Ice condenser and containing 150 gm (3.4 moles) of dimethylamine in 400 ml of anhydrous ether are added 150 gm of 12-mesh anhydrous calcium chloride and 196 gm (2.0 moles) of cyclohexanone. The reaction mixture is stirred for 64 hr at room temperature under a nitrogen atmosphere, filtered, the residue washed with ether, the ether concentrated, and the residue fractionally distilled to afford 108.3 gm (52%) of the enamine; b.p. 81°C (35 mm), n_D^{25} 1.4851, and 94 gm (0.97 mole), b.p. 156°C, n_D^{19} 1.4522, of cyclohexanone. The conversion to enamine is 83% on the basis of the ketone utilized.

By a similar method, cyclopentanone reacts with dimethylamine to afford a 56% yield (or an 87% conversion based on ketone used) of dimethylamino-1-cyclopentene, b.p. 85°–86°C (104 mm), n_D^{25} 1.4801.

2-2. Preparation of N-1-Butenylpiperazine [41]

$$
HN\text{(piperazine)}NH + n\text{-}C_3H_7\text{—CH=O} \xrightarrow[-20°C]{THF} H\text{—N}\text{(piperazine)}N\text{—CH=CH—C}_2H_5 +
$$

$$
C_3H_7\text{—CH=}\underset{\underset{C_2H_5}{|}}{C}\text{—CH=O} + \text{polymers} \qquad (13)
$$

To a flask containing 288 gm (4.0 mole) of n-butyraldehyde and cooled to −20°C is slowly added a suspension of 172 gm (2.0 moles) of piperazine in 300 ml of tetrahydrofuran over a period of 1 hr. The reaction is exothermic and the temperature climbs to 35°C for a short time. After the addition, the reaction is stirred for 1 hr at 0°C, 200 ml of THF is added, stirred for 16 hr at room temperature, filtered, the solids washed with THF and dried to afford 221.0 gm (79%), m.p. 83°–85°C. The filtrate is distilled to afford 55.0 gm (22%, based on starting n-butyraldehyde) of 2-ethyl-2-hexenal, b.p. 33°–34°C (0.4 mm), n_D^{18} 1.4542. Some polymeric residue of high molecular weight remains in the flask.

2-3. Preparation of 1-Morpholino-1-cyclohexene [20, 61]

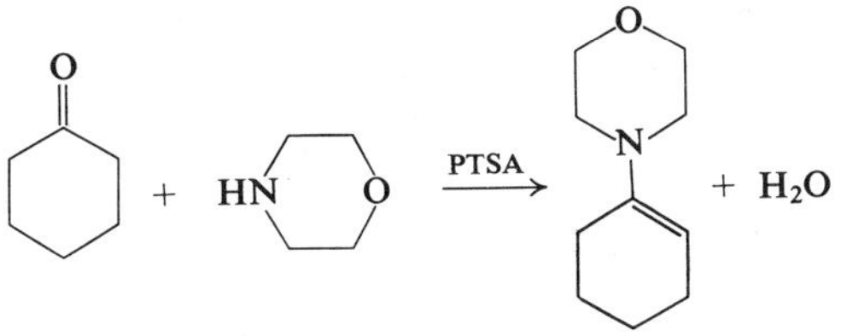

$$
\text{(cyclohexanone)} + HN \overset{O}{\underset{}{\bigcirc}} \xrightarrow{\text{PTSA}} \text{(1-morpholino-1-cyclohexene)} + H_2O \qquad (14)
$$

To a 500 ml, three-necked, round-bottomed flask equipped with a Dean and Stark trap, condenser, and stirrer is added a mixture of 52.5 gm (0.60 mole) of morpholine, 49 gm (0.5 mole) of cyclohexanone, 100 ml of toluene, and 0.5 gm of p-toluenesulfonic acid. The contents are refluxed for approximately 4 hr to remove the water of reaction. The product mixture is cooled and then distilled to afford 59 gm (71%) of the enamine, b.p. 117°–120°C (10 mm), n_D^{25} 1.5122–1.5129. Yields of enamine have been reported to vary from 71 to 80%.

2-4. Preparation of 1-Morpholino-1-isobutene [45, 46]

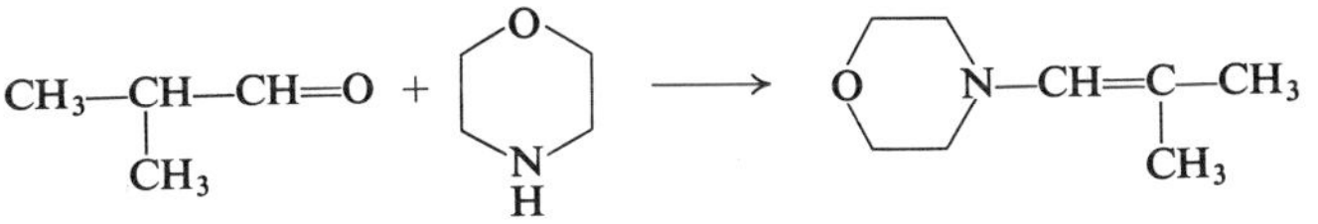

$$
CH_3\text{—}CH\text{—}CH\text{=}O + \underset{\underset{H}{N}}{\overset{O}{\bigcirc}} \longrightarrow O\bigcirc N\text{—}CH\text{=}C\text{—}CH_3 \qquad (15)
$$
$$
\underset{CH_3}{|} \qquad\qquad\qquad\qquad\qquad\qquad \underset{CH_3}{|}
$$

To an apparatus similar to that described in Preparation 2-3 and containing 43.5 gm (0.5 mole) of morpholine is slowly added 72 gm (1.0 mole) of iso-butyraldehyde. The reaction mixture is refluxed for about 3 hr until 9 ml of water separates and is then distilled to afford 66 gm (93.5%), b.p. 56°–57°C (11 mm), n_D^{20} 1.4670.

The following enamines are prepared by a similar method: (a) N,N-di-isobuten-1-yl-piperazine, 89%, b.p. 66°–67°C (1.0 mm), m.p. 35°–37°C; (b) N-isobutene-1-yl-N-methyl-cyclohexylamine, 75%, b.p. 96°C (19 mm), n_D^{20} 1.4740; and (c) N-isobuten-1-yl-N-methylaniline, 87%, b.p. 50°–53°C · (0.1 mm), n_D^{20} 1.5578.

2-5. Preparation of N,N-Dimethylisobutenylamine [53]

$$
CH_3\text{—}CH\text{—}CH\text{=}O + (CH_3)_2NH \longrightarrow (CH_3)_2N\text{—}CH\text{=}C(CH_3)_2 \qquad (16)
$$
$$
\underset{CH_3}{|}
$$

To a cooled pressure autoclave are added 288 gm (4.0 moles) of isobutyralde-hyde, 150 gm of potassium carbonate, 500 ml of xylene, and 200 gm (4.4 moles) of dimethylamine. The autoclave is sealed, and rocked, or stirred for 4 hr at 100°C, cooled, vented, and opened. The liquid is distilled to afford 122 gm of a forerun, b.p. $53°-87°C$, and then 216 gm (55%) of the enamine, b.p. $87°-88°C$, n_D^{20} 1.4221. Less xylene and potassium carbonate gave poor results. In addition, if the distillation is not carried out immediately after the reaction, lower yields of the enamine are possible.

2-6. Preparation of Methyl(5-dibutylamino)-4-ethyl-4-pentenoate [33a]

$$C_2H_5-\underset{\underset{CH=O}{|}}{CH}-(CH_2)_2-COOCH_3 + (C_4H_9)_2NH \xrightarrow{-H_2O} C_2H_5\underset{\underset{CH-N(C_4H_9)_2}{||}}{C}-(CH_2)_2COOCH_3$$

To a flask equipped with a Dean & Stark trap is added a solution of 50.0 gm (0.316 mole) of methyl 4-formylhexanoate and 46.0 gm (0.365 mole) of dibutylamine in 100 ml of dry benzene. (CAUTION: Benzene is considered a carcinogen and toluene can be substituted in its place.) The solution is refluxed for 2 hr to give 5.5 ml of water (96.6% theory). The product is recovered by distillation and gives 61.8 gm (72.6%), b.p. $108°-110°C$ (0.2 mm), n_D^{25} 1.4532. The ir of a film indicated 1747 cm^{-1} ($>C=O$) and 1660 cm^{-1} ($>C=C<$). The NMR in CCl$_4$ indicated $\delta 3.49$ (S, 3; CH$_3$O), $5.12 \left(s, 1, -CH=C{\overset{\displaystyle /}{\underset{\displaystyle \backslash}{}}} \right)$.

B. Condensation of Organometallic Reagents with Ketones

$$\underset{\underset{R}{|}}{\overset{(CH_2)_n}{\overbrace{}}}\!\!N\!-\!\!C\!=\!O + R'MgX \longrightarrow \left[\underset{\underset{R}{|}}{\overset{(CH_2)_n}{\overbrace{}}}\!\!N\!\!\overset{OMgX}{\underset{R'}{<}} \right] \xrightarrow{H^+}$$

$$\underset{\underset{R}{|}}{\overset{(CH_2)_n}{\overbrace{}}}\!\!N\!-\!R' + H_2O + some \quad \underset{\underset{R}{|}}{\overset{(CH_2)_n}{\overbrace{}}}\!\!N\!\!\overset{R'}{\underset{R'}{<}} \qquad (17)$$

Lukes [62] and Craig [63] first reported that N-methyl lactams with five- and six-membered rings react with Grignard reagents to yield 1-methyl-2-alkyl-pyrrolines ($n = 1$) and 1-methyl-2-alkylpiperideines ($n = 2$) In some cases, some 2,2-dialkylated bases are formed as by-products. This method provides a route to cyclic enamines [64] difficult to obtain by other methods. For example, 1-methyl-2-phenyl-1-azacyclohept-2-ene is synthesized via the reaction of N-methylcaprolactam with phenylmagnesium bromide, whereas the mercuric acetate oxidation (see Section 4) of 1-methyl-2-phenyl-1-azacyclo-heptane gives the open-chain amino ketone [29, 65–68] (Eqs. 18, 19).

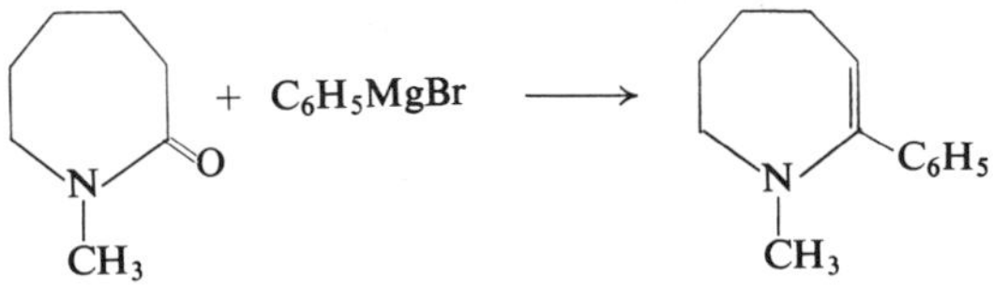

$$\text{(lactam)} + C_6H_5MgBr \longrightarrow \text{(enamine-}C_6H_5\text{)} \qquad (18)$$

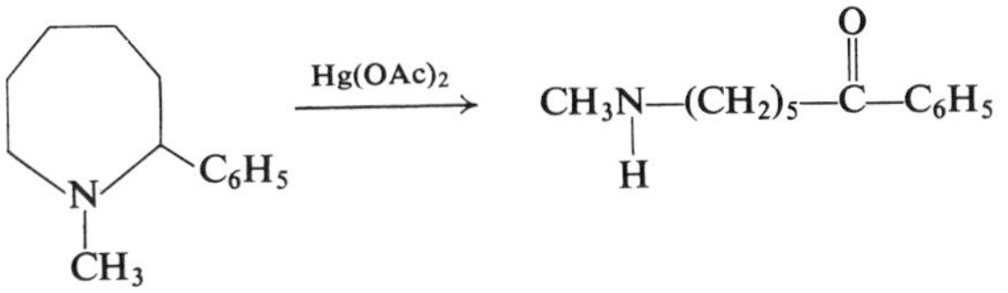

$$\text{(ring-}C_6H_5\text{)} \xrightarrow{\text{Hg(OAc)}_2} CH_3N-(CH_2)_5-\overset{\displaystyle O}{\overset{\|}{C}}-C_6H_5 \qquad (19)$$

$$\overset{\qquad}{\underset{H}{}}$$

The larger-ring lactams yield mainly acyclic amino ketones as the sole product [69–74].

2-7. *Preparation of N-Methyl-2-methylpyrroline* [75]

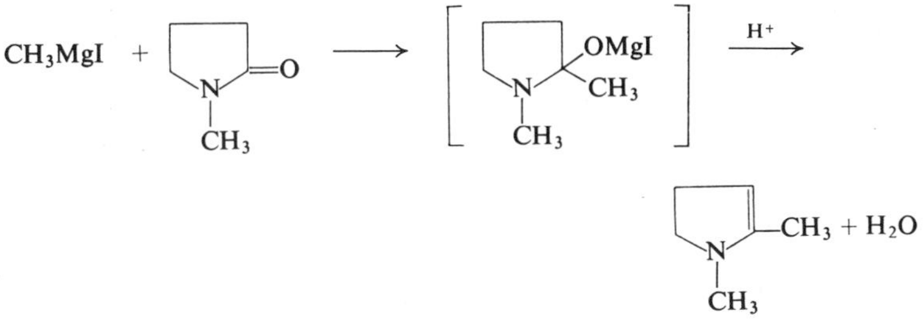

$$CH_3MgI + \text{(N-methyl-2-pyrrolidone)} \longrightarrow \left[\text{(OMgI, CH_3 intermediate)} \right] \xrightarrow{H^+}$$

$$\text{(N-methyl-2-methylpyrroline)} -CH_3 + H_2O \qquad (20)$$

To a flask containing 2 moles of methylmagnesium iodide in ether is added dropwise 99 gm (1.0 mole) of *N*-methyl-2-pyrrolidone dissolved in 100 ml of ether. The reaction mixture is allowed to stand overnight, then hydrolyzed with an equivalent of hydrochloric acid, the aqueous layer separated and made alkaline with sodium hydroxide. Steam distillation of the aqueous layer yields the product, which is dried over potassium hydroxide and then redistilled to afford 48.5 gm (50%) *N*-methyl-2-methylpyrroline, b. p. 130°–131°C.

The accompanying tabulation shows the *N*-methyl-2-R-pyrrolines prepared by a similar method.

R	B.p., °C (mm)	Yield (%)
Phenyl	105–109 (11)	70
n-Butyl	86–87 (30)	54
n-Propyl	176–177	50
Ethyl	148–149	53

2-8. Preparation of N-Methyl-2-[6'-methoxynaphthyl-(2')]-pyrideine [76]

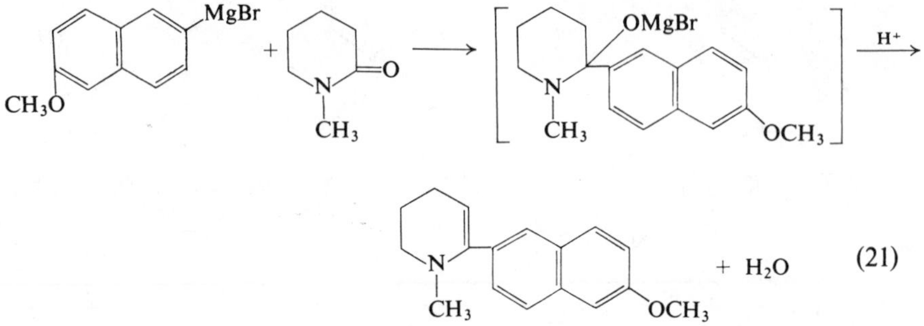

$$+ H_2O \qquad (21)$$

To a dry flask containing 8 gm (0.329 atom) of magnesium turnings, 150 ml of dry ether, and a crystal of iodine is added dropwise a solution of 50 gm (0.212 mole) of 2-bromo-6-methoxynaphthalene in 150 ml of dry benzene. The reaction mixture is refluxed on a water bath and 2 ml of ethyl bromide is added every hour for 5 hr. After this time, most of the magnesium has reacted, the flask is cooled, and a solution of 37 gm (0.328 mole) of 1-methyl-2-piperidone in 100 ml of ether is added dropwise. After the addition, the reaction mixture is refluxed for 3 hr, cooled, decomposed with dilute hydrochloric acid, the solid hydrochloride of the product, after filtration, reacts with warm aqueous sodium hydroxide, is extracted with ether, dried, and distilled to afford 11.0 gm (20.6%), b.p. 175°C (2.0 mm).

The preparation of fulvene enamines by the reaction of sodium cyclopentadienide with dimethyl sulfate complexes of amides or lactams has recently been reported by several investigators [77, 78] (Eq. 22).

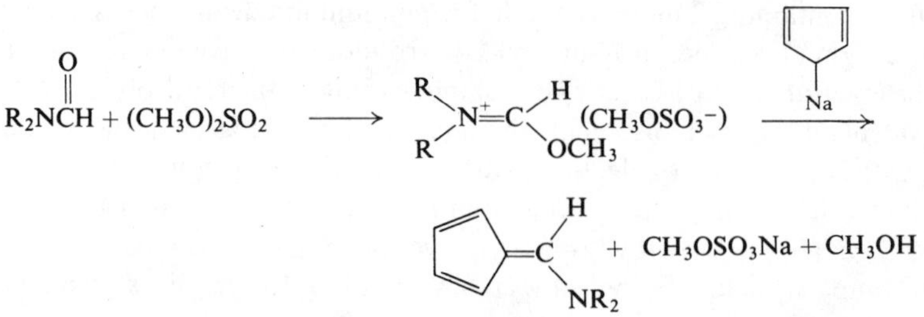

$$+ \; CH_3OSO_3Na + CH_3OH \qquad (22)$$

2-9. Preparation of 1-Methyl-2-cyclopentadienylidene-2,3,4,5-tetrahydropyrrole [79]

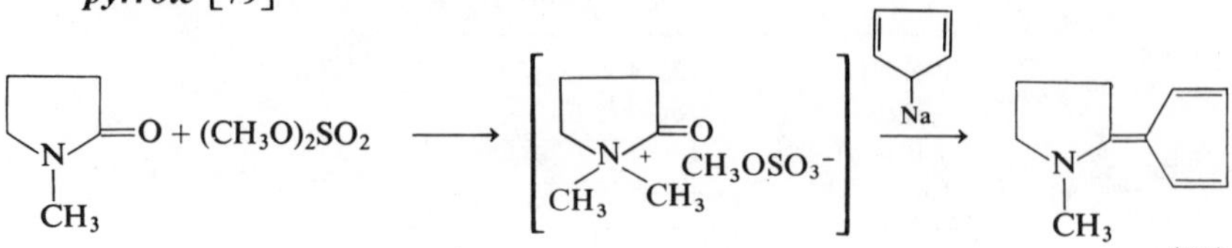

$$(23)$$

To a stirred flask containing a suspension of 4.8 gm (0.2 mole) of sodium hydride in 50 ml of THF under a nitrogen atmosphere is added dropwise 13.3 gm (0.2 mole) of cyclopentadiene. In a separate flask are mixed 19.8 gm (0.2 mole) of N-methyl-2-pyrrolidone and 25.2 gm (0.2 mole) of dimethyl sulfate. (NOTE: Separately N-methyl-2-pyrrolidone and dimethyl sulfate should be anhydrous and should be purified by distillation before reacting with each other to form the complex.) The latter mixture is heated for 20 min on a steam bath, cooled, and then added dropwise to the cyclopentadienylsodium solution at $-5°C$. After the addition, the reaction mixture is stirred for 2 hr, filtered, the filtrate concentrated, and the residue is obtained as a brown oil. On cooling, the brown oil solidifies and is recrystallized from cyclohexane to afford pale-yellow needles weighing 14.0 gm (50%), m.p. $100°-101°C$.

C. Reaction of Acetylenes with Secondary Amines

It has been postulated that vinylamine is an intermediate in the catalytic condensation of ammonia with acetylene which leads to either acetonitrile or 2-methyl-5-ethylpyridine, depending on the reaction conditions [80] (Eq. 24).

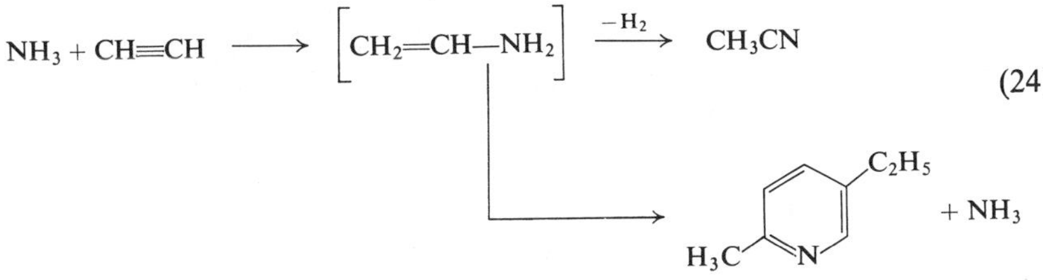

$$NH_3 + CH{\equiv}CH \longrightarrow \left[CH_2{=}CH{-}NH_2\right] \xrightarrow{-H_2} CH_3CN \tag{24}$$

$$+ NH_3$$

$$R^1R^2NH + CH{\equiv}CH \longrightarrow [R^1R^2N{-}CH{=}CH_2] \xrightarrow{CH{\equiv}CH} CH_3{-}\underset{\underset{NR^1R^2}{|}}{CH}{-}C{\equiv}CH \tag{25}$$

Primary amines react with acetylene to give resinous products. Most secondary amines react in the presence of copper acetylide to give 3-amino-1-butynes as described in Eq. (25) [81]. The R^1R^2 combinations in R^1R^2NH are shown in the accompanying tabulation.

R^1	R^2	R^1	R^2
CH_3	CH_3	$C_6H_5CH_2$	H
C_2H_5	C_2H_5	C_6H_{11}	H
$n\text{-}C_4H_9$	H	C_4H_8	C_4H_8
C_6H_5	H	C_5H_{10}	C_5H_{10}
C_6H_5	CH_3	C_6H_{12}	C_6H_{12}
C_6H_5	C_2H_5		

Some 3-amino-1-butynes have been reported to rearrange in the vapor phase to the isomeric 2-amino-1,3-butadienes or when heated from 10° to 30°C above their boiling point [81] (Eq. 26). Similarly, 1,4-bis(dialkylamino)-

$$CH_3—CH—C\equiv CH \longrightarrow CH_2=C—CH=CH_2 \qquad (26)$$
$$\overset{|}{NR_2} \qquad\qquad\qquad \overset{|}{NR_2}$$

butynes rearrange at 300°–450°C over chrome alumina catalysts to yield 1,4-bis(dialkylamino)-1,3-butadienes [81] (Eq. 27).

$$(CH_3)_2N—CH_2—C\equiv C—CH_2—N(CH_3)_2 \longrightarrow$$

$$(CH_3)_2N—CH=CH—CH=CH—N(CH_3)_2 \quad (27)$$

In addition, 3-phenylamino-1-butyne rearranges to give some 1-phenyl-pyrroline as a by-product [81] (Eq. 28).

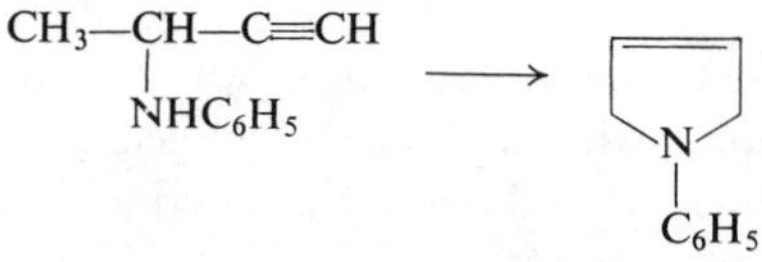

$$CH_3—CH—C\equiv CH \longrightarrow \qquad\qquad (28)$$
$$\overset{|}{NHC_6H_5}$$

Tertiary amines react with acetylene under pressure to give quaternary vinylammonium hydroxides [82a, b] (Eq. 29).

$$R_3N + H_2O + CH\equiv CH \rightarrow [R_3N—CH=CH_2]^+OH^- \qquad (29)$$

Reppe has shown that secondary amines of low basicity, such as carbazole, diphenylamine, indole, imidazole, and benzimidazole, and amides such as pyrrolidone react with acetylene in the presence of strong alkali to give vinyl derivatives [81, 83, 84a, b, c] (Eq. 30). As described by Reppe, these reactions

$$R_2NH \xrightarrow[KOH]{CH\equiv CH} R_2NCH=CH_2 \qquad (30)$$

are not recommended as laboratory procedures because special safety equipment is required when working with acetylene under pressure. A modified laboratory procedure using carbazole has been described [84c].

The addition of ammonia or primary and secondary amines to acetylenes is greatly enhanced if electron-attracting groups are attached to or conjugated with the triple bond. This reaction may be formulated as in Eq. (31).

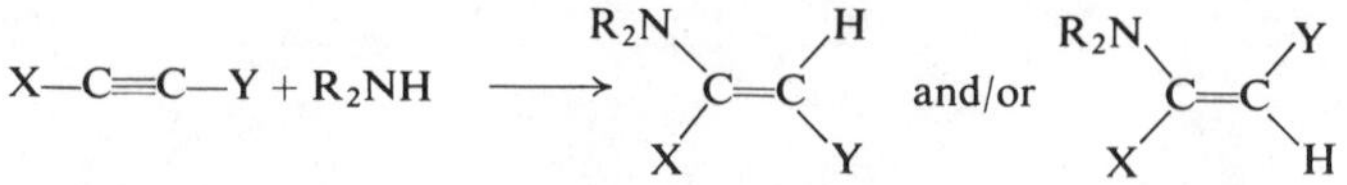

$$X—C\equiv C—Y + R_2NH \longrightarrow \quad \overset{R_2N}{\underset{X}{\diagdown}}C=C\overset{H}{\underset{Y}{\diagup}} \quad \text{and/or} \quad \overset{R_2N}{\underset{X}{\diagdown}}C=C\overset{Y}{\underset{H}{\diagup}} \qquad (31)$$

$$\text{(cis)} \qquad\qquad\qquad \text{(trans)}$$

$$X = SO_2—R' \; [85–89], \quad —SOR' \; [88], \quad \overset{O}{\underset{\|}{C}}—O—R' \; [87–95], \quad \overset{O}{\underset{\|}{C}}—R' \quad [88, 94, 95]$$

$$Y = H \text{ or alkyl}$$

Tables III, IV, and V describe some examples of this reaction.

Dolfini [89] found that, when equimolar ratios of aziridine (ethylenimine) reacts with dimethyl acetylene dicarboxylate in methanol at room temperature, a 67% yield of the trans and a 33% yield of the cis product are obtained. In dimethyl sulfoxide solvent, 95% of the cis and 5% of the trans ester are obtained (Eq. 32).

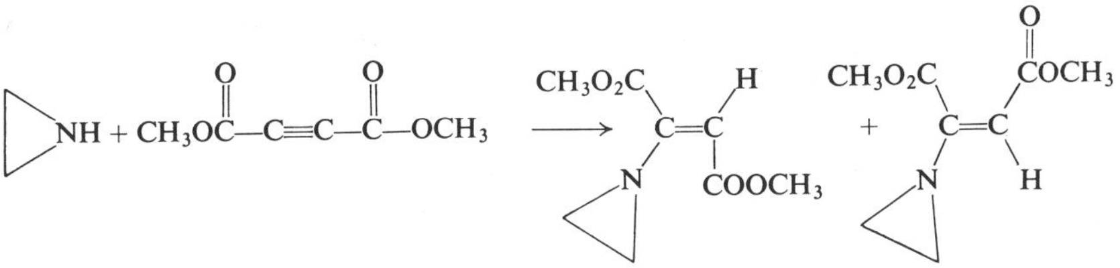

Solvent: CH₃OH	67%	33%	(32)
DMSO	5%	95%	

As seen in Tables III–V, dialkylamines react with the substituted terminal acetylenes to give only *trans*-aminovinyl products. Primary aliphatic amines react with both ethyl propiolate and 1-ethylsulfonyl-1-propyne to give mixtures of cis and trans products and with *p*-tolylsulfonylacetylene to give only trans products. Ethylenimine reacts with ethyl propiolate and 1-ethylsulfonyl-1-propyne to give mixtures of cis and trans products. Aniline reacts with *p*-tolylsulfonylacetylene to give a mixture of cis and trans products. The solvent has a great effect on the cis- and trans ratio when ethylenimine is used.

Additional information about the effect of amine structure and solvent on the steric course of the reaction of amines with acetylenecarboxylic esters has recently been reported by Huisgen [93].

It is interesting to note that Padwa and Hamilton [92] reported that *cis*-2,3-diphenylaziridine reacts with dimethyl acetylene dicarboxylate in benzene to give only the cis product (Eq. 33). Some additional examples of this reaction

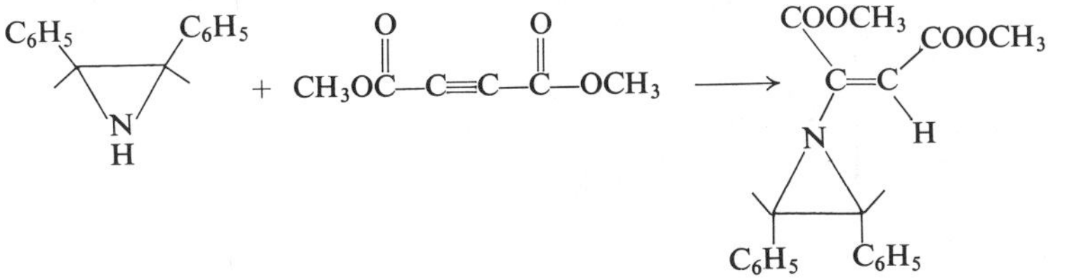

TABLE III[a]

REACTION OF RR'NH WITH $HC \equiv CSO_2C_6H_4CH_3\text{-}p$

R	R'	Solvent	Reaction time (hr)	Temp. (°C)	Yield (%)	M.p. (°C)	Configuration (%) cis	Configuration (%) trans
CH_3	CH_3	CH_2Cl_2	2	−75	73	134–135	—	100
C_2H_5	C_2H_5	C_2H_5OH	4	25–30	90	79–81	—	100
$n\text{-}C_3H_7$	H	C_2H_5OH	4	25–30	85	76–77	—	100
$i\text{-}C_3H_7$	$i\text{-}C_3H_7$	C_2H_5OH	4	25–30	91	140–141	—	100
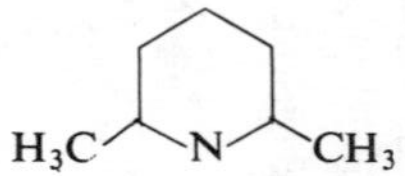 2,6-dimethylpiperidine (H₃C–N–CH₃ ring)		C_2H_5OH	4	25–30	88	114–116	—	100
C_6H_5	H	C_2H_5OH	20	25–30	80	Wide	15	85
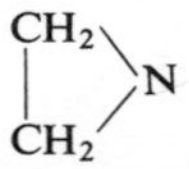		C_6H_5	4	0	89	88–89	100	—
		C_2H_5OH	4	0, 25–30	64		100	—
aziridine (CH₂–N–CH₂ ring)		CH_2Cl_2	4	0	84		100	—

[a] Tables III–V are reprinted from W. E. Truce and D. G. Brady, *J. Org. Chem.* **31**, 3543 (1966). Copyright 1966 by The American Chemical Society. Reprinted by permission of the copyright owner.

114

TABLE IV

Reaction of RR'NH with HC≡CCO₂C₂H₅

R	R'	Solvent	Reaction time (hr)	Temp. (°C)	Yield (%)	B.p., °C (mm Hg)	Configuration (%)	
							cis	trans
CH₃	CH₃	C₂H₅OH	3	0	74	90–91 (2.3), lit.[a] 97–98 (0.5)	—	100
		C₆H₆	3	0	71	—	—	100
C₂H₅	C₂H₅	C₂H₅OH	4	25–30	84	97–98 (1.3), lit.[b] 129–130 (18)	—	100
		C₆H₆	4	25–30	85	—	—	100
n-C₃H₇	H	C₂H₅OH	4	25–30	87	53–60 (2)	61	39
		C₆H₆	4	25–30	70	92–95 (5)	64	36
t-C₄H₉	H	C₂H₅OH	4	25–30	84	90–92 (4)	88	12
$\begin{array}{c}CH_2\\ \big|\\ CH_2\end{array}\!\!N$		C₂H₅OH	4	25–30	81	81–86 (7), lit.[c] 89–95 (12)	54	46
		C₆H₆	4	25–30	80	89–93 (7), lit.[c] 98–103 (12)	10	90

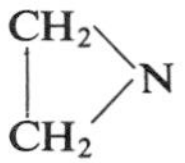

[a] J. Decombe, Ann. Chem. 18, 108 (1932).
[b] F. Straus and W. Voss, Ber. 59B, 1681 (1926).
[c] J. E. Dolfini, J. Org. Chem. 30, 1298 (1965).
[d] Product isolated after distillation is pure cis compound.

TABLE V

Reaction of RR'NH with CH₃C≡CSO₂C₂H₅

R	R'	Solvent	Reaction time (hr)	Temp. (°C)	Yield (%)	M.p. or b.p., °C (mm Hg)	Configuration (%)	
							cis	trans
C₂H₅	C₂H₅	C₂H₅OH	4	25–30	74	143–145 (0.5)	—	100
n-C₃H₇	H	C₂H₅OH	4	25–30	74	119–121 (0.4)	45	55
t-C₄H₉	H	C₂H₅OH	4	25–30	75	96–100	32	68
$\begin{array}{c}CH_2\\ \big|\\ CH_2\end{array}\!\!N$		C₂H₅OH	4	25–30	73	109–113 (0.4)	70	30
		C₆H₆	4	25–30	80	115–118 (0.3)	18	82

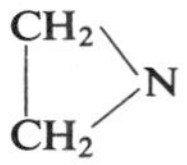

are shown in Eqs. (34)–(36) [96–98].

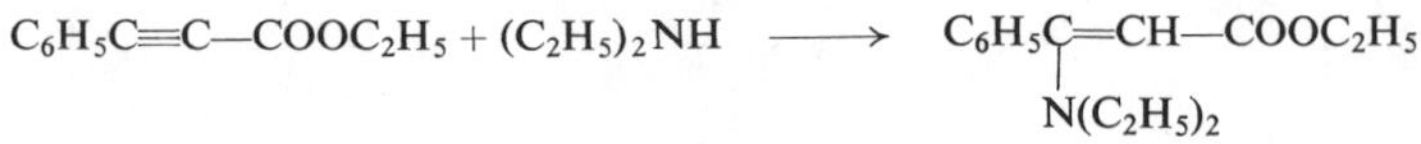

$$C_6H_5C\!\equiv\!C\!-\!COOC_2H_5 + (C_2H_5)_2NH \longrightarrow C_6H_5\underset{\underset{N(C_2H_5)_2}{|}}{C}\!=\!CH\!-\!COOC_2H_5 \quad (34)$$

$$R\underset{\underset{O}{\|}}{C}\!-\!C\!\equiv\!C\!-\!R' + R_2NH \longrightarrow R\underset{\underset{O}{\|}}{C}\!-\!CH\!=\!\underset{\underset{NR_2}{|}}{C}\!-\!R' \quad (35)$$

$$C_6H_5\underset{\underset{O}{\|}}{C}\!-\!C\!\equiv\!C\!-\!COOCH_3 \xrightarrow{C_6H_5NH_2} C_6H_5\underset{\underset{O}{\|}}{C}\!-\!CH\!=\!\underset{\underset{NHC_6H_5}{|}}{C}\!-\!COOCH_3 \quad (36)$$

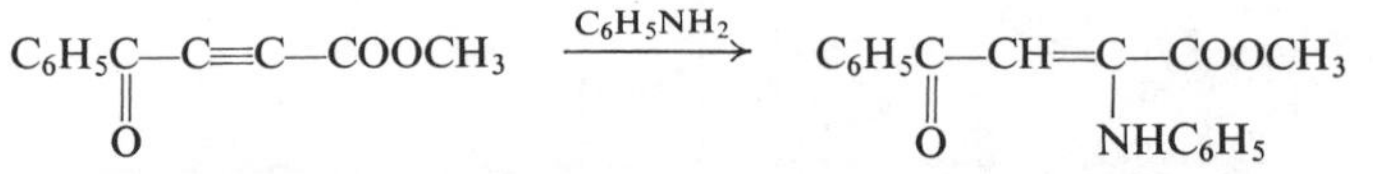

The type of addition described above takes place preferably at the triple bond even if a similarly activated double bond is also present in the molecule [98].

2-10. *Preparation of trans-Ethyl β-Dimethylaminoacrylate* [87]

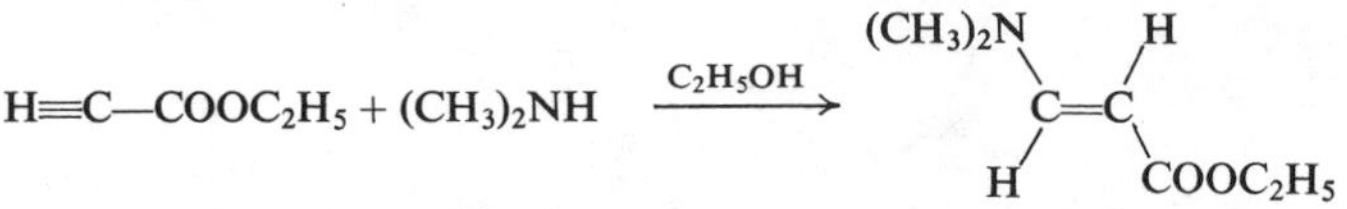

$$H\!\equiv\!C\!-\!COOC_2H_5 + (CH_3)_2NH \xrightarrow{C_2H_5OH} \begin{array}{c} (CH_3)_2N \quad\quad H \\ \diagdown\quad\quad\diagup \\ C\!=\!C \\ \diagup\quad\quad\diagdown \\ H\quad\quad COOC_2H_5 \end{array} \quad (37)$$

To a 100 ml three-necked flask equipped with a mechanical stirrer, condenser, and an addition funnel are added 2.0 gm (0.020 mole) of ethyl propiolate and 25 ml of absolute ethanol. The stirred solution is cooled to 0°C and then 3.0 gm (0.067 mole) of dimethylamine is slowly added. The reaction mixture is stirred for 2.5 hr at 0°C, concentrated under reduced pressure, and then distilled to afford 2.1 gm (74%), b.p. 90°–91°C (2.3 mm).

Benzene may be used in place of ethanol to give essentially the same results (see Table IV).

2-11. *Preparation of trans-2-Dibenzylamino-1-phenylsulfonylpropene* [85]

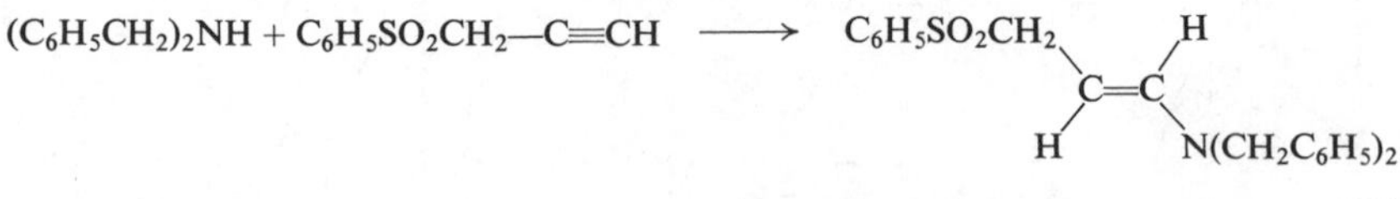

$$(C_6H_5CH_2)_2NH + C_6H_5SO_2CH_2\!-\!C\!\equiv\!CH \longrightarrow \begin{array}{c} C_6H_5SO_2CH_2 \quad\quad H \\ \diagdown\quad\quad\diagup \\ C\!=\!C \\ \diagup\quad\quad\diagdown \\ H\quad\quad N(CH_2C_6H_5)_2 \end{array}$$

$$(38)$$

To a flask containing 5.0 gm (0.0278 mole) of 3-phenylsulfonylpropyne is added a solution of 5.47 gm (0.0278 mole) of dibenzylamine in 100 ml of benzene. The resulting solution is kept at 20°C for 15 min, then concentrated [40°C

(10 mm)] to give a residue which, on addition of petroleum ether (b.p. 30°–60°C), gives 10.0 gm (95%), m.p. 104°C (recrystallized from benzene–petroleum ether.

Dienamines and enamine orthoformates are produced by the reaction of DMF acetals and ethynyl alcohols catalyzed by a trace of pivalic acid [85a].

3. ELIMINATION REACTIONS

One of the earliest preparations of an enamine was via the elimination reaction which involved the pyrolysis of hydroxyethyltrimethylammonium hydroxide to give N-vinyldimethylamine [99] (Eq. 39).

$$HO-CH_2-CH_2-\overset{+}{N}(CH_3)_3\overset{-}{O}H \quad \xrightarrow[\text{heat}]{-H_2O} \quad CH_2{=}CH-N(CH_3)_2 \qquad (39)$$

More practical methods involve the dehydrohalogenation [100], dehydrocyanation [101], or stereospecific bimolecular β-elimination of mesitoate esters using potassium t-butoxide in dimethyl sulfoxide [102] as described in Eqs. (40)–(42).

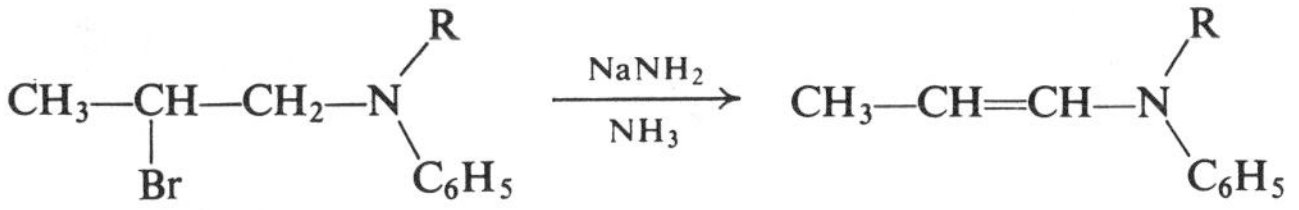

$$CH_3-\underset{\underset{Br}{|}}{CH}-CH_2-N\overset{R}{\underset{C_6H_5}{\diagup}} \quad \xrightarrow[NH_3]{NaNH_2} \quad CH_3-CH{=}CH-N\overset{R}{\underset{C_6H_5}{\diagup}} \qquad (40)$$

$$R-CH_2-\underset{\underset{CN}{|}}{\overset{\overset{R}{|}}{C}}-N\ddot{R}_2 \quad \xrightarrow[KNH_2 \text{ in } NH_3]{\text{heat or}} \quad RCH{=}\underset{}{\overset{\overset{R}{|}}{C}}-NR_2 \qquad (41)$$

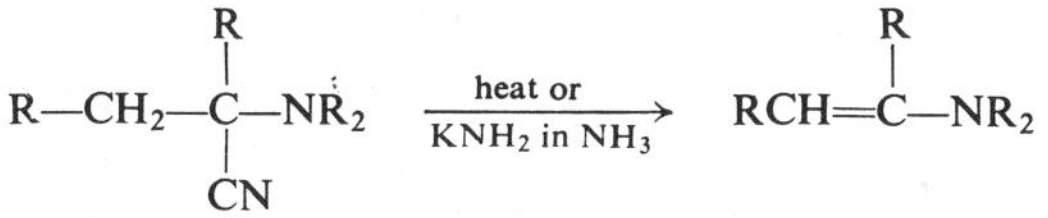

$$\text{Eq. (42)}$$

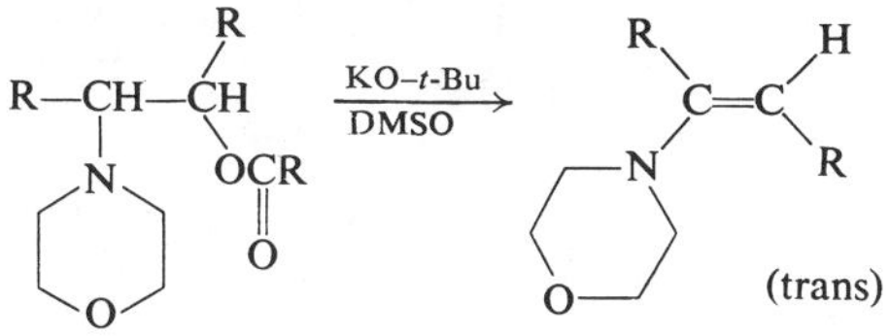

Potassium acetate has also been used for the dehydrobromination and dequaternization of 1,1,3-trimethylpyrrolidinium bromide [103] (Eq. 43).

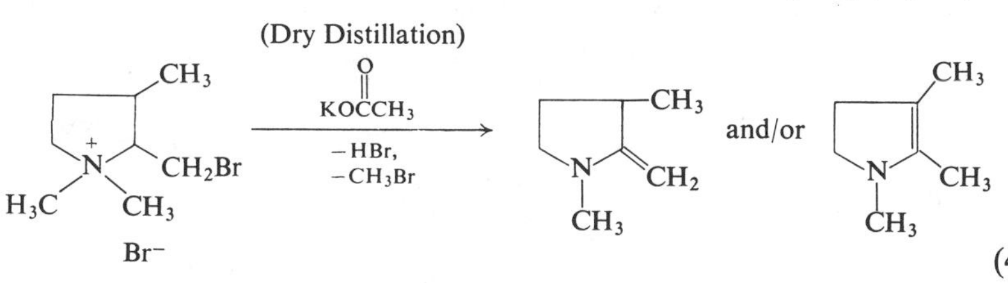

Recently Martin and Narayanan [104] reported that the α-aminonitrile 2(N,N-diethylamino)-2-phenylacetonitrile undergoes a Michael addition with acrylonitrile in the absence of solvent but in the presence of benzyltrimethylammonium hydroxide. The residue, obtained by ether extraction, is rapidly dehydrocyanated during distillation to give the enamine 4-(N,N-diethylamino)-4-phenyl-3-butenenitrile in 58% yield (Eq. 44). Whether this reaction

$$C_6H_5-\underset{\underset{CN}{|}}{\overset{\overset{N(C_2H_5)_2}{|}}{C}}-H \; + \; CH_2=CH-CN \quad \xrightarrow[38°-40°C]{C_6H_5-CH_2\overset{+}{N}(CH_3)_3OH^-} \quad \underset{N(C_2H_5)_2}{\overset{C_6H_5}{C=C}}\overset{H}{\underset{CH_2CN}{}} \; + \; HCN$$

$$(44)$$

is applicable to other α-aminonitriles is yet to be determined.

3-1. Preparation of N,N-Dimethylvinylamine [100a]

$$(CH_3)_2NCH_2CH_2Cl \xrightarrow{KO\text{-}t\text{-}Bu} (CH_3)_2NCH=CH_2 \qquad (45)$$

To a flask equipped with a stirrer and condenser is added 70.0 gm (0.485 mole) of N,N-dimethyl-N-β-chloroethylamine hydrochloride. Then an aqueous solution of potassium hydroxide is added to neutralize the amine hydrochloride to give 45.0 gm (0.42 mole) of crude (87% yield) product. The crude amine is purified by adding it to a stirred solution of 70.0 gm (0.625 mole) of potassium t-butoxide in 500 ml of purified DMF in a nitrogen atmosphere at −20°C. After 15 min, the mixture is distilled at room temperature at 2 mm Hg until approximately 100 ml of liquid is collected in a receiver cooled in a bath at −78°C. Fractional distillation of this liquid at 20 mm Hg into a receiver cooled in a bath at −78°C gives 25.0 gm (0.352 mole, 72.6%) of N,N-dimethylvinylamine. The NMR (60 MH$_2$, TMS) of the amine in DMF had the following characteristics: 3.95 (m, =CH), 6.35–6.65 (m, =CH$_2$), and 7.45 ppm (s, CH, CH$_3$). The ir (CCl$_4$) indicated the following absorption bands: 3100 (w), 2900 (m), 1730 (w), 1690 (w), 1640 (s), 1550 (w), 1430 (m), 1335 (m), 1240 (w), 1090 (m), 1050 (w), 1000 (w), 965 (m), 940 (w), 910 (w) cm^{-1}. The mass spectrum (20 eV) indicated m/c 71 (percent ion), 58, 56 (percent —CH$_3$), 45 (percent —CH≡CH), 44, 43 (percent —CH$_2$=CH$_2$).

3-2. Preparation of N-Methyl-N-propenylaniline [100]

$$CH_3-\underset{\underset{C_6H_5}{|}}{NH} + CH_3-\underset{\underset{O}{\diagup}}{CH}-CH_2 \longrightarrow CH_3-\underset{\underset{OH}{|}}{CH}-CH_2-N\overset{CH_3}{\underset{C_6H_5}{}} \xrightarrow{PBr_3}$$

$$CH_3-\underset{\underset{Br}{|}}{CH}-CH_2-N\overset{CH_3}{\underset{C_6H_5}{}} \xrightarrow[NH_3]{NaNH_2} CH_3-CH=CH-N\overset{CH_3}{\underset{C_6H_5}{}} \qquad (46)$$

(a) *Preparation of 1-(N-methyl-N-phenylamino)-2-propanol.* To an auto-clave (or Hoke pressure cylinder) containing 107.0 gm (1.0 mole) of N-methyl-aniline is added 64.9 gm (1.1 moles) of propylene oxide. The reaction mixture is heated for 7 hr at 190°C. After this time the autoclave is vented and the crude product distilled to afford 151.0 gm (91%) of 1-(N-methyl-N-phenylamino)-2-propanol, b.p. 118°C (4 mm), n_D^{25} 1.5558.

(b) *Preparation of 1-(N-methyl-N-phenylamino)-2-bromopropane.* To 39.8 gm (0.24 mole) of 1-(N-methyl-N-phenylamino)-2-propanol is added 100 gm (0.37 mole) of phosphorus tribromide. The mixture is heated to reflux for 3 hr. After this time the reaction mixture is cooled, 350 ml of benzene added, the solution is neutralized with 2 N NaOH, the benzene separated, dried, con-centrated, and the residue distilled to afford 35.6 gm (65%) of 1-(N-methyl-N-phenylamino)-2-bromopropane, b.p. 109°C (2 mm), n_D^{25} 1.5712.

(c) *Dehydrohalogenation of 1-(N-methyl-N-phenylamino)-2-bromopropane.* To 22.9 gm (0.1 mole) of 1-(N-methyl-N-phenylamino)-2-bromopropane is added a solution of 0.22 mole of sodamide in 400 ml liquid ammonia (prepared from 5 gm of sodium metal in ammonia). After 12 hr, the ammonia is evapor-ated, the product taken up in ether, concentrated, and the residue distilled to afford 12.1 gm (82%) of N-methyl-N-propenylaniline, b.p. 74°C (3–4 mm), n_D^{25} 1.5759. The product is predominantly in the trans configuration (84% trans, 16% cis) as determined by NMR spectroscopy.

3-3. *Preparation of α-Dimethylaminostilbene* [101]

$$\begin{array}{ccc}
\underset{\underset{H}{|}}{\overset{C_6H_5}{\diagdown}}\hspace{-0.5em}\text{C}\hspace{-0.5em}\underset{C_6H_5}{\overset{CN}{\diagup}}\hspace{-0.5em}\text{C}\hspace{-0.5em}\text{N(CH}_3)_2 & \xrightarrow[NH_3]{KNH_2} & \overset{C_6H_5}{\diagdown}\hspace{-0.5em}\underset{H}{\diagup}\hspace{-0.5em}\text{C}=\text{C}\hspace{-0.5em}\overset{N(CH_3)_2}{\diagup}\hspace{-0.5em}\underset{C_6H_5}{\diagdown}
\end{array} \qquad (47)$$

To a cooled flask equipped with a stirrer, dropping funnel, and condenser is added a solution of 0.1 mole of potassium amide in 500 ml of liquid ammonia. While stirring, 25 gm (0.1 mole) of α-dimethylamino-α,β-diphenylpropionitrile is added, the resulting purple-brown suspension is stirred until the ammonia has evaporated. Ether is then added, and the last traces of ammonia are removed on a steam bath. Water is then added and the ether layer separated. The water layer is extracted with ether, and the combined ether extracts concentrated under reduced pressure. The resulting oily residue is dissolved in methanol at room temperature, filtered, cooled to −78°C, and the resulting precipitate filtered to afford 18.6 gm (84%), m.p. 28°–29°C. Recrystallization from hexane and cooling to −78°C raises the melting point to 30°C.

3-4. Preparation of trans-1-(4-Morpholino)-1,2-diphenylethylene [102]

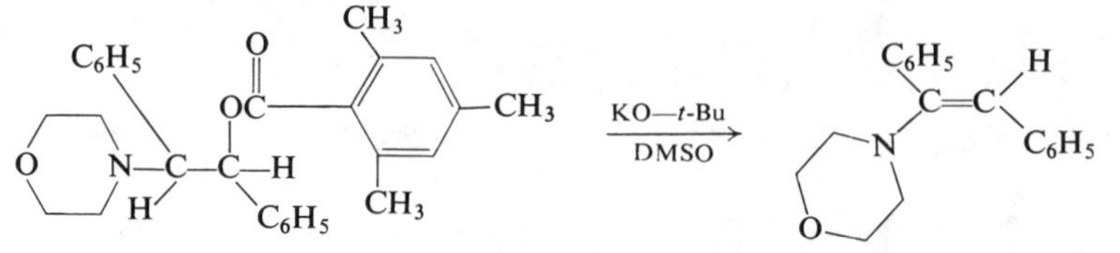

(48)

To a flask which was previously base-washed and dried is added 32 ml of 1.5 *M* potassium *t*-butoxide solution in dimethyl sulfoxide. (NOTE: The potassium *t*-butoxide (MSA Research Corp.) is sublimed under 0.1 mm pressure at 155°–160°C, transferred to a brown bottle in a dry nitrogen atmosphere, sealed with a serum cap, dry dimethyl sulfoxide added, and the resulting solution only removed with a syringe.) To the latter solution is added 1.99 gm (0.0046 mole) of the mesitoate ester of *dl-threo*-2-(4-morpholino)-1,2-diphenylethanol, the reaction is stirred for 3 hr at room temperature under a nitrogen atmosphere, and the mixture poured in 150 ml of ice-cold water. The resulting solution is extracted with ether, dried, and concentrated to afford 1.16 gm (94%) of crude enamine, m.p. 78°–81°C. The enamine is dissolved in the minimum amount of methanol at temperatures below 50°C, quickly cooled, and then filtered to afford 0.79 gm (65%), m.p. 103°–104°C.

α-Chloroenamines have been reported [102a] to be prepared by the reaction of dialkylamides first with phosgene and then with triethylamine. For example, the synthesis of 1-chloro-*N*,*N*-dimethyl-2-methylpropenylamine (b.p. 125°–133°C, 760 mm Hg):

$$(CH_3)_2\text{—}CH\text{—}\overset{\overset{\large O}{\|}}{C}NCH_3)_2 \xrightarrow[CH_2Cl_2]{COCl_2} (CH_3)_2CH\text{—}\overset{\overset{\large Cl}{|}}{C}{=}N^+(CH_3)_2Cl^-$$

$$\xrightarrow{(C_2H_5)_3N} (CH_3)_2C{=}\overset{\overset{\large Cl}{|}}{C}\text{—}N(CH_3)_2 \qquad (49)$$

(69–77%)

A good laboratory phosgenation method is described in the references to this procedure.

Some other α-chloroenamines prepared from amides are [102a]

85% (50)

45–62% (51)

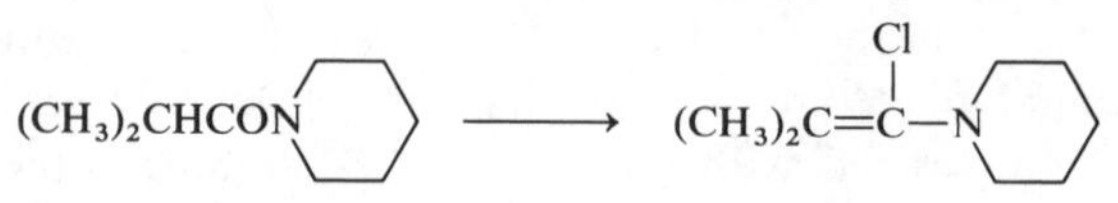

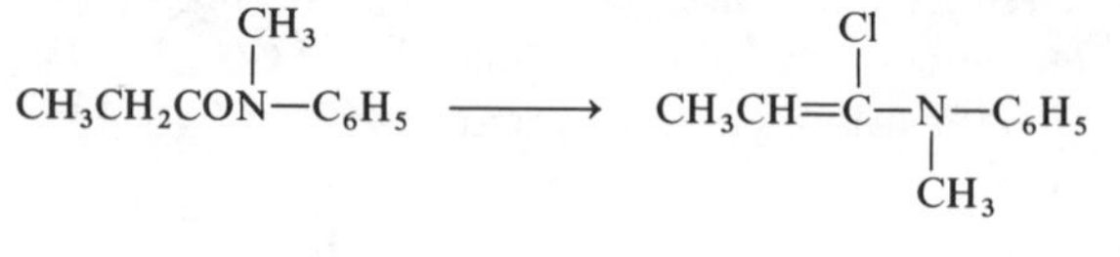

Reaction of the α-chloroenamines with Grignard reagents can be used to give alkyl or aryl substitution of the chloro group [102a]. Reaction with silver salts gives keteniminium salts that can react with olefine by [2 + 2] cycloaddition [102a]. Acetylenes and imines react in a similar manner [102a].

4. OXIDATION REACTIONS

Mercuric acetate has been used for the oxidation of amine groups to give modified alkaloid structures [105–120].

In 1955, Leonard described a general method for the oxidation of cyclic tertiary amines to give enamines. Quinolizidine was dehydrogenated using mercuric acetate in 5% aqueous acetic acid by heating on the steam bath for $1\frac{1}{2}$ hr to give 92% mercurous acetate and 60% dehydroquinolizidine.

4-1. Preparation of $\Delta^{1(10)}$-Dehydroquinolizidine [121]

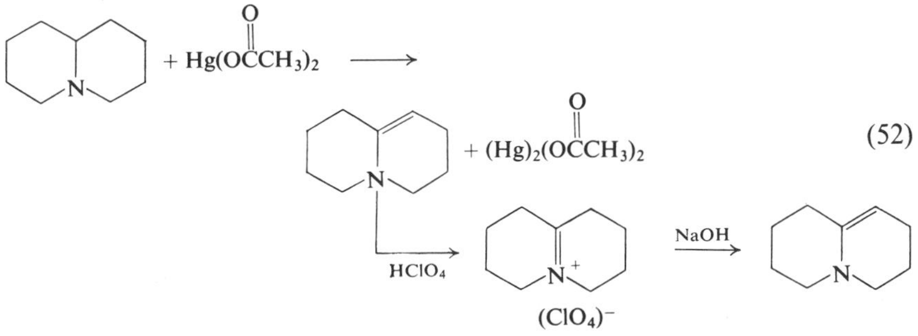

$$(52)$$

To a flask containing 20.0 gm (0.0628 mole) of mercuric acetate dissolved in 100 ml of 5% aqueous acetic acid solution is added 2.18 gm (0.0157 mole) of quinolizidine. The flask is warmed on a steam bath for $1\frac{1}{2}$ hr, then the precipitated mercurous acetate is removed by filtration (7.48 gm or 92%). The remaining mercuric ions are removed as a mercuric sulfide precipitate by saturating the reaction mixture with hydrogen sulfide. The reaction mixture is centrifuged to obtain a pale-yellow supernatant which is made basic and extracted with ether. The ether extracts are dried, treated with a solution consisting of equal volumes of 68% perchloric acid and ethanol until the ether is acid to Congo red paper. The resulting precipitate of $\Delta^{5(10)}$-dehydroquinolizidinium perchlorate is recrystallized once from ethanol to afford 2.20 gm (59%) of colorless platelets, m.p. 227°–228°C. Further recrystallization raises the melting point to 234°–235°C dec. The latter salt is dissolved in 10 ml of water, made strongly basic with 40% aqueous sodium

TABLE VI

MERCURIC ACETATE DEHYDROGENATION OF CYCLIC AMINES
TO CYCLIC ENAMINES

Starting compound	Product	Yield (%)	Ref.
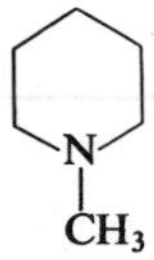	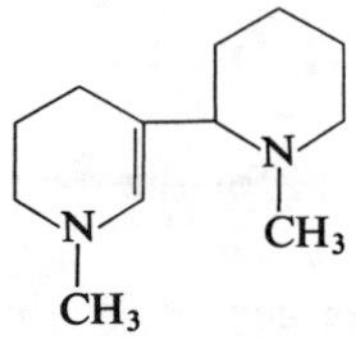	67	123
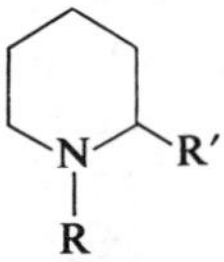	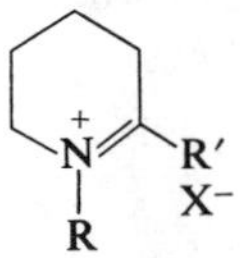	45	123
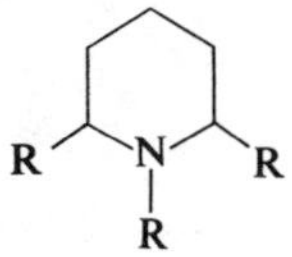	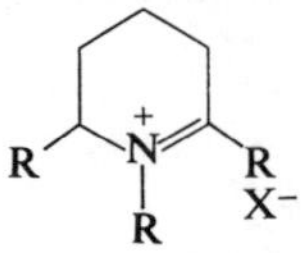	66	123
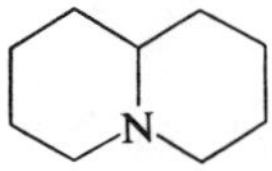	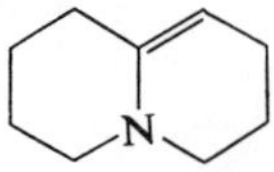	60	121, 124
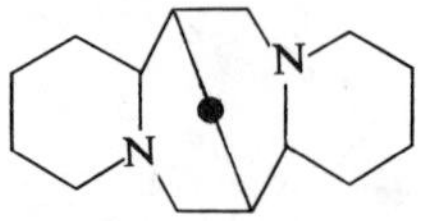	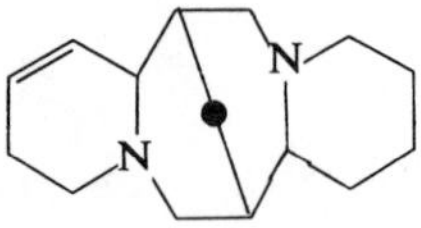	45	124
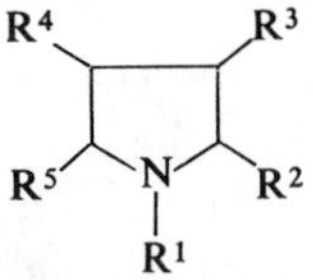	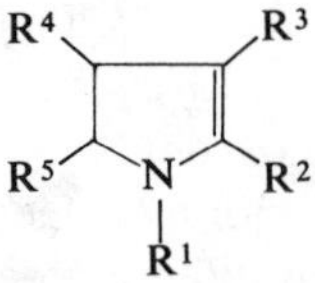	23–70	125

R^1, R^2 = alkyl, $\quad R^3 = R^4 = R^5 = H$
R^1, R^2, R^5 = alkyl, $\quad R^3 = R^4 = H$
R^1, R^3, R^4 = alkyl, $\quad R^2 = R^5 = H$

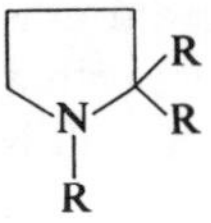	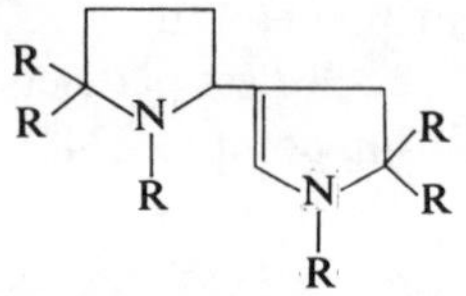	74	125

TABLE VI (*continued*)

Starting compound	Product	Yield (%)	Ref.
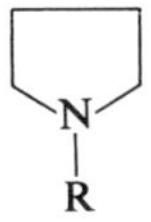	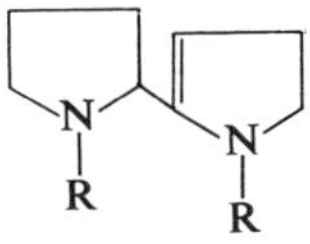	25	125
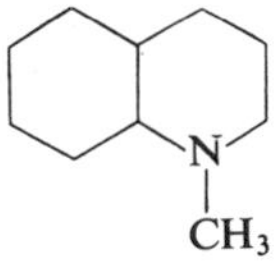	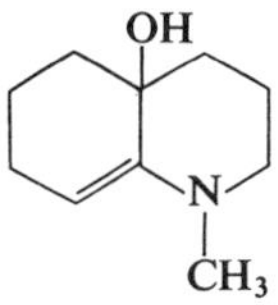	54	126
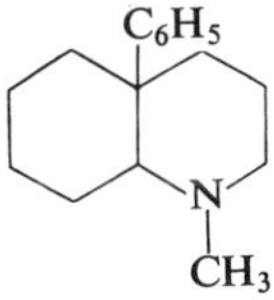	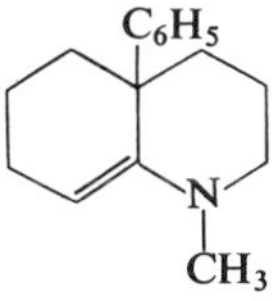	52	127
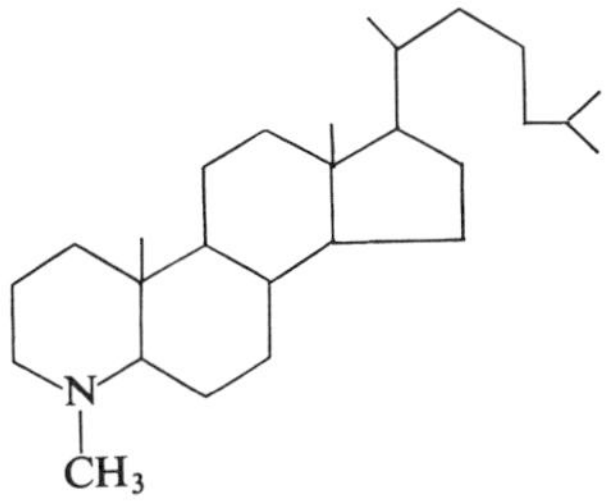	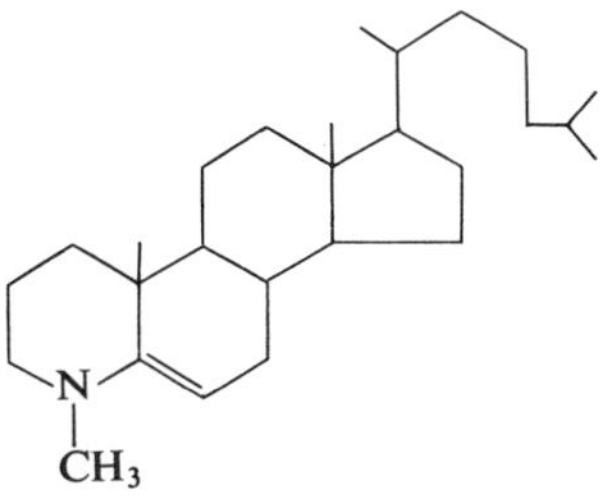	70	128
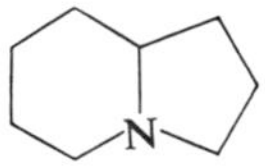	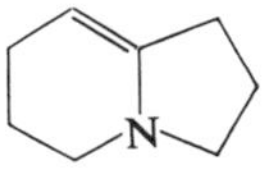	55	129

hydroxide, extracted three times with ether, dried, concentrated, and the residue is distilled under a nitrogen atmosphere to afford 1.96 gm (68%) of $\Delta^{1(10)}$-dehydroquinolizidine as a colorless oil, b.p. 80°C (18 mm), n_D^{28} 1.5116.

The infrared contains absorption bands at 1652 $\left(\diagup\!\!\!\!\diagdown C{=}C \diagdown\!\!\!\!\diagup \right)$ and 3020 cm^{-1}

$\left(\diagdown\!\!\!\!\diagup C{-}H \right)$, both of which are absent in quinolizidine.

The oxidation probably involves a mercurated complex through the electron pair on nitrogen followed by a concerted removal of a proton from the α-carbon and then cleavage of the mercury-nitrogen bond. The suggestion that a trans elimination occurs is reinforced by the evidence that yohimbine can be dehydrogenated by mercuric acetate, whereas reserpine methyl reserpate, deserpine, or pseudoyohimbine do not react [122].

After studying a large number of compounds, it was found that removal of a tertiary hydrogen α to the nitrogen atom is preferred to removal of a secondary α-hydrogen atom. The versatility of this method is illustrated by the type of compounds undergoing the mercuric acetate oxidation reaction as shown in Table VI.

Other oxidative methods have recently been reported whereby amines yield enamines either as intermediates [130] or as isolatable chloranil [131], 2,3-dichloronaphtha-1,4-quinone [131], or 2,5-dichloro-3,6-dimethoxybenzoquinone [132, 133] adducts. Benzoyl peroxide [131] and active manganese dioxide [134] have been reported as effective oxidizing agents in the reactions above. (See Eqs. 53, 54.)

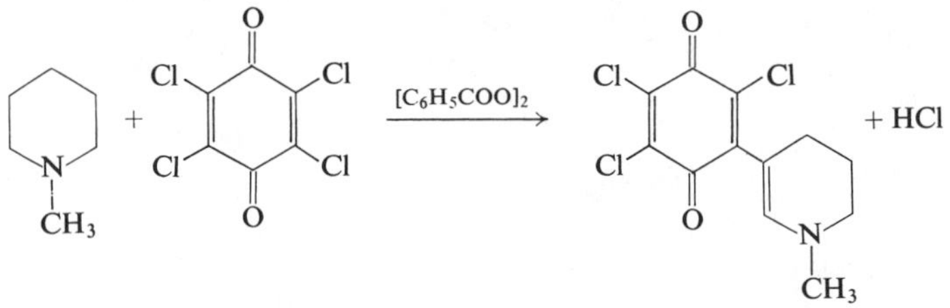

$$(53)$$

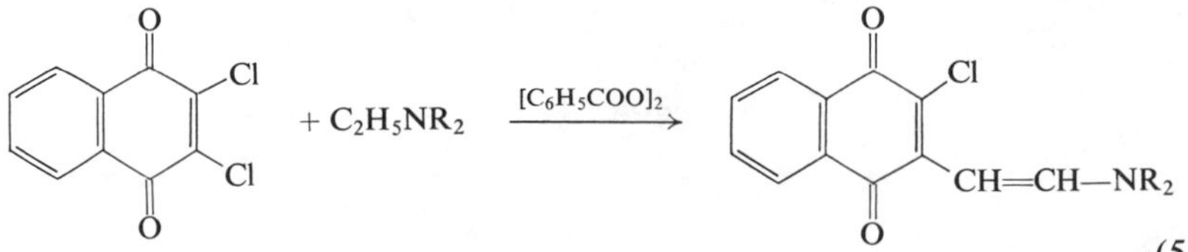

$$(54)$$

5. REDUCTION REACTIONS

The reduction methods of producing enamines from aromatic heterocyclic bases or their derivatives have thus far found only limited application [30].

The reductive route used to prepare heterocyclic enamines has the advantage of avoiding the hydroxylation reaction sometimes found in the mercuric acetate oxidation of saturated heterocyclic amines [126]. The lithium-*n*-propylamine reducing system has been used by Leonard to reduce julodine to Δ^5-tetrahydrojulolidine (66% yield) and 1-methyl-1,2,3,4-tetrahydroquinoline to a mixture of enamines (87% yield), consisting of 1-methyl-Δ^8-octahydroquinoline and 1-methyl-Δ^9-octahydroquinoline [135] (Eqs. 55, 56).

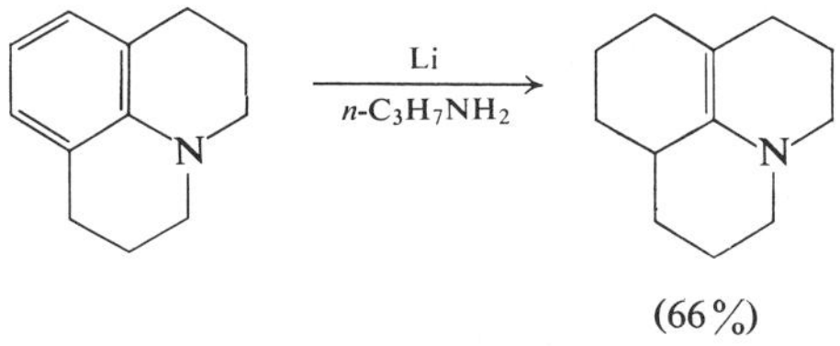

(66%)

(55)

(87%)

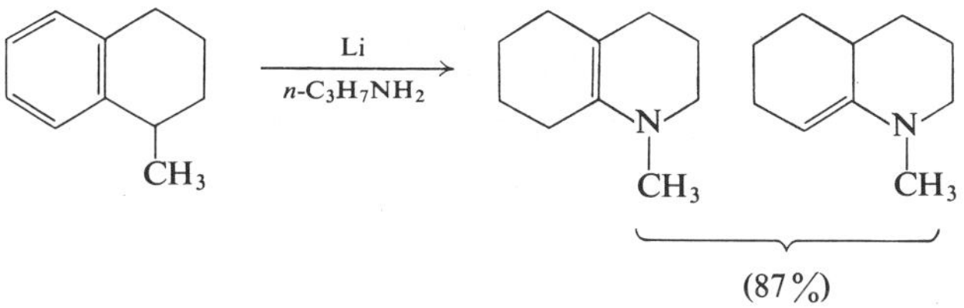

(56)

Other methods involve the use of lithium aluminum hydride to reduce N-alkyl lactams [136] (Eq. 57). By a similar method, some steroidal N-acetyl-

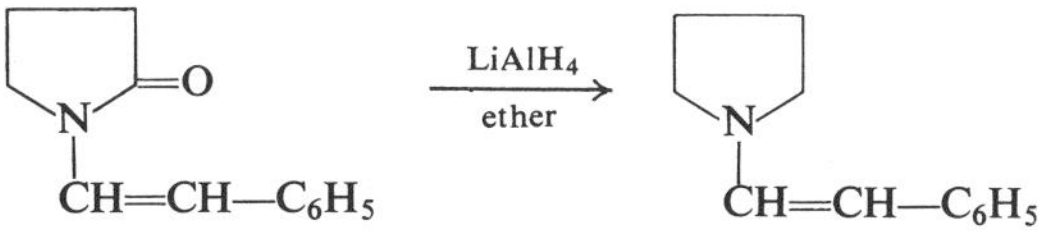

(57)

enamines can be reduced to the free enamines in 68%–90% yields [137], while others yield unstable enamines decomposing to the corresponding ketone [137]. Lithium aluminum hydride can also reduce aromatic ketene imines to stable enamines [138].

The reduction of N-alkylpiperidones with sodium in ethanol has also been reported [139a, b].

From the preparative standpoint, 1-methyl-Δ^2-piperideines are best prepared by the partial hydrogenation of quaternary pyridine salts in alkaline media [140, 141].

Other reducing agents that have been reported to afford enamines are sodium hydrosulfite [142], dialkyl aluminohydrides [143], and Grignard reagents [144] for quaternary isoquinoline salts; sodium borohydrides [145] and Grignard reagents for 3,5-dicyanopyridines; and electroreduction [146, 147] of N-methylglutarimide.

5-1. Reduction of 1-Methyl-1,2,3,4-tetrahydroquinoline to a Mixture of 1-Methyl-Δ⁸-octahydroquinoline and N-Methyl-Δ⁹-octahydroquinoline [135]

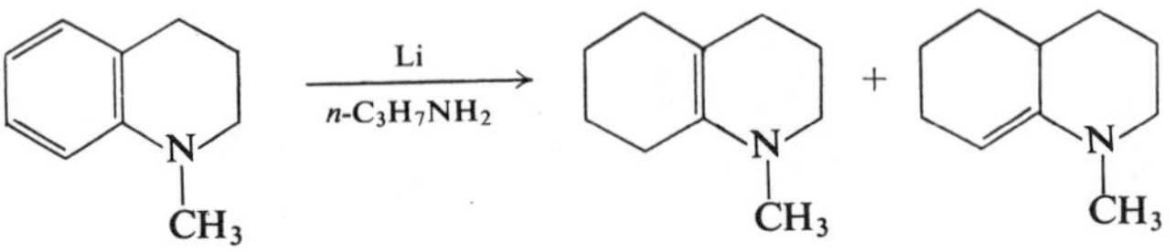

$$\xrightarrow[n\text{-C}_3\text{H}_7\text{NH}_2]{\text{Li}}$$

(58)

To a dry flask containing 415 ml of *n*-propylamine and 10.15 gm (1.45 gm-atoms) of chopped lithium wire segments is added, with stirring, 21.3 gm (0.125 mole) of 1-methyl-1,2,3,4-tetrahydroquinoline, and the mixture is stirred for 16½ hr under a nitrogen atmosphere. The unreacted lithium is removed, the excess *n*-propylamine is distilled from the flask, the semisolid residue is cooled in an ice bath, overlaid with ether, and then neutralized slowly with solid ammonium chloride. The mixture is cautiously diluted with water, the ether layer separated, the water layer extracted with ether, the ether layers combined, dried, concentrated, and the residue is distilled to afford 19.8 gm (90.5%), b.p. 60°–75°C (1.5–1.7 mm). Gas chromatography of the crude product indicates that it is 96% pure enamine. Redistillation affords 17.3 gm (79%), b.p. 63°–65°C (1.5 mm), n_D^{25} 1.5138.

The infrared spectrum of the product showed absorption at 1663 and 1642 cm^{-1}.

6. REARRANGEMENT REACTIONS

Allylamines have been reported to be isomerized with the aid of basic catalysts to enamines [147a, b]

$$R_{3-x}-N(CH_2-CH=CH_2)_x \xrightarrow{\text{Base}} R_{3-x}N(CH=C-CH_3)_x \qquad (59)$$

A dispersion of potassium amide on alumina was reported to be effective at room temperature in giving this isomerization and in most cases good yields were obtained as shown in Table VII [147a]. *N,N*-Dialkylallylamines are reported to give *cis*-enamines when isomerized in the presence of basic catalysts (K-O-*t*-Bu) [147b].

Another case of rearrangement involves the imine–enamine tautomerism.

TABLE VII

ISOMERIZATION OF $CH_2{=}CH{-}CH_2NR_2$ TO $CH_3CH{=}CHNR_2$ USING KNH_2(1.8 GM) ON ALUMINA (20 GM) AT 25°C [147a]

Substituent on $CH_2{=}CH{-}CH_2{-}$	Reaction time (min)	B.p., °C (mm Hg)	Yield (%)
$N(CH_3)_2$	30	37 (70)	70
$N(C_2H_5)_2$	30	35 (15)	94–100[a]
$N(CH_2{-}CH{=}CH_2)_2$	60[b]	28 (1)	76
N-Pyrrolidinyl	15	55 (20)	65
N-Piperidinyl	60	20 (1)	52
N-Morpholinyl	30	25 (1)	57–100[a]
N-Carbazoyl	30	160 (1)	60
$(CH_2{=}CH{-}CH_2)_2N{-}$ 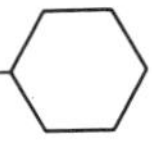	30	—	20

[a] Determined by gas–liquid chromatography using an internal standard.
[b] Reaction was carried out at 50°C.

For example, 2-(*N*-cyclohexylimino)-1,3-diphenylpropane rearranges in DMSO as shown below [147c].

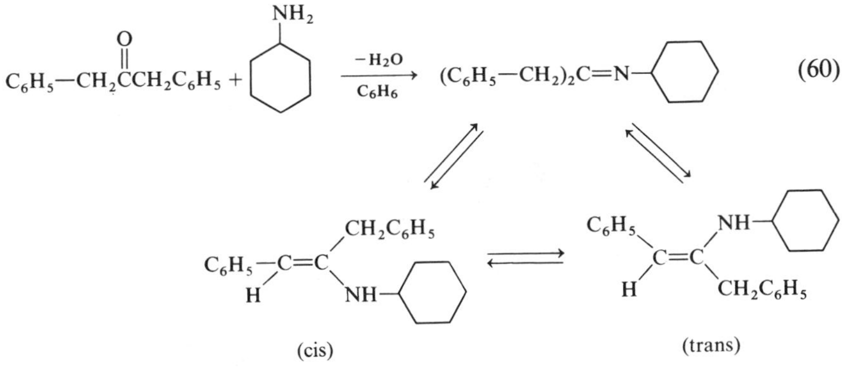

NH2
O
C6H5—CH2CCH2C6H5 +
−H2O
C6H6
(C6H5—CH2)2C=N—
CH2C6H5
C6H5—C=C
H
NH—
C6H5
NH—
C=C
H
CH2C6H5
(cis)
(trans)

(60)

7. MISCELLANEOUS METHODS

(1) Isomerization of allylamines to enamines [147a, b, 148–151].

(2) Reaction of *N*,*N*-diethyl-2,2,2-trichloroacetamide with trialkyl phosphites or triphenylphosphine to give *N*,*N*-diethyl-1,2,2-trichlorovinylamine [152–154].

(3) Preparation of enamine amides [155–157].

(4) Reaction of N-(2-bromoalkyl)-ethylamine with sodium amide in liquid ammonia to give N-ethylallenimine [158].

(5) Nucleophilic displacement reactions of some halogen-substituted phenylcyclobutenes [159–161].

(6) Halogenation of enamines to give stable halo enamines [162].

(7) Reaction of tris(dialkylamino)boranes with ketones to give enamines [163].

(8) Reaction of tris(dimethylamino)arsine or tetrakis(dimethylamino)-titanium with aldehydes or ketones to give enamines [164, 165].

(9) The enamine–immonium cation system of 9,10-dimethoxy-1,2,3,4-6,7-hexahydrobenzo[α]quinolizinium salt [166].

(10) Imido-enamine tautomery. Isolation of the polar tautomeric form of propyl β-aminovinyl ketone [167].

(11) Preparation of dienamines [168, 169].

(12) Reaction of secondary amines with allenes to give enamines [170].

(13) Preparation of silicon-containing acetylenic enamines [171].

(14) A novel synthesis of 2,4-thiophenediamines and their behavior as stable reactive enamines [172].

(15) Reaction of acetals with amines to give enamines [173].

(16) Reaction of ethyl ethoxymethylenemalonate with nitromethane in the presence of amines to afford 1-amino-2-nitroethenes [174].

(17) Reaction of methyl- and ethyl-2-cyclopentanone carboxylates with amines to give carbinolamines, enamines, and adipamides [175].

(18) Synthesis of new bicyclic enamines [176].

(19) Reaction of tetrakis(dimethylamino)methane and tris(dimethylamino-methane with fluorene to afford enamines [177].

(20) Reaction of trans-N,N-dimethyl-2-phenylcyclopropylamine with di-methylamine in methanol to give the open-chain enamine 1-dimethylamino-3-phenyl-1-propene [178].

(21) Preparation of enamines by means of the Claisen condensation [179–181].

(22) Conversion of esters of acylated aminoisobutyryl malonic acids into 3-oxo-Δ^2-pyrrolines [182–183].

(23) Conversion of isoxazolium salts by the action of alkali cyanides into 4,5-dioxo-Δ^2-pyrrolines [184, 185].

(24) Preparation of dihydropyridones by the condensation of ethyl cyano-acetate, cyanoacetamide, or malononitrile with α,β-unsaturated ketones [186].

(25) Preparation of N,N-diethyltrichlorovinylamine by the reaction of tri-ethylamine and phenyl(trichloromethyl)mercury [187].

(26) Amine-exchange reactions: Reaction of 1-(1-cyclopenten-1-yl)-piperi-dine with N-methyl-3-bromopropylamine hydrobromide [188].

(27) Preparation of N-ethylallenimine by the reaction of N-(2-bromoallyl) ethylamine with sodium amide in liquid ammonia [189, 190].

(28) Oxidative decarboxylation of N,N-dialkyl-α-amino acids by means of sodium hypochlorite to give enamines [191].

(29) Reaction of aryl hydrazines and hydrazine with dimethyl acetylene dicarboxylate to afford imine-enamine tautomers [192].

(30) Rearrangement of vinylogous urethanes to give cyclic enamines [193, 194].

(31) The enamine-imine tautomerism in the indolenine system [195].

(32) Reaction of chloramines and acetylenes to give chloroenamine intermediates [196].

(33) Preparation of N-methyl-2-cyclopentadienylidene-1,2-dihydropyridine from cyclopentadienylsodium and 2-iodopyridine methiodide [197].

(34) Preparation of enamine 2,4,6,8-nonanetetraones by the reaction of pyrrolidine with acylpyrones [198].

(35) Preparation of cyclic dienamines [199].

(36) Addition of diphenylthiirane 1,1-dioxide to enamines [200].

(37) Preparation of heterocyclic enamines by the intramolecular condensation of N-(5-oxohexyl) nitrogen heterocycles [201].

(38) Reaction of 1-bromo- and 1-chloro-2-(4-morpholino)-1,2-diphenylethene with phenyllithium in benzene to give 2-(4-morpholino)-1,2-diphenylethene [202].

(39) Preparation of bicyclic enamines and a dieneamine [203a, b, c].

(40) Reaction of aryl dichloroisonitriles ($ArN{=}CCl_2$) with dialkylamines [204].

(41) Catalytic reaction of tertiary alkylamines with olefins to give enamines [205].

(42) Reaction of aminoacetonitrile with ethyl acetoacetate to give ethyl β-cyanomethylaminocrotonate [206].

(43) Preparation of 1-N-arylamino-1,3-dienes [207].

(44) Conversion of ynamines to enamines [208].

(45) Reaction of propargyl alcohols with amine acetals [209].

(46) Preparation of cyclic-β-enamino esters [210].

(47) An endocyclic enamino synthesis [211].

(48) Preparation of enamides and enecarbamates [212].

(49) Preparation of enamines by addition of Grignard reagents of N,N-dimethylformamide [213].

REFERENCES

1. G. Wittig and H. Blumenthal, *Chem. Ber.* **60**, 1085 (1927).
2. G. Stork, R. Terrell, and J. Szmuszkovicz, *J. Amer. Chem. Soc.* **76**, 2029 (1954).
3. G. Stork and H. K. Landesman, *J. Amer. Chem. Soc.* **78**, 5128, 5129 (1956).
4. G. Optiz, H. Hellmann, and H. W. Schubert, *Ann. Chem.* **623**, 117 (1959).
5. D. F. Starr, H. Bulbrook, and R. M. Hixon, *J. Amer. Chem. Soc.* **54**, 3971 (1932).
6. N. J. Leonard and A. G. Cook, *J. Amer. Chem. Soc.* **81**, 5627 (1959).

7. N. J. Leonard, P. D. Thomas, and V. W. Gash, *J. Amer. Chem. Soc.* 77, 1552 (1955).
8. R. Adams and J. E. Mahan, *J. Amer. Chem. Soc.* 64, 2588 (1942).
9. N. J. Leonard, A. S. Hay, R. W. Fulmer, and V. W. Gash, *J. Amer. Chem. Soc.* 77, 2139 (1955).
10. R. L. Hinman, *Tetrahedron* 24, 185 (1968).
11a. E. J. Stamhuis, W. Maas, and H. Wynberg, *J. Org. Chem.* 30, 2160 (1965).
11b. P. Y. Sollenberger and R. B. Martin, *J. Amer. Chem. Soc.* 92, 4261 (1970).
12a. G. Stork, A. Brizzolara, H. K. Landesman, J. Szmuszkovicz, and R. Terrell, *J. Amer. Chem. Soc.* 85, 207 (1963).
12b. M. E. Kuehne and T. Garbacik, *J. Org. Chem.* 35, 1555 (1970).
13. G. Domschke, *Chem. Ber.* 99, 939 (1966) and earlier papers.
14. R. L. Augustine, *J. Org. Chem.* 28, 581 (1962).
14a. S. K. Malhotra, J. J. Hustynek, and A. F. Lundin, *J. Amer. Chem. Soc.* 90, 6565 (1968).
15. G. Opitz and A. Griesinger, *Ann. Chem.* 665, 101 (1963).
16. R. L. Peterson, J. L. Johnson, R. P. Holysz, and A. C. Ott, *J. Amer. Chem. Soc.* 79, 1115 (1957).
17. R. J. Parfitt, *J. Chem. Soc.* 140 (1967).
18. R. H. Hasek, P. G. Gott, and J. C. Martin, *J. Org. Chem.* 31, 1931 (1966).
19. S. Hünig, *Angew. Chem.* 71, 312 (1959).
20. S. Hünig, E. Benzing, and E. Lücke, *Chem. Ber.* 90, 2833 (1957).
21. S. Hünig, and M. Salzwedel, *Chem. Ber.* 99, 823 (1966).
22. C. D. Gutsche and D. M. Bailey, *J. Org. Chem.* 28, 607 (1963).
22a. W. G. Dauben and A. P. Kozikowski, *J. Amer. Chem. Soc.* 96, 3664 (1974).
23. R. G. Hiskey and G. L. Southard, *J. Org. Chem.* 31, 3582 (1966).
24. G. Laban and R. Mayer, *Z. Chem.* 8, 165 (1968).
24a. A. Couture, R. Dubiez, and A. Lablache-Combier, *J. Org. Chem.* 49, 714 (1984).
25. M. Colonna, *Univ. Studi Trieste Fac. Sci., Ist. Chim.* 30, 1 (1961).
25a. S. F. Dyke, "The Chemistry of Enamines," Cambridge Univ. Press, London, 1973.
26. H. Huang and C. C. Chen, *Hua Hsueh Tung Pao* [7] 411 (1964).
27. Y. Nomura, *Yuki Goesi Kagaku Kyokaishi* 19, 801 (1961).
28. J. Szmuszkovicz, *Advan. Org. Chem.* 4, 1 (1963).
29. K. Blaha and O. Cervinka, *Advan. Heterocycl. Chem.* 6, 147 (1966).
30. A. G. Cook, ed., "Enamines, Synthesis, Structure and Reactions." Dekker, New York and London, 1969.
31. C. Mannich and H. Davidsen, *Chem. Ber.* 69B, 2106 (1936).
32. M. E. Herr and F. W. Heyl, *J. Amer. Chem. Soc.* 74, 3627 (1952).
33. F. W. Heyl and M. E. Herr, *J. Amer. Chem. Soc.* 75, 1918 (1953).
33a. G. Kalaus, P. Gyö, L. Szaho, and C. Szantay, *J. Org. Chem.* 43, 5017 (1978).
34. M. E. Herr and F. W. Heyl, *J. Amer. Chem. Soc.* 75, 5927 (1953).
35. F. W. Heyl and M. E. Herr, *J. Amer. Chem. Soc.* 77, 488 (1955).
36. R. Dulou, E. Elkik, and A. Veillard, *Bull. Soc. Chim. Fr.* 1967 (1960).
37. A. Mondon, *Chem. Ber.* 92, 1461 (1959).
38. R. Jacquier, C. Petrus, and F. Petrus, *Bull. Soc. Chim. Fr.* 2845 (1966).
39. S. Hünig, E. Benzig, and E. Lücke, *Chem. Ber.* 90, 2834 (1957).
40. J. L. Johnson, M. E. Herr, J. C. Babcock, A. E. Fonkin, J. E. Stafford, and F. W. Heyl, *J. Amer. Chem. Soc.* 78, 430 (1956).
41. S. R. Sandler and M. L. Delgado, *J. Polym. Sci. Part* A-1 7, 1373 (1969).
42. C. Djerassi and B. Tursch, *J. Org. Chem.* 27, 1041 (1962).
42a. M. R. Ellenberger, D. A. Dixon, and W. E. Farneth, *J. Amer. Chem. Soc.* 103, 5377 (1981).

42b. K. Taguchi and F. W. Westheimer, *J. Org. Chem.* **36**, 1570 (1971).
43. E. P. Blanchard, Jr., *J. Org. Chem.* **28**, 1397 (1963).
44. W. A. White and H. Weingarten, *J. Org. Chem.* **32**, 213 (1967).
45. E. Benzing, *Angew. Chem.* **71**, 521 (1959).
46. E. Benzing, U.S. Pat. 3,074,940 (1963).
47. R. G. Kostianovskij and V. F. Bystrov, *Izv. Akad. Nauk SSSR Otd. Khim. Nauki* 171 (1963); R. G. Kostianovskij and O. A. Panshin, *ibid.* 182 (1963).
48. E. L. Patmore and H. Chafetz, *J. Org. Chem.* **32**, 1254 (1967).
49. A. G. Cook and C. R. Schulz, *J. Org. Chem.* **32**, 473 (1967).
50. A. G. Cook, W. C. Meyer, K. E. Ungrodt, and R. H. Mueller, *J. Org. Chem.* **31**, 14 (1966).
51. E. J. Eisenbraun, J. Osiecki, and C. Djerassi, *J. Amer. Chem. Soc.* **80**, 1261 (1958).
52. W. D. Gurowitz and M. A. Joseph, *J. Org. Chem.* **32**, 3289 (1967).
53. K. C. Brannock and R. D. Burpitt, *J. Org. Chem.* **26**, 3576 (1961).
54. L. A. Paquette, *J. Org. Chem.* **29**, 2851 (1964).
55. M. E. Kuehne, *J. Amer. Chem. Soc.* **81**, 5400 (1959).
56. A. T. Blomquist and E. J. Moriconi, *J. Org. Chem.* **26**, 3761 (1961).
57. L. Birkofer and C. D. Barnikel, *Chem. Ber.* **91**, 1996 (1958).
58. F. A. Buiter, J. H. S. Weiland, and H. Wynberg, *Rec. Trav. Chim.* **83**, 1160 (1964).
59. R. Jacquier, C. Petrus, and F. Petrus, *Bull. Soc. Chim. Fr.* 2845 (1966).
60. E. P. Blanchard, Jr., *J. Org. Chem.* **28**, 1397 (1963).
61. S. Hünig, E. Lücke, and W. Brenninger, *Org. Syn.* **41**, 65 (1961).
62. R. Lukes, *Collect. Czech. Chem. Commun.* **2**, 531 (1931).
63. L. C. Craig, *J. Amer. Chem. Soc.* **55**, 295 (1933).
64. J. Lee, A. Ziering, S. D. Heineman, and L. Berger, *J. Org. Chem.* **12**, 885 (1947).
65. O. Cervinka and L. Hub, *Tetrahedron Lett.* 463 (1964).
66. R. Lukes, V. Dudek, and L. Novotny, *Collect. Czech. Chem. Commun.* **24**, 1117 (1959).
67. N. J. Leonard and F. P. Hauck, Jr., *J. Amer. Chem. Soc.* **79**, 5279 (1967).
68. R. Lukes and O. Cervinka, *Collect. Czech. Chem. Commun.* **26**, 1893 (1961).
69. R. Lukes and K. Smolek, *Collect. Czech. Chem. Commun.* **11**, 506 (1939).
70. R. Lukes, V. Dudek, O. Sedlakova, and I. Koran, *Collect. Czech. Chem. Commun.* **26**, 1105 (1961).
71. R. Lukes and J. Dobas, *Collect. Czech. Chem. Commun.* **15**, 303 (1950).
72. R. Lukes and J. Malek, *Collect. Czech. Chem. Commun.* **16**, 23 (1951).
73. R. Lukes and L. Karliekova, *Collect. Czech. Chem. Commun.* **26**, 2245 (1961).
74. O. Cervinka and L. Hub, *Collect. Czech. Chem. Commun.* **30**, 3111 (1965).
75. L. C. Craig, *J. Amer. Chem. Soc.* **55**, 295 (1933).
76. J. Lee, A. Ziering, S. D. Heineman, and L. Berger, *J. Org. Chem.* **12**, 885 (1947).
77. K. Hafner, K. H. Vöpel, G. Ploss, and C. König, *Org. Syn.* **47**, 52 (1967).
78. J. H. Crabtree and D. J. Bertelli, *J. Amer. Chem. Soc.* **89**, 5384 (1967).
79. J. H. Crabtree and D. J. Bertelli, *J. Amer. Chem. Soc.* **89**, 5389 (1967).
80. W. Reppe, H. Krzikalla, and E. Woldan, Ger. Pat. 853,399 (1942).
81. W. Reppe, *Ann. Chem.* **596**, 1 (1955).
82a. W. Reppe and A. Magin, Ger. Pat. 860,058 (1942).
82b. G. Gardner, V. Kerrigan, J. D. Rose, and B. C. L. Weedon, *J. Chem. Soc.* 789 (1949).
83. W. Reppe, *Ann. Chem.* **601**, 81 (1956).
84a. W. Wolff, Ger. Pat. 636,213 (1935).
84b. E. Keyssner and W. Wolff, Ger. Pat. 642,424 (1935).
84c. S. R. Sandler, *Chem. Ind.* **134** (1973); U.S. Pat. 3,679,700 (1972).

85. C. J. M. Stirling, *J. Chem. Soc.* 5863 (1964).
85a. K. A. Parker, R. W. Kosley, Jr., S. L. Buchwald, and J. P. Petritis, *J. Amer. Chem. Soc.* **98**, 7104 (1976).
86. R. C. Pink, R. Spratt, and C. J. M. Stirling, *J. Chem. Soc.* 5714 (1965).
87. W. E. Truce and D. G. Brady, *J. Org. Chem.* **31**, 3543 (1966).
88. C. H. McMullen and C. J. M. Stirling, *J. Chem. Soc. B* 1217 (1966).
89. J. E. Dolfini, *J. Org. Chem.* **30**, 1298 (1965).
90. E. Winterfeldt and H. Preuss, *Angew. Chem.* **77**, 679 (1965).
91. K. Herbig, R. Huisgen, and H. Huber, *Chem. Ber.* **99**, 2546 (1966).
92. A. Padwa and L. Hamilton, *Tetrahedron Lett.* 4363 (1965).
93. R. Huisgen, B. Giese, and H. Huber, *Tetrahedron Lett.* 1883 (1967).
94. C. H. McMullen and C. J. M. Stirling, *J. Chem. Soc. B* 1221 (1966).
95. C. H. McMullen and C. J. M. Stirling, *J. Chem. Soc. B* 1217 (1966).
96. A. W. Nineham and R. A. Raphael, *J. Chem. Soc.* 118 (1949).
97. E. R. H. Jones, T. Y. Shen, and M. C. Whiting, *J. Chem. Soc.* 236 (1950).
98. K. Bowden, E. A. Braude, E. R. H. Jones, and B. C. L. Weedon, *J. Chem. Soc.* 45 (1946).
99. K. H. Meyer and H. Hopf, *Chem. Ber.* **54**, 2277 (1921).
100. M. Riviere and A. Lattes, *Bull. Soc. Chim. Fr.* 2539 (1967).
100a. P. L.-F. Chang and D. C. Dittmer, *J. Org. Chem.* **34**, 2791 (1969).
101. C. R. Hauser, H. M. Taylor, and T. G. Ledford, *J. Amer. Chem. Soc.* **82**, 1786 (1960).
102. M. E. Munk and Y. K. Kim, *J. Org. Chem.* **30**, 3705 (1965).
102a. B. Haveany, A. DeKoker, M. Cens, A. R. Sidani, J. Toye, and L. Ghosez, *Org. Syn.* **59**, 26 (1979).
103. R. Lukes and J. Pliml, *Collect. Czech. Chem. Commun.* 322 (1966).
104. C. F. Martin and V. L. Narayanan, *J. Org. Chem.* **26**, 2127 (1961).
105. J. Tafel, *Chem. Ber.* **25**, 1619 (1892).
106. A. Reissert, *Chem. Ber.* **27**, 2244 (1894).
107. J. Gadamer, *Arch. Pharm.* **253**, 274 (1915).
108. N. V. Subba Rao and T. R. Seshadri, *Proc. Indian Acad. Sci.* **11A**, 23 (1940).
109. H. Legerlotz, *Arch. Pharm.* **256**, 123 (1918).
110. J. Gadamer and H. Kollmar, *Arch. Pharm.* **261**, 153 (1923).
111. H. Dieterle and P. Dickens, *Arch. Pharm.* **264**, 257 (1926).
112. K. Winterfeld, *Arch. Pharm.* **266**, 299 (1928).
113. K. Winterfeld and C. Rauch, *Arch. Pharm.* **272**, 273 (1934).
114. L. Marion and N. J. Leonard, *Canad. J. Chem.* **29**, 355 (1951).
115. A. R. Battersby and H. T. Openshaw, *J. Chem. Soc.* S67 (1949).
116. A. R. Battersby, H. T. Openshaw, and H. C. S. Wood, *Experientia* **5**, 114 (1949).
117. P. Karrer and O. Ruttner, *Helv. Chim. Acta* **33**, 291 (1950).
118. R. N. Hazlett and W. E. McEwen, *J. Amer. Chem. Soc.* **73**, 2578 (1951).
119. H. T. Openshaw and H. C. S. Wood, *J. Chem. Soc.* 391 (1952).
120. R. F. Tietz and W. E. McEwen, *J. Amer. Chem. Soc.* **75**, 4945 (1953).
121. N. J. Leonard, A. S. Hay, R. W. Fulmer, and V. W. Gash, *J. Amer. Chem. Soc.* **77**, 439 (1955).
122. F. L. Weisenborn and P. A. Diassi, *J. Amer. Chem. Soc.* **78**, 2022 (1956).
123. N. J. Leonard and F. P. Hauck, Jr., *J. Amer. Chem. Soc.* **79**, 5279 (1957).
124. N. J. Leonard, P. D. Thomas, and V. W. Cash, *J. Amer. Chem. Soc.* **77**, 1552 (1955).
125. N. J. Leonard and A. G. Cook, *J. Amer. Chem. Soc.* **81**, 5627 (1959).
126. N. J. Leonard, L. E. Miller, and P. D. Thomas, *J. Amer. Chem. Soc.* **78**, 3463 (1956).
127. C. F. Koelsch and D. L. Ostercamp, *J. Org. Chem.* **26**, 1104 (1961).
128. J. McKenna and A. Tulley, *J. Chem. Soc.* 945 (1960).
129. M. G. Reinecke and L. R. Kray, *J. Org. Chem.* **31**, 4215 (1966).

130. S. S. Rawalay and H. Shechter, *J. Org. Chem.* **32**, 3129 (1967).
131. D. M. Gardner, R. Helitzer, and D. H. Rosenblatt, *J. Org. Chem.* **32**, 1115 (1967).
132. D. Buckley, S. Dunstan, and H. B. Henbest, *J. Chem. Soc.* 4901 (1957).
133. D. Buckley, S. Dunstan, and H. B. Henbest, *J. Chem. Soc.* 4880 (1957).
134. H. B. Henbest and A. Thomas, *J. Chem. Soc.* 3032 (1957).
135. N. J. Leonard, C. K. Steinhardt, and C. Lee, *J. Org. Chem.* **27**, 4027 (1962).
136. A. H. Boehme and G. Berg, *Chem. Ber.* **99**, 2127 (1966).
137. W. Fritsch, J. Schmidt-Thome, H. Ruschig, and W. Haede, *Chem. Ber.* **96**, 68 (1963).
138. C. L. Stevens and R. J. Gasser, *J. Amer. Chem. Soc.* **79**, 6057 (1959).
139a. L. Ruzicka, *Helv. Chim. Acta* **4**, 475 (1921).
139b. R. Lukes and J. Kovar, *Chem. Listz* **48**, 404 (1954).
140. C. Schöpf, G. Herbert, R. Rausch, and G. Schröder, *Angew. Chem.* **69**, 391 (1957).
141. E. Wenkert, K. G. Dave, and F. Haglid, *J. Amer. Chem. Soc.* **87**, 5461 (1965).
142. P. Karrer, *Helv. Chim. Acta* **21**, 223 (1938).
143. W. P. Hewmann, *Angew. Chem.* **70**, 401 (1958).
144. R. Lukes and J. Kuthan, *Collect. Czech. Chem. Commun.* **26**, 1845 (1961).
145. J. Kuthan and E. Janeckova, *Collect. Czech. Chem. Commun.* **28**, 1654 (1964).
146. R. Lukes and K. Kovar, *Chem. Listy.* **48**, 692 (1954).
147. R. Lukes and M. Smetackova, *Collect. Czech. Chem. Commun.* **5**, 61 (1931).
147a. A. J. Hubert, *J. Chem. Soc. C* 2048 (1968).
147b. J. Sauer and H. Prahl, *Tetrahedron Lett.* **25**, 2863 (1966).
147c. R. A. Clark and D. C. Parker, *J. Amer. Chem. Soc.* **93**, 7257 (1971).
148. A. J. Hubert, *J. Chem. Soc. C* 2048 (1968).
149. M. Riviere and A. Lattes, *Bull. Soc. Chim. Fr.* 2539 (1967).
150. J. Sauer and H. Prahl, *Tetrahedron Lett.* 2863 (1966).
151. G. T. Martirosyan, M. G. Indzhikyan, E. A. Grigoryan, and A. T. Babayan, *Arm. Khim. Zh.* **20**, 275 (1967).
152. A. J. Speziale and R. C. Freeman, *Org. Syn.* **41**, 21 (1961).
153. A. J. Speziale and L. R. Smith, *J. Amer. Chem. Soc.* **84**, 1868 (1962).
154. A. J. Speziale and R. C. Freeman, *J. Amer. Chem. Soc.* **82**, 903 (1960).
155. H. Bohme and G. Berg, *Chem. Ber.* **99**, 2127 (1966).
156. C. Djerassi, N. Crossley, and M. A. Kielczewski, *J. Org. Chem.* **27**, 1112 (1962).
157. N. S. Crossley, C. Djerassi, and M. A. Kielczewski, *J. Chem. Soc.* 6253 (1965).
158. A. T. Bottini and J. D. Roberts, *J. Amer. Chem. Soc.* **79**, 1462 (1957).
159. E. F. Jenny and J. Druey, *J. Amer. Chem. Soc.* **82**, 3111 (1960).
160. Y. Kitahara, M. C. Caserio, F. Scardiglia, and J. D. Roberts, *J. Amer. Chem. Soc.* **82**, 3106 (1960).
161. H. W. Whitlock, Jr. and G. L. Smith, *Tetrahedron Lett.* 1389 (1965).
162. S. J. Huang and M. V. Lessard, *J. Amer. Chem. Soc.* **90**, 2432 (1968).
163. P. Nelson and A. Pelter, *J. Chem. Soc.* 5142 (1965).
164. H. Hirsch, *Chem. Ber.* **100**, 1289 (1967).
165. H. Weingarten and W. A. White, *J. Org. Chem.* **31**, 4041 (1966).
166. Y. Ban and O. Yonemitsu, *Chem. Pharm. Bull.* **8**, 653 (1960).
167. J. Dabrowski, *Bull. Acad. Sci. Ser. Sci. Chim. Geol. Geograph.* **7**, 93 (1959).
168. D. Pocar, G. Bianchetti, and P. D. Croce, *Gazz. Chim. Ital.* **95**, 1220 1(965).
169. A. T. Babayan, M. G. Indzhikyan, and G. B. Bagdasaryan, *Dokl. Akad. Nauk SSSR* **133**, 1334 (1960).
170. P. M. Greaves and S. R. Landor, *Chem. Commun.* 322 (1966).
171. A. A. Petrov, I. A. Maretina, and K. S. Mingaleva, *Zh. Obshch. Khim.* **35**, 1720 (1965).
172. J. P. Chapp, Abstr. paper Nat. ACS Meeting, Houston, Texas, Feb. 22–27, 1970, Div. Org. Chem., ORGN No. 13.

173. J. Hoch, *C. R. Acad. Sci. Paris* **200**, 938 (1935).
174. C. D. Hurd and L. T. Sherwood, Jr., *J. Org. Chem.* **13**, 471 (1948).
175. F. C. Pennington and W. D. Kehret, *J. Org. Chem.* **32**, 2034 (1967).
176. G. N. Walker and D. Alkalay, *J. Org. Chem.* **32**, 2213 (1967).
177. H. Weingarten and N. K. Edelmann, *J. Org. Chem.* **32**, 3293 (1967).
178. A. L. Burger, C. S. Zirngible, C. S. Davis, C. Kaiser, and C. L. Zirkle, Abstr. ACS Meeting, 139th, St. Louis, Missouri, p. 1N, March, 1961.
179. K. N. Büchel and F. Korte, *Chem. Ber.* **97**, 2453 (1962).
180. F. Korte, K. H. Büchel, H. Mäder, G. Römer, and H. H. Schulze, *Chem. Ber.* **95**, 2424 (1962).
181. H. Hasse and A. Wieland, *Chem. Ber.* **93**, 1686 (1960); K. Langheld, *Chem. Ber.* **42**, 392 (1909).
182. S. Gabriel, *Chem. Ber.* **46**, 1358 (1913).
183. E. Immendorfer, *Chem. Ber.* **48**, 612 (1915).
184. O. Mumm and H. Münchmeyer, *Chem. Ber.* **43**, 3345 (1910).
185. O. Mumm and H. Hornhardt, *Chem. Ber.* **70**, 1930 (1937).
186. E. P. Kohler, A. Graustein, and D. R. Merrill, *J. Amer. Chem. Soc.* **44**, 2536 (1922).
187. D. Seyferth, M. E. Gordon, and R. Damrauer, *J. Org. Chem.* **32**, 469 (1967).
188. R. F. Parcell and F. P. Hauck, Jr., *J. Org. Chem.* **28**, 3468 (1963).
189. M. G. Ettlinger and F. Kennedy, *Chem. Ind.* (*London*) 166 (1956).
190. A. T. Bottini and J. D. Roberts, *J. Amer. Chem. Soc.* **79**, 1462 (1957).
191. E. E. van Tamelen, Y. B. Haarstad and R. L. Orvis, *Tetrahedron* **24**, 687 (1968).
192. N. D. Heindel, P. D. Kennewell, and M. Pfan, *J. Org. Chem.* **35**, 79 (1970).
193. F. Korte, *Angew. Chem.* **74**, 184 (1962).
194. F. Korte, H. Dürbeck, and G. Weisgerber, *Chem. Ber.* **100**, 1305 (1967).
195. L. J. Dolby and S. Sakai, *J. Amer. Chem. Soc.* **86**, 5362 (1964).
196. R. S. Neale, *J. Org. Chem.* **32**, 3263 (1967).
197 J. A. Berson, E. M. Evleth, and Z. Hamlet, *J. Amer. Chem. Soc.* **87**, 2887 (1965).
198. J. F. Stephen and E. Marcus, *J. Org. Chem.* **35**, 258 (1970).
199. L. A. Paquette and R. W. Begland, *J. Amer. Chem. Soc.* **88**, 4685 (1966).
200. M. H. Rosen and G. B. Bonet, Abstr. paper presented at Nat. ACS Meeting, 159th Houston, Texas, Feb. 22–27, 1970, Div. Org. Chem., ORGN No. 11.
201. R. M. Wilson and F. DiNinno, Jr., Abstr. paper presented at Nat. ACS Meeting, 159th, Houston, Texas, Feb. 22–27, 1970, Div. Org. Chem., ORGN No. 19.
202. S. J. Huang and M. V. Lessard, *J. Org. Chem.* **35**, 1204 (1970).
203a. A. G. Cook, S. B. Herscher, D. J. Schultz, and J. A. Burke, *J. Org. Chem.* **35**, 1550 (1970).
203b. J. F. Stephen and E. Marcus, *J. Org. Chem.* **34**, 2535 (1969).
203c. A. G. Cook, W. M. Kosman, T. A. Hecht, and W. Koehn, *J. Org. Chem.* **37**, 1565 (1972).
204. E. Kuhle and L. Eue, Fr. Pat. 1,256,873 (1961).
205. H. Lehmkuhland and D. Reinehr, *J. Organometal. Chem.* **55**, 215 (1973).
206. J. P. Guthrie and F. Jordan, *J. Amer. Chem. Soc.* **94**, 9132 (1972).
207. P. J. Jessup, C. B. Petty, J. Roos, and L. Overman, *Org. Syn.* **59**, 1 (1979).
208. T. Eicher and M. Urban, *Chem. Ber.* **113**, 408 (1980).
209. K. A. Parker, J. J. Petraitis, R. W. Kosley, Jr., and S. L. Buchwald, *J. Org. Chem.* **47**, 389 (1982).
210. J.-P. Ceierier, E. Deloisy, G. Lhommet, and P. Maitte, *J. Org. Chem.* **44**, 3089 (1979).
211. P. H. Merrell, V. L. Goedken, and D. H. Busch, *J. Amer. Chem. Soc.* **92**, 7590 (1970).
212. T. Shono, Y. Matsumura, K. Tsubata, Y. Sugihara, S.-I. Yamane, T. Kanazawa, and T. Aoki, *J. Amer. Chem. Soc.* **104**, 6697 (1982).
213. C. Hansson and B. Wickberg, *J. Org. Chem.* **38**, 3074 (1973).

CHAPTER 5 / **YNAMINES**

1. INTRODUCTION

Alkynylamines or ynamines have the structure shown in (I). The resonance

$$R-C\equiv C-NR_2$$
(I)

of this structure suggests that the nitrogen atom or the β-carbon atom can react as a nucleophilic center (Eq. 1).

$$R-C\equiv C-\overset{..}{N}R_2 \longleftrightarrow R\overset{-}{C}=C=\overset{+}{N}R_2$$
(1)

Attempts to synthesize ynamines with primary amino groups have resulted in producing the tautomeric nitriles [1–3]. See, for example, Eq. (2). On the

$$F-C\equiv CH \xrightarrow{\text{NaNH}_2/\text{NH}_3} [H_2N-C\equiv CH] \longrightarrow CH_3C\equiv N$$
(2)

basis of this evidence the ynols should exist in the form of the tautomeric ketenes.

Although ynamines were thought to have been prepared earlier [4–6], they were not characterized until Zaugg *et al.* reported the first ynamine, *N*-(1-propynyl)phenothiazine (II) [7].

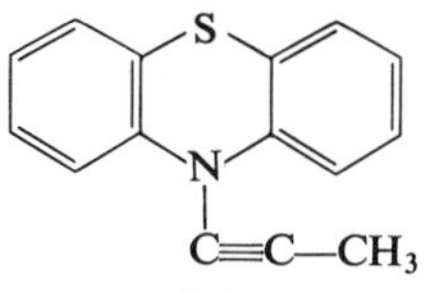

(II)

The chemistry of ynamines is relatively brief, simply because further research remains to be done on this novel functional group.

The most important methods of synthesizing ynamines are based on the reactions shown in Scheme 1.

135

SCHEME 1

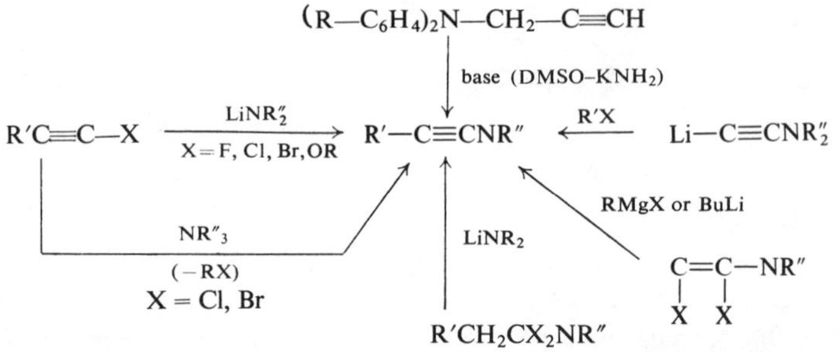

$$(R-C_6H_4)_2N-CH_2-C\equiv CH$$

$$\downarrow \text{base (DMSO-KNH}_2)$$

$$R'C\equiv C-X \xrightarrow[X=F, Cl, Br, OR]{LiNR''_2} R'-C\equiv CNR'' \xleftarrow{R'X} Li-C\equiv CNR''_2$$

$$\underset{\substack{(-RX) \\ X = Cl, Br}}{NR''_3} \qquad \underset{R'CH_2CX_2NR''}{LiNR_2} \qquad \underset{\substack{RMgX \text{ or BuLi} \\ C=C-NR'' \\ X \quad X}}{}$$

Most ynamines are water-clear liquids with a slight amine odor. They are stable and can be vacuum-distilled without decomposition. Some have now become available commercially.*

The chemistry of the reactions of ynamines is under active investigation. A review by Viehe describes some of the reactions of ynamines [8]. Some of them are shown in Eq. (3).

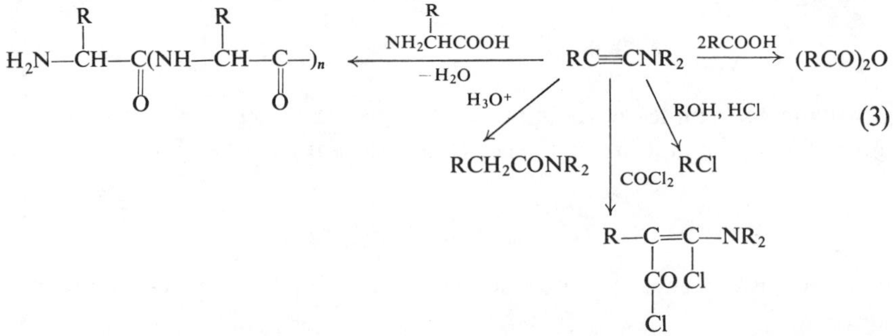

$$H_2N-\underset{R}{\overset{|}{CH}}-\underset{O}{\overset{\|}{C}}(NH-\underset{R}{\overset{|}{CH}}-\underset{O}{\overset{\|}{C}}-)_n \xleftarrow[-H_2O]{\underset{R}{\overset{|}{NH_2CHCOOH}}} RC\equiv CNR_2 \xrightarrow{2RCOOH} (RCO)_2O$$

$$\underset{RCH_2CONR_2}{\overset{H_3O^+}{\swarrow}} \quad \underset{\underset{COCl_2}{}}{\downarrow} \overset{ROH, HCl}{\underset{RCl}{\searrow}} \qquad (3)$$

$$R-\underset{\underset{Cl}{\overset{|}{CO}}}{\overset{|}{C}}=\underset{Cl}{\overset{|}{C}}-NR_2$$

More recent reports describe the Diels–Alder reactions of ynamines [8a, b], cyclo additions with isocyanates [8c], carbon dioxide [8d], sulfenes [8e], N-sulfonylamines [8f], thietenes [8g], and α,β-unsaturated sulfones [8h]. Some other interesting reactions reported are the photochemical condensation with naphthoquinones [8i] and the ynamine Claisen rearrangement [8j, k].

2. CONDENSATION REACTIONS

The condensation methods available for the preparation of ynamines are summarized in reactions (4) and (5).

* Fluka, A. G., Buchs, Switzerland has 1-diethylaminopropyne available in 99% purity, b.p. 130°–132° n_D^{20} 1.444.

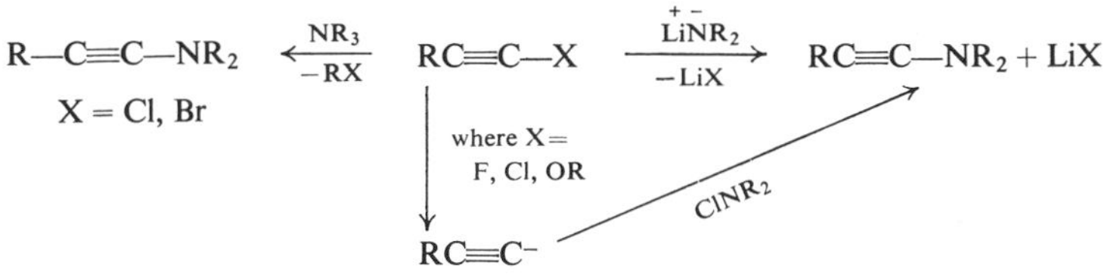

$$R—C{\equiv}C—NR_2 \xleftarrow[-RX]{NR_3} RC{\equiv}C—X \xrightarrow[-LiX]{\overset{+\ -}{LiNR_2}} RC{\equiv}C—NR_2 + LiX \quad (4)$$

$X = Cl, Br$

where X =
F, Cl, OR

$$RC{\equiv}C^-$$

$ClNR_2$

$$(R_2N)_2C{=}CHCl \xrightarrow{LiNR_2} R_2N—C{\equiv}C—NR_2 \quad (5)$$

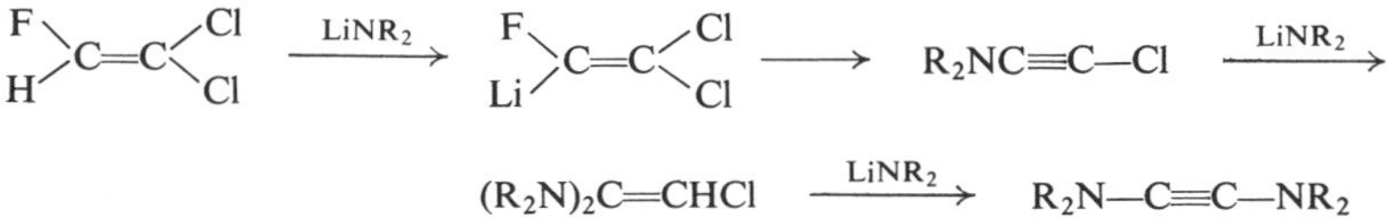

Fluoroacetylenes undergo nucleophilic substitution by lithium dialkyl-amides to give the ynamine, whereas chloroacetylenes are relatively unreactive. Chloroacetylenes may be made reactive by having electron-withdrawing groups present, as in phenylchloroacetylene [8] (Eq. 6).

$$C_6H_5—C{\equiv}C—Cl + LiN(CH_3)_2 \rightarrow C_6H_5C{\equiv}C—N(CH_3)_2 \quad (6)$$

Chlorocyclohexenyl acetylene reacts in a similar fashion with lithium diethylamide [8] (Eq. 7).

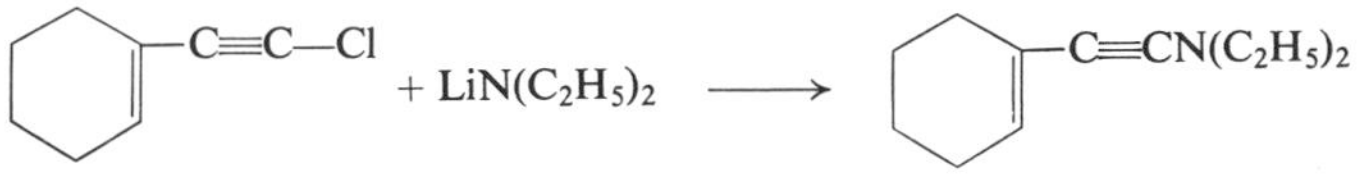

Trimethylamine also reacts with chloro- or bromoacetylenes containing electron-withdrawing groups to give ynamines [9, 10a, b]. Equation (8) gives an example. The presence of electron-donating groups such as the t-butyl group

$$C_6H_5C{\equiv}C—X + N(CH_3)_3 \xrightarrow[55°C]{40\ hr} C_6H_5C{\equiv}C—N(CH_3)_2 + CH_3X \quad (8)$$

$$X = Cl\ or\ Br$$

requires longer reaction times at elevated temperatures (Eq. 9).

$$(CH_3)_3C—C{\equiv}C—Cl + N(CH_3)_3 \xrightarrow[135°C]{60\ hr} (CH_3)_3C—C{\equiv}C—N(CH_3)_2 \quad (9)$$

$$(44\%)$$

Haloalkynes have been reported to react with bridgehead amines to give ynamines [10d].

$$R'C{\equiv}CX + R_3N \longrightarrow (R'C{\equiv}CNR_3)^+X^- \xrightarrow{R_3N} RC{\equiv}CNR_3 + R_3N^+X^- \quad (9a)$$

Examples of bridgehead amines are 1,4-diazobicyclo-[2,2,2]octane, bru-cine, dihydrobrucine, and quinilidene.

Lithium dimethylamine in ether does not react well with *t*-butyl chloro-acetylene to give the ynamine; however, the reaction appears to go easily if the related fluoroacetylene is used or if a polar solvent (hexamethylphosphor-amide should not be used since it is carcinogenic) [8] is used in place of ether.

2-1. Preparation of N,N-Dimethylaminophenylacetylene [10a]

$$C_6H_5C≡C—Cl + N(CH_3)_3 \longrightarrow C_6H_5C≡C—N(CH_3)_2 + CH_3Cl \qquad (10)$$

To a stainless steel autoclave is added 21.7 gm (0.368 mole) of trimethyl-amine and 15.0 gm (0.11 mole) of 1-chloro-2-phenylacetylene.* The autoclave is sealed and heated to 55°C for 40 hr. After this time the autoclave is cooled to room temperature, vented, and the reaction mixture is extracted with an-hydrous petroleum ether. The solvent is removed under reduced pressure and the residue is distilled to afford 7.0 gm (44%) of a light-brown oil, b.p. 90°C (40 mm). Redistillation of this product affords 5 gm (31%), b.p. 70°C (1 mm), n_D^{25} 1.5849; NMR δ 2.65 (s, 6, CH_3), 7.25 (m, 5, Ar—H).

1-Chloro-2-phenylacetylene has recently been reported to be useful in preparing 1-phenyl-2-(1-pyrrolidinyl)acetylene as shown in Eq. (11) [10e].

$$C_6H_5—C≡C—Cl + \underset{Li}{\boxed{N}} \longrightarrow \underset{\underset{C≡C—C_6H_5}{|}}{\boxed{N}} \qquad (11)$$

Other examples of the trialkylamine reaction [11] are given in Eqs. (11a) and

$$(C_2H_5O)_2\overset{O}{\overset{\|}{P}}—C≡C—Cl + N(C_2H_5)_3 \xrightarrow[\substack{2. C_6H_6, \\ 10\ min\ at \\ 80°C}]{1.\ 25°C} (C_2H_5O)_2\overset{O}{\overset{\|}{P}}C≡C—N(C_2H_5)_2 + C_2H_5Cl \qquad (11a)$$

$$C_6H_5C≡C—C≡C—Br + N(CH_3)_3 \longrightarrow C_6H_5C≡C—C≡C—N(CH_3)_2 + CH_3Br \qquad (12)$$

(12). Several additional examples of reaction (12) as reported by Dumont [12] are given in Table I.

The diynamines are very unstable and decompose rapidly even at 0°C. They can be purified by low-temperature column chromatography using alumina.

Acetylenic ethers react with lithium dialkylamides to give ynamines in good yields [13] (Eq. 13). Examples of this reaction are described in Table II.

$$R'C≡COR + LiNR_2'' \xrightarrow[110°-120°C]{} R'C≡C—NR_2'' \qquad (13)$$

---

* 1-Chloro-2-phenylacetylene is prepared by the reaction of phenylacetylene with benzenesulfonyl chloride in the presence of sodamide [10b].

TABLE I

REACTION OF TRIALKYLAMINES WITH HALOACETYLENES [13]

$$R—(C\equiv C)_2—Br \xrightarrow{R'_3N} R—(C\equiv C)_2—\overset{+}{N}R_3' \xrightarrow{R_3'N} R—(C\equiv C)_2—NR_2' \quad (14)$$
$$\quad\quad (A) \quad\quad\quad\quad\quad\quad (B) \quad\quad\quad\quad\quad\quad (C)$$

R	R' or R₃'N	Compound B (yield %)	Compound C (yield %)	Compound C, m.p. (°C)
CH_3	CH_3	31	Polymerized	Oil
C_6H_5	CH_3	70	50	50
C_6H_5	C_2H_5	60	20	Oil
C_6H_5	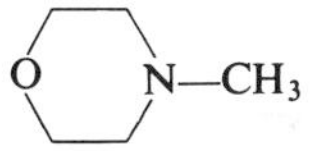 O⟨ ⟩N—CH₃	14	30	Oil

Propynyl dialkylamines are produced in low yields (see Table II) when the stoichiometric amount of reagents are used. When an excess of lithium dialkylamide is used, partial isomerization of the product to allenyl amines, $CH_2=C=CHNR_2'$ [13], takes place.

TABLE II

REACTION OF ACETYLENIC ETHERS WITH AN EQUIVALENT AMOUNT OF LITHIUM DIALKYL AMIDES TO GIVE YNAMINES [13]

$$R—C\equiv C—OC_2H_5 + LiNR_2' \rightarrow R—C\equiv C—NR_2' + LiOC_2H_5 \quad (15)$$

R	R'	B.p., °C (15 mm)	Yield (%)
CH_3	C_3H_7	57–88	14
C_2H_5	C_3H_7	71–72	67
C_2H_5	 N⟨ ⟩	78–79	55
C_3H_7	C_2H_5	60–61	68
$n-C_4H_9$	C_2H_5	75–76	65

2-2. General Method for Preparation of Ynamines by Reaction of Acetylenic Ethers with Lithium Dialkylamides [13]

The lithium dialkylamides were prepared by adding at 20°C the stoichiometric amount of amine to ether solutions of methyl- or butyllithium.

The 1-alkynyl (0.55 mole) is added in 5 min to a solution of 0.50 mole of lithium dialkylamide in 500 ml of ether at room temperature. The ether is removed by distillation and an exothermic reaction starts at a bath temperature of about 80°C. Heating is continued for an additional 30 min at 110°–120°C and then the reaction products are distilled at 15 mm Hg pressure using a short Vigreux column. The heating bath is raised to 170°C and the last traces of products are removed by distillation at 1 mm Hg pressure. In all cases the distillation receiver should be cooled to −80°C. The entire distillates are combined and fractionated through a 30 cm Widmer column. Some typical results of using this method are shown in Table II.

Ficini and Barbara [14, 15] prepared lithium aminoacetylides and allowed them to react with alkyl bromides or tosylates in hexamethylphosphoramide to give good yields of substituted ynamines (Eq. 16). This method shows promise of being a general route to ynamine derivatives.

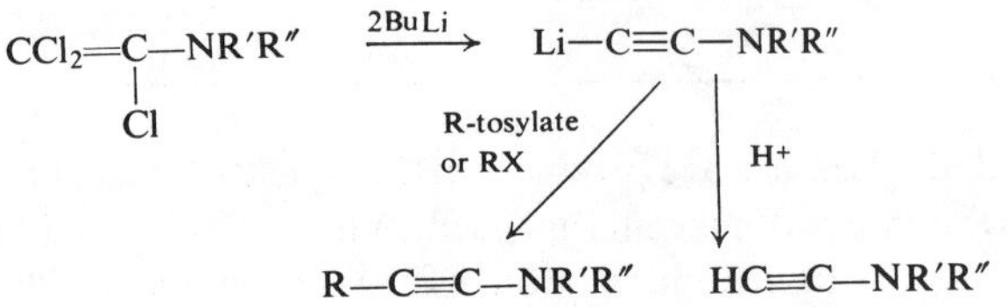

$$CCl_2{=}C{-}NR'R'' \xrightarrow{2\,BuLi} Li{-}C{\equiv}C{-}NR'R''$$

(with R-tosylate or RX → $R{-}C{\equiv}C{-}NR'R''$ and H^+ → $HC{\equiv}C{-}NR'R''$)

(16)

Examples of the condensation of lithium N-methyl-N-phenylaminoacetylide with alkyl halides is described in Table III.

TABLE III

REACTION OF LITHIUM N-METHYL-N-PHENYLAMINOACETYLIDE WITH VARIOUS REAGENTS IN HEXAMETHYLPHOSPHORAMIDE TO GIVE SUBSTITUTED YNAMINES

$$Li{-}C{\equiv}C{-}\underset{\underset{CH_3}{|}}{N}{-}C_6H_5 + RX \rightarrow R{-}C{\equiv}C{-}\underset{\underset{CH_3}{|}}{N}{-}C_6H_5$$

(17)

RX	R	Yield (%)	B.p., °C (mm Hg)	$n_D^{(t)}$
CH₃I	CH₃	65	96 (6.0)	1.5740 (23)
C₃H₇Br	C₃H₇	61	70 (0.06)	1.5523 (24)
n-C₄H₉Br	n-C₄H₉	76	88 (0.1)	1.5464 (22)
CH₃—CH₂—CH—CH₃ (Br)	CH₃—CH₂—CH—CH₃	38[a]	83 (0.1)	1.5468 (21)
CH₂=CH—CH₂Br	CH₂=CH—CH₂—	75[b]	96 (0.6)	1.5670 (21)

[a] Approximately 25% of the starting material undergoes elimination to the olefin. This is true for cyclic and tertiary alkyl halides.

[b] Some of the product decomposed on being distilled.

2-3. Preparation of N-Methyl-N-phenylaminohexyne [15]

(a) General preparation of N-methyl-N-phenyllithium aminoacetylide and related acetylides.

$$CCl_2{=}C{-}N{-}C_6H_5 + 2n\text{-BuLi} \longrightarrow Li{-}C{\equiv}C{-}N{-}C_6H_5 \qquad (18)$$
$$\overset{|}{Cl}\ \overset{|}{CH_3} \qquad\qquad\qquad\qquad\qquad \overset{|}{CH_3}$$

To a three-necked flask equipped with a stirrer, condenser, thermometer, and dropping funnel is added an ether solution of 1.0 mole of n-butyllithium. Then a 50% ether solution of N-methyl-N-phenyltrichloro-1,1,2-enamine [16], b.p. 87°C (0.1 mm), n_D 1.5859 (25), is added dropwise at −10°C under a nitrogen atmosphere until 0.42 mole has been added. The reaction is allowed to warm to room temperature in about 2 hr. The solution now possesses an intense absorption band at 2040–2050 cm^{-1}, which is characteristic of the lithium acetylide group.

(b) Condensation of N-methyl-N-phenyllithium aminoacetylide with n-butyl bromide.

$$Li{-}C{\equiv}C{-}N{-}C_6H_5 + n\text{-}C_4H_9Br \longrightarrow n\text{-}C_4H_9{-}C{\equiv}C{-}N{-}C_6H_5 \qquad (19)$$
$$\overset{|}{CH_3} \qquad\qquad\qquad\qquad\qquad\qquad \overset{|}{CH_3}$$

To the three-necked flask equipped as above and containing 0.42 mole of N-methyl-N-phenyllithium aminoacetylide is added dropwise 88 gm (0.65 mole) of n-butyl bromide in 160 ml of hexamethylphosphoramide.* The reaction is exothermic and the ether refluxes. At the end of the addition the ether is removed by distillation and the temperature is raised and kept at 75°C for 2 hr or until the absorption band at 2050 cm^{-1} disappears. The product is obtained by distilling it directly from the reaction medium under reduced pressure to afford 58 gm (75%), b.p. 88°C (0.1 mm), n_D^{22} 1.5464.

2-4. Preparation of N,N-Diethylcarbomethoxyethynylamine [17]

$$Cl_2C{=}C{-}N{-}(C_2H_5)_2 + 2n\text{-BuLi} \longrightarrow [Li{-}C{\equiv}C{-}N(C_2H_5)_2]$$
$$\overset{|}{Cl}$$
$$\xrightarrow[\displaystyle \underset{\text{ClCOCH}_3}{\overset{\text{O}}{\overset{\|}{}}}]{} CH_3OC{-}C{\equiv}CN(C_2H_5)_2 \qquad (20)$$

To a flask containing 5.0 gm (23 mmoles) of N,N-diethyl-1,2,2-trichloro-vinylamine [16] cooled to −15°C is added dropwise at −10°C under a nitrogen atmosphere 50 mmoles of n-butyllithium in hexane (diluted with ⅓ volume of dry ether). After the addition the mixture is kept at room temperature for 45

---

* CAUTION: Hexamethylphosphoramide is carcinogenic and other polar solvents would be preferable.

min. The mixture is cooled again to −10°C and then 2.16 gm (23 mmoles) of methyl chloroformate in 5 ml of dry ether is added dropwise at −10°C. The reaction is again raised to room temperature and kept there for 45 min. The reaction mixture is centrifuged and the centrifugate is distilled under reduced pressure to afford 2.5 gm (70%) of the ynamine, b.p. 91°C (2.5 mm). The product showed characteristic absorptions in the ir at 2200, 1695 cm^{-1}, and NMR (neat with a TMS standard) δ 1.20 (t, 6H), 2.97 (q, 4H), 4.35 (s, 3H).

Corey and Cane [17a] reported that n-butyllithium in tetramethylethylenediamine reacts to metalates α,β-ynamines to form the lithium derivative, which may undergo alkylation upon treatment with a variety of halides. For example,

$$CH_3C{\equiv}CNR_2 \xrightarrow[\text{TMED}]{\text{n-BuLi}} LiCH_2C{\equiv}CNR_2 \xrightarrow{\text{R'X}} R'{-}CH_2C{\equiv}CNR_2 \quad (21)$$

$$R' = CH_2{=}CH{-}CH_2Br \quad \text{or} \quad (CH_3)_3SiCl \quad \text{etc.}$$

t-Butyllithium metalates acetonitrile to form the dilithio derivative, which can be alkylated by reaction with halides. For example, when using an excess of base on Li_2C_2HN for reaction with triethylchlorosilane in THF [17b]:

$$CH_3CN + 3t\text{-BuLi} \xrightarrow[-78°C]{\text{ether}} [LiC{\equiv}CNLi_2] \xrightarrow[\substack{(CH_3)_3SiCl \\ THF}]{\text{t-BuLi}}$$

$$(CH_3)_3SiC{\equiv}CN[Si(CH_3)_3]_2 + [(CH_3)_3Si]_2C{=}C{=}NSi(CH_3)_3 \qquad (22)$$

$$(20\%) \qquad\qquad\qquad (80\%)$$

3. ELIMINATION REACTIONS

Some elimination reactions which yield ynamines involve the dehalogenation or dehydrohalogenation of substituted amines or amides as shown in Eq. (23).

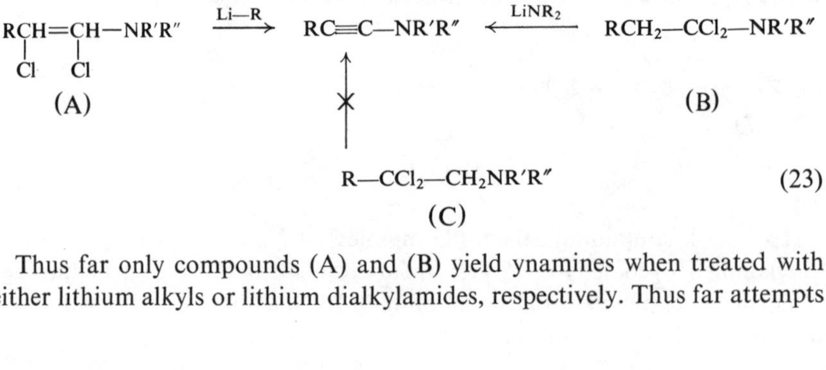

$$\begin{matrix} RCH{=}CH{-}NR'R'' \\ \;\;|\;\;\;\;\;\;| \\ \;Cl\;\;\;\;Cl \\ (A) \end{matrix} \xrightarrow{\text{Li}{-}R} RC{\equiv}C{-}NR'R'' \xleftarrow{\text{LiNR}_2} RCH_2{-}CCl_2{-}NR'R''$$

$$(B)$$

$$R{-}CCl_2{-}CH_2NR'R'' \qquad (23)$$

$$(C)$$

Thus far only compounds (A) and (B) yield ynamines when treated with either lithium alkyls or lithium dialkylamides, respectively. Thus far attempts to use compound (C) have led to other products [9]. Since little information has been published about compound (C), further research is required before it can be ruled out.

As described in Section 2, n-butyllithium reacts with α,β-dichloroenamines to yield lithium ynamines [15] (Eq. 24).

$$Cl-\underset{\underset{Cl}{|}}{C}=\underset{\underset{Cl}{|}}{C}-NR'R'' \longrightarrow Li-C\equiv C-NR'R'' \qquad (24)$$

The preparation of N-methyl-N-phenyllithium aminoacetylide is an example of this method (see Eq. 18).

As a dehydrohalogenating reagent, phenylmagnesium bromide is not as effective as are lithium dialkylamides, but in hexamethylphosphoramide it reacts with α,β-dichloroenamines to give a 35% yield of the corresponding ynamine [18] (Eq. 25).

$$\underset{\underset{Cl}{|}}{HC}=\underset{\underset{Cl}{|}}{C}-N(C_6H_5)_2 + C_6H_5-MgBr \xrightarrow[-40°C]{} Cl-C\equiv C-N(C_6H_5)_2 \qquad (25)$$

Buijle $et\ al.$ [19] reported that ynamines can be obtained by the dehydrohalogenation of α,α-dihaloamides. The α,α-dihaloamides can be obtained by the reaction of phosgene with tertiary amides [20]. An example of the use of this

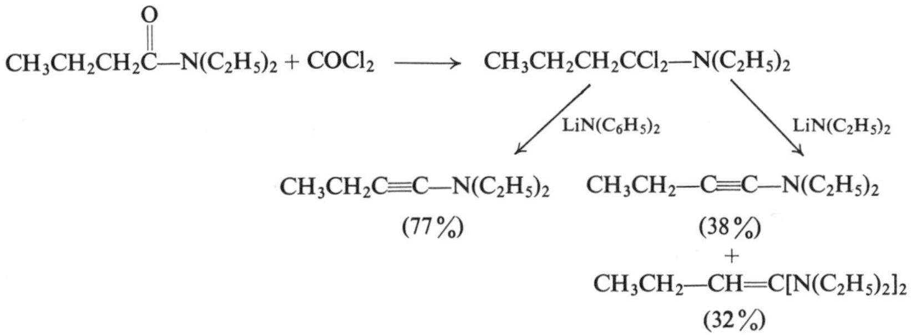

$$CH_3CH_2CH_2\overset{\overset{O}{\|}}{C}-N(C_2H_5)_2 + COCl_2 \longrightarrow CH_3CH_2CH_2CCl_2-N(C_2H_5)_2$$

$$\swarrow LiN(C_6H_5)_2 \qquad \searrow LiN(C_2H_5)_2$$

$$CH_3CH_2C\equiv C-N(C_2H_5)_2 \qquad CH_3CH_2-C\equiv C-N(C_2H_5)_2$$
$$(77\%) \qquad\qquad (38\%)$$
$$+$$
$$CH_3CH_2-CH=C[N(C_2H_5)_2]_2$$
$$(32\%)$$

$$(26)$$

method is shown in Eq. (26). Other examples with more details are shown in Table IV.

Ynamines can also be produced by α or β eliminations in the reaction of dihaloalkenes or trihaloalkenes with alkali metal amides [10, 21] (Eqs. 28–31).

TABLE IV

REACTION OF α,α-DIHALOAMIDES WITH LITHIUM DIALKYLAMIDES TO GIVE YNAMINES [19]

$$R'CH_2—CCl_2—NR_2'' + LiNR_2 \rightarrow R'C{\equiv}CNR_2'' \qquad (27)$$

R'	R''	R	B.p., °C (mm Hg)	Yield (%)
CH$_3$	C$_2$H$_5$	C$_2$H$_5$	60–62 (90)	39
		C$_6$H$_{11}$		58
	C$_5$H$_{10}$	CH$_3$	72–73 (15)	12
	C$_5$H$_{10}$	C$_2$H$_5$	—	39
	C$_5$H$_{10}$	C$_3$H$_7$	—	0
	C$_5$H$_{10}$	C$_6$H$_{11}$	—	46
C$_2$H$_5$	C$_2$H$_5$	C$_2$H$_5$	50–51 (30)	38
	C$_2$H$_5$	C$_6$H$_{11}$	—	77
n-C$_4$H$_9$	C$_2$H$_5$	C$_2$H$_5$	77–78 (15)	29
H	C$_2$H$_5$	C$_2$H$_5$	—	a

a Determined from infrared spectrum of crude reaction product in solution.

$$\underset{F}{\overset{F}{>}}C{=}C\underset{Cl}{\overset{Cl}{<}} + 3LiN(C_2H_5)_2 \xrightarrow[-80°C]{ether} (C_2H_5)_2N—C{\equiv}C—N(C_2H_5)_2 \qquad (28)$$
$$(60\%)$$

$$t\text{-Bu}—\overset{H}{\underset{Br}{C}}—\overset{Br}{\underset{H}{C}}—F + 3LiN(C_2H_5)_2 \longrightarrow t\text{-Bu}—C{\equiv}C—N(C_2H_5)_2 \qquad (29)$$
$$(48\%)$$

$$C_6H_5CH{=}C\underset{F}{\overset{Cl}{<}} + 2LiN(C_2H_5)_2 \xrightarrow{-80°C} C_6H_5C{\equiv}C—N(C_2H_5)_2 \qquad (30)$$
$$(86\%)$$

$$p\text{-CH}_3—C_6H_4—S—\underset{Cl}{\overset{|}{C}}{=}CHCl + 2LiN(C_2H_5)_2 \xrightarrow{-80°C}$$
$$p\text{-CH}_3—C_6H_4S—C{\equiv}C—N(C_2H_5)_2 \qquad (31)$$
$$(70\%)$$

Strobach [21a] and Viehe [21b] reported that ynamines are produced by the reaction of lithium dialkylamides with 1,1-difluoroethylenes.

$$RCH{=}CF_2 + 2LiN(C_2H_5)_2 \longrightarrow RC{\equiv}CN(C_2H_5)_2 \qquad (32)$$
$$R = \text{alkyl or aryl}$$

In a related manner, ketene S,N-acetals also react with alkali metal amides to give ynamines [19] (Eq. 30). Some examples are shown in Table V.

TABLE V

YNAMINES FROM KETENE S,N-ACETALS AS DESCRIBED
IN EQUATION (33) [19]

R'	R$_2''$	Base	B.p., °C (mm Hg)	Yield (%)
C$_6$H$_5$	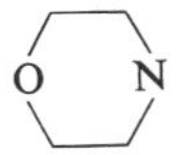 (morpholino ring) O N	LiN(C$_2$H$_5$)$_2$	99–100 (0.01)	40
		NaNH$_2$	—	50
		NaNH$_2$—C$_5$H$_{11}$N	—	50
H	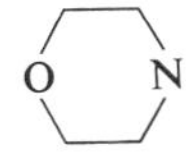 (morpholino ring) O N	NaNH$_2$—C$_5$H$_{11}$N	—	a

[a] Its presence was ascertained only from the solution ir spectrum.

$$R'CH=\underset{\underset{SR}{|}}{C}-NR_2'' \xrightarrow{\text{base}} R'C\equiv C-NR_2'' \qquad (33)$$

3-1. Preparation of 1-Chloro-2-N,N-diphenylaminoacetylene [18]

$$HC=\underset{\underset{Cl}{|}}{\underset{}{C}}-N(C_6H_5)_2 \xrightarrow[\substack{40°C,\ HMP \\ 2.\ NH_4Cl+H_2O}]{1.\ C_6H_5-MgBr} Cl-C\equiv C-N(C_6H_5)_2 + HC\equiv C-N(C_6H_5)_2$$

(Cl on the left carbon)

$$(34)$$

To a flask containing 20.1 gm (0.076 mole) of 1,2-dichloro-1-N,N-diphenyl-
aminoethylene in 100 ml of ether is added 0.1 mole of phenylmagnesium
bromide dissolved in ether and 0.12 mole of hexamethylphosphoramide. The
reaction is kept at 40°C for 2 hr and then added to an ice–ammonium chloride
solution. On removal of the ether, the product, 5.3 gm (42%), ir 2220 cm^{-1}
(C≡C), is isolated. The product on recrystallization from hexane yields white
crystals, m.p. 105°C. In addition, 1.2 gm (35%) of N,N-diphenylamino-
acetylene is isolated, b.p. 97°C (0.1 mm), n_D^{23} 1.6188. The Latter ynamine can
be obtained as a crystalline solid on cooling, m.p. not reported.

4. OXIDATION REACTIONS

The air oxidation of phenylacetylene and secondary amines in the presence
of cupric acetate in benzene solution yields ynamines [22]. This reaction requires only catalytic amounts of cupric salts and gives high conversions in less

than 30 min when the Cu^{+2}/phenylacetylene ratio is only 0.02. Only 1,4-diphenylbutadiyne is produced if the stoichiometric amount of cupric ion is used in the absence of oxygen. The yield of ynamine can be increased from 45 to 90% if the stoichiometric amounts of a reducing agent such as hydrazine are continuously added during the course of the reaction. The use of primary amines under similar conditions yields the acetamide derivative.

4-1. Preparation of N,N-Dimethylamino-2-phenylacetylene [22]

$$\text{Ph}-C{\equiv}CH + HN(CH_3)_2 \xrightarrow[C_6H_6]{Cu(OAc)_2,\ O_2}$$

$$\text{Ph}-C{\equiv}C-N(CH_3)_2 + \text{Ph}-C{\equiv}C-C{\equiv}C-\text{Ph} + H_2O \quad (35)$$

To a flask containing 2.0 gm (0.01 mole) of cupric acetate in 25 ml (0.38 mole) of dimethylamine in 100 ml of benzene at 5°C are added dropwise simultaneously a solution of 5.1 gm (0.05 mole) of phenylacetylene in 100 ml of benzene over 30 min and a stream of oxygen (1.0 ft^3/hr). The oxygen stream is added for 30 min more after the phenylacetylene addition has been completed. The copper ions are precipitated by adding 100 ml of ice water. The organic layer is separated, dried, and concentrated under reduced pressure. A gas chromatograph of the crude (column 10 ft × ½ in. 410 gum rubber) showed two peaks: N,N-dimethylamino-2-phenylacetylene and 1,4-diphenylbutadiyne. The ynamine was obtained in 40% yield as determined by reaction of the crude mixture with dilute hydrochloric acid and isolating the resultant N,N-dimethyl-aminophenylacetylene. The ynamine has characteristic ir absorption bands at 2205 and 2235 cm^{-1} and NMR absorption bands at δ 7.14 (4.0 protons) and 2.73 (6.05 protons).

5. REARRANGEMENT REACTIONS

An attempt to alkylate phenothiazine with propargyl bromide using sodium hydride in dimethylformamide afforded a 70% yield of N-(1-propynyl)phenothiazine instead of the expected N-(2-propynyl)phenothiazine [7]. This reaction (Eq. 36) constituted the first synthesis of an ynamine [7]. Other diarylamines such as diphenylamine react in a similar manner [12]. In both cases the propargylamine can be isolated and separately rearranged to the ynamine using dimethyl sulfoxide and potassium amide at ambient temperatures [12].

Viehe also reported that propargylamines can be rearranged to ynamines by contacting with a dispersion of potassium metal or potassium amide on alumina [12a]. (See Table VI.)

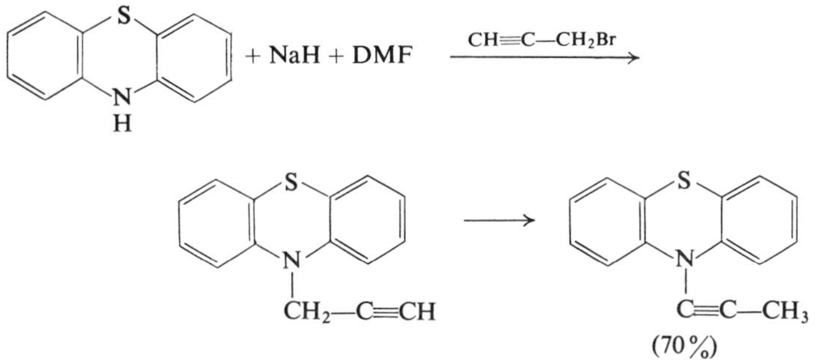

(36)

(70%)

5-1. Preparation of N-(1-Propynyl)phenothiazine [7]

To a flask maintained with a constant nitrogen atmosphere and containing 7.2 gm (0.3 mole) of sodium hydride in 600 ml of anhydrous dimethylformamide (DMF) is added portionwise 60 gm (0.3 mole) of phenothiazine. The reaction mixture is warmed to 50°C for an additional 2 hr and then raised to 70°C while a solution of 35.7 gm (0.3 mole) of propargyl bromide in 50 ml of DMF is added dropwise. The reaction mixture is heated for 2 hr at 70°C and kept overnight at room temperature. The solvent is removed by distillation at reduced pressure, and the residue is added to water. The insoluble product is extracted into ether, washed with water, dried, and concentrated by distillation to give 52 gm (87%) of an oily crude product. The oil solidifies when hexane is added, or when the oil (52 gm) in benzene is passed over a column of alumina, to give 48 gm (81%) of colorless crystals, m.p. 95°–96°C.

Other related rearrangements have been reported to occur by heating diarylaminoacetylenic tertiary alcohols with potassium hydroxide (Eq. 37) [12]. (See Table VII.)

(37)

More recently it has been reported that a dispersion of potassium amide on alumina is an active catalyst for the isomerization of N,N-dialkylprop-2-ynylamines into allenamines and ynamines [23]. For example, using potassium

TABLE VI

REARRANGEMENT OF PROPARGYLAMINES TO YNAMINES [12a]

Example	Propargylamine	Rearrangement catalyst	Ynamine	B.p. (mm Hg)	Yield (%)
1	$(C_2H_5)_2N—CH_2—C{\equiv}CH$	$Al_2O_3/KNH_2/K$	$CH_3C{\equiv}CN(C_2H_5)_2$	127–130 (760)	60
2	$(C_2H_5)_2N—CH_2—C{\equiv}CH$	$Al_2O_3 + KNH_2$	$CH_3C{\equiv}CN(C_2H_5)_2$	127–130 (760)	43
3	$(C_2H_5)_2N—CH_2—C{\equiv}CH$	$Al_2O_3 + KNH_2$	$CH_3C{\equiv}CN(C_2H_5)_2$	127–130 (760)	30
4	$(C_2H_5)_2N—CH_2—C{\equiv}CH$	$Al_2O_3 + KNH_2$	$CH_3C{\equiv}CN(C_2H_5)_2$	127–130 (760)	34
5	$(C_2H_5)_2N—CH_2—C{\equiv}CH$	$Al_2O_3 + K$	$CH_3C{\equiv}CN(C_2H_5)_2$	127–130 (760)	13
6	$(C_2H_5)_2N—CH_2—C{\equiv}CH$	$Al_2O_3 + KNH_2$	$CH_3C{\equiv}CN(C_2H_5)_2$	127–130 (760)	20
7	$(C_2H_5)_2N—CH_2—C{\equiv}CH$	$Al_2O_3 + KNH_2$	$CH_3C{\equiv}CN(C_2H_5)_2$	127–130 (760)	65
8	$(CH_3)_2N—CH_2—C{\equiv}CH$	$Al_2O_3 + KNH_2$	$CH_3C{\equiv}CN(CH_3)_2$	90–100 (760)	25
9	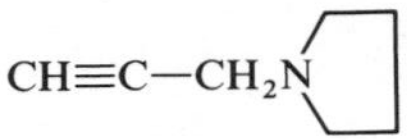 $CH{\equiv}C—CH_2N$ (pyrrolidinyl)	$Al_2O_3 + KNH_2$	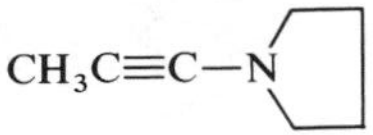 $CH_3C{\equiv}C-N$ (pyrrolidinyl)	—	35
10	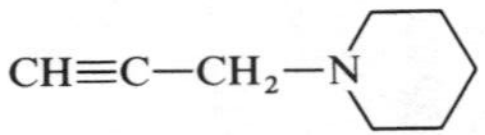 $CH{\equiv}C—CH_2—N$ (piperidinyl)	$Al_2O_3 + KNH_2$	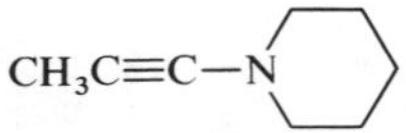 $CH_3C{\equiv}C—N$ (piperidinyl)	—	38
11	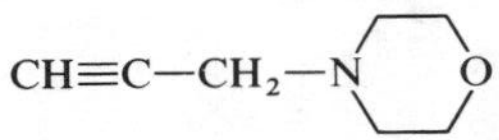 $CH{\equiv}C—CH_2—N$ (morpholinyl)	$Al_2O_3 + KNH_2$	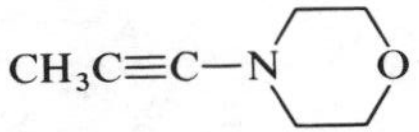 $CH_3C{\equiv}C—N$ (morpholinyl)	—	25

TABLE VII

REARRANGEMENT OF DISUBSTITUTED AMINO DIACETYLENIC
TERTIARY ALCOHOLS TO YNAMINES [12]

Starting propargylamines	Diacetylenic alcohols		Ynamine	
	Yield (%)	M.p. (°C)	Yield (%)	M.p. (°C)
N-Phenylmethylpropargylamine	60	53	50	62
N-Propargyldiphenylamine	70	74	38	68

amide/alumina 0.003 mole/ml ratio, an amine/catalyst mole ratio of 4.0, and an amine/benzene (vol./vol.) ratio of 2.0 yields 60% of the ynamine from N,N-diethylprop-2-ynylamine [23]. The reaction mixture is refluxed and the course of the reaction is followed by ir. The reaction is stopped when the ir gives the highest absorption at 4.5 μ and no absorption at 3.0 μ (0.025 mm cell) (Eq. 38).

$$HC\equiv C-CH_2NR_2 \rightarrow CH_2=C=CH-NR_2 \rightarrow CH_3-C\equiv C-NR_2 \quad (38)$$
$$(ir = 3.0\ \mu\ -C\equiv CH) \quad (ir = 5.2\ \mu\ \rangle C=C=C\langle) \quad (ir = 4.5\ \mu\ -C\equiv C-)$$

6. MISCELLANEOUS REACTIONS

(1) Conversion of aminochloroethylenes to ynamines [24].

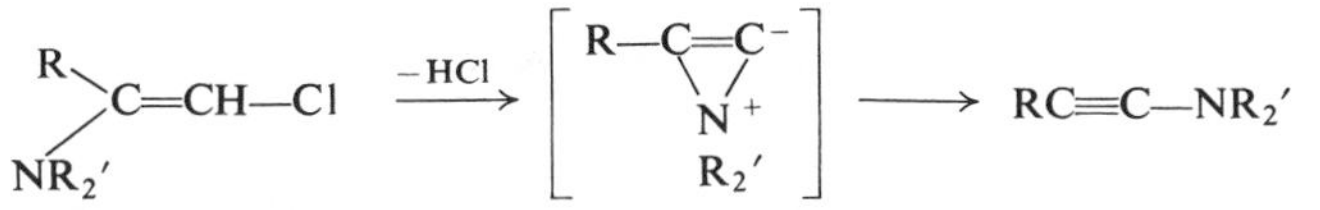

$$\begin{matrix} R \\ \rangle C=CH-Cl \\ NR_2' \end{matrix} \xrightarrow{-HCl} \left[\begin{matrix} R-C\equiv C^- \\ \diagdown / \\ N^+ \\ R_2' \end{matrix} \right] \longrightarrow RC\equiv C-NR_2' \quad (39)$$

(2) Synthesis of alkenylynamines [25].

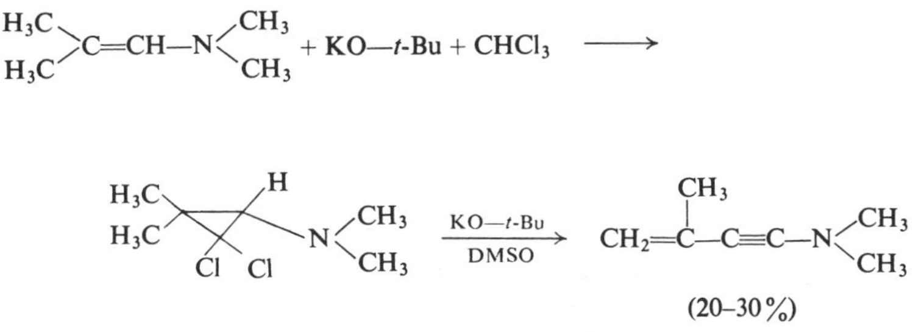

$$\begin{matrix} H_3C \\ \rangle C=CH-N\langle \\ H_3C \end{matrix} \begin{matrix} CH_3 \\ \\ CH_3 \end{matrix} + KO-t\text{-Bu} + CHCl_3 \longrightarrow$$

$$\begin{matrix} H_3C & H \\ \diagdown / \\ H_3C & \\ \bigtriangleup \\ Cl\ Cl \end{matrix} N\langle \begin{matrix} CH_3 \\ CH_3 \end{matrix} \xrightarrow[DMSO]{KO-t\text{-Bu}} CH_2=\overset{CH_3}{\underset{|}{C}}-C\equiv C-N\langle \begin{matrix} CH_3 \\ CH_3 \end{matrix} \quad (40)$$
$$(20-30\%)$$

5. Ynamines

(3) Ynamine thioethers from dichloroethylene thioethers [26].

$$\underset{H}{\overset{Cl}{\diagdown}}C=C\underset{SR''}{\overset{Cl}{\diagup}} \quad \xrightarrow{\text{3LiNRR'}} \quad RR'N-C\equiv C-SR'' \quad \xleftarrow{\text{3LiNRR'}} \quad \underset{Cl}{\overset{Cl}{\diagdown}}C=C\underset{SR''}{\overset{H}{\diagup}}$$

(41)

REFERENCES

1. I. J. Rinke, *Rec. Trav. Chim. Pays-Bas* **39**, 704 (1920); **46**, 268 (1927); **48**, 960 (1928).
2. R. A. Raphael, "Acetylenic Compounds In Organic Synthesis," p. 61. Butterworth, London and Washington, D.C., 1955.
3. H. G. Viehe and E. Franchimont, *Chem. Ber.* **95**, 319 (1962).
4. J. Bode, *Ann. Chem.* **267**, 268 (1892).
5. E. Ott, G. Dittus, and H. Weissenburger, *Chem. Ber.* **76**, 84 (1943).
6. F. Moulin, *Helv. Chim. Acta.* **34**, 2416 (1951).
7. H. E. Zaugg, L. R. Swett, and G. R. Stone, *J. Org. Chem.* **23**, 1389 (1958).
8. H. G. Viehe, *Angew. Chem. Int. Ed.* **6**, 767 (1967).
8a. A. E. Baydar, G. V. Boyd, P. F. Lindley, and F. Watson, *J. Chem. Soc. Chem. Commun.* 178 (1979).
8b. J. P. Freeman and R. C. Grabiak, *J. Org. Chem.* **41**, 3970 (1976).
8c. J. V. Piper, M. Allard, M. Faye, L. Hamel, and V. Chow, *J. Org. Chem.* **42**, 4261 (1977).
8d. J. Ficini and J. Pouliquen, *J. Amer. Chem. Soc.* **93**, 3295 (1971).
8e. D. R. Eckroth and G. M. Love, *J. Org. Chem.* **34**, 1136 (1969).
8f. J. A. Kloek and K. L. Leschinsky, *J. Org. Chem.* **45**, 721 (1980).
8g. L. A. Paquette, R. W. Houser, and M. Rosen, *J. Org. Chem.* **35**, 905 (1970).
8h. J. J. Eisch, J. E. Galle, and L. E. Hallenbeck, *J. Org. Chem.* **47**, 1610 (1982).
8i. M. E. Kuehne and H. Linde, *J. Org. Chem.* **37**, 4031 (1972).
8j. P. A. Bartlett and W. F. Hahne, *J. Org. Chem.* **44**, 882 (1979).
8k. H. G. Viehe, *in* "Chemistry of Acetylenes" (H. G. Viehe, ed.), chap. 12, Dekker, New York, 1969.
9. H. G. Viehe and N. Reinstein, *Angew. Chem.* **76**, 537 (1964).
10a. R. E. Harmon, C. V. Zenarosa, and S. K. Gupta, *J. Org. Chem.* **35**, 1936 (1970).
10b. R. Truchet, *Ann. Chim. (Paris)* **26**, 309 (1931).
10c. A. D. deWit, W. P. Tromysenaars, M. L. Pennings, D. N. Reinhardt, S. Harkeman, and G. J. van Hummel, *J. Org. Chem.* **46**, 172 (1981).
10d. J. I. Dickstein and S. I. Miller, *J. Org. Chem.* **37**, 2175 (1972).
11. H. G. Viehe, S. I. Miller, and J. I. Dickstein, *Angew. Chem.* **76**, 537 (1964); *Angew. Chem. Int. Ed. Engl.* **3**, 582 (1964).
12. J. L. Dumont, *C. R. Acad. Sci. Paris* **261**, 1710 (1965).
12a. H. G. Viehe and A. J. Hubert, U.S. Pat. 3,439,038 (1969).
13. P. P. Montijn, E. Harryvan, and L. Brandsma, *Rec. Trav. Chim. Pays-Bas* **83**, 1211 (1964).
14. J. Ficini and C. Barbara, *Bull. Soc. Chim. Fr.* 871 (1964).
15. J. Ficini and C. Barbara, *Bull. Soc. Chim. Fr.* 2787 (1965).
16. A. J. Speziale and L. R. Smith, *J. Amer. Chem. Soc.* **84**, 1868 (1962).
17. M. E. Kuehne and R. J. Sheeran, *J. Org. Chem.* **33**, 4406 (1968).
17a. E. J. Corey and D. E. Cane, *J. Org. Chem.* **35**, 3405 (1970).
17b. G. A. Gornowicz and R. West, *J. Amer. Chem. Soc.* **93**, 1714 (1971).

18. J. Ficini, C. Barbara, J. Colodny, and A. Dureault, *Tetrahedron Lett.* 943 (1968).
19. R. Buijle, A. Halleux, and H. G. Viehe, *Angew. Chem.* **78**, 593 (1966).
20. H. Eilingsfeld, M. Seefelder, and H. Weidinger, *Chem. Ber.* **96**, 2671 (1963).
21. H. G. Viehe, *Angew. Chem.* **75**, 638 (1963).
21a. D. R. Strobach, *J. Org. Chem.* **36**, 1438 (1971).
21b. H. G. Viehe, U.S. Pat. 3,369,047 (1968).
22. L. I. Peterson, *Tetrahedron Lett.* 5357 (1968).
22a. L. L. Peterson, U.S. Pat. 3,499,928 (1970).
23. A. J. Hubert and H. G. Viehe, *J. Chem. Soc. C* 228 (1968).
24. H. G. Viehe and S. Y. Delavarenne, unpublished results described in H. G. Viehe, *Angew. Chem. Int. Ed. Engl.* **6**, 767 (1967).
25. T. C. Shields, W. E. Billups and A. N. Kurtz, *Angew. Chem.* **80**, 193 (1968).
26. S. Y. Delavarenne and G. H. Viehe, *Tetrahedron Lett.* 4671 (1969).

1. INTRODUCTION

The direct synthesis of ureas from amines with carbonyl derivatives is not always convenient because of the pressure equipment that is necessary. On the industrial production scale this is usually no problem.

The most practical laboratory methods involve the condensation of amines with either ureas, isocyanates, or isocyanate derivatives. The use of difunctional reactants yields polymers (Eq. 1).

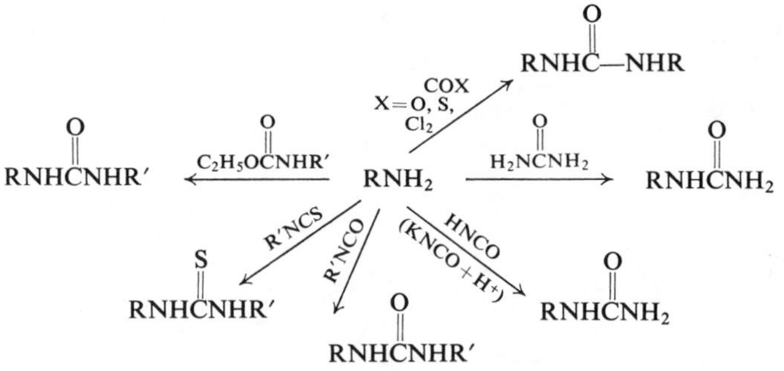

$$
\begin{array}{ccc}
 & & \overset{\displaystyle O}{\underset{\displaystyle \parallel}{RNHC}}\!\!-\!\!NHR \\
 & \overset{COX}{\underset{Cl_2}{X=O,\,S,}}\!\!\nearrow & \\
\overset{O}{\underset{\parallel}{RNHCNHR'}} \xleftarrow{C_2H_5OCNHR'} RNH_2 \xrightarrow{H_2NCNH_2} \overset{O}{\underset{\parallel}{RNHCNH_2}} & & (1) \\
\overset{R'NCS}{\swarrow}\;\;\overset{R'NCO}{\downarrow}\;\;\overset{(KNCO+H^+)}{\searrow}\;\overset{HNCO}{} & & \\
\overset{S}{\underset{\parallel}{RNHCNHR'}}\quad \overset{O}{\underset{\parallel}{RNHCNHR'}}\quad \overset{O}{\underset{\parallel}{RNHCNH_2}} & &
\end{array}
$$

If isocyanates are available, they may be allowed to react with water to yield symmetrical ureas (Eq. 2).

$$
2RNCO + H_2O \longrightarrow \overset{O}{\underset{\parallel}{RNHCNHR}} + CO_2 \qquad (2)
$$

Urea itself is the starting material for the preparation of various important linear and cyclic urea derivatives. For example, the reaction of formaldehyde with urea leads to 1,3-dimethylolurea. Heating this product converts it into a polymeric methyleneurea resin of wide industrial importance in adhesives and coatings. These resins are called U/F or urea–formaldehyde resins.

Substituted urea compositions also find use in a wide variety of other applications such as in the pharmaceutical [1b] and herbicidal [1c] areas.

2. CONDENSATION REACTIONS

A. Reactions of Amines with Urea

One of the most general methods of preparing urea derivatives in the laboratory involves the condensation of primary amines with urea. The reaction does not proceed as well with secondary amines. The availability of urea gives this method an advantage (Eq. 3).

$$H_2N\overset{O}{\overset{\|}{C}}NH_2 + RNH_2 \text{ (or } R\overset{+}{N}H_3X^-) \longrightarrow RNH\overset{O}{\overset{\|}{C}}NH_2 + NH_4X \text{ or } NH_3 \qquad (3)$$

Polyureas are obtained when a diamine is condensed with urea (Eq. 4).

$$H_2N\overset{O}{\overset{\|}{C}}NH_2 + H_2NRNH_2 \longrightarrow \text{+}RNH\overset{O}{\overset{\|}{C}}NH\text{+}_n + 2NH_3 \qquad (4)$$

2-1. Preparation of p-Ethoxyphenylurea [1a]

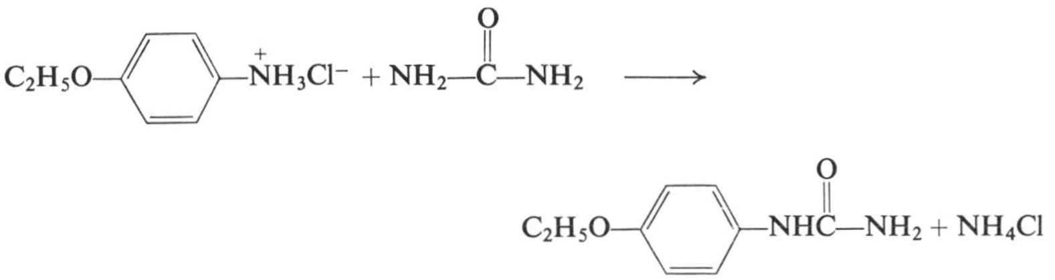

$$C_2H_5O-\langle\rangle-\overset{+}{N}H_3Cl^- + NH_2-\overset{O}{\overset{\|}{C}}-NH_2 \longrightarrow$$

$$C_2H_5O-\langle\rangle-NH\overset{O}{\overset{\|}{C}}-NH_2 + NH_4Cl \qquad (5)$$

To a 1 liter flask equipped with a condenser and mechanical stirrer is added a mixture of 87.0 gm (0.50 mole) of p-phenetidine hydrochloride and 120.0 gm (2.0 moles) of urea. To the stirred mixture is added a solution prepared from 200 ml of water, 4.0 ml of concentrated hydrochloric acid, and 4.0 ml of glacial acetic acid. The solution is heated by means of an oil bath and refluxed for $1\frac{1}{2}$ hr while the product is precipitating. At the end of the reaction period the entire contents appear to have solidified. The reaction is cooled, the solid is broken up, suspended in water, filtered, and dried to obtain 740–810 gm (82–90%) of a pale-yellow crude solid, p-ethoxyphenylurea [2–4]. Recrystallization from boiling water gives an 80% recovery of material, m.p. 173°–174°C. The by-product is sym-di(p-ethoxyphenyl)urea [(p-C$_2$H$_5$OC$_6$H$_4$NH)$_2$-CO] which is insoluble in boiling water. Prolonged boiling of p-ethoxyphenylurea in water will slowly convert it into sym-di(p-ethoxyphenyl)urea. Heating sym-di(p-ethoxyphenyl)urea with urea, ammonium carbamate, ammonium carbonate, or ethanol and ammonia converts it into p-ethoxyphenylurea again [5, 6].

More recent literature indicates that N,N-dialkylureas can be prepared in good yield by heating urea in a solvent (o-dichlorobenzene or xylene) to 120°–135°C while adding the amine at atmospheric pressure [6b] as described in Table I.

TABLE I

Preparation of N,N-Dialkylureas by the Reaction of an Alkylamine and Urea in an Organic Solvent

	Moles					$\begin{matrix} O \\ \parallel \\ R_2NCNH_2 \end{matrix}$	
Alkylamine	Alkylamine	Urea	Solvent (liters)	Temp. (°C)	Yield (%)	m.p. (°C)
$(CH_3)_2NH$	24	20	o-Dichlorobenzene (2.5)	135	97	182–183
$(C_2H_5)_2NH$	3.0	3	Xylene (0.5)	130–135	83	67
$(n\text{-}C_3H_7)_2NH$	3.0	3	Xylene (0.75)	130–135	80	69
CH_3NH_2	3.9	2.0	Xylene (0.4)	120–125	75	96
$C_2H_5NH_2$	2.75	2.0	Xylene (0.5)	120–125	83	86
$C_3H_7NH_2$	2.05	2.0	Xylene (0.55)	120–125	81	101
$n\text{-}C_5H_{11}NH_2$	2.0	2.0	Xylene	120–125	88	94

Another process utilizing no solvent has been reported for reacting dimethylamine (1 molar excess) with urea at 127°C under anhydrous conditions to give dimethylurea (m.p. 180°–185°C) [6c].

2-2. Preparation of Poly(4-oxyheptamethyleneurea) [7]

$$NH_2CNH_2 + H_2N-(CH_2)_3-O-(CH_2)_3-NH_2 \xrightarrow[\text{heat}]{N_2}$$

$$\quad\quad\quad\quad\quad -[(CH_2)_3-O-(CH_2)_3-NHC-NH]_n + 2NH_3 \quad (6)$$

To a test tube with a side arm are added 7.5 gm (0.125 mole) of urea and 16.5 gm (0.125 mole) of bis(γ-aminopropyl) ether. A capillary tube attached to a nitrogen gas source is placed on the bottom of the tube and the temperature is raised to 156°C and kept there for 1 hr, during which time ammonia is evolved. The temperature is raised to 231°C for 1 hr and then to 255°C for an additional hour. At the end of this time a vacuum is slowly applied to remove the last traces of ammonia. (CAUTION: Frothing may be a serious problem if the vacuum is applied too rapidly.) The polymer is cooled, the test tube broken, and the polymer isolated. The polymer melt temperature is approximately

190°C, and the inherent viscosity is approximately 0.6 in *m*-cresol (0.5% concentrated, 25°C).

Symmetrical diaryl- or dialkylureas are prepared in good yield by heating the respective primary amines and ureas in the dry state at 160°C [8] or by boiling an aqueous or anhydrous solution of the amine hydrochloride with urea [9, 10a, b]. The reaction is thought to involve the steps shown in Eqs. (7)–(10).

$$NH_2\overset{\overset{\displaystyle O}{\|}}{C}NH_2 \rightleftharpoons NH_4NCO \text{ (or } NH_3 + HNCO) \qquad (7)$$

$$RNH_2 + NH_4NCO \rightleftharpoons RNH\overset{\overset{\displaystyle O}{\|}}{C}\!-\!NH_2 + NH_3 \qquad (8)$$

$$RNH\overset{\overset{\displaystyle O}{\|}}{C}NH_2 \rightleftharpoons RNCO + NH_3 + H_2O \qquad (9)$$

$$RNH_2 + RNCO \longrightarrow RNH\overset{\overset{\displaystyle O}{\|}}{C}NHR \qquad (10)$$

This process may be interrupted in the case of the reaction of aniline with urea to give phenylurea in 55% yield and *sym*-diphenylurea (carbanilide) in 40% yield [11a].

Another reported process involves the reaction of urea with an alkylamine in the presence of the trialkylamine at 150°–70°C under pressure to give high yields of the 1,3-dialkylurea [11b]. Heating under pressure (21 bars) at 205°C for 4 hr of a 1:4 molar ratio of urea and alkylamine and continuously removing unreacted alkylamine and ammonia gives a 99% yield of 1,3-disubstituted urea [11c].

2-3. *Preparation of Phenylurea and sym-Diphenylurea* [11a]

$$C_6H_5\overset{+}{N}H_3Cl^- + NH_2\overset{\overset{\displaystyle O}{\|}}{C}NH_2 \xrightarrow{-NH_4Cl}$$

$$C_6H_5NH\overset{\overset{\displaystyle O}{\|}}{C}NH_2 \xrightarrow{C_6H_5\overset{+}{N}H_3Cl^-} (C_6H_5NH)_2CO + NH_4Cl \qquad (11)$$

To a flask equipped with a reflux condenser and mechanical stirrer are added 39.0 gm (0.30 mole) of aniline hydrochloride, 19.0 gm (0.32 mole) of urea, and 150 ml of water. The solution is boiled for 1½ hr while crystalline carbanilide is separating out. The mixture is filtered, the carbanilide washed with 100 ml of water, and the combined filtrates are chilled to remove the phenylurea, which crystallizes out. The filtrate is again refluxed for $1\frac{1}{2}$–2 hr while carbanilide separates out. The process [13] is repeated two or three times, and in each case the phenylurea is collected by chilling the filtrate. The mother liquor is finally evaporated to half its volume to yield additional carbanilide and phenylurea from the filtrate. The total amount of carbanilide obtained is 12–13 gm (38–40%), m.p. 235°C (recrystallized from alcohol, 100 ml/2.5 gm product). The crude phenylurea contains a little carbanilide which is less soluble in boiling water. Carbanilide separates out first and is filtered off while the solution is still hot. On further cooling, colorless, stout needles or flakes of phenylurea precipitate to yield 21–23 gm (52–55%), m.p. 147°C.

Heating aniline, instead of aniline hydrochloride, with urea produces predominantly carbanilide [8].

Unsymmetrically disubstituted ureas, such as phenyl dialkylureas, have also been reported to be made by the reaction of diphenylurea and dialkylurea at 420°–460°C [11d]. Unsymmetrical arylalkylureas can also be made by reacting a phenylurea with an alkylamine to liberate ammonia [11e]. The example cited in the latter reference is 1-chlorophenyl-3-dimethylurea.

B. Reaction of Amines with Cyanic Acid

Amines also react with cyanic acid produced in the decomposition of nitrourea to give 70–95% yields of alkylureas [13–18] (Eqs. 12 and 13). This

$$\underset{\substack{\| \\ O}}{NH_2CNH_2} + HNO_3 \longrightarrow \underset{\substack{\| \\ O}}{NH_2-C-\overset{+}{N}H_3NO_3^-} \longrightarrow$$

$$\underset{\substack{\| \\ O}}{NH_2CNHNO_2} \xrightarrow{heat} HNCO + N_2O + H_2O \qquad (12)$$

$$HNCO + R^1R^2NH \longrightarrow \underset{\substack{\| \\ O}}{R^1R^2NHCNH_2} \qquad (13)$$

$$R^1 = H$$
$$R^2 = alkyl$$
$$R^2 = HOCH_2CH_2$$
$$R^1 = R^2 = alkyl$$

reaction is also possible with the aid of cyanic acid produced from sodium or potassium cyanate and an acid.

2-4. Preparation of p-Bromophenylurea [19]

$$p\text{-Br}—C_6H_4—NH_2 + HNCO \longrightarrow p\text{-Br}—C_6H_4—NH\overset{\overset{\displaystyle O}{\|}}{C}NH_2 \qquad (14)$$

CAUTION: Use a well-ventilated hood.

To a flask equipped with a stirrer, condenser, and dropping funnel is added a solution of 86 gm (0.5 mole) of p-bromoaniline dissolved in 240 ml of glacial acetic acid and 480 ml of water at 35°C. The dropwise addition of a solution of 65 gm of sodium cyanate dissolved in 450 ml of water at 35°C is begun and continues until a white crystalline precipitate of the product is evident. At this point the remaining sodium cyanate solution is quickly added to the vigorously stirring reaction mixture. As the product begins to separate, the temperature rises to 50°–55°C. The thick suspension of the product is stirred for an additional 10 min, allowed to remain at room temperature for 2–3 hr, diluted with 200 ml of water, filtered, washed with water, and dried to yield 95–100 gm (88–93%) of p-bromophenylurea. The product is recrystallized from aqueous ethanol (12 ml ethanol and 3 ml water/gm crude product) to give a 65% recovery of white prisms, m.p. 225°–227°C. Prolonged heating of the product will convert it into di-(p-bromophenyl)urea.

p-Ethoxyphenylurea may also be prepared by this method in 95% yield by using 0.5 mole of the amine, 500 ml of water, 75 ml of acetic acid, and 1.0 mole of sodium cyanate [19].

Variation of the aromatic substituents from o-, m-, p-chloro, ethoxy, methoxy, methyl, or phenyl has little effect on the yield of products usually in the 54–96% range [19].

p-Ethoxyphenylurea (Preparation 2-1) has also been prepared by the reaction of p-potassium cyanate or p-phenetidine hydrochloride [20] or p-phenetidine acetate [21]. p-Bromophenylurea has also been reported to be prepared by the reaction of potassium cyanate on p-bromoaniline hydrochloride [22, 23a].

Isobutylurea is reported to be prepared by the reaction of an aqueous solution of isobutylamine hydrochloride and potassium cyanate. Evaporating this mixture and then extracting with hot acetone gave the product, which is also recrystallized from acetone to give a 55% yield of crystalline product, m.p. 140.5°–141.5°C [23b].

sec-Butylurea is prepared in a similar manner from sec-butylamine hydrochloride to give a crystalline product, m.p. 169°–170°C [23b].

Recently, ethyl-, allyl-, and phenylurea have been prepared by a similar method in 78–83% yield by reaction of RNH_2 and KCNO [23c]. 3-Ureidopyrrolidines can also be prepared by this method [23d].

C. Reaction of Amines with Isocyanates

Another general method of preparing ureas involves the reaction of ammonia or amines with isocyanates [24] (Eq. 15).

$$RNH_2 + R'NCO \longrightarrow R'NH\overset{\overset{\displaystyle O}{\|}}{C}NHR \qquad (15)$$

R,R′ = H, alkyl, or phenyl

p-Ethoxyphenylurea has been reported to be prepared by the reaction of ammonia on p-ethoxyphenyl isocyanate [25].

In the reaction of diamines with isocyanates, the use of less than the equivalent quantities of isocyanates and isothiocyanates yields mono- and disubstituted ureas and recovered diamine. The distribution of the products is dependent on the reactivity of the reagent, on concentration, and on the nature of the solvent [26] (Eq. 16).

$$H_2N(CH_2)_xNH_2 + RNCO \longrightarrow$$

$$H_2N(CH_2)_xNH_2 + RNHCONH(CH_2)_xNH_2 + RNHCONH(CH_2)_xNHCONHR \quad (16)$$

(I)   (II)   (III)

2-5. *Preparation of N-Cyclohexyl-N-(1-methyl-2-propynyl)-N′-p-chlorophenylurea* [27]

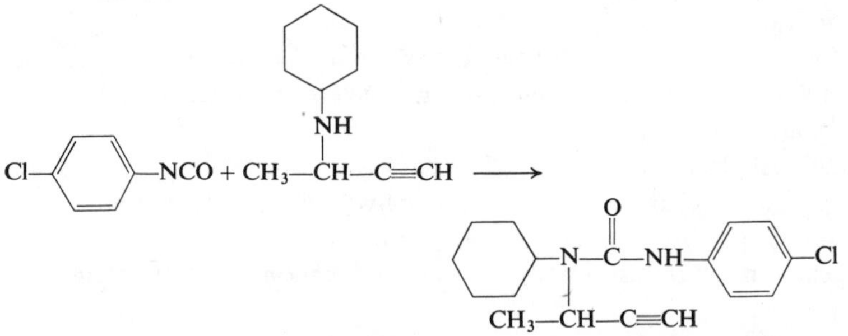

(17)

To a flask containing 17.0 gm (0.11 mole) of p-chlorophenyl isocyanate in 30 ml of ether is added 13.8 gm (0.11 mole) of 3-cyclohexylamino-1-butyne over a period of $\frac{3}{4}$ hr at 25°–35°C. An additional 300 ml of ether is added to the gelatinous slurry, and the mixture is stirred for 5 hr at room temperature. Filtration yields 11 gm of colorless crystals. An additional 20 gm of crude solid is obtained by concentrating the ether layer. The total product is recrystallized from petroleum ether (30°–60°C) to obtain 18.6 gm (60%), m.p. 107°–109°C.

In the absence of basic catalysts, propargylamines react with isocyanates to give ureas. However, in the presence of basic catalysts, 4-methylene-2-oxazolidinones and 4-methylene-2-imidazolidinones are obtained directly or through the urea derivative [27] (Eq. 18).

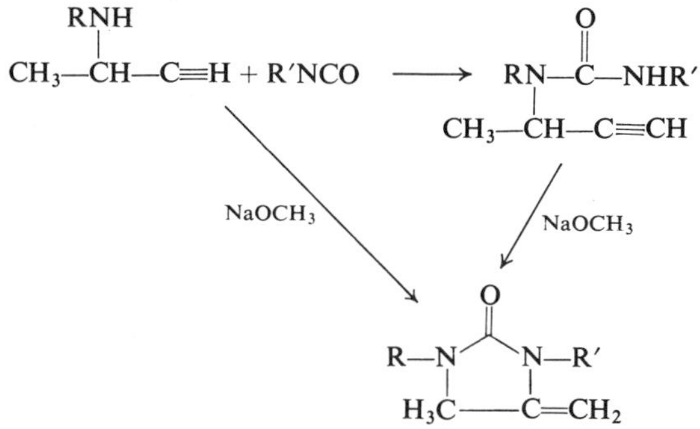

$$
\begin{array}{c}
\overset{\displaystyle RNH}{\underset{\displaystyle |}{}} \\
CH_3-CH-C\!\equiv\!H + R'NCO \longrightarrow \overset{\displaystyle O}{\overset{\displaystyle \|}{RN-C-NHR'}} \\
\end{array}
$$

$$
CH_3-CH-C\!\equiv\!CH
$$

NaOCH₃   NaOCH₃

$$
\underset{\displaystyle H_3C\text{------}C\!=\!CH_2}{R-N\diagdown N-R'}
$$

(18)

The relative reaction rates of several amines with phenyl isocyanate in ether at 0°C have been determined. Aniline is about half as reactive as ethyl-, *n*-propyl-, *n*-butylamine, and *n*-amylamine is between eight and ten times as reactive as ammonia [28].

α-Naphthyl isocyanate has been reported to be a good reagent for aliphatic amines and amides [29]. In general, the reaction is carried out in a test tube and the reaction may have to be catalyzed by 1 or 2 drops of triethylamine dissolved in ether. The reaction of α-naphthyl isocyanate with alcohols and phenols is described later, in Chapter 10.

α-Naphthyl isocyanate derivatives of amines and amides, their properties and formulas are listed in Table II.

TABLE II

α-Naphthyl Isocyanate Derivatives of Amines and Amides[a] [29]

Amine or Amide	Urea		$(C_2H_5)_3N$ catalyst added
	M.p. (°C)	Formula	
Methylamine	196–197	$C_{12}H_{12}ON_2$	Yes
Ethylamine	199–200	$C_{13}H_{14}ON_2$	Yes
Dimethylamine	158–159	$C_{13}H_{14}ON_2$	Yes
Diethylamine	127–128	$C_{15}H_{18}ON_2$	Yes
Benzylamine	202–203	$C_{18}H_{16}ON_2$	Yes
Isoamylamine	131–132	$C_{16}H_{20}ON_2$	Yes
Diisoamylamine	94–95	$C_{21}H_{30}ON_2$	Yes
Diisobutylamine	118–119	$C_{19}H_{26}ON_2$	Yes
Di-*n*-propylamine	92–93	$C_{17}H_{22}ON_2$	Yes
Acetamide	211–212	$C_{13}H_{12}O_2N_2$	No
Acetanilide	116–117	$C_{19}H_{16}O_2N_2$	No

[a] Reprinted from H. E. French and A. F. Wirtel, *J. Amer. Chem. Soc.* **48**, 1736 (1926) Copyright 1926 by the American Chemical Society. Reprinted by permission of the copyright owner.

The reaction of diamines with diisocyanates yields polyureas (Eq. 19).

$$H_2NR—NH_2 + R'(NCO)_2 \longrightarrow \overset{H}{\underset{}{\text{—[N}}}—R—NH\overset{O}{\overset{\|}{C}}—\underset{H}{N}—R'—\overset{H}{N}—\overset{O}{\overset{\|}{C}}—]_n^- \quad (19)$$

2-6. Preparation of Poly(decamethyleneurea) [30a]

$$H_2N—(CH_2)_{10}—NH_2 + OCN—(CH_2)_{10}—NCO \longrightarrow$$

$$+(CH_2)_{10}NH—\overset{O}{\overset{\|}{C}}—NH+_n \quad (20)$$

In a three-necked, round-bottomed flask equipped with stirrer, dropping funnel, condenser, and drying tubes is placed, under a nitrogen atmosphere, 19.0 gm (0.11 mole) of freshly distilled decamethylenediamine in 39 ml of distilled m-cresol. While vigorously stirring, 24.8 gm (0.11 mole) of decamethylene diisocyanate is added dropwise over a 10 min period. The dropping funnel is washed with 10 ml of m-cresol, and this is added to the reaction mixture. The temperature of the reaction mixture is raised to 218°C for 5 min, cooled to room temperature, and then poured into 1.5 liters of methanol while stirring vigorously. The polyurea which separates as a white solid is filtered, washed with ethanol in a blender, and dried at 60°C under vacuum to give 38–40 gm (90–95%), polymer melt temperature 210°C, inherent viscosity in m-cresol 8.3 (0.5% concentration at 25°C).

Ureas with heterocyclic substituents are prepared by the reaction of heterocyclic amines with aryl or alkyl isocyanates [23d, 30b]. The pyridyl ureas possess excellent anti-inflammatory properties and the 3-ureidopyrrolidines have novel pharmacological properties (analgesic, central nervous system, and psychopharmacologic activities).

D. Reaction of Amines with Isothiocyanates

In contrast to isocyanates, isothiocyanates are relatively unreactive toward hydroxyl-containing compounds; thus they are not even affected by aqueous media [31]. Their reaction with hydroxyalkyl thioureas has been reported [32]. With free 2-mercaptoethylamine, 1 mole of phenyl isothiocyanate gave the N-substituted product, 1-(2-mercaptoethyl)-3-phenyl-2-thiourea, and 2 moles gave the N,S-disubstituted product. Isothiocyanates, regardless of the relative amounts, reacted with 2-mercaptoethylamine to give the N,S-disubstituted products. Cleavage with silver nitrate affords the 1-(2-mercaptoethyl-3-alkylureas [33a].

Mono- and disubstituted thioureas have been reported to be prepared in 59–74% yield by the reaction of amines (R'NH$_2$) with isothiocyanates

(RNCS), where R′ and R are allyl, H; Ph, H; allyl, Ph; —Me, Ph; Bu, Ph; Ph, Ph; and allyl, allyl [23c]. 1,1-Dimethyl-3-alkyl thioureas are prepared by the reaction of alkyl isothiocyanates with dimethylamine [33b].

2-7. Preparation of 1-(2-Mercaptoethyl)-3-phenyl-2-thiourea [33a]

$$HSCH_2CH_2\overset{+}{N}H_3Cl^- \xrightarrow{\quad NaOH \quad}$$

$$HSCH_2CH_2NH_2 \xrightarrow{\quad C_6H_5NCS \quad} C_6H_5NH\overset{\displaystyle S}{\overset{\displaystyle \|}{C}}NHCH_2CH_2SH \qquad (21)$$

To a flask are added with stirring 6.8 gm (0.06 mole) of 2-mercaptoethyl-amine hydrochloride in 35 ml of 95% ethanol, 2.5 gm (0.06 mole) of sodium hydroxide dissolved in 4 ml of water, and then 6.8 gm (0.05 mole) of phenyl isothiocyanate. The temperature rises to about 39°C and then slowly drops to room temperature. After 2 hr at room temperature the mixture is poured into 200 ml of ice water, the solid is filtered, washed with water, and dried to give 9.7 gm (91%), m.p. 105°–114°C of crude 1-(2-mercaptoethyl)-3-phenyl-2-thiourea. The product is recrystallized twice from ethanol to give crystals melting at 113°–116.5°C.

Alkyl isothiocyanates are produced by the reaction of alkylamines with a mixture of ammonium hydroxide and carbon disulfide, followed by reaction with sodium nitrite (in situ generated nitrous acid, pH not less than 5.0) at 45°C. The alkyl isothiocyanate can be either isolated or reacted with aqueous ammonium hydroxide to give the corresponding alkyl thiourea [33g].

Allyl and methallyl isothiocyanates can be prepared by the isomerization of allyl and methallyl thiocyanates and then reacted with amines to give allyl or methallyl thioureas [33c]. The isomerization of alkyl thiocyanates to alkyl isothiocyanates has also been reported [33e].

E. Reaction of Amines with Urethanes and Carbamates

Alkylureas can be prepared by the reaction of carbamic acid esters with primary and sterically unhindered secondary aliphatic amines [33f].

$$R_1R_2NH + R_3NH\overset{\displaystyle O}{\overset{\displaystyle \|}{-C}}OR_4 \xrightarrow[110°-160°C]{\substack{catalyst \\ (C_4H_9)_2SnO}} R_1R_2N\overset{\displaystyle O}{\overset{\displaystyle \|}{C}}NHR_3 + R_4OH \qquad (22)$$

Some typical examples of this process are shown in Table III [33f].
A typical procedure used in Table III is shown in Preparation 2-8.

2-8. Preparation of n-Hexylurea [33f]

$$n\text{-}C_6H_{13}NH_2 + NH_2\overset{\displaystyle O}{\overset{\displaystyle \|}{C}}OCH_3 \longrightarrow n\text{-}C_6H_{13}NH\overset{\displaystyle O}{\overset{\displaystyle \|}{C}}NH_2 + CH_3OH \qquad (23)$$

To a flask is added 99.0 gm (1.0 mole) of *n*-hexylamine, 90.0 gm (1.2 moles) of methylcarbamate, 1.0 gm of dibutyl tin oxide, and 100 ml of xylene. The mixture is stirred and heated at 120°–130°C while the methanol is removed by distillation through a fractionating column. On concentration the product is isolated and obtained in 96 % yield (138.2 gm). The product is recrystallized once from xylene–naphtha solvent: m.p. 109.5°–110°C (19.23 % nitrogen was found).

TABLE III

PREPARATION OF ALKYLUREAS BY THE REACTION OF
AMINES WITH CARBAMATE ESTERS [33f]

Amine (R$_1$NH$_2$)	R$_3$	Yield (%)	M.p. (°C)
n-C$_4$H$_9$	H	58	96–97
n-C$_6$H$_{13}$	H	96	109.5–110.0
Cyclo-C$_6$H$_{13}$	H	90	194–196
C$_6$H$_5$CH$_2$	H	85	150–151
Cyclo-C$_6$H$_{11}$	*n*-C$_3$H$_7$	88	106–107
Cyclo-C$_6$H$_{11}$	*n*-C$_4$H$_9$	91	115–116.5
n-C$_6$H$_{13}$	C$_6$H$_5$	85	72–73
Cyclo-C$_6$H$_{11}$	C$_6$H$_5$	99	187.5–189

Other suitable catalysts are Lewis acids such as cupric acetate, stannic chloride, stannous oxalate, aluminum alkoxide, and other alkyl tin oxides used in the amount of 0.05 to 1.0 wt.% based on the amine [33f].

A bisurethane of a diamine may be used to react with a diamine to give a polyurea (Eq. 23a).

$$H_2N-R-NH_2 + C_2H_5O\overset{O}{\overset{\|}{C}}NH-R-NH\overset{O}{\overset{\|}{C}}OC_2H_5 \longrightarrow$$

$$\overset{HO}{\underset{}{}}$$

$$+R-N\overset{\|}{C}NH\overset{}{\rightarrow}_n + 2C_2H_5OH \qquad (23a)$$

2-9. *Preparation of Poly(hexamethylene-decamethyleneurea) Copolymer* [34]

$$H_2N(CH_2)_{10}NH_2 + C_2H_5O\overset{O}{\overset{\|}{C}}-NH(CH_2)_6NH-\overset{O}{\overset{\|}{C}}OC_2H_5 \longrightarrow$$

$$+(CH_2)_{10}-NH\overset{O}{\overset{\|}{C}}NH+(CH_2)_6NH-\overset{O}{\overset{\|}{C}}-NH+_n + 2C_2H_5OH \qquad (24)$$

(a) *Preparation of hexamethylene bis(ethylurethane)*. To a three-necked flask equipped with mechanical stirrer, two dropping funnels, and condenser is added 58 gm (0.5 mole) of hexamethylenediamine in 200 ml of ether. The flask is cooled to 0°–10°C with an ice-water bath, while simultaneously from separate funnels 130 gm (1.2 moles) of ethyl chlorocarbonate, 48 gm (1.2 moles) of sodium hydroxide in 400 ml of water are added with vigorous stirring. Fifteen minutes after the addition, the solid is filtered and recrystallized from benzene–petroleum ether, m.p. 84°C (% yield *not* reported).

(b) *Polymerization of decamethylenediamine and hexamethylene bis-(ethylurethane)*. To a test tube with a capillary tube for nitrogen gas are added 12.4 gm (0.072 mole) of decamethylenediamine and 18.70 gm (0.072 mole) of hexamethylene bis(ethylurethane). The test tube is heated to 202°C while the nitrogen gas is being bubbled through the contents. After 3 hr the polymer is cooled under nitrogen and then isolated to give an inherent viscosity of 0.2–0.4 in *m*-cresol (0.5% concentration, 25°C). The polymer melt temperature is about 170°C.

Ammonia can also react with urethanes to give ureas. For example, *p*-ethoxyphenylurea can be prepared by heating *p*-ethoxyphenylurethane with ammonia to 100°–180°C [35a].

Polymers with terminal benzophenone oxime carbamate groups can be reacted with dibutylamine to give dibutylurea end groups [35b].

$$(C_6H_5)_2C{=}N{-}O\underset{\underset{O}{\|}}{C}{-}NH(R)_nNH\underset{\underset{O}{\|}}{C}{-}ON{=}C(C_6H_5)_2 \xrightarrow{\ 2(C_4H_9)NH\ }$$

$$(C_4H_9)_2NH\underset{\underset{O}{\|}}{C}{-}NH(R)_n{-}NH\underset{\underset{O}{\|}}{C}N(C_4H_9)_2 \quad (25)$$

F. Reaction of Isocyanates with Water

Symmetrical ureas can also be prepared by heating isocyanates with water [36, 37] (Eq. 26).

$$2RNCO + H_2O \longrightarrow RNH\underset{\underset{O}{\|}}{\overset{\overset{O}{\|}}{C}}NHR + CO_2 \quad (26)$$

2-10. *Preparation of 3,3′-Diisocyanato-4,4′-dimethylcarbanilide* [36]

To a flask containing a solution of 8.7 gm (0.05 mole) of 2,4-toluene diisocyanate in 200 ml of anhydrous ether is slowly added a mixture of 0.432 gm (0.024 mole) of water, 50 ml of anhydrous ether, and 0.5 ml of pyridine at room temperature. The urea derivative precipitates as a fine powder. Then 10 ml of

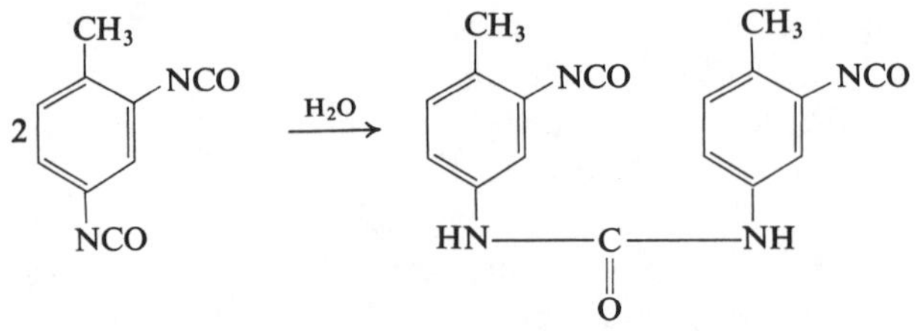

$$\text{(27)}$$

water is added and the mixture stirred for an additional $\frac{1}{2}$ hr. The product is filtered, washed several times with 1:1 anhydrous petroleum ether–diethyl ether, dried to yield the product, m.p. 172°–175°C (% yield not reported).

The intermediate isocyanates can be formed by the reaction of long-chain alkyl halides and KCNO in DMF and then hydrolyzed with the stoichiometric amount of water (in one step) [37b] to give the urea. The use of long-chain α,ω-dichloroalkanes in a similar reaction yields the corresponding polyureas. Secondary alkyl halide yielded the branched N,N'-dialkylureas [37b].

G. Direct Synthesis of Ureas and Polymers from Amines and Carbonyl Derivatives

$$2RNH_2 + COX \longrightarrow RNH\overset{\overset{\displaystyle O}{\displaystyle \|}}{C}NHR \qquad (28)$$

$$X = O, Cl_2, S$$

a. Phosgene

Primary amines react with phosgene (carbonyl chloride) to give isocyanates [38] faster than ureas at temperatures below about 100°C. For example, aniline reacts with phosgene to give phenyl isocyanate below 100°C, but carbanilide is produced when the reaction mixture is warmed at 140°C unless excess phosgene is used [39].

A few cases have been reported for the preparation of aromatic ureas from a primary amine and phosgene.

Puschin and Mitie reported that phosgene reacted with m-toluidene to give $CO(NH-C_6H_4Me)_2HCl$ and ethylenediamine gives cyclic ethyleneurea hydrochloride (2-imidizolidone hydrochloride) [40].

p-Ethoxyphenylurea has also been prepared by the reaction of p-phenetidine with phosgene in benzene or toluene and treatment of the product with ammonia [20], urethane [35], urea salts [12], acetylurea [41], and a mixture of urea and ammonium chloride [42a].

1,3-Bis(2-methoxy-5-methylphenyl)urea was recently reported to be prepared in 89% yield (m.p. 185°C) by the reaction of 4-methyl-2-aminoanisole in dry pyridine and THF with phosgene dissolved in THF at reflux temperatures [42b].

Secondary amines react with phosgene to give carbamoyl chlorides [43] which, on treatment of excess secondary or primary amine or ammonia, yields an asymmetric urea [44-46a, b]. (See also subsection 2.G.h of this chapter.)

Dialkyl amines can be reacted with phosgene in the presence of caustic to give tetraalkylureas. The use of halogenated solvent to extract the product for easier isolation has been reported [46c].

2-11. Preparation of asym-Phenylethylurea [45]

$$C_6H_5-\underset{\underset{C_2H_5}{|}}{N}-COCl + NH_3 \longrightarrow C_6H_5-\underset{\underset{C_2H_5}{|}}{N}-\overset{\overset{O}{\|}}{C}-NH_2 \qquad (29)$$

To a flask is added 15.0 gm (0.082 mole) of phenylethylcarbamyl chloride dissolved in 50 ml of benzene and 20 ml of absolute ethanol. Dry ammonia is passed through the solution until no more ammonium chloride is deposited. The solution is filtered, the filtrate evaporated, and the syrupy residue extracted with benzene. The filtered extract is evaporated again to give a faintly brown residue which crystallizes in plates to give 13.5 gm (100%), m.p. 60°C (after drying under reduced pressure for 24 hr).

Tertiary amines react with phosgene to give symmetrical tetrasubstituted ureas (Eq. 30). (See Table IV.) This reaction probably involves the intermediate formation of the carboxylchloride [47] (Eqs. 31, 32). In support of this

$$2R_3N + COCl_2 \longrightarrow R_2N-\overset{\overset{O}{\|}}{C}-NR_2 + 2RCl \qquad (30)$$

TABLE IV

REACTION OF TERTIARY AMINES WITH PHOSGENE TO GIVE
SYMMETRICAL TETRASUBSTITUTED UREAS

Tertiary amine	Symmetrical tetrasubstituted urea	M.p. (°C)	Ref.
$C_6H_5N(C_2H_5)_2$	Diethyldiphenyl-	79	44
$C_6H_5N(n\text{-Bu})_2$	Dibutyldiphenyl-	83	44
$C_6H_5N(CH_3)_2$	Dimethyldiphenyl-	122	47
$C_6H_5NCH_3(C_2H_5)$	Diethyldiphenyl-	72–3	47
$C_6H_5CH_2N(CH_3)C_6H_5$	Dimethyldiphenyl-	122	47
$C_6H_5CH_2N(C_2H_3)C_6H_5$	Diethyldiphenyl-	72	47

$$2R_3N + COCl_2 \rightarrow [R_3NCOCl_2] \longrightarrow R_2NCOCl + RCl \qquad (31)$$

$$R_2NCOCl + R_3N \longrightarrow R_2NCONR_2 + RCl \qquad (32)$$

proposed mechanism, carbamoyl chlorides react with tertiary amines at 200°C to give the symmetrical tetrasubssituted urea [47].

Polyureas may be prepared by treating diamines with one equivalent of a carbon oxy derivative as shown in Eq. (33).

$$NH_2\text{---}R\text{---}NH_2 + COX \longrightarrow$$

$$\overset{+}{N}H_3\text{---}R\text{---}NH\overset{O}{\overset{\|}{C}}X \xrightarrow{\text{heat}} \text{---}[R\text{---}NH\overset{O}{\overset{\|}{C}}\text{---}NH]_2 + H_2X \qquad (33)$$

$$X = O, S, Cl_2$$

2-12. *Preparation of Poly(hexamethyleneurea)* [48]

$$H_2N(CH_2)_6NH_2 + COCl_2 \longrightarrow \text{---}[(CH_2)_6\text{---}NH\overset{O}{\overset{\|}{C}}\text{---}NH]_n + 2HCl \qquad (34)$$

CAUTION: This reaction should be run in a well-ventilated hood.*

To a three-necked, round-bottomed flask equipped with a mechanical stirrer, dropping funnel, and condenser is added a solution of 4.95 gm (0.05 mole) of phosgene in 200 ml of dry carbon tetrachloride. The solution is vigorously stirred while the rapid addition of 5.8 gm (0.05 mole) of hexamethylenediamine and 4.0 gm (0.10 mole) of sodium hydroxide in 70 ml of water takes place. The reaction is exothermic while the polyurea forms. After 10 min, the carbon tetrachloride is evaporated off on a steam bath or with the aid of a water aspirator. The polyurea is washed several times in a blender and air-dried overnight to obtain 5.0 gm (70%), inherent viscosity 0.90 (in *m*-cresol, 0.5% concentration at 30°C), polymer melt temperature approximately 295°C.

b. CARBON DIOXIDE

In most cases the use of carbon dioxide is too impractical in the laboratory because it requires high temperatures and pressures, 100 atm [49a, b] or higher [49c-e].

More recently some reports claim the formation of N,N'-disubstituted ureas at room temperature at atmospheric pressure using amines (RNH_2, where R = Ph_2CH, cyclohexyl Ph, $PhCH_2$, Pr, or Me_2CH) and carbon dioxide in the presence of tertiary amines (Et_3N, etc.) with dicyclohexylcarbodiimide. Yields of 31-98% are claimed [49f]. When dicyclohexylcarbodiimide was replaced by $EtN{=}C{=}NCH_3$ the reaction had to be carried out under pressure [49f].

---

* We recommend the presence of small amounts of ammonia vapors in the air to reduce the danger of escaping phosgene.

N,N'-Dimethylurea has been reported to be prepared by reacting methyl-amine with CO_2 at $-30°$ to $-50°C$ for 24 hr to give $MeNHCO_2NH_3Me$, which can be decomposed to $MeNHCONHMe$ by heating in an autoclave (no pressure value reported) [49g].

c. CARBONYL SULFIDE

An example of the use of carbon oxysulfide is the reported preparation of aliphatic polyureas.

The diamine first forms the thiocarbamate salt, which is then heated to form the polyurea with liberation of hydrogen sulfide [50a] (Eqs. 35, 36).

$$2KCNS + 3H_2SO_4 + 2H_2O \longrightarrow 2COS + 2H_2SO_4 + (NH_4)_2SO_4 \qquad (35)$$

$$H_2N-R-NH_2 + COS \longrightarrow$$

$$\overset{+}{H_3N}-R-NH\overset{\overset{\displaystyle O}{\|}}{C}-S^- \xrightarrow{\text{heat}} \displaystyle\{R-NH\overset{\overset{\displaystyle O}{\|}}{C}NH\}_n + H_2S \qquad (36)$$

The effect of oxygen and hydrogen peroxide on the reaction of carbonyl sulfide on primary amines is discussed in Section 3.

More recently, urea and N,N'-dialkylureas have been prepared using COS and NH_3 or RNH_2 with decalen or tetralin in the absence of catalysts [50b]. The use of propanol as a solvent has also been reported [50c].

Zeolites have also been reported as catalysts for the reaction of carbonyl sulfide with alkyl or arylamines to give N,N'-disubstituted ureas [50d].

d. CARBON MONOXIDE AND SULFUR

A new urea synthesis involves the reaction of amines with sulfur and carbon monoxide at 120°C in methanol solvent to give high yields of 1,3-disubstit-uted ureas [51, 52] (Eq. 37). (See Section 3 for related material.)

$$2RNH_2 + CO + S \xrightarrow[120°C]{CH_3OH} RNH-\overset{\overset{\displaystyle O}{\|}}{C}-NHR + H_2S \qquad (37)$$

$$R = H, \text{ alkyl, or aryl}$$

When $R = H$, the following mechanism, in which carbonyl sulfide is an intermediate, has been suggested. The presence of carbonyl sulfide in 30% concentration in the reaction gases has been found by infrared spectroscopy when the reaction is run at 150°C [51, 52] (Eq. 38).

Aliphatic amines give high yields of 1,3-dialkylureas. Aromatic amines react well with the aid of a tertiary amine catalyst in the absence of solvent. Triethylamine is satisfactory and is used in the molar ratio of 0.3 to 1.0 of the aromatic amine at 100°–160°C for 1.5–3 hr at CO pressures of 300–600 psig. Some of the representative results are shown in Tables V–IX.

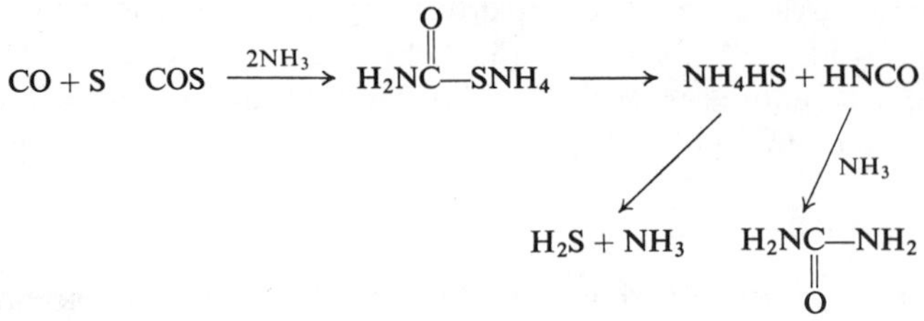

$$CO + S \quad COS \xrightarrow{2NH_3} \overset{\displaystyle O}{\underset{\displaystyle \parallel}{H_2NC}}-SNH_4 \longrightarrow NH_4HS + HNCO$$

$$H_2S + NH_3 \quad \overset{\displaystyle}{\underset{\displaystyle \parallel}{H_2NC-NH_2}} \qquad (38)$$

$$\overset{}{\underset{O}{}}$$

with NH_3 addition to give $H_2NC(=O)-NH_2$.

The preparation of 1,1-dimethyl-3-phenylurea is typical of the experimental technique used for the preparation of ureas by this new method [53].

TABLE V[a]

PREPARATION OF MONOSUBSTITUTED UREAS

$$RNH_2 + NH_3 + CO + S \rightarrow H_2NCONH_2 + RNHCONH_2 + RNHCONHR \qquad (39)$$

	Yield (%)		
RNH_2	H_2NCONH_2	$RNHCONH_2$	$RNHCONHR$
n-Octylamine	15	51	31
n-Dodecylamine	19	38	25
Aniline	52	0.7	16

[a] Tables V and VI are reprinted from R. A. Franz, F. Applegath, F. V. Morriss, F. Baiocchi, and L. W. Breed, *J. Org. Chem.* **27**, 4342 (1962). Copyright 1962 by the American Chemical Society. Reprinted by permission of the copyright owner.

TABLE VI

PREPARATION OF UNSYMMETRICAL DISUBSTITUTED UREAS

$$RNH_2 + R'NH_2 + CO + S \rightarrow RNHCONHR + RNHCONHR' + R'NHCONHR' \qquad (40)$$

		Yield (%)		
RNH_2	$R'NH_2$	RNHCONHR	RNHCONHR'	R'NHCONHR'
Ethylamine	p-Toluidine	—	17	6
n-Propylamine	Aniline	—	10	1
n-Butylamine	p-Chloroaniline	70	9	—
n-Butylamine	Aniline	—	37	—
n-Butylamine	p-Aminophenol	7	30	60
n-Amylamine	p-Aminophenol	20	35	35
n-Octylamine	p-Aminophenol	65	39	—
n-Hexylamine	N,N-Dimethyl-p-phenylenediamine	69	—	—
n-Octylamine	N,N-Dimethyl-p-phenylenediamine	62	—	—

TABLE VII[b]

1,3-Diarylureas

$$2RNH_2 + CO + S \rightarrow RNH—\overset{\overset{\displaystyle O}{\|}}{C}—NHR + H_2S \qquad (41)$$

Compound prepared	Yield (%)	M.p. (°C)		Recrystallized from
		Obsd.	Lit.	
1,3-Bis(2-pyridyl)urea	18	172	175	Methanol
1,3-Bis(3,4-dichlorophenyl)urea	20	277–279	(278–280 by synthesis)	—
1,3-Bis(2-chlorophenyl)urea	29	238	235–236	Methanol
1,3-Bis(3-chlorophenyl)urea	40	241	243	Methanol
1,3-Bis(4-chlorophenyl)urea	65	292–294	289–290	—
1,3-Diphenylurea	86	234–235	234	—
1,3-Bis(4-hydroxyphenyl)urea[a]	92	240 dec.	230 dec.	—
1,3-Bis(2-methoxy-5-chlorophenyl)urea	6	241–242	—	DMF–H$_2$O
1,3-Bis(4-carboxyphenyl)urea	34	>290	>270	—
1,3-Bis(2-methyl-5-chlorophenyl)urea	25	265–267	—	DMF–H$_2$O
1,3-Di-o-tolylurea	45	263–265	256	—
1,3-Di-p-tolylurea	79	260–262	260–261	—
1,3-Bis(4-methoxyphenyl)urea	63	232–234	234	Methanol
1,3-Di-2,4-xylylurea	23	sub. 260	262	—
1,3-Di-2,5-xylylurea	41	sub. 275	sub. 285	—
1,3-Di-2,6-xylylurea	8	>300	—	DMF
1,3-Bis(4-dimethylaminophenyl)urea	3	233–236	262 dec.	—
1,3-Bis(4-carbethoxyphenyl)urea	1	215–216	—	Methanol
1,3-Di-α-naphthylurea	14	280–292	296	DMF Methanol
1,3-Di-β-naphthylurea	18	288–295	301–302	—
1,3-Bis(4-diethylaminophenyl)urea[a]	27	212–215	223–224	—
1,3-Di-4-biphenylylurea	45	310–311	315	DMS

[a] No triethylamine used.

[b] Reprinted in part from R. A. Franz, F. Applegath, F. V. Morriss, F. Baiocchi, and C. Bolze, *J. Org. Chem.* **26**, 3311 (1961). Copyright 1961 by the American Chemical Society. Reprinted by permission of the copyright owner.

6. Ureas

TABLE VIII[b]

1,3-DIALKYLUREAS

$$2RNH_2 + CO + S \rightarrow RNH—\overset{\overset{\displaystyle O}{\|}}{C}—NHR + H_2S \qquad (42)$$

Compound prepared	Yield (%)	M.p. (°C)		Recrystallized from
		Obsd.	Lit.	
1,3-Dimethylurea	88	101–103	106	Benzene–Skellysolve B[a]
1,3-Diethylurea	75	111	112	Water
1,3-Diisopropylurea	51	192	192	Water
1,3-Di-n-propylurea	69	102–103	105	Hot water
1,3-Diisobutylurea	80	128–130	134	Skellysolve B
	—	—	128	—
1,3-Di-n-butylurea	81	67–69	71	Skellysolve F[b]
1,3-Di-sec-butylurea	71	135	137–138	Skellysolve F
1,3-Di-t-butylurea	54	245	242	Methanol
1,3-Di-n-amylurea	71	79–81	88	Skellysolve F
1,3-Dicyclohexylurea	68	229–230	230	Methanol
1,3-Di-n-hexylurea	71	73–74	—	Methanol
1,3-Di-n-octylurea	84	89–90	—	Methanol
1,3-Di-t-octylurea	29	145–147	153	Petroleum ether[c]
1,3-Di-n-decylurea	70	99–100	—	Methanol
1,3-Di-n-dodecylurea	80	105–106	105.5	Benzene
1,3-Di-n-tetradecylurea	61	106–107	—	Propanol-2
1,3-Di-n-octadecylurea	71	110–111	113–114	Benzene
1,3-Bis(2-acetamidoethyl)urea	60	200–202	—	Methanol
1,3-Bis(3-methoxypropyl)urea	60	47–49	—	Ether–Skellysolve B
1,3-Difurfurylurea	73	126–129	124–127	Water
1,3-Bis(2-N-morpholinoethyl)urea	60	109–111	—	Benzene
1,3-Bis(3-isopropoxypropyl)urea	72	37–39	—	Skellysolve B
1,3-Dibenzylurea	60	169–171	169	—
1,3-Bis(α-methylbenzyl)urea	39	121–122	—	Water
		144–148	—	—
1,3-Bis(4-methoxybenzyl)urea	64	176–177	178–179	—

[a] Boiling range 60°–71°C.

[b] Boiling range 35°–60°C.

[c] Boiling point 80°–110°C.

[d] Reprinted in part from R. A. Franz, F. V. Morriss, and F. Baiocchi, *J. Org. Chem.* **26**, 3307 (1961). Copyright by the American Chemical Society. Reprinted by permission of the copyright owner.

TABLE IX[a]

PROPERTIES AND ANALYSES OF 1,1-DIALKYL-3-ARYLUREAS

Compound	Yield (%)	M.p. (°C) Obsd.	Lit.	Recrystallized from
1,1-Dimethyl-3-phenylurea	79	131–132	127–129	H_2O
1,1-Dimethyl-3-p-tolylurea	54	152–153	—	EtOH
1,1-Dimethyl-3-(2,4-xylyl)urea	81	133–136	—	$CHCl_3$
1,1-Dimethyl-3-(2,5-xylyl)urea	68	87–88	—	EtOH
1,1-Dimethyl-3-(2,6-xylyl)urea	40	132–133	—	EtOH
1,1-Dimethyl-3-(1-naphthyl)urea	78	165–166	—	MeOH
1,1-Dimethyl-3-(2-naphthyl)urea	92	203–205	210.5–210.8	EtOH
3-(o-Chlorophenyl)-1,1-dimethylurea	23	104–106	94.1–95.4	MeOH
3-(m-Chlorophenyl)-1,1-dimethylurea	57	144–146	144.3–144.8	EtOH
3-(p-Chlorophenyl)-1,1-dimethylurea	86	170–171	170–171	EtOH
3-(3,4-Dichlorophenyl)-1,1-dimethylurea	62	152–154	155	—
1,1-Dimethyl-3-(p-methoxyphenyl)urea	88	130–132	—	$MeOH–H_2O$
3-(2,5-Dimethoxyphenyl)-1,1-dimethylurea	19	52–54	—	MeOH
3-(5-Chloro-2-methoxyphenyl)-1,1-dimethylurea	18	63–66	—	Skellysolve B
1,1-Dimethyl-3-p-hydroxyphenylurea	64	194–196	—	H_2O
3-(p-Carboxyphenyl)-1,1dimethylurea	45	214–216	—	i-PrOH
3-(p-Carbethoxyphenyl)-1,1-dimethylurea	35	144–146	—	EtOH
1,1-Dimethyl-3-(p-dimethylaminophenyl)urea	93	173–174	—	EtOH
3-(2-Benzothiazolyl)-1,1-dimethylurea	17	214–215	218–220	MeOH
1,1-Di-n-butyl-3-phenylurea	75	82–83	82.7–83.0	$EtOH–H_2O$
1,1-Diisobutyl-3-phenylurea	52	104–106	—	EtOH
3-(p-Chlorophenyl)-1,1-di-n-butylurea	39	123	122	C_6H_6
3-(p-Chlorophenyl)-1,1-diallylurea	10	73–74	80–81	i-PrOH
4-Morpholinecarboxanilide	50	156–159	161.5–162	MeOH
4'-Chloro-4-morpholinecarboxanilide	66	196–200	196	MeOH
3'-Chloro-4-morpholinecarboxanilide	73	129–131	—	C_6H_6
3',4'-Dichloro-4-morpholinecarboxanilide	23	150–152	157.1–157.8	$MeOH–H_2O$

[a] Reprinted in part from R. A. Franz, F. Applegath, F. V. Morriss, F. Baiocchi, and L. W. Breed, *J. Org. Chem.* **27**, 4342 (1962). Copyright 1962 by the American Chemical Society. Reprinted by permission of the copyright owner.

2-13. Preparation of 1,1-Dimethyl-3-phenylurea [53]

$$(CH_3)_2NH + C_6H_5NH_2 + CO + S \xrightarrow[100°C]{CH_3OH} (CH_3)_2N\overset{\overset{\displaystyle O}{\|}}{C}NHC_6H_5 \qquad (43)$$
$$(79\%)$$

To a 2 liter Hoke pressure cylinder are added 20.0 gm (0.21 mole) of aniline, 13.8 gm (0.43 mole) of sulfur, 100 ml of methanol, and 82 ml of an 8.75 M (0.73 mole) aqueous dimethylamine. The cylinder is pressurized with 100 psig of carbon monoxide, and then heated for 2 hr at 100°C. The cylinder is then vented and the contents removed from the cylinder by washing with hot methanol. The combined product and methanol washings were filtered hot and evaporated to dryness from the methanol. The urea product is recrystallized from 400 ml of water to give 27.8 gm (79%) of 1,1-dimethyl-3-phenylurea, m.p. 130°–133°C.

The highest yields were obtained as a dimethylamine to aniline ratio of 2.4:1.

Other unsymmetrical ureas are obtained by starting with a dialkylamine and a primary aromatic amine to give 1,1-dialkyl-3-arylureas (Eqs. 44–48) as shown in Table IX.

$$S + CO \rightleftharpoons COS \qquad (44)$$

$$R_2NH + COS \rightleftharpoons R_2NCOSH \cdot R_2NH \qquad (45)$$

$$RNH_2 + COS \rightleftharpoons RNHCOSH \cdot RNH_2 \rightleftharpoons RNCO + RNH_2 \cdot H_2S \qquad (46)$$

$$RNCO + RNH_2 \rightleftharpoons RNHCONHR \qquad (47)'$$

$$RNCO + R_2NH \rightleftharpoons R_2NCONHR \qquad (48)$$

e. CARBON MONOXIDE

Carbon monoxide can be used to convert amines to ureas by a catalyzed carboxylation using mild conditions. Some typical catalysts are selenium (and oxygen as a coreagent) and silver acetate. The use of palladium catalysts and manganese decacarbonyl requires more vigorous conditions (high pressures and temperatures) as described in Table X.

f. CARBONYL SELENIDE

Related to the selenium-catalyzed carbon monoxide carbonylation of amines to give ureas is the reaction of amine with carbonyl selenide to give ureas [53b]. The reaction takes place at room temperature.

g. CARBON DISULFIDE

Thioureas are prepared by reaction of amines with CS_2 by first forming the dithiocarbamate amine salt, which on subsequent heating splits off hydrogen sulfide. The reaction can be carried

$$2RNH_2 + CS_2 \longrightarrow RNH_3^- \underset{\underset{S}{\parallel}}{S}CNHR \xrightarrow[\Delta]{-H_2S} RNH\underset{\underset{S}{\parallel}}{C}NHR \qquad (49)$$

out in the presence of water and is usually carried out under pressure at 50°–150°C [53c–e]. Processes carried out at atmospheric pressure using alcohols as solvents [53f] and catalyzed by activated carbon [53g] have also been reported. Mixtures of amines have been used to give mixed dialkyl thioureas that are more easily pourable because they remain liquid for a long period and crystallize only slowly [53h].

h. CARBAMOYL CHLORIDES

Carbamoyl chlorides prepared by reacting phosgene with secondary amines can be reacted with other secondary amines to give substituted ureas.

$$R_2NH + COCl_2 \xrightarrow{-HCl} R_2NCOCl \xrightarrow[base]{R_2NH} R_2N\overset{\overset{O}{\parallel}}{C}NR_2' \qquad (50)$$

2-14. Preparation of 1-(p-Chlorophenyl)-1,3,3-trimethylurea [53j]

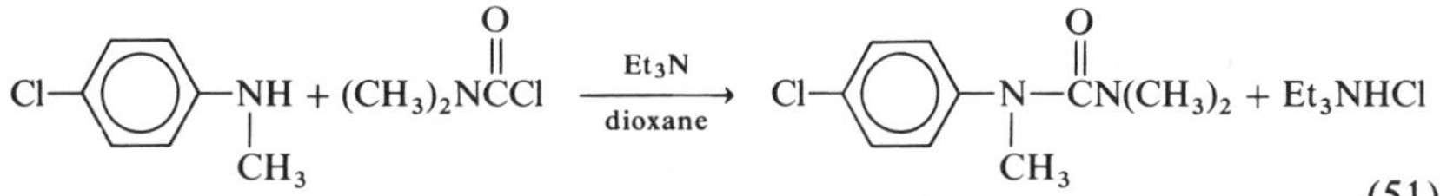

$$(51)$$

To a flask containing 16.1 gm (0.15 mole) of N,N-dimethylcarbamoyl chloride in 15 gm of dioxane and heated to 30°–60°C is added a solution of 17 gm (0.12 mole) of N-methyl-p-chloroaniline, 13.2 gm of triethylamine, and 25 gm of dioxane over a period of 2 hr. After standing overnight at room temperature the resulting mixture is heated at 77°–85°C. for 2.5 hr, then cooled, and poured into water. The product 1-(p-chlorophenyl)-1,3,3-tri-methylurea, a water-insoluble oil, is extracted into ether and the ether extracts washed successively with 1 N hydrochloric acid, 5% sodium bicarbonate, and fresh water. The ether extract is dried over sodium sulfate and distilled to yield 11.1 gm of product (43.5%), b.p. 100.5°–102.0°C (0.24 mm Hg), n_D^{25} 1.5572.

The preparation of tetramethylurea by the reaction of dimethylamine with N,N-dimethylcarbamoyl chloride in the presence of lime has also been described [53k].

TABLE X

Catalyzed Reaction of Carbon Monoxide with Amines to Give Ureas

Amine (mole)	Reagent other than CO	Catalyst (gm)	Solvent	Reaction conditions			Yield (%)	Product	Ref.
				Temp. (°C)	Time (hr)	Pressure (atm)			
$n\text{-}C_4H_9NH_2$ (0.1)	O_2 (9 ml/min.)	Se Amorphous (0.40)	$\dfrac{\text{THF}}{100\ ml}$	20	4	1 atm.	95–99	$n\text{-}C_4H_9NHCNH_2$ (with =O)	[a]
$H_2NCH_2CH_2NH_2$ (0.1)	O_2 (9 ml/min.)	Se Amorphous (0.40)	$\dfrac{\text{THF}}{100\ ml}$	20	4	1 atm.	14	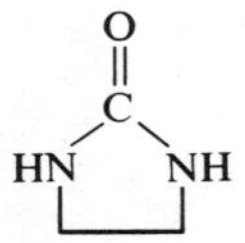 imidazolidinone	[a]
$C_6H_5NH_2$	O_2 (9 ml/min.)	Se Amorphous (0.40)	$\dfrac{\text{THF}}{100\ ml}$	20	4	1 atm.	74	$C_6H_5-CH_2NHCNH_2$	[a]
$H_2NCH_2CH_2NH_2$	O_2 (9 ml/min.)	Se Amorphous (0.40)	$\dfrac{\text{THF}}{100\ ml}$	60		50	98	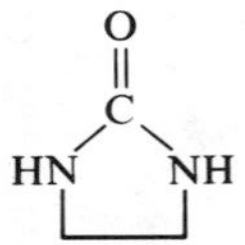 imidazolidinone	[a]

174

Amine		Catalyst	Solvent	Temp. (°C)	Time (hr)	Pressure	Yield (%)	Product	Ref.
$n\text{-}C_4H_9NH_2$	—	Se Amorphous (0.40)	THF	25	5.5	1 atm.	97	$n\text{-}C_4H_9NHCNH_2$ ($\overset{O}{\overset{\|}{}}$)	[b]
$n\text{-}C_4H_9NH_2$	—	AgOAc (10 mmole)	$(C_2H_5)_3N$ (7 ml)	25	15.5	84	~100	$n\text{-}C_4H_9NHCNH_2$ ($\overset{O}{\overset{\|}{}}$)	[c]
$CH_2{=}CH{-}CH_2NH_2$	—	AgOAc (10 mmole)	$(C_2H_5)_3N$ (7 ml)	25	15.5	84	73	$CH_2{=}CH{-}CH_2NHCNH_2$ ($\overset{O}{\overset{\|}{}}$)	[c]
$C_6H_5NH_2$	—	$PdCl_2$ (1 gm)		180	20	98	25	$C_6H_5{-}NHCNHC_6H_5$ ($\overset{O}{\overset{\|}{}}$)	[d]
$Cyclo\text{-}C_6H_{13}NH_2$ (1.33)	—	$Mn_2(CO)_{10}$ (2.26 gm)	Heptane 100 ml	180–200	12	130	40	$C_6H_{13}NHCNHC_6H_{13}$ ($\overset{O}{\overset{\|}{}}$)	[e]

[a] N. Sonoda, T. Yasuhara, K. Kondo, T. Ikeda, and S. Tsutsumi, *J. Amer. Chem. Soc.* **93**, 6344 (1971).
[b] Asahi Chem. Ind. Co. Ltd. and Chiyoda Chem. Engineering and Construction Co. Ltd., Brit. Pat. 1,275,702 (1972); *Chem. Abstr.* **77**, 151527t (1972).
[c] F. Tsuda, Y. Isegawa, and T. Saegusa, *J. Org. Chem.* **37**, 2670 (1972).
[d] J. Tsuji and N. Iwamoto, Jap. Pat. 6890 (1967); *Chem. Abstr.* **67**, 11325u (1967).
[e] F. Calderazzo, U.S. Pat. 3,316,297 (1967).

175

i. CARBONATES

The reaction of primary amines with cyclic and acyclic carbonates is reported to give ureas [53l, m].

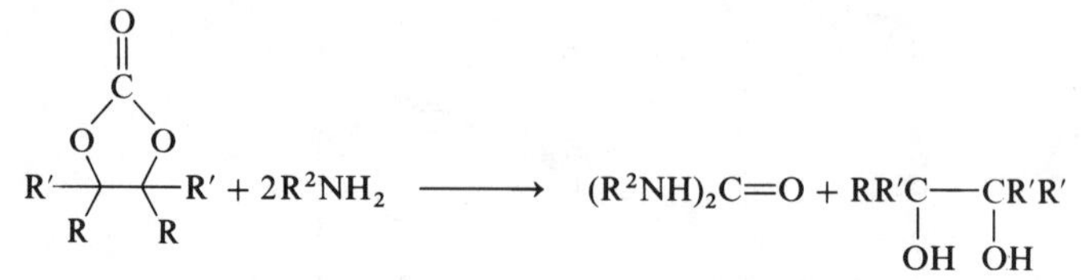

$$R' \overset{O}{\underset{R}{\diagdown}} \overset{\parallel}{C} \overset{O}{\underset{R}{\diagup}} R' + 2R^2NH_2 \longrightarrow (R^2NH)_2C{=}O + RR'C{\underset{OH}{|}}{-}CR'R'{\underset{OH}{|}} \qquad (52)$$

$$(C_6H_5O)_2C{=}O + 2RNH_2 \longrightarrow (RNH)_2C{=}O + 2C_6H_5OH \qquad (53)$$

H. Reaction of Urea with Carbonyl Derivatives

Urea reacts with a wide variety of reagents to give substituted linear and heterocyclic urea derivatives. Some of these reactions are summarized in Eq. (54) (p. 177).

Although the reactions of Wilson [60a] and Hurwitz and Auten [60b] are not concerned with the carbonyl reactions of urea, they are shown here because of the interesting 2-imidazolidinones they form.

In addition, symmetrical diphenylurea may be metalated to give the dianion which then may be condensed with active halogen compounds as shown in Eq. (55). This metallation does not work when applied to urea, phenylurea,

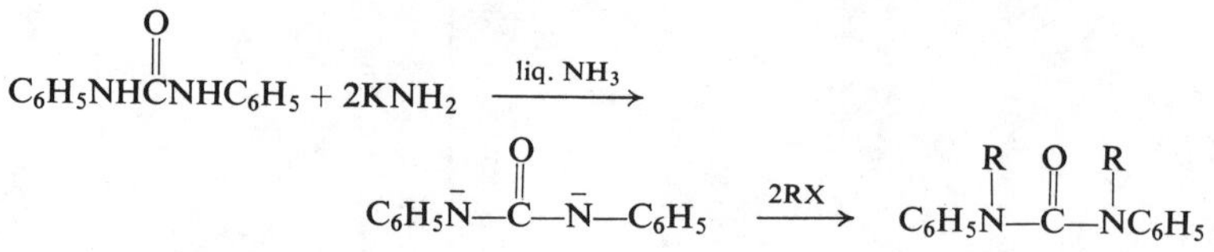

$$C_6H_5NH\overset{O}{\overset{\parallel}{C}}NHC_6H_5 + 2KNH_2 \xrightarrow{\text{liq. NH}_3}$$

$$C_6H_5\bar{N}{-}\overset{O}{\overset{\parallel}{C}}{-}\bar{N}{-}C_6H_5 \xrightarrow{2RX} C_6H_5\overset{R}{\overset{|}{N}}{-}\overset{O}{\overset{\parallel}{C}}{-}\overset{R}{\overset{|}{N}}C_6H_5 \qquad (55)$$

benzoylurea, or acetylurea [64].

2-15. Preparation of 1,3-Dihydroxymethylurea [65]

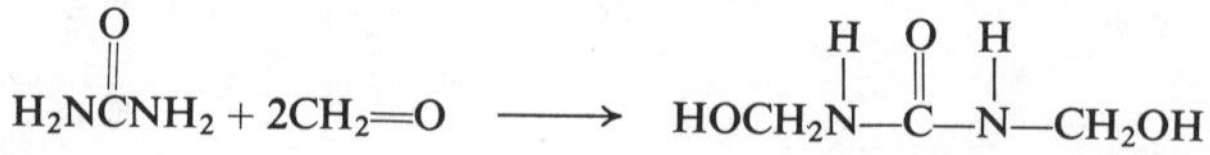

$$H_2N\overset{O}{\overset{\parallel}{C}}NH_2 + 2CH_2{=}O \longrightarrow HOCH_2\overset{H}{\overset{|}{N}}{-}\overset{O}{\overset{\parallel}{C}}{-}\overset{H}{\overset{|}{N}}{-}CH_2OH \qquad (56)$$

To a one-necked flask containing a magnetic stirrer bar is added 81.2 gm (1.0 mole) of 37% formaldehyde at pH 8.0. All at once, 30.0 gm (0.5 mole) of urea is added. The reaction is slightly exothermic and the flask is placed in a water bath at 30°C. The reaction mixture is vigorously stirred by means of a

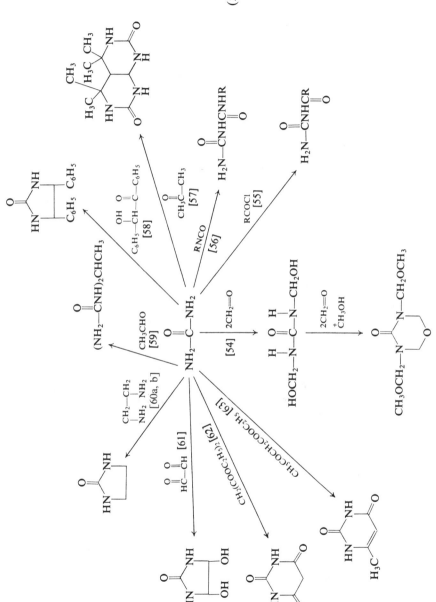

(54)

magnetic stirrer, and within 5 min the reaction becomes exothermic to 50°C. The solution is cooled to 25°C and, after standing for 18 hr dimethylolurea precipitates as a white solid. The product is washed with ethanol and dried under reduced pressure to give 30.0 gm (50%), m.p. 123°–126°C (reported 126°C) [66].

Urea can also react with more than 2 moles of formaldehyde to give the difficult to isolate tri- and tetramethylolureas [67a]. More data are needed to confirm these results. Urea also reacts with formaldehyde to give urea–formaldehyde resins [67b].

2-16. Preparation of 4,5-Dihydroxy-2-imidazolidinone [67a, b, 65].

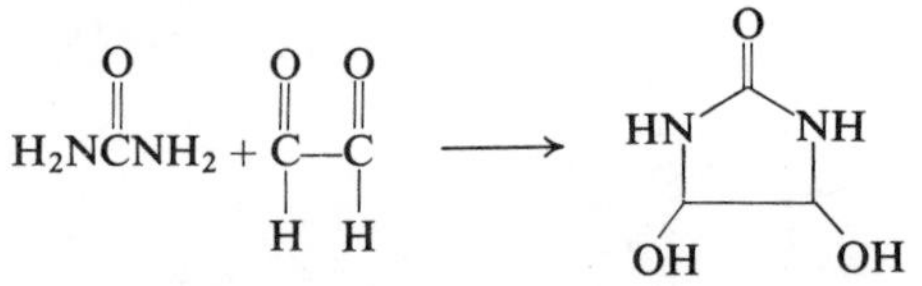

$$\text{H}_2\text{NCNH}_2 + \underset{\text{H}\ \ \ \text{H}}{\text{C}-\text{C}} \longrightarrow \text{(4,5-dihydroxy-2-imidazolidinone)} \qquad (57)$$

To a resin kettle is added 864 gm (6.0 moles) of 40% glyoxal. The pH is adjusted to approximately 7–8. The addition of 570 gm (9.0 moles) of urea gives an exothermic reaction. The resulting solution is stirred for 5 hr and then approximately half of the water is removed under reduced pressure. While concentrating, the product precipitates to give 317.0 gm, m.p. 125°–127°C. The filtrate is concentrated again, filtered, and the process repeated several times to yield an additional 132 gm of product. Total yield 64%. The by-product contains unreacted urea and other products.

Heating 4,5-dihydroxy-2-imidazolidinone with urea under acidic conditions yields tetrahydroimidazo[4,5-d]imidazole-2,5-diones (glycolurils) [68] (Eq. 58). (See Table XI.)

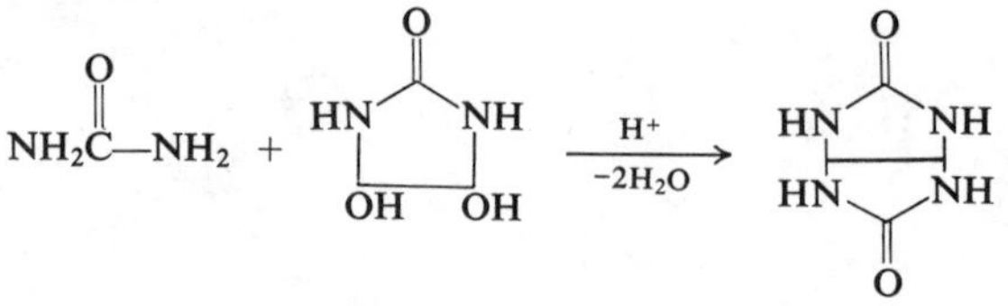

$$\text{NH}_2\text{C}-\text{NH}_2 + \text{(4,5-dihydroxy-2-imidazolidinone)} \xrightarrow[-2\text{H}_2\text{O}]{\text{H}^+} \text{(glycoluril)} \qquad (58)$$

2-17. Preparation of 2-Imidazolidinone [60a]

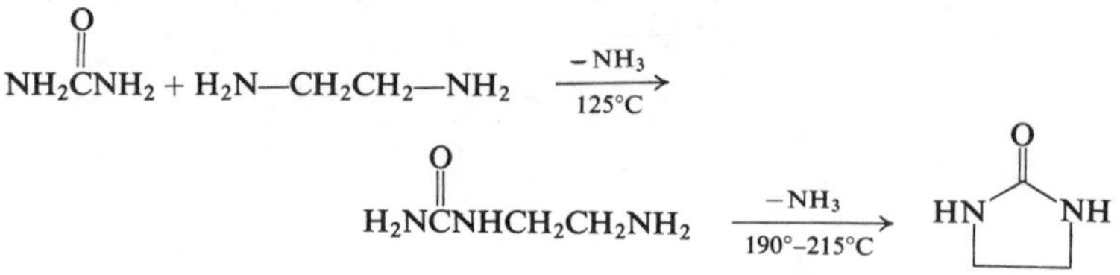

$$\text{NH}_2\overset{\text{O}}{\overset{\|}{\text{C}}}\text{NH}_2 + \text{H}_2\text{N}-\text{CH}_2\text{CH}_2-\text{NH}_2 \xrightarrow[125°\text{C}]{-\text{NH}_3}$$

$$\text{H}_2\text{N}\overset{\text{O}}{\overset{\|}{\text{C}}}\text{NHCH}_2\text{CH}_2\text{NH}_2 \xrightarrow[190°-215°\text{C}]{-\text{NH}_3} \text{(2-imidazolidinone)} \qquad (59)$$

TABLE XI[m]

PREPARATION OF SUBSTITUTED GLYCOLURILS

Product glycoluril	II[a] (moles)	Urea (moles)	H_2O (ml)	Concd. HCl (ml)	Hr at 100°C	M.p. (°C)	Yield (%)
1,3-Dimethyl-	0.35	0.4[b]	180	3.5	1	254–256	27
1-Isopropyl-	0.5	0.5[c]	200	5	0.5	248–249	51
1,3-Diisopropyl-	0.1	0.1[d]	250	3	0.5	312–313	71
1-(n-Butyl)-	0.5	0.8[f]	300	10	0.25	267–268[g]	32
1-Benzyl-	0.2	0.2[h]	300	5	1	283–284	65
1-Phenyl-	0.25	0.25[i]	500	5	0.5	300	82
1,4- and/or 1,6-Dibenzyl-	0.05[l]	0.1[h]	100	3	0.3	303–305	20
1-Carboxymethyl-	0.2	0.21[j]	80	10	0.25	226	43
1,4- and/or 1,6-Bis(carboxy-methyl)-	0.1[e]	0.21[j]	30	5	0.2	257	14
1-(Carbethoxy)methyl-	0.2	0.21[k]	100[l]	10	0.25	217–218	41

[a] 4,5-Dihydroxy-2-imidazolidinone (glyoxal monoureide).   [b] 1,3-Dimethylurea.   [c] 1-Isopropylurea.   [d] 1,3-Diisopropylurea.   [e] 30% commercial glyoxal solution.   [f] n-Butylurea.   [g] Brit. Pat. 783,051 (1957) reports m.p. 249°C.   [h] 1-Benzylurea.   [i] Phenylurea.   [j] Hydantoic acid.   [k] Ethyl hydantoate.   [l] Ethanol.   [m] Reprinted in part from F. B. Slezak, H. Bluestone, T. A. Magee, and J. H. Wotiz, *J. Org. Chem.* **27**, 2181 (1962). Copyright 1962 by the American Chemical Society. Reprinted by permission of the copyright owners.

To a resin kettle are added 257 gm (3.0 moles) of 70% ethylenediamine and 60 gm (1.0 mole) of urea and the mixture is refluxed for a period of 5 hr to a temperature of 125°C. Excess ethylenediamine is then distilled off up to a temperature of 170°C over a period of about 6 hr. Except for a slight appearance of cloudiness at 190°–215°C, the reaction mixture remains clear in appearance throughout the reaction. On cooling, the product obtained is 85.0 gm (99%), m.p. 129°–131°C (recrystallized from ethanol). Approximately 0.7% of insoluble, infusible material is obtained. Using less ethylenediamine yields more polymeric and infusible, insoluble material. When 1 mole of pure ethylenediamine per mole of urea is used, the yield is only 51% of 2-imidazolidone and the remaining material is chloroform-soluble (20 gm) and insoluble-infusible (16 gm).

The use of a solvent such as ethyleneglycol has also been reported [69].

2-Imidazolidinone has earlier been reported to be prepared from diethyl carbonate and urea [70] and from urea and phosgene [71].

The preparation of 2-imidazolidinethione (ethylenethiourea) has similarly been reported to be prepared from ethanolic ethylenediamine and carbon disulfide [72]. The reaction is exothermic, and carbon disulfide is added slowly in portions. The product melts at 197°–198°C. Substituting propylenediamine in place of ethylenediamine produces an 80% yield of colorless crystals of 2-thio-3,4,5,6-tetrahydropyrimidine, m.p. 207°–208°C, after recrystallizing from water [73].

2-18. *Preparation of Barbituric Acid* [62]

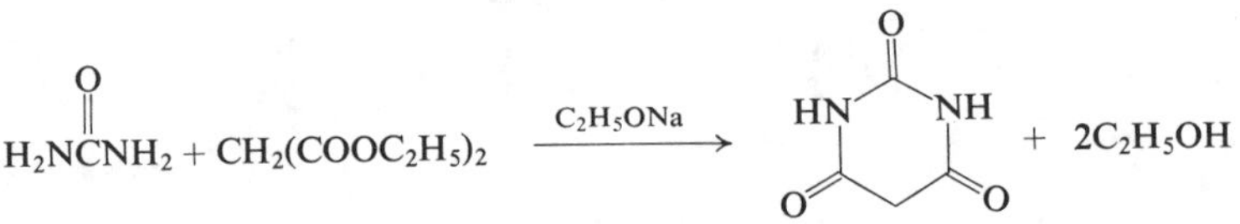

$$H_2NCNH_2 + CH_2(COOC_2H_5)_2 \xrightarrow{C_2H_5ONa} \text{(barbituric acid ring)} + 2C_2H_5OH \quad (60)$$

To a round-bottomed flask equipped with a magnetic stirring bar, a reflux condenser and a drying tube and 250 ml of absolute ethanol is added 11.5 gm (0.5 gm-atom) of finely cut sodium. To this is added a solution of 80 gm (0.5 mole) of diethyl malonate and 30 gm (0.5 mole) of urea in 250 ml of hot absolute ethanol. The mixture is stirred vigorously while the contents are refluxed for 7 hr. During the reaction a white solid separates and, after the 7 hr of refluxing, 500 ml of warm water (50°C) is added along with 45 ml of concentrated hydrochloric acid. The resulting clear solution is filtered, cooled overnight in an ice bath, the resulting solid filtered, washed with 50 ml of cold water, and dried at 105°–110°C to yield 46–50 gm (72–78%), m.p. 247°–248°C [74].

Barbituric acid may also be prepared by condensing urea with malonyl chloride [75] or with malonic acid in the presence of acetic anhydride [76a].

The reaction of thiourea with $RCOCH_2COOC_2H_5$ in the presence of $NaOC_2H_5$ has been reported to give thiouracels [76b].

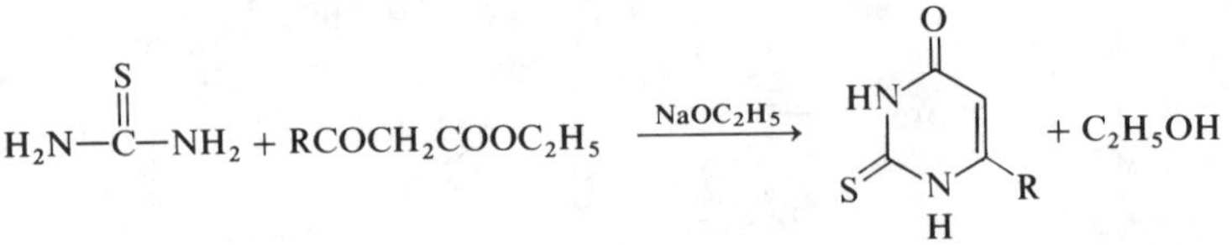

$$H_2N-\underset{S}{\overset{S}{C}}-NH_2 + RCOCH_2COOC_2H_5 \xrightarrow{NaOC_2H_5} \text{(thiouracil ring)} + C_2H_5OH \quad (61)$$

I. Reactions of Olefins or Alcohols with Urea to Produce Alkylureas

Urea can be reacted with olefins or alcohols to form alkylureas under specific conditions.

Strong acid favors the reaction of olefins and alcohols with urea to give mono- and dialkylureas. Particularly effective are tertiary olefins such as isobutylene [76c, d] and *t*-amylene [76e, f].

Tertiary alcohols such as *t*-butyl alcohol give *t*-butylurea [76g]. Primary and secondary alcohols also have been reported to react to give alkylureas [76h].

3. OXIDATION REACTIONS

Few oxidative procedures exist for the preparation of ureas. The reaction of phenetidine hydrochloride with a mixture of sodium cyanide and hydrogen peroxide has been reported to give p-ethoxyphenylurea [77-78] (Eq. 62).

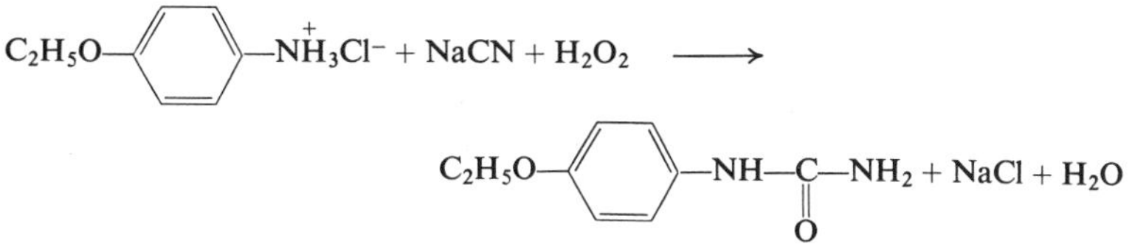

$$C_2H_5O\!-\!\!\left<\!\!\bigcirc\!\!\right>\!\!-\!\overset{+}{N}H_3Cl^- + NaCN + H_2O_2 \longrightarrow$$

$$C_2H_5O\!-\!\!\left<\!\!\bigcirc\!\!\right>\!\!-\!NH\!-\!\underset{\underset{O}{\|}}{C}\!-\!NH_2 + NaCl + H_2O \quad (62)$$

Recently ureas have been obtained by the oxidation of carbohydrazide (oxidative carbamylation) [78] with 4 moles of iodine in the presence of 10 moles of amine (or 2 moles of the amine and 8 moles of triethylamine) using dimethylacetamide as solvent (Eq. 63). (See Table XII.)

$$NH_2NHCONHNH_2 + 4I_2 + 10RNH_2 \longrightarrow$$

$$RNHCONHR + 8R\overset{+}{N}H_3Cl^- + 2N_2 \quad (63)$$

TABLE XII[e]

Ureas Obtained *via* Oxidative Carbamylation

Urea	Yield (%)	M.p. (°C)
1,3-Diphenyl-	88	238–239[a]
1,3-Dibenzyl-	84	169–170[a]
1,3-Dicyclohexyl-	87	216–218[b]
1,3-Dibutyl-	61	70[b]
1-Benzyl-3-phenyl-	90	171–172[c]
1-Cyclohexyl-3-phenyl-	89	150[d]

[a] M.p. 241°–242°C for the 1,3-diphenylurea and m.p. 170°C for the 1,3-dibenzylurea are given by P. A. Boivin, W. Bridgeo, and J. L. Boivin, *Canad. J. Chem.* **32**, 242 (1954).

[b] M.p. 216°–217°C for the 1,3-dicyclohexylurea and m.p. 67°–68°C for the 1,3-dibutylurea are given by E. Junod, *Helv. Chim. Acta* **35**, 1667 (1952).

[c] M.p. 172°C is given by R. N. Lacey, *J. Chem. Soc.* 845 (1954).

[d] M.p. 149°–150°C is given by W. B. Bennet, J. H. Saunders, and E. E. Hardy, *J. Amer. Chem. Soc.* **75**, 2101 (1953).

[e] Reprinted from Y. Wolman and P. M. Gallop, *J. Org. Chem.* **27**, 1902 (1962). Copyright 1962 by the American Chemical Society. Reprinted by permission of the copyright owner.

A similar procedure using 4-phenylsemicarbazide hydrochloride with 2 moles of iodine and 6 moles of the amine (or 1 mole of the amine and 5 moles of triethylamine) gave 90% yields of unsymmetrical ureas [78] (Eq. 64).

$$RNHCONHNH_3Cl + 2I_2 + 6R'NH_2 \longrightarrow$$
$$RNHCONHR' + 4R'\overset{+}{N}H_3I^- + R'\overset{+}{N}H_3Cl^- + N_2 \qquad (64)$$
$$R = C_6H_5$$

The method is described here only for the sake of information as more work is required to ascertain the preparative value and the general applicability of this procedure [78].

3-1. General Procedure for Preparation of Symmetrical and Unsymmetrical Ureas by the Oxidative Carbamoylation Method [79]

(a) Symmetrical Ureas.   To an ice-cooled flask containing 900 mg (10 mmole) of carbohydrazide and 100 mmole of the amine dissolved in 20 ml of dimethylacetamide is added 10 gm (40 mmole) of solid iodine with stirring. After 5 min the iodine excess is removed with a dilute solution of sodium thiosulfate and the product precipitated with 80 ml of water. The urea is filtered, air-dried, and recrystallized.

(b) Unsymmetrical Ureas.   To an ice-cooled flask containing 1.86 gm (10 mmole) of 4-phenylsemicarbazide hydrochloride and 60 mmole of the amine dissolved in 20 ml of dimethylacetamide is added 5.0 gm (20 mmole) of solid iodine with stirring. After 5 min the excess of iodine is removed with a dilute solution of sodium thiosulfate. The urea product is precipitated with 80 ml of water, filtered, air-dried, and recrystallized.

Recently a method has been reported to involve the reaction of primary amines with COS in methanol at 25°C. The mixture is oxidized with oxygen for 4 hr at 21°C at 3.5 kg/cm^2 oxygen in an autoclave, using $CuSO_4$ or $CoCl_2$ as oxidation catalyst, to give *sym*-ureas. The intermediate $RNHCOS^-$ RNH_3^+ is postulated [79].

Earlier it was reported that hydrogen peroxide and cumene hydroperoxide improve the yields of 1,3-diarylureas from the reaction of weakly basic aromatic amines with carbonyl sulfide at elevated temperatures [80]. The results are summarized in Table XIII.

3-2. Preparation of 1,3-Diphenylurea [80]

$$2C_6H_5NH_2 + COS \xrightarrow{H_2O_2} C_6H_5{-}NH{-}\overset{\overset{\displaystyle O}{\|}}{C}{-}NH{-}C_6H_5 + S + H_2O \qquad (65)$$

To a heavy-walled, 2 liter suction flask is added a suspension of 9.3 gm (0.10

TABLE XIII[a]

PEROXIDE-INFLUENCED REACTION OF AMINES WITH CARBONYL SULFIDE[a] [80]

Amine used	Peroxide used	Solvent used	Time (hr)	Temp. (°C)	Product	Yield (%)
Aniline	H_2O_2	H_2O	24	25	1,3-Diphenylurea	70
	CH[b]	CH_3OH	72	5	1,3-Diphenylurea	80[c]
	DPB[d]	CH_3OH	24	25	1,3-Diphenylurea	0
p-Methoxyaniline	H_2O_2	H_2O	72	25	1,3-Di-p-methoxyphenylurea	88
p-Chloroaniline	H_2O_2	H_2O	72	25	1,3-Di-p-chlorophenylurea	14
p-Nitroaniline	H_2O_2	H_2O	72	25	No product	0
	H_2O_2	CH_3OH	120	25	No product	0
	CH[b]	CH_3OH	120	25	No product	0
Benzylamine	CH[b]	CH_3OH	96	5	1,3-Dibenzylurea	89
	H_2O_2	H_2O	24	25	1,3-Dibenzylurea	75
n-Hexylamine	H_2O_2	H_2O	48	25[f]	1,3-Di-n-hexylurea	66
n-Decylamine	H_2O_2	H_2O/CH_3OH	48	5	1,3-Di-n-decylurea	76
2-Aminoethanol	CH[b]	CH_3OH	96	25	2-Oxazolidinone	23
Ethylenediamine	H_2O_2	H_2O	120	25	2-Imidazolidinone	6
1,2-Propanediamine	H_2O_2	H_2O	120	25	4-Methyl-2-imidazolidinone	94
p-Aminophenol	H_2O_2	CH_3OH	24	5	1,3-Di-p-hydroxyphenylurea	32
2,4,6-Trichloroaniline	H_2O_2	CH_3OH	96	25	No product	0
o-Phenylenediamine[e]	H_2O_2	CH_3OH	96	25	2-Benzimidazolol	82
2-Aminodiphenyl	H_2O_2	CH_3OH	96	25	No product	0

(continued)

TABLE XII[g] (continued)

Amine used	Peroxide used	Solvent used	Time (hr)	Temp. (°C)	Product	Yield (%)
1-Naphthylamine	CH[b]	CH₃OH	96	25	1,3-Di-1-naphthylurea	42
2-Naphthylamine	H₂O₂	H₂O	96	25	1,3-Di-2-naphthylurea	17
	CH[b]	CH₃OH	96	25	1,3-Di-2-naphthylurea	92
o-Aminophenol	CH[b]	CH₃OH	96	25	2-Benzoxazolol	89

[a] All these reactions were run essentially as described for the formation of 1,3-diphenylurea for the aromatic amines and 1,3-dibenzylurea for the aliphatic amines. The products were identified by comparison with separately synthesized authentic specimens.

[b] Cumene hydroperoxide.

[c] Yield in this case based on the addition of a sodium methoxide in a molar equivalent quantity to the amine. Without the sodium methoxide, the yield after 96 hr was only 46%.

[d] Dibenzoyl peroxide.

[e] Midwest Research Institute reports a 29% yield of benzimidazolone-2 by heating o-phenylenediamine with carbonyl sulfide at 225°C for 11 hr.

[f] When this reaction was run at 5°C, the yield of 1,3-di-n-hexylurea was only 43%.

[g] Reprinted from F. Baiocchi, R. A. Franz, and L. Horwitz, J. Org. Chem. 21, 1547 (1956). Copyright 1956 by the American Chemical Society. Reprinted by permission of the copyright owner.

mole) of aniline in 150 ml of water containing 10 ml (0.088 mole) of 30% hydrogen peroxide. The flask is placed under reduced pressure and then pressurized to 5 psi with carbonyl sulfide. A snug-fitting rubber stopper and the clamped rubber tubing are sufficient to hold the pressure for 24 hr. The solid product precipitates and is isolated to yield 9.0 gm, m.p. 231°–232°C.

The coprecipitating sulfur is removed by suspending the product for 2 min in a boiling solution of 7.0 gm of sodium sulfide and 7.0 gm of sodium hydroxide in 50 ml of water. After cooling, filtering, washing with cold water, and drying, the product was isolated in 70% yield (7.5 gm), m.p. 235°–236°C.

4. MISCELLANEOUS METHODS

(1) Reaction of α-isocyanato ester with ammonia or primary amines to give hydantoin esters [81].

(2) Synthesis of carbon-14 labeled 3-(p-chlorophenyl)-1,1-dimethylurea [82].

(3) Hydrolysis of cyanamides to give alkylureas [83, 84] (Eq. 66).

$$NH_2\!-\!C\!\equiv\!N + H_3O^+ \longrightarrow NH_2\overset{\displaystyle O}{\overset{\displaystyle \|}{C}}\!-\!NH_2 \qquad (66)$$

(4) The reaction of isocyanates with amides to give acylureas [56].

(5) Direct acylation of urea [55, 85, 86].

(6) Preparation of cyanuric acid from urea [87].

(7) Addition of aryl isocyanates to benzaldehyde and benzophenone anils [88].

(8) Addition of aryl isocyanates to N-substituted amidines at elevated temperatures [89].

(9) Certain polyureas may be made directly from the diisocyanate in the presence of dimethyl sulfoxide and a carboxylic acid [90] (Eq. 67).

$$RNCO + R'COOH \longrightarrow R\!-\!NH\overset{\displaystyle O}{\overset{\displaystyle \|}{C}}O\overset{\displaystyle O}{\overset{\displaystyle \|}{C}}R' \xrightarrow{\ DMSO\ }$$
$$RNH_2 + R'CO_2CH_2SCH_3 + CO_2 \qquad (67)$$
$$\xrightarrow{\ RNCO\ } (R\!-\!NH)_2CO$$

(10) Perfluoroalkyl isocyanates react with stoichiometric quantities of primary amines at low temperatures to yield perfluoroalkylureas, and with

secondary amines to yield perfluoroacyl amidines [91] (Eq. 68).

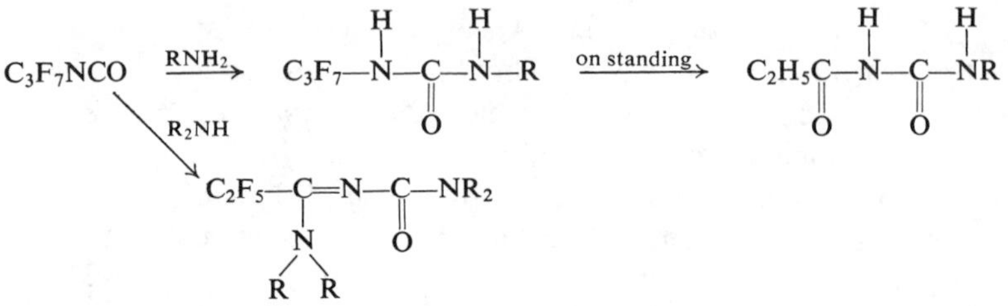

$$C_3F_7NCO \xrightarrow{RNH_2} C_3F_7-\overset{\overset{H}{|}}{N}-\underset{\underset{O}{\|}}{C}-\overset{\overset{H}{|}}{N}-R \xrightarrow{\text{on standing}} C_2H_5\underset{\underset{O}{\|}}{C}-\overset{\overset{H}{|}}{N}-\underset{\underset{O}{\|}}{C}-NR \quad (68)$$

$$\searrow R_2NH$$

$$C_2F_5-\underset{\underset{\underset{R \quad R}{/ \, \backslash}}{N}}{C}=N-\underset{\underset{O}{\|}}{C}-NR_2$$

(11) Reaction of urea with alky halides [92].

(12) Reaction of urea with halogens [93].

(13) Reaction of urea with nitrous acid [94].

(14) Preparation of N-ethylene-N'-phenylthiourea from phenyl isothiocyanate and ethylenimine [95].

(15) Preparation of sym-diphenylureas by the Curtius rearrangement of m- and p-substituted benzazides in toluene [96].

(16) Reaction of phenyl isocyanates with phenylsulfonamides, and the alkaline decomposition to the diphenylureas [97].

(17) Preparation of diarylureas by the reaction of amines with potassium O-ethyl monothiocarbonate and dicarbethoxydisulfide [98].

(18) Carbonylation of amines to form ureas [99].

(19) Addition of amine–boranes to isocyanates to form urea–borene complexes [100].

(20) Reaction of amines with silicon tetraisocyanate and silicon tetraisothiocyanate to give ureas and thioureas, respectively [101a, b].

(21) Direct fluorination of urea [102].

(22) Preparation of selenoureas [103].

(23) Preparation of 1-butyl-3-p-tolylsulfonylurea by the reaction of butylcarbamoyl azide with sodium p-toluene sulfinate [104].

(24) Synthesis of C-3 ureido steroids [105].

(25) Preparation of cyclic urea derivatives by the intramolecular condensation of 1,1,3,3-tetrakis(2-chloroethyl)urea [106].

(26) Reaction of 1-alkyl-3-chloroacetylurea with dialkylamines to give 1-alkyl-3-(dialkylglycyl)ureas [107].

(27) Reduction of the product of 2-butanone (methyl ethyl ketone) with urea to give sec-butylurea [108].

(28) Reaction of dibutylamine with carbon tetrachloride in the presence of hydrogen sulfide under pressure (autoclave) to give dibutylthiourea in almost quantitative yield [109].

(29) Reaction of urea with ethylene oxide in the presence of DMF to give a polyol [110].

(30) Reduction of N-methylolureas to N-methylureas [111].
(31) N,N′-Disubstituted thioureas from formaldimines, sulfur, and primary amines [112].
(32) Preparation of (1H-tetrazol-5-yl)ureas [113].

REFERENCES

Note that Frdl. is an abbreviation for the following source where the patent reference can be found: Friedländers Fortschritte der Teerfarbenfabrikation, Berlin.

1. D. M. Spatz and B. Cross, U.S. Pat. 4,490,165 (1984).
1a. F. Kurzer, *Org. Syn. Coll. Vol.* **4**, 52 (1963).
1b. I. Sakano, T. Yokoyama, S. Kajiya, T. Minami, Y. Okazaki, H. Tokuda, H. Kawazura, M. Kumajura, T. Nakano, and A. Awaya, U.S. Pat. 4,490,393 (1984).
2. J. D. Riedel, German Pat. 76,596 [*Frdl.* **4**, 1268 (1894–1897)].
3. N. I. Volynkin, *Zh. Obshch. Khim.* **27**, 483 (1957); *Chem. Abstr.* **51**, 15437 (1957).
4. C. J. Berlinerblau, German Pat. 63,485 [*Frdl.* **3**, 906 (1890–1894)].
5. J. D. Riedel, German Pat. 73,083 [*Frdl.* **3**, 907 (1890–1894)].
6a. J. D. Riedel, German Pat. 77,310 [*Frdl.* **4**, 1271 (1894–1897)].
6b. K. Fincleisen, R. Freimuth, and K. Wagner, U.S. Pat. 4,310,692 (Jan. 12, 1982).
6c. P. E. Throckmorton, S. Freyard, and D. Grote, U.S. Pat. 3,937,727 (Feb. 10, 1976).
7. E. I. du Pont de Nemours, Brit. Pat. 530,267 (Dec. 9, 1940).
8. T. L. Davis and H. W. Underwood, Jr., *J. Amer. Chem. Soc.* **44**, 2595 (1922).
9. T. L. Davis and K. C. Blanchard, *J. Amer. Chem. Soc.* **45**, 1817 (1923).
10a. F. Kurzer, *Org. Syn.* **31**, 11 (1951).
10b. J. Goering, M. Tauscher, and E. A. Khon, Ger. Offen. 2,411,009 (Sept. 11, 1975); *Chem. Abstr.* **84**, 16795h (1976).
11a. T. L. Davis and K. C. Blanchard, *Org. Syn. Coll. Vol.* **1**, 453 (1941).
11b. T. Sano, Japan Pat. 71 38,774 (Nov. 15, 1971); *Chem. Abstr.* **76**, 13870c (1972).
11c. O. Woerz, U. Block, R. Fischer, and G. W. Rotermand, Ger. Offen. 2,844,962 (April 30, 1980); *Chem. Abstr.* **93**, 185800x (1980).
11d. E. E. Gilbert and G. J. Sorma, U.S. Pat. 2,729,677 (Jan. 3, 1956).
11e. W. W. Thompson, U.S. Pat. 2,673,877 (Mar. 30, 1954).
12. J. D. Riedel, German Pat. 76,596 [*Frdl.* **4**, 1268 (1894–1897)].
13. T. L. Davis and K. C. Blanchard, *J. Amer. Chem. Soc.* **51**, 1790 (1929).
14. E. Biilmann and A. Klit, *Chem. Ber.* **63**, 2205 (1930).
15. R. W. Charlton and A. R. Day, *J. Org. Chem.* **1**, 552 (1937).
16. P. P. T. Sah and C. H. Kav, *Rec. Trav. Chim.* **58**, 460 (1939).
17. J. S. Buck and C. W. Ferry, *J. Amer. Chem. Soc.* **58**, 854 (1936).
18. J. S. Buck, W. S. Ide, and R. Baltzly, *J. Amer. Chem. Soc.* **64**, 2233 (1942).
19. F. Kurzer, *Org. Syn. Coll. Vol.* **4**, 49 (1963).
20. C. J. Berlinerblau, *J. Prakt. Chem.* [2], **30**, 103 (1844).
21. A. Sonn, German Pat. 399,889 (1924) [*Chem. Zentralbl.* II, **95**, 1513 (1924)].
22. A. S. Wheeler, *J. Amer. Chem. Soc.* **51**, 3653 (1929).
23a. J. R. Scott and J. B. Cohen, *J. Chem. Soc.* **121**, 2034 (1922).
23b. A. E. Dixon, *J. Chem. Soc.* **67**, 556 (1895).
23c. N. A. Ivanov, R. V. Viasova, V. A. Gancharova, and L. N. Smirnov, *Izv. Vyssh. Uchebn. Zaved. Khim. Khim. Tekhnol.* **19**(7), 1010 (1976); *Chem. Abstr.* **85**, 177037y (1976).
23d. G. C. Helsley and W. J. Welstead, Jr., U.S. Pat. 3,424,760 (Jan. 28, 1969).

24. J. H. Saunders and R. Slocombe, *Chem. Rev.* **43**, 203 (1948); J. H. Saunders and K. C. Frisch, "Polyurethanes—Chemistry and Technology." Wiley (Interscience), New York, 1962.
25. P. P. T. Sah and H. C. Chang, *Chem. Ber.* **69**, 2762 (1936).
26. O. Stoutland, L. Helgen, and C. L. Agre, *J. Org. Chem.* **24**, 818 (1959).
27. N. Shachat and J. J. Bagnell, Jr., *J. Org. Chem.* **28**, 991 (1963).
28. T. L. Davis and F. Ebersole, *J. Amer. Chem. Soc.* **56**, 885 (1934).
29. H. E. French and A. F. Wirtel, *J. Amer. Chem. Soc.* **48**, 1736 (1926).
30a. E. I. du Pont de Nemours, Brit. Pat. 535,139 (March 31, 1941).
30b. K. Thiele and G. Steinmetz, U.S. Pat. 3,404,152 (Oct. 1, 1968).
31. M. L. Moore and F. S. Crossley, *Org. Syn.* **3**, 599 (1955).
32. F. B. Davis, R. Q. Brewster, I. L. Malm, A. W. Miller, R. V. Maneval, and J. A. Sultzaberger, *J. Amer. Chem. Soc.* **47**, 1981 (1925).
33a. A. F. Ferris and B. A. Schutz., *J. Org. Chem.* **28**, 3140 (1963).
33b. R. F. Pratt and T. C. Bruice, *J. Amer. Chem. Soc.* **94**, 2823 (1972).
33c. H. A. Bruson and J. W. Eastes, *J. Amer. Chem. Soc.* **59**, 2011 (1937).
33d. D. W. Emerson, *J. Chem. Educ.* **48**, 81 (1971).
33e. P. A. S. Smith and D. W. Emerson, *J. Amer. Chem. Soc.* **82**, 3076 (1964).
33f. P. Adams, U.S. Pat. 3,161,676 (Dec. 15, 1964).
33g. P. J. Worth, Jr., and A. L. DiNapoli, U.S. Pat. 3,637,788 (Jan. 25, 1972).
34. E. I. du Pont de Nemours, Brit. Pat. 528,437 (Oct. 29, 1940).
35a. J. D. Riedel, German Pat. 77,420 [*Frdl.* **4**, 1269 (1894–1897)].
35b. A. W. Levine and J. Fech, Jr., *J. Org. Chem.* **37**, 1500 (1972); *ibid.* **37**, 2455 (1972).
36. J. A. Parker, J. J. Thomas, and C. L. Zeisse, *J. Org. Chem.* **22**, 594 (1957).
37a. N. H. Gold, M. B. Frankel, G. B. Linden, and K. Klager, *J. Org. Chem.* **27**, 334 (1962).
37b. W. Gerhardt, *J. Prakt. Chem.* **38**(1–2), 77 (1968); *Chem. Abstr.* **69**, 35357y (1968).
38. I. G. Farbenindustrie A. G. Brit. Pat. 462,182 (1937); R. J. Slocombe, E. E. Hardy, and J. H. Saunders, U.S. Pat. 2,480,089 (Aug. 23, 1949).
39. H. Shingu, T. Nishimura, and T. Takegami, *Yuki Gosei Kagaku Kyokai Shi* **15**, 140 (1957).
40. N. A. Puschin and R. V. Mitic, *Ann. Chem.* **532**, 300 (1957).
41. J. D. Riedel, German Pat. 79,718 [*Frdl.* **4**, 1270 (1894–1897)].
42a. N. E. Loginov and T. V. Polyanskii, USSR Pat. 65,779 (1946); *Chem. Abstr.* **40**, 7234 (1946).
42b. D. J. Crain, I. B. Dicker, M. Lauer, C. B. Knobler, and K. N. Trueblood, *J. Amer. Chem. Soc.* **106**, 7150 (1984).
43. T. W. Price, *J. Chem. Soc.* 3230 (1926).
44. A. Wahl, *Bull. Soc. Chem.* [5], **1**, 244 (1934).
45. E. N. Abrahart, *J. Chem. Soc.* 1273 (1936).
46a. S. Ozaki and T. Nagoya, *Bull. Chem. Soc. Jap.* **30**, 444 (1957).
46b. W. E. Coyne and J. W. Cusic, *J. Med. Chem.* **10**, 541 (1967).
46c. H. Bahad, U.S. Pat. 3,681,457 (Aug. 1, 1972).
47. R. P. Lastovskii, *J. Appl. Chem. (USSR)* **19**, 440 (1946).
48. E. L. Wittbecker, U.S. Pat. 3,816,879 (1957).
49a. G. D. Buckley and N. H. Ray, U.S. Pat. 2,550,767 (1951).
49b. L. H. Cook, U.S. Pat. 3,301,897 (Jan. 31, 1967).
49c. D. Luetzow, N. Neth, U. Wagner, and K. Volkaner, U.S. Pat. 4,178,309 (Dec. 11, 1979).
49d. BASF A.-G., Japan Kokai Tokkyo Koho 79/39,116 (Mar. 6, 1979); *Chem. Abstr.* **90**, 186418a (1979).
49e. W. Michelitsch and F. Hofmann, German Pat. 1,768,256 (Aug. 9, 1973); *Chem. Abstr.* **79**, 146024 (1973).

49f. H. Ogura, K. Takeda, R. Tokue, and K. Takanori, *Synthesis* (5), 394 (1978); *Chem. Abstr.* **89**, 42726j (1978).

49g. S. Morino, M. Sakai, I. Kashiki, A. Suzuki, and M. Miki, *Hokkaido Daigaku Suisanga Kubu Kenkyu Iho* **29**(1), 75 (1978); *Chem. Abstr.* **90**, 54439b (1979).

50a. G. J. M. Van der Kirk, H. G. J. Overmars, and G. M. Van der Want, *Rec. Trav. Chem.* **74**, 1301 (1955).

50b. K. Buechler, F. Kaess, H. Kronacher, K. Lienhard, and L. Strassberger. German Pat. 1,232,572 (Jan. 19, 1967); *Chem. Abstr.* **66**, 65136f (1967).

50c. J. Strickrodt and G. Hoffmann, German Pat. 1,468,398 (Jan. 18, 1970); *Chem. Abstr.* **73**, 76697k (1970).

51. R. A. Franz and F. Applegath, *J. Org. Chem.* **26**, 3304 (1961).

52. F. Applegath, M. D. Barnes, and R. A. Franz, U.S. Pat. 2,857,430 (1958) to Monsanto Chem. Co.

53a. R. A. Franz, F. Applegath, F. V. Morris, F. Baiocchi, and L. W. Breed, *J. Org. Chem.* **27**, 4341 (1962).

53b. N. Somoda and S. Tsuksumi, Jap. Pat. 71/11,492 (Mar. 24, 1971); *Chem. Abstr.* **75**, 88153j (1971).

53c. S. Matsuoka and Y. Kawaoka, Jap. Kokai 74/36,627 (Apr. 5, 1974); *Chem. Abstr.* **81**, 77505y (1974).

53d. V. G. Degtyar, A. A. Zalivina, and N. M. Rybal'chenko, USSR Pat. 186,440 (Aug. 9, 1963); *Chem. Abstr.* **66**, 115,337v (1967).

53e. W. W. Levis, Jr., and E. A. Waipert, U.S. Pat. 3,168,560 (Feb. 2, 1965).

53f. J. Strickrodt and G. Hoffmann, Brit. Pat. 1,173,521 (Dec. 10, 1969); *Chem. Abstr.* **72**, 54821y (1970).

53g. H. Kersten, G. Heinrichs, G. Meyer, and D. Laudien, Ger. Offen. 2,015,010 (Nov. 25, 1971); *Chem. Abstr.* **76**, 45778v (1972).

53h. J. L. Eaton, U.S. Pat. 3,288,684 (Nov. 29, 1966).

53j. N. E. Searle, U.S. Pat. 2,704,245 (Mar. 15, 1955).

53k. M. L. Weakley, U.S. Pat. 3,597,477 and U.S. Pat. 3,597,478 (1971).

53l. G. Hamprecht, K. Fischer, and O. Woerz, Ger. Offen. 2,756,409 (June 21, 1979); *Chem. Abstr.* **91**, 74220x (1979).

53m. S. Wawzonek, *Org. Prep. Proced. Int.* **8**(4), 197 (1976); *Chem. Abstr.* **85**, 123313s (1976).

54. M. T. Beachem, J. C. Oppelt, F. M. Cowen, P. D. Schickedantz, and D. V. Maier, *J. Org. Chem.* **28**, 1876 (1963).

55a. R. W. Stoughton, H. L. Dickison, and O. G. Fitzhugh, *J. Amer. Chem. Soc.* **61**, 408 (1939).

55b. R. Richter, B. Tucker, and H. Ulrich, *J. Org. Chem.* **43**, 4150 (1978).

56. P. F. Wiley, *J. Amer. Chem. Soc.* **71**, 1310 (1949).

57. T. Inoi, T. Okamoto, and Y. Koizumi, *J. Org. Chem.* **31**, 2700 (1966).

58. B. B. Corson and E. Freeborn, *Org. Syn.* **2**, 231 (1943).

59. Y. Ogata, A. Kawasaki, and N. Okumura, *J. Org. Chem.* **30**, 1636 (1965).

60a. A. L. Wilson, U.S. Pat. 2,517,750 (1950).

60b. M. D. Hurwitz and R. W. Auten, U.S. Pat. 2,613,212 (1952).

61. H. Pauly and H. Sauter, *Chem. Ber.* **63B**, 2063 (1930).

62. J. B. Dickey and A. R. Gray, *Org. Syn. Coll. Vol. 2*, 60 (1943).

63. J. D. Donleavy and M. A. Kise, *Org. Syn. Coll. Vol. 2*, 422 (1943).

64. D. R. Bryant, S. D. Work, and C. R. Hauser, *J. Org. Chem.* **29**, 235 (1964).

65. Author's Laboratory (S.R.S.).

66. R. E. Kirk and D. F. Othmer, *Encycl. Chem. Techn.* **14**, 468 (1955).

67a. Iliceto, *Ann. Chim.* (*Rome*) **43**, 625 (1953).

67b. B. Meyer, "Urea-Formaldehyde Resins," Addison-Wesley, Reading, Mass., 1979.

68. F. B. Slezak, H. Bluestone, T. A. Magee, and J. H. Wotiz, *J. Org. Chem.* **27**, 2181 (1962).

69. R. L. Wayland, Jr., U.S. Pat. 2,825,732 (1958).
70. E. Fischer and H. Koch, *Annu. Chem.* **232**, 227 (1886).
71. N. A. Paschinand and R. V. Mitic, *Annu.* **532**, 300 (1937).
72. C. F. H. Allen, C. O. Edens, and J. Van Allan, *Org. Syn. Coll. Vol.* **3**, 394 (1955), and earlier references cited.
73. J. Van Allan, *Org. Syn. Coll. Vol.* **3**, 395 (1955).
74. A. Michael, *J. Prakt. Chem.* (2), **35**, 456 (1887).
75. E. Grimaux, *C. R. Acad. Sci. Paris* **87**, 757 (1878).
76a. H. Biltz and H. Wittek, *Chem. Ber.* **54**, 1035 (1921).
76b. E. Baeuerlein and R. Kiehl, *Justus Liebigs Ann. Chem.* (4), 675 (1978); *Chem. Abstr.* **89**, 109356j (1978).
76c. D. E. Morris and F. R. Smith, U.S. Pat. 3,847,981 (Nov. 12, 1974).
76d. H. C. Brown, U.S. Pat. 2,548,585 (Apr. 10, 1951).
76e. T. F. Rutledge, U.S. Pat. 3,280,186 (Oct. 18, 1966).
76f. J. C. Ambalang and G. M. Massie, U.S. Pat. 2,849,488 (Aug. 26, 1958).
76g. L. I. Smith and O. H. Emersin, *Org. Syn. Coll. Vol.* **3**, 151 (1955).
76h. K. G. Tyshchuk and I. P. Tsukervanik, Jr., *Tashkent Politekh Inst.* 17 (1968); *Chem. Abstr.* **73**, 109456m (1970).
77. J. D. Riedel, German Pat. 313,965 [*Frdl.* **13**, 1049 (1916–1921)].
78. Y. Wolman and P. M. Gallop, *J. Org. Chem.* **27**, 1902 (1962).
79. J. E. Anderson and C. E. Parish, Belg. Pat. 663,137 (1965).
80. Z. Baiocchi, R. A. Franz, and L. Horwitz, *J. Org. Chem.* **21**, 1546 (1956).
81. A. C. Smith, Jr., and C. C. Unruh, *J. Org. Chem.* **22**, 442 (1957).
82. N. E. Searle and H. E. Cupery, *J. Org. Chem.* **19**, 1622 (1954).
83. F. H. S. Curd, D. G. Davey, and D. N. Richardson, *J. Chem. Soc.* 1732 (1949).
84. F. H. S. Curd, D. G. Davey, D. N. Richardson, and R. B. Ashworth, *J. Chem. Soc.* 1745 (1949).
85. Austrian Pat. 186,643 (1956) to Donen-Pharamazie G.m.b.H.
86. R. W. Stoughton, *J. Org. Chem.* **2**, 514 (1938).
87. D. Bundrit, Brit. Pat. 570,715 (1945).
88. R. Richter, *Chem. Ber.* **102**, 938 (1969).
89. R. Richter and W. P. Trautwein, *Chem. Ber.* **102**, 931 (1969).
90. W. R. Sorenson, *J. Org. Chem.* **24**, 978 (1959).
91. R. L. Dannley, D. Yamashiro, and R. G. Taborsky, *J. Org. Chem.* **24**, 1706 (1959).
92. H. Finkbeiner, *J. Org. Chem.* **30**, 2861 (1965).
93. H. B. Conahoe and C. A. Van der Werf, *Org. Syn. Coll. Vol.* **4**, 157 (1963).
94. F. Arndt, *Org. Syn. Coll. Vol.* **2**, 461 (1943).
95. Y. Iwakura and A. Nabeya, *Nippon Kagaku Zasshi* **77**, 773 (1956).
96. Y. Yukawa and Y. Tsuno, *J. Amer. Chem. Soc.* **79**, 5530 (1957).
97. O. Baeyer and E. Caner, French Pat. 993,465 (1951).
98. C. N. V. Namburg and K. Kapoor, *Vikram* **3**, 11 (1959).
99. D. M. Fenton, U.S. Pat. 3,277,061 (1966).
100. R. H. Cragg and N. N. Greenwood, *J. Chem. Soc. A* 961 (1967).
101a. R. G. Neville and J. J. McGee, *Org. Syn.* **45**, 69 (1965).
101b. R. G. Neville, *J. Org. Chem.* **23**, 937 (1958).
102. V. Grakauskas and K. Baum, *J. Amer. Chem. Soc.* **92**, 2096 (1970).
103. D. L. Klayman and R. J. Shine, *J. Org. Chem.* **34**, 3549 (1969).
104. V. J. Bauer, W. J. Fanshawe, and S. R. Safir, *J. Org. Chem.* **31**, 3440 (1966).
105. A. Yagi, J. Liang, and D. K. Fukushima, *J. Org. Chem.* **32**, 713 (1967).
106. J. A. Settepani and G. R. Pettit, *J. Org. Chem.* **35**, 843 (1970).

107. P. Truitt and J. T. Witkowski, *J. Med. Chem.* **13**, 574 (1970).
108. J. K. Simons, U.S. Pat. 2,673,859 (Mar. 30, 1954).
109. R. T. Wragg and C. E. Kendall, U.S. Pat. 3,299,130 (Jan. 17, 1967).
110. T. Uno, G. Katagiri, and K. Shimomura, Jap. Pat. 70/11, 130 (Apr. 22, 1970); *Chem. Abstr.* **73**, 76699n (1970).
111. J. Auerbach, M. Zamore, and S. M. Weinreb, *J. Org. Chem.* **41**, 725 (1976).
112. Badische Anilin- und Soda-Fabrik A. G., Fr. Pat. 1,585,353 (Jan. 16, 1970); *Chem. Abstr.* **74**, 41958q (1971).
113. G. H. Denny, E. J. Cragoe, Jr., C. S. Rooney, J. P. Springer, J. M. Hirshfield, and J. A. McCauley, *J. Org. Chem.* **45**, 1662 (1980).

CHAPTER 7 / **PSEUDOUREAS**

1. INTRODUCTION

In 1894 the first oxygen ether of a urea, a pseudourea, was quantitatively prepared by the reaction of ethanol with diphenylcarbodiimide at 180°C in a sealed tube [1] (Eq. 1). Shortly thereafter it was shown that sodium ethylate

$$C_6H_5-N{=}C{=}N-C_6H_5 + C_2H_5OH \longrightarrow \underset{\underset{OC_2H_5}{|}}{C_6H_5-N{=}\overset{\overset{H}{|}}{C}-N-C_6H_5} \quad (1)$$

reacts with the hydrogen chloride–diphenylcarbodiimide addition product at lower temperatures, without the use of sealed tubes, to give the same pseudourea [2] (Eq. 2). It was also shown that the pseudoureas mentioned can be

$$\underset{\underset{Cl}{|}}{C_6H_5-N{=}\overset{\overset{H}{|}}{C}-N-C_6H_5} + NaOC_2H_5 \longrightarrow \underset{\underset{OC_2H_5}{|}}{C_6H_5-N{=}\overset{\overset{H}{|}}{C}-N-C_6H_5} \quad (2)$$

made readily at lower temperatures by the sodium ethoxide-catalyzed reaction of diphenylcarbodiimide with ethanol [3, 4].

In 1874 Claus [5] reported that thiourea reacts with ethyl iodide or bromide to give compounds whose structures were first suggested to be pseudothioureas by Bernthsen and Klinger [6] (Eq. 3). In place of thiourea, symmetrical

$$CH_3CH_2I + NH_2\overset{\overset{S}{\|}}{C}NH_2 \longrightarrow \left[\underset{\underset{SCH_2CH_3}{|}}{H_2\overset{+}{N}{=}C-NH_2}\right] I^- \quad (3)$$

disubstituted or monosubstituted thioureas can be effectively used in the same reaction [7].

Although thiourea readily reacts with alkyl halides or a mixture of hydrogen halides and alcohols to give pseudothiourea salts, urea fails to react under similar conditions [8, 9] (Eq. 4).

$$ROH + H_2N-\overset{\overset{S}{\|}}{C}-NH_2 + HX \longrightarrow (H_2\overset{+}{N}{=}\underset{\underset{SR}{|}}{C}-NH_2)X^- \qquad (4)$$

$$RX + H_2N\overset{\overset{S}{\|}}{C}-NH_2$$

Werner [10] later found that urea reacts with methyl sulfate to give 2-methyl-pseudourea, but the isolation of the product from the syrup is difficult and is usually done via the picrate. The reaction is usually carried out at about 40°C (7 hr) because at temperatures above 100°C the reaction is too violent [11] (Eq. 5).

$$H_2N-\overset{\overset{O}{\|}}{C}-NH_2 + (CH_3O)_2SO_2 \longrightarrow HN{=}\underset{\underset{OCH_3}{|}}{C}-NH_2 + CH_3HSO_4 \qquad (5)$$

Janus [11] has recently shown that urea can also be smoothly methylated with methyl toluene-p-sulfonate at 80°–90°C to give good yields of crystalline 2-methylpseudourea toluene-p-sulfonate.

Stieglitz and McKee [12] in an extension of their earlier work reported that 2-methylpseudourea can also be prepared by the addition of methanol to cyanamide in the presence of hydrogen chloride (Eq. 6).

$$NH_2CN + CH_3OH + HCl \longrightarrow NH_2{-}\underset{\underset{OCH_3}{|}}{C}{=}NH \qquad (6)$$

Recently cyanates have been also shown to react with amines to give pseudoureas [13].

The best methods for preparing pseudoureas are based on these earlier methods; they are summarized in Eqs. (7)–(13).

$$ROH + R'N{=}C{=}NR'' \xrightarrow[[14-16]]{} R'N{=}\underset{\underset{OR}{|}}{C}-NHR'' \qquad (7)$$

$$ROH + NH_2CN + HCl \xrightarrow[[12, 17-22]]{} \overset{-}{Cl}H_2\overset{+}{N}{=}\underset{\underset{OR}{|}}{C}-NH_2 \qquad (8)$$

$$ROH + AgNHCN + HCl \xrightarrow[[23a, b]]{} \overset{-}{Cl}H_2\overset{+}{N}{=}\underset{\underset{OR}{|}}{C}-NH_2 \qquad (9)$$

$$R'OH + N{\equiv}C{-}N{\overset{R''}{\underset{R'''}{\big\langle}}} \xrightarrow[[12, 17-22]]{R'ONa} HN{=}\underset{\underset{OR'}{|}}{C}{-}N{\overset{R''}{\underset{R'''}{\big\backslash}}} \qquad (10)$$

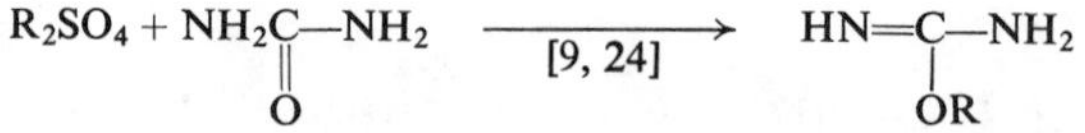

$$R_2SO_4 + NH_2C-NH_2 \xrightarrow[{[9, 24]}]{} HN=C-NH_2 \qquad (11)$$

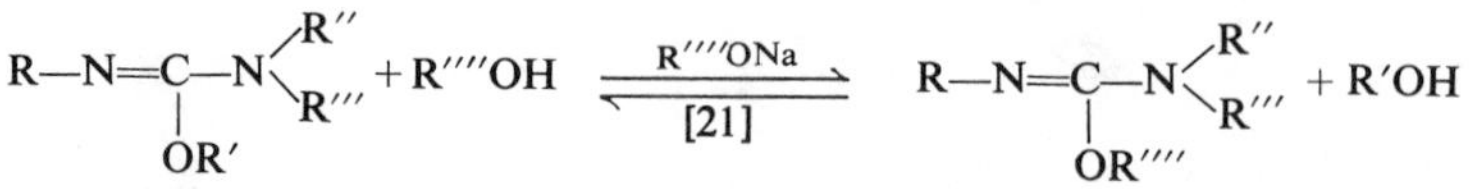

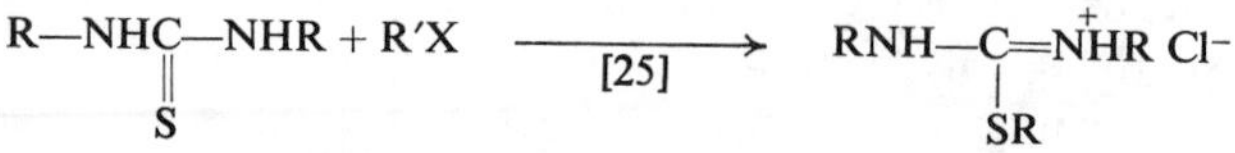

with O below the first structure and OR below the product.

$$R-N=C-N\begin{smallmatrix}R''\\R'''\end{smallmatrix} + R''''OH \underset{[21]}{\overset{R''''ONa}{\rightleftharpoons}} R-N=C-N\begin{smallmatrix}R''\\R'''\end{smallmatrix} + R'OH \quad (12)$$

with OR' and OR'''' below respectively.

$$R-NHC-NHR + R'X \xrightarrow[{[25]}]{} RNH-C=\overset{+}{N}HR \ Cl^- \qquad (13)$$

with S below the reactant and SR below the product.

Pseudoureas undergo some of the additional reactions [14, 14a, 16, 26] shown in Scheme 1.

Some of the commercial uses that pseudoureas have found are: herbicides [27, 28]; inhibitors for discoloration of color photographic layers [29]; monomers for the preparation of some novel copolymers [30, 31] and dyes [32, 33]. A survey of the recent decennial indexes of *Chemical Abstracts* should be made to see the large number of pseudoureas reported for varied uses. Pseudothioureas have also been found useful as radiation protective agents [34].

SCHEME 1

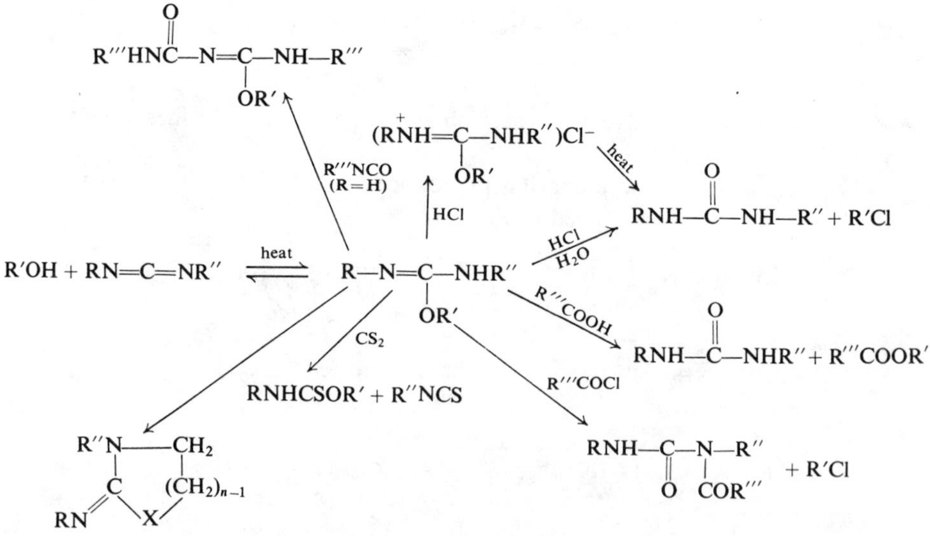

If $R' = (CH_2)_n-Cl$, then $X = O$;

if $R' = (CH_2)_n-NHR$, then $X = NR$.

To date no general review of pseudoureas or pseudothioureas has appeared, but there is a recent review [14] which is limited to O,N,N'-trisubstituted pseudoureas. Another review on the use of pseudoureas for esterification and alkylation reactions is worth consulting [14a].

NOMENCLATURE

Chemical Abstracts uses the name pseudourea rather than isourea, which was widely used in the older literature. The numbering of substituents is shown in (I). An analogous system is used with pseudothioureas.

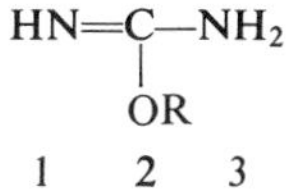

$$HN\!\!=\!\!\underset{\underset{\underset{1\quad\ \ 2\quad\ \ 3}{OR}}{|}}{C}\!\!-\!\!NH_2$$

(I)

2. CONDENSATION REACTIONS

The most important methods of preparing pseudoureas involve: (a) the condensation of alcohols with carbodiimides or cyanamides in the presence of inorganic bases or salts; (b) the condensation of methyl sulfates or sulfonates with urea; and (c) the condensation of alkyl halides with thiourea to give 2-alkylpseudothiourea. The pseudourea and pseudothioureas may react with a variety of compounds that react with either amino or active hydrogen groups to give substituted pseudoureas or pseudothioureas. In some cases heterocyclic compounds are formed.

A. Condensation of Carbodiimides with Alcohols

The reaction of alcohols with diarylcarbodiimides [16, 35, 36] is catalyzed by sodium alkoxides but not by triethylamine. Recently Schmidt and Moosmüller [37, 38] found that cuprous chloride (0.075 mole/100 moles dialkylcarbodiimide) catalyzes the reaction of primary, secondary, and tertiary alcohols with aliphatic carbodiimides to give *O,N,N'*-trialkylpseudoureas. Several examples of the use of these methods are described in Tables I and II.

In the condensation of diols, halogenated alcohols, amino alcohols, cyclic hydroxy ethers, or other bifunctional hydroxy compounds with carbodiimides, 5-, 6-, and 7-membered 1,3-*O-N*- or 1,3-*N,N*-heterocyclics are obtained [14].

Allyl alcohols such as linalool have also been reported to react with carbodiimides such as diisopropylcarbodiimide [14b].

2-1. Preparation of 2-Methyl-1,3-diisopropylpseudourea [37]

$$\underset{\underset{CH_3}{|}}{\overset{\overset{CH_3}{|}}{HC}}-N=C=N-\underset{\underset{CH_3}{|}}{\overset{\overset{CH_3}{|}}{CH}} \quad + CH_3OH \xrightarrow{\text{CuCl}} \underset{\underset{CH_3}{|}}{\overset{\overset{CH_3}{|}}{HC}}-NH-\underset{\underset{OCH_3}{|}}{C}=N-\underset{\underset{CH_3}{|}}{\overset{\overset{CH_3}{|}}{CH}} \quad (14)$$

To a dry flask containing 31.6 gm (0.25 mole) of diisopropylcarbodiimide and 8.0 gm (0.25 mole) of methanol is added 0.05 gm of cuprous chloride. After a short time the reaction mixture gets hot and becomes alkaline. After stirring for 4 hr, the product is distilled under reduced pressure to give 38.7 gm (97.8%), b.p. 54°C (10 mm) (bath temp. = 80°C), n_D^{20} 1.4379, d_4^{20} 0.88045.

This pseudourea reacts with oxalic acid and mercuric chloride to give addition compounds useful for identification purposes.

2-2. Preparation of 2-n-Butyl-1,3-di-p-tolylpseudourea [35]

$$CH_3-\!\!\!\!\bigcirc\!\!\!\!-N=C=N-\!\!\!\!\bigcirc\!\!\!\!-CH_3 + CH_3CH_2CH_2CH_2OH \xrightarrow{CH_3CH_2CH_2CH_2ONa}$$

$$CH_3-\!\!\!\!\bigcirc\!\!\!\!-NH-\underset{\underset{OCH_2CH_2CH_2CH_3}{|}}{C}=N-\!\!\!\!\bigcirc\!\!\!\!-CH_3 \quad (15)$$

To a flask equipped with a stirrer and a condenser with a drying tube is added 25 ml (0.274 mole) of n-butanol followed by 1.0 gm (0.0435 mole) of sodium metal. After the sodium has reacted, 8.9 gm (0.04 mole) of di-p-tolylcarbodiimide in 15 ml of ether is added rapidly. The mixture becomes warm and after 15 min it is diluted with ether, washed three times with water, dried, and the ether removed under reduced pressure. The residue is a viscous oil which crystallizes on prolonged suction to give 11.8 gm (100%).

The product is recrystallized with petroleum ether (b.p. 30°–60°C) by keeping the solution at −20°C to afford clusters of fine needles, m.p. 34°C.

B. Reaction of Cyanamides with Alcohols

Cyanamide and substituted cyanamides react with alcohols to give pseudoureas. Cyanamide reacts with alcohols in the presence of anhydrous hydrogen chloride, whereas disubstituted cyanamides require slightly more than an equivalent of sodium alkoxide for reaction with the alcohol. (See Tables III and IV.) Catalytic amounts of sodium alkoxides are sufficient to induce the reaction of a cyanamide with primary, secondary, or tertiary alcohols [21]

TABLE I

Preparation of Pseudoureas by the Reaction of Carbodiimides with an Alcohol Catalyzed by 0.1 Equivalent of Sodium Alkoxide[a]

$$R^1OH + R^2N{=}C{=}NR^3 \rightarrow R^2N{=}C{-}NR^3 \overset{\displaystyle H}{\underset{\displaystyle OR^1}{|}} \qquad (16)$$

R^1	R^2	R^3	Yield (%)	B.p., °C (mm)	n_D^{25} (or m.p., °C)	d_4^{25}
$n\text{-}C_4H_9$	$n\text{-}C_4H_9$	$n\text{-}C_4H_9$	65	58 (0.001)	1.4478	—
C_5H_{11} (cyclic)	$n\text{-}C_4H_9$	$n\text{-}C_4H_9$	73	106 (0.001)	1.4715	—
C_6H_5[b]	$n\text{-}C_4H_9$	$n\text{-}C_4H_9$	44	—	1.5068	0.9832
CH_3	C_6H_{11} (cyclic)	C_6H_{11} (cyclic)	89	85 (0.025)	1.4945	—
C_6H_{11} (cyclic)	C_6H_{11} (cyclic)	C_6H_{11} (cyclic)	60	—	(85)	—
$n\text{-}C_4H_9$	C_6H_5	C_6H_5	85	80 (0.005)	1.5088	0.9823
$n\text{-}C_4H_9{-}CH{-}CH_2$ $\vert$ C_2H_5	C_6H_5	C_6H_5	41	120 (0.001)	—	—

[a] Data taken from S. E. Forman, C. A. Erickson, and H. Adelman, *J. Org. Chem.* **28**, 2653 (1963).

[b] Sodium alkoxide was ineffective as a catalyst but saturated hydrogen chloride gave the product as shown.

TABLE II[h]
O,N,N'-TRIALKLYLPSEUDOUREAS (I) FROM MONOHYDRIC ALCOHOLS AND DIALKYLCARBODIIMIDES

ROH	R'[f]	R"[f]	Yield (%)	B.p., °C/mm Hg (m.p. °C)	n_D^{20}	d_4^{20}	Ref.
Methanol	i-Pr	i-Pr	98	54/10	1.4379	0.88045	a
	n-Bu	c-Hex	93	132–134/11	1.4772	0.9446	a
	c-Hex	c-Hex	94	162–163/11 (32–33)	1.4945[g]	—	a,b
	Me	t-Bu	90	51–53.5/11	1.4460	0.9053	c
Ethanol	Me	t-Bu	85	58–60/10	1.4438	0.8894	a
	i-Pr	i-Pr	94	64–65/10	1.4340	0.8649	a
2-Ethoxyethanol	i-Pr	i-Pr	88	103–105/10	1.4414	0.9071	a
2-Diethylaminoethanol	i-Pr	i-Pr	92	123/11	1.4509	0.8833	c
n-Butanol	i-Pr	i-Pr	81	87–89/10	1.4397	0.8582	b
	n-Bu	n-Bu	65	58/0.001	1.4478[g]	—	a
1,2-O-Isopropylideneglycerol	i-Pr	i-Pr	99	86–88/0.006	1.4516	0.9745	a
2-Nitro-2-methyl-1-propanol	i-Pr	i-Pr	73	(53)	—	—	a
n-Octanol	i-Pr	i-Pr	93	140–141/10	1.4458	0.8569	a
Furfuryl alcohol	i-Pr	i-Pr	88	65–67/0.002	1.4764	0.9887	d
Tetrahydrofurfuryl alcohol	i-Pr	i-Pr	97	72–74/0.003	1.4649	0.9674	d
Isopropyl alcohol	i-Pr	i-Pr	93	66–67/10	1.4298	0.8508	a
Cyclohexanol	i-Pr	i-Pr	92	115–117/10	1.4632	0.9144	a
	n-Bu	n-Bu	65	58/0.001	1.4478[g]	—	b

Menthol	i-Pr	i-Pr	88	75–77/0.002	1.4584	0.8395	d
Borneol	i-Pr	i-Pr	82	84–86/0.002	1.4726	0.9343	d
t-Butanol	i-Pr	i-Pr	64	67–68/9	1.4264	0.8377	a
t-Amyl alcohol	i-Pr	i-Pr	51	81.5–82.5/10	1.4331	0.8491	a
	Me	t-Bu	68	80–82/10	1.4431	0.8750	a
N-Methyl-N-(2-hydroxyethyl)aniline	i-Pr	i-Pr	97	123/0.01	1.5213	0.9830	e
N-Methyldiethanolamine	i-Pr	i-Pr	86	128.5–130/0.001	1.4663	0.9337	e
N-(2-Hydroxyethyl)aniline	i-Pr	i-Pr	99	122.5/0.001	1.5248	0.9948	e
2-Phenoxyethanol	i-Pr	i-Pr	97	97–98/0.005	1.5010	0.9960	e
Citronellol	i-Pr	i-Pr	97	102–103/0.01	1.4597	0.8703	e
Cholesterol	i-Pr	i-Pr	94	(83.5–84)	—	—	e
Allyl alcohol	i-Pr	i-Pr	98	75–76/10	1.4053	0.8829	e
Benzyl alcohol	i-Pr	i-Pr	98	92–93/0.01	1.5001	0.9620	e
p-Nitrobenzyl alcohol	i-Pr	i-Pr	71	(42)	—	—	e
Quinine	i-Pr	i-Pr	72	(110–111)	—	—	e

[a] E. Schmidt and F. Moosmüller, Ann. Chem. 597, 235 (1955).

[b] S. E. Forman, C. A. Erickson, and H. Adelman, J. Org. Chem. 28, 2653 (1963).

[c] W. Carl, Dissertation, Universität München, 1960.

[d] E. Schmidt and W. Carl, Ann. Chem. 639, 24 (1961).

[e] E. Däbritz, Dissertation, Universität München, 1963; E. Schmidt, E. Dübritz, and K. Thulke, Ann. Chem. 685, 161 (1965).

[f] The abbreviations are: i-Pr = isopropyl; c-Hex = cyclohexyl; n-Bu = n-butyl; Me = methyl.

[g] n_D^{25}.

[h] Reprinted from E. Däbritz, Angew. Chem. Int. Ed. Engl. 5, 470 (1966) with permission of the publishers.

(Eq. 17). When both R′ groups are secondary alkyl groups, alkaline catalysts cannot be used to prepare the pseudourea; however, the use of anhydrous

$$RO^- + N\!\equiv\!C\!-\!N\overset{\displaystyle R'}{\underset{\displaystyle R'}{\diagup}} \rightleftharpoons \;\; ^-N\!=\!C\!-\!N\overset{\displaystyle R'}{\underset{\displaystyle R'}{\diagup}} \tag{17}$$

hydrogen chloride gives the pseudouronium salt. When secondary and primary alkyl groups are present, alkaline catalysts are effective [25].

Monoalkylcyanamide reacts with alcohols only in the presence of anhydrous hydrogen chloride [21].

2-3. Preparation of 2-n-Butyl-3,3-di-n-butylpseudourea [21]

$$n\text{-}C_4H_9OH + N\!\equiv\!C\!-\!N\overset{\displaystyle n\text{-}C_4H_9}{\underset{\displaystyle n\text{-}C_4H_9}{\diagup}} \xrightarrow{n\text{-}C_4H_9ONa} HN\!=\!C\!-\!N\overset{\displaystyle n\text{-}C_4H_9}{\underset{\displaystyle n\text{-}C_4H_9}{\diagup}} \tag{18}$$

$$O\!-\!n\text{-}C_4H_9$$

To a flask containing 405 gm (5.47 moles) of n-butanol is added 25.3 gm (1.1 gm-atom) of sodium metal in small portions. After the reaction is complete, 154.0 (1.0 mole) of N,N-di-n-butylcyanamide is added with sufficient benzene to keep the refluxing pot temperature at approximately 100°C. The mixture is refluxed for 3 hr, glacial acetic acid is added to neutralize the base, the mixture is washed with water, the organic layer is separated, dried, and then distilled to afford 194.0 gm (85%), b.p. 117°C (2.0 mm), n_D^{25} 1.4500, d_4^{25} 0.8871.

This procedure is essentially the same for the other pseudoureas shown in Table III using either 1.1 or 0.1 equivalent of sodium alkoxide.

Other references to related procedures can be found in McKee [39] and Elderfield and Green [40].

1-Cyanopyrrolidine has been reported to react with allylic alcohols when catalyzed by potassium hydride in THF to form allylic pseudoureas. These pseudoureas can rearrange on heating to form allylic amines [14b]. Propargyl alcohols react with 1-cyanopyrrolidine at 0°–25°C to give acetylenic pseudoureas in crude yields of 85%. On heating the propargylic pseudoureas pseudoureas are converted to 2-pyridones [14c–e].

Donetti and co-workers have reported using this reaction for the initial conversion of disubstituted cyanamide to the pseudourea, which was then hydrolyzed easily to the secondary amine [14f, g].

TABLE III[a]

Preparation of Pseudoureas by the Reaction of Cyanamides with Alcohols[b] in the Presence of Sodium Alkoxides

$$R^1OH + N{\equiv}C{-}N{\diagup}^{R^2}_{\diagdown R^3} \xrightarrow{R^1ONa} H{-}N{=}\underset{OR^1}{\overset{}{C}}{-}N{\diagup}^{R^2}_{\diagdown R^3} \qquad (19)$$

R^1	R^2	R^3	Catalyst R^1ONa, equivalents	Yield (%)	B.p., °C (mm Hg)	n_D^{25} (or m.p., °C)	d_4^{25}
$n\text{-}C_4H_9$	$n\text{-}C_4H_9$	$n\text{-}C_4H_9$	1.1	85	117 (2)	1.4500	0.8871
CH_3	$C_6H_5CH_2$	$C_6H_5CH_2$	1.1	62	128 (0.01)	(51–53)	—
C_2H_5	$C_6H_5CH_2$	$C_6H_5CH_2$	1.1	57	133 (0.01)	1.5652	1.0823
C_2H_5	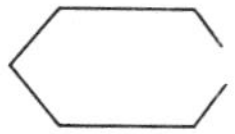 (cyclohexane ring spanning R^2–R^3)		1.1	66	109 (15)	1.4794	1.0015
$CH_2{=}CH{-}CH_2$	$CH_2{=}CHCH_2$	$CH_2{=}CH{-}CH_2$	1.1	58	94 (5)	1.4809	0.9555
$n\text{-}C_4H_9$	$CH_2{=}CH{-}CH_2$	$CH_2{=}CH{-}CH_2$	1.1	43	100 (5)	1.4796	0.9005
CH_3	$CH_2{=}\underset{CH_3}{C}{-}CH_2$	$CH_2{=}\underset{CH_3}{C}{-}CH_2$	1.1	73	71 (0.7)	1.4730	0.9414
C_2H_5	$CH_2{=}\underset{CH_3}{C}{-}CH_2$	$CH_2{=}\underset{CH_3}{C}{-}CH_2$	1.1	83	73 (0.5)	1.4665	—
$n\text{-}C_4H_9$	$CH_2{=}\underset{CH_3}{C}{-}CH_2$	$CH_2{=}\underset{CH_3}{C}{-}CH_2$	1.1	82	102 (2.2)	1.4684	0.9134
$C_2H_5{-}\underset{CH_3}{CH}$	$CH_2{=}\underset{CH_3}{C}{-}CH_2$	$CH_2{=}\underset{CH_3}{C}{-}CH_2$	1.1	74	59 (0.005)	1.4656	0.9056

TABLE III (continued)

R^1	R^2	R^3	Catalyst R^1ONa equivalents	Yield (%)	B.p., °C (mm Hg)	n_D^{25} (or m.p., °C)	d_4^{25}
$n\text{-}C_4H_9\text{-}CH(C_2H_5)\text{-}CH_2$	$CH_2=C(CH_3)\text{-}CH_2$	$CH_2=C(CH_3)\text{-}CH_2$	1.1	57	110 (0.2)	1.4698	0.9043
$CH_2=C(CH_3)\text{-}CH_2$	$CH_2=C(CH_3)\text{-}CH_2$	$CH_2=C(CH_3)\text{-}CH_2$	1.1	18	87 (1)	1.4731	0.9136
$n\text{-}C_4H_9\text{-}OCH_2CH_2$	$CH_2=C(CH_3)\text{-}CH_2$	$CH_2=C(CH_3)\text{-}CH_2$	1.1	68	128 (1.8)	1.4692	0.9409
$n\text{-}C_4H_9(OCH_2CH_2)_2$	$CH_2=C(CH_3)\text{-}CH_2$	$CH_2=C(CH_3)\text{-}CH_2$	1.1	72	140 (0.6)	1.4723	0.9626
$n\text{-}C_3H_7$	$n\text{-}C_3H_7$	$n\text{-}C_3H_7$	0.1	72	108 (10)	1.4450	0.8939
$CH_2=C(CH_3)\text{-}CH_2$	$n\text{-}C_3H_7$	$n\text{-}C_3H_7$	0.1	62	118 (12)	1.4569	0.9052
$n\text{-}C_3H_7$	$n\text{-}C_4H_9$	$n\text{-}C_4H_9$	0.1	72	79 (0.15)	1.4474	0.8825
$n\text{-}C_4H_9$	$n\text{-}C_4H_9$	$n\text{-}C_6H_5$	0.1	68	117 (0.15)	1.4520	0.8131
	$n\text{-}C_4H_9$	C_6H_5	0.1	76	102 (0.1)	1.5040	0.9758
	$(CH_3)_2\text{-}C(NO_2)\text{-}CH_2$	$n\text{-}C_4H_9$	0.1	70	83 (0.02)	1.4650	1.0126
CH_3	 (cyclic ring structure)		0.1	90	146 (0.001)	1.4710	—
C_6H_{11} (cyclic)	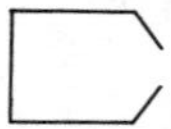 (cyclic ring structure)		0.1	89	77 (0.001)	1.4980	1.032

[a] Data taken from S. E. Forman, C. A. Erickson, and H. Adelman, *J. Org. Chem.* **28**, 2653 (1963).
[b] This reaction does not work for phenol, furfuryl alcohol, or t-butyl alcohol.

TABLE IV[a]

PREPARATION OF PSEUDOUREAS BY THE REACTION OF CYANAMIDES WITH ALCOHOLS[b] IN THE PRESENCE OF ANHYDROUS HYDROGEN CHLORIDE [2, 4]

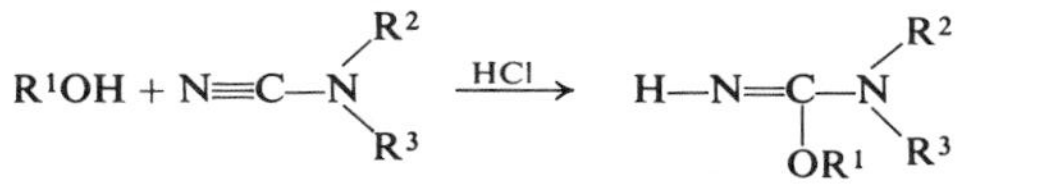

$$R^1OH + N\equiv C-N{<}^{R^2}_{R^3} \xrightarrow{HCl} H-N=C{<}^{OR^1}_{N{<}^{R^2}_{R^3}} \qquad (20)$$

R^1	R^2	R^3	Yield (%)	B.p., °C (mm Hg)	n_D^{25} (or m.p., °C)	d_4^{25}
$n\text{-}C_4H_9$	$i\text{-}C_3H_7$	$i\text{-}C_3H_7$	39	62 (0.2)	1.4491	0.9005
C_6H_{11} (cyclic)	C_6H_{11} (cyclic)	C_6H_{11} (cyclic)	14	115 (0.003)	(80)	—
$n\text{-}C_4H_9-\underset{\underset{C_2H_5}{\|}}{CH}-CH_2$	H	C_6H_5	37	95 (0.004)	1.5156	1.0068
$n\text{-}C_4H_9-\underset{\underset{C_2H_5}{\|}}{CH}-CH_2$	H	$n\text{-}C_4H_9$	20	86 (0.005)	1.4591	0.9062
$n\text{-}C_4H_9$	H	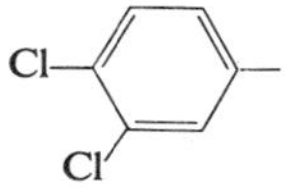 (3,4-dichlorophenyl)	44	—	—	—

[a] Data taken from S. E. Forman, C. A. Erickson, and H. Adelman, J. Org. Chem. 28, 2653 (1963).
[b] This method does not work for phenol.

203

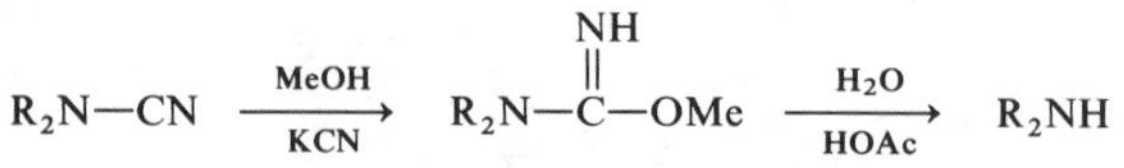

$$R_2N\text{---}CN \xrightarrow[\text{KCN}]{\text{MeOH}} R_2N\text{---}\overset{\overset{\displaystyle NH}{\|}}{C}\text{---}OMe \xrightarrow[\text{HOAc}]{\text{H}_2\text{O}} R_2NH \qquad (21)$$

2-4. Preparation of 2-n-Butyl-3,3-diisopropylpseudourea [21]

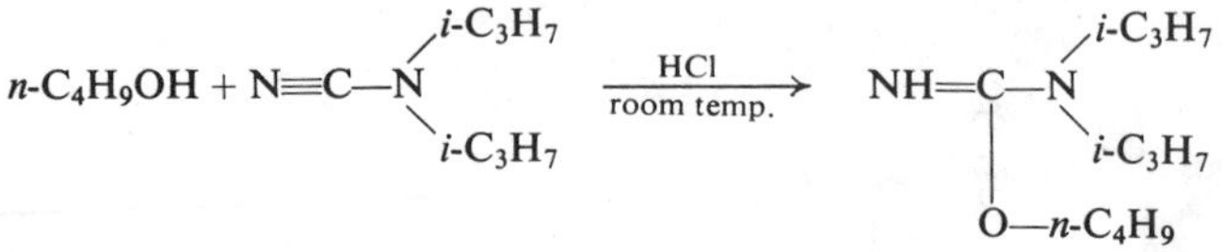

$$n\text{-}C_4H_9OH + N\!\!\equiv\!\!C\text{---}N\!\!\begin{array}{l}{}^{i\text{-}C_3H_7}\\ {}_{i\text{-}C_3H_7}\end{array} \xrightarrow[\text{room temp.}]{\text{HCl}} \underset{O\text{---}n\text{-}C_4H_9}{NH\!\!=\!\!C\text{---}N}\!\!\begin{array}{l}{}^{i\text{-}C_3H_7}\\ {}_{i\text{-}C_3H_7}\end{array} \qquad (22)$$

To a flask containing 250 ml (2.74 moles) of n-butanol is added 37.8 gm (0.3 mole) of N,N-diisopropylcyanamide. Dry hydrogen chloride gas is slowly added, while the temperature is kept at 25°C, to saturate the reaction mixture (1.0–1.1 equivalents of hydrogen chloride are satisfactory) [41]. After standing for 1–15 days the mixture is made alkaline with sodium hydroxide, extracted with benzene, and distilled to afford 23.3 gm (39%), b.p. 62°C (0.2 mm), n_D^{25} 1.4491, d_4^{25} 0.9005.

The other isomers shown in Table IV were prepared by a similar procedure.

Basterfield and co-workers [41] suggest that the use of a large excess of hydrogen chloride gas slows the rate of reaction. The use of large amounts of alcohol and temperatures in the range 40°–50°C speeds the rate of reaction.

Other compounds prepared by a similar method are found in Kurzer and Lawson [42], Hunsdiecker and Vogt [43], Battegay [44, 45], and Puetzer [46].

C. O-Alkylation of Urea

Werner [10] suggested that, on heating, urea tautomerizes to the pseudourea structure which, on reaction with methyl sulfate, allows the isolation of 2-methylpseudourea (Eq. 23). The product was difficult to crystallize and it was isolated as the picric acid derivative.

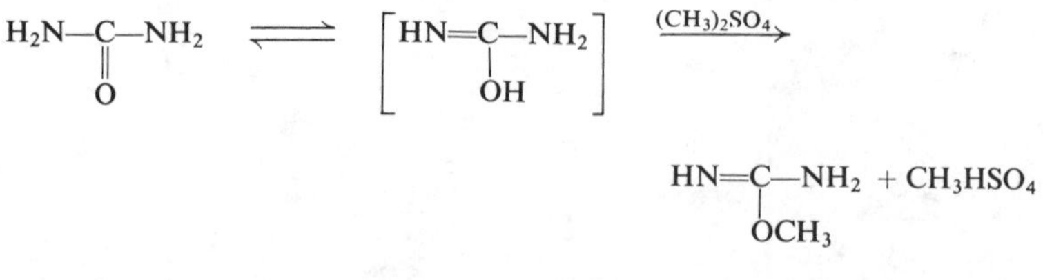

$$H_2N\text{---}\underset{O}{\overset{\|}{C}}\text{---}NH_2 \;\rightleftharpoons\; \left[\underset{OH}{HN\!\!=\!\!C\text{---}NH_2}\right] \xrightarrow{(CH_3)_2SO_4}$$

$$\underset{OCH_3}{HN\!\!=\!\!C\text{---}NH_2} + CH_3HSO_4 \qquad (23)$$

Recently Piasek and Urbanski [47] gave further proof of the pseudourea structure of urea by showing that diazomethane reacts at room temperature

with urea dissolved in methanol or ethanol to give 2-methyl- or 2-ethylpseudo-
urea, respectively, in 11–12 % yield.

No alkylation of urea occurred when the reaction was carried out in a less
polar solvent than ethyl ether.

Janus [11] has also showed that urea reacts with methyl sulfate or methyl-
p-toluenesulfonate at room temperature to give the 2-alkylpseudoureas. At
temperatures of 100°C or more the reaction is too violent; a reaction temperature of 30°–60°C is preferred.

2-5. *Preparation of 2-Methylpseudourea* [11]

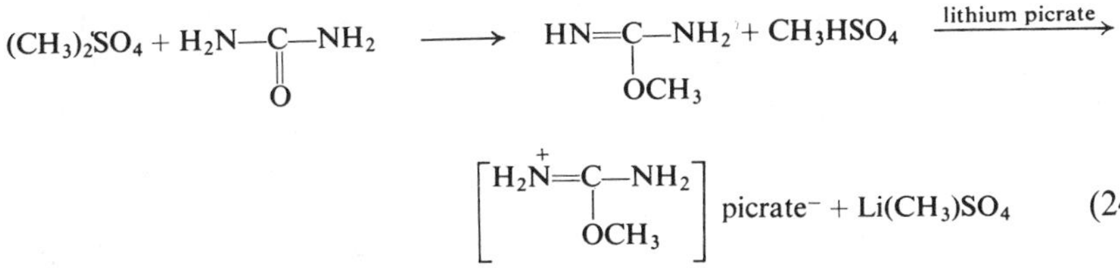

$$(CH_3)_2SO_4 + H_2N-\underset{\underset{O}{\|}}{C}-NH_2 \longrightarrow HN\!=\!\underset{\underset{OCH_3}{|}}{C}-NH_2 + CH_3HSO_4 \xrightarrow{\text{lithium picrate}}$$

$$\left[H_2\overset{+}{N}\!=\!\underset{\underset{OCH_3}{|}}{C}-NH_2\right] picrate^- + Li(CH_3)SO_4 \qquad (24)$$

To a flask containing 60 gm (1.0 mole) of urea is added 126 gm (1.0 mole) of
methyl sulfate, and the stirred reaction mixture is warmed for 7 hr at 40°C.
Stirring is continued until the reaction mixture is clear. The O-methylisourea is
isolated as the picrate salt by adding 350–400 ml of 2 M lithium picrate in 300
ml of ethanol. The resulting mixture is boiled on the steam bath for 5 min.
Most of the picrate dissolves but precipitates on slowly cooling to 0°C. The
picrate is filtered, washed with ethanol, and air-dried to afford 182 gm (60%),
m.p. 177°C.

A similar methylation when carried out at 18°C requires 5 days for completion (3 days to become clear). Carrying out the reaction at 60°C requires only
about 1 hr.

NOTE: Pure 2-methylpseudourea has been prepared by Piasek and
Urbanski [47] and Stieglitz and McKee [12] (m.p. 44°–45°C).

Reaction of *sym*-dimethylurea in dry diglyme with sodium hydride while
refluxing until gas evolution ceases followed by reaction with benzyl chloride
gives N,N'-dimethyl-O-benzylisourea [11a].

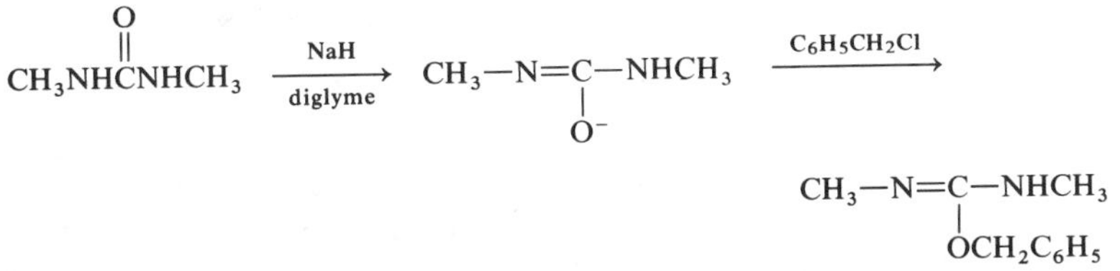

$$CH_3NH\overset{\overset{O}{\|}}{C}NHCH_3 \xrightarrow[\text{diglyme}]{NaH} CH_3-N\!=\!\underset{\underset{O^-}{|}}{C}-NHCH_3 \xrightarrow{C_6H_5CH_2Cl}$$

$$CH_3-N\!=\!\underset{\underset{OCH_2C_6H_5}{|}}{C}-NHCH_3 \qquad (25)$$

D. S-Alkylation of Thiourea

As for urea, thiourea has been shown to tautomerize to the pseudothiourea structure [48] (Eq. 26).

$$
\underset{\substack{\|\\ \text{S}}}{H_2N-C-NH_2} \rightleftharpoons \underset{\substack{|\\ \text{SH}}}{H_2N-C=NH} \tag{26}
$$

Before the synthesis of the pseudoureas was published, Bernthsen and Klinger [6] reported a pseudothiourea synthesis involving the reaction of thioureas with alkyl halides. This reaction was briefly reviewed by Dains [16] and Stieglitz [49, 50], and it found many commercial applications [51–53]. The preparation of isothiouronium salts by the direct action of thiourea and halogen acids on alcohols (primary, secondary, and tertiary) was reported by Stevens [8] and further developed by Johnson and Sprague [54, 55] (Eq. 27).

$$
\underset{\substack{\|\\ \text{S}}}{H_2N-C-NH_2} + RX \xrightarrow{} \underset{\substack{|\\ \text{SR}}}{H_2N-C=\overset{+}{N}H_2X^-} \tag{27}
$$

$$\text{ROH} + \text{HX}$$

Polymeric pseudothioureas are made by reacting linear polythioureas with alkylating agents [56]. Thioureas may also be alkylated by alkyl sulfates, halohydrins [57, 58], sulfonates [59], alkyl halide derivatives [60–64], and certain unsaturated acids and their esters [65].

Applications of these methods to various synthetic objectives have been reported: the preparation of organosilicon-containing pseudothiuronium salts [66], mercaptans from alcohols [9, 67], 2-alkylpseudothiuronium picrates for the estimation of tertiary alcohols [68, 69], and alkylsulfonyl chlorides from pseudothioureas by chlorination of pseudothiourea–HCl salts [55]. (CAUTION: May cause an explosion [70].)

Thiourea can also be S-alkylated by the acid-catalyzed reaction of unsaturated compounds with thiourea. For example, thiourea has been reported to add with acid catalysis to acrylonitrile and acrylamides. This reaction is a route to the synthesis of S-(β-carboxamidoethyl)isothiuronium salts, which can be hydrolyzed with cold sodium hydroxide to β-mercaptopropionitrile and β-mercaptopropionamides [12a, b].

$$
S=C(NH_2)_2 \xrightarrow{HCl} [HS=C(NH_2)_2]^+Cl^- \xrightarrow[C_2H_5OH]{CH_2=CH-CN} \underset{\substack{\|\\ \text{NH}_2{}^+\text{Cl}^-}}{H_2N-C-SCH_2CH_2CN} \tag{27a}
$$

$$
\underset{\substack{|\\ \text{CH}_3}}{\overset{CH_2}{\underset{}{\diagdown}}} \underset{C}{\diagup} \overset{CH_3}{} + S=C(NH_2)_2 \xrightarrow{HBr} \tag{27b}
$$

The aqueous acid-catalyzed hydrolysis of arylmethylisothioureas was recently reported to take place by an SN_1 mechanism [12b, c].

Alkylpseudothiourea salts (or alkylisothiourea salts) have been used for anesthetic agents [12d].

2-6. *Preparation of 1-Phenyl-3-(2-cyanoethyl)-2-methylpseudothiuronium Iodide* [71]

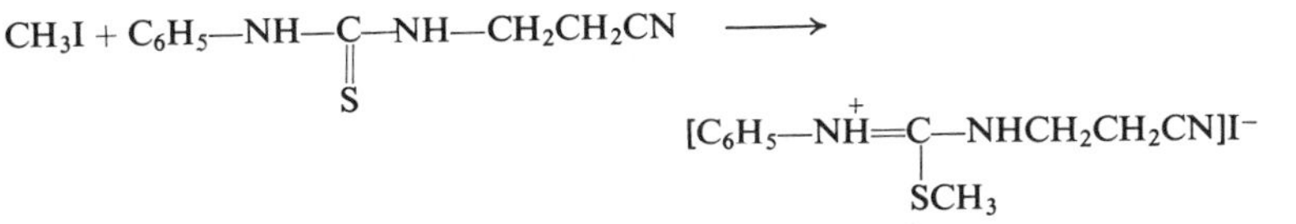

$$CH_3I + C_6H_5-NH-\underset{\underset{S}{\|}}{C}-NH-CH_2CH_2CN \longrightarrow$$

$$[C_6H_5-N\overset{+}{H}=\underset{\underset{SCH_3}{|}}{C}-NHCH_2CH_2CN]I^- \quad (28)$$

To a refluxing solution of 9.73 gm (0.047 mole) of 1-phenyl-3-(2-cyanoethyl)-thiourea in 75 ml of absolute methanol is added dropwise 7.4 gm (0.052 mole) of methyl iodide over a 5 min period. After refluxing for 3 hr the solution is evaporated to dryness under reduced pressure to give an oil. The oil is triturated with 50 ml of acetone to afford 15.1 gm (93%). Three recrystallizations from 3:1 absolute ethanol–ether solution raised the melting point from 152°–159°C to 165°–167°C (picrate derivative, m.p. 139°–141°C).

E. Reaction of Various Reagents with Pseudoureas to Give Substituted Pseudoureas

The active hydrogens of pseudoureas react with various reagents, such as sodium hypochlorite [72], acyl halides [73], sulfonyl halides [19], isocyanates [26], isothiocyanates [26, 74] cyanamide, [75] and esters [76], to give linear or cyclic derivatives as shown in Scheme 2.

SCHEME 2

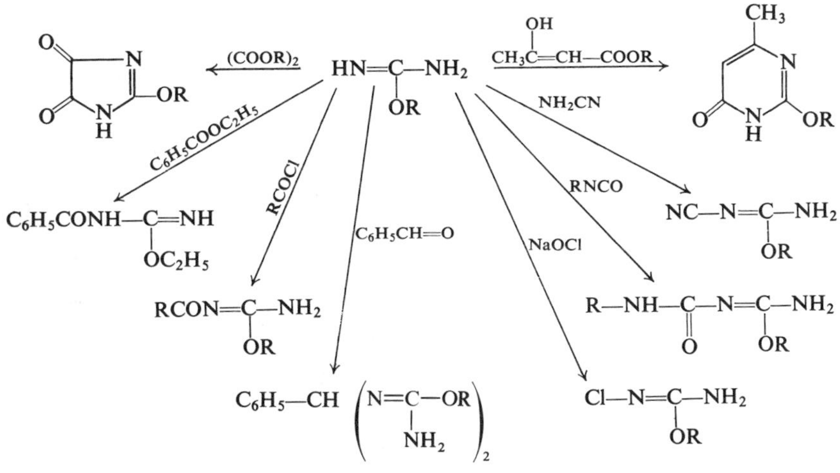

TABLE V[a]

TRANSALKYLATION OF THE 2-METHYL OR 2-ETHYL GROUP OF A TRISUBSTITUTED PSEUDOUREA WITH 0.1 EQUIVALENT OF SODIUM ALKOXIDE ($R_2 = CH_3$ OR C_2H_5)

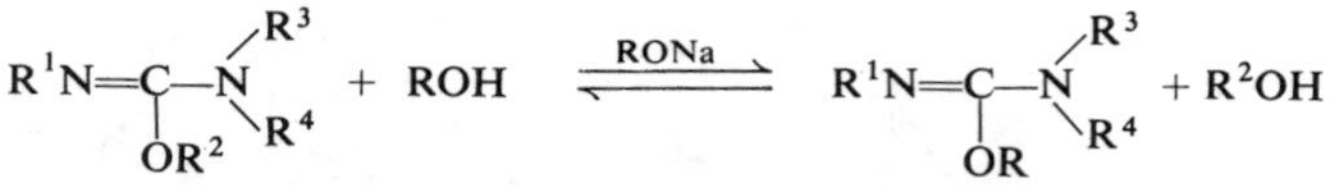

$$R^1N{=}C\genfrac{}{}{0pt}{}{\diagup N R^3R^4}{\diagdown OR^2} + ROH \xrightarrow{RONa} R^1N{=}C\genfrac{}{}{0pt}{}{\diagup N R^3R^4}{\diagdown OR} + R^2OH \quad (29)$$

R^1	R	R^3	R^4	Yield (%)	B.p., °C (mm)	n_D^{25} (or m.p., °C)	d_4^{25}
H	$n\text{-}C_4H_9\text{—}\overset{}{\underset{C_2H_5}{CH}}\text{—}CH_2$	$n\text{-}C_4H_9$	$n\text{-}C_4H_9$	89	94 (0.025)	1.4535	—
H	$i\text{-}C_4H_9\text{—}\overset{CH_3}{\underset{CH_3}{C}}\text{—}CH_2$	$n\text{-}C_4H_9$	$n\text{-}C_4H_9$	75	74 (0.002)	1.4538	0.8810
H	C_6H_{11} (cyclic)	$n\text{-}C_4H_9$	$n\text{-}C_4H_9$	74	94 (0.001)	1.4692	0.9320
H	$(CH_3)_3\text{—}C\text{—}CH_2$	$CH_2{=}\overset{CH_3}{\underset{}{C}}\text{—}CH_2$	$CH_2{=}\overset{CH_3}{\underset{}{C}}\text{—}CH_2$	86	83 (0.5)	1.4623	0.8992
H	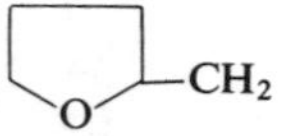 (tetrahydrofurfuryl ring with CH_2)	$CH_2{=}\overset{CH_3}{\underset{}{C}}\text{—}CH_2$	$CH_2{=}\overset{CH_3}{\underset{}{C}}\text{—}CH_2$	82	83 (0.004)	1.4883	1.0095
C_6H_{11} (cyclic)	$n\text{-}C_4H_9\text{—}\overset{}{\underset{C_2H_5}{CH}}\text{—}CH_2$	H	C_6H_{11} (cyclic)	96	128 (0.025)	1.4815	—

[a] Data taken from S. E. Forman, C. A. Erickson, and H. Adelman, *J. Org. Chem.* **28**, 2653 (1963).

Trisubstituted pseudoureas undergo a transalkylation reaction where R^1 or $R^3 = H$. 2-Methyl or 2-ethyl groups are replaced by alkyl groups of higher molecular weight on reaction with the corresponding alcohol in the presence of sodium alkoxide catalysts [21] (Eq. 30). The lower-boiling alcohol is removed by distillation. This is a useful method for preparing novel 2-alkyl derivatives, as seen in Table V. When $R^3 = R^4 =$ secondary alkyl groups, the reaction fails with basic or acidic catalysts. When $R^3 =$ secondary alkyl and $R^4 =$ primary alkyl, the reaction proceeds readily. With two secondary alkyl groups on different nitrogen atoms, the transalkylation reaction proceeds

$$R^1N=C-N\begin{smallmatrix}R^3\\R^4\end{smallmatrix} + ROH \xrightarrow{RONa} R^1N=C-N\begin{smallmatrix}R^3\\R^4\end{smallmatrix} + R^2OH \qquad (30)$$
$$\quad\ \ OR^2 \qquad\qquad\qquad\qquad\qquad OR^2$$

readily under basic conditions. 1,2,3,4-Tetrasubstituted pseudoureas did not transalkylate in the presence of an equivalent or excess sodium alkoxide [21].

2-7. Transalkylation of Pseudourea: Preparation 2-Cyclohexyl-3,3-di-n-butyl-pseudourea [21]

$$HN=C-N\begin{smallmatrix}n\text{-}C_4H_9\\n\text{-}C_4H_9\end{smallmatrix} + \text{(cyclohexanol)} \rightleftharpoons HN=C-N\begin{smallmatrix}n\text{-}C_4H_9\\n\text{-}C_4H_9\end{smallmatrix}$$
$$\quad OC_2H_5$$

$$+ \ C_2H_5OH \qquad (31)$$

To a flask containing 500 ml (4.75 moles) of cyclohexanol is added 2.3 gm (0.1 gm-atom) of sodium metal and, after the reaction is complete, 190.0 gm (1.0 mole) of o-ethyl-N,N-di-n-butylpseudourea is added. Benzene is added in an amount sufficient to hold the pot temperature at about 100°C while the mixture is distilled through a 120×35 cm column packed with 5 mm glass helices. The mixture is kept at a reflux ratio of 20:1 to remove the ethanol as an azeotrope with the benzene. When no more ethanol distills, the sodium salt is neutralized with glacial acetic acid and the mixture is distilled to afford 180.6 gm (74%), b.p. 94°C (0.001 mm), n_D^{25} 1.4692 d_4^{25} 0.9320.

3,3-Dibutyl-2-(2,2,4-trimethylpentyl)pseudourea, an insect repellent, is reported to be prepared by the base-catalyzed transalkylation reaction of 2,2,4-trimethylpentyl alcohol with 2-methyl-3,3-dibutylpseudourea [21a–c].

3. MISCELLANEOUS METHODS

(1) Preparation of 1,2,3,4-tetrasubstituted pseudoureas by the reaction of alkyl-dialkylaminochloroformamidines [21].

(2) Reaction of N-(4-chlorophenylimino)chloroformic acid 2,4-dichloro-phenyl ester with dialkylamines at room temperature in benzene solution to give N-(4-chlorophenyl)-2-(2,4-dichlorophenyl)-1,3-dimethylpseudourea. Other pseudoureas were prepared similarly [27, 77].

(3) Photodecomposition of 1,4-dimethyltetrazolinone [78].

(4) Preparation of selenium analogs of 2-aminoethylpseudothiouronium salts [79].

(5) Preparation of 2-(6-purinyl)-2-pseudothioureas [80].

(6) Reaction of cyanates with primary and secondary amines to give 2-substituted pseudoureas [13].

(7) Reaction of mercaptans with cyanamide to give pseudothioureas [81].

(8) Preparation of arylsulfonyl 2-substituted pseudothioureas by the reaction of arylsulfonyl halides with pseudothioureas [82, 83].

(9) Reaction of ω-(phthalimidooxy)alkyl bromides with pseudothiourea to give 2-[ω-(aminooxy)alkyl]pseudothiouronium salts [84].

(10) Reaction of thiourea with 2-haloacetamides and 2-haloacetates to give pseudothiouronium salts and pseudothiohydantoins [85].

(11) Polymeric pseudoureas [86].

(12) O-Acylisoureas by rearrangement of acylureas [87].

(13) Thiophosphorylisothiourea derivatives [88].

REFERENCES

1. F. Lengfeld and J. Stieglitz, *Chem. Ber.* **27**, 926 (1894).
2. F. Lengfeld and J. Stieglitz, *Amer. Chem. J.* **17**, 112 (1892).
3. F. Lengfeld and J. Stieglitz, *Chem. Ber.* **27**, 1283 (1894).
4. F. Lengfeld and J. Stieglitz, *Chem. Ber.* **28**, 1004 (1895).
5. A. Claus, *Chem. Ber.* **7**, 226 (1874).
6. A. Bernthsen and H. Klinger, *Chem. Ber.* **11**, 492 (1878).
7. E. A. Werner, *J. Chem. Soc.* **57**, 283 (1890).
8. H. P. Stevens, *J. Chem. Soc.* **81**, 79 (1902).
9. G. G. Urquhart, J. W. Gates, Jr., and R. Connor, *Org. Syn.* **21**, 36 (1941).
10. E. A. Werner, *J. Chem. Soc.* 923 (1914).
11. J. W. Janus, *J. Chem. Soc.* 3551 (1955).
11a. W. S. Wadsworth, Jr., *J. Org. Chem.* **34**, 2994 (1969).
12. J. Stieglitz and R. H. McKee, *Chem. Ber.* **33**, 1517 (1900).
12a. L. Bauer and T. L. Welsh, *J. Org. Chem.* **26**, 1443 (1961).
12b. D. R. Flanagan and A. P. Simonelli, *J. Org. Chem.* **41**, 3118 (1976).
12c. D. R. Flanagan and A. P. Simonelli, *J. Org. Chem.* **41**, 3114 (1976).
12d. B. M. Regan, U.S. Pat. 3,448,198 (1969).

13. Farbenfabriken Bayer, Neth. Appl. 6,406,887 (1964).
14. E. Dabritz, *Angew. Chem. Int. Ed. Engl.* **5**, 470 (1966).
14a. L. J. Mathias, *Synthesis* 561 (1979).
14b. S. Tsuboi, P. Stronquist, and L. E. Overman, *Tetrahedron Lett.* (15), 1145 (1976).
14c. L. E. Overman and S. Tsuboi, *J. Amer. Chem. Soc.* **99**, 2813 (1977).
14d. L. E. Overman, S. Tsuboi, J. P. Roos, and G. F. Taylor, *J. Amer. Chem. Soc.* **102**, 747 (1980).
14e. L. E. Overman and J. P. Roos, *J. Org. Chem.* **46**, 811 (1981).
14f. A. Donetti, A. Omodei-Sole, and A. Mantegani, *Tetrahedron Lett.* (39), 3327 (1969).
14g. A. Donetti and E. Bellora, *J. Org. Chem.* **37**, 3352 (1972).
15. E. Schmidt, M. Seefelder, R. C. Jannin, W. Striewsky, and H. von Martius, *Ann. Chem.* **571**, 83 (1951).
16. F. B. Dains, *J. Amer. Chem. Soc.* **21**, 136 (1899).
17. F. Kurzer and A. Lawson, *Org. Syn.* **4**, 645 (1963).
18. R. H. McKee, *Amer. Chem. J.* **26**, 209, 245 (1901).
19. S. Basterfield and E. C. Powell, *Can. J. Res.* **1**, 261 (1929).
20. E. H. Cox and S. H. Raymond, Jr., *J. Amer. Chem. Soc.* **63**, 300 (1941).
21. S. E. Forman, C. A. Erikson, and H. Adelman, *J. Org. Chem.* **28**, 2653 (1963).
21a. H. H. Incho and S. E. Dorman, U.S. Pat. 3,266,979 (1966).
22. F. C. Schaefer and G. A. Peters, *J. Org. Chem.* **26**, 412 (1961).
23a. J. Stieglitz and R. H. McKee, *Chem. Ber.* **33**, 810 (1900).
23b. R. H. McKee, *Amer. Chem. J.* **26**, 244 (1901).
24. E. A. Werner, *J. Chem. Soc.* 927 (1914).
25. F. B. Dains and W. C. Thompson, *Univ. Kansas Sci. Bull.* **13**, 118 (1922).
26. J. A. Snyder, U.S. Pat. 2,780,535 (1957).
27. E. Kuehle, L. Eue, and O. Bayer, Ger. Pat. 1,137,000 (1962).
28. N. E. Searle, U.S. Pat. 2,849,306 (1958).
28a. H.-L. M. Chin, U.S. Pat. 4,400,529 (1983).
29. R. A. Jeffreys, A. N. Davenport, and D. G. Saunders, U.S. Pat. 3,095,302 (1963).
30. J. A. Price, U.S. Pat. 2,654,725 (1953).
31. J. A. Price, U.S. Pat. 2,655,493 (1953).
32. E. E. Renfrew and D. I. Randall, U.S. Pat. 2,645,641 (1953).
33. T. Chadderton and R. Thornton, U.S. Pat. 2,599,371 (1952).
34. M. Furnica and G. Furnica, *Acad. Repub. Pop. Rom. Stud. Cercet. Biochem.* **4**, 387 (1961).
35. H. G. Khorana, *Canad. J. Chem.* **32**, 227 (1954).
36. J. Stieglitz, *Chem. Ber.* **28**, 573 (1895).
37. E. Schmidt and F. Moosmüller, *Ann. Chem.* **597**, 235 (1955).
38. E. Schmidt, F. Moosmüller, and R. Schnegg, Ger. Pat. 956,499 (1957).
38a. E. Schmidt, E. Däbritz, K. Thulke, and E. Grassmann, *Ann. Chem.* **685**, 161 (1965).
39. R. H. McKee, *Amer. Chem. J.* **26**, 209 (1901); **36**, 208 (1906); **42**, 1 (1909).
40. R. C. Elderfield and M. Green, *J. Org. Chem.* **17**, 431 (1952).
41. S. Basterfield, F. B. S. Rodman, and J. W. Tomecko, *Can. J. Chem.* **17B**, 390 (1939).
42. F. Kurzer and A. Lawson, *Org. Syn. Coll. Vol.* **4**, 645 (1963).
43. H. Hunsdiecker and E. Vogt, Brit. Pat. 426,508 (1935).
44. M. Battegay, U.S. Pat. 2,039,514 (1936).
45. M. Battegay, Fr. Pat. 736,175 (1933).
46. B. Puetzer, U.S. Pat. 2,156,193 (1939).
47. Z. Piasek and T. Urbanski, *Tetrahedron Lett.* 723 (1962).
48. A. Tkac, *Chem. Zvesti* **3**, 332 (1949).
49. F. Lengfeld and J. Stieglitz, *Chem. Ber.* **14**, 1490 (1881); **15**, 1314 (1882); **21**, 962, 1857 (1888).

50. F. Lengfeld and J. Stieglitz, *Chem. Ber.* **21**, 962, 1857 (1888).
51. A. F. Crowther, F. H. S. Curd, J. A. Hendry, and F. L. Rose, Brit. Pat. 603,070 (1948).
52. Wellcome Foundation Ltd., Belg. Pat. 626,015 (1963).
53. F. K. Velichko, B. J. Keda, and S. D. Polikvrpova, *Zh. Obshch. Khim.* **34**, 2356 (1964).
54. T. B. Johnson and J. M. Sprague, *J. Amer. Chem. Soc.* **58**, 1348 (1936).
55. J. M. Sprague and T. B. Johnson, *J. Amer. Chem. Soc.* **59**, 1837 (1937).
56. I. G. Farbenindustrie A-G., Belg. Pat. 449,460 (1943).
57. P. L. DuBrow, U.S. Pat. 3,093,666 (1963).
58. F. Arndt, *Chem. Ber.* **54**, 2236 (1921).
59. D. Klamann and F. Drahowzal, *Monatsh. Chem.* **83**, 463 (1952).
60. N. J. Kartinos, U.S. Pat. 3,116,327 (1963).
61. D. I. Randall, U.S. Pat. 2,619,492 (1952).
62. F. Lodge and J. Wardleworth, U.S. Pat. 2,655,515 (1953).
63. F. Lodge and J. Wardleworth, Brit. Pat. 696,679 (1953).
64. J. W. Bolger, U.S. Pat. 3,124,595 (1964).
65. H. Behringer and P. Zillikens, *Ann. Chem.* **574**, 140 (1951).
66. S. Nozakura, *Nippon Kagaku Zasshi* **75**, 958 (1954).
67. R. L. Frank and A. V. Smith, *J. Amer. Chem. Soc.* **68**, 2103 (1946).
68. L. Schotte and S. Verbel, *Acta. Chem. Scand.* **7**, 1357 (1953).
69. L. Long, Jr., R. C. Clapp, F. H. Bissett, and T. Hasselstrom, *J. Org. Chem.* **26**, 85 (1961).
70. K. Folkers, A. Russell, and R. W. Bost, *J. Amer. Chem. Soc.* **63**, 3530 (1941).
71. M. E. Kreling and A. F. McKay, *Canad. J. Chem.* **40**, 143 (1962).
72. J. Goerdeler and F. Bechlars, *Chem. Ber.* **88**, 843 (1955).
73. Bayer & Co., Aust. Pat. 72,300 (1916).
74. W. M. Bruce, *J. Amer. Chem. Soc.* **26**, 119, 419 (1904).
75. K. Kawans and K. Odo, *Yuki Gasei Kayaku Kyokai Ski* **24**, 955 (1966).
76. S. Basterfield, A. E. Baughen, and I. Bergsteinsson, *Trans. Roy. Soc. Can. III* **33**, 115 (1936).
77. E. Kuehle, L. Eue, Ger. Pat. 1,138,039 (1962).
78. W. S. Wadsworth, Jr., *J. Org. Chem.* **34**, 2994 (1969).
79. S. H. Chu and H. G. Mautner, *J. Org. Chem.* **27**, 2899 (1962).
80. C. Temple, Jr., and J. A. Montgomery, *J. Org. Chem.* **31**, 1417 (1966).
81. Schering-Kahlbaum A.-G., Brit. Pat. 296,782 (1927).
82. E. H. Cox and J. M. Sprague, U.S. Pat. 2,557,633 (1951).
83. E. Toyoshima, S. Tanaka, and K. Hashimoto, Japan. Pat. 28,263 (1965).
84. L. Bauer and K. S. Suresh, *J. Org. Chem.* **28**, 1604 (1963).
85. A. J. Speziale, *J. Org. Chem.* **23**, 1231 (1958).
86. S. Pundak and M. Wilchek, *J. Org. Chem.* **46**, 808 (1981); M. Wilchek, T. Oka, and Y. J. Topper, *Proc. Natl. Acad. Sci. U.S.A.* **72**, 1055 (1975).
87. W. T. Brady and R. A. Owens, *J. Org. Chem.* **42**, 3220 (1977).
88. L. Kuruc, V. Konecny, and S. Truchlik, Czech. Pat. 172,676, (May 15, 1978); *Chem. Abstr.*, **90**, 54501r (1979).

1. INTRODUCTION

Several reviews have covered the preparation and derivatives of semicarb-azides [1, 2]. In the recent literature the nomenclature follows the numbering

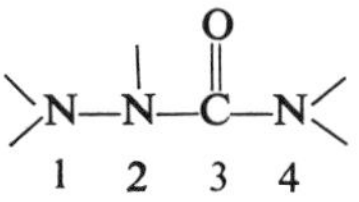

$$\text{>N—N—C(=O)—N<}$$
$$\text{1 \quad 2 \quad 3 \quad 4}$$

(I)

system in (I) and, the older literature, the system in (II).

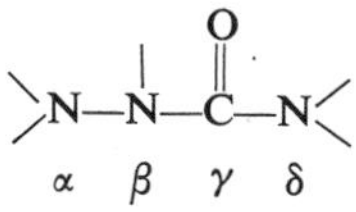

$$\text{>N—N—C(=O)—N<}$$
$$\alpha \quad \beta \quad \gamma \quad \delta$$

(II)

The methods of preparation of semicarbazides starting with hydrazine are summarized in Scheme 1.

SCHEME 1

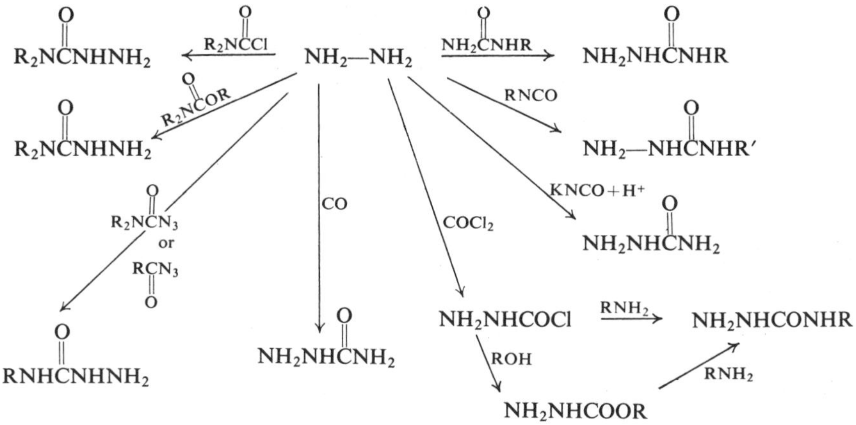

213

Chemical Abstracts now names the semicarbazides as hydrazinecarboxamide using the following numbering system:

$$\underset{2\quad\;1\qquad\;N}{H_2N-NH-\overset{\overset{\displaystyle O}{\|}}{C}-NH_2}$$

In Scheme 1 the most useful laboratory method for the preparation of semicarbazides involves the reaction of hydrazines with ureas or isocyanate derivatives (organic and inorganic).

Other substituted semicarbazides are prepared by analogous methods using alkyl- or arylhydrazines and similarly substituted derivatives (for example, Eq. 1).

$$RNHNH_2 + C_6H_5-NCS \longrightarrow C_6H_5NH\overset{\overset{\displaystyle S}{\|}}{C}-\overset{\overset{\displaystyle R}{|}}{N}-NH_2 \qquad (1)$$

Semicarbazide, like other hydrazines, reduces Fehling solution and, unless substituted in the 1-position, undergoes the typical hydrazine reactions with esters, carbonyl compounds and halides, etc. On heating, it decomposes to hydrazine and biurea.

SCHEME 2

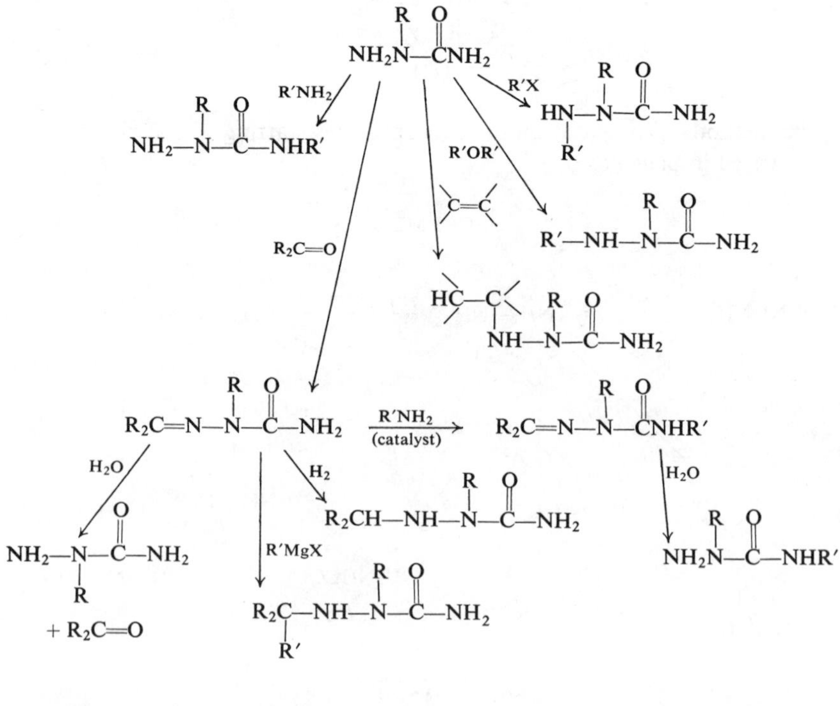

One of the most important uses of semicarbazide is as a reagent for identifying aldehydes and ketones [3a]. Many semicarbazones have higher melting points and lower solubilities than the corresponding phenylhydrazones. The carbonyl compound is easily recovered by hydrolysis in boiling water. Semicarbazones have also recently been reported to have antihypertensive properties [3b]. In addition, semicarbazides have been reported to be heat stabilizers for thermoplastic polyesters [3c].

Semicarbazide or its derivatives undergo addition reactions which lead to substituted semicarbazides; they are shown in Scheme 2.

CAUTION: Semicarbazides should be handled with care since, like many hydrazine derivatives, they may be mutagenic and carcinogenic [3d].

2. CONDENSATION REACTIONS

A. Reaction of Hydrazines with Urea

The reaction of hydrazine or a substituted hydrazine with urea is a very convenient and useful method of preparing semicarbazides in good yields (Eq. 2). Table I gives some results of this reaction, and Table II some results of the reaction of substituted ureas with hydrazine to give semicarbazides.

$$R^1R^2N-NHR^3 + NH_2CONR^4R^5 \longrightarrow NH_3 + R^1R^2N-\underset{\underset{R^3}{|}}{N}-CO-NR^4R^5 \quad (2)$$

The preparation of phenylsemicarbazide from aniline and hydrazine hydrate is one of the earliest reports of this method [4]. Some more recent examples are reported for the preparation of 2-*p*-cymyl-4-semicarbazide [5], phenylthiosemicarbazides [6], and 1-isonicotinylsemicarbazide [7].

TABLE I

REACTION OF UREA AND HYDRAZINES TO GIVE SEMICARBAZIDE

Starting hydrazine	1-Semicarbazide product	Ref.
Hydrazine	Semicarbazide	8–10
Phenyl-	1-Phenyl-	11–14
m-Tolyl-	1-*m*-Tolyl-	15
1-Benzyl-1-phenyl-	1-Benzyl-1-phenyl-	16
2-Nitro-5-chlorophenyl-	(2-Nitro-5-chlorophenyl-)-	17
Cymyl-	Cymyl-	18
4-Sulfamidophenyl-	(4-Sulfamidophenyl)-	19

TABLE II

REACTION OF SUBSTITUTED UREAS WITH HYDRAZINE TO GIVE
SEMICARBAZIDES

Substituted urea	4-Semicarbazide product	Ref.
Phenyl-	Phenyl-	20, 21
m-Tolyl-	*m*-Tolyl-	22
p-Tolyl-	*p*-Tolyl-	23
4-Bromophenyl-	(4-Bromophenyl)-	24
4-Nitrophenyl-	(4-Nitrophenyl)-	15
3-Nitrophenyl-	(3-Nitrophenyl)-	25
2,4-Dinitrophenyl-	(2,4-Dinitrophenyl)-	15
4-Biphenyl-	(4'-Biphenyl)-	26
2-Naphthyl-	(2-Naphthyl)-	27
β-Naphthyl-	(β-Naphthyl)-	28
1,1-Diphenyl-	Diphenyl-4,4-	29
Phenyl- (with phenylhydrazine)	Diphenyl-1,4-	6
p-Cymyl- (with phenylhydrazine)	(*p*-Cymyl)-2-phenyl-1-	5

2-1. Preparation of 4-Phenylsemicarbazide [21]

$$NH_2NH_2 \cdot H_2O + C_6H_5NHCONH_2 \xrightarrow[-NH_3]{H_2O} C_6H_5-NH\overset{O}{\overset{\|}{C}}NH-NH_2 \qquad (3)$$

To a flask equipped with a reflux condenser are added 68 gm (0.5 mole) of phenylurea and 120 ml (1.0 mole) of 42% hydrazine hydrate solution. The mixture is refluxed for 12 hr, filtered hot using decolorizing charcoal, and then concentrated to 100 ml. On cooling, the crystalline product separates and it is washed with the minimum amount of water. The filtrate is concentrated to 25 ml and another crop of crystals is collected. The total yield of the crude product is 47–52 gm, m.p. 100°–115°C. The product is white but turns brown on drying. The product is purified by dissolving in 200 ml of hot absolute alcohol and then treating with 250 ml of concentrated hydrochloric acid. The hydrochloride precipitates and is filtered, washed with alcohol, and dried to yield 46–48 gm, m.p. approximately 215°C. The hydrochloride is dissolved in three times its weight of boiling water. The solution is filtered, heated with approximately 100 gm (0.25 mole) of 10% sodium hydroxide solution. The free base separates, and after cooling the crystalline product is filtered to yield 28–30 gm (37–40%).

In an improved method, 34 gm (0.25 mole) of phenylurea, 25.0 gm (0.50 mole) of 100% hydrazine hydrate, and 25 ml of absolute ethanol are refluxed for 24 hr. This procedure eliminates the purification procedure as no unreacted phenylurea remains. The yields are also higher [21].

2-2. Preparation of 1-Isonicotinylsemicarbazide [7]

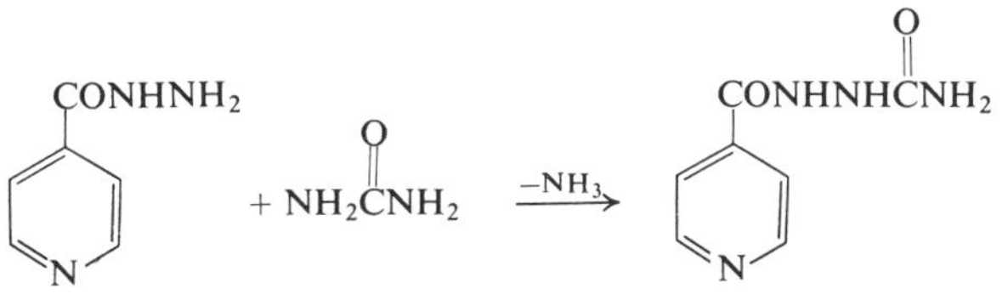

$$(4)$$

A mixture of 13.6 gm (0.1 mole) of isonicotinic acid hydrazide, 16.6 gm (0.28 mole) of urea, and 40 ml of water is refluxed for 8 hr while ammonia is being evolved. The solid product which separates on cooling is dissolved in 10% sodium hydroxide solution, treated with charcoal, filtered, and acidified with 10% hydrochloric acid to pH 6.0. The white crystalline product that separates is purified by recrystallization from water to afford 17.0 gm (94%) of colorless shining needles, m.p. 244°C dec. ([30]: m.p. 243°C dec.).

2-3. Preparation of 1-Benzoyl-4-phenylsemicarbazide [7]

$$C_6H_5CONHNH_2 + C_6H_5NHCNH_2 \xrightarrow{-NH_3} C_6H_5CONHNHCNHC_6H_5 \qquad (5)$$

A mixture of 6.0 gm (0.05 mole) of phenylurea, 6.0 gm (0.044 mole) of benzoylhydrazine, and 15 ml of water is refluxed for 6 hr. On cooling, a crystalline solid separates which after recrystallization from water gives 6.0 gm (47%) of white needles, m.p. 212°–214°C.

2-4. Preparation of Thiosemicarbazides [6]

$$C_6H_5NHNH_2 + RNHCNH_2 \xrightarrow{-NH_3} C_6H_5NHNHCNHR \qquad (6)$$

A mixture of 10 gm of the respective thiourea, suspended in alcohol, and a slight excess of phenylhydrazine is refluxed for 6 hr and then poured into water. Examples of some of the products are shown in Table III.

The thioureas were prepared by boiling an alcoholic solution of the aniline, carbon disulfide, and sodium hydroxide for 14 hr. The product, diaryl thioureas, were recrystallized from benzene.

8. Semicarbazides

TABLE III[g]

THE PREPARATION OF THIOSEMICARBAZIDES FROM PHENYLHYDRAZINE
AND sym-DIARYLTHIOUREAS

Thiosemicarbazide	Formula	M.p. (°C)
1-Phenyl-4-(4-bromophenyl)-	$C_{13}H_{12}N_3BrS$	179–180
1-Phenyl-4-(2,5-dibromophenyl)-[a]	$C_{13}H_{11}N_3Br_2S$	188
1-Phenyl-4-(2,4-dibromophenyl)-[b]	$C_{13}H_{11}N_3Br_2S$	177–178
1-Phenyl-4-(o-tolyl)-[c]	$C_{14}H_{15}N_3S$	170–171
1-Phenyl-4-(m-tolyl)-[d]	$C_{14}H_{15}N_3S$	173–174
1-Phenyl-4-(α-naphthyl)-[e]	$C_{17}H_{15}N_3S$	192–193
1-Phenyl-4-(3-nitrophenyl)-[f]	$C_{13}H_{12}O_2N_4S$	172

[a] The composition was also checked by analysis for bromine. *Anal.* Subs. 0.1534: AgBr, 0.1429. Calcd. for $C_{13}H_{11}N_3Br_2S$: Br, 39.90. Found: Br, 39.65.

[b] Composition checked for halogen content. *Anal.* Subs. 0.1559: AgBr, 0.1454. Calcd. for $C_{13}H_{11}N_3Br_2S$: Br, 39.90. Found: Br, 39.69.

[c] Obtained previously by interaction of phenylhydrazine and o-tolyl mustard oil by A. E. Dixon [*J. Chem. Soc.* **57**, 258 (1890)], who reported a melting point of 162°–163°C. Here a solution of 10 gm of the required thiourea and 5 gm of phenylhydrazine in 250 cm³ of benzene was evaporated at its boiling point to a volume of about 25 cm³ and allowed to cool.

[d] Reported by P. K. Bose and D. C. Ray-Chaudhury [*J. Ind. Chem. Soc.* **4**, 261 (1927)], who prepared it by the action of m-tolyl mustard oil on phenylhydrazine. Here prepared by the method given above for the ortho isomeride.

[e] Checked by analysis because the product obtained by A. E. Dixon [*J. Chem. Soc.* **61**, 1019 (1892)] through the action of phenylhydrazine on α-naphthyl mustard oil and supposed to be identical with this one was recorded as melting at 183°C.

[f] Kjeldahl method was modified as directed by J. Milbauer [*Z. Anal. Chem.* **42**, 728 (1903)] in order to avoid loss of nitrogen.

[g] Reprinted from L. C. Raiford, and W. T. Daddow, *J. Amer. Chem. Soc.* **53**, 1552 (1931). Copyright 1931 by the American Chemical Society. Reprinted by permission of the copyright owner.

B. Reaction of Hydrazines with Cyanates

$$RNHNH_2 + \overset{+}{M}\overset{-}{N}CX \longrightarrow RNHNHCN\overset{-}{H}M^+ \xrightarrow{HCl}$$
$$\underset{X}{\overset{\|}{}}$$

$$RNHNHCNH_2 + \overset{+}{M}Cl^- \qquad (7)$$
$$\underset{X}{\overset{\|}{}}$$

$$X = O, S$$

The reaction of hydrazines with sodium or potassium cyanate or thiocyanate gives semicarbazides or thiosemicarbazides in good yields. Monosubstituted alkylhydrazines give 2-semicarbazides (Eq. 8).

$$R—NHNH_2 + N\equiv C—OH \longrightarrow H_2N—\underset{R}{N}—CONH_2 \qquad (8)$$

TABLE IV

REACTION OF ALKYL- OR ARYLHYDRAZINES WITH CYANATES

Hydrazine substituent	Semicarbazide product	Ref.
Methyl-	2-Methyl-	31
Amyl-	2-Amyl-	32
2,4-Dimethylbenzyl-	2-(2,4-Dimethylbenzyl)-	33
1,2-Diisopropyl-	1,2-Diisopropyl-	34
4-Methoxyphenyl-	1-(4-Methoxyphenyl)-	35
1,1-Diphenyl-	1-(1,1-Diphenyl)-	36

In contrast, monosubstituted arylhydrazines react with cyanates to yield
1-semicarbazides (Eq. 9), probably as a result of the decreased basicity of the

$$C_6H_5-NHNH_2 + N{\equiv}C-OH \longrightarrow C_6H_5-NHNHC-NH_2 \qquad (9)$$
$$\underset{O}{\|}$$

nitrogen atom adjacent to the phenyl group.

Examples of semicarbazides prepared from alkyl- or aryl-substituted hydrazines are shown in Table IV, and of the reaction of hydrazine derivatives
with HOCN or HSCN in Table V.

TABLE V

REACTION OF HYDRAZINE DERIVATIVES WITH HOCN OR HSCN

Substituted hydrazine or derivative	Reagents	Semicarbazide, thiosemicarbazide, or semicarbazone	Ref.
$C_6H_5SO_2NHNH_2$	$NaOCN + HCl$	$C_6H_5-SO_2NHNHCONH_2$ m.p. 218°C dec.	37
$(H_2NNH)_2C{=}S$	$KSCN + HCl$	$H_2NC-NHNHC-NHNH_2$ (with two S above the C's) m.p. 191°–192°C	38
$HOCH_2CH_2NHNH_2$	$KOCN + HCl$	$NH_2-NHCNH_2$ (O above C), CH_2CH_2OH m.p. 110°C	39
$C_6H_5NHN{=}C(CH_3)_2$	$KOCN + CH_3COOH$	$C_6H_5N(CONH_2)N{=}C(CH_3)_2$	40
NH_2NH_2	NH_4SCN	$NH_2-C-NHNH_2$ (S above C) m.p. 180°–182°C	41

2-5. Preparation of Benzenesulfonylsemicarbazide [37]

Method A.

$$C_6H_5SO_2Cl + H_2NNH_2 \longrightarrow C_6H_5SO_2NHNH_2 \xrightarrow[\text{HCl}]{\text{KOCN}}$$

$$C_6H_5SO_2NHNH\overset{\text{O}}{\underset{\|}{C}}NH_2 \quad (10)$$

To a flask are added 17.2 gm (0.1 mole) of benzenesulfonylhydrazide, 100 ml of water, and 8.2 ml of concentrated hydrochloric acid. The solution that forms is filtered and then treated with 9.8 gm (0.12 mole) of sodium cyanate in 100 ml of water. A crystalline precipitate forms and, after stirring for 1 hr, an additional 8.2 ml of concentrated hydrochloric acid is added to destroy the excess of sodium cyanate. The crystalline product is washed with water, ethanol, dried, and recrystallized from acetic acid to yield 17.0 gm (79%), m.p. 218°C dec.

Method B.

$$(NH_2NH_2)_2 \cdot SO_4 + 2KOCN + H_2O \longrightarrow 2NH_2NH\overset{\text{O}}{\underset{\|}{C}}NH_2$$

$$NH_2NH\overset{\text{O}}{\underset{\|}{C}}NH_2 + C_6H_5SO_2Cl \longrightarrow C_6H_5SO_2NHNH\overset{\text{O}}{\underset{\|}{C}}NH_2 \quad (11)$$

To a flask containing 176.6 gm (1.0 mole) of benzenesulfonyl chloride in 1 liter of water is added dropwise a freshly prepared solution of semicarbazide prepared from 81.0 gm (1.0 mole) of potassium cyanate in 200 ml of water and 81.0 gm (0.51 mole) of dihydrazine sulfate in 200 ml of water. After half the semicarbazide solution has been added, a solution of 53.0 gm (0.50 mole) of sodium carbonate in 200 ml of water is added concurrently with the remaining semicarbazide solution. After standing for a few hours at room temperature, the product is filtered, washed with water, and dried to give 156 gm (73%), m.p. 218°C dec. (from glacial acetic acid).

By similar methods, the R-SO$_2$NHNHCONH$_2$ compounds listed in the accompanying tabulation have been prepared [37].

R	M.p., °C
p-CH$_3$C$_6$H$_4$-	236
CH$_3$-	194
C$_2$H$_5$-	149–151
n-C$_4$H$_9$-	188
n-C$_8$H$_{17}$-	182–183
C$_6$H$_5$CH$_2$-	230
β-Naphthyl-	214

2-6. Preparation of Thiosemicarbazide [41]

$$NH_4SCN + H_2NNH_2 \cdot H_2O \longrightarrow H_2NNH\overset{\overset{S}{\|}}{C}-NH_2 \qquad (12)$$

To a flask are added 200 gm (2.6 moles) of ammonium thiocyanate, 59 gm (1.0 mole) of 85% hydrazine hydrate, and 25 ml of water. The solution is refluxed for 3 hr under nitrogen, cooled, the sulfur filtered, and the filtrate allowed to cool for 10–18 hr. The product crystallizes slowly on standing and is filtered. The crude product, on recrystallization from a 1:1 water–ethanol mixture, gives 46 gm of thiosemicarbazide, m.p. 180°–182°C. Evaporation of the mother liquor to half its volume and then cooling give an additional 5 gm of product. The total yield is 57%.

C. Reaction of Hydrazine with Isocyanates

Hydrazines react with isocyanates to give either 1,4- or 2,4-disubstituted semicarbazides (see Scheme 1). Alkylhydrazines react in a similar fashion to produce only the 2,4-dialkylsemicarbazides [41, 42] (Eq. 13).

$$RNCO + R'-\overset{\overset{H}{|}}{N}-NH_2 \longrightarrow RNH\overset{\overset{}{C}}{\underset{\overset{}{O}}{\|}}-\overset{\overset{}{N}}{\underset{\overset{}{NH_2}}{|}}-R' \qquad (13)$$

Reaction of diisocyanates with hydrazine yields polysemicarbazides [43, 44].

Busch and Frey [45] noted earlier that 2,4- and 1,4-substituted semicarbazides are always formed when using arylhydrazines, the product depending on the reaction solvent used. The 1,4-substituted compounds are obtained in absolute ether, benzene, and neutral solvents. In alcoholic solution containing acetic or formic acid the 2,4-isomer is obtained in 25% yield along with the 1,4-substituted isomer.

Isothiocyanates react with substituted aromatic hydrazines at low temperature or in acid media to give 2,4-substituted thiosemicarbazides. At elevated temperatures 1,4-thiosemicarbazides are produced [46–48a] (Eqs. 14, 15).

$$RNHNH_2 + C_6H_5NCS \xrightarrow{\text{Low Temp.}} R-\overset{\overset{}{N}CS-NHC_6H_5}{\underset{\overset{}{NH_2}}{|}} \qquad (14)$$

$$RNHNH_2 + C_6H_5-NCS \xrightarrow{\text{Elev. Temp.}} RNHNHCSNHC_6H_5 \qquad (15)$$

The hydrazine group reacts preferentially with isocyanates even when hydroxyl groups are present as for the case reported in which *N,N*-bis(2-hydroxyethyl)hydrazine reacts with methoxymethyl isocyanate to give $(HOCH_2CH_2)_2NNHCONHCH_2OCH_3$ [48b].

TABLE VI

REACTION OF SUBSTITUTED HYDRAZINES WITH ALKYL OR ARYL
ISOCYANATES

Substituted hydrazine	Isocyanate	Semicarbazide product	Ref.
Hydrazine	Methyl	4-Methyl-	31
Methyl-	Methyl	2,4-Dimethyl-	31
Phenyl-	Methyl	1-Phenyl-4-methyl-	48
Methyl-	Phenyl	2-Methyl-4-phenyl-	49
Phenyl-	4-Biphenyl	1-Phenyl-4-(4-biphenyl)-	42a

Examples of the reaction of substituted hydrazines with aryl or alkyl iso-cyanates are given in Table VI.

1-Acetyl-1-methylhydrazine has also been reported to react with phenyl isocyanate to give 1-acetyl-1-methyl-4-phenylsemicarbazide in 86% yield [42b].

2-7. Preparation of 4-(4-Diphenyl)semicarbazide [42a]

$$NH_2NH_2 \cdot H_2O + p\text{-}C_6H_5\text{—}C_6H_4\text{—}NCO \longrightarrow p\text{-}C_6H_5\text{—}C_6H_4\text{—}NH\overset{\overset{\displaystyle O}{\|}}{C}\text{—}NHNH_2 \tag{16}$$

To a mortar cooled with ice is added 10 gm (0.20 mole) of hydrazine hydrate, and then 4.5 gm (0.023 mole) of finely divided 4-diphenyl isocyanate is added in small portions while mixing. The reaction is exothermic and a thick white paste is formed. The paste is cooled and then filtered, washed with water, boiling alcohol, and dried to afford 3.8 gm (73%) of colorless leaflets, m.p. 250°–260°C dec.

2-8. Preparation of 1,1-Dimethyl-4-phenylthiosemicarbazide [50]

$$\begin{array}{c}H_3C\\H_3C\end{array}\!\!\!\diagdown N\text{—}NH_2 + C_6H_5NCS \longrightarrow \begin{array}{c}H_3C\\H_3C\end{array}\!\!\!\diagdown N\text{—}NH\overset{\overset{\displaystyle S}{\|}}{C}\text{—}NH\text{—}C_6H_5 \tag{17}$$

To a flask are added 6.0 gm (0.10 mole) of 1,1-dimethylhydrazine, 13.5 gm (0.10 mole) of phenyl isothiocyanate, and 50 ml of ethanol. The contents is refluxed for 4 hr and then concentrated to $\frac{1}{3}$ the original volume. After cooling, the product crystallizes and is filtered to afford 12.6 gm (65%). Recrystalliza-tion from ethanol yields a product with m.p. 188°C.

2-9. Preparation of 4-(3,4-Dichlorophenyl)semicarbazide [51]

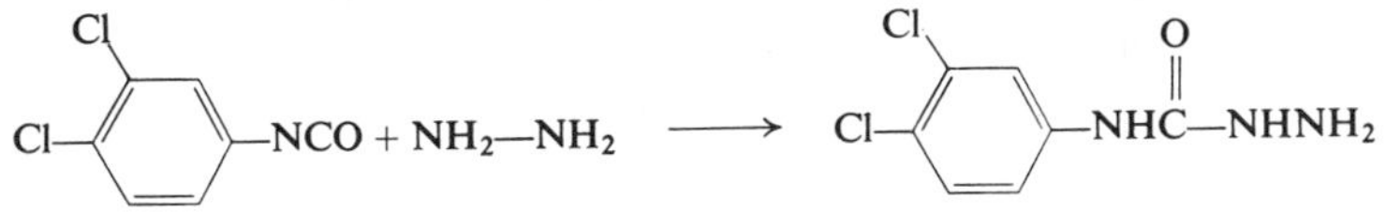

To a solution of 298 gm (9.3 moles) of hydrazine in ether is slowly added with stirring and cooling 215 gm (1.15 moles) of 3,4-dichlorophenyl isocyanate in 2 liters of ether while keeping the temperature at 20°C. After the addition the ether layer is decanted from the oily layer. On dilution of the oily layer with water is obtained 237.5 gm of a solid. Recrystallization from ethanol and filtration to remove the insoluble 1,6-bis(3,4-dichlorophenyl)biurea afford 108.6 gm of a solid, m.p. 175°–177°C. An additional recrystallization from 700 ml of ethyl acetate affords 77.3 gm (30%), m.p. 173°175°C, ir 3.00, 3.10 μ (NH), 5.90 μ (C=O).

2-10. Preparation of Poly-2,4-toluenesemicarbazide [43]

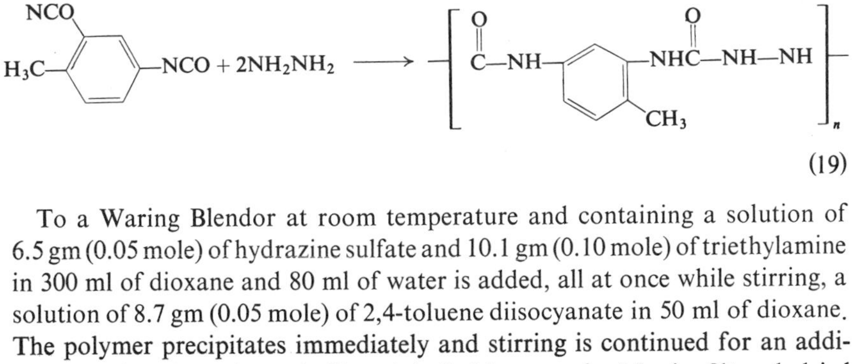

To a Waring Blendor at room temperature and containing a solution of 6.5 gm (0.05 mole) of hydrazine sulfate and 10.1 gm (0.10 mole) of triethylamine in 300 ml of dioxane and 80 ml of water is added, all at once while stirring, a solution of 8.7 gm (0.05 mole) of 2,4-toluene diisocyanate in 50 ml of dioxane. The polymer precipitates immediately and stirring is continued for an additional 5 min. The polymer is filtered, boiled in water for 15 min, filtered, dried under reduced pressure at 70°C to afford a quantitative yield of polymer, inherent viscosity 0.31 (sulfuric acid), polymer melt temperature 306°C. The polymer is soluble in formic acid from which a film can be cast.

D. Condensation Reactions Involving Semicarbazides

The basic amino group of the 1-position in semicarbazide or thiosemicarbazide may be used to react by a substitution reaction with activated halides [52], ethers [51], hydroxy [53], phenoxy [54], and amino groups [55] to yield substituted 1-semicarbazides or thiosemicarbazides. In addition, the amino group of the 1-position may add to electron-deficient double bonds [56].

Formaldehyde and other aldehydes may add to all the available free NH groups to give methylol, alkylol, or polymeric products under basic conditions [57]. Aldehydes or ketenes usually give semicarbazone derivatives, and these in turn are used analytically to identify the purity or structure of a known aldehyde [3].

2-11. Preparation of 2,4-Dinitronaphthyl-1-semicarbazide [52]

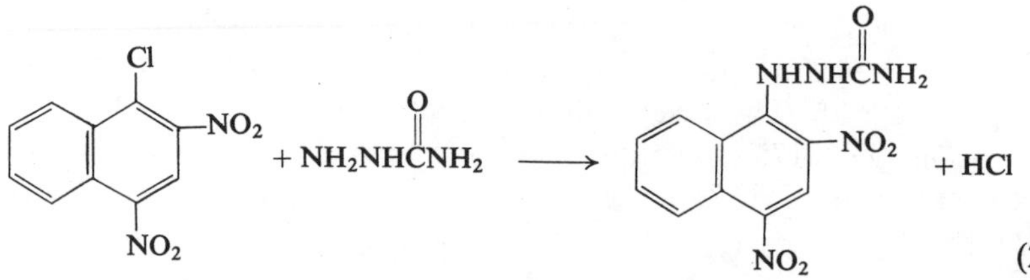

$$(20)$$

To a flask containing 8.8 gm (0.079 mole) of semicarbazide hydrochloride in 100 ml of water, 200 ml alcohol, and 158.5 ml of 0.5 N alcoholic sodium ethylate solution is added 10 gm (0.04 mole) of 1-chloro-2,4-dinitronaphthalene dissolved in 1 liter of ethanol. The red solution is refluxed for 10 min, allowed to stand overnight, and filtered to afford 5.0 gm (43.1 %) of 2,4-nitronaphthyl-1-semicarbazide (bright yellow), m.p. 185°–187°C dec.

2-12. Preparation of Semicarbazidomethylenemalononitrile [51]

$$C_2H_5OCH=C(CN)_2 + H_2N-NH\overset{\overset{\displaystyle O}{\|}}{C}NH_2 \longrightarrow H_2N\overset{\overset{\displaystyle O}{\|}}{C}NHNHCH=C(CN)_2 + C_2H_5OH$$

$$(21)$$

To a flask containing 11.1 gm (0.10 mole) of semicarbazide hydrochloride in 75 ml of water is added 8.2 gm (0.1 mole) of sodium acetate. To the latter solution is added 12.2 gm (0.10 mole) of ethoxymethylenemalononitrile in 175 ml of ethanol. On standing the clear solution deposits a tan solid. Filtration after 1½ hr affords 5.5 gm (36 %), m.p. 169°C (earlier reported [58] >300°C). The infrared spectrum has the following characteristic absorption bands: 2.91, 3.08 μ (NH), 4.51 μ (CN), 5.96 μ (C=O), 6.12 μ (amide); NMR: τ −40 (s, 1, NH), 1.28, 1.40 (singlets, 0.3H and 0.7H) (NHCH=C), 3.83 (bs, 2, NH$_2$).

In a related manner, substituted 4-semicarbazides and 4-thiosemicarbazides can react with other ethoxyvinyl derivatives as shown in Scheme 3 and Table VII.

TABLE VII[c,d] [51]

SEMICARBAZIDO- AND THIOSEMICARBAZIDOMETHYLENEMALONATES, AND
SEMICARBAZIDO- AND THIOSEMICARBAZIDO-2-CYANOACRYLATES

Compd.	M.p. (°C)	Yield (%)	Formula	Ir spectra		
				NH	C=O	CN
1a	190–192	97[a]	$C_{10}H_{17}N_3O_5$	3.00, 3.10	5.89, 6.01	—
1b	161–162	95	$C_{15}H_{19}N_3O_5$	3.00, 3.10	5.85 sh, 5.95, 6.10	—
1c	178–180	96	$C_{15}H_{17}Cl_2N_3O_5$	3.05, 3.15	5.85, 5.95, 6.01	—
2a	160–162	84	$C_{11}H_{19}N_3O_4S$	3.08, 3.24	5.87, 6.04	—
2b	151–153	66	$C_{12}H_{19}N_3O_4S$	3.08, 3.22	5.84, 6.05	—
2c	129–131	74[b]	$C_{15}H_{19}N_3O_4S$	3.10, 3.20	5.85, 6.03	—
3a	180–181	85[a]	$C_8H_{12}N_4O_3$	3.02, 3.10	5.85 sh, 5.95, 6.01	4.51
3b	158–161	85	$C_{13}H_{14}N_4O_3$	3.07	5.85 sh, 5.97	4.51
3c	161–163	72	$C_{13}H_{12}Cl_2N_4O_2$	3.00, 3.08	5.85 sh, 5.91, 6.05	4.51
4a	141–142	90	$C_9H_{14}N_4O_2S$	3.02, 3.20	5.85 sh, 5.92	4.51
4b	139–141	68	$C_{10}H_{14}N_4O_2S$	3.20	5.85, 5.90 sh	4.51
4c	137–138	28	$C_{13}H_{14}N_4O_2S$	3.20	5.85	4.51

[a] Reaction in 50% aqueous ethanol.
[b] Reaction in ethanol–THF, and solution cooled on ice to precipitate product.
[c] All reactions were allowed to stand overnight before isolating the product.
[d] Scheme 3 and Table VII are reprinted from R. K. Howe and S. C. Bolluyt, *J. Org. Chem.* **34**, 1713 (1969). Copyright 1969 by the American Chemical Society. Reprinted by permission of the copyright owner.

SCHEME 3 [51]

$$\underset{\text{RNHCNHNH}_2}{\overset{\overset{\text{X}}{\|}}{}} + \text{EtOCH=C(Y)CO}_2\text{Et} \xrightarrow{\text{EtOH}} \underset{\text{RNHCNHNHCH=C(Y)CO}_2\text{Et}}{\overset{\overset{\text{X}}{\|}}{}}$$

1a, X = O, R = CH₃,	Y = CO₂Et
b, X = O, R = C₆H₅,	Y = CO₂Et
c, X = O, R = 3,4-Cl₂C₆H₃,	Y = CO₂Et
2a, X = S; R = Et,	Y = CO₂Et
b, X = S; R = allyl,	Y = CO₂Et
c, X = S, R = C₆H₅,	Y = CO₂Et
3a, X = O, R = CH₃,	Y = CN
b, X = O, R = C₆H₅,	Y = CN
c, X = O, R = 3,4-Cl₂C₆H₃,	Y = CN
4a, X = S, R = Et,	Y = CN
b, X = S, R = allyl,	Y = CN
c, X = S, R = C₆H₅,	Y = CN

3. REACTION OF CARBAMIC ACID DERIVATIVES TO GIVE SEMICARBAZIDES

$$R^1-\underset{\underset{R^2}{|}}{N}H + COCl_2 \longrightarrow R^1\underset{\underset{R^2}{|}}{N}COCl \xrightarrow{R^3NH_2} R^1-\underset{\underset{R^2}{|}}{N}-CONHR^3 \quad (22)$$

$$R^1 = R^4NH$$
$$R^2, R^3, R^4 = \text{alkyl or aryl}$$

Phosgene reacts with phenylhydrazine hydrochloride, and the product is directly treated with aniline to afford 1,4-diphenylsemicarbazide [59] (Eq. 23).

$$C_6H_5-NHNH_2 + COCl_2 \longrightarrow C_6H_5-NHNHCOCl \xrightarrow{2C_6H_5-NH_2}$$

$$C_6H_5-NHNHCONHC_6H_5 + C_6H_5-\overset{+}{N}H_3Cl^- \quad (23)$$

Carbamyl chlorides, available by the reaction of phosgene on secondary amines, can react with hydrazines to give good yields of 1,1-dialkylsemicarbazides [31] (Eq. 24).

$$R_2NH + COCl_2 \longrightarrow R_2NCOCl \xrightarrow{H_2N-NH_2} R_2N\underset{\underset{O}{\|}}{C}-NHNH_2 \quad (24)$$

Alkyl carbamates prepared from the carbamyl chlorides also react with amines to give semicarbazides [15, 17] (Eq. 25).

$$C_6H_5-NHNHCOCl + ROH \longrightarrow C_6H_5-NHNHCOOR \xrightarrow{C_6H_5-NH_2}$$

$$C_6H_5-NHNH\underset{\underset{O}{\|}}{C}NHC_6H_5 + ROH \quad (25)$$

Examples of the products of the reaction of carbamyl chlorides and derivatives with hydrazines are listed in Table VIII.

TABLE VIII

REACTION OF CARBAMYL CHLORIDES AND DERIVATIVES WITH HYDRAZINES

Carbamyl derivative	Hydrazine	Product semicarbazide	Ref.
$(CH_3)_2NHCOCl$	NH_2NH_2	4,4-Dimethyl-	31
$(CH_3)_2NHCOCl$	$C_6H_5NHNH_2$	1-Phenyl-4,4-dimethyl-	60
$(C_6H_5)_2NHCOCl$	$C_6H_5-NHNH_2$	1,4,4-Triphenyl-	61
$C_2H_5OCNH_2$ $\underset{O}{\|\|}$	$C_6H_5NHNH_2$	1-Phenyl-	61
Benzoxazolane	NH_2NH_2	4-(2-Hydroxyphenol)-	62

4. REDUCTION METHODS

The reduction methods involve primarily the reduction of semicarbazones using various catalysts. However, since semicarbazones are not readily available, the reduction of nitroureas or guanidines to semicarbazides may be more useful if these compounds can be made readily. Nitroureas are reduced with either zinc dust and hydrochloric acid [63] or electrolytically, employing various electrodes made of iron [64], tin [65], lead [66], copper, nickel, or lead and mercury in hydrochloric or sulfuric acid solution [67, 68] (Eq. 26).

$$NH_2CONHNO_2 + H_2SO_4 + 6H \text{ (electrolytically reduced)} \longrightarrow$$

$$NH_2CONH\overset{+}{N}H_3HSO_4^- + 2H_2O \quad (26)$$

Semicarbazones are reduced by sodium amalgam in alcohol [69, 70], sodium in alcohol [71], sodium hydrosulfite [72], or catalytic hydrogenation using either platinum [73, 74], platinum oxide [75, 76], or Raney nickel [71]. The method involving sodium hydrosulfite is simple and is preferable to the other methods (Eq. 27).

$$R_2C{=}N{-}\underset{R}{N}{-}CONR_2 \xrightarrow{\text{(H)}} R_2CH{-}NH{-}\underset{R}{N}{-}CONR_2 \quad (27)$$

4-1. Preparation of 1-(2,3,5-Trimethyl-4-hydroxyphenyl)semicarbazide [72]

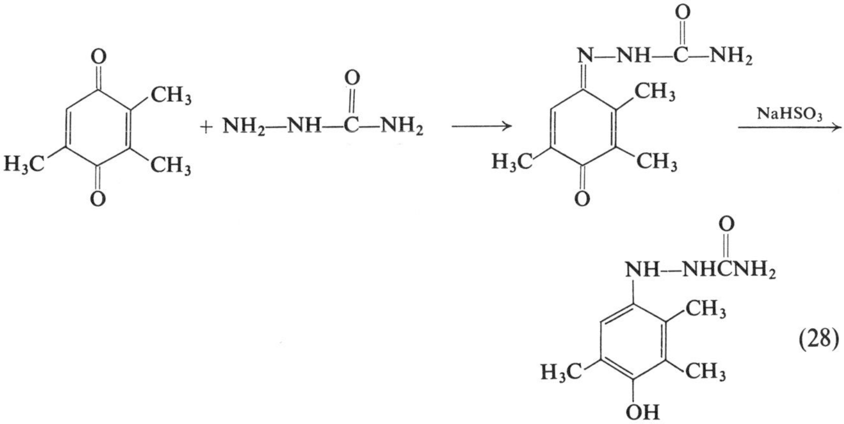

$$(28)$$

To a flask containing 40 ml of ethanol and 60 ml of water are added 7.5 gm (0.67 mole) of semicarbazide hydrochloride and 10 gm (0.67 mole) of 2,3,5-trimethylquinone. After standing for a few hours at room temperature, the yellow semicarbazone precipitates and is filtered to afford 12.7 gm (78%). Recrystallization from ethanol gives a crystalline product, m.p. 252°–253°C dec.

To another flask are added 1.22 gm (0.0059 mole) of the semicarbazone and 35 ml of alcohol. The mixture is refluxed and then 5.0 gm (0.048 mole) of sodium hydrosulfite in 25 ml of water is slowly added. The mixture immediately becomes a colorless solution which on concentration yields 1.12 gm (91 %) of the semicarbazide, m.p. 194°C dec. On exposure of the product to air in the solid form or in solution, oxidation occurs to yield the yellow semicarbazone, m.p. 250°–251°C dec.

4-2. Preparation of Fenchylsemicarbazide [77]

$$+ \text{Pt} + 2\text{H}_2 \longrightarrow \tag{29}$$

To a pressure bottle are added 43 ml of 10% chloroplatinic acid, 0.5 gm of gum arabic, and 10 ml of water, 50 ml of methyl alcohol, 1.0 ml of hydrochloric acid, and 10.0 gm (0.052 mole) of fenchylsemicarbazone. The mixture is shaken at over 2 atm hydrogen pressure until the theoretical amount of hydrogen is absorbed. The colloidal reaction mixture is added to acetone and the solution is concentrated under reduced pressure. The residue is precipitated with ammonium hydroxide. Recrystallization of the product from ethyl acetate affords 10.1 gm (100%), m.p. 181°C.

TABLE IX[d] [77]

HYDROGENATION EXPERIMENTS

Semicarbazide formed	Semicarbazone (gm)	Methyl alcohol (cm³)	Water (cm³)	HCl[a]	Pt. sol. (cm³)	Time (hr)	Yield (%)
Fenchyl-	10	50	10	Calc.	43	24	100
Carvomenthyl-	50	100	200	½ calc.	5	5	80
Hexahydrophenyl-	50	100	200	Trace	5	3	80
Menthyl-	50	100	200	½ calc.	5	2.75	80
i-Propyl-	50	100	200	Calc.[b]	5	5	76
Benzyl-	25	100	200	Calc.[c]	5	2.5	50
Bornyl-	10	20	40	½ calc.	10	20	100

[a] Concentrated hydrochloric acid was used to keep the volume of solution a minimum.

[b] The acid was added in three equal portions, one at the beginning and the other two when the hydrogen absorbed equal ⅓ and ⅔, respectively, of the amount required.

[c] This semicarbazone was not isolated, but benzaldehyde and semicarbazide hydrochloride, in calculated amounts, were added to the reduction mixture.

[d] Reprinted from E. J. Path and J. R. Bailey, J. Amer. Chem. Soc. 45, 3004 (1923). Copyright 1923 by the American Chemical Society. Reprinted by permission of the copyright owner.

Related semicarbazides as described in Table IX have also been produced by a similar technique.

5. MISCELLANEOUS METHODS

(1) Reaction of carbon monoxide with hydrazine to give semicarbazide [78].

(2) Reaction of hydroxylamine with *o*-tolylurea to give *o*-tolylsemicarbazide [79].

(3) Reaction of semicarbazides with aryl esters of cyanic acid to give semicarbamid-carbamic acid esters [80].

(4) Reaction between diiodoacetylene and semicarbazide to give hydrazodicarboxamide [81].

(5) Reaction of ethyl carbazinates with hydrazine to give carbohydrazides (4-aminosemicarbazides) [82, 83].

(6) Reaction of acyl azides with hydrazines to give semicarbazides [84–86].

(7) Reaction of chlorine with urea to give semicarbazide or hydrazinedicarboxamide [87, 88].

(8) Reaction of alkyl iodides and thiosemicarbazides to form *S*-alkylthiosemicarbazides [89].

(9) Reaction of hydrazine hydrate and phenylurethanes [4].

(10) Reaction of α-naphthyl isocyanate and hydrazides to yield 1-acyl-4-arylsemicarbazides [90].

(11) Preparation of polymers containing urethane and semicarbazide groups [91].

(12) Preparation of silyl-substituted semicarbazides from the reaction of trimethylsilyl isocyanate and hydrazine or derivatives [92].

(13) Bromination of unsaturated semicarbazides [93].

(14) Chloroacetylation of 4-arylsemicarbazides [94].

(15) Preparation of semicarbazidylphthalalides and use as plant growth regulators [95].

(16) Preparation of cyclic semicarbazide derivatives such as 3,4-dihydro-4-methyl-1*H*-1,3,4-benzothiazepine-2,5-dione [96].

(17) The catalyzed isomerization of (Z)- to (E)-benzaldehyde semicarbazone by aliphatic and aromatic thiols [97].

(18) Condensation of semicarbazide with diacetone acrylamide–acrylic acid copolymers [98].

(19) Metal complexes of benzaldehyde semicarbazone [99].

(20) Preparation of formaldehyde semicarbazone [100].

REFERENCES

1. G. D. Byrkit and G. A. Michalek, *Ind. Eng. Chem.* **42**, 1862 (1950).
2. L. Peyron, *Bull. Soc. Chim. Fr.* D12–26 (1954); *Chem. Abstr.* **48**, 5112 (1954).

3a. R. L. Shriner and R. C. Buson, "The Systematic Identification of Organic Compounds," 3d ed., pp. 229–231, 262–266, Wiley, New York, 1948.

3b. J. D. Warren, D. L. Woodward, and R. T. Hargreaves, *J. Med. Chem.* **20**(11), 1520 (1977).

3c. K. Sugie, N. Saiki, S. Kurisu, K. Shoji, Jap. Kokai Tykkyo Koho 79/21,454 (Feb. 17, 1979); *Chem. Abstr.* **91**, 5938a (1979).

3d. H. Shimiau, K. Hayashi, and N. Takemura, Nippon Eiseigaku Azsshi **33**(3), 474 (1978); *Chem. Abstr.* **90**, 133751a (1979).

4. T. Curtius and A. Burkhardt, *J. Prakt. Chem.* **58**, 205 (1898).

5. A. S. Wheeler and J. G. Park, *J. Amer. Chem. Soc.* **51**, 3079 (1929).

6. L. C. Raiford and W. T. Daddow, *J. Amer. Chem. Soc.* **53**, 1552 (1931).

7. S. Dutta, B. P. Das, B. K. Paul, A. K. Acharyya, and U. P. Basu, *J. Org. Chem.* **33**, 858 (1968).

8. T. Curtius and K. Heidenreich, *Chem. Ber.* **27**, 57 (1898).

9. T. Curtius and K. Heidenreich, *J. Prakt. Chem.* **52**, 465 (1895).

10. S. M. Mistry and P. C. Guha, *J. Indian Chem. Soc.* **7**, 793 (1930).

11. A. Andraga, *Anal. Quin. Farm.* 15 (1941).

12. H. Milrath, *Monatsh. Chem.* **29**, 337 (1908).

13. K. H. Slotta and K. R. Jacobi, *Z. Anal. Chem.* **77**, 344 (1929).

14. J. M. Das Gapta, *J. Indian Chem. Soc.* **10**, 111 (1933).

15. French Pat. 349,962 (June 6, 1904) to Bayer Co.

16. H. Milrath, *Monatsh. Chem.* **29**, 909 (1908).

17. A. A. Mangini and C. Deliddo, *Gazz. Chim. Ital.* **66**, 343 (1936).

18. A. S. Wheeler and C. T. Thomas, *J. Am. Chem. Soc.* **51**, 3135 (1929).

19. M. Amorosa, *Farm. Sci. Tec.* 3 389 (1948).

20. P. P. T. Sah, *J. Chin. Chem. Soc.* (*Peiping*) **2**, 32 (1934).

21. A. S. Wheeler, *Org. Syn.* **1**, 450 (1941).

22. P. P. T. Sah, *J. Chin. Chem. Soc.* (*Peiping*) **4**, 187 (1936).

23. P. P. T. Sah and H. H. Lei, *J. Chin. Chem. Soc.* (*Peiping*) **2**, 167 (1934).

24. M. Busch and G. Haase, *J. Prakt. Chem.* **115**, 186 (1927).

25. A. S. Wheeler and T. T. Walker, *J. Amer. Chem. Soc.* **47**, 2792 (1925).

26. P. P. T. Sah and I. S. Kao, *Rec. Trav. Chim.* **54**, 253 (1935).

27. P. P. T. Sah and S. H. Chiang, *J. Chin. Chem. Soc.* (*Peiping*) **4**, 496 (1936).

28. P. P. T. Sah and S. H. Chiang, *J. Chin. Chem. Soc.* (*Peiping*) **4**, 501 (1936).

29. B. Toschi, *Gazz. Chim. Ital.* **44**, 443 (1914).

30. T. Vitali and S. Sardella, *Chimica* **7**, 229 (1952); *Chem. Abstr.* **47**, 6414 (1953).

31. C. Vogelsang, *Rec. Trav. Chim.* **62**, 5 (1943).

32. K. A. Taipale and P. V. Usachev, *J. Russ. Phys. Chem. Soc.* **62**, 1241 (1930).

33. T. Curtius, *J. Prakt. Chem.* **85**, 137, 393 (1904).

34. H. L. Lochte, W. A. Noyes and J. R. Bailey, *J. Amer. Chem. Soc.* **44**, 2556 (1922).

35. W. Borsche, W. Müller, and C. A. Bodenstein, *Ann. Chem.* **472**, 201 (1929).

36. K. Michaelis, *Chem. Ber.* **41**, 1427 (1908).

37. U. S. Rubber Co., Brit. Pat. 896,497 (May 16, 1962).

38. E. S. Scott and L. F. Audrieth, *J. Org. Chem.* **19**, 742 (1953).

39. G. Gever, C. O'Keefe, G. Drake, F. Ebetino, J.Michels, and K. Hayes, *J. Amer. Chem. Soc.* **77**, 2277 (1955).

40. R. C. Goodwin and J. R. Bailey, *J. Amer. Chem. Soc.* **46**, 2827 (1924).

41. E. S. Scott, E. E. Zeller, and L. F. Audrieth, *J. Org. Chem.* **19**, 749 (1953).

42a. V. M. Gelderen, *Rec. Trav. Chim.* **52**, 979 (1933).

42b. N. P. Peet, S. Sunder, and R. J. Cregge, *J. Org. Chem.* **41**, 2733 (1976).

43. J. Farago, U.S. Pat. 2,980,651 (April 18, 1961).

44. S. Petersen, *Ann. Chem.* **562**, 205 (1949).
45. M. Busch and R. Frey, *Chem. Ber.* **36**, 1362 (1903).
46. W. Marchwald, *Chem. Ber.* **25**, 3098 (1892).
47. M. Busch and H. Holzmann, *Chem. Ber.* **34**, 320 (1901).
48a. C. C. Pacilly, *Rec. Trav. Chim.* **55**, 101 (1936).
48b. H. D. Winkelmann, H. Oertel, and N. Weimann, Ger. Offen. 2,542,449 (Apr. 7, 1977); *Chem. Abstr.* **87**, 24675w (1977).
49. M. Busch, E. Opfermann, and H. Walterh, *Chem. Ber.* **37**, 2318 (1904).
50. R. S. Levy, *Mem. Poudres* **40**, 429 (1958).
51. R. K. Howe and S. C. Bolluyt, *J. Org. Chem.* **34**, 1713 (1969).
52. H. W. Talen, *Rec. Trav. Chim.* **47**, 345 (1938).
53. A. Doucte, *C. R. Acad. Sci. Paris* **177**, 1120 (1923).
54. P. C. Guha and S. P. Mukherjee, *J. Indian Inst. Sci.* **28A**, 63 (1946).
55. H. Mazonrewith, *Bull. Soc. Chim. Fr.* 1183 (1924).
56. D. E. Worral and F. Benington, *J. Amer. Chem. Soc.* **62**, 493 (1940).
57. F. L. Johnston, U.S. Pat. 2,308,696 (Jan. 19, 1943).
58. M. J. Kamlet, *J. Org. Chem.* **24**, 714 (1959).
59. T. Lieser and G. Nischk, *Chem. Ber.* **82**, 527 (1949).
60. E. Jolles and B. Bini, *Gazz. Chim. Ital.* **68**, 510 (1938).
61. N. A. Valyashko and I. T. Depeshko, *Zh. Obsch. Khim.* **20**, 479 (1950).
62. J. D. Bower and F. F. Stephens, *J. Chem. Soc.* 325 (1951).
63. J. Thiele and C. Heuser, *Ann. Chem.* **288**, 312 (1895).
64. V. C. Sprefer and E. Briner, *Helv. Chim. Acta.* **32**, 215 (1949).
65. H. J. Backer, *Rev. Trav. Chem.* **31**, 25 (1912).
66. A. W. Ingersoll, L. J. Bircher, and M. M. Brubaker, *Org. Syn.* **1**, 485 (1941).
67. L. J. Bircher, A. W. Ingersoll, B. F. Armendt, and G. Cook, *J. Amer. Chem. Soc.* **47**, 393 (1925).
68. M. Levi and I. Pesheva, *Khim. Ind. (Sofia)* **39**, 203 (1967).
69. S. Kessler, and H. Rupe, *Chem. Ber.* **45**, 26 (1912).
70. H. Rupe and E. Destreicher, *Chem. Ber.* **45**, 30 (1912).
71. J. Gole, *Bull. Soc. Chem. Fr.* 894 (1949).
72. L. I. Smith and W. M. Schubert, *J. Amer. Chem. Soc.* **70**, 2656 (1948).
73. D. Neighbors, A. L. Foster, S. M. Clark, J. E. Miller, and J. R. Bailey, *J. Amer. Chem. Soc.* **44**, 1557 (1922).
74. K. A. Taipale and A. S. Smirnov, *Chem. Ber.* **56B**, 1794 (1923).
75. C. S. Marvel and D. B. Glass, *J. Amer. Chem. Soc.* **60**, 1051 (1938).
76. E. L. May and E. Mosettig, *J. Org. Chem.* **14**, 1137 (1949).
77. E. J. Poth and J. R. Bailey, *J. Amer. Chem. Soc.* **45**, 3004 (1923).
78. G. D. Buckley and N. H. Ray, *J. Chem. Soc.* 1156 (1949).
79. H. H. Lei, P. P. T. Sah, and C. Shih, *J. Chin. Chem. Soc. (Peiping)* **3**, 246 (1935).
80. E. Grigat and R. Pütter, *Chem. Ber.* **98**, 2619 (1965).
81. A. T. d'Arcangels, *Rev. Fac. Cienc. Quim.* **18**, 81 (1943).
82. P. C. Guha and M. A. Hyd, *J. Indian Soc.* **7**, 933 (1930).
83. E. P. Nesinov and P. S. Pel'kis, *Dopov. Akad. Nauk Ukr. RSR* 1080 (1962).
84. M. O. Forster, *J. Chem. Soc.* **95**, 433 (1909).
85. R. Stolle, N. Nieland, and M. Merkle, *J. Prakt. Chem.* **116**, 192 (1927).
86. E. Oliveri-Mandala, *Gazz. Chim. Ital.* **44-I**, 662 (1914).
87. C. P. Riley, Jr., R. Strauss, and W. P. Terhorst, U.S. Pat. 3,238,226 (March 1, 1966).
88. C. P. Riley, Jr., R. Strauss, and W. P. Terhorst, Brit. Pat. 1,063,893 (April 5, 1967).
89. E. Lieber and G. B. L. Smith, *Chem. Rev.* **25**, 213 (1939).

90. V. S. Misra and R. S. Varma, *J. Indian Chem. Soc.* **40**, 799 (1963).

91. H. Oertel, H. Rinke, and W. Thoma, Ger. Pat. 1,123,467 (Feb. 6, 1962).

92. U. Wannaget and C. Kreuger, *Monatsh. Chem.* **94**, 63 (1963).

93. F. Gozzo, P. M. Boschi, and A. Longari, Ger. Offen. 2,750,813 (May 18, 1978); *Chem. Abstr.* **89**, 109506h (1978).

94. Y. V. Svetkin, A. N. Minlibaeva, A. M. Vinokurov, and R. Z. Biglova, Jr., *Organ Khim.* (3), 121 (1975); *Chem. Abstr.* **86**, 189362 (1977).

95. J. L. Kirkpatrick and R. N. Patel, U.S. Pat. 4,272,279 (June 9, 1981).

96. N. P. Peet and S. Sunder, *J. Org. Chem.* **40**, 1909 (1975).

97. P. R. Conlon and J. M. Sayer, *J. Org. Chem.* **44**, 262 (1979).

98. W. F. DeWinter, D. M. Timmerman, A. H. DeCat, and G. A. Kockelbergh, Ger. Offen. 2,728,072 (Jan. 12, 1978); *Chem. Abstr.* **89**, 34124w (1978).

99. C. B. Mahto, *J. Indian Chem. Soc.* **57**(5), 553 (1980); *Chem. Abstr.* **93**, 178666e (1980).

100. M. Pomerantz, S. Bittner, and S. B. Khader, *J. Org. Chem.* **47**, 2217 (1982).

1. INTRODUCTION

The application of carbodiimide to problems of organic synthesis is of recent origin. The preparation and study of this functional group has been gaining interest because of its application as a condensing agent. The first review of carbodiimides was published in 1953 [1a], and a more recent one appeared in 1967 [1b]. Other reviews are also worth considering [2–5]. Substituted carbodiimides have been applied to the synthesis of amidines, anhydrides, amides, carboxylic acids, sulfonic acids, esters, and other functional groups [2]. The synthesis of peptide linkages with the aid of carbodiimides has also attracted a great amount of interest [2a]. The use of carbodiimides to help prepare immobilized enzymes is increasing [2b].

The earliest preparations of carbodiimides involved the reaction of mercuric or lead oxide with thioureas [6–10].

The most practical methods for the preparation of carbodiimides are summarized in Scheme 1. The carbodiimide structure may be viewed as a dehydrated urea.

SCHEME 1

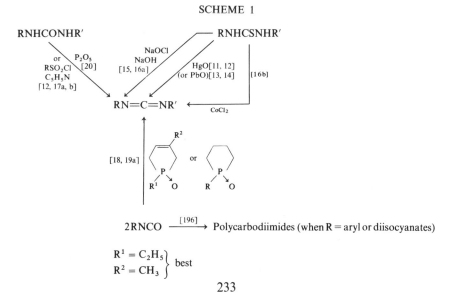

$$2\,RNCO \xrightarrow{[196]} \text{Polycarbodiimides (when R = aryl or diisocyanates)}$$

$$\left.\begin{array}{l} R^1 = C_2H_5 \\ R^2 = CH_3 \end{array}\right\} \text{ best}$$

233

The structure of N,N'-disubstituted carbodiimide is distinct from the corresponding disubstituted cyanamides and is supported by chemical and physical evidence (I, II).

$$RN{=}C{=}NR'$$

$$\begin{matrix} R \\ \\ R' \end{matrix}{\Large >}N{-}C{\equiv}N$$

Carbodiimide   Cyanamide

(I)   (II)

At room temperature aliphatic and aromatic carbodiimides are liquid or solid. They usually can be purified by distillation under reduced pressure or by recrystallization to yield neutral products. On prolonged standing they polymerize to yield basic products. The liquid carbodiimides are less stable than the solid ones. The stability of carbodiimides increases with the degree of branching of the alkyl substituents attached to the nitrogen atom: $RCH_2 < R_2CH < R_3C$. For example, diethylcarbodiimide [14, 21] polymerizes in a few days, whereas diisopropyl- and dicyclohexylcarbodiimides [14] are stable for several months. Unsaturated substituents also cause a marked decrease in stability [22].

In the aromatic series, diphenylcarbodiimide [23] and di-p-iodophenyl-carbodiimide [23] polymerize in a few days, whereas di-p-tolylcarbodiimide when obtained in pure crystalline form is stable for several months [24, 25]. Di-p-dimethylaminophenylcarbodiimide is stable for three years [25].

2. ELIMINATION REACTIONS

A. Catalytic Conversion of Isocyanates into Carbodiimides

$$2RNCO \longrightarrow RN{=}C{=}NR + CO_2 \qquad (1)$$

Hofmann in 1885 [26] reported that heating of isocyanate esters in the absence of catalysts gave products which analyzed correctly for carbodiimides.

Prolonged heating of free isocyanates in the absence of catalysts has been reported recently to give poor yields of carbodiimides unless a slow stream of nitrogen was passed through the boiling isocyanate [18]. In the presence of phosphorus catalysts [18, 19, 27–39], aromatic carbodiimides are obtained in high yield under mild conditions from isocyanates but not isothiocyanates [18]. Aliphatic isocyanates react more slowly, but improved yields are obtained in high-boiling solvents. Under the latter conditions the carbodiimides may be contaminated by isocyanuric acid derivatives, complex products, or polymeric carbodiimides [35]. For reasons not well understood, diphenylmethyl isocyanate fails to give the carbodiimide, giving only the isocyanuric acid derivative [35].

Some examples of the catalytic conversion of isocyanates into carbodiimides are shown in Table I.

In the case of aromatic isocyanates, electron-withdrawing groups increase the rate of carbodiimide formation in proportion to their electron-withdrawing power. Electron-releasing groups tend to inhibit the reaction. p-Nitrophenyl isocyanate reacts almost explosively to yield the carbodiimide when catalyzed [5]. o-Chlorophenyl isocyanate reacts seven times as fast as the o-methyl analog [33].

Phosphoramidate anion also reacts with isocyanates to give carbodiimides [40, 41]. However, the procedure is limited by the tendency of many isocyanates, especially the straight-chained aliphatic ones, to polymerize in the basic medium [42].

TABLE I[a]

CATALYTIC REACTION OF ISOCYANATES TO CARBODIIMIDES

$$2\text{RNCO} \longrightarrow \text{RN}=\text{C}=\text{NR} + CO_2 \qquad (2)$$

Substituent R	Temp. (°C)	Time (hr)	Solvent	Carbodi-imides (yield %)	B.p.,°C n_D^t	M.p. (°C)
C_6H_5-	50	2.5	Benzene	94	119–121 (0.4–0.5) n_D 1.6372 (25)	—
$o\text{-}CH_3C_6H_4-$	50–55	95	None	87	118–122 (0.1–0.15) n_D 1.6230 (25)	—
$m\text{-}CH_3C_6H_4-$	25	16	None	98	151 (1.1)	—
$p\text{-}CH_3C_6H_4-$	140	4	Xylene	88	165 (2.4)	56.6–57.4
$o\text{-}CH_3O\text{—}C_6H_4-$	25	16	None	92	—	73.5–74.5
$p\text{-}CH_3O\text{—}C_6H_4-$	140	4	Xylene	99	—	52–53
$m\text{-}Cl\text{—}C_6H_4-$	25	2	None	99	—	42–43
$p\text{-}Cl\text{—}C_6H_4-$	25	2	None	91	—	56–57
$o\text{-}NO_2\text{—}C_6H_4-$	50	$\frac{1}{6}$	Cyclohexane	95	—	93–96
$m\text{-}NO_2C_6H_4-$	60	$\frac{1}{6}$	None	91	—	157–159
$p\text{-}NO_2\text{—}C_6H_4-$	60–62	4	CCl$_4$	52	—	166–167
$\alpha\text{-}C_{10}H_7$	140	4	Xylene	99	—	91–91.5
$n\text{-}C_4H_9$	110–114	15	None	60	82 (8)	—
$2\text{-}C_8H_7$	140–145	$\frac{1}{2}$	None	89	108–109 (1.5)	—

[a] Data taken from T. W. Campbell, J. J. Monagle, and V. S. Foldi, *J. Amer. Chem. Soc.* **84**, 3673 (1962).

Other compounds were screened by Monagle [32] for catalytic activity by reaction with phenyl isocyanate to form diphenylcarbodiimide. Traces of impurities, especially in the isocyanate, had a large effect on the yield. The data reported in Tables II and III give the best of several results obtained with each compound.

It has been suggested that the results in Tables II and III are in agreement with the mechanism proposed in Eqs. (3) and (4) [32].

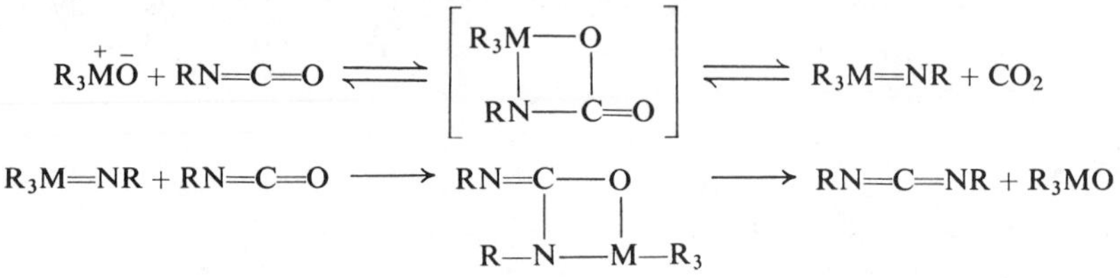

$$R_3\overset{+}{M}\overset{-}{O} + RN{=}C{=}O \rightleftharpoons \left[\begin{array}{cc} R_3M{-}O \\ | \quad | \\ RN{-}C{=}O \end{array} \right] \rightleftharpoons R_3M{=}NR + CO_2 \qquad (3)$$

$$R_3M{=}NR + RN{=}C{=}O \longrightarrow \begin{array}{c} RN{=}C{-}O \\ | \quad | \\ R{-}N{-}M{-}R_3 \end{array} \longrightarrow RN{=}C{=}NR + R_3MO \qquad (4)$$

M = Group VB and Group VIB elements

In agreement with these mechanism, it was shown [43] that, when ^{18}O-enriched phosphine oxides were substituted as catalysts, the evolved CO_2 contained significant amounts of ^{18}O. Compounds with high dipole moments

TABLE II[b] [32]

GROUP VB AND VIB OXIDES AS CATALYSTS FOR
CARBODIIMIDE FORMATION[a]

Catalyst	Amount (moles)	Temp. (°C)	Time (hr)	Yield (%)
Triphenylphosphine oxide	0.004	115–120	2	5
	0.004	160–162	3	27
	0.004	160–162	8	56
Triphenylarsine oxide	0.004	70–75	1	85
Triphenylstibine oxide	0.004	115–120	2	41
Tri(m-nitrophenyl) phosphine oxide	0.004	160–164	3	16
Tri(m-nitrophenyl) arsine oxide	0.004	90	3	85
Triethylphosphine oxide	0.006	122–125	3	64
Pyridine N-oxide	0.027	110–123	1.5	33
Dimethylsulfoxide	0.006	162–164	2.5	13
Diphenylsulfoxide	0.004	145	7	0
Diphenylsulfone	0.004	145	5	0
Trimethylamine oxide	0.004	111–126	1.5	0
1-Phenylarsolidine 1-oxide	0.002	99–108	1	0
4-Nitropyridine N-oxide	0.007	160	1.5	0

[a] Phenyl isocyanate (0.45 mole) used in each case.
[b] Reprinted from J. J. Monagle, *J. Org. Chem.* **27**, 3851 (1962). (Copyright 1962 by the American Chemical Society. Reprinted by permission of the copyright owner.)

TABLE III[b] [32]

PHOSPHORUS COMPOUNDS AS CATALYSTS FOR CARBODIIMIDE FORMATION[a]

Catalyst	Formula	Amount (moles)	Temp. (°C)	Time (hr)	Yield (%)
Hexamethylphosphoramide	$[(CH_3)_2N]_3PO$	0.008	113–135	1.6	27
Dilauryllaurylphosphonamide	$C_{12}H_{25}P(O)(NHC_{12}H_{25})_2$	0.0004	162–163	102	14.4
Triphenylphosphoramide	$(C_6H_5NH)_3PO$	0.004	141–200	20	66.6
Diethyl phosphoramidate	$(C_3H_5O_2)P(O)NH_2$	0.004	140–200	19	77.6
Triethyl phosphate	$(C_2H_5O)_3PO$	0.004	150–175	3	32
Triallyl phosphate	$(CH_2=CHCH_2O)_3PO$	0.004	142–164	3	4.8
Triphenyl phosphate	$(C_6H_5O)_3PO$	0.04	160–190	21.5	62.8
Bis(β-chloroethyl) vinylphosphonate	$CH_2=CHP(O)(OCH_2CH_2Cl)_2$	0.005	147–191	1	88
	$CH_2=CHP(O)(OCH_2CH_2Cl)_2$	0.004	160–164	4	8
Bis(β-chloroethyl-β-chloroethylphosphonate	$ClCH_2CH_2P(O)(OCH_2CH_2Cl)_2$	0.005	145–172	5	20
Ethylene phenylthiophosphonate	$(CH_2O)_2P(S)C_6H_5$	0.004	162	16.5	20

[a] Phenyl isocyanate (0.45 mole) used in each case.

[b] Reprinted from J. J. Monagle, J. Org. Chem. 27, 3851 (1962). (Copyright 1962 by the American Chemical Society. Reprinted by permission of the copyright owner.)

or the highest negative charge on the oxygen of $R_3M^+O^-$ should exhibit the best catalytic activity. This may account for the fact that triphenylarsine oxide (5.50*D*) is better than triphenylphosphine oxide (4.31*D*) as a catalyst. Other elements such as nitrogen (as in pyridine *N*-oxide) are not capable of forming a pentacovalent bond but may help to polarize the isocyanate group so that it reacts with another isocyanate group to give the carbodiimide product. A compound similar to $R_3M{=}NR$ had been prepared earlier by Staudinger and Meyer [44], who showed that it reacts with isocyanate groups to give carbodiimides (Eqs. 5, 6).

$$R_3P + R'N_3 \longrightarrow R_3P{=}NR' + N_2 \tag{5}$$

$$R_3P{=}NR' \xrightarrow{CO_2} R_3PO + R'NCO$$

$$R_3P{=}NR' \xrightarrow{R''NCO} R''N{=}C{=}NR' + R_3PO \tag{6}$$

Other more recent examples of the catalytic conversion of isocyanates and isothiocyanates to carbodiimide are shown in Tables IV and V.

The use of reusable polymer catalysts is shown in Table V. They have the advantage of being easily separated from the product and then used again.

2-1. Preparation of Diphenylcarbodiimide [18]

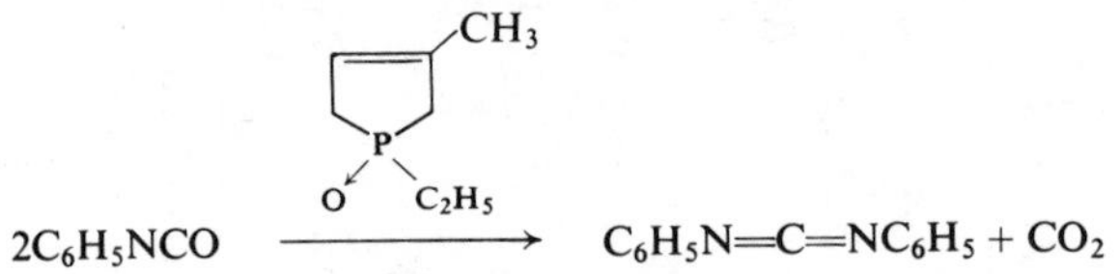

CH3
P
O
C2H5
2C6H5NCO
C6H5N=C=NC6H5 + CO2

(7)

(*a*) *Catalyst formation. Preparation of 1-ethyl-3-methyl-3-phospholine 1-oxide.*

CH3
C2H5PCl2 + CH2=C—CH=CH2
CH3
P
Cl
Cl
C2H5
H2O
CH3
P
O
C2H5

(8)

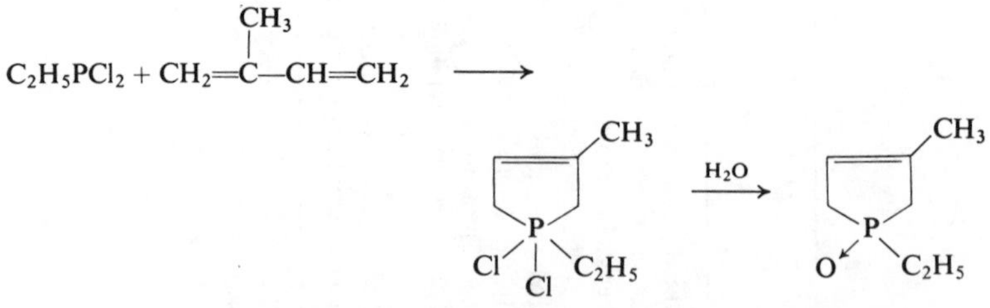

To a 2 liter, four-necked flask equipped with a spiral condenser topped with a Dry Ice condenser, 1 liter dropping funnel with pressure-equalizing arm, and

TABLE IV

MORE RECENT REPORTS ON THE CATALYTIC CONVERSION OF ISOCYANATES OR ISOTHIOCYANATES TO CARBODIIMIDES

Catalyst	RNCO/RNCS	Temp. (°C)	Time (hr)	Yield (%)	Product	B.p., °C (mm Hg)	n_D (°C)	Ref.
$(C_6H_5)_3As$	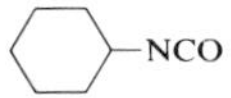 cyclohexyl–NCO	140–148	1.7	83.5	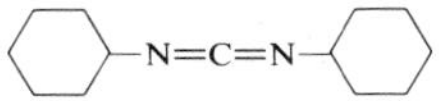 cyclohexyl–N=C=N–cyclohexyl	95–97 (0.1)	—	[a]
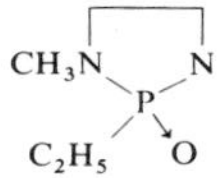 $CH_3N{-}P{-}O$, C_2H_5	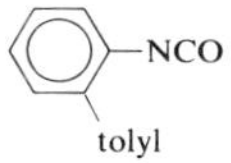 tolyl–NCO (tolyl)	Reflux	—	84.0	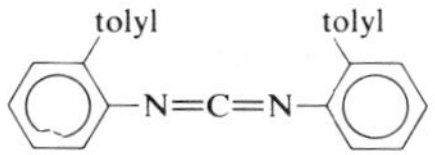 tolyl–N=C=N– (tolyl)	128–130 (0.3)	—	[b]
KOt-Bu	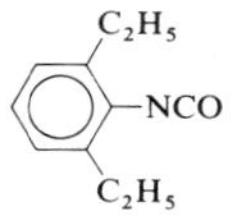 C_2H_5 substituted –NCO C_2H_5	200–250	1.5	90.1	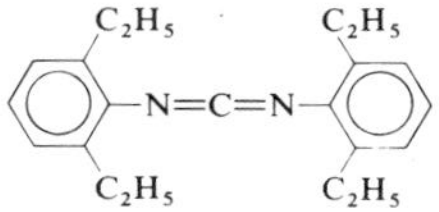 C_2H_5 –N=C=N– C_2H_5	194–197 (0.5)	1.5912 (23.5)	[c]
$Fe(CO)_5$, $W(CO)_6$ or $Mo(CO)_6$	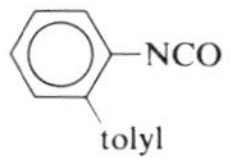 –NCO (tolyl)	180–250	0.6	85.0	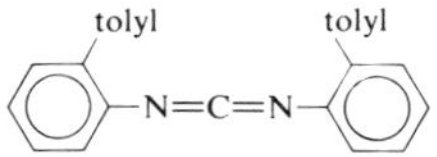 –N=C=N– (tolyl)	130–132 (0.1)	—	[d]
$Ti(OiPr)_4$	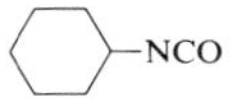 cyclohexyl–NCO	100–102	4.5	87	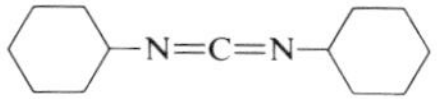 cyclohexyl–N=C=N–cyclohexyl	—	—	[e]
$Zr(OiPr)_4$	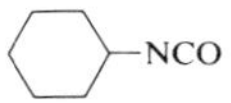 cyclohexyl–NCO	180	10	64.3	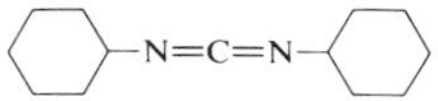 cyclohexyl–N=C=N–cyclohexyl	—	—	[e]
$[Bu_3Sn]_2O$	$CH_2{=}CH{-}CH_2NCS$	100 (90 mm Hg)	7	10	Polymerized diallylcarbodiimide	—	—	[f]
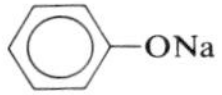 –ONa	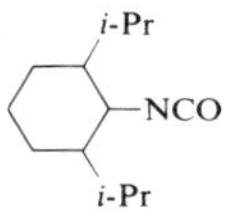 i-Pr substituted –NCO i-Pr	180–200	4	95	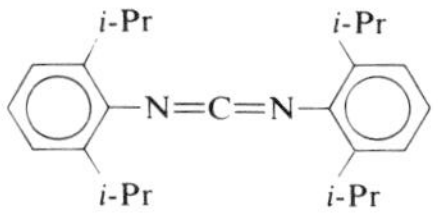 i-Pr –N=C=N– i-Pr	152–160 (0.05) (m.p. 47–49.5)	—	[g]

(continued)

TABLE IV (*continued*)

MORE RECENT REPORTS ON THE CATALYTIC CONVERSION OF ISOCYANATES OR ISOTHIOCYANATES OF CARBODIIMIDES

Catalyst	RNCO/RNCS	Temp. (°C)	Time (hr)	Yield (%)	Product	B.p., °C (mm Hg)	n_D (°C)	Ref.
$C_6H_5-\overset{O}{\overset{\|\|}{P}}(NH_2)_2$	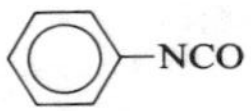 C_6H_5-NCO	170–180	14	98	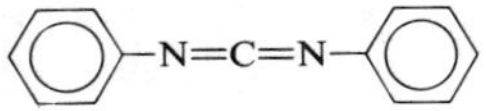 $C_6H_5-N=C=N-C_6H_5$	—	—	h
$(ClCH_2)_3P \rightarrow O$	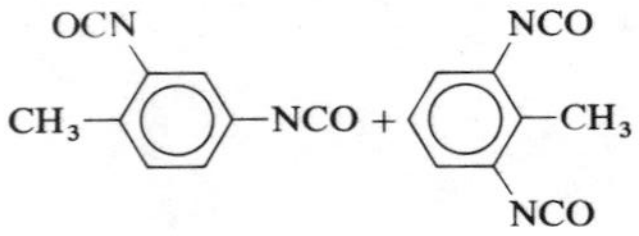 toluene diisocyanate mixture (CH$_3$ ring with NCO groups)	80	2		Liq. carbodiimide containing organic polyisocyanate	—	—	i
Camphene phenyl phosphine oxide	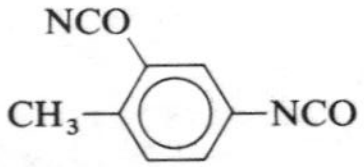 CH$_3$-substituted aryl diisocyanate	125	1	100	Polymeric carbodiimide	—	—	j
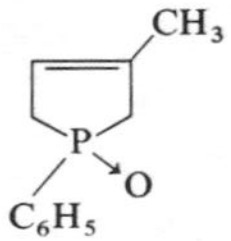 methyl phospholene phenyl oxide	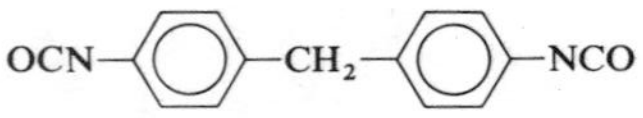 $OCN-C_6H_4-CH_2-C_6H_4-NCO$	25–130	1.5	100	Polymeric carbodiimide			k
n-BuP$(C_6H_5)_3$P$^+$ Br$^-$	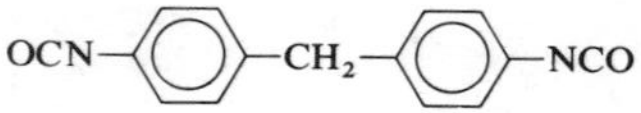 $OCN-C_6H_4-CH_2-C_6H_4-NCO$	200	3	100	Polymeric carbodiimide			l
$\overset{O}{\overset{\|\|}{CH_3C}}NH_2$	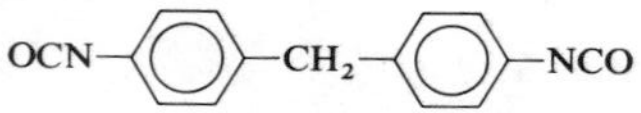 $OCN-C_6H_4-CH_2-C_6H_4-NCO$	225	1	100	Polymeric carbodiimide			m
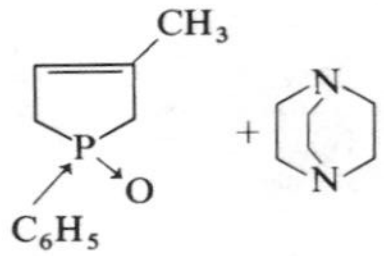 piperazine + methyl phospholene phenyl oxide	 $CH_3-C_6H_4-NCO$	115	8	77	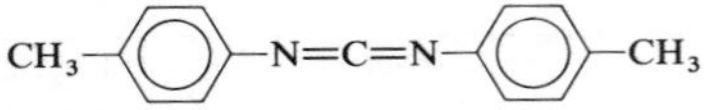 $CH_3-C_6H_4-N=C=N-C_6H_4-CH_3$	—	—	n

240

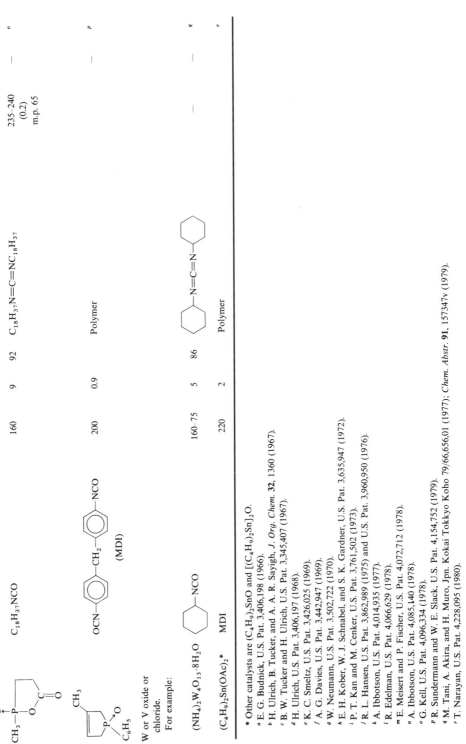

Isocyanate	Temp	%	Product	b.p./m.p.	Ref.
$C_{18}H_{37}NCO$	160	9	$C_{18}H_{37}N=C=NC_{18}H_{37}$	235–240 (0.2) m.p. 65	o
OCN–C₆H₄–CH₂–C₆H₄–NCO (MDI)	200	0.9	Polymer	—	p
cyclohexyl–NCO	160–75	5	86	(dicyclohexylcarbodiimide) —	q
MDI	220	2	Polymer	—	r

CH₃–P(=O) (phospholane oxide structure)

CH₃ / phospholene oxide with C_6H_5

W or V oxide or chloride.

For example:

$(NH_4)_2W_4O_{13}\cdot8H_2O$

$(C_4H_9)_2Sn(OAc)_2$*

241

* Other catalysts are $(C_4H_9)_2SnO$ and $[(C_4H_9)_2Sn]_2O$.

[a] E. G. Budnick, U.S. Pat. 3,406,198 (1966).
[b] H. Ulrich, B. Tucker, and A. A. R. Sayigh, J. Org. Chem. 32, 1360 (1967).
[c] B. W. Tucker and H. Ulrich, U.S. Pat. 3,345,407 (1967).
[d] H. Ulrich, U.S. Pat. 3,406,197 (1968).
[e] K. C. Smeltz, U.S. Pat. 3,426,025 (1969).
[f] A. G. Davies, U.S. Pat. 3,442,947 (1969).
[g] W. Neumann, U.S. Pat. 3,502,722 (1970).
[h] E. H. Kober, W. J. Schnabel, and S. K. Gardner, U.S. Pat. 3,635,947 (1972).
[i] P. T. Kan and M. Cenker, U.S. Pat. 3,761,502 (1973).
[j] R. L. Hansen, U.S. Pat. 3,862,989 (1975) and U.S. Pat. 3,960,950 (1976).
[k] A. Ibbotson, U.S. Pat. 4,014,935 (1977).
[l] R. Edelman, U.S. Pat. 4,066,629 (1978).
[m] E. Meisert and P. Fischer, U.S. Pat. 4,072,712 (1978).
[n] A. Ibbotson, U.S. Pat. 4,085,140 (1978).
[o] G. Kell, U.S. Pat. 4,096,334 (1978).
[p] R. Sundermann and W. E. Slack, U.S. Pat. 4,154,752 (1979).
[q] M. Tani, A. Akira, and H. Muro, Jpn. Kokai Tokkyo Koho 79/66,656,01 (1977); Chem. Abstr. 91, 157347v (1979).
[r] T. Narayan, U.S. Pat. 4,228,095 (1980).

TABLE V

Polymer Catalysts for the Conversion of Isocyanates to Carbodiimides

Polymer catalyst	RNCO/RNCS	Temp. (°C)	Time (hr)	Yield (%)	Product	B.p., °C (mm Hg)	n_D	Ref.
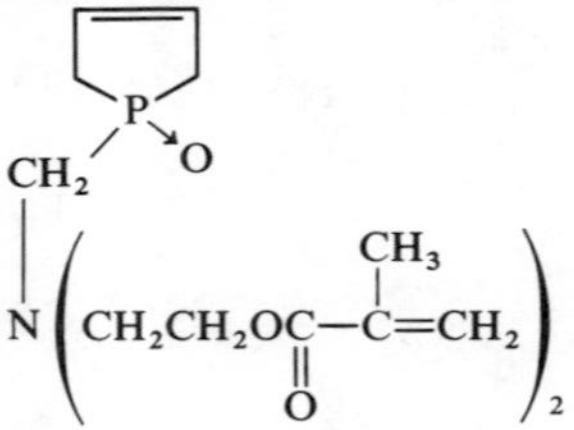	C_6H_5NCO in hexene	60	9.25	100	$C_6H_5N=C=NC_6H_4$	—	—	[a]
Polymer of above monomer is catalyst for above entry	C_6H_5NCO	155–160	2.5	—	$C_6H_5N=C=N-C_6H_5$	—	—	[b]

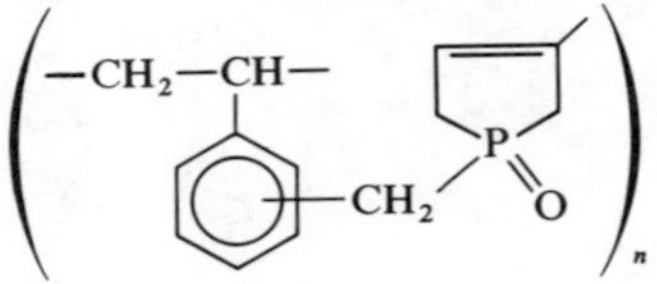

242

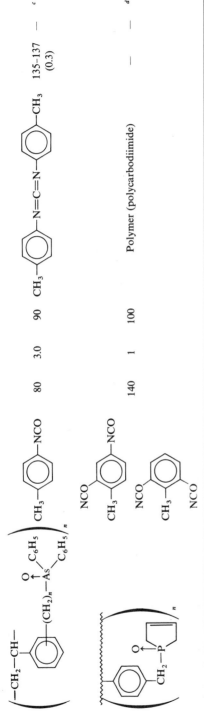

	80	3.0	90	135–137 (0.3) [c]
	140	1	100	— — [d]

Polymer (polycarbodiimide)

[a] C. P. Smith, U.S. Pat. 4,105,643 (1978).
[b] C. P. Smith, U.S. Pat. 4,105,642 (1978). C. P. Smith, U.S. Pat. 4,068,055 (1978).
[c] C. P. Smith, U.S. Pat. 4,137,386 (1979).
[d] W. Schäfer, K. Wagner, and H.-D. Block, U.S. Pat. 4,344,855 (1982).

243

a magnetic stirrer are added 1 gm of copper stearate, 780 gm (5.96 moles) of dichloroethylphosphine [45], and, from the dropping funnel, 447 gm (6.56 moles) of freshly distilled isoprene. The stirred solution is refluxed under a nitrogen atmosphere for 42 hr, cooled, allowed to stand for 2 days, and then refluxed an additional 5 days. The excess isoprene is removed by distillation and 850 ml of water is added dropwise while cooling the stirred solution in an ice bath. The dark-brown aqueous solution is transferred to a 5 liter flask and then 1250 ml of 30% sodium hydroxide solution is gradually added so that the solution obtains a pH of 8. The mixture is filtered and the aqueous solution is continuously extracted with chloroform for 12 days. The chloroform is dried and then distilled to yield a residue which on vacuum distillation affords 435 gm (51%), b.p. 115°–120°C (1.2 mm). The product is further purified by oxidation at 50°C with an excess of 3% hydrogen peroxide for 6 hr. The aqueous layer is continuously extracted with benzene, dried, and then concentrated to give a residue which distills as follows: b.p. 115°–119°C (1.2–1.3 mm), n_D^{25} 1.5050.

(*b*) *Carbodiimide formation.* A solution consisting of 50 ml of anhydrous benzene, 54 gm (0.45 mole) of phenyl isocyanate, and 0.20 gm of 1-ethyl-3-methyl-3-phospholine-1-oxide catalyst is refluxed for approximately 2.5 hr or until carbon dioxide evolution ceases. The benzene is removed under reduced pressure and the residue on distillation affords 41.1 gm (94%) of an oil, b.p. 119°–121°C (0.40–0.45 mm), n_D^{25} 1.6372. The product is stable for several weeks at 0°C. At room temperature it gradually solidifies to a mixture of trimer and polymer.

2-2. *Preparation of 4,4′-Dinitrodiphenylcarbodiimide* [29a]

$$2O_2N{-}{-}NCO \xrightarrow{\quad\quad}$$

$$O_2N{-}{-}N{=}C{=}N{-}{-}NO_2 + CO_2 \qquad (9)$$

p-Nitrophenyl isocyanate (21.0 gm, 0.128 mole) is melted and treated with 0.03 gm of 1-ethyl-3-methyl-3-phospholine-1-oxide catalyst at 60°C. The reaction takes place with great vigor and all the carbon dioxide is evolved in 1–2 min. The mixture begins to solidify to a crystalline mass and the last traces of carbon dioxide are removed under reduced pressure. The yellow carbodiimide that is obtained weighs 18 gm (100%), m.p. 165°–170°C. Recrystallization from petroleum ether–chloroform affords a sample of m.p. 164°–166°C.

2-3. Preparation of a Polycarbodiimide from Methylenebis(4-phenyl Isocyanate) [29a]

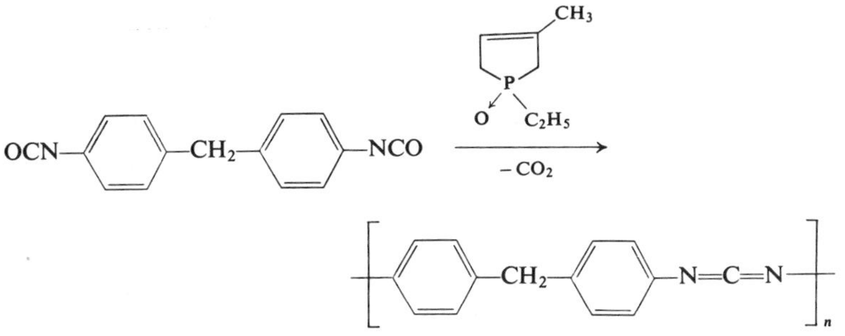

$$OCN-\!\!\!\bigcirc\!\!\!-CH_2-\!\!\!\bigcirc\!\!\!-NCO \xrightarrow[- CO_2]{}$$

$$\left[-\!\!\!\bigcirc\!\!\!-CH_2-\!\!\!\bigcirc\!\!\!-N\!=\!C\!=\!N-\right]_n \qquad (10)$$

To a flask containing 20 gm (0.08 mole) of methylenebis(4-phenyl isocyanate) in 150 ml of xylene is added 0.03 gm of 1-ethyl-3-methyl-3-phospholine-1-oxide catalyst. The solution is refluxed for 4 hr, during which time carbon dioxide is being evolved. The xylene solvent is removed under reduced pressure to afford a nearly quantitative yield of the solid polycarbodiimide. The latter polymer was pressed at 250°C to afford a clear, crystalline, orientable film with a tenacity of about 50,000 psi, an initial modulus of 410,000 psi, and a 20% elongation at 25°C.

2-4. Preparation of Di(2,6-diisopropylphenyl)carbodiimide [29b]

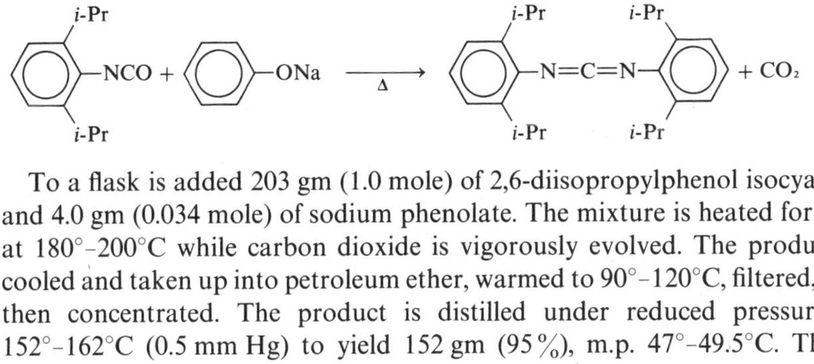

$$i\text{-}Pr\text{-}\bigcirc\text{-}NCO + \bigcirc\text{-}ONa \xrightarrow{\Delta} i\text{-}Pr\text{-}\bigcirc\text{-}N\!=\!C\!=\!N\text{-}\bigcirc + CO_2 \qquad (11)$$

To a flask is added 203 gm (1.0 mole) of 2,6-diisopropylphenol isocyanate and 4.0 gm (0.034 mole) of sodium phenolate. The mixture is heated for 4 hr at 180°–200°C while carbon dioxide is vigorously evolved. The product is cooled and taken up into petroleum ether, warmed to 90°–120°C, filtered, and then concentrated. The product is distilled under reduced pressure at 152°–162°C (0.5 mm Hg) to yield 152 gm (95%), m.p. 47°–49.5°C. The ir indicates a strong band between 2130 and 216 cm^{-1} which is characteristic of the $-N\!=\!C\!=\!N-$ group. Analysis for $C_{25}H_{34}N_2$, m.w. 362.54; calcd. C, 82.82; H, 9.45; N, 7.73; found C, 82.15; H, 9.56; and 8.16.

B. Dehydration of N,N'-Disubstituted Ureas to Carbodiimides

$$R\text{---}NHC\text{---}NHR' \longrightarrow RN\!=\!C\!=\!NR' + H_2O \qquad (12)$$
$$\underset{O}{\overset{\|}{}}$$

The reagents used to effect the dehydration of N,N'-dicyclohexylurea to N,N'-dicyclohexylcarbodiimide in 82% yield is p-toluenesulfonyl chloride in pyridine as solvent and base [17] (Eq. 13). Methylene chloride may also be

$$(C_6H_{11}NH)_2CO + CH_3-C_6H_4-SO_2Cl + 2C_5H_5N \longrightarrow$$

$$C_6H_{11}-N=C=N-C_6H_{11} + CH_3C_6H_4-SO_3H \cdot C_5H_5N + C_5H_5N \cdot HCl \quad (13)$$

effectively used to replace part of the pyridine as a solvent [12]. The method described in Eq. (13) has the advantage over desulfurization procedures in that large quantities of expensive mercuric oxide and long reaction times are not required. In addition, the products are not contaminated with sulfur-containing impurities.

Other reagents that have been used to effect the dehydration of ureas to carbodiimides are: alkyl chloroformates (for thioureas) [46], phosgene [47], phosphorus oxychloride [48], and phosphorus pentachloride [38, 49]. Dichlorocarbodiimides have been shown to be intermediates when $COCl_2$ or PCl_5 reacts with ureas (Eq. 14). If R^1 and R^3 are H, then the intermediate

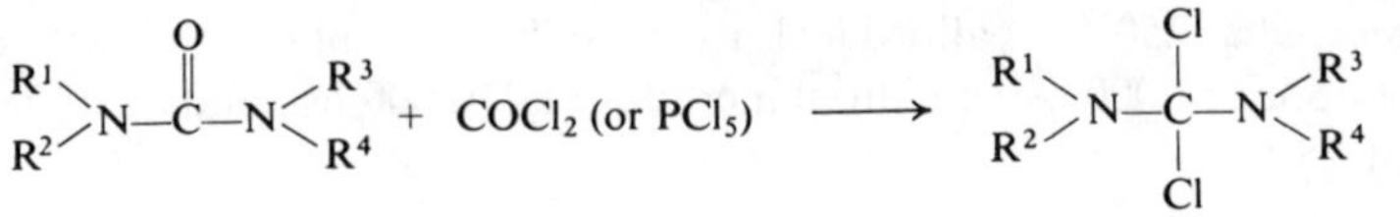

$$\begin{array}{c} O \\ \parallel \\ {R^1 \atop R^2}\!\!>\!\!N-C-N\!\!<\!\!{R^3 \atop R^4} \end{array} + COCl_2 \text{ (or } PCl_5) \longrightarrow \begin{array}{c} Cl \\ \mid \\ {R^1 \atop R^2}\!\!>\!\!N-C-N\!\!<\!\!{R^3 \atop R^4} \\ \mid \\ Cl \end{array} \quad (14)$$

probably has the structure shown in (III), which readily hydrolyzes to carbodiimides [49].

$$[R-NHCCl=NHR']^+ Cl^-$$

(III)

Table VI gives the melting points of several dichlorocarbodiimides obtained from substituted ureas and phosgene [49].

Recently [20] it has been reported that N,N'-disubstituted ureas dehydrate easily with phosphorus pentoxide and pyridine to give the corresponding carbodiimide in good yields (Eq. 15). (See Table VII.)

TABLE VI

DICHLOROCARBODIIMIDES FROM THE REACTION
OF PHOSGENE AND SUBSTITUTED UREAS [49]

Dichlorocarbodiimides	M.p. (°C)
N,N'-Diphenyl-	123–125
N,N'-Dimethyl-	138–143
N-t-Butyl-	110–113
Tetramethyl-	110–112

TABLE VII

DEHYDRATION OF *N*,*N*'-Substituted Ureas by P_2O_5–PYRIDINE
TO CARBODIIMIDES [20]

Urea substituent, R—NHCNHR, R = ‖ O	Urea (moles)	P_2O_5 (moles)	Pyridine (ml)	Temp. (°C)	Time (min)	Yield (%)[a]	B.p., °C (mm Hg)	M.p. (°C)
Cyclohexyl-	0.088	0.70	700	116	145	76	143 (3.5)	34–35
Phenyl-	0.033	0.246	400	116	145	53	110 (0.2)	—
p-Ethoxyphenyl-	0.01	0.085	200	116	180	86	—	46–47[b]
p-Chlorophenyl-	0.10	0.70	750	116	145	56	—	53–54[c]

[a] Carbodiimides possess a strong characteristic infrared absorption band in the 4.6–4.8 μ region.

[b] Recrystallized from *n*-hexane.

[c] Recrystallized from petroleum ether.

$$RNH-\underset{\underset{O}{\|}}{C}-NHR' \xrightarrow[\text{pyridine}]{P_2O_5} R-N=C=N-R' \qquad (15)$$

Table VIII gives examples of recent reports on preparing carbodiimides by the dehydration of *N*,*N*'-disubstituted ureas and thioureas.

2-4. Preparation of Dicyclohexylcarbodiimide Using p-Toluenesulfonyl Chloride–Pyridine [17]

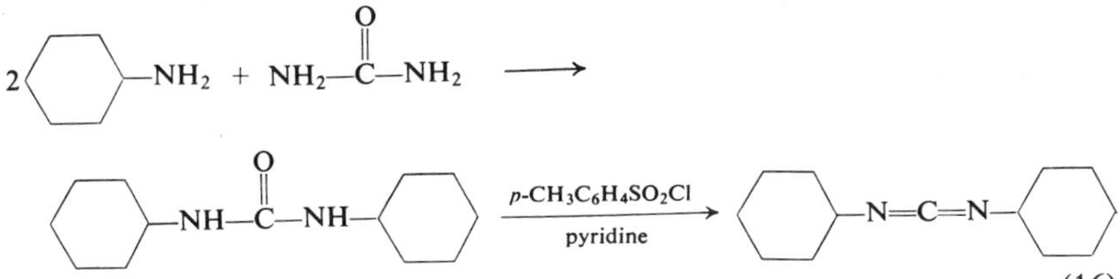

$$\text{(16)}$$

(a) *Preparation of dicyclohexylurea.* A solution of 60 gm (1.0 mole) of urea, 240 gm (2.4 moles) of cyclohexylamine, and 480 ml of isoamyl alcohol is refluxed for 20 hr, cooled, filtered, washed with ether, and dried at 80°C to afford 200 gm (89%), m.p. 234°–235°C (recrystallized from absolute ethanol).

(b) *Preparation of dicyclohexylcarbodiimide.* A stirred mixture consisting of 200 gm (0.90 mole) of *N*,*N*'-dicyclohexylurea, 600 ml of pyridine, and 300 gm

TABLE VIII

RECENT REPORTS ON THE PREPARATION OF CARBODIIMIDE BY THE DEHYDRATION OF N,N'-DISUBSTITUTED UREAS AND THIOUREAS

Urea or thiourea $R-NHC-NHR'$, $\parallel X$	Dehydrating agent	Temp. (°C)	Time (hr)	Yield (%)	B.p., °C (mm Hg)	Ref.
$R=R'=O-CH_3-C_6H_4$, $X=S$	$COCl_2$	78-80	2.5	79.1	150-155 (1.5)	a
(4-methylcyclohexyl) $R'=R'$, 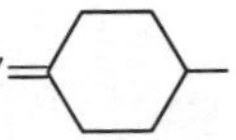 $X=S$	Cyanuric chloride	15-20	0.5	79.1	—	b
$R=R'=C_6H_5$, $X=S$	Ph_3PBr_2 (in place of Br, Cl or I can be used)	80	3	89.1	—	c
$R=(CH_3)_3C-$, $R'=CH_3-$, $X=S$	$COCl_2$	2-8	2.75	65.3	124-125 (760)	d
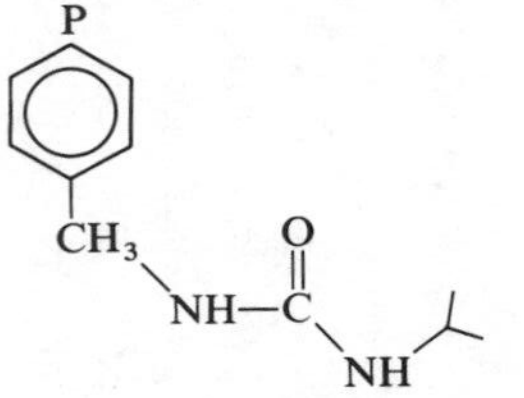 (structure)	$p\text{-}CH_3C_6H_4SO_2Cl$	35	50		Polymer	e

P=Polymer

COCl$_2$ 120–170   0.25   30–61   —[g]   [f]

[a] A. A. R. Sayigh and H. Ulrich, U.S. Pat. 3,301,895 (1967); see also [83a–c] in this chapter.

[b] S. Furumoto, *Nippon Kagaku Zasshi* **92**(11), 1005 (1971); *Chem. Abstr.* **76**, 153720d (1972); S. Furumoto, Japan Kokai, 73/15,824 (1973); *Chem. Abstr.* **79**, 65892t (1973).

[c] G. Heinrichs, H. Maegerlein, and G. Meyer, Brit. Pat. 1,252,707 (1971); *Chem. Abstr.* **78**, 15763k (1973); C. Palomo and R. Mestres, *Synthesis* (5), 373 (1981); C. Palomo and R. Mestres, *Bull. Soc. Chim. Fr.* (9–10), pt. 2, 361 (1981); J. C. Calere and C. N. Palomo, Span. Pat. 481,581 (1980); *Chem. Abstr.* **93**, 70177g (1980).

[d] H. Ulrich, B. Tucker, and A. A. R. Sayigh, *J. Amer. Chem. Soc.* **94**, 3484 (1972).

[e] N. M. Weinshenker, C. M. Shen, and J. Y. Wong, *Org. Synth.* **56**, 95 (1977).

[f] W. J. Schnabel and R. M. Early, U.S. Pat. 4,263,221 (1981).

(1.57 mole) of *p*-toluenesulfonyl chloride is heated at 70°C for 1 hr. The contents are added to 1500 ml of ice water and then extracted with ether. The ether extract is washed with water, dried over magnesium sulfate, and then concentrated to yield a residue which on distillation under reduced pressure affords 152 gm (82%), b.p. 148°–152°C (11 mm), m.p. 35°C.

2-5. *Preparation of N,N'-Dicyclohexylcarbodiimide Using Phosphorus Pentoxide–Pyridine* [20]

$$\text{[cyclohexyl]}-NHCNH-\text{[cyclohexyl]} + P_2O_5 \xrightarrow{\text{pyridine}} \text{[cyclohexyl]}-N=C=N-\text{[cyclohexyl]} \tag{17}$$

A stirred mixture of 19.7 gm (0.088 mole) of *N,N'*-dicyclohexylurea (see Preparation 2-4 for synthesis of dicyclohexylurea), 100 gm (0.70 mole) of phosphorus pentoxide, 175 gm of sand, and 700 ml of pyridine is refluxed for 145 min. After 30 min, further stirring is not possible and the mixture is filtered. The solids are extracted with 100 ml of hot pyridine and then the combined pyridine solutions are concentrated under reduced pressure to yield an oil. The oil is extracted with two 100 ml portions of boiling petroleum ether (b.p. 60°–80°C), followed by 100 ml of ether. The combined extract is washed with water, dried, filtered, and concentrated under reduced pressure to give 17.4 gm of an oil which on distillation affords 13.7 gm (76%), b.p. 143°C (3.5 mm), m.p. 34°–35°C.

3. OXIDATION REACTIONS

A. Desulfurization of Thioureas with Metal Oxides and Other Metallic Compounds

$$RNHCNHR' + HgO \longrightarrow RN=C=NR' + HgS + H_2O \tag{18}$$

As a result of the readily available substituted thioureas (see Chapter 6) the desulfurization of the *N,N'*-disubstituted derivatives by mercuric oxide is one of the best methods for the preparation of carbodiimides [11]. The preferred solvents are ether, benzene, and acetone. Toluene and other high-boiling solvents favor the formation of by-products such as aniline, isothiocyanates, and guanidines [50]. The presence of drying agents such as $CaCl_2$ [51], Na_2SO_4 [52], or $MgSO_4$ [53] remove the water formed so that it will be unable to add to the carbodiimide to form urea. Water may also be removed by azeotroping it off using a Dean-Stark trap [23]. However, for most aliphatic carbodiimides the presence of water causes no urea and dehydrating agents are often not employed [54, 55].

Other catalysts have been employed but they are not so effective as mercuric oxide. Some of the other catalysts are: PbO [8, 10, 56], PbO—S— (or Se) [57], ZnO [58], PbCO₃, Pb(NO₃)₂, and PbCl₂ [59]. The quantity and the method of preparation of the PbO influence the yields of products [60].

Cyclic carbodiimides have been prepared where $n = 5$ or greater [61] (see structures IV and V).

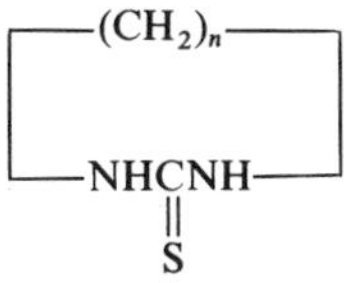

$$\text{(IV)} \qquad\qquad \text{(V)}$$

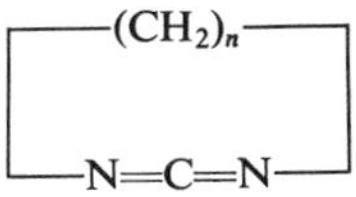

In the preparation of aromatic carbodiimides the presence of substituents in the benzene ring affects the yields of the product and the by-product iso-thiocyanate [60].

Examples of the preparation of carbodiimides by the action of metallic compounds are shown in Table IX.

Some other reports on the preparation of carbodiimides by the oxidation of disubstituted thioureas are given in Table X.

3-1. *Preparation of 1-Cyclohexyl-3-[2-morpholinyl-(4)-ethyl]carbodiimide* [54]

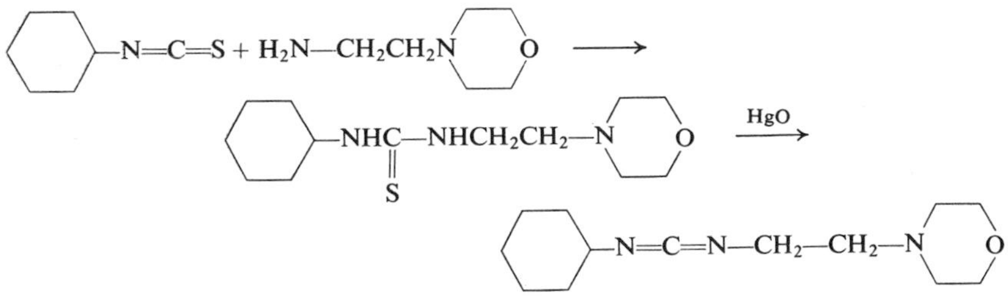

$$(19)$$

(a) *Preparation of 1-cyclohexyl-3-[2-morpholinyl-(4)-ethyl]thiourea.* To 500 ml of ether are added 19.4 gm (0.138 mole) of cyclohexyl isothiocyanate and 18.0 gm (0.138 mole) of N-(2-aminoethyl)morpholine. The solution is refluxed for 10 min, cooled, filtered, and the crystalline solid dried to yield 36.0 gm (96%), m.p. 128°–129°C.

(b) *Preparation of 1-cyclohexyl-3-[2-morpholinyl-(4)-ethyl]carbodiimide.* To 50 ml of acetone are added 4.0 gm (0.0147 mole) of the thiourea prepared

9. Carbodiimides

TABLE IX

PREPARATION OF CARBODIIMIDES BY THE ACTION OF METALLIC
OXIDES OR OTHER COMPOUNDS ON THIOUREAS OR ISOCYANATES

Starting thiourea or isocyanate	Metallic oxide	Yield (%)	Carbodiimide B.p., °C (mm Hg)	Carbodiimide M.p. (°C)
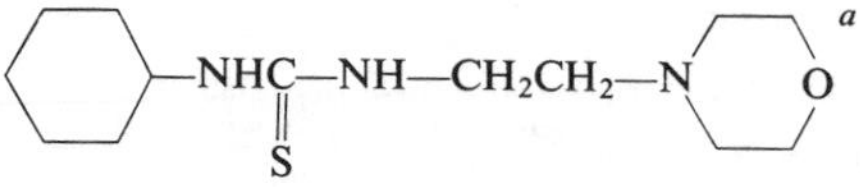 (cyclohexyl)—NHC(=S)—NH—CH$_2$CH$_2$—N(morpholine)[a]	HgO	70	145 (0.2)	—
C$_2$H$_5$NH—C(=S)—NHC$_2$H$_5$ [b]	HgO	—	35–40 (1.0)	—
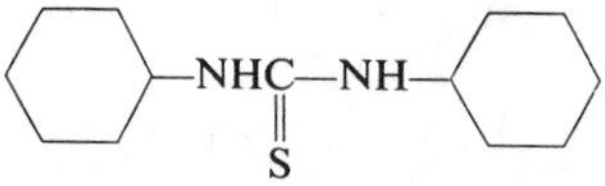 (cyclohexyl)—NHC(=S)—NH—(cyclohexyl) [c]	Ag$_2$O	91	141–144 (6)	—
o-CH$_3$C$_6$H$_4$NCO [d]	Fe(CO)$_5$	85	—	—
RC$_6$H$_4$NHC(=S)—NHC$_6$H$_4$—R [e]				
R = H	PbO	53	163–165 (11)	—
R = p-CH$_3$	PbO	96	—	53
R = m-CH$_3$	PbO	81	60 (1.5)	—
R = p-(CH$_3$)$_2$N	PbO	96	—	88
R = p-Cl	PbO	83	—	54
R = m-NO$_2$	PbO	30	—	154
R = m-CN	PbO	50	—	137

[a] J. C. Sheehan and J. J. Hlavka, *J. Org. Chem.* **21**, 439 (1956).
[b] R. P. Parker and R. S. Long, U.S. Pat. 2,749,498 (1949).
[c] B. D. Wilson, U.S. Pat. 3,236,882 (1966).
[d] H. Ulrich, B. Tucker, and A. A. R. Sayigh, *Tetrahedron Lett.* 1731 (1967).
[e] S. Hunig, H. Lehmann, and G. Grimmer, *Ann. Chem.* **579**, 77 (1953).

above and 6.0 gm (0.0277 mole) of yellow mercuric oxide. The mixture is refluxed for 6 hr, mercuric sulfide is removed by filtration, and then another 6.0 gm (0.0277 mole) of mercuric oxide is added. The suspension is refluxed for an additional 6 hr, filtered, the filtrate concentrated under reduced pressure, and on distillation of the residue the yield is 2.4 gm (70%), b.p. 145°C (0.2 mm).

TABLE X

RECENT REPORTS ON THE PREPARATION OF CARBODIIMIDES BY THE OXIDATION OF
DISUBSTITUTED THIOUREAS

Catalyst	Thiourea RNHCNHR' $\overset{\parallel}{S}$	Temp. (°C)	Time (hr)	Yield (%)	B.p., °C, (mm Hg)	M.p. (°C)	Ref.
PbO + S	R=R'= 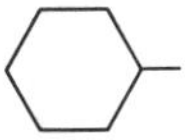 ⬡	70	8	94.5	133–134 (2–3)	—	a
PbO + S	R=R'=i-Pr	72	7.5	—	144–144.8 (760)	—	a
PbO + S	R=R'=o-tolyl	72	2	90.6	169–174 (4.5)	—	a
Ag₂O	R=R'= 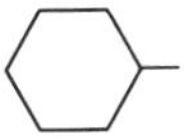 ⬡	50–60	18	91.0	141–144 (6)	—	b
PbO	R=R'=i-Pr	100	1.25	95	—	—	c
NaNH₂	R=R'= 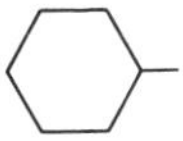 ⬡	110	4	45	123–126 (0.7)	—	d
NaNH₂	R= 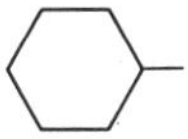 ⬡ R'=t-bu	110	3.5	65	126–128	—	d

[a] W. R. Ruby, U.S. Pat. 3,201,463 (1965).
[b] B. D. Wilson, U. S. Pat. 3,236,882 (1966).
[c] G. T. White and K. B. Mullin, U.S. Pat. 3,352,908 (1967).
[d] P. Schlack and G. Keil, U.S. Pat. 3,320,309 (1967).

3-2. Preparation of Diphenylcarbodiimide [60]

$$C_6H_5NHCNHC_6H_5 + PbO \xrightarrow{\text{acetone}} C_6H_5N=C=N-C_6H_5 + PbS + H_2O \quad (20)$$
$$\overset{\parallel}{S}$$

To a flask are added 20.0 gm (0.088 mole) of diphenylthiourea, 0.4 gm of sulfur, 200 ml of acetone, and 35 gm (0.157 mole) of finely dispersed lead oxide. The mixture is refluxed for 40 min, cooled, filtered, and the filtrate concentrated to 60 ml under reduced pressure. The latter filtrate is cooled to −70°C in order to separate the unreacted thiourea, which in this case amounts to 5.0 gm (0.022

mole). The acetone is removed under reduced pressure and the residue is extracted with petroleum ether (b.p. 30°–40°C). The petroleum ether is also removed under reduced pressure to yield 9.0 gm (53%) of diphenylcarbodiimide, b.p. 163°–165°C (11 mm).

B. Desulfurization of Thioureas with Alkaline Hypochlorite

The oxidation of N,N'-dialkylthioureas to the corresponding carbodiimides can also be accomplished in excellent yields by alkaline hypochlorites below 0°C [15, 16, 62–64]. The use of excess hypochlorite converts the sulfur into sulfate ion. This process has the main advantage of using inexpensive reagents and thus can be applied to the large-scale production of carbodiimides (Eq. 21).

$$\underset{\overset{\|}{S}}{RNHCNHR'} + 4NaOCl + 2NaOH \longrightarrow$$

$$RN{=}C{=}NR' + 4NaCl + Na_2SO_4 + 2H_2O \qquad (21)$$

This reaction is also applicable to aromatic thioureas as reported by Lee and Wragg [64b] but the yields were low. For example, di-p-tolylcarbodiimide was prepared in 25% yield, m.p. 54°–55°C, by reacting 1.0 mole of the appropriate thiourea with 4 moles of 12% NaOCl, 4 moles of sodium dioxide (25%) in ether (200 ml) at 5°C [64b].

3-3. *Preparation of N,N'-Di[admantyl-(1)]carbodiimide* [64]

$$\underset{\overset{\|}{S}}{R-NHCNHR} + NaOCl + Na_2CO_3 \longrightarrow RN{=}C{=}N{-}R \qquad (22)$$

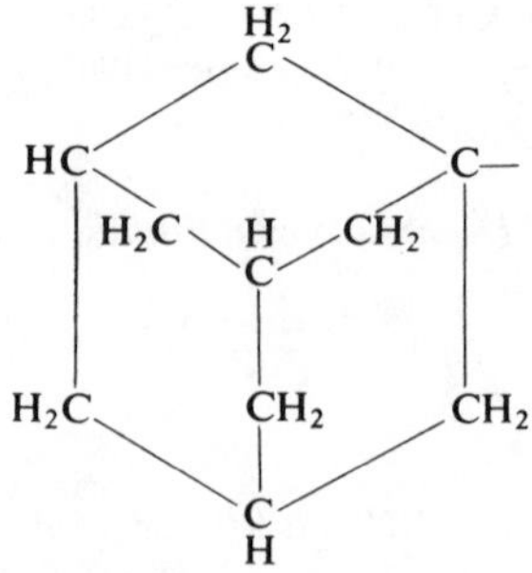

To a stirred mixture consisting of 3.5 gm (0.011 mole) of N,N'-di[adamantyl-(1)]thiourea in 50 ml of methylene chloride, 2.2 gm (0.03 mole) of sodium hypochlorite, and 1.5 gm (0.014 mole) of sodium carbonate is added 0.08 gm

(0.81 mole) of cuprous chloride. The reaction is allowed to remain standing overnight without stirring, and the next day the two layers are separated. The water layer is extracted with 10 ml of methylene chloride and then the combined methylene chloride layers are washed with water and dried. The methylene chloride solvent is removed under reduced pressure to give a residue which is then purified by crystallization from dimethylformamide to afford 2.5 gm (73%), m.p. 340°–343°C.

4. MISCELLANEOUS METHODS

(1) Pyrolysis of tetrazoles to carbodiimides [65a, b].

(2) Reaction of phosphinimines with isothiocyanates or isocyanates to give carbodiimide [66].

(3) Pyrolysis of amidino dichlorides (from urea and phosgene) to carbodiimide [67].

(4) Reaction of N,N'-disubstituted imino chlorides with base to yield carbodiimides [68].

(5) Fission of S-alkyl-N,N'-diphenylisothioureas into mercaptan and diphenylcarbodiimide [69].

(6) Isocyanate–carbodiimide exchange reaction [70].

(7) Reaction of isocyanide dichlorides with primary amine hydrochlorides in an inert solvent at 180°C [71].

(8) Reaction of N-substituted trichlorophosphazenes with phenyl isocyanate [72].

(9) Thermal decomposition of 1,2,4-oxadiazetidines to carbodiimides [73].

(10) Pyrolysis of 4,5-diphenyl-1,2,3,5-thiaoxadiazole 1-oxide at 100°C yields diphenylcarbodiimide and sulfur dioxide [74].

(11) Reaction of aliphatic N-acylimino chloride with thiourea in the presence of triethylamine [75].

(12) Reaction of N,N'-diphenylthiourea with 2,4-dimethylphenyl cyanate to yield diphenylcarbodiimide and O-(2,4-dimethylphenyl)thiocarbamate. Monosubstituted thioureas yield the cyanamide [76].

(13) Conversion of isocyanates into carbodiimides with isopropyl methylphosphonofluoridate as catalyst [77].

(14) Preparation of carbodiimides by heating isocyanates with 1,2,3-trimethyl-2-oxo-1,3,2-diazaphopholidine or analogs [78].

(15) Catalytic conversion of isocyanates into carbodiimides with the aid of 1,3-dimethyl-1,3,2-diazaphospholidine 2-oxides and 1,3-dimethylhexahydro-1,3,2-diazaphosphorine 2-oxides [79].

(16) Carbodiimides by the catalytic action of 1-oxo-1-alkoxyphospholenes on alkyl or aryl isocyanates [80].

(17) Carbodiimides by the pyrolysis of 4,5-disubstituted 1-oxo-1,2,3,5-thiaoxadiazoles [81].

(18) Polycarbodiimide foams from polyisocyanates and 0.001–1% by weight of phospholine or phospholidine [82].

(19) Reaction of 1,3-disubstituted thioureas with phosgene [83a–c].

(20) Elimination of mercaptans from isothioureas [84a–c].

(21) Reaction of azides with isocyanides [85].

(22) Pyrolysis of pyridine imidoyl-N-imines [86].

(23) Isomerization of N-aryl-1-aziridinecarboximodoyl chlorides to N-(2-chloroalkyl)-N-arylcarbodiimides [87].

(24) Carbodiimides from olefinic cyanamide and *tert*-butyl hypochlorite [88].

(25) Conversion of isocyanide dichloride to carbodiimides [89].

(26) Synthesis of N,N'-dicyclohexylcarbodiimide from N,N'-dicyclohexyl-formamidine [90].

(27) Synthesis of carbodiimide from N,N'-disubstituted thioureas by reaction with 1-chlorobenzotriazole or trichlorocyanuric acid [91].

(28) Reaction of iminophosphoranes with CS_2 to give carbodiimides [92].

(29) Reaction of 2-(trifluoromethyl)-3,3-difluorooxaziridine with trimethylsilicon cyanide $(CH_3)_3SiCN$ to give 1-(trifluoromethyl-3-trimethylsilyl)-carbodiimide [93].

(30) Synthesis of unusual carbodiimides (e.g., nine-membered ring carbodiimides) [94].

(31) Addition of sulfonyl isocyanates to carbodiimides to give sulfonyl carbodiimides [95].

(32) Dehydration of amidoximes [96].

(33) Reaction of azide with isocyanide using iron carbonyl catalyst to give carbodiimides [97].

(34) Reaction of p-tolylisocyanate with $Ba[(C_4H_9)_2B\ (OC_4H_9)_2]_2$ catalyst to give p-tolylcarbodiimide [98].

(35) Oxidation of carbene palladium (II) complex with silver oxide [99].

REFERENCES

1a. H. G. Khorana, *Chem. Rev.* **53**, 145 (1953).

1b. F. Kurzer and K. Douraghi-Zadeh, *Chem. Rev.* **67**, 107 (1967).

2a. J. R. Schaeffer, *Org. Chem. Bull.* **2**, 33 (1961).

2b. K. Mosbach, P.-O. Larsson, and C. Lowe, *in* "Methods in Enzymology" (K. Mosbach, ed.), vol. XLIV, chapter on immobilized coenzymes, pp. 859–864, Academic Press, New York, 1976.

3. W. Neumann and P. Fischer, *Angew. Chem. Int. Ed. Engl.* **1**, 621 (1962).

4. B. V. Bocharov, *Usp. Khim.* **34**, 488 (1965).

5. K. Nowak and I. Z. Siemion, *Wiad. Chem.* **14**, 327 (1960).

6. W. Weith, *Chem. Ber.* **6**, 1389 (1873).

7. R. Rotter, *Monatsh. Chem.* **47**, 355 (1926).

8. F. Zetzche and A. Fredrich, *Chem. Ber.* **73**, 1114 (1940).

9. Farbenfabriken Bayer Akt.-Ges., Ger. Pat. 129,417 (1950).

10. I. W. Herzog, *Angew. Chem.* **33**, 140 (1920).

11. H. G. Khorana, *J. Chem. Soc.* 2081 (1952).

12. J. C. Sheehan, P. A. Cruickshank, and G. L. Boshart, *J. Org. Chem.* **26**, 2525 (1961).

13. W. Weith, *Chem. Ber.* **8**, 1530 (1875).

14. E. Schmidt, W. Striewsky, and F. Hitzler, *Ann. Chem.* **560**, 222 (1948).

15. E. Schmidt and R. Schnegg, U.S. Pat. 2,656,383 (1953).

16a. E. Schmidt and R. Schnegg, Ger. Pat. 823,445 (1951).

16b. A. A. R. Sayigh and H. Ulrich, U.S. Pat. 3,301,895 (1967).

17a. G. Amiard and R. Heymes, *Bull. Soc. Chim. Fr.* 1360 (1956).

17b. J. C. Sheehan and P. A. Cruickshank, *Org. Syn.* **48**, 83 (1968).

18. T. W. Campbell, J. J. Monagle, and V. S. Foldi, *J. Amer. Chem. Soc.* **84**, 3673 (1962).

19a. W. B. McCormack, *Org. Syn.* **43**, 73 (1963).

19b. R. Sundermann and W. E. Slack, U.S. Pat. 4,154,752 (1979).

20. C. L. Stevens, G. H. Spinghal, and A. B. Ash, *J. Org. Chem.* **32**, 2895 (1967).

21. H. Staudinger and E. Hauser, *Helv. Chim. Acta.* **4**, 861 (1921).

22. E. Mulder, *Chem. Ber.* **6**, 655 (1873).

23. F. Zetzsche, E. Luescher, and H. E. Meyer, *Chem. Ber.* **71B**, 1088 (1938).

24. F. Zetzsche, H. E. Meyer, H. Overbeck, and H. Lindler, *Chem. Ber.* **71B**, 1516 (1938).

25. F. Zetzsche, H. E. Meyer, H. Overbeck, and W. Nerger, *Chem. Ber.* **71B**, 1512 (1938).

26. A. W. Hofmann, *Chem. Ber.* **18**, 765 (1885).

27. W. J. Balon, U.S. Pat. 2,853,518 (1958).

28. T. W. Campbell and J. J. Monagle, *Org. Syn.* **43**, 31 (1963).

29a. T. W. Campbell and J. J. Monagle, *J. Amer. Chem. Soc.* **84**, 1493 (1962).

29b. W. Neumann, U.S. Pat. 3,502,722 (1970).

30. T. W. Campbell and J. Verbanc, U.S. Pat. 2,853,473 (1958).

31. E. Dyer and R. E. Reed, *J. Org. Chem.* **26**, 4677 (1961).

32. J. J. Monagle, *J. Org. Chem.* **27**, 3851 (1962).

33. J. J. Monagle, T. W. Campbell, and H. B. McShane, Jr., *J. Amer. Chem. Soc.* **84**, 4288 (1962).

34. J. J. Monagle and R. H. Nace, U.S. Pat. 3,056,835 (1962).

35. W. Neumann and P. Fischer, *Angew. Chem. Int. Ed. Engl.* **1**, 621 (1962).

36. W. Neumann, E. R. Mueller, H. Logemann, H. Marzolph, and F. Moosmüller, Belg. Pat. 618,389 (1962).

37. K. C. Smeltz, U.S. Pat. 3,157,662 (1964).

38. H. Ulrich and A. A. R. Sayigh, *J. Chem. Soc.* 5558 (1963).

39. W. B. McCormack, U.S. Pat. 2,663,737 (1953).

40. W. S. Wadsworth, Jr., and W. D. Emmons, *J. Amer. Chem. Soc.* **84**, 1316 (1962).

41. W. S. Wadsworth, Jr., and W. D. Emmons, *J. Org. Chem.* **29**, 2816 (1964).

42. V. E. Shashoura, W. Sweeney, and R. F. Tielz, *J. Amer. Chem. Soc.* **82**, 866 (1960).

43. J. J. Monagle and J. V. Mengenhauser, *J. Org. Chem.* **31**, 2321 (1966).

44. H. Staudinger and J. Meyer, *Helv. Chim. Acta* **2**, 636 (1919).

45. M. S. Kharasch, E. V. Jensen, and S. Weinhouse, *J. Org. Chem.* **14**, 429 (1949).

46. R. F. Coles and H. A. Levine, U.S. Pat. 2,942,025 (1960).

47. J. S. Jochims, *Chem. Ber.* **98**, 2128 (1965).

48. H. Walther, Ger. (East) Pat. 22,437 (1961).

49. H. Eilingsfeld, M. Seefelder, and H. Weidinger, *Angew. Chem.* **72**, 836 (1960).

50. J. Pump and E. G. Rochow, *Z. Anorg. Allg. Chem.* **330**, 101 (1964).
51. R. Rotter and E. Schandy, *Monatsh. Chem.* **58**, 245 (1931).
52. E. Schmidt, F. Hitzler, and E. Lahde, *Chem. Ber.* **71**, 1933 (1938).
53. G. D. Meakin and R. J. Moss, *J. Chem. Soc.* 993 (1957).
54. J. C. Sheehan and J. J. Hlavka, *J. Org. Chem.* **21**, 439 (1956).
55. E. Schmidt and W. Striewsky, *Chem. Ber.* **74**, 1285 (1941).
56. N. Zinin, *Jahresber. Fortsch. Chem. Verw. Wiss.* 628 (1852).
57. F. Zetzsche and W. Nerger, *Chem. Ber.* **73**, 467 (1940).
58. R. F. Coles, U.S. Pat. 2,946,819 (1960).
59. J. C. Sheehan, U.S. Pat. 3,135,748 (1964).
60. H. Hunig, H. Lehman, and G. Grimmer, *Ann. Chem.* **579**, 77 (1953).
61. H. Behringer and H. Meier, *Ann. Chem.* **607**, 72 (1957).
62. Farbenfabriken Bayer, Brit. Pat. 685,970 (1953).
63. E. Schmidt, M. Seefelder, R. G. Jennen, W. Striewsky, and M. von Martius, *Ann. Chem.* **571**, 83 (1951).
64a. H. Stetter and C. Wulff, *Chem. Ber.* **95**, 2302 (1962).
64b. T. Lee and R. Wragg, *J. Appl. Polym. Sci.* **14**, 115 (1970).
65a. R. A. Olofson, W. R. Thompson, and J. S. Michelman, *J. Amer. Chem. Soc.* **86**, 1865 (1964).
65b. D. M. Zimmerman, R. A. Olofson, Abstr. Meeting Nat. Amer. Chem. Soc., 158th, New York, Sept. 7–12, 1969, paper in the Org. Chem. Sec. No. 133.
66. L. Horner and H. Hofmann, *Angew. Chem.* **68**, 473 (1956).
67. S. Neubauer, M. Seefelder, and H. Weidinger, *Chem. Ber.* **97**, 1232 (1964).
68. H. Eilingsfeld, M. Seefelder, and H. Weidinger, *Angew. Chem.* **72**, 836 (1960).
69. P. Schlack and G. Keil, *Ann. Chem.* **661**, 164 (1963); P. Schlack and G. Keil, Ger. Pat. 1,173,460 (1964).
70. W. Neumann and P. Fischer, *Angew. Chem. Int. Ed. Engl.* **1**, 621 (1962).
71. Farbenfabriken Bayer Akt.-Ges., Ger. Pat. 1,149,712 (1963).
72. H. Ulrich and A. A. R. Sayigh, *J. Org. Chem.* **30**, 2779 (1965).
73. C. K. Ingold, *J. Chem. Soc.* 87 (1924).
74. P. Rajagopalan and H. U. Daeniker, *Angew. Chem.* **75**, 91 (1963).
75. K. Hartke and J. Bartulin, *Angew. Chem.* **74**, 214 (1962).
76. E. Grigat and R. Putter, *Chem. Ber.* **98**, 1168 (1965).
77. J. O. Appleman and V. J. DeCarlo, *J. Org. Chem.* **32**, 1505 (1967).
78. E. Daebritz and H. Herlinger, Fr. Pat. 1,469,946 (Feb. 17, 1967).
79. H. Ulrich, B. Tucker, and A. A. R. Sayigh, *J. Org. Chem.* **32**, 1360 (1967).
80. K. Hunger, *Tetrahedron Lett.* 5929 (1966).
81. P. Rajagopalan and B. G. Advani, *J. Org. Chem.* **30**, 3369 (1965).
82. H. Nohe, R. Platz, and E. Wegner, Belg. Pat. 657,835 (1965).
83a. H. Ulrich and A. A. R. Sayigh, *Angew. Chem. Int. Ed. Engl.* **5**, 704 (1966).
83b. H. Ulrich, B. Tucker, and A. A. R. Sayigh, *Tetrahedron* **22**, 1565 (1966).
83c. R. K. Gupta and C. H. Stammer, *J. Org. Chem.* **33**, 4368 (1968).
84a. W. Will, *Chem. Ber.* **14**, 1485 (1881).
84b. A. F. Ferris and B. A. Schutz, *J. Org. Chem.* **28**, 71 (1963).
84c. C. G. McCarty, J. E. Parkinson, and D. M. Wieland, *J. Org. Chem.* **35**, 2067 (1970).
85. T. Saegusa, Y. Ito, and T. Shimizu, *J. Org. Chem.* **35**, 3995 (1970).
86. A. R. Katritsky, P.-L. Nie, A. Dondoni, and D. Tassi, *J. Chem. Soc. Perkin Trans.* **1**(8), 1961 (1959).
87. D. A. Tomalia, T. J. Giacobbe, and W. A. Sprenger, *J. Org. Chem.* **36**, 2142 (1971).
88. L. deVries, U.S. Pat. 3,769,344 (1973).

89. E. Kühle, U.S. Pat. 3,231,610 (1966).
90. D. Furomoto, *Yuki Gosei Kagaku Kyokai Shi* **34**(7), 499 (1976); *Chem. Abstr.* **85**, 176906u (1976).
91. S. Furomoto, *Yuki Gosei Kagaku Kyokai Shi* **32**(9), 727 (1974); *Chem. Abstr.* **82**, 125361t (1975).
92. P. Molina, M. Alajarin, A. Arques, and J. Saez, *Synth. Commun.* **12**(7), 573 (1982).
93. W. Y. Lam and D. D. DesMarteau, *J. Amer. Chem. Soc.* **104**, 4034 (1982).
94. R. R. Hiatt, M-J. Shaio and F. Georges, *J. Org. Chem.* **44**, 3265 (1979).
95. H. Ulrich, B. Tucker, F. A. Stuber, and A. A. R. Sayigh, *J. Org. Chem.* **34**, 2250 (1969).
96. J. H. Boyer and P. J. A. Frintz, *J. Org. Chem.* **35**, 2449 (1970).
97. T. Saegusa, Y. Ito, and T. Shimizu, *J. Org. Chem.* **35**, 3995 (1970).
98. K. C. Smeltz, U.S. Pat. 3,157,662 (1964).
99. Y. Ito, T. Hira, and T. Saegusa, *J. Org. Chem.* **40**, 2981 (1975).

CHAPTER 10 / *N*-CARBAMATES (URETHANES)

1. INTRODUCTION

The electronic structure of the isocyanate group indicates the following possible resonance structure (Eq. 1):

$$R\overset{..}{\underset{..}{N}}-\overset{+}{C}=\overset{..}{\underset{..}{O}} \leftrightarrow R-\overset{..}{N}=C=\overset{..}{\underset{..}{O}} \leftrightarrow R-\overset{+}{N}=C-\overset{..}{\underset{..}{O}}\colon \qquad (1)$$

(I)

The usual reactions with isocyanates involve the reaction of active hydrogen with the nitrogen of the isocyanate group as is shown in Eq. 2.

$$R-N{=}C{=}O + HA \longrightarrow RNH-\overset{\displaystyle O}{\overset{\|}{C}}-A \qquad (2)$$

Some of these products are stable, while others can be decomposed easily to the starting material or other products. The reaction of isocyanate groups with alcohols is discussed in detail in this chapter (Eq. 3). Reference is also made to the preparation of polymers using difunctional starting materials.

$$R-NCO + R'OH \longrightarrow RNH-\overset{\displaystyle O}{\overset{\|}{C}}OR' \qquad (3)$$

The synthesis of O-alkyl carbamates [1a] $NH_2\overset{\|}{\underset{O}{C}}-OR$, is discussed in Chapter 11. This chapter mainly describes *N*-alkyl carbamates, $RNH\overset{\|}{\underset{O}{C}}OR$, as derived from the condensation reactions of isocyanates. The conversion of O-alkyl carbamates to *N*-alkyl carbamates is briefly described.

$$NH_2\overset{\|}{\underset{O}{C}}OR + RNH_2 \xrightarrow{R_2N} RNH\overset{\|}{\underset{O}{C}}OR + NH_3 \qquad (4)$$

$$NH_2\overset{\|}{\underset{O}{C}}OR + RCH{=}CH_2 \xrightarrow{H^+} RCH_2CH_2NH\overset{\|}{\underset{O}{C}}OR \qquad (5)$$

The chemistry of N,N-dialkyl carbamates is referred to briefly in Section 4, as are other methods for the preparation of N-alkyl carbamates.

CAUTION: The handling of isocyanates should be done with great care because of their known toxicity hazard. The hazards and chemistry of methyl isocyanate have recently been reported [1b]. Toluene diisocyanate (TDI), methylene diisocyanate (MDI), and methyl isocyanate (MIC) are flammable, very reactive, and have in common low maximum allowable concentrations for employee exposure by the Occupational Safety and Health Administration (OSHA). The exposure limits, averaged over an 8 hr period, are 0.02 ppm each for MDI and MIC and only 0.005 ppm for TDI [1b].

2. CONDENSATION REACTIONS

A. Reaction of Isocyanates with Hydroxy Compounds

For the most part, the preparation of monomeric and polymeric carbamates (urethanes), semicarbazides, and ureas consists of condensation reactions of isocyanates with alcohols, hydrazines, or amines. The synthesis of ureas and semicarbazides are described in Chapters 6 and 8, respectively.

Isocyanates readily react with primary alcohols at $25°–50°C$, whereas secondary alcohols react 0.3 as fast, and tertiary alcohols approximately 0.005 as fast as the primary [2]. Thus, the effect of steric hindrance shows a pronounced effect on these reactions. For example, triphenylcarbinol has been reported to be completely unreactive [3a, b]. Other tertiary alcohols such as t-butanol may react under uncatalyzed conditions with isocyanates to give olefin formation, as shown in Eq. (6), using phenyl isocyanate [4, 5].

$$2C_6H_5NCO + (CH_3)_3C\!-\!OH \rightarrow C_6H_5NHCONHC_6H_5 + CO_2 + (CH_3)_2C\!\!=\!\!CH_2 \qquad (6)$$

The olefins formed under these conditions using other tertiary alcohols tend to follow the Hofmann rule [6a]. However, tertiary alcohols and phenols [6b] can be made to react with isocyanates to give urethanes when they are catalyzed by acids or bases such as pyridine [6b], triethylamine, sodium acetate, boron trifluoride etherate, hydrogen chloride, or aluminum chloride [7a] and lithium alkoxide [7b]. Table I indicates the results of preparing phenylurethanes of tertiary alcohols using 2 or 3 gm of the alcohol with an equivalent amount of phenylisocyanate in the presence of 0.1 gm of anhydrous sodium acetate for 4–5 hr on a steam bath. In each case the reaction product was contaminated with some diphenylurea and unreacted phenyl isocyanate. Recrystallization or distillation under reduced pressure yielded the pure product.

TABLE I[a]

PREPARATION OF PHENYLURETHANES OF TERTIARY ALCOHOLS [7a]

Phenylurethane of	M.p. (°C) (uncorr.)	Formula
Dimethylbutylcarbinol	62–63	$C_{14}H_{21}NO_2$
Diphenylmethylcarbinol	124–125	$C_{21}H_{19}NO_2$
Triethylcarbinol	61–61.5	$C_{14}H_{21}NO_2$
Methylethylbenzylcarbinol	83.5–84	$C_{18}H_{21}NO_2$
Diethylbenzylcarbinol	96–96.5	$C_{19}H_{23}NO_2$

[a] Reprinted from D. S. Tarbell, R. C. Mallatt, and J. W. Wilson, *J. Amer. Chem. Soc.* **64**, 2229 (1942). (Copyright 1942 by the American Chemical Society. Reprinted by permission of the copyright owner.)

Using 9.2 mmoles each of *o*-cresol and α-naphthyl isocyanate in 3.00 ml of purified ligroin, the effect of catalysts was tested. The sealed tubes were heated in a 65°C bath and after a definite time the solid urethane derivative was isolated, washed, dried, and weighed. The results (see Fig. 1) show that triethylamine is the most effective amine, boron trifluoride etherate being the most effective acid tested at that time [7a].

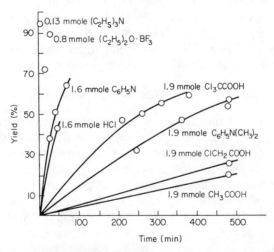

FIG. 1. Effect of catalysts on the rate of reaction of *o*-cresol and α-naphthyl isocyanate [7]. Reprinted from D. S. Tarbell, R. D. Mallatt, and J. W. Wilson, *J. Amer. Chem. Soc.* **64**, 2229 (1942). (Copyright 1942 by the American Chemical Society. Reprinted by permission of the copyright owner.)

2-1. Preparation of the Phenylurethane of Ethanol [8]

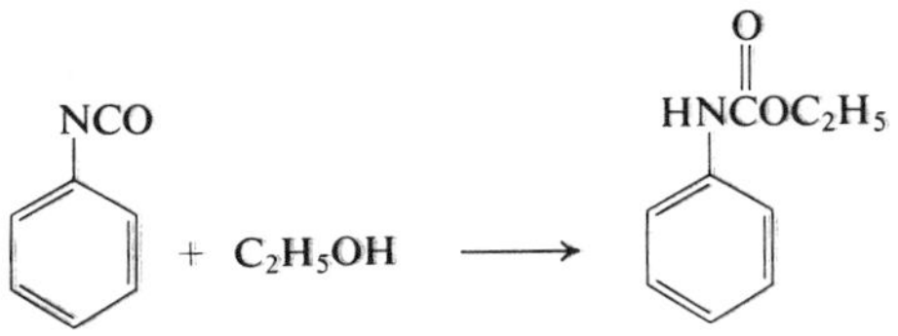

$$\text{(7)}$$

To a dry Erlenmeyer flask are added 29.6 gm (0.25 mole) of phenyl isocyanate, 1 drop of pyridine, and 11.5 gm (0.25 mole) of absolute ethanol. The reaction mixture, after standing for 5–10 min, becomes hot and is cooled and stirred. After the reaction appears to be complete, the contents are heated to

TABLE II[c]

COMPARISON OF THE CATALYTIC ACTIVITY OF SOME ORGANOTIN COMPOUNDS WITH THAT OF OTHER CATALYSTS[a] [9]

Catalyst	Catalyst (mole %)	Relative activity at 1.0 (mole %)
None		1.0[b]
N-Methylmorpholine	1.0	4
Triethylamine	1.0	8
N,N,N′,N′-Tetramethyl-1,3-butanediamine	1.0	27
1,4-Diazabicyclo[2,2,2]octane	1.0	120
Stannous chloride	0.10	2,200
Stannic chloride	0.10	2,600
Ferric acetylacetonate	0.01	3,100
Tetra-n-butyltin	1.0	160
Tetraphenyltin	1.0	9
Tri-n-butyltin acetate	0.001	31,000
Dimethyltin dichloride	0.001	78,000
Di-n-butyltin diacetate	0.001	56,000
Di-n-butyltin dichloride	0.001	57,000
Di-n-butyltin dilaurate	0.001	56,000
Di-n-butyltin dilaurylmercaptide	0.001	71,000
Bis(2-ethylhexyl)tin oxide	0.001	35,000
Di-n-butyltin sulfide	0.001	20,000
2-Ethylhexylstannonic acid	0.001	30,000

[a] Phenyl isocyanate–butanol reaction, both 0.25 M in dioxane at 70°C.

[b] $K_0 = 0.37 \times 10^{-4}$ l mole^{-1} sec^{-1} at 70°C.

[c] Reprinted from F. Hostettler and E. F. Cox, *Ind. Eng. Chem.* **52**, 609 (1960). (Copyright 1960 by the American Chemical Society. Reprinted by permission of the copyright owner.)

TABLE III[m]

Tertiary Amine Catalytic Activities in the Reaction of Phenyl
Isocyanate with 1-Butanol in Toluene at 39.69°C[l] [11]

Catalyst	Catalytic activity	pK_a
N-Methylmorpholine[a]	1.00	7.41[b]
N-Ethylmorpholine[a]	0.68	7.70[b]
Ethyl morpholinoacetate	0.21	5.2[c]
Dimorpholinomethane	0.075	7.4[c]
N-(3-Dimethylaminopropyl)morpholine	2.16	
Triethylamine[d]	3.32	10.65[e]
N-Methylpiperidine[a]	6.00	10.08[e]
N,N,N′,N′-Tetramethyl-1,3-propanediamine	4.15	9.8
N,N-Dimethyl-*N′,N′*-diethyl-1,3-propanediamine	3.10	—
N,N,N′,N′,N″-Pentamethyldiethylenediamine	3.47	9.4[f]
N,N,N′,N′-Tetraethylmethanediamine	0.085	10.6[c]
Bis(2-diethylaminoethyl)adipate[g]	1.00	8.6[c]
Bis(2-dimethylaminoethyl)adipate[g]	1.92	8.8[c]
N,N-Dimethylcyclohexylamine[g,h]	6.00	10.1[c]
N,N-Diethylcyclohexylamine[h]	0.70	10.0[c]
N-Methyl-*N*-octylcyclohexylamine[g]	2.00	9.8[c]
N-Methyl-*N*-dodecylcyclohexylamine[g]	1.90	—
N-Methyl-*N*-(2-ethylhexyl)cyclohexylamine[g]	0.16	9.6[c]
N-Methyldicyclohexylamine[g]	0.16	
1,4-Diazabicyclo[2.2.2]octane[i]	23.9	5.4[j]
1,2-Dimethylimidazole[i]	13.9	
Quinine[d]	11.3	7.8[c]
Pyridine[d]	0.25	5.29[k]

[a] From Union Carbide Chemicals Co.   [b] H. K. Hall, *J. Phys. Chem.* **60**, 63 (1956).   [c] Determined in this work [11].   [d] From Distillation Products, Ind.   [e] H. K. Hall, *J. Amer. Chem. Soc.* **79**, 5441 (1957).   [f] R. Rometsch, A. Marxer, and K. Miescher, *Helv. Chim. Acta* **34**, 1611 (1951).   [g] From Naugatuck Chemical Co.   [h] From E. I. du Pont de Nemours & Co.   [i] From Houdry Process Corp.   [j] pK_b, A. Farkas and K. G. Flynn, *J. Amer. Chem. Soc.* **82**, 642 (1960).   [k] H. H. Jaffe and G. O. Doak, *J. Amer. Chem. Soc.* **77**, 4441 (1955).   [l] Catalytic activity of *N*-methylmorpholine is taken as 1.00. The amines were compared at equal amine equivalents which was about 0.0300 *N*. The isocyanate and alcohol concentrations were about 0.100 *M*.   [m] Reprinted from J. Burkus, *J. Org. Chem.* **26**, 779 (1961). (Copyright 1961 by the American Chemical Society. Reprinted by permission of the copyright owner.)

60°–70°C for ½ hr. The contents are cooled and a few drops are placed on a watchglass to get a seed crystal. The seed crystal is added to the rest of the flask and, on stirring the contents, the urethane precipitates to yield 40.0 gm (97%), m.p. 55°C.

Other examples of the catalytic activity of amines and other catalysts [9] are given in Tables II and III, which show that their activity is not dependent on base strength [10a].

In the reaction of 2,4-toluene diisocyanate the 4-position isocyanate group reacts first and then the one at the 2-position [12]. Some representative examples of the reactivity of diisocyanates with 2-ethylhexanol are shown in Fig. 2.

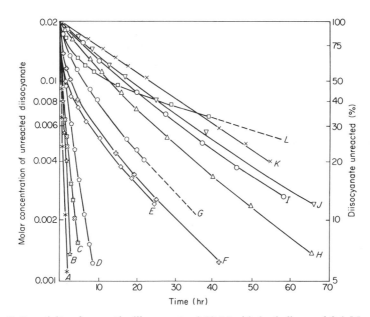

FIG. 2. Reactivity of aromatic diisocyanates 0.02 M with 2-ethylhexanol 0.4 M and diethylene glycol adipate polyester in benzene at 28°C. (A) 1-Chloro-2,4-phenylene diisocyanate. (B) m-Phenylene diisocyanate. (C) p-Phenylene diisocyanate. (D) 4,4′-Methylene bis(phenyl isocyanate). (E) 2,4-Tolylene diisocyanate. (F) Tolylene diisocyanate (60%, 2,4-isomer, 40% 2,6-isomer). (G) 2,6-Tolylene diisocyanate. (H) 3,3′-Dimethyl-4,4′-biphenylene diisocyanate (0.002 M) in 0.04 M 2-ethylhexanol. (I) 4,4′-Methylene bis(2-methylphenyl isocyanate). (J) 3,3′-Dimethoxy-4,4′-biphenylene diisocyanate. (K) 2,2′,5,5′-Tetramethyl-4,4′-biphenylene diisocyanate. (L) 80% 2,4- and 20% 2,6-isomer of tolylene diisocyanate with diethylene glycol adipate polyester (hydroxyl No. 57, acid No. 1.6, and average molecular weight 1900). Reprinted from M. E. Bailey, V. Kirss, and R. G. Spaunburgh, *Ind. Eng. Chem.* **48**, 794 (1956). (Copyright 1956 by the American Chemical Society. Reprinted by permission of the copyright owner.)

Cation exchange (free tertiary amine or quaternary ammonium groups) resin catalysts such as Amberlyst A-21 (Rohm & Haas) have been reported to catalyze the reaction of alkyl, allyl, or aryl isocyanates with hydroxy groups or compounds [10b].

Polyurethane resins are usually prepared by the reaction of a long-chain diol with an excess of the diisocyanate to obtain a "prepolymer" with terminal isocyanate groups [13a] (Eq. 8). The "prepolymer" can react separately with

R(NCO)$_2$ + HO———〰〰〰〰———OH ⟶

OCN—R—NHCOO〰〰〰〰OCONH—R—NCO   (8)

diols or diamines of low molecular weight to cause further chain extension or polymerization (curing of the prepolymer). On reaction with water, the "prepolymer" can also be used to give foams, amines or tin compounds being used as catalysts in the foaming process. The properties of the polyurethane resin are controlled by the choice of the diisocyanate and polyol. The stiffness of the aromatic portion of the diisocyanate may be increased by using 1,5-naphthalene diisocyanate in place of 2,4-toluene diisocyanate. In addition, the flexibility may be increased by using a polyether polyol derived from propylene oxide or tetrahydrofuran. Using aromatic polyols and triols would lead to chain stiffness and crosslinking in the case of the triols. The use of aromatic diamines leads to a polyurea of greater rigidity than a polyether polyurethane. Some examples of the reactivity of diols with a representative diisocyanate are shown in Table IV.

TABLE IV

REACTIVITY OF DIOLS WITH *p*-PHENYLENE DIISOCYANATE
AT 100°C [14]

Diol type	Relative order of reactivity
Polyethylene adipate	100
Polytetrahydrofuran	17–54
1,4-Butanediol	15
1,4-*cis*-Butenediol	7
1,5-Bis(β-hydroxyethoxy)naphthalene	4
1,4-Butynediol	1[a]

[a] Rate of reaction = 0.6×10^{-4} liter/mole-sec.

2-2. Preparation of a Polyurethane Prepolymer [15a]

$$\text{(structure)} \quad + \quad 2 \,\text{(2,4-toluene diisocyanate)} \quad \longrightarrow$$

$$\text{(polyurethane prepolymer structure)} \tag{9}$$

To a 500 ml dry, nitrogen-flushed resin kettle equipped with a mechanical stirrer, thermometer, condenser with drying tube, dropping funnel, and heating mantle are added 34.8 gm (0.20 mole) of 2,4-toluene diisocyanate and 0.12 gm (0.000685 mole) of *o*-chlorobenzoyl chloride. Polyoxypropylene glycol (Dow Chemical Co. P-1000, mol. wt. 1000) is added dropwise over a 2 hr period at 65° ± 3°C until 100.0 gm (0.1 mole) has been added. Stirring at 65°C is maintained for an additional 2 hr and then the reaction is cooled. The final product is a clear viscous liquid; NCO calc. 6.23%; found 6.27%.

The use of hydroxyalkylmethacrylates on reaction with diisocyanates leads to polymerizable urethanes [15b].

The reaction of α-naphthyl isocyanate with alcohols has been reported to be a convenient analytical method for the preparation of solid derivatives [16–18]. In addition, the by-product dinaphthylurea is very insoluble in hot ligroin (b.p. 100°–120°C). The urethanes are readily soluble in hot ligroin, and on cooling the solution they recrystallize to sharp-melting solids. It is recommended that two recrystallizations be performed to obtain substances for analysis. Primary alcohols react well without the need for heating the reaction mixture. Secondary alcohols require additional heat, and the yields of urethane oft are smaller than when primary alcohols are used. Tertiary alcohols other than *t*-butyl [17] or *t*-amyl [17] were not able to react under the conditions used. Table V lists some representative alcohols and their α-naphthylurethane derivatives.

TABLE V[a]

α-Naphthlyurethane Derivatives of Alcohols [16]

Alcohol	α-Naphthylurethane, m.p. (°C)
Methyl	124
Benzyl	134.5
Cinnamyl	114
Phenylethyl	119
Lauryl	80
n-Amyl	68
Furfuryl	129–130
m-Xylyl	116
o-Methoxybenzyl	135–136
Ethylene glycol	176
Trimethylene glycol	164
Glycerol	191–192
Ethylene–bromohydrin	86–87
Trimethylene–bromohydrin	73–74
Ethylene–chlorohydrin	101
Trimethylene–chlorohydrin	76
Phenylmethyl carbinol	106
Phenylethyl carbinol	102
Menthol	119
Borneol	127
Isoborneol	130
Cholesterol	160
Benzoin	140
Diphenyl carbinol	135–136
Cyclohexanol	128–129
2-Methylcyclohexanol	154–155
3-Methylcyclohexanol	122
4-Methylcyclohexanol	159–160
Methylhexyl carbinol	63–64
Diethyl carbinol	71–73
Triphenyl carbinol	No reaction
Diethylmethyl carbinol	No reaction
Citronellol	No reaction

[a] Reprinted in part from V. T. Bickel and H. E. French, *J. Amer. Chem. Soc.* **48**, 747 (1926). (Copyright 1926 by the American Chemical Society. Reprinted by permission of the copyright owner.)

2-3. General Method for the Preparation of α-Naphthylurethanes [16]

$$ROH + \underset{\text{NCO}}{\overset{}{\bigcirc\!\bigcirc}} \longrightarrow \underset{\text{HN—C—OR}}{\overset{\overset{\text{O}}{\|}}{\bigcirc\!\bigcirc}} \qquad (10)$$

To a small flask or test tube are added 2–3 gm of the alcohol and a slight stoichiometric excess of the α-naphthyl isocyanate. If the reactants do not react immediately to give a solid product, heat is applied. The contents of the flask are extracted with boiling ligroin (b.p. 100°–120°C) which dissolves the urethane but leaves the insoluble dinaphthylurea.

In the case of cinnamyl alcohol, menthol, borneol, isoborneol, cholesterol, and benzoin, the alcohol is dissolved in approximately 10 ml of ligroin before it reacts with α-naphthyl isocyanate.

The urethanes are generally recrystallized from ligroin with the exception of glycerol and ethylene glycol, which are recrystallized from alcohol.

Phenols but not polyhydroxybenzenes also react with α-naphthyl isocyanate to give α-naphthylurethanes. In some cases in which solid derivatives were difficult to obtain, the addition of 1 or 2 drops of triethylamine solution in ether caused a solid derivative to form rapidly. Table VI [19a] gives several phenols and the respective melting points for their α-naphthylurethane derivatives. Again in this case the by-product dinaphthylurea has the advantage that it is insoluble in boiling ligroin, whereas diphenylurea (from phenyl isocyanate) is soluble and causes problems in separating it from the urethane derivative. The experimental procedure is similar to that described in Preparation 2-3.

B. Reaction of Carbonates with Amines

N-Alkyl carbamates can be prepared by the reaction of dialkyl carbonates with alkylamines. Delepine and Schving earlier reported that *N*-methyl methylcarbamate could be prepared by the reaction of methylamine with dimethyl carbonate [19b, f]. Dimethylamine reacted similarly to give *N*-dimethyl methylcarbamate [19b, c].

$$(CH_3)_2NH + CH_3O\underset{\overset{\|}{O}}{C}—OCH_3 \longrightarrow (CH_3)_2—\underset{\overset{\|}{O}}{C}—OCH_3 + CH_3OH \quad (11)$$

Aromatic carbonates react similarly with aliphatic amines in the presence or absence of solvents [19d].

TABLE VI[c]

α-NAPHTHYLURETHANE DERIVATIVES OF SOME
REPRESENTATIVE PHENOLS [19]

Phenolic compound	Urethane, m.p. (°C)	Catalyst $(C_2H_5)_3N$ added
m-Cresol	127–128	No
p-Cresol	146	No
Thymol	160	No
Carvacrol	116	No
o-Nitrophenol	112–113	Yes
m-Nitrophenol	167	No
p-Nitrophenol	150–151	No
o-Chlorophenol	120	Yes
m-Chlorophenol	157–158	Yes
p-Chlorophenol	165–166	No
o-Bromophenol	128–129	Yes
p-Bromophenol	168–169	Yes
2,4,6-Tribromophenol	153	Yes
2-Chloro-5-hydroxytoluene	153–154	Yes
4-Hydroxy-1,2-dimethylbenzene	141–142	Yes
4-Hydroxy-1,3-dimethylbenzene	134–135	Yes
2-Hydroxy-1,4-dimethylbenzene	172–173	Yes
Resorcinol–monomethyl ether	128–129	Yes
Guiacol	118	No
Eugenol	122	No
Isoeugenol	149–150	Yes
Orcinol[a]	160	Yes
Resorcinol	No reaction	—
Hydroquinol	No reaction	—
Catechol	No reaction	—
Pyrogallol	No reaction	—
α-Naphthol	152	Yes
β-Naphthol	156–157	Yes
1-Nitro-2-naphthol[a]	128–129	Yes
o-Aminophenol[b]	201	No

[a] Reacts with great difficulty and gives low yields of the urethanes.

[b] Reacts readily at room temperature.

[c] Reprinted in part from H. E. French and A. F. Wirtel, *J. Amer. Chem. Soc.* **48**, 1736 (1926). (Copyright 1926 by the American Chemical Society. Reprinted by permission of the copyright owner.)

In the absence of amines ureas can be used to react with alkyl carbonates under elevated temperatures and pressures when catalyzed by metal salts or oxides. For example, N,N''-diphenylurea (1.0 mole) reacts with diethylcarbonate (6.0 moles), at 180°C in a small pressure vessel when catalyzed by lead oxide (0.8 gm) to give a 98% yield of N-phenyl-O-ethylurethane [19e].

C. N-Alkylation of Alkyl Carbamates

The reaction of alkyl carbamates with amines to give N-alkyl carbamates has been reported to be catalyzed by tertiary amines [19f]. Aromatic amines such as aniline have been reported to react with methyl carbamate in methanol at 190°C when catalyzed by zinc chloride [19g].

The reaction of olefins (such as isobutylene with alkyl carbamates) is reported to be catalyzed by acids [19h, i].

Ethyl carbamate can also be substituted on the nitrogen by means of condensations involving formaldehyde [19i].

$$C_2H_5O\overset{\overset{\displaystyle O}{\|}}{C}NH_2 + CH_2{=}O \quad CH_3{-}C_6H_4{-}SO_2Na \xrightarrow[70°-75°C]{\overset{\overset{\displaystyle O}{\|}}{H{-}COH}}$$

$$CH_3{-}C_6H_4SO_2CH_2NH\overset{\overset{\displaystyle O}{\|}}{C}OC_2H_5 + H_2O + NaO\overset{\overset{\displaystyle O}{\|}}{C}H \quad (12)$$

3. MISCELLANEOUS REACTIONS OF ISOCYANATES AND ALCOHOLS

(1) Reaction of α-hydroxy esters with isocyanates [20].
(2) Ring formation on the reaction of glycidol with phenyl isocyanate [21].
(3) Reaction of ethanolamine with isocyanates [22].
(4) Reaction of phenols with isocyanates [23a, b].
(5) Reaction of acrylyl isocyanate with alcohols [23c].

4. MISCELLANEOUS METHODS OF PREPARING URETHANES (N-CARBAMATES)

(1) Reaction of inorganic cyanates with organic halides in the presence of alcohols [24].

(2) Reaction of 1,4-dichloro-2-butene with sodium cyanate and 1,4-butanediol to give poly(tetramethylene-2-butene-1,4-dicarbamate) [25].

(3) Direct fluorination of N-substituted carbamates [26].

(4) Reaction of dimethylcarbamoyl chloride with *p*-methoxyphenol to give *p*-methoxyphenol *N,N*-dimethylcarbamate [27].

(5) Reaction of ethyl chloroformate with amines to give ethyl alkylcarbamates [28].

(6) Reaction of phenyl chloroformate with amines [29].

(7) Reaction of ethyl chloroformate with allylamine to give ethyl allylcarbamate [30].

(8) Reaction of alcohols with the inner salt of methyl (carboxysulfamoyl)-triethylammonium hydroxide to give methyl *N*-alkyl carbamates [31].

(9) Reaction of nitro compounds with carbon monoxide and an alcohol in the presence of catalysts to give urethanes [32].

(10) Preparation of benzoxazolone [33].

(11) Carbonylation of amines to carbamates [34].

(12) Fluorination of carbamates [35].

(13) Preparation of *N*-chlorocarbamates [36].

(14) Carbamates by the oxidative alkoxycarbonylation of amines [37].

(15) Preparation of vinyl carbamates [38].

(16) Carboalkoxylation and carboxylation of amines to give carbamates [39].

(17) Vicinal oxyamination of olefins by *N*-chloro-*N*-metallocarbamates [40].

(18) Anodic methoxylation of amines to give α-methoxycarbamates [41].

REFERENCES

1a. P. Adams and F. A. Baron, *Chem. Rev.* **65**, 567 (1965).
1b. W. Worthy, *Chem. Eng. News*, pp. 27–33 (Feb. 11, 1985).
2. T. L. Davis and J. M. Farnum, *J. Amer. Chem. Soc.* **56**, 883 (1934).
3a. E. Knovenagel and A. Schurenberg, *Ann. Chem.* **297**, 148 (1897).
3b. V. T. Bickel and H. E. French, *J. Amer. Chem. Soc.* **48**, 747 (1926).
4. G. Karmas, U.S. Pat. 2,574,484 (Nov. 13, 1951).
5. J. H. Saunders and R. J. Slocombe, *Chem. Rev.* **43**, 201 (1948).
6a. W. J. Bailey and F. Cesare, Meeting Amer. Chem. Soc., April 1959.
6b. A. Segal, G. Lavers, and B. L. Van Duuren, *J. Chem. Eng. Data* **15**, 337 (1970).
7a. D. S. Tarbell, R. C. Mallatt, and J. W. Wilson, *J. Amer. Chem. Soc.* **64**, 2229 (1942).
7b. W. J. Bailey and J. R. Griffith, *J. Org. Chem.* **43**, 2690 (1978).
8. S. R. Sandler, unpublished data (1970).
9. F. Hostettler and E. F. Cox, *Ind. Eng. Chem.* **52**, 609 (1960).
10a. A. Farkas and K. G. Flynn, *J. Amer. Chem. Soc.* **82**, 642 (1960).
10b. D. W. Peck, Ger. Offen. 2,461,129 (July 3, 1975); *Chem. Abstr.* **83**, 113128e (1975).
11. J. Burkus, *J. Org. Chem.* **26**, 779 (1961).
12. M. E. Bailey, V. Kirss, and R. G. Spaunburgh, *Ind. Eng. Chem.* **48**, 794 (1956).
13a. J. H. Saunders and K. C. Frisch, "Polyurethanes—Chemistry and Technology," Wiley (Interscience), New York, 1962.
13b. H. Ulrich, in *Kirk–Othmer Encyclopedia of Chemical Technology* **23**, 576 (1983).
14. W. Cooper, R. W. Pearson, and J. Drake, *Ind. Chem.* **36**, 121 (1960).
15a. S. R. Sandler and F. R. Berg, *J. Appl. Polym. Sci.* **9**, 3909 (1965).

15b. R. H. Leitheiser and J. J. Szwarc, U.S. Pat. 4,078,015 (Mar. 7, 1978).

16. V. T. Bickel and H. E. French, *J. Amer. Chem. Soc.* **48**, 747 (1926).

17. C. Neuberg and E. Kansky, *Biochem. Z.* **20**, 446 (1909).

18. R. L. Shriner and R. C. Fuson, "The Systematic Identification of Organic Compounds," 3rd ed. Wiley, New York, 1948.

19a. H. E. French and A. F. Wirtel, *J. Amer. Chem. Soc.* **48**, 1736 (1926).

19b. M. Delepine and F. Schving, *Soc. Chim. Fr.* **7**, 894 (1910).

19c. A. Akoyunoglou and M. Calvin, *J. Org. Chem.* **28**, 1484 (1963).

19d. U. Romano, U.S. Pat. 4,097,676 (June 27, 1978).

19e. H.-J. Boysch, H. Krimm, and W. Richter, U.S. Pat. 4,381,404 (Apr. 26, 1983).

19f. S. J. Falcone and J. J. McCoy, U.S. Pat. 4,336,402 (June 22, 1982); *Chem. Abstr.* **97**, 144414p (1982).

19g. P. Heitkämper, K. König, R. Fauss, and K. Findeisen, U.S. Pat. 4,388,238 (June 14, 1983).

19h. F. Merger and G. Nestler, Ger. Offen. DE 3,204,711 (Aug. 18, 1983); *Chem. Abstr.* **100**, 5886m (1984).

19i. G. Nestler and F. Merger, Ger. Offen. DE 3,233,309 (Mar. 8, 1984); *Chem. Abstr.* **101**, 6645q (1984).

19j. A. M. van Leusen and J. Strating, *Org. Synth.* **57**, 951 (1977).

20. R. F. Rekker, A. C. Faber, D. H. E. Tom, H. Verheur, and W. T. Natta, *Rec. Trav. Chim.* **70**, 113 (1951).

21. Y. I. Wakura and Y. Taneda, *J. Org. Chem.* **24**, 1992 (1959).

22. L. Knorr and P. Rössler, *Chem. Ber.* **36**, 1278 (1903).

23a. S. Petersen, *Ann. Chem.* **562**, 205 (1949).

23b. E. A. Shilov and M. N. Bogdanov, *Zh. Obshch. Khim.* **18**, 1060 (1948); *Chem. Abstr.* **43**, 1351h (1929).

23c. S. F. Reed, Jr., U.S. Pat. 4,001,191 (Jan. 4, 1977).

24. P. A. Argabright, D. H. Rider, and R. Sieck, *J. Org. Chem.* **30**, 3317 (1965).

25. S. Ozaki, *Polym. Lett.* **5**, 1053 (1967).

26. V. Grakauskas and K. Baum, *J. Org. Chem.* **34**, 2840 (1969).

27. E. Lustig, W. R. Benson, and N. Duy, *J. Org. Chem.* **32**, 851 (1967).

28. A. Donetti and E. Morazzi-Uberti, *J. Med. Chem.* **13**, 747 (1970).

29. D. A. Scola and J. S. Adams, Jr., *J. Chem. Eng. Data* **15**, 347 (1970).

30. J. L. Brewbaker and H. Hart, *J. Amer. Chem. Soc.* **91**, 711 (1969).

31. E. M. Burgess, H. R. Penton, Jr., E. A. Taylor, and W. M. Williams, *Org. Synth.* **56**, 40 (1977).

32. R. Becker, C. Rasp, G. Stammann, and J. Grolig, U.S. Pat. 4,339,592 (July 13, 1982); T. Takeuchi, M. Nishi, T. Irie, and H. Ryuto, U.S. Pat. 4,469,882 (Sept. 4, 1984).

33. Y. Querou, U.S. Pat. 4,309,549 (Jan. 5, 1982) and U.S. Pat. 4,309,550 (Jan. 5, 1982).

34. D. Moy, U.S. Pat. 4,242,520 (Dec. 30, 1980) and U.S. Pat. 4,258,201 (Mar. 24, 1981); H. S. Kesling, Jr., U.S. Pat. 4,251,667 (Feb. 17, 1981); K. Kondo, N. Sonoda, and S. Tsutsumi, *Chem. Lett.* (5), 373 (1972); *Chem. Abstr.* **77**, 87801y (1972); G. Stammann, R. Becker, J. Grolig, and H. Waldmann, Eur. Pat. Appl. EP 54,218 (June 23, 1982); *Chem. Abstr.* **97**, 198005z (1982).

35. V. Grakauskas and K. Baum, *J. Org. Chem.* **34**, 2840 (1969).

36. S. C. Czapf, H. Gottlieb, G. F. Whitfield, and D. Swern, *J. Org. Chem.* **38**, 2555 (1973).

37. S. Furuoka, M. Chono, and M. Kohno, *J. Org. Chem.* **49**, 1460 (1984).

38. P. J. Stang and G. H. Anderson, *J. Org. Chem.* **46**, 4585 (1981).

39. Y. Kita, J.-I. Haruta, H. Tagawa, and Y. Tamura, *J. Org. Chem.* **45**, 4579 (1980).

40. E. Herranz and K. B. Sharpless, *J. Org. Chem.* **45**, 2710 (1980).

41. T. Shono, Y. Matsumura, and S. Kashimura, *J. Org. Chem.* **48**, 3338 (1983).

CHAPTER 11 / *O*-CARBAMATES

1. INTRODUCTION

O-Carbamates (structure I) represent a class of compounds distinct from those discussed in Chapter 10, which are *N*-carbamates (substituents on the nitrogen atom as in structure II). In this chapter only carbamates with varying substituents on the oxygen atom are discussed.

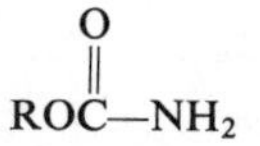

$$\underset{\text{(I) } O\text{-carbamate}}{ROC\!\!-\!\!NH_2} \qquad \qquad \underset{\text{(II) } N\text{-carbamate}}{ROC\!\!-\!\!NHR}$$

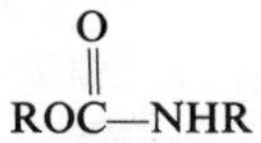

The recent industrial applications of a carbomates such as those reported as a tranquilizer [1] (structure III), a crease-resistant agent (when reacting with formaldehyde) in the textile industry [2], a solvent [3], hair conditioners [4], a plasticizer [5], and a fuel additive [6] have stimulated interest in the synthesis of various *O*-carbamates.

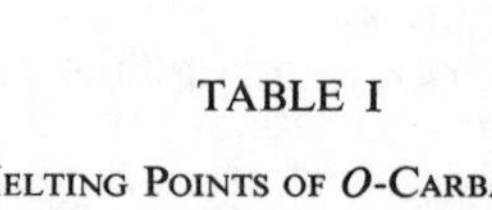

$$\underset{C_3H_7 \quad CH_2OCONH_2}{\overset{CH_3 \quad CH_2OCONH_2}{\diagdown \underset{\diagup}{C}\diagup}}$$

(III)   Meprobamate, a tranquilizer

The *O*-carbamates are usually solids; some representative examples are shown in Table I.

TABLE I

MELTING POINTS OF *O*-CARBAMATES

Carbamate	M.p. (°C)
Methyl	54.2
Ethyl	48.2
n-Propyl	60
Isopropyl	95
n-Butyl	53
Lauryl	81–82
n-Octyl	67
n-Stearyl	94–95

The most important methods of preparing O-carbamates are shown in Eq. (1).

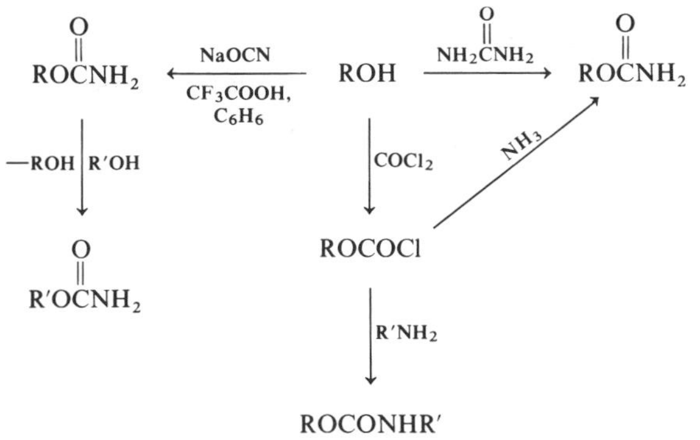

$$
\begin{array}{ccccc}
 & & \overset{O}{\underset{\|}{\text{ROCNH}_2}} & & \\
\overset{O}{\underset{\|}{\text{ROCNH}_2}} & \xleftarrow[\substack{\text{CF}_3\text{COOH,} \\ \text{C}_6\text{H}_6}]{\text{NaOCN}} & \text{ROH} & \xrightarrow{\text{NH}_2\text{CNH}_2} & \overset{O}{\underset{\|}{\text{ROCNH}_2}} \\
\Big\downarrow{\text{—ROH}\,\big|\,\text{R'OH}} & & \Big\downarrow{\text{COCl}_2} & \nearrow^{\text{NH}_3} & \\
\overset{O}{\underset{\|}{\text{R'OCNH}_2}} & & \text{ROCOCl} & & \\
 & & \Big\downarrow{\text{R'NH}_2} & & \\
 & & \text{ROCONHR'} & &
\end{array}
\tag{1}
$$

2. CONDENSATION REACTIONS

A. Reactions of Alcohols with Urea

The reaction of primary alcohols with urea gives carbamates when the reaction is carried out at 115°–150°C [7–10] (Eqs. 2, 3). Since 150°C is the temperature for the optimum dissociation of urea to cyanic acid and ammonia, lower-boiling alcohols (methyl, ethyl, and propyl) must be heated under pressure. Refluxing urea and n-butanol at 115°–120°C requires a 40-hr reaction time to give a 75 % yield of butyl carbamate [8].

$$
\overset{O}{\underset{\|}{\text{NH}_2\text{CNH}_2}} \;\underset{\text{heat}}{\rightleftharpoons}\; \text{HNCO} + \text{NH}_3
\tag{2}
$$

$$
\text{HNCO} + \text{ROH} \;\longrightarrow\; \text{RO}\underset{\underset{O}{\|}}{-}\text{CNH}_2
\tag{3}
$$

In order to shorten the reaction time, various heavy metal salts (zinc, lead, and manganese acetates) of weak organic acids, zinc or cobalt and tin chlorides are added to the reaction mixture [11]. For example, refluxing an uncatalyzed mixture of 3 moles of isobutyl alcohol and urea for 150 hr at 108°–126°C gives a 49 % yield of the carbamate. Adding lead acetate or cobalt chloride to the same reaction lowers the reaction time to 75 hr, at which point an 88–92 % yield is obtained. In another example, ethylene glycol (1 mole) and urea (2 moles) are heated for 3 hr at 135°–155°C with Mn(OAc)_2 to give a 78 % yield of the diurethane [11]. The commercial production of butyl carbamate uses catalytic quantities of cupric acetate [12].

The effects of catalysts on the yields and rates of formation of carbamates from alcohols and urea are shown in Tables II and III. The catalytic effectiveness of BF_3 has also been reported [13].

Apparently a systematic study has not been made to evaluate the comparative effectiveness of various metals as catalysts for the reaction of urea with alcohols to give carbamates. The reaction of urea with tertiary alcohols and phenols fails to give carbamates. In concentrated sulfuric acid at 20°–25°C tertiary alcohols alkylate urea in 31–33% yields [14, 15]. For the preparation of tertiary alkyl, secondary alkyl, and phenolic carbamates see Section C.

2-1. Preparation of Methyl Carbamate [12]

$$
\underset{\text{NH}_2\overset{\displaystyle\text{O}}{\overset{\|}{\text{C}}}\text{NH}_2}{} + \text{CH}_3\text{OH} \quad \xrightarrow[130°C]{\text{Cu(OAc)}_2} \quad \text{CH}_3\text{O}\overset{\displaystyle\text{O}}{\overset{\|}{\text{C}}}\text{NH}_2 + \text{NH}_3 \tag{4}
$$

TABLE II[a]

CATALYTIC EFFECT ON THE YIELDS AND RATES OF FORMATION OF CARBAMATES
FROM ALCOHOLS AND UREA

Reactants				Time	Temp.	Carbamate
Urea (moles)	Alcohol (moles)	Catalyst	(gm)	(hr)	(°C)	(yield %)
	Benzyl alcohol					
1.0	3.0	O	—	5	175–185	56
1.0	2.0	Zn(OAc)$_2$	4	8	150–160	87
1.0	2.0	SnCl$_4$	4	8	150–160	92
1.0	2.0	conc. H$_2$SO$_4$	1	5	175–185	0
1.0	2.0	NH$_4$HSO$_4$	6	5	175–185	0
1.0	2.0	Glycerin	8	5	175–185	0
	Isobutyl alcohol					
1.0	3.0	O	—	150	108–126	49
1.0	3.0	Pb(OAc)$_2$	6	75	108–126	88
1.0	3.0	COCl$_2$	5	75	108–126	92
1.0	1.95	85% H$_3$PO$_4$	5	12	80–85	0
1.0	1.95	HCl gas		10	70–75	0
	Methyl- Cyclohexanol					
1.0	3.02	O	—	4	160–205	18
1.0	3.02	O	—	6	150–160	56
1.0	3.02	Zn(OAc)$_2$	7	5	100–105	83
1.0	3.02	85% H$_3$PO$_4$	7	5	160–185	0

[a] Data taken from Paquin [11].

To a flask are added 60 gm (1.0 mole) of urea, 4.0 gm (0.022 mole) of cupric acetate, and 0.5 ml of methanol. The mixture is heated to 130°C, and 32 gm (1.0 mole) of methanol is then added dropwise over a 3 hr period while ammonia is evolved completely. The residue in the flask weighs 31.1 gm (41%) and has a melting point of 54°C. The advantage of this catalyst is that it allows the reaction to be carried out at atmospheric pressure. Other examples are described in Table IV.

Ethyl, butyl, and 2-methoxyethyl carbamates are prepared similarly.

2-2. Preparation of Ethyl Carbamate [13]

$$C_2H_5OH + NH_2\overset{\overset{\text{O}}{\|}}{C}NH_2 \xrightarrow{BF_3} C_2H_5O\overset{\overset{\text{O}}{\|}}{C}NH_2 + NH_3 \qquad (5)$$

To a tared flask containing 600.6 gm (10 moles) of urea dissolved in 2073.2 gm of ethanol at 70°C is added, via a gas addition tube, boron trifluoride until 678.2 gm (4.3 moles) is absorbed. The ethanol is removed by distillation until the pot temperature reaches 100°C, then the reaction mixture is filtered. The solids, which consist of BF_3-NH_3 complex, are washed with ethanol and then the ethanol is added to the original filtrate. The ethanol is then distilled off again as above and filtered again to give more BF_3-NH_3 solids. The total BF_3-NH_3 solids weigh 535 gm. This solid is washed with hot ethanol and the total alcohol washing and remaining ethanol filtrates cooled to give 255 gm of $BF_3-4(NH_2CONH_2)$ complex, m.p. 96°–98°C. The combined filtrates are freed of alcohol by distillation, extracted with benzene, and the extract on distillation under reduced pressure affords 380 gm (99%) of $C_2H_5OCONH_2$, b.p. 115°C (100 mm), m.p. 48°–50°C. The tacky residue weighs 150 gm.

An example of the uncatalyzed reaction of n-butanol with urea is described in Preparation 2-3, which requires a 30 hr reaction time for completion. The more convenient catalyzed reactions are shown in Preparations 2-4 and 2-5. Preparation 2-4 is preferred because of its simplicity (also cost) and use of no added solvents which may complicate recovery. Other examples of useful catalyzed reactions are shown in Tables II, III, and IV.

2-3. Preparation of n-Butyl Carbamate [8]

$$NH_2\overset{\overset{\text{O}}{\|}}{C}NH_2 + n\text{-}C_4H_9OH \longrightarrow n\text{-}C_4H_9O\overset{\overset{\text{O}}{\|}}{C}NH_2 + NH_3 \qquad (6)$$

To a flask in a hood is added 970 gm (13.1 moles) of n-butanol. The n-butanol is stirred and warmed to 100°C while urea (180 gm, 3 moles) is added in small portions. The reaction is exothermic, and the temperature is maintained

USE OF MISCELLANEOUS CATALYSTS FOR THE FORMATION OF O-CARBAMATES BY THE REACTION OF UREA WITH ALCOHOLS

Catalyst* (gm)	Alcohol (moles)	Urea (moles)	Time (hr)	Temp. (°C)	O-Carbamate yield (%)	Ref.
Cu_2O-CuO 0.5	$C_6H_5CH_2OH$ 1.0	0.5	3.5	130	77.1	[a]
—	$C_2H_5OCH_2CH_2OH$ 3.0	1.7	7.0	190 (7 bars)	96	[b]
Ni-containing ion exchange resin 18	$CH_3OCH_2CH_2OH$ 9.0	5.0	7	149	98.5 (b.p. 96°–98°C at 0.1 mm Hg)	[c]
(From Amberlite 200 and $NiSO_4$) H_3PO_4 588	CH_3OH 53	6.0	0.66	138–140	88.2	[d]
$AlCl_3$ 15	Dry $CH_3OCH_2CH_2OH$ 10	5.0	5.0	130–147	98.5	[e]
ZnO 1.2%	n-C_4H_9OH 2–4	1.0	9	100–150	94	[f]
Aluminosilicate 14	CH_3OH	1.0	—	160 (40 atm)	42	[g]
PbO_2 4.0	CH_3OH 1.0	1.0	3	130	50.6 (m.p. 53°–54°C)	[h]
$Cu(OAc)_2$ 4.0	C_1–C_5 alcohols 1.0	1.0	3	130	31.5	[h]
5% ZnO + 0.3 mole H_3PO_4	C_1–C_5 alcohols 9.0	1.0	0.25	140	57.8–63.5	[i]
Benzene hexachloride 97	CH_3OH 3.0	1.0	4	160	82	[j]
$(CH_3O)_3P$ 1.8	n-BuOH 3.0	1.0	20	Reflux	85	[k]
Diazobicyclo[2.2.2]octane (triethylene diamine) 32	$CH_3OCH_2CH_2OH$ 20	10.0	9	Reflux	71	[l]

CS$_2$	CH$_3$OH		1.0	—	180	80.7	[m]
	14						
CS$_2$	C$_2$–C$_5$ alcohols		1.0	—	180	60–80	[m]
	14						
HNO$_3$	C$_5$–C$_9$ Alcohols		1.0 in DMF		130–140	25–45	[n]
Polyethylene glycol 200	C$_2$H$_5$OH		3.0		140	53	[o]
	26				(250 psig)		
100	C$_4$–C$_6$		1.0	3–4	120–172	83–95 (C$_4$)	[p]
						69–75 (C$_6$)	
Zn(OAc)$_2$	1.0						
10							
Dibutyltindilaurate	n-BuOH	4.0	1.0	17	Reflux	57	[q]
9.2							

*Note: In some cases the catalyst is also a reactant, as in the case of benzene hexachloride. In one case the catalyst is a polar solvent which is recovered (as, for example, polyethylene glycol 200).

[a] K. Kozlowski, A. Goraczko, S. Musial, and J. Doda, Pol. PL. 117,711 (Dec. 31, 1982); Chem. Abstr. 99, 53397k (1983).

[b] O. Mattner, F. Merger, N. Gerhard, and F. Towae, Ger. Offen. DE 3,200,559 (July 21, 1983); Chem. Abstr. 99, 104818h (May 29, 1983).

[c] T. Dockner and H. Petersen, U.S. Pat. 4,156,784 (May 29, 1979).

[d] J. D. Slater and W. J. Culbreth, U.S. Pat. 3,554,730 (Jan. 12, 1971).

[e] J. G. DeMarco, U.S. Pat. 3,725,464 (Apr. 3, 1973).

[f] M. O. Robeson, U.S. Pat. 3,574,711 (Apr. 13, 1971).

[g] T. P. Vishnyakova, Y. M. Puushkin, T. A. Faltynek, I. A. Golubeve, V. V. Federov, and G. F. Shuleika, Khim. Tekhnol. Topl. Masel 16(7), 10 (1971); Chem. Abstr. 75, 129311e (1971).

[h] H. Hatas, Japan Pat. 70/23,536 (Aug. 7, 1970); Chem. Abstr. 73, 109312m (1970). See also S. Bernfest and J. Halpern, U.S. Pat. 3,013,064 (Mar. 10, 1960). See also ref. [12] in this chapter.

[i] T. P. Vishnyakova, Y. M. Paushkin, T. A. Faltynek, I. A. Golubova, A. G. Liakumovich, Y. I. Michurov, Neftepererab. Neftekhim (Moscow) (1), 27 (1969); Chem. Abstr. 70, 86993f (1969).

[j] M. Taguchi, M. Aramaki, and K. Nagano, Japan Pat. 68/22,290 (Sept. 25, 1968); Chem. Abstr. 70, 57221x (1969).

[k] T. Fujita, Y. Aida, S. Shimpo, and T. Nakamura, Japan Pat. 69/05,370 (May 6, 1969); Chem. Abstr. 71, 3012g (1969).

[l] H. G. Goodman, Jr., and C. A. Duprez, U.S. Pat. 3,449,406 (June 10, 1969); Chem. Abstr. 71, 38388m (1969).

[m] T. P. Vishnyakova, I. A. Golubeva, and T. W. Faltynek, Neftepererab. Neftekhim. (Moscow) (3), 31 (1969); Chem. Abstr. 71, 85008x (1969).

[n] R. Zielinski, Zesz. Nauk. Akad. Ekom Poznanin Ser. I 88, 103 (1981); Chem. Abstr. 98, 161220y (1983). See also L. Guerci, Gi. Chim. Ind. Appl. 4, 60 (1921); Chem. Abstr. 16, 2481 (1922).

[o] J. Levy, U.S. Pat. 3,871,259 (Jan. 27, 1959).

[p] B. A. Dadashev, O. G. Seidbekova, and G. G. Alimamedov, Azerb. Khim. Zh. (4), 21 (1965); Chem. Abstr. 64, 8028 (1966).

[q] A. Ibbotson and H. J. Twitchett, Brit. Pat. 984,084 (Feb. 24, 1965); Chem. Abstr. 62, 14510 (1965).

TABLE IV

PREPARATION OF *O*-ALKYL CARBAMATES BY THE PROCESS OF PREPARATION 2-4 [12]

Urea (moles)	Cupric acetate (g)	Alcohol (moles)	Temp. (°C)	Time (hr)	Yield product (%)	M.p. (°C)
		Ethanol				
1.0	4.0	1.0	130	3	60	49
		n-Butanol				
1.0	2.0	1.1	118–160	4	71	52
		Methyl cellosolve[a]				
1.0	2.0	1.1	120–160	4	50	46.8

[a] $CH_3OCH_2CH_2OH$.

below the melting point of urea so that the urea dissolves and does not settle as a molten layer. The solution is refluxed for 30 hr while ammonia escapes from the top of the condenser. The reaction mixture is distilled until the temperature reaches 150°C. On cooling, the residue solidifies and is then boiled with 1 liter of ligroin (b.p. 60°–90°C). The mixture is filtered. The remaining solids are refluxed with ligroin and then filtered. Approximately 12–18 gm (9–14%) of almost pure cyanuric acid is obtained from the ligroin-insoluble material. Distillation of the combined dry ligroin filtrate up to 150°C affords a residue which on distillation at reduced pressure gives 266 gm (76%), b.p. 108°–109°C (14 mm), m.p. 53°–54°C.

2-4. *Preparation of n-Butyl Carbamate* [12]

$$H_2N\overset{\overset{\displaystyle O}{\|}}{C}NH_2 + n\text{-}C_4H_9OH \xrightarrow{Cu(OAc)_2} n\text{-}C_4H_9O\overset{\overset{\displaystyle O}{\|}}{C}NH_2 + NH_3 \qquad (7)$$

To a flask was added 60 gm (1 mole) of urea, 100 ml of *n*-butanol (1.1 mole), and 2.0 gm (1.01 mmole) of cupric acetate. The contents were heated to reflux (118°–160°C) for 4 hr until 99.5% of the theoretical amount of ammonia was evolved. After the reaction 83.2 gm (71%) of *n*-butyl carbamate was obtained having a melting point of 52°C.

2-5. *Preparation of n-Butyl Carbamate* [13b]

$$H_2N\overset{\overset{\displaystyle O}{\|}}{C}NH_2 + n\text{-}C_4H_9OH \xrightarrow[\substack{\text{triethylene} \\ \text{diamine}}]{DMF} n\text{-}C_4H_9O\overset{\overset{\displaystyle O}{\|}}{C}NH_2 + NH_3 \qquad (8)$$

To a flask were added 296.0 gm (4.0 mole) of *n*-butanol, 180.0 gm (3.0 mole) of urea, 180.0 gm (2.5 mole) of dimethylformamide, and 20. 4 gm

(0.19 mole) of triethylenediamine (diazobicyclo[2.2.2]octane). The contents were refluxed for 6 hr until the theoretical amount of ammonia was evolved. During the reflux period it was observed that only a small amount of ammonium carbamate formed in the condenser. The reaction mixture was distilled under reduced pressure to give 221.0 gm (64% yield) of product.

B. Reactions of Alkyl Chloroformates with Ammonia

The ammonolysis of alkyl chloroformates affords an excellent laboratory method for the preparation of carbamates [16–20]. However, since phosgene is involved in the preparation of the chloroformate ester, great caution must be exercised. It is mandatory that phosgene be used in a well-ventilated hood. Where possible, the urea–alcohol method of Section A should be used because of the more favorable weight relationship of urea compared to phosgene. Assuming 100% reaction, the weight relationships given in Eqs. (9), (10), and (11) would hold.

Section B:

$$CH_3OH + COCl_2 \longrightarrow CH_3OCOCl + HCl \qquad (9)$$
$$32 \text{ gm} \quad 99 \text{ gm} \qquad 94.5 \text{ gm} \quad 36.5 \text{ gm}$$

$$CH_3OCOCl + 2NH_3 \longrightarrow CH_3OCONH_2 + NH_4Cl \quad (10)$$
$$94.5 \text{ gm} \quad 34 \text{ gm} \qquad 75 \text{ gm} \quad 53.5 \text{ gm}$$

Section A:

$$CH_3OH + NH_2CONH_2 \longrightarrow CH_3OCONH_2 + NH_3 \qquad (11)$$
$$32 \text{ gm} \quad 60 \text{ gm} \qquad 75 \text{ gm} \quad 17 \text{ gm}$$

It should be noted that the method of Section A gives ammonia by-product which can be used to prepare more urea. However, Section B gives ammonium chloride, which requires costly conversion into free ammonia.

2-6. Preparation of 2-Ethylbutyl Carbamate [19]

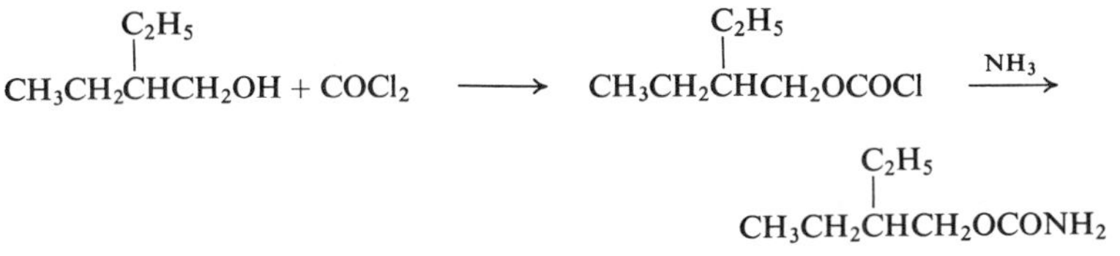

$$\underset{\substack{| \\ CH_3CH_2CHCH_2OH}}{\overset{C_2H_5}{}} + COCl_2 \longrightarrow \underset{\substack{| \\ CH_3CH_2CHCH_2OCOCl}}{\overset{C_2H_5}{}} \xrightarrow{NH_3}$$

$$\underset{\substack{| \\ CH_3CH_2CHCH_2OCONH_2}}{\overset{C_2H_5}{}} \quad (12)$$

To a flask situated in a well-ventilated hood is added 1 liter of toluene. The toluene is cooled to 5°C with an ice bath and then phosgene is passed into it until 200 gm (2.0 mole) has been absorbed. To this phosgene solution is added 182 gm (1.8 mole) of 2-ethyl-1-butanol with rapid stirring. The reaction is exothermic and the temperature rises to 35°C while hydrogen chloride and some phosgene are evolved and passed through a potassium hydroxide trap. The reaction mixture is stirred for an additional 18 hr, then a dry nitrogen stream is passed through the solution for 1 hr in order to remove excess phosgene. The solution is poured, with rapid stirring, into 400 ml of concentrated aqueous ammonia, cooled to 5°C. The toluene layer is separated, concentrated under reduced pressure, and cooled in an ice bath to precipitate 195 gm (75%), m.p. 81°C (corr.) of 2-ethylbutyl carbamate.

Other carbamates such as ethyl-*n*-propyl, isopropyl, 3-butyl, isobutyl, *sec*-butyl, *n*-amyl, isoamyl, 2-methylbutyl, 1-ethylpropyl, and 2-ethylhexyl carbamate were prepared in a manner similar to that described here for 2-ethylbutyl carbamate in 55–76% yields [19]. Benzyl carbamate has also been reported to be prepared by this method in 91–94% yield [17].

2-7. *Preparation of 3-Methyl-1-butynyl-3-carbamate* [17b]

$$CH_3CHC{\equiv}CH + COCl_2 \xrightarrow[\substack{\text{ether} \\ -10°C}]{\text{quinoline}} \underset{OH}{} \; \overset{C{\equiv}CH}{\underset{O}{CH_3{-}\overset{|}{CH}{-}O\overset{\|}{C}Cl}} \xrightarrow{NH_3}$$

$$\overset{C{\equiv}CH}{\underset{O}{CH_3{-}\overset{|}{CH}{-}O\overset{\|}{C}NH_2}} \quad (13)$$

To a vigorously stirred flask containing 70 gm (1.0 mole) of 1-butyne-3-ol and 99 gm (1.0 mole) of phosgene in 400 ml of ether at $-10°C$ is added 129 gm (1.0 mole) of quinoline dissolved in 140 ml of absolute ether over a 4 hr period. After allowing the mixture to stand for 16 hr at $-6°C$ the precipitated quinoline hydrochloride is filtered off by suction and the ether solution added dropwise to 750 ml of ether at $-5°C$ while ammonis gas is passed through it. After the reaction is complete the mixture is transferred to a separatory funnel containing 100 ml of water and 2000 ml of ether. The aqueous ammonium chloride solution is separated and the ether layer washed several times with water, then several times with 10% sulfuric acid, and finally with water. The resulting ether solution is dried over sodium sulfate, concentrated at atmospheric pressure, and the residue distilled under reduced pressure to give 50 gm of product (49% yield), b.p. 108°–109°C (12 mm Hg), m.p. 48°–50°C.

Tables V and VI give the preparation of several chloroformates and their conversion into carbamates with aqueous ammonia.

TABLE V[a]

PREPARATION OF CHLOROFORMATES AND CONVERSION INTO
CARBAMATES USING AQUEOUS AMMONIA

	ROCH$_2$CH$_2$OH + COCl$_2$ → ROCH$_2$CH$_2$OCOCl			ROCH$_2$CH$_2$OCOCl + NH$_3$ → ROCH$_2$CH$_2$OCONH$_2$			
R	Yield (%)	B.p., °C (mm Hg)	n_D^t	M.p. (°C)	B.p., °C (mm Hg)	n_D^t	Yield (%)
CH$_3$	93	58.7 (13)	1.4163 (25)	46.8	—	—	13.3
C$_2$H$_5$	77	67.2 (14)	1.4169 (25)	62.2	—	—	39.0
CH$_3$(CH$_2$)$_3$	91	93.0–93.5 (14)	1.4241 (25)	—	132.2–132.4 (2.5)	—	63.5
(C$_6$H$_5$)$_2$CH	72	87.0–87.5 (8.5)	1.4261 (25)	—	133–134 (2.5)	—	63.5

[a]Data taken from Porai-Koshits and Remizov [16].

C. Reactions of Sodium Cyanate with Alcohols in the Presence of Acids

Werner [21] reported that ethanol reacted with an aqueous solution of sodium cyanate and hydrochloric acid to give a 56% yield of ethyl carbamate. Similar results were obtained with potassium cyanate [22, 23]. Tertiary alcohols were reported earlier to react with potassium cyanate in acetic acid to give dehydration or rearrangement but no carbamate [24]. Marshall [25] and others [26, 27] described a method whereby tertiary acetylenic alcohols reacting *in situ* generated cyanic acid from a mixture of anhydrous sodium cyanate in trichloroacetic acid (see Table VII). Loev [28a, b, c] reported that modifying Marshall's procedure by using trifluoroacetic acid affords *t*-butyl carbamates in over 90% yields. The beneficial effect of trifluoroacetic acid does not appear to be solely related to increased acid strength as hydrochloric acid or methanesulfonic acid gives only traces of carbamate when used in place of trifluoroacetic acid under similar conditions. The role of trifluoroacetic acid in the mechanism of this reaction requires further investigation. When the reaction is run in benzene and methylene chloride, better yields are obtained than from similar reactions run in ether, tetrahydrofuran, or carbon tetrachloride.

Surprisingly it was reported that when potassium cyanate is substituted for sodium cyanate the yields of carbamates are reduced to less than 5%. The reason for this drastic effect is not known at this time. In addition, the use of other alkali or alkaline metal cyanates in this reaction has not been investigated. The Loev [28a, b, c] procedure appears applicable to the synthesis of carbamates from primary, secondary, and tertiary alcohols (2 hr reaction time affords 60-90% yield), cyclic and acyclic 1,3-diols, phenols,

TABLE VI

Preparation of O-Alkyl Carbamates by the Reaction of Ammonia with Chloroformates [17b]

Alcohol (moles)	Phosgene (moles)	$\overset{O}{\overset{\|}{NH_2CCl}}$ (moles)	Ether solvent (ml)	Amine	Product	B.p. °C (mm)
3-Methyl-1-pentyne-3-ol 0.43	—	0.43	250	—	$H_2N-\overset{O}{\overset{\|}{C}}O\overset{CH_3}{\underset{C_2H_5}{C}}C\equiv CH$	120–121 (12 mm Hg) (m.p. 56°–58°C)
3-Methyl-1-butyne-3-ol 0.72	0.75	—	250	Quinoline 0.75 NH_3 excess	$H_2N\overset{O}{\overset{\|}{C}}O\overset{CH_3}{\underset{CH_3}{C}}C\equiv CH$ (34%)	111 (12 mm) (m.p. 108°C)
1-Ethynylcyclohexanol-1 0.5	0.5	—	150	Quinoline 0.5 NH_3 excess	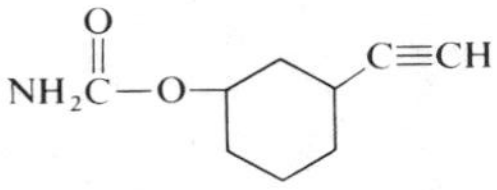 (53%)	108–110 (3 mm) (m.p. 94°–95°C)
3-Methyl-1-Pentyne-3-ol	0.5	—	150	Quinoline 0.5 CH_3NH_2 0.75	$CH_3NH\overset{O}{\overset{\|}{C}}O\overset{CH_3}{\underset{CH_3}{C}}C\equiv CH$	111–112 (14 mm) (m.p. 53°–54°C)
1-Butyne-3-ol 1.0	1.0	—	540	Quinoline 1.0 NH_3 excess	$H_2N\overset{O}{\overset{\|}{C}}O\overset{}{\underset{CH_3}{C}}H-C\equiv CH$	108–109 (12 mm) (m.p. 48°–50°C)

TABLE VII

PREPARATION OF THE CARBAMATES OF 3-METHYL-1-PENTYNE-3-OL [25]

Alcohol 3-methyl-1-pentyne-3-ol	Moles		Solvent (ml)	CCl₃COOH (moles)	O-Alkyl carbamate		
	KOCN	NaOCN			Yield (%)	B.p. (°C)	M.p. (°C)
			Dioxane				
0.4	0.4	—	80	0.4	—	120–121	53.5–55 (from cyclohexane)
0.8	—	0.5	—	0.5	65	—	52–53.5 (from cyclohexane and pet ether)
22.0	—	10	—	10	80 crude 66 recryst.	—	53–55 (from cyclohexane)

oximes, aldoximes, and ketoximes, and primary, secondary, and tertiary mercaptans. Carbamates could not be obtained from diphenylethylcarbinol (dehydrated to 1,1-diphenylethylene) or trichloro- and trifluoromethylcarbinols.

Although earlier investigators [24, 29, 30] reported that cyanic acid reacts with alcohols to give allophonates, Loev [28a, b, c] never detected any of these in his procedure. The only by-product was trifluoroacetamide, which did not present a difficulty because of its high solubility in water and organic solvents.

Table VIII illustrates the production of carbamates by the reaction of sodium cyanate and alcohols in the presence of trifluoroacetic acid.

2-8. *Preparation of t-Butyl Carbamate* [28a]

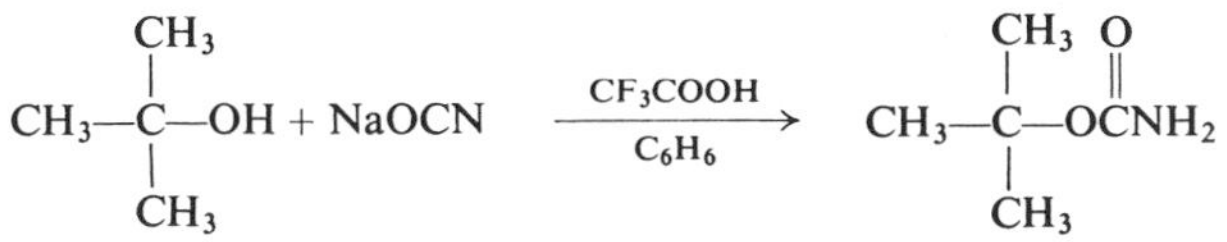

$$CH_3-\underset{\underset{CH_3}{|}}{\overset{\overset{CH_3}{|}}{C}}-OH + NaOCN \xrightarrow[C_6H_6]{CF_3COOH} CH_3-\underset{\underset{CH_3}{|}}{\overset{\overset{CH_3}{|}}{C}}-O\overset{\overset{O}{\|}}{C}NH_2 \qquad (14)$$

CAUTION: Use a hood.

To a stirred mixture of 7.4 gm (0.1 mole) of *t*-butyl alcohol and 13.0 gm (0.2 mole) of sodium cyanate in 50 ml of benzene is slowly added 15.5 ml (0.21 mole) of trifluoroacetic acid. The reaction is exothermic and some gas bubbles are formed. The reaction is stirred for 3 hr, then 15 ml of water is added, the organic layer separated, dried, the solvent removed under reduced pressure at

TABLE VIII[a]

SYNTHESIS OF CARBAMATES VIA THE REACTION OF SODIUM CYANATE AND ALCOHOLS IN THE PRESENCE OF TRIFLUOROACETIC ACID

Alcohol	CF_3COOH (moles)	NaOCN (moles)	Solvent (ml)	Time (hr)	Temp. (°C)	Product carbamate	Yield (%)	M.p. (°C)	Recryst. solvent
n-BuOH	0.21	0.20	C_6H_6 100	2	30	n-BuOCONH₂	73	53–54	H₂O
t-BuOH	0.21	0.20	100	2	30	t-BuOCONH₂	92	107–108	Hexane
CH≡C—CH₂OH	0.21	0.20	Ether 100	3	30	CH≡C—CH₂OCONH₂	60	47–50	Benzene
C₆H₅OH	0.21	0.20	100	48	30	C₆H₅OCONH₂	62	145–148	H₂O
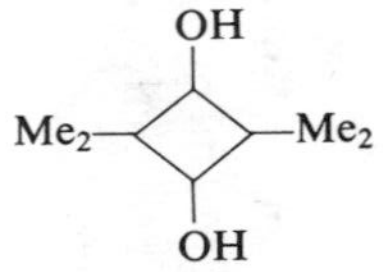 Me₂ cyclobutane-OH,OH	0.21	0.20	100	18	30	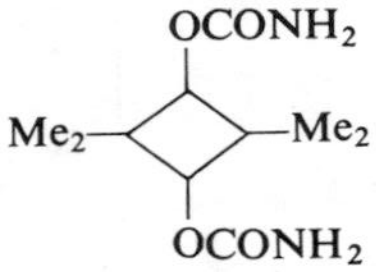 Me₂ cyclobutane OCONH₂, OCONH₂	45	175–185	H₂O
n-BuSH	0.21	0.20	100	3	30	n-BuSCONH₂	65	98–100	Hexane
t-BuSH	0.21	0.20	100	4	30	t-BuSCONH₂	50	92–95	Hexane
Me / Et₂C—SH	0.21	0.20	CH_2Cl_2 100	18	30	Me / Et₂C—SCONH₂	25	45–47	Hexane
=NOH (cyclohexanone)	0.21	0.20	100	18	30	=NOCONH₂ (cyclohexanone)	90	94–96	H₂O

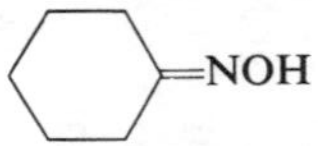

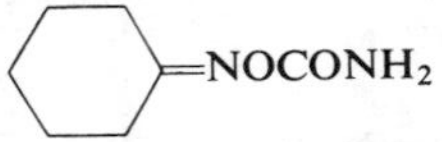

[a] Data taken from Loev and Kormendy [28a].

a pot temperature of 40°–50°C; the resulting residue solidifies to afford 10.7 gm (92%), m.p. 98°–101°C. Recrystallization of the entire product from water affords 8 gm (69%), m.p. 107°–108°C.

Cyanuric acid (HOCN) has been reported to be generated by reaction with sodium cyanate and trichloroacetic acid and then bubbled into aromatic alcohols (HOXPh) dissolved in trichloroacetic acid to give carbamates in 87% yield [28d].

D. Transesterification of Carbamates

Carbamates have been prepared by heating ethyl carbamate with a higher-boiling alcohol in the presence or absence of catalysts [31–33]. Aluminum isopropoxide has been reported [34] to be an excellent catalyst for the interchange reaction between ethyl carbamate and benzyl alcohol. The interchange reaction is also effective for N-alkyl carbamates as well as unsubstituted carbamates [35]. This catalyst is effective in preparing mono- and dicarbamates in excellent yields from primary and secondary alcohols and diols. Other effective catalysts are: dibutyltin dilaurate [36], dibutyltin oxide [37], sulfuric acid or p-toluenesulfonic acid [31], and sodium metal (reacts with alcohols to give the alkoxide catalyst) [33].

Examples of the transesterification reaction of ethyl carbamates and alcohols are given in Table IX.

TABLE IX[a]

TRANSESTERIFICATION REACTION OF ETHYL CARBAMATES
AND ALCOHOLS

Alcohol	Mole ratio of ethyl carbamate/ alcohol	Catalyst	Temp. (°C)	Time (hr)	Yield (%)	M.p. (°C)
Isobutyl	1/6	O	110–120	103	—	—
	1/6	H_2SO_4[e]	110–120	19	87 (56)[b]	63–64
	1/6	$(n\text{-Bu})_3$N	110–120	8	—	—
sec-Butyl	1/6	H_2SO_4	105–110	16	37	92–93
t-Butyl	1/6	H_2SO_4	85–90	—	—	—
Benzyl	1/1.5	O	190–230	19	70 (53)[b]	85–86
	1/1.5	H_2SO_4	145–240	5	9	82–84
	1/1[c]	Al (iso-PrO)$_3$	130–140[d]	5–10	86	86–87

[a] Data taken from Gaylord and Sroog [31] except where noted.
[b] Recrystallized product.
[c] Data from Kraft [34].
[d] Bath temperature.
[e] Approximately 2 ml concentrated H_2SO_4/0.5 mole ethyl carbamate used.

Tertiary alcohols and phenols do not undergo the transesterification reaction of carbamates with acidic or basic catalysts [38].

2-9. Preparation of Benzyl Carbamate [34]

$$C_2H_5O\overset{\overset{O}{\|}}{C}NH_2 + C_6H_5CH_2OH \xrightarrow{\text{Al (iso-PrO)}_3} C_6H_5CH_2O\overset{\overset{O}{\|}}{C}NH_2 + C_2H_5OH \qquad (15)$$

To a 250 ml three-necked flask equipped with a thermometer and a 20 cm distillation column filled with glass beads are added 44.5 gm (0.5 mole) of ethyl carbamate, 54.1 gm (0.5 mole) of benzyl alcohol, and 60 ml of toluene. The reaction flask is heated with an oil bath at 110°–125°C in order to remove any water in the reagents. The bath is cooled to 100°C, and 2.0 gm (0.01 mole) of aluminum isopropoxide is added all at once. The reaction is heated with the oil bath set at 130°–140°C in order to remove about 50 ml of the ethanol–toluene azeotrope at 77°C. The residue is recrystallized from toluene to yield 64.5 gm (85.5%), m.p. 86°–87°C.

NOTE: Substituting sodium methoxide for aluminum isopropoxide gave poor yields.

3. MISCELLANEOUS METHODS

(1) Ammonolysis of diethyl carbonate to ethyl carbamate [39].

(2) Preparation of tertiary alkyl esters of carbamic acid by the ammonolysis of mixed phenyl alkyl carbonates to yield alkyl carbamate and phenol [24, 40, 41].

(3) Reaction of urea and ethylene oxide to give aminoethyl carbamates [42].

(4) Reaction of carbamoyl chloride (NH_2COCl) with alcohols to yield carbamates [43].

(5) Reaction of alkyl oxamates with bromine and sodium ethoxide [44].

(6) Reaction of silver carbamate with alkyl halides [45].

(7) Hydrolysis of cyanates [46].

(8) Reaction of polyvinyl alcohol with urea to give polyvinyl carbamates [11, 47, 48].

(9) Preparation of mixtures of *O*-alkyl carbamates and dialkyl imidocarbonates by the reaction of alcohols with cyanogen and water in the presence of dipolar, aprotic solvents such as acetonitrile and an acid catalyst such as hydrochloric acid [49].

(10) Preparation of *O*-alkyl carbamates by the acid-catalyzed reaction of cyanogen chloride with aliphatic alcohols [50].

(11) Preparation of O-alkyl carbamates by the reaction of thiocyanogen trichloride with alcohols [51].

REFERENCES

1. F. M. Berger and B. J. Ludwig, U.S. Pat. 2,724,720 (1955).
2a. R. L. Arcenaux, J. G. Frick, H. O. Reid, and G. A. Gantreaux, *Amer. Dyest. Rep.* **50**, 37 (1961).
2b. G. C. Tesoro, U.S. Pat. 3,687,891 (1972).
3. Berkeley Chemical Corp., Liquithane, Bulletin, 1960.
4. I. H. Updegraff and A. Contras, U.S. Pat. 2,937,966 (1960).
5. R. T. Pollock, U.S. Pat. 2,438,452 (1948).
6. P. Kirby and T. Owen, Brit. Pat. 961,569 (1964).
7. R. A. Jacobson, *J. Amer. Chem. Soc.* **60**, 1742 (1938).
8. T. L. Davis and S. C. Lane, *Org. Syn. Coll. Vol.* **1**, 140 (1941).
9. A. E. A. Werner, *J. Chem. Soc.* **113**, 622 (1918).
10. F. M. Meigs, U.S. Pat. 2,197,479 (1940).
11. M. Paquin, *Z. Naturforsch.* **1**, 518 (1946).
12. S. Beinfest, P. Adams, and J. Halpern, U.S. Pat. 2,837,561 (1958).
13a. F. J. Sowa, U.S. Pat. 2,834,799 (1958).
13b. H. G. Goodman, Jr., and C. A. Duprezi, U.S. Pat. 3,449,406 (June 10, 1969).
14. L. I. Smith and O. H. Emerson, *Org. Syn. Coll. Vol.* **3**, 151 (1955).
15. M. T. Harvey and S. Caplan, U.S. Pat. 2,247,495 (1941).
16. H. G. Ashburn, A. R. Collett, and C. L. Lazzell, *J. Amer. Chem. Soc.* **60**, 2933 (1938).
17a. H. E. Carter, R. L. Frank, and H. W. Johnson, *Org. Syn. Coll. Vol.* **3**, 167 (1955).
17b. K. Junkmann and H. Pfeiffer, U.S. Pat. 2,816,910 (Dec. 17, 1957).
18. G. A. Kirkhgof and R. Y. Astrova, *Khim. Farm. Prom.* 282 (1933); *Chem. Abstr.* **28**, 3718 (1934).
19. W. M. Kraft and R. M. Herbst, *J. Org. Chem.* **10**, 483 (1945).
20a. J. Thiele and F. Dent, *Ann. Chem.* **302**, 246 (1898).
20b. F. D. Chattaway and R. Sakoens, *J. Chem. Soc.* **117**, 708 (1920).
20c. K. Junkmann, U.S. Pat. 2,816,910 (Dec. 17, 1957).
21. E. A. Werner, *Sci. Proc. Roy. Dublin Soc.* **24**, 209 (1947).
22. O. Folin, *Amer. Chem. J.* **19**, 336 (1897).
23. J. H. Barnes, M. V. A. Chepman, P. A. McCrea, P. G. Marshall, and P. A. Walsh, *J. Pharm. Pharmacol.* **13**, 39 (1961).
24. W. M. McLamore, S. Y. P'An, and A. Bavley, *J. Org. Chem.* **20**, 1379 (1955).
25. P. G. Marshall, J. H. Barnes, and P. A. McCrea, U.S. Pat. 2,814,637 (1957).
26. W. Stuehmer and S. Funke, U.S. Pat. 2,878,158 (1959).
27. J. H. Barnes, *J. Pharm. Pharmacol.* **13**, 39 (1961).
28a. B. Loev and M. F. Kormendy, *J. Org. Chem.* **28**, 3421 (1963).
28b. M. C. Flores and B. Loev, U.S. Pat. 3,072,710 (Jan. 8, 1963); *Chem. Abstr.* **59**, 3833 (1963).
28c. B. Loev, M. F. Kormendy, and M. G. Goodman, *Org. Syn. Coll. Vol.* **5**, 162 (1973).
28d. I. Nagayama, A. Maruyama, S. Yamada, S. Nakai, and T. Kutzuma, Japan Kokai 72/16,433 (Sept. 1, 1972); *Chem. Abstr.* **77**, 139649b (1972).
29. W. J. Close and M. A. Spielman, *J. Amer. Chem. Soc.* **75**, 4055 (1953).
30. K. W. Blohm and E. I. Becker, *Chem. Rev.* **51**, 471 (1952).
31. N. G. Gaylord and C. C. Sroog, *J. Org. Chem.* **18**, 1632 (1953).
32. G. Heilner, Ger. Pat. 551,777 (1932).
33. M. Metayer, *Bull. Soc. Chim. Fr.* 802 (1951).

34. W. M. Kraft, *J. Amer. Chem. Soc.* **70**, 3569 (1948).
35. R. Hazard, J. Cheymal, P. Charbrier, A. Skera, and E. Echefralaire, *Bull. Soc. Chim. Fr.* 2087 (1961).
36. W. Kern, K. J. Ranterkus, W. Weber, and W. Herz, *Makromol. Chem.* **57**, 241 (1962).
37. P. Adams and F. A. Baron, *Chem. Rev.* **65**, 570 (1965).
38. N. G. Gaylord, *J. Org. Chem.* **25**, 1874 (1960).
39. A. Cahours, *Ann. Chem.* **56**, 266 (1845).
40. K. Hafner and W. Kaiser, *Tetrahedron Lett.* **32**, 2185 (1964).
41. Societe des Laboratoires Labaz, Brit. Pat. 802,557 (1959).
42. W. F. Touisignant and T. Houtmann, Jr., U.S. Pat. 2,842,523 (1958).
43. L. Gatterman, *Ann. Chem.* **244**, 29 (1888).
44. J. Miliotis, *Prakt. Akad. Athenon* **10**, 190 (1935).
45. U. Iwase, Japan. Pat. 17,655 (1963).
46. D. Martin, *Angew. Chem.* **76**, 303 (1964).
47. I. Sakurada, A. Nakajima, and K. Shibatani, *J. Polym. Sci. A* **2**, 3545 (1964).
48. F. Masuo and T. Watanabe, Japan. Pat. 13,886 (1960).
49. J. L. Greene, U.S. Pat. 4,301,087 (Apr. 17, 1981).
50. R. Fuks and M. A. Marinus, *Bull. Soc. Chim. Belg.* **82**(1–2), 23 (1973); *Chem. Abstr.* **78**, 158870d (1973).
51. R. G. R. Bacon and R. S. Irwin, *J. Chem. Soc.* **50**, 79 (1960); *Chem. Abstr.* **55**, 11345 (1961).

1. INTRODUCTION

This chapter is concerned with compounds containing the structure RR'C=NR". These compounds are known as imines, azomethines, anils, or, more commonly, Schiff bases [1].

A wide variety of compounds having the primary amino group condense with carbonyl compounds to give $\diagup$C=N— compounds by the elimination of water. The latter reactions are usually acid catalyzed because protonation of the carbonyl group enhances its reactivity toward nucleophilic attack by the amino group on the carbonyl carbon atom (Eq. 1).

$$\begin{array}{c} R \\ \diagdown \\ \diagup C{=}O + H_2N{-}R' \\ R' \end{array} \longrightarrow \begin{array}{c} R \\ \diagdown \\ \diagup C{=}N{-}R' + H_2O \\ R' \end{array} \tag{1}$$

R = alkyl, aryl, H; R' = alkyl or aryl

$$R' = alkyl, aryl, -NH_2, -NH-alkyl, -NH\overset{\overset{\displaystyle O}{\|}}{C}NH_2, OH, H$$

It will be noted that compounds in which R' is —NH$_2$ or —NH alkyl are hydrazones. When R' is OH, the product is an oxime. Such compounds are treated elsewhere in this work. To do otherwise would only lead to confusion. Unsaturated heterocyclic compounds, such as isoquinoline, which do have

the —C=N— feature, ketenimines, $\begin{array}{c} R \\ \diagdown \\ \diagup C{=}C{=}N{-}R, \\ R' \end{array}$ semicarbazones,

azines, and carbodiimides, are also not considered in this chapter. Since the structure R—CH$\diagdown$$\overset{}{\underset{NH}{\diagup}}CH_2$ is frequently termed an ethylenimine derivative, these compounds are briefly discussed here [2].

In addition to the reaction of amino groups with carbonyl compounds, another useful method for the preparation of imines employs the reduction or condensation of nitriles with other reagents (Eq. 2).

291

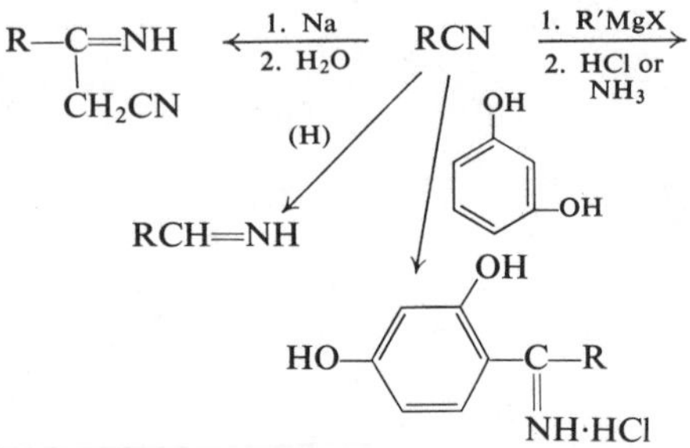

$$\text{R—C=NH} \xleftarrow[\text{2. H}_2\text{O}]{\text{1. Na}} \text{RCN} \xrightarrow[\substack{\text{2. HCl or} \\ \text{NH}_3}]{\text{1. R'MgX}} \text{R'C=NH}$$

(2)

It is interesting to note that several long-chain dipolar Schiff bases have physical properties which impart to them liquid crystal behavior [3, 4]. Most of these imines are disubstituted N-benzylideneanilines (structure 1) which

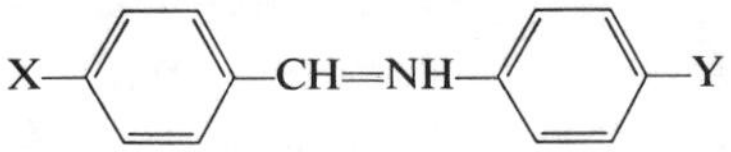

$$X—\bigcirc—CH=NH—\bigcirc—Y$$

(I)

allow themselves to be arranged in a parallel orientation with each other in the liquid melt, giving smectic and nematic mesomorphic states before reaching the melting point or isotropic liquid state. Thus the melting-point range of many of these types of compounds are broad, as seen in Table XI, owing to the appearance of these mesomorphic transition states. As a result of the continuing interest in liquid crystal behavior, many types of Schiff bases have been prepared [4].

TABLE I

NOMENCLATURE USED TO DESCRIBE THE VARIOUS IMINES OF STRUCTURE RR'C=NR"

Name	Source	Substituents
Imine	*Chemical Abstracts*	R" = hydrogen (H)
Amine[a]	*Chemical Abstracts*	R" = alkyl (R) or aryl (Ar) group
Aldimine	Common usage	R = R or Ar; R' = H
Ketimine	Common usage	R, R' = R or Ar
Schiff base	Common usage	R = Ar; R' = H; R" = R or Ar
Anils	Common usage	R, R'=R, Ar, H; R"=phenyl or substituted phenyl

[a] That is, "ylideneamine" systems.

General review articles on the chemistry of imines are by Sprung [7] and by Layer [8]. A review of papers by Hine and co-workers will be of interest for their study of the internal acid catalysis of imine formation on reaction of acetone and other carbonyl compounds with N,N-dialkyl diamines [9].

The geometric isomerism of imines is a subject of some controversy. Layer [8] stated that both E and Z forms are possible but that there may be rapid interconversion between the isomers. On the other hand, Karabatsos has indicated that in the case of aldimines, only the E isomer formed [10]. Upon dehalogenation of N-chloro-N-fluoroalkylamines such as CF_3CF_2NClF with mercury, N-fluoroimines such as $CF_3CF=NF$ form only in the Z form [11].

Imine derivatives have long found great use in analytical chemistry and are now being actively investigated in the medicinal and polymer chemistry areas.

The nomenclature of imines is highly variable. For example, one recent article identified a compound as (3-methoxy-3-methyl-2-butylidene)isopropylamine, i.e. an -ylideneamine system [5]. Another recent article in the same journal talks of compounds such as cinnamaldehyde N-isopropylimine—an aldehyde-imine system [6]. The "ylideneamine" system is probably the most systematic one, although the synthetic chemist would probably opt for the latter method, which shows the carbonyl compound and the amine from which the Schiff base may be prepared. Table I lists various nomenclature systems found in the literature.

Examples of some representative substituted imines are shown in Tables II through IX along with references to their preparations. Other compounds are discussed in greater detail below.

TABLE II[a]

ALKYL—CH=N—ALKYL

Compound	B.p., °C (mm Hg)	Yield (%)	Ref.
$CH_3CH=NCH_3$	27	55	[a]
$CH_3CH=NC_2H_5$	48	77	[a]
$CH_3CH=NC_3H_7$	74	69	[a]
$CH_3CH=NC_3H_7\text{-}i$	59	69	[a]
$CH_3CH=NC_4H_9$	102	70	[a]
$CH_3CH=N\langle\bigcirc\rangle$	54 (18)	76	[a]
$CH_3CH=NCH_2C_6H_5$	94 (21)	18	[a]
$CH_3CH_2CH=NCH_3$	53	77	[a]
$CH_3CH_2CH=NC_2H_5$	74	81	[a]
$CH_3CH_2CH=NC_3H_7$	101	78	[a]
$CH_3CH_2CH=NC_4H_9\text{-}sec$	111	75	[a]
$CH_3CH_2CH=NC_4H_9\text{-}i$	144	78	[a]
$CH_3CH_2CH=NCH_2CH=CH_2$	102	48	[a]
$CH_3CH_2CH_2CH=NCH_3$	81	76	[a]
$CH_3CH_2CH_2CH=NC_2H_5$	102	84	[a]
$CH_3CH_2CH_2CH=NC_3H_7$	125	78	[a]
$CH_3CH_2CH_2CH=NC_3H_7\text{-}i$	112	82	[a]
$CH_3CH_2CH_2CH=NC_4H_9\text{-}i$	166	74	[a]
$CH_3CH_2CH_2CH=NCH_2CH=CH_2$	128	66	[a]
$CH_3CH_2CH_2CH=NCH_2C_6H_5$	101 (16)	40	[a]
$CH_3CH_2CH_2CH=NC(CH_3)_2C(CH_3)_3$	60 (2.5)	93	[a]
$(CH_3)_2CHCH=NCH_3$	70	82	[a]
$(CH_3)_2CHCH=NC_2H_5$	90	80	[a]
$(CH_3)_2CHCH=NC_3H_7$	115	77	[a]
$(CH_3)_2CHCH=NC_4H_9\text{-}sec$	124	79	[a]

(continued)

TABLE II (continued)

Compound	B.p., °C (mm Hg)	Yield (%)	Ref.
$(CH_3)_2CHCH=N$〈⟩	82 (26)	87	a
$(CH_3)_2CHCH=NCH_2C_6H_5$	105 (15)	85	a
$(CH_3)_2CHCH=NCH_2CH=CH_2$	117	82	a
$(CH_3)_2CHCH_2CH=NC_3H_7$	130–139	64	a
$(CH_3)_2CHCH_2CH=NC_4H_9$	93 (100)	67	a
$C_6H_{13}CH=NCH_3$	160	66	a
$C_6H_{13}CH=NC_2H_5$	175	70	a
$C_6H_{13}CH=NC_3H_7$	195	66	a
$C_6H_{13}CH=NCH_2CH=CH_2$	86 (20)	61	a
$C_7H_{15}CH=NC_2H_5$	89 (20)	60	a
$C_8H_{17}CH=NCH_3$	94 (22)		a
$C_9H_{19}CH=NCH_3$	103 (15)	60	a
$CH_3CH=CHCH=NCH_3$	114	24	a
$CH_3CH=CHCH=NC_3H_7$	137	23	a
$CH_3CH=CHCH=NC_3H_7\text{-}i$	128	65	a
$CH_3CH=CHCH=NCH_2CH=CH_2$	140	32	a
$CH_3CH=CHCH=N$〈⟩	64–77 (4)	29	—
$CH_3CH=CHCH=NC(CH_3)_2C(CH_3)_3$	108–111	89	b
$CH_3CH_2CH_2CH=C(C_2H_5)CH=NC(CH_3)_2C(CH_3)_3$	141–148 (17)	—	b
$(CH_3)_3CCH_2CH(CH_3)CH_2CH=NC(CH_3)_2C(CH_3)_3$	90–94 (0.2)	82	b
$[CH_3CH_2CH_2C(CH_3)_2N=CH_2-]_2$	90 (1)	—	b
$[C_6H_5C(CH_3)_2N=CH_2-]_2$	180–200 (1)	—	b
$[(CH_3)_3CN=CH_2-]_2$	90–100 (25)	—	b
$(CH_3)_2CCH=NCH_2C(CH_3)_3$	90 (104)	68	c

[a] R. Tiollais, *Bull. Soc. Chim. Fr.* 708 (1947).

[b] M. D. Hurwitz, U.S. Pat. 2,582,128; *Chem. Abstr.* 46, 8146 (1952).

[c] H. S. Mosher and E. J. Blanz, Jr., *J. Org. Chem.* 22, 445 (1957).

[d] Reprinted from R. W. Layer, *Chem. Rev.* 63, 4891 (1963). (Copyright 1963 by the American Chemical Society. Reprinted by permission of the copyright owner.)

2. CONDENSATION REACTIONS

A. Condensation of Amines with Carbonyl Compounds

Aliphatic and aromatic aldehydes condense with aliphatic and aromatic primary amines to form *N*-substituted imines. Reaction conditions vary widely, such as passing a gaseous amine through a melt of a solid ketone [12, 13]; reaction of amines with aldehydes without solvent or catalyst [14];

TABLE III[c]

$(Alkyl)_2$—C=N—Alkyl

Compound	B.p., °C (mm Hg)	Yield (%)	Ref.
$(CH_3)_2C=NC_3H_7$	107	84	[a]
$(CH_3)_2C=NCH_2CH=CH_3$	113	70	[a]
$(CH_3)_2C=NC_4H_9\text{-}sec$	116	97	[a]
$(CH_3)_2C=NCH_2CH(CH_3)_2$	122	85	[a]
$(CH_3)_2C=N\langle\hexagon\rangle$	181	95	[a]
$CH_3CH_2(CH_3)C=NC_3H_7$	129	96	[a]
$CH_3CH_2(CH_3)C=NCH_2CH=CH_2$	135	86	[a]
$CH_3CH_2(CH_3)C=NC_4H_9\text{-}sec$	134	86	[a]
$(CH_3)_2CHCH_2(CH_3)C=NC_3H_7$	163	88	[a]
$(CH_3)_2CHCH_2(CH_3)C=NCH_2CH=CH_2$	110 (100)	99	[a]
$(CH_3)_2CHCH_2(CH_3)C=NC_4H_9\text{-}sec$	109	96	[a]
$(CH_3)_2CHCH_2(CH_3)C=NC_4H_9\text{-}i$	175	96	[a]
$[(CH_3)_2CHCH_2]_2C=NC_3H_7$	134 (100)	81	[a]
$[(CH_3)_2CHCH_2]_2C=NC_3H_7\text{-}i$	126 (100)	62	[a]
$[(CH_3)_2CHCH_2]_2C=NC_4H_9\text{-}sec$	145 (100)	75	[a]
$CH_3CH_2CH_2(CH_3)C=NCH(CH_3)CH_2CH(CH_3)_2$	110 (50)	—	[b]
$CH_3CH_2CH_2CH_2CH_2(CH_3)C=NCH_2CH_2OH$	88 (6.5)	—	[b]
$[(CH_3)_2CHCH_2]_2C=NCH_2CH_2OH$	93 (60)	—	[b]
$\langle\hexagon\rangle{=}NCH_2CH_2OH$	109 (50)	—	[b]

[a] K. Langheld, *Chem. Ber.* **42**, 2360 (1909).

[b] V. E. Haury, U.S. Pat. 2,421,937; *Chem. Abstr.* **41**, 5892 (1947).

[c] Reprinted from R. W. Layer, *Chem. Rev.* **63**, 489 (1963). (Copyright 1963 by the American Chemical Society. Reprinted by permission of the copyright owner.)

azeotropic removal of water in a low-boiling solvent in the preparation of some aldimines [15]; treatment of aldehydes with amines in solvents such as ether, tetrahydrofuran, or benzene in the presence of a molecular sieve [6, 16–19]; reaction of aqueous amines with ketones [20]; reactions catalyzed by potassium or sodium hydroxide solutions at low temperatures [15, 21–24]; reactions in the presence of base under heat and pressure [15]; dehydrations in the presence of acidic reagents such as p-toluenesulfonic acid [25, 26]; methanesulfonic acid [15], or titanium tetrachloride [15, 27, 5]. In the case of systems where subsequent cyclization might take place between an imino function and other parts of the molecule, the use of a cationic or an anionic surfactant, in a nonaqueous system, is said to reduce intramolecular cyclization and enhance the rate of reaction [28]. Aldehydes react with silazanes in

TABLE IV[b]

ARYL—CH=N—ARYL[a]

Compound	B.p., °C (mm Hg)	M.p. (°C)
$C_6H_5CH=NC_6H_5$	—	52
$C_6H_5CH=NC_6H_4CH_3\text{-}p$	—	33–35
$C_6H_5CH=NC_6H_4Cl\text{-}p$	—	56–58
$C_6H_5CH=NC_6H_4Br\text{-}p$	—	66–67
$C_6H_5CH=NC_6H_4Cl\text{-}o$	325 (775)	—
$C_6H_5CH=NC_6H_4CH_3\text{-}m$	315 (775)	—
$C_6H_5CH=NC_6H_4CH_3\text{-}o$	307 (775)	—
$o\text{-}HOC_6H_4CH=NC_6H_4Br\text{-}p$	—	112
$p\text{-}CH_3OC_6H_4CH=NC_6H_4OCH_3\text{-}p$	—	142
$p\text{-}ClC_6H_4CH=NC_6H_5$	—	66
$p\text{-}HOC_6H_4CH=NC_6H_5$	—	51
$(CH_3)_3C_6H_2CH=NC_6H_5$	—	56
$(CH_3)_3C_6H_2CH=NC_6H_4Cl\text{-}p$	—	74
$p\text{-}ClC_6H_4CH=NC_6H_4Cl\text{-}p$	—	111
$o\text{-}ClC_6H_4CH=NC_6H_4Cl\text{-}p$	—	68
$m\text{-}NO_2C_6H_4CH=NC_6H_4Cl\text{-}p$	—	84
$p\text{-}NO_2C_6H_4CH=NC_6H_4Cl\text{-}p$	—	128

[a] A. Hantzsch, *Chem. Ber.* **34**, 822 (1901).

[b] Reprinted from R. W. Layer, *Chem. Rev.* **63**, 489 (1963). (Copyright 1963 by the American Chemical Society. Reprinted by permission of the copyright owner.)

the presence of a ruthenium catalyst to form imines. Ketones, under similar conditions, yield enamines [29] (cf. Eqs. 3 and 4).

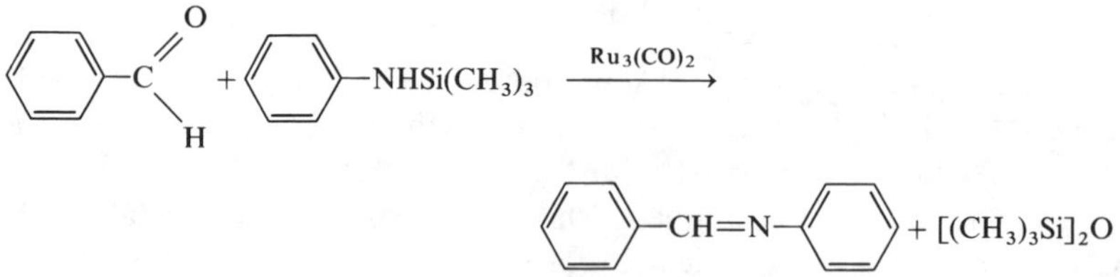

$$\text{(3)}$$

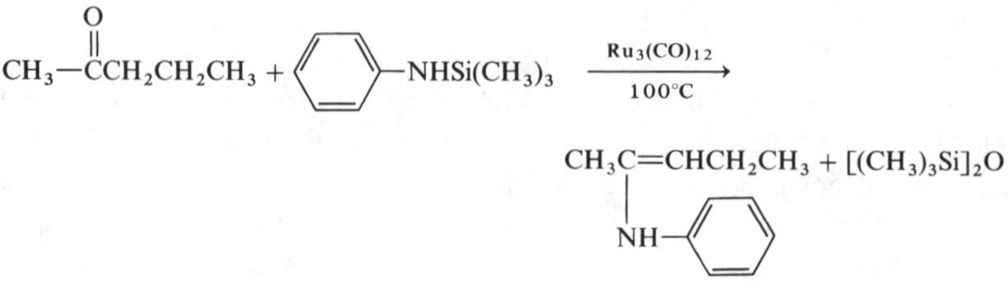

$$\text{(4)}$$

TABLE V[d]

RR′C=NH (R, R′ = ALKYL OR ARYL)

Compound		B.p., °C (mm Hg)	M.p. (°C)	Yield (%)	Ref.
R	R′				
2-Cyclohexylethyl	sec-Butyl	101 (2)	—	56	a
2-Cyclohexylpropyl	sec-Butyl	107 (1)	—	43	a
2-Cyclohexylbutyl	sec-Butyl	121 (1)	—	45	a
2-Cyclohexylpentyl	sec-Butyl	130 (1)	—	60	a
m-Tolyl	Isopropyl	227 (740)	—	—	b
p-Tolyl	Isopropyl	228 (740)	—	—	b
o-Tolyl	t-Butyl	235 (740)	—	—	b
m-Tolyl	t-Butyl	238 (740)	—	—	b
p-Tolyl	t-Butyl	235 (740)	—	—	b
Phenyl	Ethyl	101 (13)	—	—	c
Phenyl	Propyl	99 (8)	—	—	c
Phenyl	Isobutyl	113 (12)	—	—	c
Phenyl	Cyclohexyl	136 (5)	—	—	c
Phenyl	Phenyl	127 (3)	—	—	c
Phenyl	o-Tolyl	136 (4)	—	—	c
Phenyl	p-Tolyl	147 (5)	37	—	c
Phenyl	α-Naphthyl	182 (4)	68–69	—	c

[a] P. L. Pickard and T. L. Talbert, *J. Org. Chem.* **26**, 4886 (1961).
[b] P. L. Pickard and G. W. Polly, *J. Amer. Chem. Soc.* **76**, 5169 (1954).
[c] C. Moureau and G. Mignonac, *C. R. Acad. Sci. Paris* **156**, 1801 (1913).
[d] Reprinted from R. W. Layer, *Chem. Rev.* **63**, 489 (1963). (Copyright 1963 by the American Chemical Society. Reprinted by permission of the copyright owner.)

TABLE VI[b]

ALKYL—CH=N—ARYL[a]

Compound	M.p. (°C)
$CH_3CH_2CH=NC_6H_5$	103.4
$(CH_3)_2CH=NC_6H_5$	140
$CH_3CH_2CH_2CH_2CH=NC_6H_5$	97
$(CH_3)_2CHCH_2CH=NC_6H_4CH_3\text{-}p$	99

[a] W. Miller and J. Plochl, *Chem. Ber.* **25**, 2020 (1892).
[b] Reprinted from R. W. Layer, *Chem. Rev.* **63**, 489 (1963). (Copyright 1963 by the American Chemical Society. Reprinted by permission of the copyright owner.)

TABLE VII[b]

CYCLIC ALIPHATIC KETONE ANILS[a]

Compound	B.p., °C (mm Hg)	M.p. (°C)
Cyclohexanone anil	140 (19)	—
Methone anil	160 (12)	—
Dihydrocarvone anil	170 (15)	—
Pulegone anil	142 (12)	—
Carvone anil	181 (17)	—
Camphor anil	—	13.5
2-Cyclohexylidene cyclohexanone anil	210 (14)	—
Methone cyclohexanone (p-toluidine)	178 (16)	—
Methone cyclohexanone (p-anisidine)	—	62.0

[a] A. Rahman and M. O. Farooq, *Rec. Trav. Chim.* **73**, 423 (1954).
[b] Reprinted from R. W. Layer, *Chem. Rev.* **63**, 489 (1963). (Copyright 1963 by the American Chemical Society. Reprinted by permission of the copyright owner.)

TABLE VIII[d]

(ALKYL)(ALKYL OR ARYL)—C=N—ARYL

Compound	B.p., °C (mm Hg)	M.p. (°C)	Yield (%)	Ref.
$(CH_3)_2C{=}NC_6H_5$	86 (13)	23.5	—	a
$(C_2H_5)_2C{=}NC_6H_5$	117 (25)	—	75	b
$(C_3H_7)_2C{=}NC_6H_5$	130 (17)	—	76	b
$(C_3H_7)_2C{=}NC_6H_4CH_3\text{-}o$	140 (18)	—	84	b
$(C_3H_7)_2C{=}NC_6H_4CH_3\text{-}m$	144 (20)	—	74	b
$(C_3H_7)_2C{=}NC_6H_4CH_3\text{-}p$	144 (17)	—	63	b
$C_9H_{19}(CH_3)C{=}NC_6H_5$	206 (24)	—	60	b
⬡=NC₆H₅	157 (30)	—	65	b
$C_6H_5(CH_3)C{=}NC_6H_5$	175 (13)	—	—	c
$C_6H_5(CH_3)C{=}CHC(C_6H_5){=}NC_6H_5$	250 (13)	98–99	—	c
$(C_6H_5)_2C{=}NC_6H_5$		113	—	c
$(C_6H_5)_2C{=}NC_4H_4CH_3\text{-}p$	225 (15)	—	—	c
$(C_6H_5)_2C{=}NC_{10}H_7\text{-}\alpha$		138	—	c
$C_6H_5C(\alpha\text{-}C_{10}H_7){=}NC_6H_5$		94	—	c

[a] M. Sekiya and T. Fujita, *J. Pharm. Soc. Jap.* **71**, 941 (1951); *Chem. Abstr.* **46**, 3983 (1952).
[b] J. Hoch, *C.R. Acad. Sci. Paris* **199**, 1428 (1934).
[c] G. Reddelein, *Chem. Ber.* **46**, 2172, 2718 (1913).
[d] Reprinted from R. W. Layer, *Chem. Rev.* **63**, 489 (1963). (Copyright 1963 by the American Chemical Society. Reprinted by permission of the copyright owner.)

In general, aliphatic imines (C_5–C_{10}) are obtained in good yield by acid-catalyzed azeotropic dehydration. However, they are unstable and must be used directly after their distillation [4].

Tertiary aliphatic and aromatic aldehydes at room temperature react readily and nearly quantitatively with amines to give the imines without the aid of catalysts [1]. Primary aliphatic aldehydes tend to give polymeric materials with amines as a result of the ease of their aldol condensation [8]. The use of low temperatures and potassium hydroxide favors the formation of the imine product [21, 22]. Secondary aliphatic aldehydes readily form imines with amines with little or no side reaction [30].

Schiff bases from substituted and unsubstituted benzaldehyde and aliphatic or aromatic amines are stable [14, 31–36]. Benzaldehyde with substituents such as nitro, dialkylamino, hydroxyl, methoxyl, or halo have been used [31, 33].

Preparation 2-1 is an example of the preparation of a ketimine from a gaseous amine and a molten ketone without catalyst or solvent.

2-1. Preparation of Benzophenone N-methylimine [12, 13]

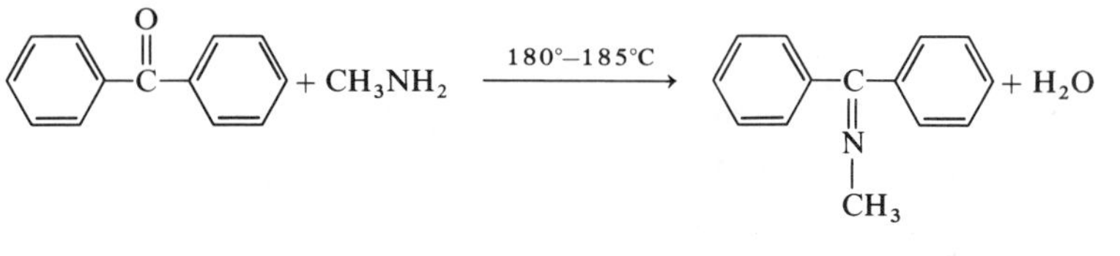

$$\text{(5)}$$

In a well-ventilated hood, in a suitable reaction vessel 66.5 gm (0.35 mole) of benzophenone is melted and heated to 180°–185°C. A stream of methylamine is passed through the melt for about 10 hr. The reaction is considered over when the evolution of water has stopped. The oily flask residue is allowed to cool and dissolved in ether. The ether solution is then extracted with several portions of ice-cold 2 N hydrochloric acid. Each portion of the acid extract is made alkaline as quickly as possible with 40 % aqueous sodium hydroxide. The product oil is extracted with ether. The ether solution is dried over anhydrous phosphorus pentoxide, filtered, and freed of solvent on a rotary evaporator. The pure imine is finally distilled under reduced pressure at 126°–128°C/2.5 mm. Yield: 55.3 gm (82 %).

By a similar procedure 4-methylbenzophenone was reacted with methylamine. Yield: 68 %; b.p. 140°–142°C/2.4 mm.

The preparation of benzalaniline is an example of an aldimine preparation without any catalyst and without extraneous solvents.

2-2. Preparation of Benzalaniline (Benzylideneaniline) [14]

$$C_6H_5CH{=}O + C_6H_5NH_2 \longrightarrow C_6H_5CH{=}N{-}C_6H_5 + H_2O \qquad (6)$$

To a 500 ml, three-necked flask containing 106 gm (1.0 mole) of benzaldehyde is added 93 gm (1.0 mole) of aniline while stirring rapidly. The exothermic reaction starts immediately as water begins to separate. After 15 min the mixture is poured into a beaker containing 165 ml of vigorously stirred 95% ethanol. Crystallization begins in about 5 min, and after 10 min the mixture is placed in an ice bath for $\frac{1}{2}$ hr. The semisolid is filtered by suction and air-dried to yield 152–158 gm (84–87%) of benzalaniline, m.p. 52°C. From concentrating the filtrate under reduced pressure an additional 10 gm, m.p. 51°C, is obtained. Further recrystallization may be made from 95% ethanol.

Weingarten et al. [15] outlined five methods of preparing ketimines. According to these authors, with increasing complexities of the ketones and amines, increasingly more vigorous reaction conditions are required. Table IX lists the structures and synthetic methods used. Some of these procedures will be illustrated below.

TABLE IX

PREPARATION OF KETIMINES OF INCREASING COMPLEXITY[a]

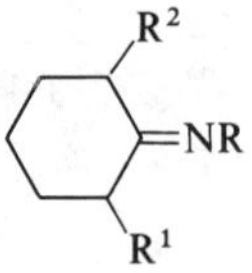

R^1	R^2	R	Method[b]	Yield[c] (%)	C=N/ C=O[d]	Bp (mm). °C	n_D^{24}
H	H	CH_3	A	80		70–74 (35)	1.4772[c]
H	H	$(CH_3)_2CH$	A	80		70–74 (28)	1.4615
CH_3	H	CH_3	A	<20[f]		72–74 (25)	
CH_3	CH_3	H	B	55[f]	0.78	61–65 (10)	
CH_3	CH_3	CH_3	B	e	0.71	61–70 (10)	
CH_3	CH_3	C_2H_5	B	<10[f]	0.26	61–70 (10)	
$(CH_3)_2C$	cis-CH_3 trans-CH_3	H	B	50	0.48	93–94 (12)	1.4694
CH_3	H	$CH_2{=}CHCH_2$	C	74		76–78 (10)	1.4796
CH_3	H	$(CH_3)_2CH$	C	30	1.42	59–60 (8)	1.4564
C_2H_5	H	$CH_2{=}CHCH_2$	C	66	2.6	88–91 (10)	1.4795
CH_3	CH_3	$CH_2{=}CHCH_2$	C	<30[f]	0.27	75 (10)	
H	H	$CH_3OCH_2CH_2$	D	60		72–75 (3)	1.4741
CH_3	H	$CH_3OCH_2CH_2$	D	79	3.6	98–100 (15)	1.4672
C_2H_5	H	$CH_3OCH_2CH_2$	D	69		113 (13)	1.4684
$(CH_3)_2CH$	H	$CH_3OCH_2CH_2$	D	79	6.0	112–114 (10)	1.4690
CH_3	CH_3	$CH_3OCH_2CH_2$	D	47	2.6	95 96 (10)	1.4666
CH_3	H^g	C_2H_5	E	53	5.4	81 (14)	1.4642

TABLE IX (continued)

R¹	R²	R	Method[b]	Yield[c] (%)	C=N/ C=O[d]	Bp (mm). °C	n_D^{24}
$(CH_3)_2C$	H	C_2H_5	E	56	4.5	94–97 (12)	1.4665
CH_3	CH_3	$(CH_3)_2CH$	E[h]	79	13.5	65–71 (10)	1.4554
$(CH_3)_3C$	H	$CH_3OCH_2CH_2$	E[h]	61	3.0	59–61 (1)	1.4680
$(CH_3)_3C$	cis-CH_3 trans-CH_3	CH_3	E[h]	10	1.0	95 (15)	1.4702
$(CH_3)_2CH$	$(CH_3)_2CH$	CH_2=$CHCH_2$	E	70	3.8	133–134 (14)	1.4763
2,3-		(structure) =NCH_3	E	28		92 (15)	1.4935
3,4-		(structure) =$NCH(CH_3)_2$	E[j]	67		98 (17)	1.4690

[a] Table IX is reprinted in part with permission from J. P. Chupp and W. White, *J. Org. Chem.* **32**, 3246 (1967). Copyright 1967 by The American Chemical Society.

[b] Method A, dehydration in the presence of 85% KOH; method B, reaction under pressure with KOH pellets; method C, azeotropic dehydration with pentane; method D, azeotropic dehydration with benzene, catalyzed by methylsulfonic acid; method E, reaction in the presence of $TiCl_4$.

[c] Unless otherwise indicated, yield represents conversion to pure distilled imine, based on ketone charged.

[d] Absorption intensities measured by infrared at about 5.9–6.2 and 5.8–6.1 μ, respectively. Although differing in magnitude, the molar absorption intensities of the cyclohexanones at these wavelengths are approximately two times those of imines.

[e] R. N. Blomberg and W. F. Bruce [U.S. Patents 2,700,681 and 2,700,682 (1955)] give n_D^{20} 1.4747.

[f] Product contained appreciable ketone on distillation and no attempt was made to obtain pure imine. Yield estimated from infrared, NMR, or vapor phase chromatography.

[g] Starting ketone is 2,3-dimethylcyclohexanone.

[h] Method B gave no imine.

[i] Method D gave no imine.

[j] Method A gave no imine.

2-3. Preparation of N-Allyl-2-ethylcyclohexylideneamine [15]

$$\text{(structure)} =O + CH_2=CH-CH_2-NH_2 \xrightarrow[-H_2O]{\text{pentane}} \text{(structure)} =N-CH_2CH=CH_2 + H_2O$$

(7)

In a hood in a 500 ml Erlenmeyer flask fitted with a Dean–Stark trap topped with a condenser and a magnetic stirrer, 37 gm (0.3 mole) of 2-ethylcyclohexanone is mixed with 50 ml of allylamine and 200 ml of pentane.

The mixture is heated with stirring for several hours to remove water by azeotropic distillation. When the water evolution has stopped, the pentane is distilled off and the residual imine is distilled at 88°–91°C/10 mm. Yield: 32.9 gm. Instead of pentane, 2-methylbutane has also been used in this method.

The removal of water formed during the synthesis of imines may be accomplished with a molecular sieve. Preparation 2-4 illustrates the method.

2-4. Preparation of N-(5-Methyl-5-hexenylidene)benzylamine [17]

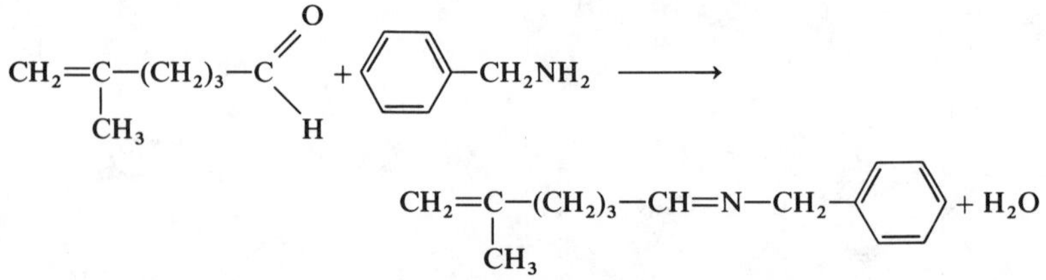

$$CH_2{=}C{-}(CH_2)_3{-}C\overset{O}{\underset{H}{\diagdown}} \ + \ \langle\bigcirc\rangle{-}CH_2NH_2 \longrightarrow$$
$$\underset{CH_3}{|}$$

$$CH_2{=}\underset{\underset{CH_3}{|}}{C}{-}(CH_2)_3{-}CH{=}N{-}CH_2{-}\langle\bigcirc\rangle + H_2O \quad (8)$$

A solution of 0.55 gm (4.9 mmole) of 5-methyl-5-hexenal and 0.5 gm (4.7 mmole) of benzylamine in 20 ml of benzene is treated with 10 gm of 4-Å molecular sieve for 1 hr. (CAUTION: Benzene is carcinogenic. Toluene should be evaluated as a replacement.) The solvent is then removed under reduced pressure, leaving 1 gm of crude imine. Infrared (benzene) v_{CN} 1680 cm^{-1}.

Other imines prepared by similar procedures are those formed by reaction of (+)-(R)-citronellal and benzylamine, with (−)-(S)-α-phenylethylamine [17].

In THF, 2-pyridinecarboxaldehyde has been reacted with a number of branched primary amines with 5-Å molecular sieves in excellent yields [18].

An example of an imine synthesis using potassium hydroxide is Preparation 2-5.

2-5. Preparation of Butyledinepropylamine [22]

$$CH_3CH_2CH_2NH_2 + CH_3CH_2CH_2CH{=}O \longrightarrow$$

$$CH_3CH_2CH_2CH{=}N{-}C_3H_7 + H_2O \quad (9)$$

To an ice-cooled, 250 ml, three-necked, round-bottomed flask equipped with reflux condenser, stirrer, and dropping funnel is added 23.6 gm (0.4 mole) of n-propylamine. Butyraldehyde (28.8 gm, 0.4 mole) is added dropwise at 0°C over a period of 2 hr. After the addition, 15 min is allowed to elapse and potassium hydroxide flakes are added. In about 10 min two layers appear and the organic layer is separated. The organic layer is added to a flask containing some crushed potassium hydroxide flakes and is placed in the refrigerator for a few hours. The dried material is distilled to yield 31.6 gm (70%), b.p. 120°–124°C, n_D^{20} 1.4149.

The physical constants of several other imines prepared by a similar procedure are shown in Table X. The aldimines listed in the table can be obtained only if certain precautions are strictly observed [22]. The method of Emerson *et al.* [38] could not be extended satisfactorily and the method described in Preparation 2-5 is a modification of the one described by Chancel [39] for propylidenepropylamine. The reaction is best carried out by adding the aldehyde to the amine, without a solvent, at 0°C. When the order of addition is reversed, the yields are much lower. Potassium hydroxide is added at the end in order to remove the water formed during the reaction. The use of other drying agents such as potassium carbonate or magnesium sulfate failed to yield aldimines on distillation. The aldimines should always be distilled from fresh potassium hydroxide to yield water-white products.

TABLE X

PHYSICAL CONSTANTS OF SEVERAL ALDIMINES PREPARED BY PROCEDURE 2-5

Aldimine	Yield (%)	B.p.[a] (°C)	n_D^{20}	d_4^{20}
Ethylidenepropylamine[b]	72	74–81	1.4006	0.7342
Ethylidenebutylamine	83	98–106	1.4098	0.7513
Propylideneethylamine[a]	70	70–76	1.4004	0.7353
Propylidenebutylamine	78	118–127	1.4153	0.7601
Butylideneethylamine	52	100–108	1.4082	0.7558
Butylidenepropylamine	70	120–124	1.4149	0.7611
Butylideneisopropylamine	64	100–111	1.4063	0.7436
Butylidenecyclohexylamine[e]	60	78–88 (20 mm)	1.4564	0.8475
Isobutylidenepropylamine[b]	76	108–114	1.4087	0.7456
Isoamylidenepropylamine	64	130–139	1.4170	0.7615
Isoamylidenebutylamine	67	90–96 (100 mm)	1.4217	0.7687

[a] The boiling range given is that of the usable fractions; the bulk of the material in each case boiled over a much narrower range.

[b] L. Henry, *C.R. Acad. Sci. Paris*, **120**, 837 (1895), reported the following boiling points: ethylidenepropylamine, 75°–77°C; propylidene–ethylamine 75°–78°C, butylidenepropylamine, 117°–118°C.

[c] A. Skita and G. Pfeil, *Ann. Chem.* **485**, 152 (1931), obtained a boiling point of 85°–87°C (16 mm Hg) for butylidene–cyclohexylamine.

[d] Reprinted from K. N. Campbell, A. H. Sommers, and B. K. Campbell, *J. Amer. Chem. Soc.* **66**, 82 (1944). (Copyright 1944 by the American Chemical Society. Reprinted by permission of the copyright owner.)

The aldimines are unstable and should be used within a few hours after their distillation; otherwise polymeric products are obtained.

Aliphatic ketones react more slowly than aldehydes with amines to form imines (see Table VII). Higher reaction temperatures, longer reaction times, and the removal of water aid in giving high yields of imines (80–95%). Sterically hindered ketones react slowly. Methyl ketones require mild acid catalyst and are more prone to aldol condensation by-products than are methylene ketone [8].

Aromatic ketones react even more slowly than aliphatic ketones and require strong acid catalysts and higher temperatures to effect the isolation of good yields of imine. Acetophenone reacts with aniline at reflux temperature in the presence of aniline hydrochloride to give the imine [40] (see Table VIII).

In general, the weaker the amine, the slower is the reaction rate with any given carbonyl compound.

The high-pressure preparation of Weingarten et al. [15] involves reaction of ketones with ammonia and gaseous amines over KOH pellets in an autoclave heated to 260°C for 60 hr. Under these conditions, in a 1.4 liter autoclave, pressures of 6400 psi develop.

The use of acid catalysts in the preparation of imines is quite common. For example, the benzylideneanilines of Table XI are prepared by refluxing a mixture of 0.010 mole each of the appropriate para-substituted benzaldehyde and aniline in 150 ml of benzene containing 0.1 gm of benzenesulfonic acid for 2–4 hr. A Dean–Stark trap is used to collect the water and then the solvent is removed under reduced pressure. The residue is recrystallized from hexane or other suitable solvent to give 47–95% yields. The nonmesomorphic compounds are recrystallized to constant melting point, whereas the nematic materials are recrystallized until the nematic-isotropic transition temperatures are constant and reversible [37].

It is interesting to note that the reaction of α,β-unsaturated aldehydes with amines proceeds in ether solution at room temperature without catalysis. On the other hand, α,β-unsaturated ketones react with amines in benzene solution at reflux temperatures upon catalysis with zinc chloride [6].

The use of titanium tetrachloride as an acid catalyst for the preparation of ketimines is particularly attractive with complex imines and hindered ketones. The quantities of reactants are determined by the equation:

$$2R\overset{\overset{\displaystyle O}{\|}}{C}-R + 6RNH_2 + TiCl_4 \longrightarrow 2R\overset{\overset{\displaystyle N}{\|}}{\underset{}{C}}{}^{\nearrow R}-R + 4RNH_2HCl + TiO_2 \quad (10)$$

Generally, slight excesses of the amine and the titanium tetrachloride are used for best results [15].

2-6. Preparation of N-Isopropyl-N-(2,6-dimethylcyclohexylidene Amine [15]

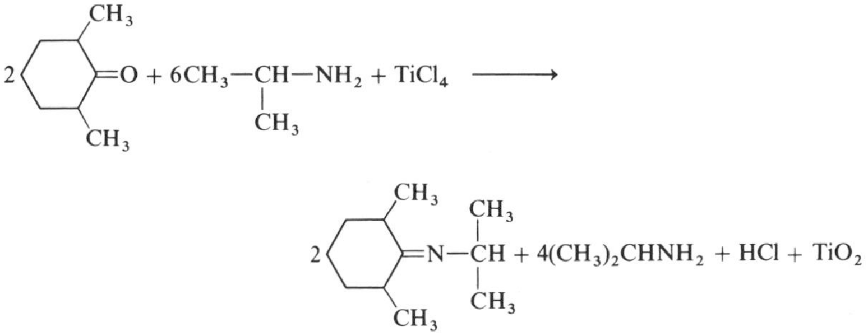

$$2 \underset{CH_3}{\overset{CH_3}{\diagup}} \!\!\!\! =\!\! O + 6CH_3\!\!-\!\!CH\!\!-\!\!NH_2 + TiCl_4 \longrightarrow$$
$$\underset{CH_3}{}$$

$$2 \underset{CH_3}{\overset{CH_3}{\diagup}} \!\!\!\! =\!\! N\!\!-\!\!\underset{CH_3}{\overset{CH_3}{CH}} + 4(CH_3)_2CHNH_2 + HCl + TiO_2 \quad (11)$$

(CAUTION: $TiCl_4$ reacts exothermically with ether. Also, since $TiCl_4$ hydrolyzes readily in moist air, equipment should be carefully protected against moisture and an efficient fume hood must be used.)

In a hood, in a suitable apparatus, a solution of 63 gm (0.5 mole) of 2,6-dimethylcyclohexanone and 88.5 gm (1.5 moles) of isopropylamine in 300 ml of ether is prepared. To this solution, with stirring, 200 ml of a solution of 48 gm (0.25 mole) of titanium tetrachloride in pentane is added over a period of approximately 0.75 hr. The reaction mixture is allowed to come to room temperature with stirring over an additional hour. Then the mixture is heated to reflux for 1.5 hr and allowed to stand at room temperature overnight.

The ether solution is filtered. The salt cake is slurried with ether several times. The ether extracts are combined with the filtrate. The ether–pentane solution of the product is evaporated to near dryness behind a safety shield. (CAUTION: The usual precautions of evaporating ether to dryness must be observed.) The residue is distilled under reduced pressure. After a small forerun (less than 1 gm), the product distilled at 69°–71°C/10 mm. Yield: 65.5 gm (79 %); infrared maximum at 6.09 μ (C=N).

By similar techniques N-arylketimines have been prepared from aceto-phenone and anilines [27] as well as α-halogenated ketimines of the type [5]:

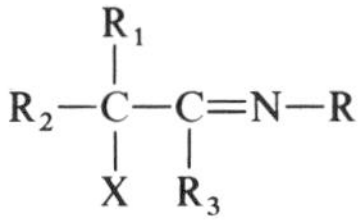

$$R_2\!\!-\!\!\underset{X}{\overset{R_1}{\underset{|}{\overset{|}{C}}}}\!\!-\!\!\underset{R_3}{\overset{|}{\underset{|}{C}}}\!\!=\!\!N\!\!-\!\!R$$

X = Cl or Br
R_1 = Me, Ph
R_2 = 4
R_3 = Me, i-Pr, Et
R = i-Pr, Me, cyclophenyl, allyl, t-butyl

TABLE XI

SUBSTITUTED BENZYLIDENEANILINES (SCHIFF BASES)

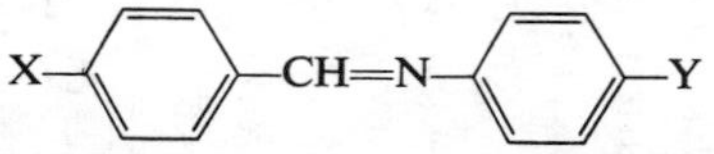

X—[C₆H₄]—CH=N—[C₆H₄]—Y

X	Y	Smectic range (°C)	Nematic range (°C)	Calcd. C	Calcd. H	Calcd. N	Found C	Found H	Found N
CH₃O	CH:CHCO₂C₄H₉	71–95	95–113[a]	74.75	6.87	4.15	74.78	6.90	4.27
C₇H₁₅O	CH:CHCO₂C₄H₉	102–138	—	76.92	8.37	3.32	76.88	8.42	3.52
CH₃CO₂	OCH₃	—	112–117[b]	71.37	5.61	5.20	71.80	5.65	5.32
CH₃CONH	OCH₃	—	188[c,d]	71.62	6.01	10.44	71.75	6.03	10.36
C₂H₅O	CO₂C₂H₅	—	94[e] (87)[f,g]	72.71	6.44	4.71	72.75	6.44	4.77
CH₃O	CONH₂	176–95	—	70.86	5.51	11.02	70.60	5.68	11.20
CH₃O	CN	—	105–118[h]	76.25	5.12	11.86	76.51	5.61	11.93
CH₃CO₂	CN	—	157–158[c]	72.69	4.58	10.60	72.65	4.54	10.50
Ph	OCOCH₃	—	155–182	79.95	5.39	4.45	80.18	5.44	4.58
CH₃O	OCOCH₃	—	83–110[i]	71.36	5.62	5.20	71.42	5.57	5.17
CH₃O	COCH₃	—	124.5[c,j]	75.87	5.97	5.53	75.82	5.53	5.50
C₂H₅O	COCH₃	—	123[e] (118)[f]	76.38	6.41	5.24	76.20	6.49	5.22
n-C₄H₉O	COCH₃	85–97	98–111	77.26	7.17	4.74	77.66	7.13	4.66

n-C₈H₁₇O	COCH₃	73–116	117–119	78.59	8.32	3.99	78.88	8.61	4.06
CH₃	CN	—	130ᶜ	81.77	5.49	12.72	81.97	5.69	12.40
H	CN	—	71.5–2ᶜ	81.55	4.85	13.59	81.52	4.86	13.67
HO₂C	CN	259–305	306–18ᵏ	71.99	4.04	11.19	72.12	4.10	11.05
Cl	CN	—	169–70ᶜ,ᵏ	69.86	3.77	11.64	69.79	3.93	11.40
(CH₃)₂N	CN	—	183ᶜ	77.08	6.05	16.86	77.34	6.29	16.51
O₂N	CN	—	190ᶜ	66.92	3.61	16.73	66.84	3.78	16.80
NC	OCOCH₃	—	164–165ᶜ	72.69	4.58	10.60	73.05	4.74	10.78
CH₃CO₂	SCH₃	—	116–117ᶜ	67.60	5.28	4.92	67.91	5.38	4.95
CH₃S	OCOCH₃	—	100ᵉ(82)ᶠ	67.60	5.28	4.92	67.84	4.98	5.00
CH₃S	SCH₃	—	143–145ᶜ	66.20	5.54	5.18	66.43	5.73	5.24
CH₃S	OCH₃	—	142ᶜ,ˡ	—	—	—	—	—	—
n-C₄H₉O	SCH₃	(83.5)ᵐ	99ᵉ(94)ᶠ	72.20	7.03	4.67	72.53	6.88	4.79
n-C₄H₉O	OCH₃	—	111ᵉ(106)ᶠ	76.60	7.43	4.95	76.80	7.30	5.15

ᵃ D. Vorlander, *Chem. Ber.* **41**, 2033 (1908) reports a mesomorphic range of 112°–118°C. ᵇ P. Hansen (Diss. Halle (1907)) reports a nematic range of 58°–76°C. ᶜ Not mesomorphic. ᵈ Recrystallized from benzene. ᵉ Melts to isotropic liquid at this temperature. ᶠ Monotropic (isotropic liquid → nematic) transition temperature. ᵍ H. Sackmann and D. Demus, *Z. Phys. Chem. (Leipzig)* **224**, 177 (1963). No transition temperatures are given. Although Castellano *et al.* observed one mesophase, Sackmann and Demus report the presence of a smectic phase. ʰ A. Frohlich (Diss. Halle (1910)) reports a nematic range of 103°–113.5°C. ⁱ P. Hansen (Diss. Halle (1907)) reports a nematic range of 81.5°–108°C. ʲ M. Guia and E. Bagiella, *Gazz. Chim. Ital.*, **51**, II, 116 (1921). ᵏ Recrystallized from ethanol. ˡ A. Hantzsh *et al.*, *Chem. Ber.* **34**, 832 (1901). ᵐ Monotropic (nematic → smectic) transition temperature. ⁿ Reprinted from J. A. Castellano, J. E. Goldmacher, L. A. Barton, and J. S. Kane, *J. Org. Chem.* **33**, 3501 (1968). (Copyright 1968 by the American Chemical Society. Reprinted by permission of the copyright owner.)

Primary aliphatic aldehydes such as $RCH_2CH{=}O$ react with ammonia to give addition compounds called "aldehyde ammonias." These compounds can be converted into the starting materials again, or they may lose water to give imines which then polymerize. Secondary aliphatic aldehydes such as $R_2CHCH{=}O$ react with ammonia by a different course. For example [41, 42], isobutyraldehyde reacts with ammonia to give the products shown in Eq. (12).

$$3(CH_3)_2CHCH{=}O + 2NH_3 \longrightarrow (CH_3)_2CH{-}CH[N{=}CHCH(CH_3)_2]_2 + H_2O$$

$$\downarrow \text{heat} \qquad\qquad (12)$$

$$(CH_3)_2C{=}CHN{=}CH{-}CH(CH_3)_2 + NH_3$$

Aromatic aldehydes react to give diimine derivatives as in Eq. (13).

$$3\,ArCH{=}O + 2\,NH_3 \xrightarrow{-3H_2O} ArCH{\Big\langle}{\begin{array}{l} N{=}CHAr \\ N{=}CHAr \end{array}} \qquad (13)$$

Monochloroamine reacts with substituted benzaldehydes to form ald-chlorimines [43].

Related to the imines are their alkylated salts, the ternary iminium salts, of structure

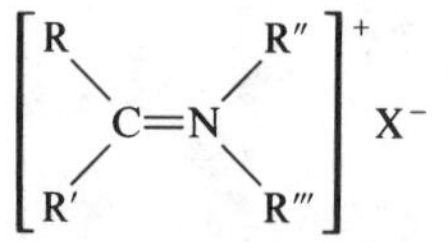

$$\left[\begin{array}{c} R \\ \diagdown \\ \diagup \\ R' \end{array} C{=}N \begin{array}{c} R'' \\ \diagup \\ \diagdown \\ R''' \end{array} \right]^{+} X^{-}$$

A simple synthesis of such compounds involves the direct reaction of a perchlorate salt of a secondary amine such as pyrrolidine or morpholine with a ketone such as acetone. With complex ketones azeotropic dehydration with benzene may be required [44].

The above discussion demonstrates that the formation of Schiff base may proceed readily in many cases under relatively mild conditions. It is therefore not surprising that biochemistry, biomedical research, immunochemistry, etc. have found uses for the reaction of amines (for example, those found in amino acids or in proteins) with aldehydes, particularly glutaraldehyde, $HCO(CH_2)_3CHO$. It may be expected that the reaction of polyfunctional amino compounds such as proteins and a dialdehyde such as glutaraldehyde will give rise to complex products after initial Schiff base formation (cf. [23, 24, 26, 28], i.a., for some examples of the reaction of polyfunctional reagents and of subsequent cyclization reactions).

By reacting the enzyme carboxypeptidase A with glutaraldehyde in a sodium veronal buffer at pH 7.5, the enzyme was insolubilized by cross-linking. It was postulated that the reaction involved the ε-amino group of the lysine segments of the enzyme but that the reaction proceeded beyond the formation of cross-links via Schiff base formation [45].

Enzymes have also been "immobilized" on polymeric substrates, often with considerable retention of the biochemical activity of the specific enzyme under study. In effect, this leads to polymers with surfaces that may act as enzymes. Hornby and Goldstein [46] mention nylon tubes which have been partially acid-hydrolyzed to free a few amino groups. The tube is then treated with glutaraldehyde followed by coupling of the enzyme—urease in this case—by reaction with free aldehyde groups.

In an article on the use of glutaraldehyde as a coupling agent for proteins and peptides, Reichlin [47] rules out simple Schiff base formation upon reaction of aldehydes with protein because of the stability of the cross-links to acid hydrolysis. Be that as it may, many techniques have used reactions of glutaraldehyde with proteins. For example, in [47] a procedure is given for coupling adrenocorticotropic hormone (ACTH) to bovine serum albumin (BSA) at pH 7.0 with glutaraldehyde as well as a method for coupling glucagon to rabbit serum albumin at pH 10. Proteins have also been coupled to polyacrylamide. In many cases, the biological activity of the protein is retained after coupling with glutaraldehyde to another substrate.

B. Condensation Reactions Involving Organometallic Compounds

Grignard reagents react with nitriles to produce ketimines. The intermediate addition compounds are isolated preferably with anhydrous ammonia [48–50] or anhydrous hydrogen chloride [48–50] to give 50–90% yields of the ketimines. Some ketimines are easily hydrolyzed to ketones and thus must be stored under anhydrous conditions [51]. Other ketimines are not easily hydrolyzable and are thus easily isolated [50]. For example, 2,2,6-trimethylcyclohexylcyanide and phenylmagnesium bromide give a ketimine stable toward hydrolysis [52]. In addition, t-butyl-o-tolylketimine is stable toward hydrolysis [50]. The latter two examples suggest that sterically hindered imines may help to stabilize those ketimines toward hydrolysis. Further research in this area is required before sound conclusions based on steric hindrance can be made.

Aliphatic nitriles also react with aliphatic Grignard reagents to give high yields of the ketimine when the complex reaction product is slowly decomposed with anhydrous methanol rather than anhydrous ammonia or aqueous decompositions [53].

2-7. Preparation of Diethylketimine [53]

$$C_2H_5CN + C_2H_5MgBr \longrightarrow \overset{\overset{\displaystyle NMgBr}{\|}}{CH_3CH_2C}-C_2H_5 \tag{14}$$

$$\overset{\overset{\displaystyle NMgBr}{\|}}{CH_3CH_2-C}-C_2H_5 \xrightarrow{CH_3OH} \overset{\overset{\displaystyle NH}{\|}}{C_2H_5-C}-C_2H_5 + MgBrOCH_3 \tag{15}$$

To a flask containing a solution of 0.5 mole of ethylmagnesium bromide in 300 ml of ether is added dropwise 16 gm (0.285 mole) of propionitrile. The mixture is refluxed for 4 hr and then decomposed with 2.0 moles of anhydrous methanol. The ketimine is isolated as the hydrochloride by treatment of the filtrate with hydrogen chloride, and then regenerated with anhydrous ammonia to yield on distillation 12.2 gm (50.5%), b.p. 86.5°C (730 mm), n_D 1.4626 (20°C); hydrochloride, m.p. 104°C.

Several other ketimines prepared by a similar method are shown in Table XII. The general procedure for the preparation is described in Procedure 2-8.

2-8. General Procedure for Ketimine Preparation [53]

Ketimines were prepared by the following general procedure; variations from the method are described separately. A Grignard reagent–nitrile complex

TABLE XII[f]

KETIMINES PREPARED BY THE METHANOL MODIFICATION

$$R-\overset{\overset{\displaystyle NH}{\|}}{C}-R'$$

R	R'	Yield (%) By CH₃OH	Yield (%) Other	B.p.,[a] °C (mm Hg)	n_D^{20}	d_D^{20}	Hydro-chloride, m.p.,[a] °C
Ethyl	Ethyl	50.5	—	86.5 (730)	1.4626	0.8523	104
n-Propyl	Isopropyl	38.0	—	58 (90)	1.4006	0.7973	—
sec-Butyl	o-Tolyl	86.0	77[b]	83 (4)	1.5239	0.9446	143–144
Phenyl	Phenyl	82.0	60[c]	127 (3.5)	1.6191	1.0849	139
Ethyl	2-Pyridyl	40.9	—	192.5 (730)	1.5578	1.5491	143
o-Tolyl	2-Thienyl	61.0	57[d]	163 (4)	1.6262	1.1464	205
2-Thienyl	5-Acridyl	5.0	—	172[e]	—	—	167

[a] Uncorrected.

[b] P. L. Pickard and S. H. Jenkins, J. Am. Chem. Soc. 75, 5899–5901 (1953).

[c] C. Moureu and G. Mignonac, C.R. Acad. Sci. Paris 156, 1801–6 (1913).

[d] J. O. Snyder, Ph.D. Thesis, University of Oklahoma, 1954.

[e] Melting point.

[f] Preparation 2-8 and Table XII are reprinted from R. L. Pickard and T. L. Tolbert, J. Org. Chem. 26, 4886 (1961). (Copyright 1961 by the American Chemical Society. Reprinted by permission of the Copyright owner.)

was prepared by the dropwise addition or 0.45 mole of nitrile to a stirred Grignard reagent prepared from 0.50 mole of halide and 0.51 gm-atom of magnesium turnings in 300 ml of anhydrous ether, followed by an 8–12 hr reflux. After cooling to room temperature, the stirred complex was decomposed by the dropwise addition of 3 moles of anhydrous methanol. Reaction was vigorous; in every case, completion of the decomposition gave a slurry of white, easily filtered, crystalline solid, although at intermediate stages the mixture was sometimes gummy. Immediately after the decomposition, which required 20–40 min, the slurry was filtered and the filtrate was distilled.*

The preparation of diphenyl ketimine is given in considerable detail in *Organic Syntheses* [54].

The chlorine groups of *N*-chloroimines [55, 56] and *C*-chloro-*N*-benzylideneanilines [57] react with Grignard reagents in fair to good yields to give the corresponding imines as described in Eqs. (16) and (17).

$$RC{=}NR + R'MgX \longrightarrow RC{=}NR + MgXCl \qquad (16)$$
$$\underset{Cl}{|} \qquad\qquad\qquad \underset{R'}{|}$$

$$R_2C{=}N{-}Cl + R'MgX \longrightarrow R_2C{=}NR' + RCN + RCl + MgX_2 \qquad (17)$$

Oximes react with Grignard reagents to give the amine of the Grignard reagent with the ketimine as the secondary product as shown in Eq. (18) [58].

$$ArCH{=}N{-}OH + RMgX \longrightarrow ArC{=}NH + ArCHNHR + MgXOH \qquad (18)$$
$$\underset{R}{|}$$

Titanium alkyls react with aryl- or alkylnitriles to give imines after hydrolysis [59].

Recently, Richey and Erickson [60] found that primary amines react under mild conditions with organolithium compounds to form imines and α-substituted primary amines. Equation (18) outlines a scheme by which such compounds may be formed:

$$
\begin{array}{c}
\text{RCH}_2\text{NH}_2 \xrightarrow{\text{R'Li}} \text{RCH}_2\text{NHLi} \qquad \text{RCH}{=}\text{NH} \qquad \text{R}{-}\underset{\text{R'}}{\underset{|}{\text{CHNH}_2}} \\[2em]
\text{R'Li} \big\downarrow \qquad\qquad \underset{\text{R'Li}}{\searrow}{-}\text{LiH} \qquad\qquad \uparrow \qquad\qquad \text{R'Li} \big\updownarrow \text{ROH} \qquad (19) \\[2em]
\text{RCH}_2\text{NLi}_2 \xrightarrow{-\text{LiH}} \text{RCH}{=}\text{NLi} \xrightarrow{\text{R'Li}} \underset{\text{R'}}{\underset{|}{\text{RCHNLi}_2}}
\end{array}
$$

This reaction procedure for the preparation of imine needs to be evaluated further since the experimental section of this paper deals only with the preparation of millimolar quantities. Table XI lists some typical constants of imines prepared by this and other methods.

Lithium aldimines may be prepared by treating an isocyanide like 1,1,3,3-tetramethylbutyl isocyanide with *tert*-butyllithium and *n*-butyllithium. Lithium aldimides are capable of reacting with a variety of electrophilic reagents, e.g., alkyl halides, carbon dioxide, ethyl chloroformate, benzaldehyde, other carbonyl compounds, and allyl and benzyl halides [61, 62]. The products are α-hydroxy imines, which usually undergo an Almadori rearrangement, ultimately leading to α-aminoketones. Lithium aldimides add 1,4 to methyl acrylate but 1,2 to acrolein. Hydrolysis of these compounds gives the corresponding carbonyl compounds. The reaction of lithium aldimines with allyl bromide results in an E,Z mixture of α,β-unsaturated imines (cf. Table XIII).

Sodium dialkylamine derivatives may cause the formation of iminonitriles (cf. Eq. 2). Under certain circumstances, cyclic iminonitriles may form (Eq. 20) [63-68].

$$CN-(CH_2)_4-CN + NaNR_2 \longrightarrow \begin{array}{c} \text{=NH} \\ \text{-CN} \end{array} \tag{20}$$

The base-catalyzed condensation of two nitriles to imines is sometimes called the Thorpe reaction. Preparation 2-9 is an application of this process.

TABLE XIII

PYSICAL PROPERTIES OF MISCELLANEOUS IMINES

	M.p. °C	B.p. °C/mm	Ref.
N,N'-(p-Phenylenedimethylidyne)dianiline	162–163	—	16
N,N'-(p-Phenylenedimethylidyne)bis(methylamine)	92	—	16
N,N'-(p-Phenylenedimethylidyne)bis(ethylamine)	51	—	16
N,N'-(p-Phenylenedimethylidyne)bis(butylamine)	—	143/0.03	16
N-Undecyl-2-picolinimine	—	125–150/0.2	19
2,2,4,4-tetramethylpentan-3-imine	—	64–109	60
2,2-Dimethyl-3-phenyl-3-imine	—	46–47/0.35	60
Diphenylmethylimine	—	123–139/0.9	60
Diphenylmethylidenediphenylamine	155–157	—	60
2-[5-(Carbomethoxy)-2,2-dimethyl-3-phenylidene]-2,4,4-trimethylpentane	—	123–125/0.2	62
(E,Z)-3-[(2,2-Dimethyl-4-ene-3-hexylidene)amino]-2,4,4-trimethylpentane	—	74/0.55	62

2-9. *Preparation of 3-Iminobutyronitrile [64] (Thorpe Reaction)* [65, 66]

$$CH_3CN + Na \xrightarrow[-CH_4, NaCN]{\substack{C_6H_6 \\ \text{reflux temp.}}} \underset{\overset{\|}{CH_3C-CH_2CN}}{\overset{NNa}{}} \xrightarrow[-NaOH]{H_2O} \underset{\overset{\|}{CH_3C-CH_2CN}}{\overset{NH}{}} \quad (21)$$

In a three-necked flask equipped with a long reflux condenser, stirrer, and dropping funnel is placed 78 gm (3.9 gm-atom) of powdered sodium in 800 ml of dry benzene. Acetonitrile (246 gm, 6.0 moles) is slowly added over a period of $2\frac{1}{2}$ hr at a rate to maintain the reaction mixture boiling. Once the reaction starts, it is very exothermic and some cooling may be necessary. Methane is evolved, and care should be taken to prevent loss of benzene.

After the addition of acetonitrile the mixture is refluxed for 3 hr or longer until, in cooling, the sodium salt or 3-iminobutyronitrile and sodium cyanide crystallize from the reaction mixture. The product salts are filtered, suspended in 1 liter of ether, and water is slowly added until the salts dissolve. The ether layer is separated, dried, and the ether removed by atmospheric distillation. After a few hours the residue crystallizes to yield 121 gm (49.7%) of crude product. Recrystallization from benzene yields 91 gm (37.4%), m.p. 63°–71°C. Von Meyer [67] has reported that this compound exists in a stable form with m.p. 50°–54°C and in a labile form, m.p. 79°–84°C.

3. REDUCTION REACTIONS

Lithium aluminum hydride in tetrahydrofuran has been found to reduce aromatic nitriles to give an amine and to give an imine which is formed from the addition of the amine to a nonisolatable intermediate imine followed by elimination of ammonia [69] (Eq. 22). This is simpler than catalytic hydrogenation of nitriles [70], which gives poor yields of imines.

$$C_6H_5CN \xrightarrow{LiAlH_4} C_6H_5CH_2NH_2 + C_6H_5CH_2N{=}CHC_6H_5 + NH_3 \quad (22)$$
$$(58\%) \qquad\qquad (30\%)$$

Nitriles can also be reduced to imines with stannous chloride in ethyl acetate containing hydrogen chloride [71]. The imines are isolated as the stannous chloride salt.

β-Nitrostyrenes can be reduced with lithium aluminum hydride below 0°C, then hydrolyzed to the phenylacetaldimine (16% yield) with a 20% aqueous solution of potassium sodium tartrate [72] (Eq. 23).

$$C_6H_5CH{=}CHNO_2 \xrightarrow[-40°C]{-30° \text{ to}} C_6H_5CH{=}CH-NH_2 \quad (23)$$

$$\Updownarrow$$

$$C_6H_5CH_2CH{=}NH_2$$

Oximes of aliphatic and aromatic ketones are catalytically reduced in the presence of hydrogen to ketimines. Acetophenone oxime gives the imine in 30% yield [73]. This reaction is not of practical value and gives much poorer yields than the direct reaction of ammonia or amines with the carbonyl compounds.

4. OXIDATION REACTIONS

Hydroperoxides and peroxides oxidize primary and secondary aliphatic amines to imines. Thus t-butyl hydroperoxide oxidizes 4-methyl-2-pentyl-amine to 2-(4-methylpentylidene)-4-methyl-2-pentylamine in 66% yield [74]. Di-t-butyl peroxide reacts in a similar manner [74]. However, this reaction is

$$\text{RNHCHR'R''} + \text{R''OOH} \longrightarrow \text{RN=CR'R''} + \text{R''OH} \tag{24}$$

not suggested for preparative use because further work is required to ascertain its value for the general preparation of amines (see Eq. 24).

Recently, complex aromatic amines with reagents such as nitrobenzene, silver oxide, and potassium ferricyanide [75] are oxidized to cyclic imines. The resulting imines may cyclize as indicated in Eq. (25) [75].

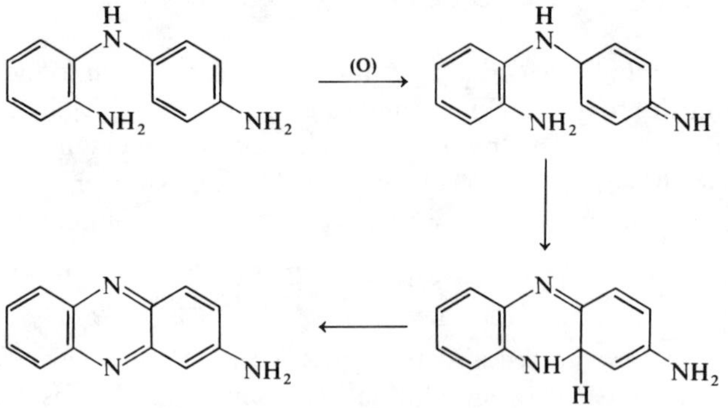

$$\tag{25}$$

5. MISCELLANEOUS METHODS

(1) Aziridines [2]. Ethylenimine is conveniently prepared in the laboratory in 34–37% yields by heating the sulfate ester of ethanolamine with aqueous alkali [76, 77]. Ethylenimine is considered a carcinogen by the FDA. The

preparations are given here only for historical interest. It is recommended that this preparation *NOT* be attempted.

5-1. *Preparation of Ethylenimine* [36] (*CAUTION*: A known carcinogen)

$$
\begin{array}{c}
H_2C-OH \\
| \\
H_2C-NH_2
\end{array}
\xrightarrow{H_2SO_4}
\begin{array}{c}
H_2COSO_2O^- \\
| \\
H_2C-NH_3^+
\end{array}
\xrightarrow{NaOH}
\begin{array}{c}
H_2C \\
\diagdown \\
| \quad > NH \\
\diagup \\
H_2C
\end{array}
\tag{26}
$$

(*a*) *Preparation of β-aminoethylsulfuric acid* [77]. Ethanolamine (6 moles, 366 gm) and 98% sulfuric acid (6 moles, 600 gm) are separately diluted with half their weight of water and cooled in an ice bath. The amine is slowly added to the acid with constant stirring in a flask also cooled in an ice bath. The mixture is boiled under reduced pressure with the aid of an aspirator. Boiling stones are used to prevent bumping. When the temperature of the liquid reaches 140°–160°C, the heating is stopped and the material solidifies. The product is not isolated, but methanol or ethanol is added in order to wash the product free from any remaining sulfuric acid or ethanolamine. The yield is approximately 761–804 gm (90–95%).

(*b*) *Preparation of ethylenimine.* To a 5 liter flask equipped with a distillation head are added 564 gm (4.0 moles) of β-aminoethylsulfuric acid and 1760 gm of 40% by weight of sodium hydroxide solution. The mixture is heated to the boiling point to initiate the reaction. The heating mantle is removed and the mixture continues to boil for several minutes. When the initial reaction subsides, heating is continued and about 500 ml of distillate is quicky collected in a cooled receiver. Approximately 450–500 gm of potassium hydroxide is gradually added to the distillate to cause the imine layer to separate as a top layer. The aqueous layer is separated and distilled through a wrapped 10 in. Vigreux column. The distillate boiling at 50°–100°C is collected in a chilled receiver and saturated with potassium hydroxide pellets. The upper layer of ethylenimine is separated and combined with the previously obtained crude product. The ethylenimine is distilled from fresh potassium hydroxide to yield 59–62 gm (34–37%), b.p. 56°–58°C. The product is stored over sodium hydroxide pellets in the refrigerator.

Ethylenimine has also been prepared for β-bromoethylamine hydrobromide by reaction with silver oxide [78], sodium methoxide [79], potassium hydroxide [80]; from β-chloroethylamine hydrochloride by reaction with sodium hydroxide [81]; from β-aminoethylsulfuric acid by reaction with sodium hydroxide [76, 77, 82–86], and by heating to 100°–300°C oxazolidone or substances yielding it [87].

Ethylenimines may also be prepared by the reaction of 1,2-dihaloalkanes with amines or ammonia. For example, ethylenimine may be prepared in the laboratory or on an industrial scale by the reaction of ethylene dichloride with ammonia [88]. The reaction with amines with α,β-dibromoketones yields imines as described in Eq. (27) [89].

$$R-\underset{\underset{Br}{|}}{CH}-\underset{\underset{Br}{|}}{CH}-COR' + R''NH_2 \longrightarrow R-\underset{\diagdown}{CH}\underset{\underset{|}{N}}{}\underset{\diagup}{CH}-COR' \qquad (27)$$
$$\underset{R''}{}$$

Aryl alkyl ketoximes react with aryl or alkyl Grignard reagents to give substituted ethylenimines [90, 91]. Additional methods for the synthesis of ethylenimines (aziridines) is found in Dermer and Ham [2]. The toxicity and carcinogenicity of the compounds are not well known. However, since ethylenimine itself is carcinogenic, the related derivatives of it, i.e., the whole class of aziridine, must be considered suspect. We do not recommend that any of the syntheses of aziridines be attempted at this time. The syntheses are discussed here primarily for reference, historical interest, and also to point out to the reader that relatively innocent appearing reagents may lead to dangerous products.

5-2. *Preparation of 2-Phenyl-2-ethylethylenimine* [90]

$$C_2H_5MgBr + C_6H_5\underset{\underset{N-OH}{\|}}{C}-CH_3 \longrightarrow$$

$$\left[\begin{array}{c} \underset{|}{C_2H_5} \\ C_6H_5\underset{\diagup N \diagdown}{C}-CH_3 \\ MgBr \quad OH \end{array} \longrightarrow \begin{array}{c} \underset{|}{C_2H_5} \\ C_6H_5\underset{\diagdown N \diagup}{C}-CH_2 \\ MgBr \end{array} \right] \xrightarrow[\substack{H_2O, \\ cold}]{NH_4Cl} \begin{array}{c} \underset{|}{C_2H_5} \\ C_6H_5\underset{\diagdown N \diagup}{C}-CH_2 \\ | \\ H \end{array} \qquad (28)$$

To a flask containing 1.0 mole of ethylmagnesium bromide prepared in 350 ml of dry ether is attached a distillation head and column. The ether is removed by distillation until 200 ml is collected, and then 200 ml of dry toluene is added. From a dropping funnel is added dropwise a solution of 34 gm (0.25 mole) of acetophenone oxime in 200 ml of dry toluene over a period of 2 hr. The oil bath around the flask is kept at 90°–95°C during the addition and for an $\frac{1}{2}$ hr afterward. The cool reaction mixture is hydrolyzed by pouring all at once into an ice–ammonium chloride solution. The water layer is extracted several times with ether, and the combined organic layers are dried over magnesium sulfate. The residue remaining after the ether and toluene are removed is distilled under reduced pressure to yield 14.4 gm (37%) of 2-phenyl-2-ethylethylenimine, b.p. 85°–86°C (7.0 mm), n_D^{20} 1.5318. Yields as high as 60% and as low as 20% have been obtained by this procedure; thus further preparative modifications may lead to more consistently higher yields.

Ethylenimine derivatives can be made to undergo nucleophilic substitution reactions to give a wide variety of substituted ethylenimines.

Several examples of reactions to prepare substituted ethylenimines are shown in Table XIV.

(2) Phenols are also known to be able to condense with nitriles to give phenolic ketimines as shown in Eq. (2) [92]. The reaction between the phenol and nitrile is carried out in ether solution when hydrogen chloride gas is added to the saturation point. Less reactive phenols also require the presence of zinc chloride.

(3) Reaction of nitroso compounds with active methylene groups to give imines [93, 94].

(4) Reaction of metal amides of primary amines with aromatic ketones to give imines [95].

(5) Diethyl ketals react with alkyl- or arylamines. Aromatic amines give better yields of imines [94].

(6) Nitrones react with potassium cyanide to give C-cyanoimines [97].

(7) Phenyl isocyanate reacts with p-dimethylaminobenzaldehyde at 100°C to give the imine in quantitative yield [98].

(8) Olefins react with hydrazoic acid in sulfuric acid to give imines [99].

(9) Tertiary alcohols or halides react with hydrazoic acid in sulfuric acid to give imines. Under the same conditions, benzhydrol gives N-benzylideneaniline in 90% yield [100].

(10) Secondary nitroalkanes react with primary amines to give imines [101].

(11) Alkylidene triphenylphosphoranes react with nitroso compounds to give imines [102].

(12) 1-Butylperfluoroisobutylene reacts with primary amines to give imines [103].

(13) Ketones as well as imines, in the presence of an acid catalyst, react with ketimines to exchange the higher- for the lower-boiling ketone [104].

(14) Diimines of α-diketones—reaction of aromatic aldimines with sodium in DMSO at 20°C (72 hr) [105].

(15) Fluorination of Schiff bases [106].

(16) Reaction of iodine azide with olefins followed by reduction leads to a stereospecific synthesis of aziridines [107].

(17) Preparation of benzylidene hydrazides (anticancer agents) [108].

(18) Preparation of poly(Schiff bases) (polyazomethines) [109-111].

(19) Addition of aziridines to ketones via titanium chloride catalysis [112].

TABLE XIV

SUBSTITUTED ETHYLENIMINES

Reaction	Product	B.p., °C (mm Hg)	$n_D^{(t)}$	Yield (%)	Ref.
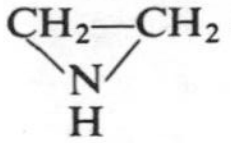 CH₂—CH₂ + NaOCl (N—H)	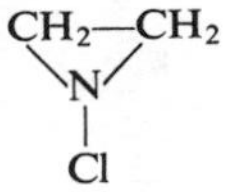 CH₂—CH₂ N—Cl	37.5–38.3 (245)	1.4433 (20)	80	[a]
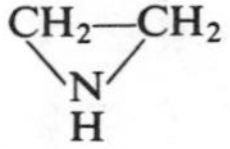 CH₂—CH₂ + CH₃COCl (N—H)	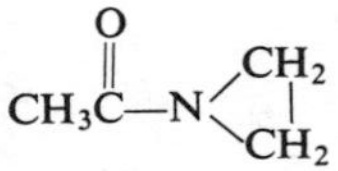 CH₃C(=O)—N(CH₂CH₂)	38–39 (17)	1.4378 (20)	30	[b]
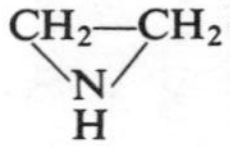 CH₂—CH₂ + C₂H₅OCOCl (N—H)	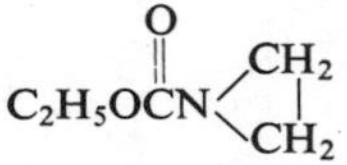 C₂H₅OC(=O)N(CH₂CH₂)	60–63 (21)	—	60–65	[c]
2 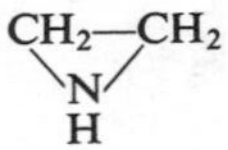 CH₂—CH₂ + COCl₂ (N—H)	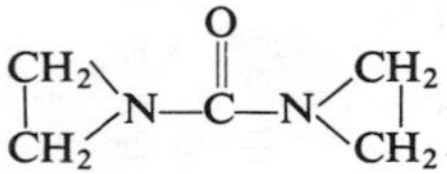 (CH₂CH₂)N—C(=O)—N(CH₂CH₂)	55–56 (0.4) m.p. 39–41°C	—	72	[c]
3 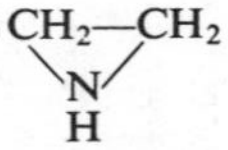 CH₂—CH₂ + POCl₃ (N—H)	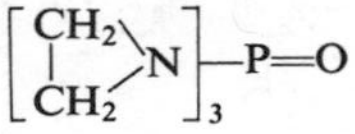 [(CH₂CH₂)N]₃P=O	90–91 (0.3) m.p. 41°C	—	90–91	[b,c]

[a] A. F. Graefe and R. E. Mayer, *J. Amer. Chem. Soc.* **80**, 3939 (1958).
[b] A. H. Bestian, J. Henya, A. Bauer, G. Ehlers, B. Hirsekorn, T. Jacobs, W. Noll, W. Weibezahn, and F. Römer, *Ann. Chem.* **566**, 210 (1950).
[c] S. Skorokhodov, *Zh. Obshch. Khim.* **31**, 3626 (1961).

(20) Reaction of glyoxal with aromatic primary amines [113].

(21) Dehydrofluorination of N-chloro-N-fluoroalkylamine [11].

(22) Formation of anti-Bredtimines by photolysis of 1-azidobicyclo-[2.2.2]octane and related bicyclic compounds [114].

(23) Cycloaddition of tert-butylcyanoketene to isocyanides [115].

(24) Synthesis of diaziridinimines [116].

(25) Formation of thioimino esters [117].

(26) Alkylation of imines to form iminium salts [118].

(27) Oxidation of imino ethers [119, 120].

REFERENCES

1. H. Schiff, *Ann. Chem.* **131**, 118 (1864); *Ann. Chem. (Suppl.)* **3**, 343 (1864–1865).
2. O. C. Dermer and G. E. Ham, "Ethylenimine and Other Aziridines—Chemistry and Applications," Academic Press, New York, 1969.
3. G. W. Gray, "Molecular Structure and the Properties of Liquid Crystals," Academic Press, New York, 1962.
4. S. L. Arora, T. R. Taylor, and J. L. Fergason, *J. Org. Chem.* **35**, 170 (1970).
5. N. DeKimpe, P. Sulmon, R. Verhé, L. DeBuyck, and N. Schamp, *J. Org. Chem.* **48**, 4320 (1983).
6. W. T. Brady and C. H. Shieh, *J. Org. Chem.* **48**, 2499 (1983).
7. M. M. Sprung, *Chem. Rev.* **26**, 297 (1940).
8. W. R. Layer, *Chem. Rev.* **43**, 489 (1963).
9. J. Hine and W.-S. Li, *J. Org. Chem.* **40**, 2622 (1975); J. Hine and F. A. Via, *J. Org. Chem.* **42**, 1972 (1977); J. Hine, J. P. Ziegler, and M. Johnston, *J. Org. Chem.* **44**, 3540 (1979); J. Hine, W.-S. Li, and J. P. Ziegler, *J. Amer. Chem. Soc.* **102**, 4403 (1980); J. Hine and Y. Chou, *J. Org. Chem.* **46**, 649 (1981).
10. G. J. Karabatsos and S. S. Lande, *Tetrahedron* **24**, 3907 (1968).
11. A. Sekiya and D. D. Des Marteau, *J. Org. Chem.* **46**, 1277 (1981).
12. C. R. Hauser and D. Lednicer, *J. Org. Chem.* **24**, 46 (1959).
13. R. L. Ehrhardt, G. Gopalakrishnan, and J. L. Hogg, *J. Org. Chem.* **48**, 1586 (1983).
14. L. A. Bigelow and H. Eatough, *Org. Syn. Coll. Vol.* **1**, 80 (1941).
15. H. Weingarten, J. P. Chupp, and W. White, *J. Org. Chem.* **32**, 3246 (1967).
16. T. Kajimoto and J. Tsuji, *J. Org. Chem.* **48**, 1685 (1983).
17. G. Demailly and G. Solladie, *J. Org. Chem.* **46**, 3102 (1981).
18. S. E. Dinizo and D. S. Watt, *J. Amer. Chem. Soc.* **97**, 6900 (1975).
19. J. H. Babler and B. J. Invergo, *J. Org. Chem.* **46**, 1937 (1981).
20. A. H. Sommers and S. E. Aaland, *J. Org. Chem.* **21**, 484 (1956).
21. R. Tiollias, *Bull. Soc. Chim. Fr.* **14**, 708 (1947).
22. K. N. Campbell, A. H. Sommers, and B. K. Campbell, *J. Amer. Chem. Soc.* **66**, 82 (1944).
23. W. A. Mosher and S. Piesch, *J. Org. Chem.* **35**, 1026 (1970).
24. K. N. Sawhney and T. M. Lemke, *J. Org. Chem.* **48**, 4326 (1983).
25. R. O. Hutchins, W.-Y. Su, R. Sivakumai, F. Cistone, and Y. P. Stercho, *J. Org. Chem.* **48**, 3412 (1983).
26. F. Sannicolo, *J. Org. Chem.* **48**, 2924 (1983).
27. N. DeKimpe, R. Verhé, L. DeBuyck, and N. Schamp, *J. Org. Chem.* **46**, 2079 (1981).
28. J. Sunamoto, H. Kondo, J.-I. Kikuchi, H. Yoshinaga, and S. Takei, *J. Org. Chem.* **48**, 2423 (1983).

29. M. T. Zoekler and R. M. Laine, *J. Org. Chem.* **48**, 2539 (1983).
30. W. Miller and J. Plochl, *Chem. Ber.* **25**, 2020 (1892).
31. N. H. Cromwell and H. Hoeksema, *J. Amer. Chem. Soc.* **67**, 1658 (1945).
32. R. B. Moffett and W. M. Hoehn, *J. Amer. Chem. Soc.* **69**, 1792 (1947).
33. R. E. Lutz, P. S. Bailey, R. Porolett, Jr., J. W. Wilson, III, R. K. Allison, M. T. Clark, N. H. Leake, R. H. Jordan, R. J. Keller, III, and K. C. Nicodemus, *J. Org. Chem.* **12**, 760 (1947).
34. K. A. Jensen and N. H. Bang, *Ann. Chem.* **548**, 106 (1941).
35. E. Walker and M. Latif, *Chem. Ind. (London)* 51 (1969).
36. K. N. Campbell, C. H. Helbing, M. P. Florkowski, and B. K. Campbell, *J. Amer. Chem. Soc.* **70**, 3868 (1948).
37. J. A. Costellano, J. E. Goldmacher, L. A. Barton, and J. S. Kane, *J. Org. Chem.* **33**, 3501 (1968).
38. W. S. Emerson, S. M. Hess, and F. C. Uhle, *J. Amer. Chem. Soc.* **63**, 872 (1941).
39. F. Chancel, *Bull. Soc. Chim. Fr.* (3) **11**, 933 (1894).
40. G. Reddelien, *Chem. Ber.* **46**, 2172, 2718 (1913).
41. H. H. Strain, *J. Amer. Chem. Soc.* **52**, 820 (1930).
42. R. H. Hasek, E. U. Elam, and J. C. Martin, *J. Org. Chem.* **26**, 1822 (1961).
43. C. R. Hauser, A. G. Gillaspie, and J. W. LeMaistre, *J. Amer. Chem. Soc.* **57**, 567 (1935).
44. N. J. Leonard and J. V. Paukstelis, *J. Org. Chem.* **28**, 3021 (1963).
45. F. A. Quiocho and F. M. Richards, *Biochemistry* **5**, 4062 (1966).
46. W. E. Hornby and L. Goldstein, in "Methods in Enzymology" (K. Mosbach, ed.), vol. 44, Academic Press, New York, 1976, p. 118 ff. and references therein.
47. M. Reichlin, in "Methods in Enzymology" (K. Mosbach, ed.), vol. 70, Academic Press, New York, 1980, p. 159 ff and references therein.
48. C. Moureu and G. Mignonac, *C. R. Acad. Sci. Paris* **156**, 1801 (1913).
49. J. H. Cloke, *J. Amer. Chem. Soc.* **62**, 117 (1940).
50. P. L. Pickard and D. J. Vaughan, *J. Amer. Chem. Soc.* **72**, 876, 5017 (1950).
51. A. Lachman, *Org. Syn. Coll. Vol.* **2**, 334 (1941).
52. H. L. Lochte, J. Horeczy, P. L. Pickard, and A. D. Barton, *J. Amer. Chem. Soc.* **79**, 2012 (1948).
53. P. L. Pickard and T. L. Tolbert, *J. Org. Chem.* **26**, 4886 (1961).
54. P. L. Pickard and T. L. Tolbert, *Org. Syn.* **5**, 520 (1973).
55. J. W. LeMaistre, A. E. Rainsford, and C. R. Hauser, *J. Org. Chem.* **4**, 106 (1939).
56. K. S. Ng and H. Alper, *J. Org. Chem.* **46**, 1039 (1981).
57. M. Busch and F. Falco, *Chem. Ber.* **43**, 2557 (1910); M. Busch and M. Fleischmann, *ibid.* **43**, 2553 (1910).
58. P. Grammatickis, *C. R. Acad. Sci. Paris* **210**, 716 (1940).
59. B. A. Porai-Koshits and A. L. Remizov, *Sb. Stat. Obshch. Khim.* **2**, 1570 (1953); *Chem. Abstr.* **49**, 5367 (1955).
60. H. G. Richey, Jr., and W. F. Erickson, *J. Org. Chem.* **48**, 4349 (1983).
61. G. E. Niznik, W. H. Morrison, III, and H. M. Walborsky, *J. Org. Chem.* **39**, 600 (1974).
62. M. J. Marks and H. M. Walborsky, *J. Org. Chem.* **46**, 5405 (1981).
63. K. Ziegler, H. Ohlinger, and H. Eberle, Ger. Pat. 591,269 (1934); *Chem. Abstr.* **28**, 2364 (1934).
64. H. Adkins and G. M. Whitman, *J. Amer. Chem. Soc.* **64**, 150 (1942).
65. E. F. G. Atkinson and J. F. Thorpe, *J. Chem. Soc.* **89**, 1906 (1906).
66. N. Lees and J. F. Thorpe, *J. Chem. Soc.* **91**, 1282 (1907).
67. E. von Meyer, *J. Prakt. Chem.* (2) **52**, 84 (1895).
68. V. L. Hansley, U.S. Pat. 2,742,503 (1956); *Chem. Abstr.* **50**, 16830 (1956).

69. L. M. Soffler and M. Katz, *J. Amer. Chem. Soc.* **78**, 1705 (1956).
70. V. Grignard and R. Escourrou, *C. R. Acad. Sci. Paris* **180**, 1883 (1925); R. Juday, and H. Adkins, *J. Amer. Chem. Soc.* **77**, 4559 (1955).
71. T. Stephen and H. Stephen, *J. Chem. Soc.* 4695 (1956).
72. R. T. Gilsdorf and F. F. Nord, *J. Amer. Chem. Soc.* **72**, 4327 (1950).
73. G. Mignonac, *C. R. Acad. Sci. Paris* **170**, 936 (1920).
74. H. E. De La Mare, *J. Org. Chem.* **25**, 2114 (1960).
75. N. P. Loveless and K. C. Brown, *J. Org. Chem.* **46**, 1182 (1981).
76. C. F. H. Allen, F. W. Spangler, and E. R. Webster, *Org. Syn.* **30**, 38 (1950).
77. P. A. Leighton, W. A. Perkins, and M. L. Renquist, *J. Amer. Chem. Soc.* **69**, 1540 (1947).
78. S. Gabriel, *Chem. Ber.* **21**, 1049 (1888).
79. L. Knorr and G. Meyer, *Chem. Ber.* **38**, 3129 (1905).
80. S. Gabriel, *Chem. Ber.* **21**, 2664 (1888); S. Gabriel and R. Stelzner, *ibid.* **28**, 2929 (1895).
81. E. I. du Pont de Nemours & Co., U.S. Pat. 2,212,146 (1940); *Chem. Abstr.* **35**, 463 (1937).
82. H. Wenker, *J. Amer. Chem. Soc.* **57**, 2328 (1935).
83. G. D. Jones, A. Langsjoen, Sister M. M. C. Neumann, and J. Zomlefer, *J. Org. Chem.* **9**, 125 (1944).
84. E. J. Mills, Jr. and M. T. Bogert, *J. Amer. Chem. Soc.* **62**, 1173 (1940).
85. Brit. Pat. 460,888; *Chem. Abstr.* **31**, 4676 (1937).
86. W. A. Reeves, G. L. Drake, Jr., and C. L. Hoffpauir, *J. Amer. Chem. Soc.* **73**, 3522 (1951).
87. O. Sunden (to Stockholm Superfosfat Fabriks A/B) Swed. Pat. 148,559; *Chem. Abstr.* **50**, 2679 (1956).
88. Dow Chemical Co., Brit. Pat. 923,528 (April 10, 1963).
89. N. H. Cromwell and J. A. Caughlan, *J. Amer. Chem. Soc.* **67**, 2235 (1945); N. H. Cromwell, *ibid.* **69**, 258 (1947).
90. B. K. Campbell, E. P. Chaput, and J. F. McKenna, *J. Org. Chem.* **8**, 103 (1943).
91. K. N. Campbell, B. K. Campbell, L. G. Hess, and I. J. Schaffner, *J. Org. Chem.* **9**, 184 (1944).
92. K. Hoesch, *Chem. Ber.* **48**, 1122 (1915); K. Hoesch, *ibid.* **50**, 462 (1917); J. Houben and W. Fischer, *J. Prakt. Chem.* **123**, 89 (1929).
93. F. Barrow and F. J. Throneycraft, *J. Chem. Soc.* 769 (1939).
94. F. Ullmann and B. Frey, *Chem. Ber.* **37**, 855 (1904).
95. C. R. Hauser, W. R. Braser, P. S. Skell, S. W. Kantor, and A. E. Brodhog, *J. Amer. Chem. Soc.* **78**, 1653 (1956).
96. J. Hoch, *C. R. Acad. Sci. Paris* **199**, 1428 (1934); L. Claisen, *Chem. Ber.* **29**, 2931 (1896).
97. V. Bellairta, *Gazz. Chim. Ital.* **65**, 897 (1935); *Chem. Abstr.* **30**, 3420 (1936).
98. H. Staudinger and R. Endle, *Chem. Ber.* **50**, 1042 (1917).
99. R. Adams, *Org. React.* **3**, 324 (1949).
100. J. H. Boyer and F. C. Canter, *Chem. Rev.* **54**, 1 (1954).
101. H. L. Wehrmeister, *J. Org. Chem.* **25**, 2132 (1960).
102. U. Schollkopf, *Angew. Chem.* **71**, 260 (1959).
103. I. L. Knunyants, L. S. German, and B. L. Dyatkin, *Izv. Akad. Nauk SSSR Otd. Khim. Nauk* **221** (1960); *Chem. Abstr.* **54**, 20870 (1960).
104. V. E. Haury, U.S. Pat. 2,692,284 (1954); *Chem. Abstr.* **49**, 15946 (1955).
105. J. S. Walia, J. Singh, M. S. Chattha, and M. Satyanarayana, *Tetrahedron Lett.* 195 (1969).
106. R. F. Merritt and F. A. Johnson, *J. Org. Chem.* **32**, 416 (1967).
107. A. Hassner, G. J. Matthews, and F. W. Fowler, *J. Amer. Chem. Soc.* **91**, 5046 (1969).
108. D. W. Boykin, Jr., and R. S. Varma, *J. Med. Chem.* **13**, 583 (1970).
109. A. L. Gershuns and G. A. Kreimer, *Ukr. Khim. Zh.* **30**, 1195 (1964).
110. J. Danhaeuser and G. Manecke, *Makromol. Chem.* **84**, 238 (1965).

111. Y. A. Popov, B. E. Davydov, N. A. Kubasova, B. A. Kreutsel, and I. I. Konstantinov, *Vysokomol. Soedin.* **7**, 835 (1965).
112. S. C. Kuo and W. H. Daly, *J. Org. Chem.* **35**, 1861 (1970).
113. J. M. Kliegman and R. K. Barnes, *J. Org. Chem.* **35**, 3140 (1970).
114. T. Sasaki, S. Eguchi, T. Okano, and Y. Wakata, *J. Org. Chem.* **48**, 4067 (1983).
115. H. W. Moore and C.-C. Yu, *J. Org. Chem.* **46**, 4935 (1981).
116. G. L'abbé, A. Verbruggen, T. Minami, and S. Toppet, *J. Org. Chem.* **46**, 4478 (1981).
117. D. Anderson, P. Zinke, and S. L. Razniak, *J. Org. Chem.* **48**, 1544 (1983).
118. T. Shono, H. Hamaguchi, M. Sasaki, S. Fujita, and K. Nagami, *J. Org. Chem.* **48**, 1621 (1983).
119. A. J. Biloski and B. Ganem, *J. Org. Chem.* **48**, 3118 (1983).
120. D. H. Aue and D. Thomas, *J. Org. Chem.* **39**, 3855 (1974).

CHAPTER 13 / **AZIDES**

1. INTRODUCTION

The chemistry of organic azides has received increasing attention in recent years not only because azides are sources of nitrenes [1] but also because heterocyclic nitrogen compounds such as carbazoles, furoxans, azepines, Δ^2-triazolines, triazoles [2], tetrazoles [3], aziridines [4, 5], and azirines have been obtained by either addition or decomposition reactions of azides.

Of considerable current interest is functionalization of polymers for use in biochemical research. For example, a variety of carboxylated polymers may be transformed to the corresponding polymeric acylhydrazides. On reacting with amino groups of proteins, the hydrazides form conjugates (Eq. 1). The resultant product may be water-insoluble material that may have retained virtually all of the biochemical activities of the protein. Thus a specific enzyme may be fixed to a solid substrate [6–11].

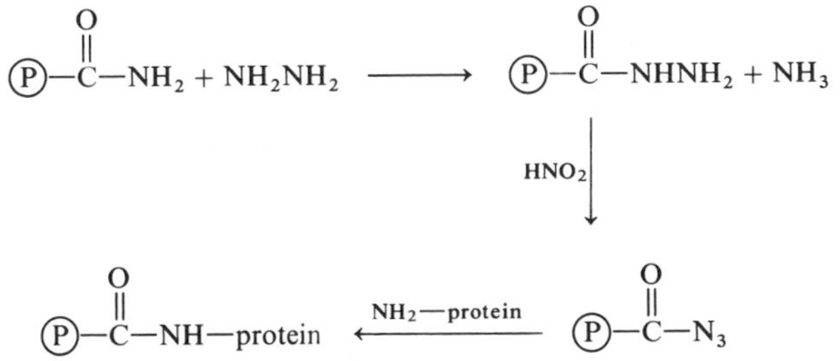

$$ \underset{\substack{\| \\ O}}{\textcircled{P}-C-NH_2} + NH_2NH_2 \longrightarrow \underset{\substack{\| \\ O}}{\textcircled{P}-C-NHNH_2} + NH_3 $$

$$ \downarrow \text{HNO}_2 \tag{1} $$

$$ \underset{\substack{\| \\ O}}{\textcircled{P}-C-NH-\text{protein}} \xleftarrow{\ NH_2-\text{protein}\ } \underset{\substack{\| \\ O}}{\textcircled{P}-C-N_3} $$

In solid-phase polypeptide syntheses, Merrifield and co-workers extended a peptide chain which had one end attached to a poly(chloromethylated styrene) resin and a free end terminated with a hydrazide by converting the hydrazide portion of the molecule to an acylazide and coupling the resin-acylazide with another amino acid [12].

References [1, 13–18] represent a selection of reviews of the preparation and chemistry of alkyl, aryl, and related azides.

The most generally useful method of preparing azide and acyl azides makes use of the displacement of other functional groups by azide ions (Eq. 2).

$$ RX + NaN_3 \longrightarrow RN_3 + NaX \tag{2} $$

X = halide, acyl halide, sulfonyl halide, bridgehead hydroxyl, etc.

323

Other useful preparations involve diazo transfer reactions. In the aromatic series diazonium salts may be reacted with sodium azides (Eq. 3).

$$\overset{+}{ArN_2}X^- \xrightarrow{\text{NaN}_3} ArN_3 \tag{3}$$

Hydrazine derivatives may be treated with nitrous acid to form azides, a reaction which is of particular value in preparing polymeric acrylazide.

Hydrazoic acid may be added to activated olefinic bonds to yield azides. Some epoxy compounds have also been reacted with sodium azide to form hydroxy azides.

Hazards and Safe Handling Practices

Although organic azides are reputed to be explosive materials, detailed information on hazards and safety precautions is sparse. Smith called acetyl azide "treacherous" [19].

The explosive hazard may be a function of the size of the molecule. For example, methyl azide is reported to be handled in a routine manner (but not in the presence of mercury) [19]. Yet, while this chapter was in preparation, Burns and Smith reported an explosion during the preparation of this very compound from dimethyl sulfate and sodium azide while sodium hydroxide was being added [20]. They attributed the explosion to the formation of hydrazoic acid during the preparation, when the pH of the reaction mixture may have dropped below 7. They therefore recommend adding to the indicator bromthymol blue to the reaction mixture. This indicator changes color from deep blue at pH 6.5 to yellow at pH 8. The rate of base addition can then be monitored readily by observing the color of the reaction mixture and maintaining the pH at 8 or higher, throughout the reaction. Since the indicator may fade during the process, additional quantities will have to be added from time to time (approximately once each hour).

Burns and Smith more recently [20a] reported an explosion during their preparation of 20 gm ethyl azide following their precautions cited above for methyl azide [20]. The explosion took place just after approximately 0.5 ml of additional indicator solution was added. They postulated that either the acidic indicator or exposure to the ground glass joint initiated the detonation. The force of the detonation of this 20 gm ethyl azide left a 1-cm deep depression in their 16-gauge stainless steel hood floor and shattered the safety glass around the hood's fluorescent lamp.

The present authors are not in a position to comment on either the validity of the hypothesis of Burns and Smith or their recommendation. Obviously, all reactions involving the preparation, use, and disposal of solvents and by-products of azides and related compounds must be carried out on a very small scale with suitable protection of personnel even if a particular reaction has been repeatedly carried out without incident. The avoidance of ground glass joints, the protection from strong light and the use of dilute solutions (as with diazo methane preparations) are additional precautions suggested by Burns and Smith [20a].

In the case of "triflyl" azide, the recommendation has been made that the compound not be allowed to be completely free of solvent and that it not be stored for any length of time [21]. Ethyl azidoformate could be distilled at about 100°C. It did not detonate until 160°C. On the other hand, the vapors of this azide are toxic, leading to vertigo, severe headaches, and sometimes vomiting [22].

Of the 1,2-diazobenzenes, the parent compound, 1,2-diazobenzene, could be detonated on an anvil with a hammer. However, the 3-methyl-4-methyl, 4-methoxy, and 4-chloro derivatives were not said to be shock-sensitive [23].

These observations show that extreme care must be exercised in the handling and preparation of azides and their derivatives.

In addition, sodium azide and hydrazoic acid must be handled safely. In working with sodium azide, the salt must not come in contact with copper, lead, mercury, silver, gold, their alloys and their compounds. All of these form sensitive explosive azides. Azide salts must not be thrown into sinks or sewers since all azide salts are highly toxic; react with acids to form explosive, toxic, and gaseous hydrazoic acid; and react with copper and lead pipes. Azides and hydrazoic acid are thought to be more toxic than cyanides and hydrogen cyanide [24].

Decontamination of rags, filter paper, solutions containing sodium azide, and apparatus which has been contaminated with sodium azide should be done by soaking, in a fume hood, with acidified sodium nitrite until the azides have been destroyed, followed by washing or other disposal [24b]. In general, the handling and disposition of sodium and other azides must be in conformity with all appropriate laws and regulations.

2. CONDENSATION REACTIONS

A. Displacement Reactions with Azide Ions

Allyl and aryl azides are commonly prepared by the nucleophilic displacement of such functional groups as halide, sulfide, phenyldiazonium, hydroxyl, nitrate, iodoxy, alkoxy, tosylate [13], aliphatic and aromatic sulfonyl chlorides [25], acyl chlorides [22], and quaternary ammonium salts [26] with sodium azide. Acetyl azide, generated *in situ* as required, has been used as an azidation agent of an alkyl halide [27]. This reaction, along with the displacement of amino groups by azide moieties using trifluoromethenesulfonyl azide [21], may be looked upon as a "diazo transfer" reaction.

While the use of silver azide has been recommended in some syntheses, in most cases this does not appear to be necessary. In view of the explosive hazards associated with heavy metal azides, use of this azide is best avoided (see Section 1A). Table I cites experimental details for the preparation of azides from the literature [28–59]. Preparation 2-1 is a typical example of the procedures used in the synthesis. In view of the general hazards of handling azides, the scale of reaction should probably be reduced considerably. The removal of unreacted alkyl halide with silver nitrate may lead to silver azide formation and should, therefore, be replaced by another procedure.

2-1. *Preparation of n-Butyl Azide* [29]

$$n\text{-}C_4H_9Br + NaN_3 \xrightarrow[CH_3OH]{H_2O} n\text{-}C_4H_9N_3 + NaBr \tag{4}$$

With suitable safety precautions, to a flask containing 34.5 gm (0.53 mole) of sodium azide in 70 ml of water and 25 ml of methanol is added 68.5 gm (0.50 mole) of *n*-butyl bromide while stirring at room temperature. The resulting mixture is heated and stirred on a steam bath for 24 hr. The bottom layer of *n*-butyl bromide disappears after this time and a top layer of crude *n*-butyl azide forms. The crude azide is separated and then treated overnight with alcoholic silver nitrate to remove traces of butyl bromide. The mixture is then filtered, washed with water, and distilled behind a safety barricade to yield 40.0 gm (90%) of *n*-butyl azide, b.p. 106.5°C (760 mm), $n_D^{29.5}$ 1.4152, $d^{29.5}$ 0.8649. [NOTE: *n*-Butyl azide and methanol form an azeotrope (b.p. 60°C) from which the azide is liberated by the addition of a saturated solution of calcium chloride.]

The use of carbitol as a reaction solvent rather than water or aqueous alcohol has been reported to improve the reaction of alkyl halides with commercial sodium azide [31]. At one time, an activation of sodium azide with hydrazine hydrate was suggested [63]. However, the use of Carbitol seems to have eliminated the need for this step. Sealed tube reactions also seem to have been common. But this must be considered hazardous, since a violent explosion has been reported when an attempt was made to seal a tube for the preparation of an alkyl azide [64]. No reactions reported in this chapter involve sealed tubes. Table II describes the conditions and azides prepared using the Carbitol solvents. Procedure 2-2 is an example of a synthesis utilizing this solvent system.

2-2. *Preparation of Pentyl Azide* [31]

$$n\text{-}C_5H_{11}I + NaN_3 \xrightarrow{Carbitol} n\text{-}C_5H_{11}N_3 + NaI \tag{5}$$

With suitable safety precautions, to a flask containing 16.9 gm (0.26 mole) of sodium azide, 300 ml of Carbitol, and 50 ml of water is added all at once, with stirring, 39.6 gm (0.20 mole) of pentyl iodide. After a few minutes the homogeneous solution is heated to 95°C and kept there for 24 hr. After cooling to room temperature, the reaction mixture is poured into 1 liter of ice water. The organic layer is separated and the water layer is extracted with two 200 ml portions of ether. The ether and organic layers are combined, dried, and concentrated. The residue is distilled under reduced pressure to afford 18.9 gm (83.6%), b.p. 77°–78°C (112 mm), n_D^{20} 1.4266.

In their paper on the preparation of α-azidovinyl ketones, L'abbe and Hassner [60] mention five azide syntheses of which three may be considered displacement reactions of alkyl halides with azide ions (Eqs. 6, 7, 9). Another process involves addition of iodine azide to a *trans-α,β*-unsaturated ketone

$$
\underset{\substack{\;\;\;\;\;\;\;| \;\;\;\; | \\ \;\;\;\;\;\;\;Br \;\; Br}}{R-\overset{\overset{O}{\|}}{C}-CH-CH-R'} \xrightarrow[\text{DMF}]{2NaN_3} \underset{\substack{| \\ N_3}}{R-\overset{\overset{O}{\|}}{C}-C=CH-R'} \tag{6}
$$

$$
\underset{\substack{\;\;\;\;\;\;\;| \\ \;\;\;\;\;\;\;Br}}{R-\overset{\overset{O}{\|}}{C}-C=CH-R'} \xrightarrow[\text{DMF}]{NaN_3,\ NH_3} \underset{\substack{| \\ N_3}}{R-\overset{\overset{O}{\|}}{C}-C=CH-R'} \tag{7}
$$

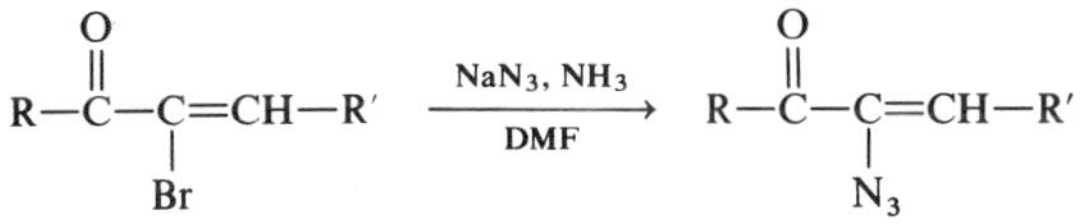

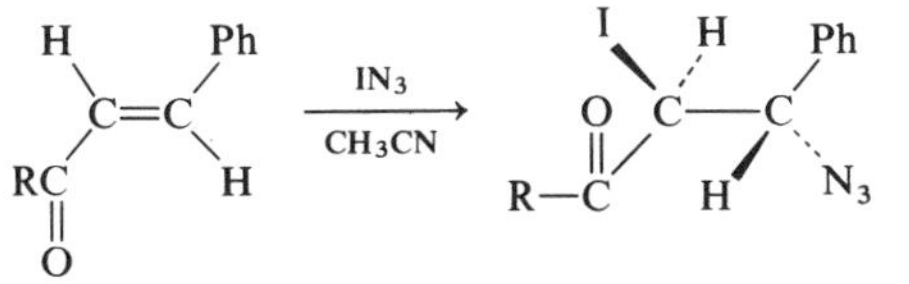

(8)

(Eq. 9). The fifth process, due to Knittel *et al.* [64a], is an aldol condensation of azidoacetophenones with substituted benzaldehydes (Eq. 10).

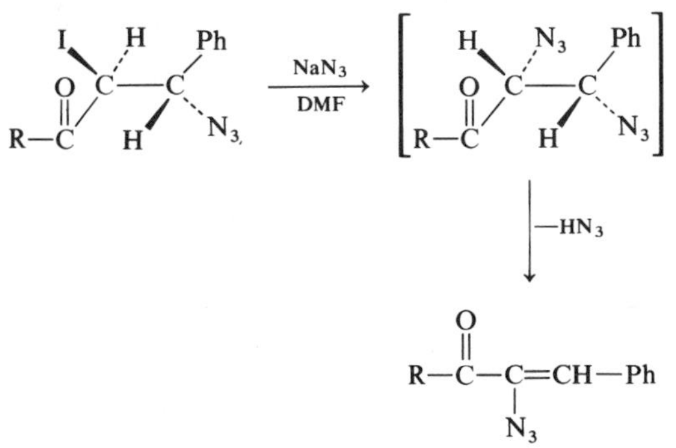

(9)

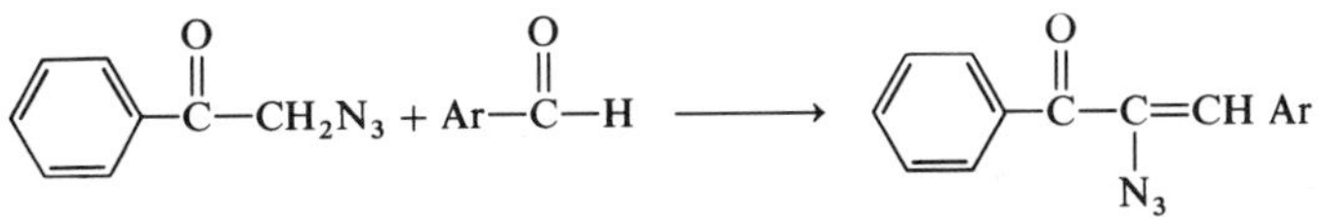

$$
\underset{\substack{| \\ N_3}}{R-\overset{\overset{O}{\|}}{C}-C=CH-Ph}
$$

(10)

TABLE I

DISPLACEMENT OF VARIOUS FUNCTIONAL GROUPS BY AZIDE ION IN POLAR SOLVENTS

Starting material	Azide	Solvent	React. temp. (°C)	React. time (hr)	Azide (yield %)	Ref.
ClCH$_2$COOC$_2$H$_5$	NaN$_3$	C$_2$H$_5$OH + H$_2$O	80	3	91	28
n-C$_4$H$_9$Br	NaN$_3$	CH$_3$OH + H$_2$O	50-60	24	90	29-30
RX, X = I, Br, R = alkyl	NaN$_3$	Carbitol	95	24	64-100	31
Bis(chloromethyl)oxetane	NaN$_3$	Diethylene glycol	90-100	30	25	32
RX, X = Br, Cl R = alkyl or alkyl aryl	Tetramethyl-guanidinium N$_3$	CH$_2$Cl$_2$	32	1¼	60-100	33
R-(Br)$_2$, R = alkylene	NaN$_3$	CH$_3$OH	56	24	60-89	34-35
RBr	NaN$_3$	C$_2$H$_5$OH + H$_2$O	80	24	70	36
9-Bromomethylfluorene	NaN$_3$	DMF	0-10	73	100	37
2,3-Dichloro-1,4-naphthoquinone	NaN$_3$	H$_2$O + CMF	25	¼	61	38
10-Bromo-1-undecene	NaN$_3$	Acetone		16	80	39
RSO$_2$C$_6$H$_4$—CH$_3$	LiN$_3$	CH$_3$OH + H$_2$O	56	24	57	32, 40
RSO$_2$Cl	NaN$_3$	C$_2$H$_5$OH + H$_2$O	0-10	1	84	41-43
R$_2$NSO$_2$Cl	NaN$_3$	C$_2$H$_5$OH + H$_2$O	32	1	90	44
RCOCl	NaN$_3$	Acetone + H$_2$O	25	1	98	45-48

	Reagent	Solvent	Temperature	Time	Yield	Ref.
$RCOCH_2Cl$	NaN_3	$C_2H_5OH + H_2O$	0–10	24	93	49–52
RN_2^+	NaN_3	H_2O	0–10	1	63–69	53–58
R_2NCOCl	NaN_3	$C_2H_5OH + H_2O$	—	—	75–80	15
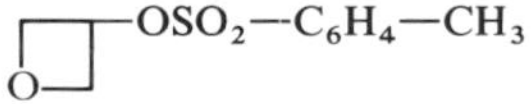 ☐—OSO_2—C_6H_4—CH_3 (oxetane)	NaN_3	Carbowax 300	120–130 (7–10 mm)	15	86	59
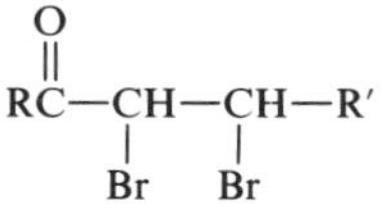 $RC(=O)$—CH—CH—R' (Br, Br)	NaN_3	DMF	R.T.			60
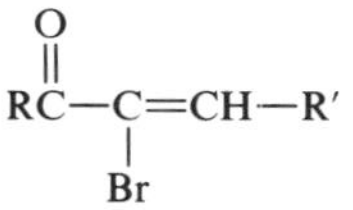 $RC(=O)$—C=CH—R' (Br)	NaN_3	DMF	R.T.			60
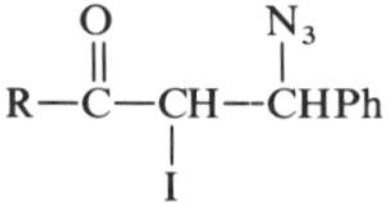 R—C(=O)—CH—CHPh (N_3, I)	NaN_3	DMF	R.T.	0.5–2		60, 61
 $ROCCl$ (O=)	NaN_3	H_2O	0	2.5	60	22, 62

TABLE II[d] [10]

PREPARATION OF ALKYL AZIDES FROM SODIUM AZIDE, ALKYL IODIDES,
OR BROMIDES, AND CARBITOL SOLVENTS AT 95°C FOR 24 HR

Azide	NaN$_3$[a] (moles)	Halide (moles)	Solvent type	ml	H$_2$O (ml)	Yield (%)	B.p., °C (mm Hg)	n_D^{20}
Propyl	0.4	0.3[b]	Methyl Carbitol	450	75	64.4	58 (357)	1.4105
Butyl	0.4	0.3[c]	Methyl Carbitol	450	75	78.3	71 (225)	1.4192
Hexyl	0.4	0.3[b]	Carbitol	450	75	86.6	85 (63)	1.4318
Heptyl	0.26	0.2[b]	Carbitol	300	50	99.6	70 (13)	1.4343
Octyl	0.26	0.2[b]	Butyl Carbitol	300	60	94.8	62 (3.3)	1.4368
Decyl	0.26	0.2[b]	Methyl Carbitol	300	60	89.2	67 (0.65)	1.4425
Cyclopentyl	0.26	0.2[c]	Carbitol	300	50	82.0	72 (77)	1.4616
Cyclohexyl	0.40	0.3[b]	Carbitol	150	50	75.2	72 (30)	1.4693

[a] Not activated.   [b] Iodide.   [c] Bromide.

[d] Reprinted from E. Lieber, T. S. Chao, and C. N. R. Rao, *J. Org. Chem.* **22**, 238 (1957). (Copyright 1957 by the American Chemical Society. Reprinted by permission of the copyright owner.)

The processes of Eqs. (6), (7), and (9) are displacement reactions using dimethyl formamide (DMF) instead of Carbitol as the reaction solvent.

Preparations of acyl azides from acid chlorides, anhydrides, or the sulfonyl chlorides are similar to those involving alkyl halides. The aliphatic members of the series are said to be generally unstable and explode on heating, whereas aroyl azides are more stable thermally and may be isolated more readily [65]. Note that in Preparation 2-3, by adding the acid chloride to a sodium azide solution, the reaction mixture may be maintained at a high pH throughout the process.

2-3. *Preparation of Ethyl Azidoformate* [22]

$$C_2H_5O-\overset{\overset{\displaystyle O}{\|}}{C}-Cl + NaN_3 \longrightarrow C_2H_5-O-\overset{\overset{\displaystyle O}{\|}}{C}-N_3 + NaCl \qquad (11)$$

With suitable safety precautions, with the reaction flask surrounded by a large ice–salt bath, to an ice-cooled aqueous solution of 35 gm (0.54 mole) of sodium azide in 190 ml of water is added dropwise and with vigorous stirring a solution of 50 gm (0.45 mole) of ethyl chloroformate in 50 ml of peroxide-free ether. Cooling and stirring are continued for 2.5 hr. The infrared spectrum of the ether layer in toluene is taken from time to time during the reaction period. The reaction is considered complete when there is no acid chloride absorption at 1798 cm^{-1}.

The ether layer is separated and preserved. The aqueous layer is extracted twice with ether. The ether extracts are combined with the original ether layer and dried over anhydrous sodium sulfate. After filtration, the ether is removed by distillation at atmospheric pressure. The residue is distilled under reduced pressure through a short Vigreux microcolumn. At 39°–41°C at approximately 30 mm, 40.5 gm of crude ethyl azidoformate is isolated. Upon redistillation at 40°C (30.5 mm), 31.1 gm (60 % yield) is isolated; $n_D^{24.8}$ 1.4180. Infrared spectrum (in CCl_4) showed the N_3 peaks at 2416 (w), 2185 (s), and 2137 (s); C=O at 1759 (s) and 1730 (vs); and C—O at 1242 cm^{-1} (vs).

A new method of preparing acyl azides of aromatic acids makes use of zinc iodide as a catalyst for the reaction of aroyl chlorides with trimethylsilyl azide at room temperature [65].

2-4. General Procedure for the Preparation of Aromatic Acid Azides [65]

$$\underset{\substack{\| \\ O}}{Ar-\overset{O}{\overset{\|}{C}}-Cl} + \underset{\substack{| \\ CH_3}}{CH_3-\overset{CH_3}{\overset{|}{Si}}-N_3} \xrightarrow{ZnI_2} Ar-\overset{O}{\overset{\|}{C}}-N_3 \qquad (12)$$

With suitable safety precautions, to a stirred dispersion, under nitrogen, of 20 mmoles of an aroyl chloride and 20 millimoles of trimethylsilyl azide in 80 ml of methylene chloride at 0°C is added 20 mg of anydrous zinc iodide. Stirring at 0°C is continued for 0.5 hr. Then the stirred mixture is allowed to warm slowly to room temperature. Stirring is continued at room temperature until the reaction has been completed (cf. Table III). The reaction mixture is poured into 100 ml of ice water. The methylene chloride layer is separated

TABLE III

PREPARATION OF AROMATIC ACID AZIDES FROM ACID
CHLORIDES AND TRIMETHYLSILYL AZIDE, CATALYZED WITH
ZINC IODIDE [65]

Substituent on Benzoyl chloride	Reaction time (hr)	M.p., °C (b.p., °C/mm Hg)	Yield (%)
4-(CH_3O—)	3	70–71	96
4-(CH_3)	3	32–33	84
3-(CH_3—)	3	^{a}	93
(H)	3	25–27	95
4-HBr—)	5	48	86
4-(Cl—)	3	39–42	95
4-(F—)	30	(72/1.0)	85
3-(F—)	34	(74–77/1.0)	94
4-(NO_2—)^b	72	64–66	83

^{a} Decomposed upon distillation to produce an isocyanate by Curtius rearrangement.

^{b} Required use of two equivalents of M_3SiN_3 and one equivalent of ZnI_2.

and preserved. The aqueous layer is extracted with two 100-ml portions of methylene chloride. The methylene dichloride solutions are combined, washed once with 5 % aqueous sodium thiosulfate, and then washed with two 100-ml portions of cold water. The organic layer is then dried over anhydrous magnesium sulfate and filtered. The solvent is evaporated off. The residue may be recrystallized or, with sufficient care, distilled. Table III gives details of starting materials, reaction times, and physical properties.

Preparation 2-5 is an example of the preparation of an aromatic sulfonyl azide from sulfonyl chloride.

2-5. *Preparation of o-Nitrobenzenesulfonyl Azide* [43]

$$\text{(13)}$$

A stirred solution of 5.0 gm (0.077 mole) of sodium azide in 75 ml of acetone is cooled to $-10°C$ while 15 gm (0.66 mole) of o-nitrobenzenesulfonyl chloride dissolved in 75 ml of acetone is added dropwise. The reaction mixture is stirred for 1 hr at $-10°C$ and for 1 hr at room temperature. The solution is filtered and diluted with 500 ml of ice water. The yellow product is removed by filtration, washed with water and dried. The product is dissolved in 150 ml of warm ethanol and freed from any insoluble material by filtration. The clear filtrate is cooled to afford 12.0 gm (80%), m.p. 71°–73°C.

Using a similar procedure, p-toluenesulfonyl azide is prepared in 65% yield, m.p. 10°–20°C [43].

Benzenesulfonyl azide has been reported [42] to decompose violently on heating, especially if it is in the dry solid state.

The preparation of azides using other acid halides follows a procedure similar to that for o-nitrobenzenesulfonyl azide. Reference [25] gives one preparation of methanesulfonyl azide, the azides of 4-bromo-, 4-methoxy-, 3-nitro-, and 4-nitrobenzenesulfonic acid, and 2-naphthalenesulfonic acid.

The reagent trifluoromethanesulfonyl azide ("triflyl azide") may be prepared rapidly by treating trifluoromethanesulfonyl anhydride at $0°C$ with aqueous sodium azide. The product separates as a water-insoluble layer. However, at least one explosion took place during this synthesis [66]. On the other hand, when the preparation was carried out in the presence of an organic solvent such as methylene chloride (evidently reaction conditions are such that no significant reaction with sodium azide and at least one of the chlorines of CH_2Cl_2 takes place), N,N-dimethylformamide, tetrahydrofuran, dimethyl sulfoxide, dioxane, acetonitrile, methanol, acetone, or aliphatic amines which are used as coreagents, no difficulty was reported [66].

2-6. Preparation of Trifluoromethansulfonyl Azide (Triflyl Azide) [66]

$$(CF_3SO_2)_2O + NaN_3 \xrightarrow[0°C]{CH_2Cl_2} CF_3SO_2N_3 + Na(CF_3SO_3) \qquad (14)$$

With suitable safety precautions, to a rapidly stirred dispersion of 8 gm (0.16 mole) of sodium azide in 20 ml of water and 25 ml of methylene chloride maintained at 0°C is slowly added 7.06 gm (0.005 mole) of trifluoromethanesulfonyl anhydride. The low temperature and stirring are continued for 2 hr. The lower, organic layer is separated and preserved. The aqueous layer is extracted with two 25-ml portions of methylene chloride. The resulting triflyl azide solution may be stored for at least 24 hr. It should not be allowed to dry out since an explosion occurred when no organic solvent was present. The ir spectrum has peaks at 4.65 and 7.1 and three peaks between 8 and 9 μ.

Zaloom and Roberts [21] suggest that residual traces of acid be removed by washing the product solution with 1 N aqueous sodium hydroxide prior to use. They assume that the yield of the reaction is 50%. The reagent solution may be used for the preparation of azides from aliphatic amines, including tertiary amines and amino acids (see below).

Acetyl azide, prepared *in situ* by adding a solution of acetyl chloride in methylene chloride at 5°C to an aqueous solution of sodium azide, has been proposed as an azidation agent under nonbasic conditions [27]. The procedure appears to be suitable for converting labile methylol amine derivatives to the corresponding azide (Eq. 15).

$$
\begin{array}{c}
R \\ \diagdown \\ \diagup \\ R
\end{array}
N{-}CH_2OH \longrightarrow
\begin{array}{c}
R \\ \diagdown \\ \diagup \\ R
\end{array}
N{-}CH_2OAc \xrightarrow{CH_3COBr}
$$

$$
\begin{array}{c}
R \\ \diagdown \\ \diagup \\ R
\end{array}
N{-}CH_2Br \xrightarrow{CH_3CON_3}
\begin{array}{c}
R \\ \diagdown \\ \diagup \\ R
\end{array}
N{-}CH_2N_3 \qquad (15)
$$

The actual example of this reaction cited in [27] is for synthesis of an analog of a high explosive, 1-(azidomethyl)-3,5,7-trinitro-1,3,5,7-tetraazacyclooctane. Whether the process is applicable to more mundane starting materials (as we have indicated in Eq. 15) will yet have to be determined.

The direct conversion of alcohols to azides appears to be of limited utility. At one time it was thought that azides could only be prepared from alcohols which formed carbonium ions readily by the reaction of hydrogen azide in trichloroacetic acid. For example, benzhydrols and trialkyl carbinols were reacted with these reagents to form azides [67]. Sulfuric acid has been substituted for trifluoroacetic acid in recent procedures. It is assumed that either acid causes carbonium ion formation prior to the reaction with azide ions [68].

Preparation 2-7 is an example of the reaction involving a triarylcarbinol.

2-7. Preparation of Triarylmethyl Azides [69]

$$(C_6H_5)_3COH + NaN_3 + H_2SO_4 \longrightarrow (C_6H_5)_3C-N_3 \qquad (16)$$

With suitable safety precautions, to an ice-cooled flask containing 5 gm (0.077 mole) of sodium azide, 5 ml of water, 20 ml of chloroform, and 5 gm (1.019 mole) of triphenylcarbinol in 50 ml of chloroform is added dropwise with stirring 5 ml of concentrated sulfuric acid. After 1 hr the mixture is neutralized with sodium hydroxide and the chloroform solution is separated, dried, and concentrated under reduced pressure to yield an oil. Recrystallization of the oil from hexane affords 4.8 gm (88%), m.p. 64°–65°C.

Sasaki and co-workers have extended this reaction in recent years to the conversion of corresponding bridgehead azides, using sodium azide and 57% aqueous sulfuric acid [70a–d, 71]. A modification of the workup procedure was suggested by Kovacic and co-workers [72, 73].

2-8. General Procedure for the Preparation of Bridgehead Azides from Alcohols [70a]

$$(17)$$

96%
1-azidoadamantane

With suitable safety precautions, 10 ml of 57% aqueous sulfuric acid is prepared and cooled in an ice bath. With rapid stirring, 10 ml of chloroform and 10 mmole of the bridgehead alcohol are added. To the resulting dispersion is added in small increments 1.30 gm (20 mmole) of sodium azide. The ratio of sodium azide to alcohol is varied with different alcohols from 2.0 to 1 to 4.3 to 1. The addition of sodium azide requires approximately 0.5 hr. The resulting mixture is allowed to warm to room temperature with stirring. Stirring is continued from 2 to 27 hr, depending on the nature of the organic starting material. Then the reaction mixture is poured into ice water. The product is extracted with four 10-ml portions of methylene chloride. The combined extracts are washed in turn with 10 ml of 5% aqueous sodium bicarbonate and 5 ml of water. The product solution is dried over anhydrous sodium sulfate and filtered. The solvent is then evaporated off under the reduced pressure of a water aspirator to afford the azide. Some of the azides may be purified further by recrystallization from aqueous methanol and by sublimation. Some oily azide products may be purified on a silica gel column, eluting with n-hexane–methylene chloride or by "Kugelrohr" distillation.

The yields are reported to be in the range 70 to 80%. Infrared spectra (neat films for oily azides, KBr pellets for solids): 2090 to 2100 cm^{-1}.

(The acidic aqueous layer from the reaction is made basic with 50% aqueous sodium hydroxide and extracted with five 10-ml portions of chloroform. The chloroform extract is dried over anhydrous sodium sulfate and evaporated to dryness. From the residue various rearrangement products may be isolated.)

Table IV outlines the specific reaction conditions used with various bridgehead alcohols.

The modification of this procedure by Kovacic and co-workers for the preparation of 1-azidomantane included increasing the reaction time to 4–4.5 hr. The purification was carried out either by dissolving the crude product in n-hexane, filtering, and evaporating the solvent off under reduced pressure [72] or by dissolving the crude product in methanol and precipitating it by cooling the solution in a Dry Ice–acetone bath, followed by sublimation [73].

B. Diazo Transfer Reactions

The synthesis of diazo compounds in which a preformed N_2 group is transferred from one reagent to another with an exchange of two hydrogens has been termed a diazo transfer reaction. The application of this process to the formation of aliphatic diazo compounds is discussed in Volume I of this edition, p. 483 ff. There, the reaction of tosyl azide with an active methylene

TABLE IV

SYNTHESIS OF BRIDGEHEAD AZIDES FROM ALCOHOLS [70a]

Alcohol	Mole ratio of NaN$_3$ to alcohol	Reaction time (hr) (including NaN$_3$ addition)	Yield of azide (%)	M.p. (°C)
1-Hydroxyadamantane	2.0	3.0	96	78.5–79[a]
1-Hydroxy-3,5-dimethyl-adamantane	3.0	2.0	72	27
1-Hydroxy-3,5,7-trimethyl-adamantane	3.0	24.0	76	15
1-Hydroxybicyclo-[3.3.1]nonane	4.3	2.0	70	20
3-Hydroxyhomoadamantane	4.0	27.0	73	97–98[b]

[a] Ref. [72].
[b] Ref. [73].

compound is used to illustrate the preparation of a diazo compound (cf. Eq. 18).

$$CH_3-S-\overset{\overset{O}{\|}}{C}-CH_2-\overset{\overset{O}{\|}}{C}-OCH_3 \xrightarrow[\text{Et}_3\text{N}]{\text{TsN}_3} CH_3-S-\overset{\overset{O}{\|}}{C}-\underset{\underset{N_2}{\|}}{C}-\overset{\overset{O}{\|}}{C}-O-CH_3 \quad (18)$$

Sasaki and co-workers [71] applied the same approach to converting bridgehead amines to bridgehead azides.

2-9. Preparation of 1-Azidobicyclo[2.2.2]octane [71]

$$\text{NH}_2 \xrightarrow[\text{NaH/THF}]{\text{TsN}_3} \text{N}_3 \quad (19)$$

With suitable safety precautions, an ice-cooled stirred dispersion of 600 mg of a 6% dispersion of sodium hydride in mineral oil (25 mmole) in 20 ml of anhydrous tetrahydrofuran, under nitrogen, is treated in turn with a solution of 125 mg (1.0 mmole) of 1-aminobicyclo[2.2.2]octane in 2 ml of THF and with a solution of 300 mg (1.52 mmole) of p-toluensulfonyl azide in 2 ml of THF. The mixture is allowed to warm to room temperature with stirring. Stirring at room temperature is continued for 2 days. To the mixture is then added, cautiously, with ice cooling, 2 ml of methanol. Then the mixture is poured over an ice-water mixture. The aqueous mixture is extracted with four 10-ml portions of peroxide-free ether. The combined extracts are washed with 20 ml of water, dried over anhydrous magnesium sulfate, and evaporated under reduced pressure. The oily residue is put on a silica gel column and eluted with n-pentane. The product is isolated as a colorless oil. Yield: 126 mg (83% of theory); H NMR (CDCl$_3$) δ 1.66 (s), mass spectrum m/Z 151 (M$^+$).

Cavender and Shiner [66] used freshly prepared triflyl azide for the diazo transfer reaction to prepare n-hexyl azide and 2,4,4-trimethyl-2-pentyl azide from the corresponding primary amines in methylene chloride solution with 2,6-lutidine as the base. This method seems preferable to the one given in Preparation 2-9 since it does not require handling sodium hydride dispersions and, perhaps more important, when trioctylmethylammonium chloride is used as a base catalyst, optically pure amino acids can be converted to azido acids without racemization. Esters of amino acids have also been used in this preparation [21].

2-10. *Preparation of Ethyl L-Azidopropanoate* [21]

$$CH_3-\underset{\underset{NH_2}{|}}{CH}-\overset{\overset{O}{\parallel}}{C}-OC_2H_5 \quad \xrightarrow[TOMA\cdot Cl]{CF_3SO_2N_3} \quad CH_3-\underset{\underset{N_3}{|}}{CH}-\overset{\overset{O}{\parallel}}{C}-OC_2H_5 \qquad (20)$$

With suitable safety precautions, to a solution of 5.73 gm (0.049 mole) of ethyl L-aminopropionate (L-alanine, ethyl ester) in 10 ml of methylene chloride is added a solution of triflyl azide in methylene chloride. (See Preparation 2-6 for the synthesis of triflyl azide. For the present experiment, four times the scale of Preparation 2-6 will be needed.) Then 0.4 gm (ca. 0.1 mmole) of trioctylmethylammonium chloride is added and the reaction mixture is stirred with a magnetic stirrer at room temperature for 10 to 15 hr. The progress of the reaction may be monitored by gas chromatography (6 ft., SE 30/Chromasorb W). Then the reaction mixture is washed with two 30-ml portions of 0.5 N citric acid. The methylene chloride layer is dried over anhydrous sodium sulfate, filtered, and evaporated under reduced pressure. The residual azide ester is distilled under reduced pressure. Yield: 4.32 gm (62% of theory); b.p. 70°–74°C (at water aspirator pressure).

C. Reactions with Aromatic Diazonium Salts

Aromatic amines may be diazotized by conventional treatments with acids and sodium nitrite. The resulting diazonium salts are then stirred with sodium azide (without the addition of copper or copper salts, which might lead to the formation of very explosive copper azide) to form the corresponding aromatic azide.

Huisgen and Ugi [57] presented evidence for the intermediate formation of a pentazene when the benzene diazonium ion reacted with the azide ion (Eq. 21).

$$ArN_2{}^+N_3{}^- \quad \longrightarrow \quad Ar-N\underset{\diagdown}{\overset{\overset{N=\!\!=\!\!N}{|\qquad|}}{\diagdown}}N \quad \longrightarrow \quad ArN_3 + N_2 \qquad (21)$$

The pentazene then decomposes to an aromatic azide and nitrogen. This dual decomposition was shown by the ^{15}N activity of the products (starting with ^{15}N in the diazonium only), 82.5% [15]N in $C_6H_5N_3$ and 17.5% in N_2.

Preparation 2-11 is an example of the preparation of an aromatic azide by the diazotization procedure. To be noted is that even an *ortho*-substituted compound undergoes the reaction.

2-11. Preparation of 2-(2-Azidophenyl)-5-methylbenzotriazole [58]

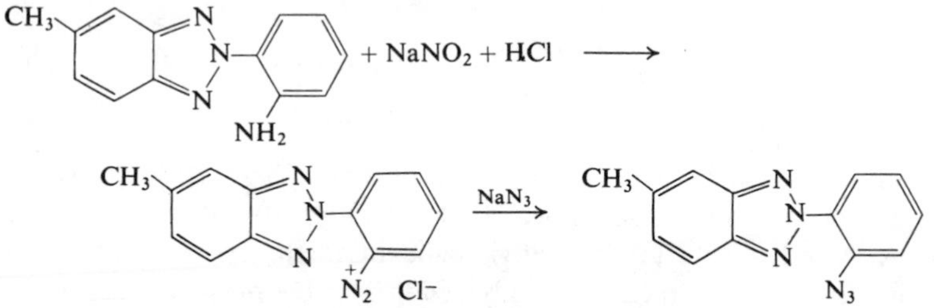

To 10 ml of 6N hydrochloric acid is added 0.28 gm (1.25 mmole) of 2-(2-aminophenyl)-5-methylbenzotriazole. The solution is cooled to 0°C and diazotized by the addition of 0.10 gm (1.45 mmole) of sodium nitrite in water. After stirring for 15 min, 0.18 gm (2.5 mmole) of sodium azide in water is added. The reaction mixture is allowed to stand for 3 hr, then filtered; the solid is washed with water and dried to afford 0.26 gm (84%) of colorless needles, m.p. 83.7°–84°C (from ethanol).

Other examples of the preparation using diazonium salts as intermediates are the preparation of methyl 4-azido-2-methoxybenzoate and 3,4-dimethoxyphenyl azide [74]; o-diazidobenzene [75]; o-diazidotoluene, related 1,2-diazidobenzene derivatives, 1,2-diazidonaphthalene [23]; o-azidoaniline, 8-azido-1-naphthylamine, 2-azido-2'-biphenylylamine, and o-iodophenyl azide [76]; and p-azidoacetophenone [77].

In the case of the preparation of azidoindoles and azidotryptophans, the indole nuclei are sensitive to strong mineral acids used in diazotization procedures. Substituting 80% aqueous acetic acid for dilute hydrochloric acid in the diazotization step led to good yields of azido derivatives [78].

2-12. General Procedure for the Preparation of Azidoindoles and Azidotryptophans [78]

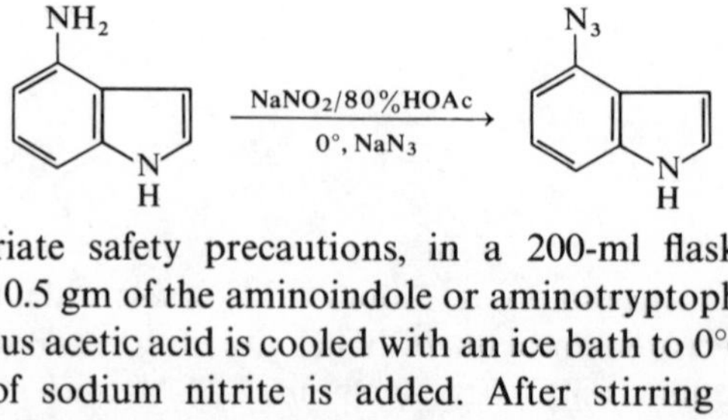

With appropriate safety precautions, in a 200-ml flask fitted with a magnetic stirrer, 0.5 gm of the aminoindole or aminotryptophan dissolved in 80% (v/v) aqueous acetic acid is cooled with an ice bath to 0°C. With stirring 1.1 equivalent of sodium nitrite is added. After stirring for 5 min, 1.1 equivalents of sodium azide dissolved in 1 ml of ice-cold water is added. The mixture is stirred at 0°C for 2.5 hr. Then the reaction system is evaporated under reduced pressure at 30°C to dryness.

A typical workup of the product is the case of the preparation of 4-azidoindole. From 0.506 gm (3.83 mmole) of 4-aminoindole a black tar is obtained. To this tar is added 25 ml of water and 25 ml of diethyl ether. The ether layer is separated and preserved. The aqueous layer is extracted with three 25-ml portions of diethyl ether. The ether layers are combined and evaporated to dryness at room temperature. The residual tar is purified by column chromatography, using chloroform as the eluant. Yield: 0.283 gm (47% of theory). The orange solid had a melting point of 68°–69°C; ir 2111 (N_3) cm^{-1}.

In the preparation of 5-azidoindole, a crude dark solid is isolated. This is purified by a crystallization procedure, m.p. 84°–85°C (88% yield), ir 2104 (N_3) cm^{-1}; and 6-azido-L-tryptophan is isolated in 54% yield, m.p. approximately 190°C dec. (evacuated, sealed tube), ir 2110 (N_3) cm^{-1}.

Diazonium salts have also been converted to azides by treatment with a variety of nitrogen compounds such as hydrazine [79], hydroxylamine [80], ammonia and bromine [81], and arylsulfonamides [82]. In his review of the Curtius reaction, Smith mentions the formation of acyl azides from aromatic hydrazides by the action of diazonium salts rather than by nitrous acid. This is a process due to Curtius and co-workers [83].

D. Reaction of Nitrous Acid with Hydrazine Derivatives

This method allows conversion of either a hydrazine or a hydrazide into an azide (Eqs. 24, 25). Phenyl azide has been prepared from phenylhydrazine and nitrous acid in 65–68% yield by this method [84].

$$RNHNH_2 + HNO_2 \longrightarrow RN_3 + 2H_2O \tag{24}$$

$$\underset{O}{\underset{\|}{R C}}\!\!-\!\!NHNH_2 + HNO_2 \longrightarrow \underset{O}{\underset{\|}{R C}}\!\!-\!\!N_3 + 2H_2O \tag{25}$$

Several substituted benzazides have also been prepared from the hydrazides shown in Table V [85].

The preparation of succinamoyl azide, in 57% yield, has been reported by interaction of succinamoyl hydrazide and a molar equivalent of nitrosyl chloride [86].

2-13. Preparation of Phenyl Azide [84]

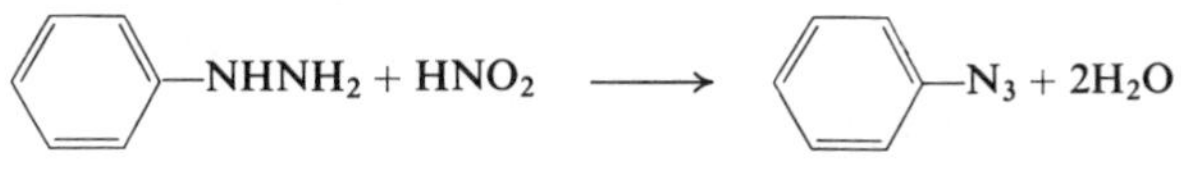

$$\text{C}_6\text{H}_5\!\!-\!\!NHNH_2 + HNO_2 \longrightarrow \text{C}_6\text{H}_5\!\!-\!\!N_3 + 2H_2O \tag{26}$$

TABLE V

MELTING POINTS[a] OF SUBSTITUTED BENZAZIDES,
BENZHYDRAZIDES, AND DIPHENYLUREAS [85]

Substituents	Azides, m.p. (°C)	Hydrazides, m.p. (°C)	sym-Diphenyl- ureas, m.p. (°C)
H	27	—	237
p-HO[b]	132	255–256	Dec.
p-CH₃O[c,d]	70–71	136	233–234
p-C₂H₅O[d]	31	128	224
p-t-C₄H₉[e]	63–65	134	281
p-CH₃[c,f]	35	117	263
p-Br[g]	47	164	244–246
p-Cl[h]	43	163	305
p-NO₂[b,i]	65	209–211	308–311
m-CH₃[i]	Liquid	97	210–211
m-CH₃O[c]	Liquid	109	
m-Br[g]	Liquid	153	260
m-NO₂[b,i]	68	154	242

[a] All melting points are uncorrected.   [b] T. Curtius, A. Struve, and R. Radenhausen, *J. Prakt. Chem.* [2] **52**, 227 (1895).   [c] C. Naegeli, A. Tyabji, and L. Conrad, *Helv. Chim. Acta* **21**, 1127 (1938). [d] P. P. T. Sah and Kwang-Shih Chang, *Chem. Ber.* **69**, 2762 (1936); R. Robinson and M. L. Tomlinson, *J. Chem. Soc.* 1524 (1934). [e] H. Yale, K. Losee, J. Martins, M. Holsing, F. M. Perry, and J. Bernstein, *J. Amer. Chem. Soc.* **75**, 1933 (1953).   [f] T. Curtius and H. Franzen, *Chem. Ber.* **35**, 3239 (1902).   [g] T. Curtius and E. Portner, *J. Prakt. Chem.* [2] **58**, 190 (1898).   [h] C. H. Kao, H. Y. Fang, and P. P. T. Sah, *J. Chin. Chem. Soc.* **3**, 137 (1935); *Chem. Abstr.* **29**, 6172 (1935).   [i] T. Curtius and O. Trachmann, *J. Prakt. Chem.* [2] **51**, 165 (1895).   [j] R. Stolle and H. P. Stevens, *ibid.* [2] **69**, 366 (1904).   [k] Reprinted from Y. Yukawa and Y. Tsuno, *J. Amer. Chem. Soc.* **79**, 5530 (1957). (Copyright 1957 by the American Chemical Society. Reprinted by permission of the copyright owner.)

To an ice-cooled (0°C to.−10°C) stirred flask containing 300 ml of water and 55.5 ml of concentrated hydrochloric acid is added dropwise over a 10 min period 33.5 gm (0.31 mole) of freshly distilled phenylhydrazine. Phenyl-hydrazine hydrochloride crystals precipitate as they are formed. Addition of 100 ml of ether at 0°C is followed by the dropwise addition (25 min) of a solution of 25 gm (0.36 mole) of sodium nitrite in 30 ml of water. At all times the reaction temperature is kept below 5°C.

The product is isolated by carrying out a steam distillation of the reaction mixture to yield 400 gm of distillate. The ether layer is separated from the distillate, and the water layer is extracted with 25 ml of ether. The combined ether layers are concentrated at 25°–30°C under reduced pressure. The residue is distilled under reduced pressure to afford 24–25 gm (65–68 %) of phenyl azide, b.p. 49°–50°C (5 mm).

CAUTION: Phenyl azide decomposes violently when heated at 80°C or above. Care must be taken that the bath temperature never exceeds 60°–70°C. The product should be stored in a cool place in brown bottles.

The preparation of acyl azides from hydrazides is of particular interest for their biochemical applications. As illustrated in Eq. (1), polymers which bear acyl azide functional groups act as substrates for binding proteins to the polymer chains. For example, a particular antibody may be bonded to a polymer while retaining the biochemical activity of the antibody. This resin may then be used to ascertain whether a particular antigen is present in a serum. By this means various diseases may be diagnosed [87]. Preparation 2-14 illustrates the preparation of a polyacrylazide from cross-linked poly-acrylamide. A similar procedure for the preparation of a polymeric acyl azide from a monodispersed poly(styrene–coacrylamide) latex is described in the patent cited in [87].

2-14. *Preparation of Cross-Linked Poly(acrylazide)* [11]

$$\left(\!\!\begin{array}{c} CH_2-CH \\ | \\ CONH_2 \end{array}\!\!\right)_n + NH_2NH_2 \longrightarrow \left(\!\!\begin{array}{c} CH_2-CH \\ | \\ CONHNH_2 \end{array}\!\!\right)_n \qquad (27)$$

$$\left(\!\!\begin{array}{c} CH_2-CH \\ | \\ CONHNH_2 \end{array}\!\!\right)_n + HNO_2 \longrightarrow \left(\!\!\begin{array}{c} CH_2-CH \\ | \\ CON_3 \end{array}\!\!\right)_n \qquad (28)$$

(a) *Preparation of poly(acrylhydrazide).* In a siliconized glass-stoppered Erlenmeyer flask, 1 gm of cross-linked polyacrylamide beads are allowed to swell overnight in an excess of distilled water equal to approximately 1.3 times the bed volume of the gel. The flask is then suspended in a constant-temperature bath maintained at 47°C. At the same time, a glass-stoppered cylinder containing six times the number of equivalents of the acrylamide in the resin of hydrazine hydrate is immersed in the constant-temperature bath. After about 45 min, the hydrazine is added to the swollen polyacrylamide, a magnetic stirrer is inserted in the flask, the flask is stoppered, and the mixture is stirred at 47°C for 7 hr.

In a fume hood, the gel is washed with 0.1 M aqueous sodium chloride on a Büchner funnel and finally by sedimentation. This washing operation is repeated until the aqueous supernatant solution is free of hydrazine. The gel is finally washed and suspended in a storage buffer at pH 7.3, which is 0.20 M sodium chloride, 0.002 M Na$_2$EDTA, 0.10 M boric acid, 0.005 M sodium hydroxide, and 5×10^{-6} M pentachlorophenol. The resin is reported to contain approximately 4 milliequivalents of hydrazide per gram of dry resin.

(b) *Preparation of poly(acrylazide)*. About 50 ml of the polymer from step (a) is washed with 0.1 M aqueous sodium chloride and 0.25 N hydrochloric acid and resuspended to a 32-ml volume with 0.25 N hydrochloric acid. The suspension is cooled to 0°C. Then 8 gm of crushed ice is added. The container is placed in an ice bath and, with efficient magnetic stirring, 4.0 ml of 1.0 M aqueous sodium nitrite is added.

About 90 sec later, the reagent to be coupled to the resin (e.g., a protein with free amino groups) is added rapidly while maintaining cooling at 0°C. Stirring is continued for the reaction period required for the specific protein involved (see [11] for typical examples).

The excess unreacted azide is reconverted into the hydrazide and then to the stable acetyl hydrazide by adding 1.5 ml of hydrazine hydrate and stirring for 0.5 to 1 hr, followed by washing on a Büchner funnel in turn with 100 ml of 0.1 M aqueous sodium chloride, 100 ml of 0.2 M aqueous sodium acetate in which 4 ml of acetic anhydride was dissolved immediately before the washing, 100 ml of 0.1 M aqueous sodium chloride, 50 to 100 ml of 2 M aqueous sodium chloride, and a storage buffer.

Inman and Diutzis [11] projected in 1969 that among the uses of such reactive polyacrylamide derivatives would be those involving separations such as antigens with antibodies which had been bound to a polymer, enzymes with substrates, and molecules having free sulfhydryl groups with mercurials bound to the resin. Indeed, this approach of binding or complexing biochemical reagents to a polymer for selective reaction with suitable materials is currently of great interest. For example, in the laboratory of one of the authors (W.K.) we are developing a variety of polymeric particles —some of highly uniform particle size ranging from 0.1 to 20 μ in diameter, some with a variety of dyes or fluorescent reagents incorporated, some with reactive functional groups on their surfaces, and some also encapsulating pigment or metal particles.

3. ADDITION REACTIONS

An azide group may be introduced into a compound by the addition of IN$_3$ [88, 89] or HN$_3$ [90–93] to an unsaturated center. A sodium azide solution in water and an acetic acid solution of acrolein were found to produce 3-azidopropanol in 63% yield [94]. Metallic azides may also be used for addition reactions with epoxy groups [95–98].

Oliveri-Mandala found that HN_3 adds to benzoquinone to give azido-hydroquinone [99]. Extension of this reaction to acetylenes led to triazoles [100] but failed entirely with cinnamic acid, fumaric acid, styrene, vinyl bromide, ethylene, and other olefinic compounds [101].

A. Addition of Azides to Olefins

Activated double bonds, such as the olefinic bonds in α,β-unsaturated carbonyl compounds, were found by Boyer [92] to add hydrazoic acid (Eq. 29).

$$R'R''C\!\!=\!\!CH\!-\!\!\underset{\underset{R}{|}}{C}H\!\!=\!\!O + HN_3 \longrightarrow R'R''\underset{\underset{N_3}{|}}{C}\!-\!CH_2\!-\!\underset{\underset{R}{|}}{C}\!\!=\!\!O \qquad (29)$$

Reaction works with

$$R = R' = R'' = H$$
$$R = OCH_3, R' = R'' = H$$
$$R = OH, R' = R'' = H$$
$$R = R' = R'' = CH_3$$

Reaction fails with

R = OH,	R' = C₆H₅,	R'' = H
R = R'' = H,	R' = C₆H₅	
R = OC₂H₅,	R' = C₆H₅,	R'' = H
R = NHCH₃,	R' = C₆H₅,	R'' = H
R = R' = C₆H₅,	R'' = H	

Some of the results and physical properties of the products are summarized in Table VI.

3-1. General Procedure for the Addition of Hydrogen Azide to Olefins [92]

To 0.2 mole of an olefin dissolved in 30 ml of glacial acetic acid is added 19.5 gm (0.3 mole) of sodium azide in 75 ml of water. The addition to acrolein and β-nitrostyrene underwent rapid addition, requiring cooling with an ice–salt bath and slow addition of the sodium azide. Addition to methyl acrylate, acrylic acid, and acrylonitrile required 1–3 days at room temperature, α-vinyl pyridine and mesityl oxide required heating for 24 hr on a steam bath. Other olefins underwent no reaction even after 7 days of heating and were recovered unchanged.

The reaction mixture is poured into water and extracted with ether. The ether layer is washed with sodium carbonate solution (except for the case of β-azido-propionic acid) and dried over magnesium sulfate. The ether is removed by passing an air stream through the ether layer. The oily azide residue is purified by distillation at reduced pressure except for β-azidopropionaldehyde and β-nitro-α-phenyl-α-azidoethane, both of which decompose on heating. Several examples using this procedure are recorded in Table VI.

TABLE VI [92]

SOME PRODUCTS, AND THEIR PHYSICAL PROPERTIES, OF ADDITION OF AZIDES TO OLEFINS

Olefin	HN$_3$ addition product	Yield (%)	B.p., °C (mm Hg)	Analyses (%) Calcd.	Found	n_D^{20a}	d_D^{26a}	MR_D Calcd.	Found	Decomp. by conc. H$_2$SO$_4$
CH$_2$=CHCHO	N$_3$CH$_2$CH$_2$CHO^b	71.0	—	Not attempted		—	—	—	—	Violent at 0°C
CH$_2$=CHCO$_2$CH$_3$	N$_3$CH$_2$CH$_2$CO$_2$CH$_3$	35.4	40–45 (1)	C, 37.20 H, 5.47	37.32 5.51	1.4408	1.139	30.0	29.8	Vigorous at 60°C
CH$_2$=CHCO$_2$H	N$_3$CH$_2$CH$_2$CO$_2$H	24.0	80 (1)	C, 31.31 H, 4.28 N, 36.51	30.90 4.15 36.76	1.4645	—	—	—	Vigorous at 60°C
CH$_3$—C(CH$_3$)=CHCOCH$_3$	N$_3$—C(CH$_3$)$_2$—CH$_2$COCH$_3$	38.0	57–58 (3)	C, 51.04 H, 7.86 N, 29.77	51.68 7.97 29.98	1.4428	0.9955	37.7	37.6	Vigorous at 25°C
C$_6$H$_5$—CH=CHNO$_2$^{c}	C$_6$H$_5$CH(N$_3$)CH$_2$NO$_2$	69.0	—	Not attempted		—	—	—	—	Violent at 0°C
CH$_2$=CHC≡N	N$_3$CH$_2$CH$_2$C≡N	17.3	64 (1)	C, 37.50 H, 4.20 N, 58.30	38.38 4.09 58.04	1.4570	1.1138	23.4	23.5	Vigorous at 60°C
C$_5$H$_4$NCH=CH$_2$^{d}	C$_5$H$_4$NCH$_2$CH$_2$N$_3$	49.5	65 (1)	C, 56.74 H, 5.44 N, 37.82	56.61 5.35 37.39	1.5289	1.1122	41.0	41.1	Vigorous at 60°C

^{a} The refractive indices for methyl β-azidopropionate and 1-α-pyridyl-2-azidoethane were determined at 25°C.

^{b} This compound decomposed on standing at room temperature or in the refrigerator and regenerated acrolein, identified by its color. An anhydrous 10% chloroform solution was successfully stored in the refrigerator for several weeks with no decomposition.

^{c} Attempts to reduce this compound using stannous chloride led only to the oxime of phenylacetaldehyde, the reduction product, under these conditions, of β-nitrostyrene.

^{d} α-Vinyl pyridine.

^{e} Reprinted from J. H. Boyer, J. Amer. Chem. Soc. 73, 5248 (1951). (Copyright 1951 by the American Chemical Society. Reprinted by permission of the copyright owner.)

Aryl azide, acyl azide, sulfonyl azides, and azidoformate add to olefins by a 1,3-dipolar cycloaddition mechanism to yield triazoline. This addition also occurs with other unsaturated systems such as α,β-unsaturated olefins, and enamines [102, 103].

The addition of cyanogen azide to hydrocarbon olefins leads to N-cyanoaziridines [104].

The addition of iodine azide, prepared *in situ*, to olefins leads to iodo azides, which, under certain reaction conditions, may dehydrohalogenate to vinyl azides. Preparation 3-2 illustrates the addition reaction to an olefin. A related reaction, also by Hassner and co-workers, is outlined in Eqs. (8) and (9) [60, 61].

3-2. *Preparation of 2-Azido-1-iodo-2-methylpropane* [105]

$$ICl + NaN_3 + (CH_3)_2C\!\!=\!\!CH_2 \quad \xrightarrow{\text{acetonitrile}} \quad CH_3\!-\!\overset{\overset{\displaystyle CH_3}{|}}{\underset{\underset{\displaystyle N_3}{|}}{C}}\!-\!CH_2I \qquad (30)$$

A flask containing 15.0 gm (0.25 mole) of sodium azide in 100 ml of aceto-nitrile is stirred and cooled in an ice bath while 18.3 gm (0.113 mole) of iodine monochloride is added over a 10–20 min period. After stirring for an additional 5–10 min, 5.6 gm (0.1 mole) of isobutylene is added and then the reaction is allowed to warm to room temperature for 20 hr. The red-brown slurry is poured into 250 ml of water, extracted with 100 ml of ether three times, washed with 150 ml of 5% sodium thiosulfate, washed four times with 250 ml portions of water, and dried over magnesium sulfate. The ether is removed at reduced pressure, leaving a slightly orange oil. Distillation behind a barricade yields 13.5 gm (60%), b.p. 68°–69°C (7 mm), n_D^{25} 1.5292.

CAUTION: Distillation is not recommended when using olefins boiling higher than hexene, because the iodo function is labile and violent decomposition of the azido function may occur when the adducts are heated.

Dehydrohalogenation of the iodo azides yields vinyl azides, and several examples have been reported [89]. (See Table VII for examples.)

Harvey and Ratts [106] also reported that vinyl azides can be prepared by the addition of azide ion to conjugated allenic esters as in Eq. (31). A terminally disubstituted allene ester or amide does not give a vinyl azide even after several hours of heating.

$$CH_2\!\!=\!\!C\!\!=\!\!\overset{\overset{\displaystyle R}{|}}{C}\!-\!COOC_2H_5 \quad \xrightarrow[\substack{THF, \\ H_2O}]{NaN_3} \quad CH_3\!-\!\overset{\overset{\displaystyle R}{|}}{C}\!\!=\!\!\underset{\underset{\displaystyle N_3}{|}}{C}\!-\!COOC_2H_5 \qquad (31)$$

$$R = H \text{ or } CH_3$$

SYNTHESES OF IODO AZIDES AND VINYL AZIDES

Olefin	β-Iodo azide	Unsaturated azide
$PhCH=CH_2$	$PhCHCH_2I$ (N$_3$ on 1-C)	$PhC=CH_2$ (N$_3$ on 1-C)
$t\text{-}BuCH_2CH=CH_2$	$t\text{-}BuCH_2CHCH_2I$ (N$_3$)	$t\text{-}BuCH_2C=CH_2$ (N$_3$)
$t\text{-}BuCH=CH_2$	$t\text{-}BuCHCH_2N_3$ (I)	$t\text{-}BuCH=CHN_3$
$CH_3CH=CHCH_3$ (cis)	$CH_3CHCHCH_3$ (threo) N$_3$ I	$CH_3C=CHCH_3$ (trans) N$_3$
$(CH_3)_2CHCH=CHCH_3$ (cis)	$(CH_3)_2CHCHCHCH_3$ (threo) I N$_3$	$(CH_3)_2CHC=CN_3$ (trans) H CH$_3$
$CH_3CH=CHCH_3$ (trans)	$CH_3CHCHCH_3$ (erythro) N$_3$ I	$CH_3C=CHCH_3$ (cis) N$_3$
$EtCH=CHEt$ (trans)	$EtCHCHEt$ N$_3$ I	$EtC=CEt$ (cis) N$_3$ H
$PhCH=CHCH_3$ (trans)	$PhCHCHCH_3$ N$_3$ I	$PhC=CCH_3$ (cis) N$_3$ H
$t\text{-}BuCH=CHCH_3$ (trans)	$t\text{-}BuCHCHCH_3$ I N$_3$	$t\text{-}BuC=CCH_3$ (cis) H N$_3$
$CH_3CH=CHCO_2Et$ (trans)	$CH_3CHCHCO_2Et$ N$_3$ I	$CH_3C=CCO_2Et$ (cis) N$_3$ H
$PhCH=CHCO_2CH_2$ (trans)	$PhCHCHCO_2CH_3$ N$_3$ I	$PhC=CCO_2CH_3$ N$_3$ H
$CH_2=CHCO_2CH_3$	$CH_2CHCO_2CH_3$ N$_3$ I $CH_2CHCO_2CH_3$ I N$_3$	$HC=CCO_2CH_3$ N$_3$ H $CH_2=CCO_2CH_3$ N$_3$
$(CH_3)_2CHCH=CH_2$	Mixture	$(CH_3)_2CHCH=CHN_3$ $(CH_3)_2CHC=CH_2$ N$_3$

^{a} All compounds gave NMR spectra consistent with their structure is indicative of high purity; the mass spectra of these compounds often show absence of parent peaks but are consistent with their structure

^{b} Reprinted from A. Hassner and F. W. Fowler, *J. Org. Chem.* **33**, 2687 (1968). (Copyright 1968 by the American Chemical Society. Reprinted by permission of the copyright owner.)

The azides of the 1,1-diphenyl series were prepared by adding hydrogen azide to the appropriate olefin. The reaction was catalyzed by trichloroacetic acid and carried out in benzene solution [107].

B. Addition of Azides to Epoxides

Sodium azide reacts with various epoxides at 25°–30°C at pH 6–7 to give azido alcohols [95–97]. The use of harsh conditions as earlier employed by Van der Werf *et al.* [97] (16–40 hr reflux in dioxane) led to the production of 1,3-diazidopropanol instead of 1-azido-3-chloro-2-propanol when starting with epichlorohydrin. Several representative examples of the conditions and products of the reaction of azides with epoxides are shown in Table VIII.

The preparation of 1-azido-2-propanol (0.49 mole) from 1-chloro-2-propanol and sodium azide (0.51 mole) in 33 % aqueous ethanol (by volume) has been reported to give a 55 % yield [9a].

The example in Preparation 3-3 illustrates the general applications of the azide epoxy reaction to large molecules.

3-3. *Preparation of 6β-Azido-5α,11α-dihydroxypregnane-3,20-dione Bis[cyclic ethylene ketal]-11-acetate* [108]

(I) $C_{27}H_{40}O_7$ (II) $C_{27}H_{41}N_3O_7$

With suitable safety precautions, 10 gm (0.021 mole) of (I) and 5.0 gm (0.077 mole) of sodium azide is added to 200 ml of dioxane and 50 ml of water. The solution is refluxed under a nitrogen atmosphere for 5 days, cooled, treated with Darco G50 (2 mg), and diluted to 500 ml with water. The solid product is filtered, washed with water, and dried at 60°C under reduced pressure to afford 9.19 gm (83 % crude), m.p. 164°–230°C. This material is adsorbed onto 275 gm of Fluorsil in methylene chloride and eluted with twenty-one 375 ml fractions of Skellysolve B over the range 5–20 % acetone. On combination, fractions 4–10 are concentrated and then recrystallized

TABLE VIII

PREPARATION OF ORGANIC AZIDE COMPOUNDS BY THE REACTION OF SODIUM AZIDE WITH EPOXIDES IN AQUEOUS SOLUTION [98]

| Reactants | | | | | | | | | | |
Epoxide (moles)	NaN$_3$ (moles)	H$_2$O (ml)	Temp. (°C)	Time (hr)	pH	Product	B.p., °C (mm Hg)	n_D^{25}	d_D^{25} (gm/ml)	Yield (%)
Propylene oxide, 1.0	1.54	250	25	24		1-Azido-2-propanol	72–74 (21)	1.4533	1.060	70
Epichlorohydrin, 6.25	7.0	1500	25–30	8–9	6–9 (use HClO$_4$) MgClO$_4$	1-Azido-2,3-propane-diol	92–95 (0.3)	1.4892	1.2581	65
Epichlorohydrin, 1.3	1.8	700	0–5	24	0.8 mole	1-Azido-3-chloro-2-propanol	44–48 (0.5)	1.4918	1.291	60–75
a	—	a	a	a	a	Glycidyl azide	37–39 (8)	1.4530	1.129	86
1-Chloro-3-nitrato-2-propanol 6.25	7.0	1500	25–30	8–9	6–9 (HClO$_4$)	1-Azido-3-nitrato-2-propanol	88 (1)	1.4851	1.3636	45–80

a Prepared from 1-azido-3-chloro-2-propanol and aqueous sodium hydroxide as described for glycidyl nitrate. See W. L. Petty and P. L. Nichols, Jr., J. Amer. Chem. Soc. **76**, 4385 (1954).

from acetone to afford 3.44 gm (32%) of (II), m.p. 184.5°–185.5°C, v_{max} 3510, 2100, 1723, and 1250 cm^{-1}.

4. MISCELLANEOUS METHODS

(1) Preparation of 2,5-diazido-1,4-benzoquinone from sodium azide, acetic acid, and 1,4-benzoquinone [109].

(2) Preparation of azidodithiocarbonic acid [110].

(3) Preparation of polyazides as cross-linking agents for polymers [111, 112].

(4) Bromination of azides [113].

(5) Formation of benzyl azide from the reaction of 1,1,4,4-tetrabenzyltetrazene with lead tetraacetate [114].

(6) Reaction of aryldiazonium fluoroborate with either isopropyl fluorocarbamate or difluoramine to give aryl azides [115].

(7) Azidoazomethine–tetrazole equilibrium [116, 117].

(8) Azides and amines from Grignard reagents and tosyl azide [118].

(9) Preparation of triphenylsilyl azide [119].

(10) Preparation of 2-azido-1,4-benzoquinones [120].

(11) Preparation of triarylmethyl azides by the reaction of nitrous acid on hydrazine derivatives [68, 121, 122].

(12) Preparation of 1,2-diazido-3-phenylethylene and 2-azido-2-phenylethyl *tert*-butyl ether from styrene, sodium azide, *tert*-butyl hydroperoxide in the presence of ferrous sulfate [123].

(13) A study of the reactivity of ethyl chlorosodiocarbamate [EtOC(O)—N$^-$Cl Na$^+$] toward various azides [124].

(14) A study of the chemistry of 2-azido-3-benzoylenamines [125].

(15) Azide formation from the reaction of aromatic nitroso compounds with hydroxylamine [126] or with hydrazoic acid [127].

(16) Competing S_{NAr} and S_{N^2} reactions with *p*-nitrobenzoate and *o*-nitrobenzoate esters with azides [128].

REFERENCES

1. G. L'abbe, *Chem. Rev.* **69**, 345 (1969).
2. F. R. Benson and W. L. Savell, *Chem. Rev.* **46**, 1 (1950).
3. F. R. Benson, *Chem. Rev.* **41**, 1 (1947).
4. J. F. Fruton, *Heterocycl. Compounds* **1**, 69 (1960).
5. P. A. Gembitskii, N. M. Loim, and D. S. Zhuk, *Usp. Khim.* **35**, 229 (1966); *Russ. Chem. Rev.* **35**, 185 (1966).
6. F. Micheel and J. Evers, *Makromol. Chem.* **3**, 200 (1949).
7. M. A. Mitz and L. J. Summaria, *Nature (London)* **189**, 576 (1961).
8. W. E. Hornby, M. D. Lilly, and E. M. Crook, *Biochem. J.* **98**, 420 (1966).

9. C. W. Wharton, E. M. Crook, and K. Brocklehurst, *Eur. J. Biochem.* **6**, 565 (1968).
10. I. Whittam, B. A. Edwards, and K. P. Wheeler, *Biochem. J.* **107**, 3 (1968).
11. J. K. Inman and H. M. Dintzis, *Biochemistry* **8**, 4074 (1969).
12. A. M. Felix and R. B. Merrifield, *J. Amer. Chem. Soc.* **92**, 1385 (1970).
13. J. H. Boyer and F. C. Canter, *Chem. Rev.* **54**, 1 (1954).
14. E. Lieber, R. L. Missin, Jr., and C. N. R. Rao, *Chem. Rev.* **65**, 377 (1965).
15. B. Toschi, *Gazz. Chim. Ital.* **44**, 447 (1914).
16. E. Miller, ed., "Methoden der Organischen Chemie (Houben-Weyl)," vol. 8, G. Thieme Verlag, Stuttgart, 1952).
17. P. A. S. Smith, "Open-Chain Nitrogen Compounds," vol. 2, Benjamin, New York, 1966.
18. E. Patai, ed., "The Chemistry of the Azido Group," Wiley-Interscience, New York, 1971.
19. Smith [17], p. 214.
20. M. E. Burns and R. H. Smith, Jr., *Chem. Eng. News* p. 2 (Jan. 9, 1984).
20a. M. E. Burns and R. H. Smith, Jr., *Chem. Eng. News* p. 2 (Dec. 16, 1985).
21. J. Zaloom and D. C. Roberts, *J. Org. Chem.* **46**, 5173 (1981).
22. W. Lwowski and T. W. Mattingly, Jr., *J. Amer. Chem. Soc.* **87**, 1947 (1965).
23. J. H. Hall and E. Patterson, *J. Amer. Chem. Soc.* **89**, 5856 (1967).
24a. Lonza, Inc., Fairlawn, New Jersey, "Sodium Azide, Determination of Hydrazoic Acid and Azide in the Atmosphere of Azide Plants" (October, 1983).
24b. Military Specification MIL-S-20552A, "Sodium Azide, Technical" (July 24, 1952).
25. M. T. Reagan and A. Nickon, *J. Amer. Chem. Soc.* **90**, 4096 (1968).
26. A. N. Nesmeyanor and M. I. Rybinskaya, *Izv. Akad. Nauk SSSR Otd. Khim. Nauk* 761, 816 (1962).
27. M. B. Frankel and D. O. Woolery, *J. Org. Chem.* **48**, 611 (1983).
28. M. O. Forster and H. E. Fierz, *J. Chem. Soc.* **93**, 72 (1908).
29. J. H. Boyer and J. Hamer, *J. Amer. Chem. Soc.* **77**, 951 (1955).
30. J. H. Boyer, F. C. Canter, J. Hamer, and R. K. Putney, *J. Amer. Chem. Soc.* **78**, 325 (1956).
31. E. Lieber, T. S. Chao, and C. N. R. Rao, *J. Org. Chem.* **22**, 238 (1957).
32. W. R. Carpenter, *J. Org. Chem.* **27**, 2085 (1962).
33. A. J. Papa, *J. Org. Chem.* **31**, 1426 (1966).
34. A. H. Sommers and J. D. Barnes, U.S. Pat. 2,769, 819 (Nov. 6, 1956).
35. A. H. Sommers and J. D. Barnes, *J. Amer. Chem. Soc.* **79**, 3491 (1957).
36. J. H. Boyer and L. R. Morgan, Jr., *J. Amer. Chem. Soc.* **81**, 3369 (1959).
37. G. Smolensky and C. A. Pryde, *J. Org. Chem.* **33**, 2411 (1968).
38. J. A. Van Allan, W. J. Priest, A. S. Marshall, and C. A. Reynolds, *J. Org. Chem.* **33**, 1100 (1968).
39. A. Oskerko, *Mem. Int. Chem. Acad. Sci. Ukr. SSR* **3**, 415 (1936).
40. D. H. R. Barton and L. R. Morgan, Jr. *J. Chem. Soc.* **622** (1962).
41. J. H. Boyer, C. H. Mack, N. Gaebel, and L. R. Morgan, Jr., *J. Org. Chem.* **23**, 1051 (1958).
42. O. C. Dermer and M. T. Edmison, *J. Amer. Chem. Soc.* **77**, 70 (1955).
43. J. E. Leffler and Y. Tsuno, *J. Org. Chem.* **28**, 902 (1963).
44. W. B. Hardy and F. H. Adams, U.S. Pat. 2,863,866 (1958).
45. J. Munch-Peterson, *Org. Syn. Coll. Vol.* **4**, 715 (1963).
46. J. B. Hendrickson and W. A. Wolf, *J. Org. Chem.* **33**, 3610 (1968).
47. C. W. MacMullen and G. R. Leader, U.S. Pat. 2,865,932 (1958).
48. E. W. Barrett and C. W. Porter, *J. Amer. Chem. Soc.* **63**, 3434 (1941).
49. J. H. Boyer and D. Straw, *J. Amer. Chem. Soc.* **75**, 2683 (1953).
50. J. H. Boyer and D. Straw, *J. Amer. Chem. Soc.* **75**, 1642 (1953).
51. J. H. Boyer and D. Straw, *J. Amer. Chem. Soc.* **74**, 4506 (1952).
52. J. H. Boyer, W. E. Krueger, and R. Modler, *J. Org. Chem.* **34**, 1987 (1969).
53. P. A. Smith and J. H. Boyer, *Org. Syn. Coll. Vol.* **4**, 75 (1963).

54. P. A. Smith and B. B. Brown, *J. Amer. Chem. Soc.* **73**, 2438 (1951).
55. P. A. Smith, G. J. W. Breen, M. K. Hajek, and D. V. C. Awang, *J. Org. Chem.* **35**, 2215 (1970).
56. G. Smolensky, *J. Org. Chem.* **26**, 4108 (1961).
57. R. Huisgen and I. Ugi, *Angew. Chem.* **68**, 705 (1956).
58. J. H. Hall, J. G. Stephanie, and D. K. Nordstrom, *J. Org. Chem.* **33**, 2951 (1968).
59. K. Baum, P. T. Berkowitz, V. Grakauskus, and T. S. Archibald, *J. Org. Chem.* **48**, 2953 (1983).
60. G. L'abbe and A. Hassner, *J. Org. Chem.* **36**, 258 (1971).
61. A. B. Belinka, Jr., A. Hassner, and J. M. Hendler, *J. Org. Chem.* **46**, 631 (1981).
62. M. Mitani, M. Takayana, and K. Koyama, *J. Org. Chem.* **46**, 2226 (1981).
63. P. A. S. Smith, *Org. React.* **3**, 382 (1946).
64. C. Grundmann and H. Haldenwanger, *Angew. Chem.* **62A**, 410 (1950).
64a. D. Knittel, H. Hemetsberger, and H. Weidmann, *Monatsh. Chem.* **101**, 157 (1970).
65. G. K. Surya Prakash, P. S. Iyer, M. Arvanaghi, and G. A. Olah, *J. Org. Chem.* **48**, 3358 (1983).
66. C. J. Cavender and V. J. Shiner, Jr., *J. Org. Chem.* **37**, 3567 (1972).
67. S. N. Eğe and K. W. Sherk, *J. Amer. Chem. Soc.* **75**, 354 (1953); A. L. Logothetis, *J. Amer. Chem. Soc.* **87**, 749 (1965).
68. C. G. Swain, C. B. Scott, and K. H. Lohmann, *J. Amer. Chem. Soc.* **75**, 136 (1953).
69. W. H. Saunders, Jr., and J. C. Ware, *J. Amer. Chem. Soc.* **80**, 3328 (1958).
70a. T. Sasaki, S. Eguchi, T. Katada, and O. Hiroaki, *J. Org. Chem.* **42**, 3741 (1977).
70b. T. Sasaki, S. Eguchi, M. Yamaguchi, and T. Esaki, *J. Org. Chem.* **46**, 1800 (1981).
70c. T. Sasaki, S. Eguchi, and T. Okano, *J. Org. Chem.* **46**, 4474 (1981).
70d. T. Sasaki, S. Eguchi, and T. Okano, *J. Org. Chem.* **49**, 444 (1984).
71. T. Sasaki, S. Eguchi, T. Okano, and Y. Wakata, *J. Org. Chem.* **48**, 4067 (1983).
72. D. Margosian and P. Kovacic, *J. Org. Chem.* **46**, 877 (1981).
73. D. Margosian, J. Speier, and P. Kovacic, *J. Org. Chem.* **46**, 1346 (1981).
74. R. A. Mustill and A. H. Rees, *J. Org. Chem.* **48**, 5041 (1983).
75. J. H. Hall, *J. Amer. Chem. Soc.* **87**, 1147 (1965).
76. L. Benati and P. C. Montevecchi, *J. Org. Chem.* **46**, 4570 (1981).
77. T. T. Ngo, C. F. Yam, H. M. Lenhoff, and J. Ivy, Jr., *Biol. Chem.* **256** (21), 11313 (1981).
78. L. L. Melhado and N. J. Leonard, *J. Org. Chem.* **48**, 5130 (1983).
79. A. Wohl and H. Schiff, *Ber. Dtsch. Chem. Ges.* **33**, 2741 (1900).
80. L. Gatterman and R. Ebert, *Ber. Dtsch. Chem. Ges.* **49**, 2117 (1916).
81. M. O. Forster, *J. Chem. Soc.* **107**, 260 (1915).
82. H. B. Schneider and H. Rager, *Monatsh. Chem.* **81**, 970 (1950).
83. P. A. S. Smith, *Org. React.* **3**, 272 (1946).
84. R. O. Lindsay and C. F. H. Allen, *Org. Syn. Coll. Vol.* **3**, 710 (1955).
85. A. Yakawa and Y. Tsuno, *J. Amer. Chem. Soc.* **79**, 5530 (1957).
86. R. A. Clement, *J. Org. Chem.* **27**, 1904 (1962).
87. L. C. Dorman, U.S. Patent 4,046,723 (Sept. 6, 1977).
88. F. W. Fowler, A. Hassner, and L. A. Levy, *J. Amer. Chem. Soc.* **89**, 2077 (1967).
89. A. Hassner and F. W. Fowler, *J. Org. Chem.* **33**, 2686 (1968).
90. G. R. Harvey and K. W. Rath, *J. Org. Chem.* **31**, 3907 (1966).
91. R. E. Schaad, U.S. Pat. 2,557,924 (1951).
92. J. H. Boyer, *J. Amer. Chem. Soc.* **73**, 5248 (1951).
93. C. H. Heathcock, *Angew. Chem.* **81**, 148 (1969).
94. J. Szmuszkovicz, M. P. Kane, L. G. Laurian, L. G. Chidester, and T. A. Scahill, *J. Org. Chem.* **46**, 3562 (1981).
95. G. Swift and D. Swern, *J. Org. Chem.* **31**, 4226 (1966).

96. J. D. Ingham, W. L. Petty, and P. L. Nichols, Jr., *J. Org. Chem.* **21**, 373 (1956).
97. G. A. Van der Werf, R. Y. Heisler, and W. E. McEwen, *J. Amer. Chem. Soc.* **76**, 1231 (1954).
98. Data taken from Ingham *et al.* [96].
99. E. Oliveri-Mandala and E. Calderas, *Gazz. Chim. Ital.* **45**, 307 (1915).
100. E. Oliveri-Mandala and A. Coppola, *Atti Accad. Lincei* **19**, 563 (1910).
101. E. Oliveri-Mandala and G. Coronna, *Gazz. Chim. Ital.* **71**, 182 (1941).
102. G. L'abbe, *Ind. Chim. Belge* **32**, 541 (1967).
103. S. P. McManus, M. Ortiz, and R. A. Abramovitch, *J. Org. Chem.* **46**, 336 (1981).
104. M. E. Hermes and F. D. Marsh, *J. Org. Chem.* **37**, 2969 (1972).
105. F. W. Fowler, A. Hassner, and L. A. Levy, *J. Amer. Chem. Soc.* **89**, 2077 (1967).
106. G. R. Harvey and K. W. Ratts, *J. Org. Chem.* **31**, 3907 (1966).
107. C. H. Grudmundsen and W. E. McEwen, *J. Amer. Chem. Soc.* **79**, 329 (1957).
108. W. J. Wechter, *J. Org. Chem.* **1**, 2136 (1966).
109. H. R. Moore, H. R. Shelden, and D. F. Shellhamer, *J. Org. Chem.* **34**, 1999 (1969).
110. G. B. L. Smith, *Inorg. Synth.* **1**, 77 (1939).
111. G. B. Field, Belg. Pat. 661,070 (1965).
112. G. B. Field, Brit. Pat. 996,350 (1965).
113. P. A. S. Smith, J. H. Hall, and R. O. Kan, *J. Amer. Chem. Soc.* **84**, 485 (1962).
114. G. Koga and J. P. Anselme, *J. Amer. Chem. Soc.* **91**, 4323 (1969).
115. K. Baum, *J. Org. Chem.* **33**, 4333 (1968).
116. C. Temple, Jr., C. L. Kussner, and J. A. Montgomery, *J. Org. Chem.* **31**, 2210 (1966).
117. J. A. Montgomery, *J. Org. Chem.* **30**, 2395 (1965).
118. P. A. S. Smith, C. D. Rowe, and L. B. Bruner, *J. Org. Chem.* **34**, 3430 (1969).
119. N. Wiberg, F. Raschig, and R. Sustmann, *Angew. Chem. Int. Ed. Engl.* **1**, 551 (1962).
120. C. G. Overberger and P. S. Yuen, *J. Amer. Chem. Soc.* **92**, 1667 (1970).
121. H. Wieland, *Chem. Ber.* **42**, 3021, 3025 (1909).
122. J. K. Senior, *J. Amer. Chem. Soc.* **38**, 2719 (1916).
123. F. Minisci, *Gazz. Chim. Ital.* **89**, 626 (1959).
124. H. H. Gibson, Jr., M. R. Macha, S. J. Farrow, and T. L. Ketchersid, *J. Org. Chem.* **48**, 2062 (1983).
125. A. P. Ahern, K. J. Dignam, and A. F. Hagarty, *J. Chem. Soc.* **'45**, 4302 (1980).
126. E. Bamberger, *Justus Liebigs Ann. Chem.* **424**, 233 (1921).
127. S. Maffli and L. Coda, *Gazz. Chim. Ital.* **88**, 1300 (1955).
128. M. W. Logue and B. H. Han, *J. Org. Chem.* **46**, 1638 (1981).

CHAPTER 14

AZO COMPOUNDS

1. INTRODUCTION

For perhaps a century, the primary interest in azo compounds has been in dye chemistry. Recently, aliphatic azo compounds, which are thermally less stable than their aromatic counterpart, have enjoyed attention as sources of free radicals for polymerization reactions. In this regard, an initiator such as α,α'-azobis(isobutyronitrile) is important for two reasons:

(1) Its thermolysis is evidently strictly first order and no complications arise because the induced decompositions so common with peroxidic initiators do not take place with this compound. This permits closer control of the formation of linear addition polymers.

(2) Since aliphatic azo compounds can, at best, act only as hydrogen abstracters in oxidation reactions, the color formation associated with oxygen addition is eliminated or reduced even in the presence of atmospheric oxygen, thus leading to polymers low in unwanted color.

With the increased research in the field of liquid crystals, some of which is based on the azoxy functional group, the methods for preparing azo compounds as intermediates for the synthesis of azoxy derivatives have taken on added importance.

Perhaps the best-known method of preparing aromatic azo compounds involves the coupling of diazonium salts with sufficiently reactive aromatic compounds such as phenols, aromatic amines, phenyl ethers, the related naphthalene compounds, and even sufficiently reactive aromatic hydrocarbons. Generally, the coupling must be carried out in media which are neutral or slightly basic or which are buffered in the appropriate pH range. The reaction may also be carried out in nonaqueous media. While some primary and secondary aromatic amines initially form an *N*-azoamine, which may rearrange to the more usual amino-*C*-azo compound, tertiary amines couple in a normal manner.

Under some conditions, phenolic ethers are dealkylated during coupling. However, the dealkylation follows the coupling step and is acid-catalyzed. Consequently, use of an excess of sodium acetate as a buffer or use of a nonaqueous medium obviates the dealkylation.

Two molecules of a diazonium salt may couple with loss of some nitrogen. In the Bogoslovskii reaction, this reaction has been developed as a means of preparing *o,o′*-dihydroxyazo compounds, which are difficult to obtain by other means. This reaction involves the use of a cuprous complex as the reaction catalyst. Self-coupling of diazonium salts also takes place in the presence of sodium sulfite.

Several intramolecular couplings of diazonium salts with ortho substituents bearing an active methylene group give rise to cinnolines, a class of cyclic azo compounds (the Borsche, von Richter, and Widman–Stoermer syntheses).

A very flexible method of preparing unsymmetrical azo compounds makes use of the condensation of *C*-nitroso compounds with amines. Thionylamines have also been condensed with substituted hydroxylamines to produce azo compounds not usually accessible by other means. Treatment of dialkylsulfuric diamides with sodium hypochlorite is one means of preparing aliphatic azo compounds. Aromatic amines and aromatic nitro compounds at high temperature produce azo compounds.

Diazonium salts may also couple with certain active methylene compounds—particularly with active methinyl compounds.

Conjugated olefinic azo compounds have been prepared by the reaction of 2 moles of a substituted hydrazine with an α-halocarbonyl compound.

Grignard reagents and diazonium zinc chloride double salts have been used to prepare azo compounds. The reaction of arylzinc chloride with a diazonium salt has also been used. Mixed aromatic aliphatic azo compounds have been prepared by reacting aliphatic zinc iodides with diazonium salts.

Olefins have been added to diazoalkanes to prepare 1-pyrazolines, a class of cyclic azo compounds.

The reductive methods of preparing azo compounds involve, as starting materials, aromatic nitro compounds, azoxy compounds, and azines.

The bimolecular reduction of nitro compounds is believed to involve reduction of some of the starting material to a nitroso compound and another portion to either a substituted hydroxylamine or an amine. These intermediates, in turn, condense to form the azo compound. The exact mechanism of the reaction requires critical study. On the one hand, reducing conditions are always on the alkaline side to prevent the benzidine rearrangement of an intermediate hydrazo compound under acidic conditions, yet it is difficult to visualize the formation of hydrazo compounds by the indicated condensation. As a practical matter, this method is of value only if symmetrically substituted azo compounds are desired.

A large variety of reducing agents have been proposed for this reduction. However, zinc and sodium hydroxide offer the most common system, and lithium aluminum hydride merits consideration. The reduction of azoxy compounds with lithium aluminum hydride has value mainly in structural determinations. Its importance in a preparative procedure is limited; normally such a reaction sequence would be a matter of putting the cart before the horse. The reduction of azines has potential value because of the accessibility of azines; unfortunately, only under specialized circumstances has it been possible simply to add the required gram-molecule of hydrogen to the structure. Usuaally, chlorine is added to an azine structure to produce dichloro azo compounds. An extension of the reaction permits the preparation of α,α'-diacyloxyazoalkanes from azines.

Among the oxidative procedures for preparing azo compounds are: oxidation of aromatic amines with activated manganese dioxide; oxidation of fluorinated aromatic amines with sodium hypochlorite; oxidation of aromatic amines with peracids in the presence of cupric ions; oxidation of hindered aliphatic amines with iodine pentafluoride; oxidation of both aromatic and aliphatic hydrazine derivatives with a variety of reagents such as hydrogen peroxide, halogens or hypochlorites, mercuric oxide, N-bromosuccinimide, nitric acid, and oxides of nitrogen.

The oxidation of hydrazine derivatives with diethyl azodicarboxylate is of particular interest because it involves direct hydrogen abstraction. The oxidation of keto hydrazones with lead tetraacetate leads to azoacetates, presumably by a free radical mechanism.

Hydrazones, in the presence of anhydrous potassium hydroxide, at a temperature below 120°C, have been isomerized to azo compounds. The reaction is usually carried out under reduced pressure by continuous distillation under carefully controlled heating conditions.

A normally stable trans-azo isomer can be converted into the cis isomer by irradiation with ultraviolet light. Separation of the isomers is carried out by column chromatography.

a. NOMENCLATURE

The variety of methods of naming azo compounds which has been in use for many years may lead to considerable confusion, especially when attempts are made to name structural formulas of highly substituted dye molecules with several azo linkages. Furthermore, in regard to the older dye literature, an intuitive interpretation of an author's intention frequently seems more productive than a detailed analysis of the system of nomenclature which he may be using.

There are about a half-dozen systems of designating azo compounds: an IUPAC system [1], an older *Chemical Abstracts* system used prior to volume 76, a newer *Chemical Abstracts* system (see *Chemical Abstracts*, Index Guide [2]), an old "diazene" system resembling the new *Chemical Abstracts* system, an old "diimine" system, and a practical system used in the dye industry.

(1) For monoazo compounds in which identical radicals are linked by the azo group, the IUPAC [1] and older *Chemical Abstracts* systems are in

general agreement that the prefix "azo" is simply added to the name of the parent molecule (Rules C-911.1 and C-912.1), e.g.,

$$CH_3-N{=}N-CH_3$$

(I) Azomethane   (II) Azobenzene

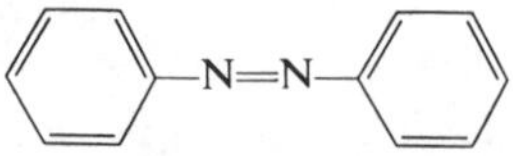

The new *Chemical Abstracts* name for this compound (II) would be diphenyldiazene (or, for indexing purposes: diazene, diphenyl-).

(2) Once substituents are introduced on even one aromatic ring, these two systems deviate. Thus structure (III) becomes azobenzene-4-sulfonic acid in

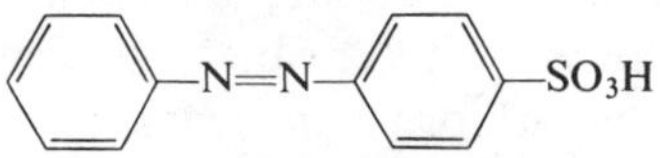

(III)

the IUPAC system (Rule C-911) and *p*-phenylazobenzenesulfonic acid in the old *Chemical Abstracts* system (IUPAC Alternate Rule C-912.3).

The new *Chemical Abstracts* name would be diazene, (phenyl)-(4-phenyl-sulfonic acid).

When the same number and kinds of substituents are carried by the two aromatic radicals, the IUPAC system simply numbers the substituents in a conventional manner, whereas the old *Chemical Abstracts* system names the compound as an assembly of identical units with the prefix "azodi-" preceding the name of the unsubstituted parent compounds. The new *Chemical Abstracts* name of compound IV may be identical to the one generated by the old rules since there are identical functional groups present which are expressible as suffixes.

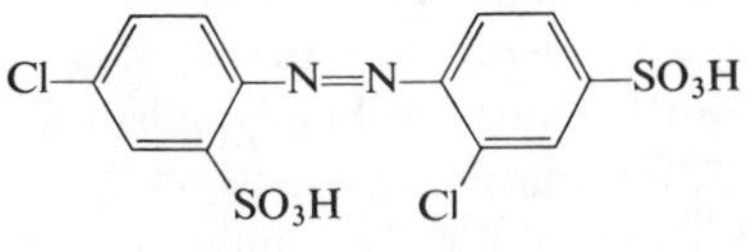

2′,4-dichloroazobenzene-2,4′-disulfonic acid
(IUPAC Rule C-911.1)
2′,4-dichloro-2,4′-azodibenzenesulfonic acid
(*Chemical Abstracts*, IUPAC Alternative Rule C-912.2)

(IV)

(3) When the two radicals attached to the azo group are derived from different parent molecules, the IUPAC system places the term "azo" between the complete names of the (substituted) parent molecules (Rule C-911.2). This system resembles the older "numbered azo bridge" system. The old *Chemical*

Abstracts system names the compound as a parent molecule RH substituted by a radical R′N=N— (Rule C-912.4). This system is particularly convenient for unsymmetrically substituted aliphatic azo compounds. The new *Chemical Abstracts* system evidently also uses this approach.

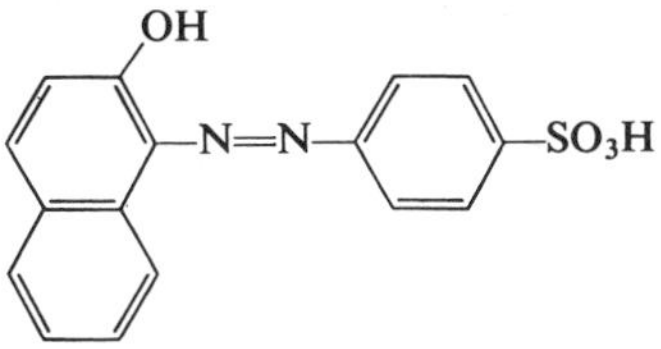

2-Hydroxynaphthalene-1-azo-4-benzenesulfonic acid
(IUPAC Rule C-911.2)
2-Hydroxynaphthalene⟨1-azo-4⟩benzenesulfonic acid
(numbered azo bridge system)
p-(2-Hydroxy-1-naphthylazo)-benzenesulfonic acid
(*Chemical Abstracts*, IUPAC Alternate Rule C-912.4).

(V)

As indicated, the old *Chemical Abstracts* name for structure VI would be ethylazocyclohexane. The newer name would be ethylphenyldiazene

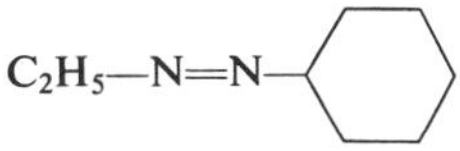

(VI)

(4) Two older systems of nomenclature name aliphatic azo compounds "diazenes" or "diimines" as in structure VII. The "diazene" system obviously is similar to the one curently advocated by *Chemical Abstracts*.

$$C_2H_5—N=N—CH_3$$

Ethyl methyl diazene   or
Ethyl methyl diimine

(VII)

(5) An interesting and practical system of nomenclature used in the dye literature denotes with a connecting arrow the compounds which may have been coupled to form an azo bridge (structure VIII). In connection with this method of naming azo dyes, it must be pointed out that, while the name usually indicates which reagent was diazotized (*p*-sulfanilic acid in this case) and with what reagent it was coupled (β-naphthol here), it does not indicate the exact location of the azo bridge in the final compound, a matter of little concern to

the dye manufacturer as long as he obtains his desired dye reproducibly. Since, in principle, the same compound may be produced by the reaction of a

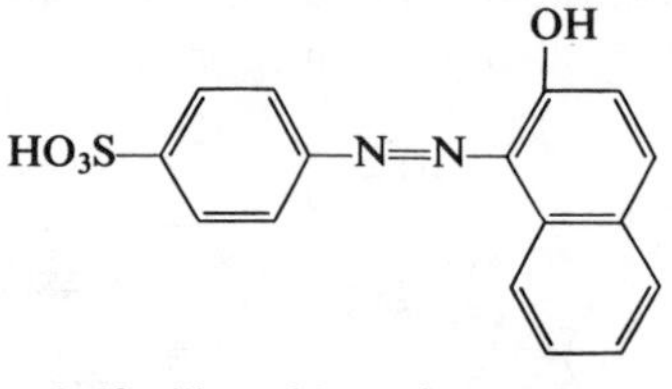

p-Sulfanilic acid → β-naphthol

(VIII)

variety of diazotized amines with different dye couplers, this system can lead to a great variety of names for the same compound, a matter which is compounded by the general use of the trivial names of the reagents.

Some recent authors working with cyclic azo compounds have found it convenient to name their compounds "diaza cycloolefins". For example, Overberger and Merkel [3] refer to a cis-3,8-dimethyl-1,2-diaza-(Z)-1-cyclooctene.

b. REVIEWS

The chemistry of azo compounds has been investigated by Overberger and co-workers since 1949. Reference [3] is the 53rd in the series. Patai edited an extensive review [4]. Representative references with particular emphasis on the dye aspects of azo compounds are [5-11].

c. SAFETY NOTE

While azo dyes have been in use for many years, only recently have questions of health hazards associated with dyes generally been raised. Derivatives of p-dimethylaminoazobenzene (DMAB, butter yellow) are known carcinogens [12, 13]. Consequently, we suggest that other azo compounds, especially aromatic azo compounds, be handled with great caution.

2. CONDENSATION REACTIONS

A. Coupling of Diazonium Salts with Aromatic Compounds

The coupling of aromatic diazonium salts with a variety of aromatic compounds is the basis of the azo dye industry. A variety of dyeing techniques are available but, fundamentally, two procedures are involved:

(1) Application of a solution of an azo dye to the textile followed by removal of the solvent.

(2) Formation of the dye in or on the textile fiber by treatment, in turn, with a coupling agent and a diazonium salt solution.

An interesting variation of the latter technique finds application in enzyme chemistry. In this procedure a tissue section is exposed to a relatively colorless derivative of β-naphthol, such as sodium β-naphthyl acid phosphate. A phosphotase enzyme reacts with this reagent (often called an "enzyme substrate"), leaving free β-naphthol behind. Subsequent treatment with a solution of a diazonium salt produces highly colored spots in the tissue section. Thus not only can the presence of phosphotase enzyme be demonstrated, but also the location of the enzyme in the tissue can be determined. The intensity and chroma of the color produced and the solubility of the azo dye in the cell materials can be varied by judicious selection of the reagents.

Generally the coupling of diazonium salt is carried out under neutral to slightly alkaline conditions. In coupling with amines or phenols, it has been demonstrated that the active species are the diazonium cation, the free amine, or the phenoxide ion [14, 15]. The fact that coupling does not require a diazonium hydroxide is demonstrated by Preparation 2-1, which is carried out in a nonaqueous medium.

2-1. *Preparation of 1-Phenylazo-2-naphthol (in a Nonaqueous Medium)* [15]

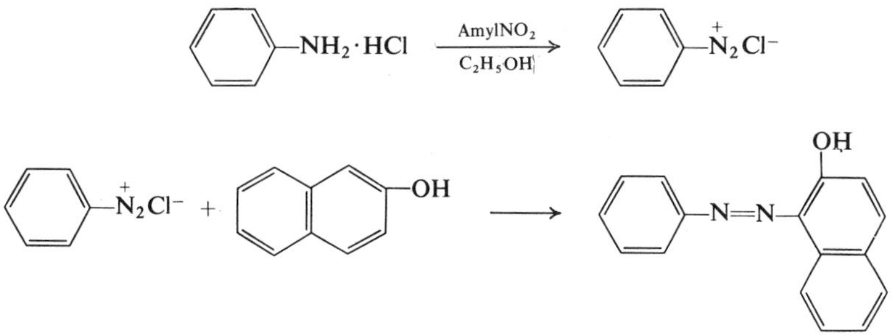

(a) *Preparation of phenyldiazonium chloride in nonaqueous medium.* In a hood, behind a shield, 3.5 gm (0.027 mole) of aniline hydrochloride is dissolved in 20 ml of absolute ethanol. To this solution is added 0.5 ml of a saturated solution of hydrogen chloride in absolute ethanol. The solution is cooled and stirred in an ice bath, and 4 gm of isoamyl nitrite is added dropwise while maintaining the reaction temperature between 0°C and 5°C. After the addition has been completed, the reaction system is allowed to warm to room temperature and to stand at room temperature for about 10 min. The diazonium salt is then precipitated by the addition of absolute ether.

CAUTION: From this point on, care must be taken never to allow the diazonium salt to become completely dry (*explosion hazard*).

In a hood, behind a shield, the salt is rapidly removed by gravity filtration, washed with 5 ml of a 1:1 (v:v) mixture of absolute ether and absolute alcohol and then with 10 ml of ether. For the next step the moist product is added to pyridine as described below.

(*b*) *Coupling reaction.* With suitable precautions, the moist diazonium salt prepared above is suspended in 25 ml of dry pyridine and chilled in an ice bath. To the stirred mixture is added a solution of 3.6 gm (0.025 mole) of β-naphthol in 25 ml of pyridine. The reaction mixture is stirred in the ice bath for 1 hr, warmed to room temperature, and then stirred at room temperature for an additional hour (deep-red solution).

Under a hood, the reaction mixture is poured cautiously, with stirring, over a mixture of 50 ml of hydrochloric acid and 300 gm of ice. After the ice has melted 600 ml of water is added and the product is removed by filtration, dried, and recrystallized from ethanol. Yield 2.5 gm (48.5% based on recovery of 0.6 gm of unreacted β-naphthol from the mother liquor), m.p. 121°–122.5°C.

A similar yield is obtained when sodium β-naphthylate is substituted for β-naphthol in the reaction above.

If the diazonium salt is derived from alkoxyanilines in which the alkoxy moiety contains heptyl, decyl, or cetyl groups and the anion is fluoroborate, perchlorate, or *p*-toluenesulfonate, the salt is soluble in organic solvents such as benzene or ether and the azo compounds derived therefrom may be prepared in nonaqueous systems (16).

2-2. *Preparation of 1-(p-n-Decyloxyphenylazo)-2-naphthol* [16]

$$\text{CH}_3(\text{CH}_2)_9\text{O} - \underset{}{\bigcirc} - \overset{+}{\text{N}}_2\,\text{X}^- \;+\; \text{(2-naphthol with OH)} \;\longrightarrow$$

$$\text{CH}_3(\text{CH}_2)_9\text{O} - \bigcirc - \text{N}{=}\text{N} - \text{(naphthol with OH)} \qquad (2)$$

In a hood, behind a shield, to a solution of 1.5 gm (0.0104 mole) of β-naphthol in 100 ml of dry benzene containing 1.7 ml of dry pyridine maintained at 55°C is added a solution of 4.3 gm (0.103 mole) of *p-n*-decyloxybenzenediazonium toluene-*p*-sulfonate in 50 ml of dry benzene maintained at 60°C. The mixture is kept overnight. Then the precipitated pyridinium toluene-*p*-sulfonate is filtered off. The filtrate is cautiously evaporated to dryness. On recrystallization of the residue from ethanol, 3.4 gm of 1-(*p-n*-decyloxyphenylazo)-2-naphthol is isolated, m.p. 73°–76°C, yield 85%. On further recrystallization the melting point may be raised to 78°–79°C.

As might be expected, the stability and reactivity of diazonium salts are affected by substituents as well as by the anion. The rate of coupling and the location of the azo bond are influenced by substituents on the coupling reagent. Some primary and secondary aromatic amines derived from benzene initially form *N*-azo compounds, which rearrange to *p*-aminoazo compounds in acidic media [8b, c]. Tertiary amines, on the other hand, behave normally. Preparation 2-3, while of a rather complex molecule, illustrates the simple techniques commonly used in coupling in an aqueous system. Note the presumed preferential tendency of coupling to take place predominantly in the para position.

2-3. *Preparation of 4,4'-[Oxybis(4-nitro-o-phenyleneazo)]bis(N,N-dimethylaniline)* [19]

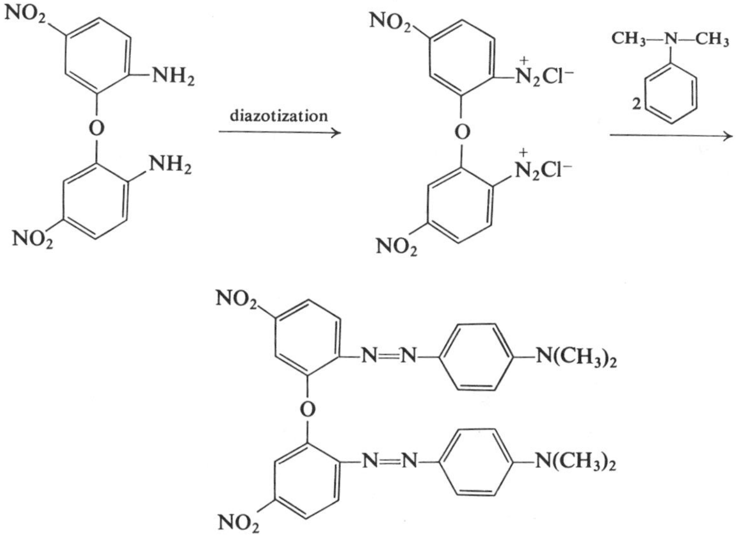

$$(3)$$

To 1.1 gm (0.0038 mole) of bis(2-amino-5-nitrophenyl) ether dissolved in 10 ml of 5 *N* hydrochloric acid and cooled to 0°–5°C by adding ice to the solution is slowly added 8 ml of a 0.1 *N* solution of sodium nitrite (0.008 mole) while maintaining the temperature between 0° and 5°C. After the addition has been completed (see Volume I, Chapter 15, Diazo and Diazonium Compounds, for notes and precautions in carrying out diazotizations), stirring is continued for 1 hr (see Wistar and Bartlett [14] on the sluggishness of some diazotizations).

Meanwhile, 1 gm (0.0083 mole) of dimethylaniline is dissolved in dilute hydrochloric acid. After the diazotization has been completed, the dimethylaniline hydrochloride solution is added to the diazonium salt solution in one portion. By use of solid sodium acetate, the pH is rapidly adjusted to approximately 5, and stirring is continued until coupling has been completed (see Bunnett·and Hoey [20] on the slow rate of some coupling reactions).

[The completeness of coupling may be checked by the "R-salt" test as follows: A few drops of the clear supernatant liquid is added to a small quantity of a solution of "R-acid" (β-naphthol-3,6-disulfonic acid) in an excess of 2 N sodium hydroxide. An intense red color is produced if unreacted diazonium salts are present. Obviously this test is satisfactory only if R-acid couples more rapidly with the diazonium salt than the coupling agent involved in the reactions and if the change to a red color is not obscured by other colors present in the reaction system.]

After the coupling has been completed, the product is removed by filtration, air-dried, and recrystallized from toluene. The red crystalline product has m.p. 217°–219°C dec. The yield is said to be quantitative.

Recently, 4′-aminobenzo-18-crown-6 was diazotized and coupled to N-n-butylaniline in an aqueous system. The yield was reported to be 9.9% [20a].

Among the solvents suggested for azo coupling reactions, aside from aqueous and specialized nonaqueous systems as mentioned above, are mixtures of water with water-soluble alcohols (e.g., methanol, ethanol, propanol), other water-soluble solvents such as tetrahydrofuran, N,N-dimethylacetamide, N,N-dimethylformamide, and organic acids such as formic, acetic, and propionic acids [21].

There are conflicting reports in the literature in regard to the coupling of diazonium salts with phenolic ethers. Whereas in many cases the expected product was isolated, in others dealkylation of the ether linkage was observed. The question naturally arises whether the cleavage of the ether link takes place before or after the coupling step. It has been shown that dealkylation takes place after coupling and that it is acid-catalyzed and can be controlled by coupling in a system buffered with an excess of sodium acetate or by working in a strictly anhydrous medium [20]. In Preparation 2-4, the ethoxy unit is retained because the reaction is carried out in a specially dried acetic acid. Presumably, in the presence of significant levels of water during the coupling step, the corresponding azonaphthol would be isolated.

2-4. Preparation of Ethyl 4-(p-Nitrophenylazo)-1-naphthyl Ether [20]

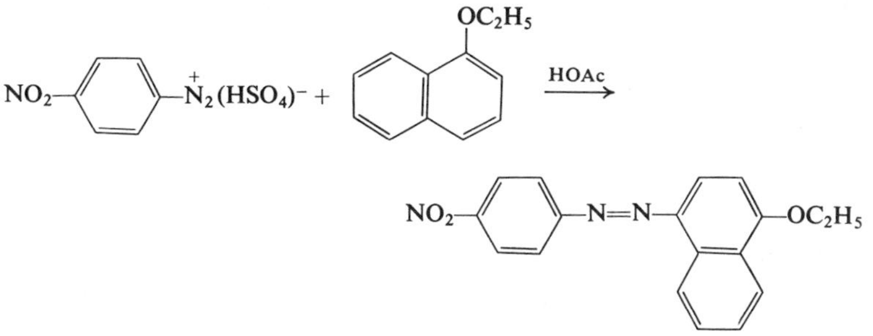

(a) *Preparation of stock acetic acid.* In a reflux apparatus protected against atmospheric moisture with calcium chloride, a solution of 60 ml of acetic anhydride and 1 liter of glacial acetic acid is refluxed for several hours. The resulting solution is cooled, stored, and dispensed without exposure to atmospheric moisture (openings to the room are protected with drying tubes filled with crushed alumina).

(b) *Diazotization.* In an apparatus protected against moisture, a cooled and stirred mixture of 6.0 gm (0.044 mole) of p-nitroaniline, 75 ml of the stock acetic acid, and 24 ml of concentrated sulfuric acid is treated with 3.4 gm (0.049 mole) of solid sodium nitrite. Stirring is continued for some time until diazotization appears to be complete. The reaction mixture is cooled to 0° to 5°C.

(c) *Coupling.* To a stirred, cooled solution of 7.2 gm (0.042 mole) of ethyl 1-naphthyl ether in 25 ml of stock acetic acid is added the cooled diazonium salt solution prepared above while maintaining the temperature with an ice-water bath. A purple precipitate forms. Stirring is continued at the ice temperature for 15 min. Then the mixture is allowed to warm to room temperature (27°C) and maintained at this temperature for 80 min.

Then the reaction mixture is poured into a mixture of 100 gm of clean ice and 300 ml of water. After the ice has melted, the deep-orange solid is collected by filtration, washed free of acids with water, air-dried, and crystallized from ethanol. The final product is a deep-red crystalline material, m.p. 154°–156°C; yield 108 gm (80%).

The melting points of some related nitrophenylazo-1-naphthalene derivatives are given in Table I.

Azo coupling of diazonium salts with aromatic (or pseudo-aromatic) hydrocarbons is possible if the coupling agent is highly substituted. For example, azo compounds have been produced from pentamethylbenzene [22], benzpyrene [23], and azulene [24].

TABLE I

Melting Points of Nitrophenylazo-1-naphthalene Derivatives [20]

Compound	M.p. (°C)
4-(p-Nitrophenylazo)-1-naphthol	287–289 (dec.)
4-(p-Nitrophenylazo)-1-naphthylamine	249–251
Methyl 4-(p-nitrophenylazo)-1-naphthyl ether	168–169.5
Ethyl 4-(p-nitrophenylazo)-1-naphthyl ether	154–156
Isopropyl 4-(p-nitrophenylazo)-1-naphthyl ether	133.5–135

A number of observations have been made in which two molecules of a diazonium salt couple with each other with loss of one molecule of nitrogen. In this reaction the azo bridge forms by displacement of one of the diazonium salt groupings by the entering group.

Perhaps the first observation of this reaction was made in connection with by-product formation of the Sandmeyer reaction in which a diazonium salt is treated with cuprous complexes to afford some azobenzene [25, 26].

By careful preparation of the catalyst system, a useful method of synthesis has been developed by Bogoslovskii [27]. The method lends itself to the preparation of o,o'-dihydroxyazo compounds which could not be prepared by other methods. Incidentally, in the Bogoslovskii reaction the product may separate as a stable copper chelate, and a step must be included in the preparation to decompose the chelated complex and free the product of residual copper salts. Preparation 2-5 illustrates the technique.

2-5. Preparation of 2,2'-Dihydroxyazobenzene (Bogoslovskii Reaction) [28]

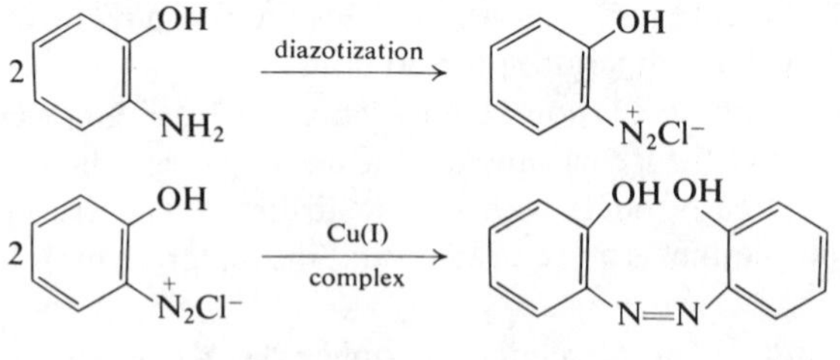

$$(5)$$

$$(6)$$

(a) Preparation of catalyst stock solution. Copper(II) sulfate pentahydrate, 28.5 gm, is dissolved in 100 ml of hot water, cooled to room temperature, and then enough concentrated aqueous ammonia is added dropwise until the soluble complex has been formed completely. To this solution is added a solution of 7 gm of hydroxylamine hydrochloride in 20 ml of water to reduce all of the copper(II) complex to the colorless copper(I) complex.

(b) *Diazotization and coupling.* To a solution of 11.0 gm (0.1 mole) of o-aminophenol and 7.0 gm (0.1 mole) of sodium nitrite in 250 ml of 5 % aqueous sodium hydroxide is dropwise added concentrated hydrochloric acid until a positive reaction with starch sodium iodide test paper is observed. The solution of the diazo compound is added rapidly with stirring to the catalyst solution contained in a large vessel. The excess foaming may be controlled by the addition of a small quantity of ether. The reaction mixture is allowed to stand at room temperature with occasional stirring for 1 hr. Then the brown solid is removed by filtration.

The resultant chelated complex is mixed with 500 ml of concentrated hydrochloric acid and warmed gently. The mixture is cooled, diluted with ice water, and filtered. The product is recrystallized three times from benzene to give 5.7 gm (53 %) of yellow-orange needles, m.p. 172°–172.7°C.

The yield may be increased to 78 % by continuous liquid–liquid extraction of the dilute hydrochloric acid filtrate solution with ether.

An old reaction patented by Lange [29] involves the coupling of naphthyl-diazonium salts with sodium sulfite to yield azonaphthalene. A more recent study of this reaction indicates that the first step of the reaction is the formation of an aryl-*syn*-diazosulfonate, which couples with another molecule of a diazonium salt and by a multicentered rearrangement ultimately affords an azo compound [30, 31]. The reaction is represented in Eq. (7). The validity of

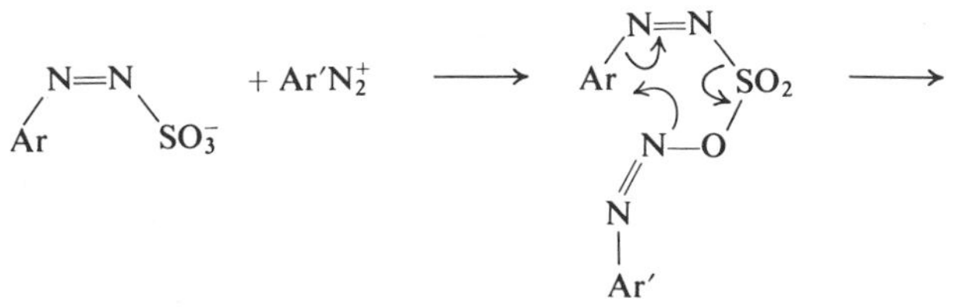

$$Ar{-}N{=}N{-}Ar' + N_2 + SO_2 \qquad (7)$$

this mechanism was demonstrated by use of heavy nitrogen [31]. The reaction permits the formation of unsymmetrical azo compounds, particularly in the naphthalene series. Unfortunately, experimental details are lacking in the more recent reports. The reaction conditions should be studied in greater detail, since diazonium salts are also reduced to arylhydrazines by sodium sulfite.

B. Coupling of Diazonium Salts with Active Methylene Compounds

Diazonium salts are capable of reacting with active methylene compounds. In the case of linear compounds, isomerization usually takes place to afford the corresponding hydrazones [32]. We believe that, since the azo-hydrazone tautomerism appears to be pH dependent (see below), it is possible that the normal reaction conditions favor formation of hydrazones.

In the case of active methinyl compounds, which, of course, lack the necessary mobile hydrogen, coupling with diazonium salts results in azo compounds.

The related Japp-Klingemann reaction, which involves addition to an active methinyl compound with subsequent loss of a group, normally leads to hydrazones [33].

The preparation of 2-phenylazo-1,3-indandione (m.p. 192°–193°C) by condensation of benzenediazonium chloride with 1,3-indandione is an example of azo formation from an active methylene compound [34]. With an active methinyl compound such as the methyl 1,3-indandione-2-carboxylate, the corresponding 2-arylazo-1,3-indandione-2-carboxylate ester has been prepared [35] (Eq. 8).

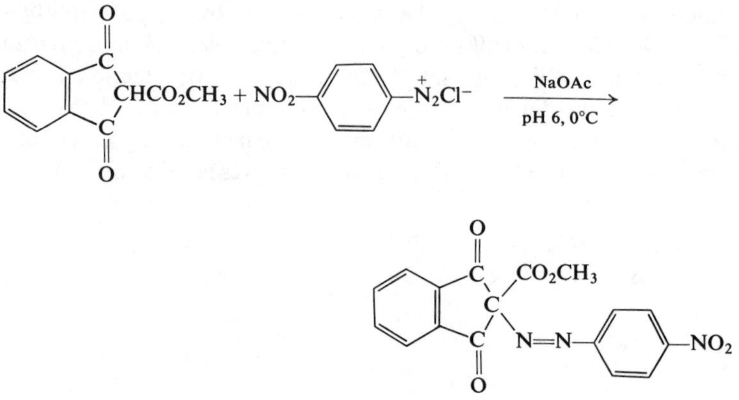

Methyl 2-(4-nitrophenylazo)-
1,3-indandione-2-carboxylate

Yield   60%
M.p.   155°C

In the patent literature a procedure is given for coupling a diazonium salt with ethyl α-methylacetoacetate to afford an azo compound which could be used as an acetylating agent because its acetyl group was readily cleaved under mild conditions [36]. In the coupling reaction, relatively weak bases such as pyridine, picolines, quinoline, and collidine are said to be effective. The example is given here for illustrations only. It is to be noted that this is a "composition of matter" patent.

2-6. *Preparation of Ethyl α-Phenylazo-α-methylacetoacetate* [36]

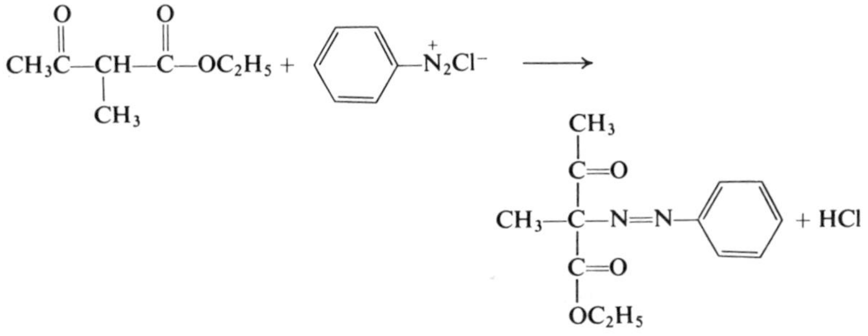

$$
\underset{\underset{CH_3}{|}}{\overset{\overset{O}{\|}}{CH_3C}-CH}-\overset{\overset{O}{\|}}{C}-OC_2H_5 + \langle\!\!\!\bigcirc\!\!\!\rangle-\overset{+}{N_2}Cl^- \longrightarrow
$$

$$
CH_3-\underset{\underset{OC_2H_5}{|}}{\overset{\overset{\overset{CH_3}{|}}{\overset{C=O}{|}}}{C}}-N{=}N-\langle\!\!\!\bigcirc\!\!\!\rangle + HCl \qquad (9)
$$

(a) *Preparation of diazonium salt.* To a solution of 13 gm (0.14 mole) of aniline dissolved in 120 ml of 5N hydrochloric acid and cooled to 0°C is slowly added a cooled, saturated solution of sodium nitrite until the equivalent of 10 gm of pure sodium nitrite has been added. This solution is kept cold until used.

(b) *Coupling reaction.* To a solution of 20 gm (0.14 mole) of ethyl α-methylacetoacetate dissolved in a mixture of 50 ml of pyridine and 150 ml of water and cooled to 0°C is added dropwise with constant stirring and while maintaining a reaction temperature of 0°C the diazonium salt solution prepared above. When the addition has been completed, a yellow oil forms on the bottom of the flask. The cold supernatant liquid is extracted with ether. The extract is combined with the yellow oil and washed in turn with very dilute ice-cold acetic acid and with several portions of ice water. The ether solution is then dried over magnesium sulfate and filtered. The filtrate is evaporated under reduced pressure, leaving a crude yellow liquid as residue. This residue is distilled under reduced pressure, and the fraction boiling between 138°C and 149°C (4 mm Hg) is collected; yield 27 gm (78 %).

Intramolecular coupling of diazonium salts with ortho substituents bearing an active methylene grouping or its equivalent gives rise to cinnolines, which may be considered a class of cyclic azo compounds. Three name reactions are cited here for reference only.

(1) The Borsche synthesis [37]

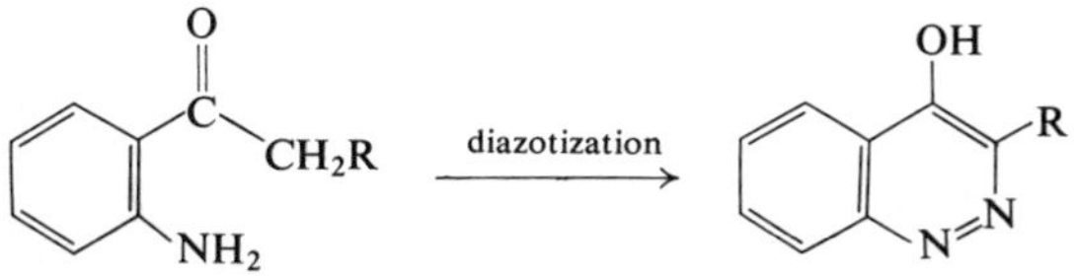

$$
\text{(structure)} \xrightarrow{\text{diazotization}} \text{(structure)} \qquad (10)
$$

(2) The von Richter synthesis [28]:

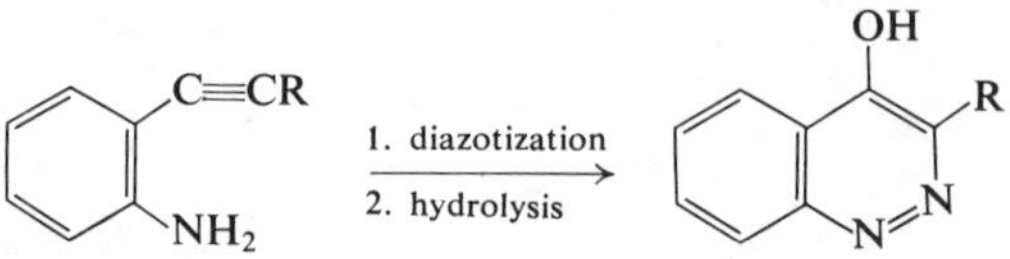

$$(11)$$

(3) The Widman–Stoermer synthesis [29, 30]:

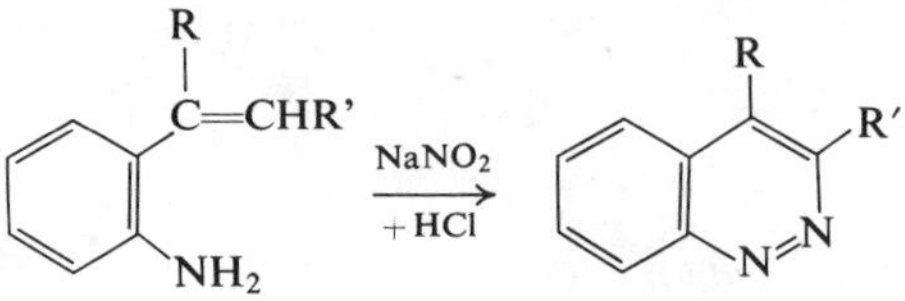

$$(12)$$

C. Condensation of Nitroso Compounds with Amines

A convenient synthetic procedure for the preparation of azo compounds, particularly unsymmetrically substituted ones, involves the reaction of aromatic nitroso compounds with aromatic amines [41–46]. The reaction is of additional interest because the replacement of amines by the corresponding hydroxylamine leads to the formation of the related azoxy compounds.

2-7. Preparation of m-Nitrophenylazobenzene [41]

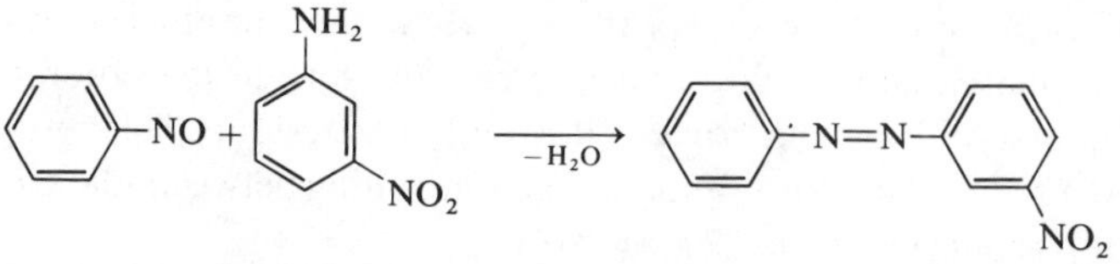

$$(13)$$

To a solution of 27.6 gm (0.2 mole) of m-nitroaniline in 240 ml of glacial acetic acid maintained at 40°C is added, with shaking, 23.3 gm (0.22 mole) of nitrosobenzene. The reaction mixture is allowed to stand in the dark for 2 days. Then the crude product is removed by filtration (31 gm, m.p. 94°C). To the mother liquor is added 80 ml of water, and another 5 gm of crude product precipitates. Greater dilution must be avoided to prevent tar formation. Yield 80%. The product may be recrystallized from ethanol, m.p. of pure m-nitrophenylazobenzene 96°C.

In this preparation the nitroso compound may be added in solution form to a solution to the amine at low temperature. The solvent may be a suitable mixture of acetic acid and ethanol as well as acetic acid alone. The product may also be extracted from a reaction mixture diluted with water by use of ether

[47]. Purification of the final product may be carried out by chromatography on an alumina column.

The reaction seems to have wide applicability and has been used even to prepare fluorinated compounds, such as 3,5-difluoro-4-N-methylaminoazobenzene [48, 49].

Although the bulk of azo chemistry involves aromatic systems, 1,1,1-trifluoromethylazomethane (b.p. 2.6°C) has been prepared by the condensation of methylamine with trifluoronitrosomethane [50]. An extension of the reaction to other aliphatic systems would be of considerable interest. A study of several benzylamines with nitrosobenzene in solvents such as diethyl ether, benzene, and dimethyl sulfoxides showed that azoxybenzenes and/or imines, possibly followed by amine exchange, give substituted imines [44].

Reaction of aliphatic amines with nitrosobenzenes is still considered to give contradictory results. Quite recently, aqueous methylamine or aqueous ethylamine formed the corresponding (phenol)azoalkanes in ether or chloroform, respectively, when reacted with nitrosobenzene. The nature of the alkyl group on NH_2- was thought to govern the distribution of the possible products [46].

The reaction of amines with nitroso compounds appears to be more rapid when carried out in the presence of strong bases than acids. An example using potassium t-butoxide in dimethyl sulfoxide–t-butanol (80:20) is given in Preparation 2-8.

2-8. Preparation of o,o,p-Trideuterioazobenzene (Azobenzene-d₃) [51]

$$(14)$$

In an apparatus through which a gentle stream of nitrogen is flowing, to a solution of 450 mg (4 mmoles) of potassium t-butoxide in 20 ml of a solvent mixture of 80% of dimethyl sulfoxide and 20% of t-butyl alcohol is added a solution of 190 mg (1.8 mmoles) of nitrosobenzene and 170 mg (1.8 mmoles) of o,o,p-trideuterioaniline in 5 ml of the same solvent solution. The reaction mixture is shaken for 5 min, then poured into 50 ml of water. The precipitate is separated by shaking the diluted reaction mixture with ether, separating the ether layer, and evaporating the ether layer to dryness. The product is recrystallized from methanol (yield 130 mg). An additional 90 mg of product is isolated by evaporation of the methanol solution. The total product of 220 mg of azobenzene-d_3 had m.p. 65°–66°C, yield 67–70% (based on two preparations).

In the case of heterocyclic amines that are difficult to diazotize (e.g., the 2-
and 4-aminopyridines), the preparation of their azo derivatives may be
carried out by condensation with nitrosoaryl compounds by forming the
sodio derivative of the amine. Alternatively, disodio derivatives of p-nitrodi-
methylaniline have been condensed with 2- and 4-aminopyridines [43].
Taylor and co-workers have prepared azo dyes by condensing nitroso
derivatives of heterocyclic compounds with aromatic amines. Table II gives
physical properties of some aryl azo heterocyclic compounds prepared by
these methods.

An example of the condensation of a nitrosoheterocyclic compound with
an aromatic amine is the preparation of 2-[(p-chlorophenylazo]-4-methyl-
pyridine.

2-9. Preparation of 2-[(p-chlorophenyl)azo]-4-methylpyridine [45]

$$(15)$$

With suitable safety precautions, in a hood, in a 100-ml Erlenmeyer flask
fitted with a magnetic stirrer, a solution of 0.244 gm (2.0 mmole) of 4-methyl-
2-nitrosopyridine in 35 ml of methylene chloride containing 2 drops of glacial
acetic acid is stirred with 0.279 gm (2.0 mmole) of p-chloroaniline at room
temperature for 12 hr. Then the solvent is removed by evaporation under
reduced pressure.

The residue is stirred with 50 ml of a solution of hexane and chloroform
(2.5:1). The mixture is filtered. The filtrate is concentrated and placed on a
short column of silica gel. The product is eluted with chloroform. The
product solution is evaporated and the residue is recrystallized from petroleum ether. Yield: 0.76 gm (78%), light yellow crystals; m.p. 93°–94°C; NMR
δ 2.50 (s, 3H, CH$_3$), 7.10–7.35 (m, 1H, H(3)), 7.35–7.75 (m, 3H), 8.03 (d, 2H,
$J = 9$ Hz), 8.62 (d, 1H, H96, $J = 4$ Hz).

Table II lists a number of other heterocyclic amine-based azo compounds
prepared by similar techniques. In this connection it should be noted that
some of the required nitroso compounds may be prepared in situ by the
oxidation of the corresponding S,S-dimethylsulfilimines with m-chloroperbenzoic acid. The S,S-dimethylsulfilimines are prepared by reacting an
aminoheterocyclic compound with dimethyl sulfide in methylene chloride at
−20°C with N-chlorosuccinimide followed by the addition of a methanolic
solution of sodium methoxide [45].

TABLE II

PROPERTIES OF SUBSTITUTED PHENYLAZO HETEROCYCLIC COMPOUNDS

Compound	M.p. (°C)	Yield (%)	Ref.
2-[(p-Dimethylaminophenyl)azo]pyridine	111–112	47.3[a], 52.3[b]	43
4-[(p-Dimethylaminophenyl)azo]pyridine	207–209	44.3[a], 53.8[b]	43
2-[(p-Dimethylaminophenyl)azo](3-methylpyridine)	158–160	45.8[a], 57.8[b]	43
4-[(p-Dimethylaminophenyl)azo](3-methylpyridine)	151–153	45.8[a], 53.5[b]	43
2-[(p-Dimethylaminophenyl)azo](5-methylpyridine)	154–147	42.5[a], 55.6[b]	43
2-[(p-Dimethylaminophenyl)azo](6-methylpyridine)	107–108	45.8[a], 61.9[b]	43
2-[(p-Dimethylaminophenyl)azo](6-methylpyridine)	210–213	31.2[a]	43
2-[(p-Chlorophenyl)azo]pyridine	115–118, 114–117	44.8[b,c]	43, 45
2-[Phenylazo]pyridine	Oil	41[b]	43
2-[(p-Methylphenyl)azo]pyridine	72–74	42.5[b]	43
2-[(p-Methylphenyl)azo]pyridine	50–52, 51–52	55.3[b], 70[c]	43–45
2-[(p-Chlorophenyl)azo]-4-methylpyridine	43–44	78[c]	45
2-[(p-Chlorophenyl)azo]pyrazine	133–134	63[c]	45
2-[(p-Methoxyphenyl)azo]-4-methylpyridine	84–85	83[c]	45
1-[(p-Methoxyphenyl)azo]isoquinoline	54–55	54[c]	45
2-[(p-Methoxyphenyl-azo]pyrimidine	105–109	32[c]	45
2-[(p-Methoxyphenyl)azo]pyrazine	116–117	64[c]	45

[a] By reaction of sodioaminopyridine.
[b] By reaction of disodium nitro derivatives.
[c] By condensation of a nitroheterocyclic compound with an amine.

D. Reaction of Thionylamines with Substituted Hydroxylamines

While the condensation of amines with nitroso compounds appears to have wide applicability in the benzene series, it seems to lead to complex dye molecules in the naphthalene series. A method has been developed using a somewhat complex reaction between thionylamines and substituted hydroxylamines to produce azo compounds derived from naphthalenes. This synthesis is of particular interest because it helped to settle the question whether true naphthylazo compounds with hydroxyl groups could exist. [52].

Although the yield of the thionylamines itself is not always high, its reaction with the hydroxylamine is one of high yield.

The mechanism of the reaction is complex. The equation for the reaction presented here is based on those of Michaelis and Petou for the preparation of simpler azo compounds [53].

In the preparation of 2-phenylazo-3-naphthol it is to be noted that the action of thionyl chloride on 2-amino-3-naphthol results in preferential attack on the amino group. This has been attributed, at least in part, to the fact that the

372 14. Azo Compounds

hydroxyl group is in an ortho position. If the reaction is to be extended to aminonaphthols which do not have ortho hydroxyl groups, provisions will probably have to be made for the protection of the hydroxyl group against attack by thionyl chloride.

The preparation itself we are presenting here is substantially as given in the cited literature. We suggest several points for investigation to improve the thionylamine synthesis.

(1) Since hydrogen chloride is generated in the thionylamine formation, we suggest that either 3 moles of amine be reacted with each mole of thionyl chloride or that 2 moles of a strong base, such as a tertiary amine, be present to scavenge the 2 moles of hydrogen chloride which form. The fact that the yield is reported to be 30% implies to us that two-thirds of the 2-amino-3-naphthol were used up as an acid scavenger.

(2) The thionyl chloride should be of the highly purified, colorless variety, which has been carefully freed from dissolved hydrogen chloride.

(3) Since both o- and p-aminohydroxy compounds are powerful reducing agents, they should be allowed to react under an inert atmosphere.

(4) The order of addition should probably be reversed, i.e., the solution of 2-amino-3-naphthol in benzene should be reasonably dilute and be added slowly to warm thionyl chloride. The HCl scavenger may either be dissolved directly in the benzene solution or added separately at an appropriate rate.

2-10. Preparation of 2-Phenylazo-3-naphthol [52]

(16)

+ amine hydrochloride

(17)

With suitable safety precautions, to a solution of 8 gm (0.05 mole) of 2-amino-3-naphthol (m.p. 225°C) in 100 ml of benzene is added 6 gm (0.03 mole) of freshly distilled thionyl chloride. The mixture is refluxed for 3 hr. The reaction mixture is then cooled and filtered through a fritted-glass funnel. The filtrate is evaporated to dryness to leave 4 gm of an olive-brown residue of the hydroxythionylnaphthylamine (yield 30%).

The crude product is dissolved in 150 ml of benzene and heated with 4.3 gm of phenylhydroxylamine. The dark-brown precipitate is filtered off. The filtrate is dried over anhydrous sodium sulfate and filtered through a layer of alumina to afford an orange-red solution. On evaporation, yellow-orange needles form. After recrystallization from ethanol, 2.2 gm of 2-phenylazo-3-naphthol is obtained (m.p. 173°C); yield 41 %.

It is interesting to note that aliphatic azo compounds have been prepared in good yield from N,N'-dialkylamides of sulfuric acid with 2 moles of sodium hypochlorite in a 1 N alkaline solution [54, 55]. The proposed course of the reaction is shown in Eq. (18). This method may well be one of the simplest procedures for synthesizing azo compounds generally.

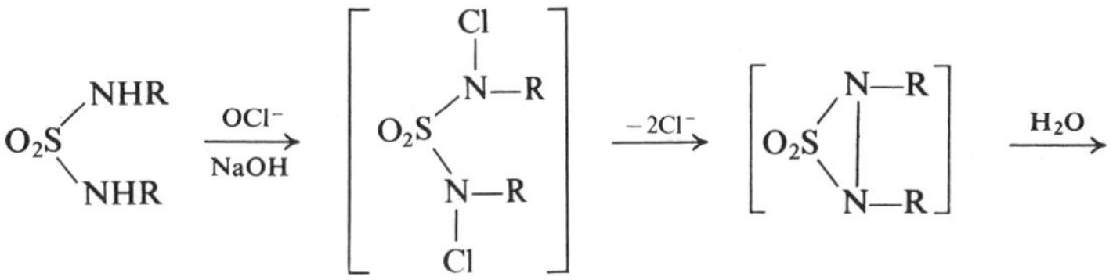

$$RNHNHR + SO_4^{2-} \xrightarrow{OCl^-} RN{=}NR \qquad (18)$$

By this method, azopropane is said to have been prepared in 100% yield, azo-n-butane in 54% yield, and azocyclohexane in 80% yield (m.p. 33°C). The identity of the product was determined by vapor-phase chromatography and by comparison with products obtained from the corresponding hydrazine derivatives on oxidation with hypochlorite ions.

E. Reaction of Aromatic Amines with Nitro Compounds

The high-temperature reaction of aromatic amines with aromatic nitro compounds in the presence of base affords primarily an azo compound [39, 40]. Because two independent laboratories have reported reasonable results with this synthesis, the procedure is given here. To be noted is that, while the reaction as described here involves 2-naphthylamine, a known carcinogenic intermediate, it is given only for reference to the procedure. Evidently, other aromatic amines also undergo the reaction.

2-11. *Preparation of 2-Phenylazonaphthalene* [56]

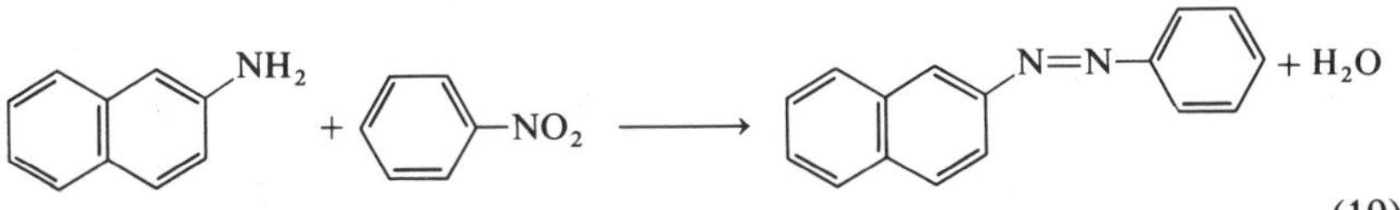

$$(19)$$

WARNING: 2-Naphthylamine is a known carcinogen. This procedure is given for reference only.

In a distillation apparatus fitted with a mechanical stirrer, thermometer, and provisions for adding solids, to an agitated mixture of 26.7 gm (0.187 mole) of 2-naphthylamine (CAUTION: carcinogenic material) and 20 gm (0.163 mole) of nitrobenzene maintained at 180°C is added slowly 17 gm of powdered sodium hydroxide over a 20 min period. After completion of the addition, heating is continued for 10 min. After cooling, the reaction mixture is treated repeatedly with dilute hydrochloric acid. The excess nitrobenzene is then separated by steam distillation. The residue from the steam distillation is treated with ethanol at 70°C to precipitate insoluble impurities which are removed by filtration. On cooling the filtrate, product crystals separate which, after filtration, are taken up in petroleum ether, leaving petroleum ether-insoluble impurities behind. The petroleum ether extract is evaporated to dryness and the residue is recrystallized from ethanol at 75°C: yield 17 gm (41%), m.p. 84°C.

By a similar procedure, 1-phenylazonaphthalene was prepared in 50% yield, m.p. 70°C. In both cases, by column chromatography, small amounts of the corresponding azoxy compounds were isolated [57].

F. Preparation from α-Halocarbonyl Compounds

Olefinic azo compounds, in which the double bond is conjugated with the azo group, have been prepared from α-halocarbonyl compounds by reaction with a substituted hydrazine to form a hydrazone which, on treatment with a base, is dehydrohalogenated and isomerized to an olefinic azo compound. The reaction may be represented by Eqs. (20) and (21).

$$
R-\overset{\overset{O}{\|}}{C}-\underset{\underset{Cl}{|}}{C}H-R' + R''-NH-NH_2 \longrightarrow R''-NH-N=\overset{\overset{Cl-CH-R'}{|}}{\underset{\underset{R}{|}}{C}} \qquad (20)
$$

$$
R''-NH-N=\overset{\overset{Cl-CH-R'}{|}}{\underset{\underset{R}{|}}{C}} \xrightarrow{\ OH^-\ } R''-N=N-\underset{\underset{R}{|}}{C}=CHR' \qquad (21)
$$

When phenylhydrazine is used as one of the reactants, the product is a mixed aromatic–aliphatic azo compound [58]. When methylhydrazine is used, a completely aliphatic azo compound is produced [59]. To be noted in the latter synthesis is that half of the methylhydrazine used in the reaction serves as base for the dehydrohalogenation step. This is an exothermic reaction.

The choice of reaction solvents appears to be quite wide. For example, when inorganic bases are employed, water and ethanol have been used [58]. When methylhydrazine is used, anhydrous conditions are usually maintained. Solvents include chlorobenzene or methylene chloride. In some cases no solvent was used.

2-12. *Preparation of Ethyl 2-Phenylazocrotonate* [58]

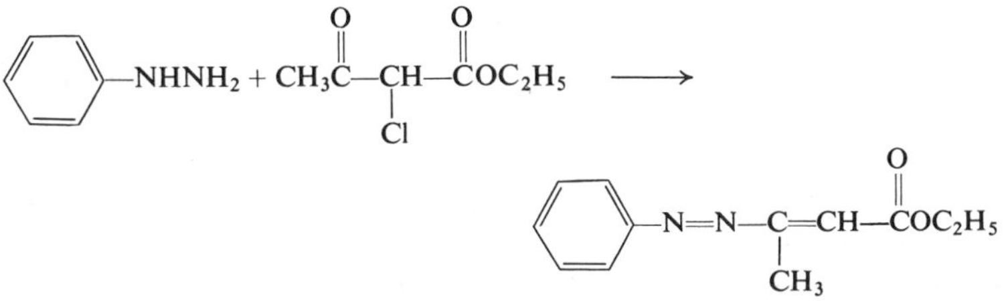

$$\text{(22)}$$

To a solution of 16.5 gm (0.1 mole) of ethyl 2-chloroacetylacetate in approximately 60 ml of ethanol, well-cooled in an ice-water bath, is added, with shaking, as quickly as possible, a mixture of 10.8 gm (0.1 mole) of phenylhydrazine dissolved in 100 ml of ethanol and 13.6 gm (0.1 mole) of crystalline sodium acetate trihydrate in 30 ml of water. The reaction mixture rapidly turns red and subsequently turns cloudy as the product begins to separate. If powdered ethyl 2-phenylazocrotonate is available from a previous preparation, inoculation with a small quantity of it facilitates formation of a beautiful red, needlelike product, m.p. 51°C (yield reported only as "good").

On prolonged storage of the mother liquor, a small amount of 1-phenyl-3-methyl-4-phenylazopyrazolone (structure IX) separates from the mother liquor.

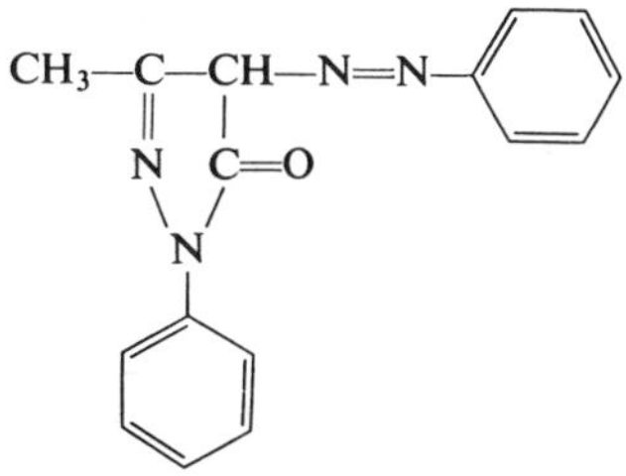

(IX) (m.p. 158°)

By a similar procedure the corresponding methyl ester has also been prepared, m.p. 46°C.

2-13. *Preparation of 2-(Methylazo)propene* [59]

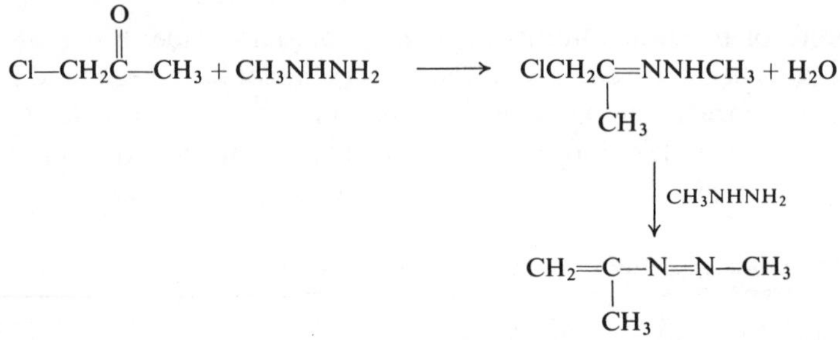

$$
\underset{\text{O}}{\text{Cl—CH}_2\overset{\displaystyle\parallel}{\text{C}}\text{—CH}_3} + \text{CH}_3\text{NHNH}_2 \longrightarrow \underset{\text{CH}_3}{\text{ClCH}_2\text{C}=\text{NNHCH}_3} + \text{H}_2\text{O}
$$

$$\downarrow \text{CH}_3\text{NHNH}_2$$

$$
\underset{\text{CH}_3}{\text{CH}_2=\text{C—N}=\text{N—CH}_3} \qquad (23)
$$

With suitable safety precautions, in a hood, to a well-stirred dispersion of 62 gm (1.35 moles) of methylhydrazine, 175 ml of chlorobenzene, and 20 gm of powdered, anhydrous sodium sulfate is added dropwise, with cooling, 61 gm (0.663 mole) of freshly distilled 1-chloro-2-propanone. After the addition has been completed, the reaction mixture is stirred for 1 hr. The moist sodium sulfate is separated by filtration and the filtrate is distilled. The fraction boiling between 60°C and 64°C is collected as product, yield 16.8 gm (31%) of a yellow liquid. On redistillation the boiling point is raised slightly to 61°–64°C; refractive index n_D^{23} 1.4300.

Table III lists yields and properties of several other olefinic azo compounds prepared by this method.

G. Reaction of Diazonium Salts with Grignard and Related Reagents

The reaction of Grignard reagents with diazonium zinc chloride double salts to afford azo compounds was initially reported in 1945 [60]. Unfortunately this work was followed by contradictory observations. The reaction was reinvestigated in 1956 [61]. It was found that the reaction proceeds most satisfactorily in a heterogeneous system rather than in solution. It was also

TABLE III

PROPERTIES OF OLEFINIC AZO COMPOUNDS [59]

Compound	Solvent	Yield (%)	B.p., °C (mm Hg)	n_D^t
2-(Methylazo)propene	Chlorobenzene	31	61–64 (atm)	1.4300[23]
1-(Methylazo)cyclohexene	Methylene chloride	61	101–105 (80)	1.4990[29]
1-(Methylazo)isobutylene	Methylene chloride	35	75–80 (200)	1.4590[26]
2-(Methylazo)-3-methyl-2-butene	No solvent	43	64–65 (22)	1.4574[18]
2-(Methylazo)-2-butene	Methylene chloride with MgSO$_4$	38	87–88 (145)	1.4591[25]

found that alkyl and aryl zinc chlorides react with diazonium fluoborates to give azo compounds and that organolithium, -cadmium, and -mercury compounds give greatly reduced yields when reacting with diazonium fluoroborates in a homogeneous reactive system. The example cited here is given for reference only, since we believe that subsequent work (see below) indicates the direction for an improved procedure.

2-14. Preparation of (4-Chlorophenylazo)benzene Using 4-Chlorophenylzinc Chloride [61]

$$\text{Cl}-\!\!\!\left\langle\!\!\bigcirc\!\!\right\rangle\!\!-\text{MgBr} + \text{ZnCl}_2 \longrightarrow \text{Cl}-\!\!\!\left\langle\!\!\bigcirc\!\!\right\rangle\!\!-\text{ZnCl} \tag{24}$$

$$\text{Cl}-\!\!\!\left\langle\!\!\bigcirc\!\!\right\rangle\!\!-\text{ZnCl} + \left\langle\!\!\bigcirc\!\!\right\rangle\!\!-\overset{+}{\text{N}}_2\text{BF}_4^- \longrightarrow \text{Cl}-\!\!\!\left\langle\!\!\bigcirc\!\!\right\rangle\!\!-\text{N}=\text{N}-\!\!\left\langle\!\!\bigcirc\!\!\right\rangle \tag{25}$$

With appropriate safety precautions, to a solution of p-chlorophenylmagnesium bromide prepared from 0.29 gm (0.012 gm-atom) of magnesium and 2.1 gm (0.011 mole) of 4-chlorobromobenzene in 30 ml of anhydrous ether is added 1.1 gm (0.011 mole) of freshly fused anhydrous zinc chloride in 50 ml of anhydrous ether. The resulting solution is transferred to an addition funnel with nitrogen and added slowly to a suspension of 1.92 gm (0.01 mole) of benzenediazonium fluoroborate (prepared by addition of a 20% aqueous solution of ammonium fluoroborate to a solution of benzene diazonium chloride, filtering the precipitate, washing the latter in turn with anhydrous methanol and anhydrous ether, and drying in a vacuum desiccator) in 50 ml of anhydrous ether with cooling in an ice-salt bath. The reaction mixture is stirred for 6 hr, then diluted by cautious addition to a mixture of diluted hydrochloric acid and ice. The neutral product fraction is extracted with ether. The ether extract is dried over anhydrous magnesium sulfate and filtered. After evaporation of the solvent, 0.914 gm of 82% pure crude product (brown solid) is isolated. The yield is estimated at 35%, on the basis of infrared absorption spectroscopy. The crude product may be recrystallized from ethanol. The melting point has been reported variously as 78°–84°C, 87°–88.5°C, and 90°–91°C. The product may also be chromatographically purified on an alumina column using cyclohexane as eluent.

By use of aliphatic zinc iodides, in toluene solution, mixed aromatic-aliphatic azo compounds have also been prepared [62]. Reported properties are

Phenylazoethane: b.p. 75°–77°C (19 mm Hg), $n_D^{28.4}$ 1.5480
2-Phenylazopropane: b.p. 85°–87°C (7 mm Hg), $n_D^{20.9}$ 1.5219

The earlier work using the reaction of Grignard reagents with diazonium zinc chloride double salts has been criticized by the observation that the reactions are normally carried out with equimolar or excess quantities of Grignard reagent to diazonium salts. Consequently, resultant azo compounds may react with additional quantities of the Grignard reagent to form hydrazo compounds and their derivatives [62–64]. By using an excess of the diazonium salt, improved yields are obtained. The generalized procedure for the preparation of the zinc chloride double salt follows.

2-15. Generalized Procedure for Preparation of Diazonium Zinc Chloride Double Salts [63]

To a mixture of 1 mole of the aromatic amine and 3–4 moles of hydro-chloric acid (diluted 1 : 1 with water) is added, at 0°C, the requisite amount of sodium nitrite dissolved in a minimum amount of water. While maintaining the diazonium salt solution at ice temperatures, 200 ml of a cold saturated solution of zinc chloride is added with occasional stirring. After allowing the mixture to stand for some time in an ice bath, the product is filtered off, washed several times with small portions of ice-cold ethanol followed by anhydrous ether. The salt is then air-dried and finally dried in a vacuum desiccator. In most cases the yield is in the range 90–98%.

2-16. Preparation of 2-Bromo-2'-methylazobenzene [(2-Bromophenylazo)-2'-toluene] [63]

$$\text{(2-Br-C}_6\text{H}_4\text{)}-N_2^+X^- + \text{(2-CH}_3\text{-C}_6\text{H}_4\text{)}-MgBr \longrightarrow \text{(2-Br-C}_6\text{H}_4\text{)}-N{=}N-\text{(2-CH}_3\text{-C}_6\text{H}_4\text{)} + MgBrX \quad (26)$$

With suitable safety precautions, in a flask equipped with an efficient stirrer, reflux condenser, an addition funnel is placed 17 gm of the zinc chloride double salt of 2-bromo-benzenediazonium chloride. Into the dropping funnel is introduced the Grignard reagent, prepared from 7 gm of 2-bromotoluene and 1.3 gm of magnesium in ether, dilute to about 40 ml with additional dry ether. At room temperature the Grignard solution is added to the rapidly stirred content of the reaction flask without additional cooling at a rate to maintain a gentle reflux (about 15 min). The reflux continues for a short time after the addition has been completed. After the reaction mixture has been diluted with ether to twice its initial volume, dilute hydrochloric acid is added cautiously with agitation, then water. The ether layer is separated and dried over calcium chloride. On evaporation of the solvent, red crystals are isolated. They may be recrystallized from ethanol with a small amount of charcoal. Yield 7 gm (64%, based on 2-bromotoluene).

Table IV gives data for other unsymmetrically substituted azobenzenes prepared by this technique.

When the zinc chloride double salts of diazotized aminocarboxylic esters were subjected to this reaction, the corresponding azo compounds were isolated with only poor to modest yields [64].

The same group of investigators also prepared a series of methoxy-substituted azo compounds [65] and a series of phenylazonaphthalenes [66] in poor to modest yields, using the reaction of appropriate Grignard reagents with diazonium zinc chloride double salts. Data for these azo compounds are given in Tables V, VI, and VII. It seems to us that it would be appropriate to reevaluate the preparation of these compounds by using the Curtin and Ursprung technique [61] of reacting arylzinc chlorides with diazonium fluoroborates.

H. Addition Reactions: Addition of Olefins to Diazoalkanes

In the course of an extensive study of the chemistry of azo compounds, Overberger *et al.* [68] prepared 1-pyrazolines, which may be considered cyclic azo compounds, by the addition of diazoalkanes to styrene. The reaction is considered a stereospecific 1,3-dipolar addition, and in the reaction between *p*-methoxyphenyldiazomethane and *p*-methoxystyrene both *trans*-3,5-bis(*p*-methoxyphenyl)pyrazoline (m.p. 129°C, dec.) and *cis*-3,5-*bis*(*p*-methoxyphenyl)pyrazoline were prepared and separated [69].

TABLE IV

PROPERTIES OF BROMOMETHYLAZOBENZENES [63]

Substituents	M.p. (°C)	Yield[a] (%)
2-Bromo-2'-methyl	88–89	64
2-Bromo-3'-methyl	44.5–45.5	40
2-Bromo-4'-methyl	60.5–61.5	60
3-Bromo-2'-methyl	49	61
3-Bromo-3'-methyl	79–79.5	52
3-Bromo-4'-methyl	119.5	65
4-Bromo-2'-methyl	53.5–54	42
4-Bromo-3'-methyl	81.5	68
4-Bromo-4'-methyl	153	68
4-Bromo-2-methyl	73–73.5	73
4-Bromo-3-methyl	68.5	74
2-Bromo-4-methyl	43.5–44	65.6

[a] Based on bromo compound converted into Grignard reagent.

TABLE V

PROPERTIES OF SUBSTITUTED METHOXYAZOBENZENES [65]

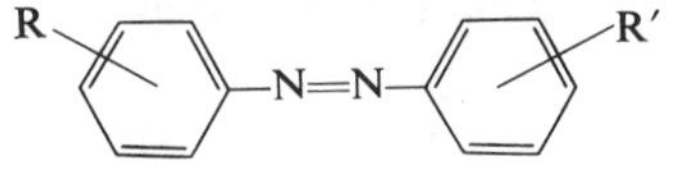

$$R-\overset{}{\underset{}{\bigcirc}}-N=N-\overset{}{\underset{}{\bigcirc}}-R'$$

Substituents		M.p. (°C)	B.p., °C (mm Hg)	Yield[a] (%)
R	R'			
2-Methyl	2'-Methoxy	73.5–75	—	18
3-Methyl	2'-Methoxy	—	176 (6)	11
4-Methyl	2'-Methoxy	56–57	—	11
2-Methyl	3'-Methoxy	—	160–165 (7)	29
3-Methyl	3'-Methoxy	—	168 (6)	20
4-Methyl	3'-Methoxy	78–79	—	14
2-Methyl	4'-Methoxy	58–59	173 (6)	8.9
3-Methyl	4'-Methoxy	56–57	171 (6)	4.0
4-Methyl	4'-Methoxy	109–110[b]	—	13
2-Methoxy	2'-Methoxy	143–144	—	1.4
3-Methoxy	2'-Methoxy	—	167 (6)	11.3
4-Methoxy	2'-Methoxy	41–42	165–172 (5)	7.3
3-Methoxy	3'-Methoxy	77–77.3	—	7.6
3-Methoxy	4'-Methoxy	41–42	—	34
4-Methoxy	4'-Methoxy	145–155	—	6.9
2-Methoxy	3'-Bromo	65–65.5	—	8.8
3-Bromo	3'-Methoxy	86–87	—	17.1
4-Methoxy	3'-Bromo	59–60	—	10.3
2-Methoxy	4'-Bromo	81.5–82.5	—	20.0
3-Methoxy	4'-Bromo	88.5–89.5	—	0.9
4-Methoxy	4'-Bromo	146–146.5	—	14.0

[a] Based on bromo compound converted into Grignard reagent.
[b] Horner and Dehner [67] report m.p. 110°–111°C.

The reaction appears to be reasonably general. Thus a series of 1-pyrazolines have been prepared according to the reaction scheme in Eq. (27), where X represents an electron-withdrawing group such as CO_2R, COR, CN, NO_2, or $\bigcirc$ [70, 71].

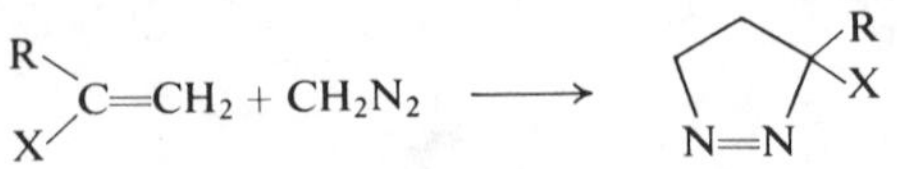

$$\underset{X}{\overset{R}{\diagdown}}C=CH_2 + CH_2N_2 \longrightarrow \underset{N=N}{\overset{R}{\diagdown}}{\diagup}X \qquad (27)$$

Preparation 2-16, *trans*-3,5-diphenyl-1-pyrazoline, demonstrates the general reaction procedure.

TABLE VI

PROPERTIES OF SUBSTITUTED PHENYLAZO-1-NAPHTHALENES [66]

Substituent, R	M.p. (°C)	Yield[a] (%)
H	60.5–67.5	22
2-Methyl	67–68	7
3-Methyl	42–43	23
4-Methyl	81–82	20
2-Methoxy	102–103	12
3-Methoxy	55–56	12
4-Methoxy	73–74.5	9
2-Bromo	103–104	3.5
3-Bromo	84–85	4.2
4-Bromo	132–133	11.4
2-Ethyl carboxylate	b	Trace
3-Ethyl carboxylate	101–102	25
4-Ethyl carboxylate	88.5–89	4
2-Carboxylic acid	141–142	Trace
3-Carboxylic acid	207	Trace
4-Carboxylic acid	234–234.5	Trace

[a] Based on bromo compound converted into Grignard reagent.
[b] B.p. 170°C (11 mm Hg).

TABLE VII

PROPERTIES OF SUBSTITUTED PHENYLAZO-2-NAPHTHALENES [66]

Substituent, R	M.p. (°C)	Yield[a] (%)
H	81.5–82.5	11
2-Methyl	70–71	12
3-Methyl	98–99.5	6
4-Methyl	126–127	7
2-Methoxy	93–94.5	11
3-Methoxy	67–68	6.4
4-Methoxy	102–103	4
2-Bromo	92.5–93	12
3-Bromo	117–118	12
4-Bromo	142.5–143	13
2-Ethyl carboxylate	(Oil)	1
3-Ethyl carboxylate	88–89	1
4-Ethyl carboxylate	112.5–113.5	Trace
2-Carboxylic acid	146–146.5	Trace
3-Carboxylic acid	220–221	Trace
4-Carboxylic acid	220	Trace

[a] Based on bromo compound converted into Grignard reagent.

2-17. Preparation of trans-3,5-Diphenyl-1-pyrazoline [68]

$$\text{(structure)} \quad \text{CH}=\text{CH}_2 + \overset{+}{\text{N}}=\text{N}-\overset{-}{\text{CH}}\text{(structure)} \longrightarrow \text{(structure)}$$

(28)

(By conventional procedures, 106 gm of benzaldehyde reacts in 600 ml of ether with 66 gm of anhydrous hydrazine to give benzalhydrazone, which is, with due precautions, directly oxidized in anhydrous ether with 300 gm of mercuric oxide to phenyldiazomethane.)

To a flask covered with aluminum foil, over a 1 hr period, a solution of 35 gm (0.336 mole) of inhibitor-free styrene in 90 ml of ether is added to 525 ml of the ether solution of phenyldiazomethane prepared as above. The reaction mixture is allowed to stand at room temperature overnight. Then approximately 200 ml of the ether is evaporated off under reduced pressure. The flask is immersed in an ice bath and stirred for a total reaction time of 36 hr. After thorough cooling in an ice bath, the precipitated product is removed by filtration and washed with ice-cold ether until all the red coloration has been taken up by the solvent. The ether washings and filtrate may again be partially evaporated and cooled to produce a further quantity of product. The total yield of crude product is 17.7 gm (23.7%), m.p. 107°–109°C. After two recrystallizations from methanol and drying in a vacuum desiccator over phosphorus pentoxide, the product consists of shiny white flakes, m.p. 109°–110°C, dec.

By a similar technique, 3,5-bis(p-chlorophenyl)-1-pyrazoline was also prepared, m.p. 120°–121°C, dec.

The cis and trans isomers may be separated by fractional precipitation, fractional crystallization, or careful fractional distillation.

3. REDUCTION REACTIONS

The nature of the azo bond is such that only a very limited number of possible functional groups have the necessary features to serve as starting materials for reductive methods of preparation. In a sense, the Bogoslovskii reaction [27, 28] may be considered a reduction of a diazonium salt by copper(I) ions. However, because the reaction resembles the other condensations of diazonium salts, its classification among the condensation reactions seems appropriate. The direct reduction of azoxy compounds as such is of minor preparative importance except as a method of identification of an azoxy compound. However, in the various bimolecular reduction procedures of aromatic nitro compounds, it has been postulated that an azoxy intermediate forms in the course of the reaction. This intermediate azoxy compound is ultimately reduced to an azo compound.

The reduction azines is of interest but requires further development.

A. Reduction of Nitro Compounds

The reduction of aromatic nitro compounds is believed to proceed to an intermediate mixture of nitroso compounds and substituted hydroxylamines which are not isolated but condense to form an azoxy compound which, in turn, is reduced to an azo compound. Contributing evidence to substantiate this mechanism is that the reduction of a mixture of two aromatic nitro compounds leads to a mixture of azo compounds consistent with that predicted if each of the nitro compounds were reduced to a nitroso compound and a hydroxylamine and these, in turn, reacted with each other in all possible combinations. This observation also implies that the bimolecular reduction of nitro compounds is practical only from the preparative standpoint for the production of symmetrically substituted azo compounds. Spectrophotometric studies of the reaction kinetics of the reduction of variously substituted nitro compounds may, however, uncover reasonable procedures for the synthesis of unsymmetrical azo compounds.

Further complications of the reduction of aromatic nitro compounds are the possibility of complete reduction to aromatic amines (which may condense with nitroso compounds to give the desired azo compounds), reduction of azo compounds to the corresponding hydrazo compounds, followed by a benzidine (or semidine) rearrangement. It is clear, therefore, that the level of reducing agent used and other reaction conditions are quite critical.

It has been pointed out that, under acidic reducing conditions, either azoxy compounds [72] or hydrazo compounds (with subsequent benzidine rearrangements) are produced [73]. Alkaline reduction yields hydrazo compounds which are converted into azo compounds by dissolved atmospheric oxygen. This oxidation becomes independent of oxygen concentration once a certain critical value of oxygen concentration is reached and the rate of reduction increases with decreasing hydrogen ion concentration [73]. It is curious that here both a reducing agent and an oxidizing agent function in the same medium, a situation reminiscent of the "redox" initiation systems found in emulsion polymerizations. The formation of hydrazo compounds by the condensation of hydroxylamines with nitroso compounds is difficult to visualize.

Among the many reducing agents suggested for the bimolecular reduction of nitro aromatic compounds to azo compounds are a sodium–lead alloy ('Drynap,' supplied by Wako Pure Chemical Co. of Japan) [74]; sodium and ethylene glycol [75, 76]; potassium hydroxide in ethanol [77]; ferrous oxide [78]; iron [79]; sodium amalgam and methanol [76]; hydrogen with a platinum oxide catalyst [80]; hydrogen with a Raney nickel catalyst [81]; and hydrazine with a palladium catalyst [82, 83]. Reference is also made in the literature to electrolytic reductions [83], reductions with alkaline carbohydrates such as glucose, dextrose, cellulose, and molasses, and reductions with silicon, stannites, arsenites, sulfides, sulfites, and stannous salts [84]. Many of these reducing systems have only limited uses in specific industrial processes.

From the standpoint of laboratory-scale preparations, probably the most commonly used method is the reduction with zinc in a sodium hydroxide medium. The use of lithium aluminum hydride also merits consideration.

The preparation of azobenzene by reduction of nitrobenzene with zinc dust and sodium hydroxide has been well described [85], except that no specific mention is made of the desirability of carrying the reaction out in the presence of air. As a matter of fact, in the older literature, mention is made of blowing air through the filtrate after the reduction has been completed and the insoluble salts have been filtered off [84]. This procedure has also been recommended more recently. Actually, air has been drawn through the product solution for as long as 6 hr to oxidize any hydrazo compound which may be present [86]. Blackadder and Hinshelwood [73] have shown that the presence of modest amounts of air toward the end of the reduction cycle should be sufficient for obtaining adequate yields. Yet Shinkai, Manabe, and co-workers [87] aerated their reduction product derived from 4′-nitrobenzo-15-crown-5 for 4 hr.

We believe that attention should also be focused on the fact that metal ions will form chelated complexes with azo compounds. Hence a vigorous posttreatment with acids should be attempted if the stability of the product permits. Usually this treatment should be carried out before the final crystallization.

The wide applicability of the zinc-sodium hydroxide reduction is illustrated here by two examples. In the first, the reduction of p-nitrostyrene, conditions evidently were sufficiently mild that the vinyl groups were not saturated [88]. The second example illustrates the reduction of a dinitro compound to produce an intramolecular azo bridge [89].

3-1. Preparation of 4,4′-Divinylazobenzene [88]

$$2O_2N{-}\langle\!\rangle{-}CH{=}CH_2 \xrightarrow[\text{NaOH}]{\text{Zn}}$$

$$CH_2{=}CH{-}\langle\!\rangle{-}N{=}N{-}\langle\!\rangle{-}CH{=}CH_2 \quad (29)$$

In a 250 ml, four-necked flask fitted with mechanical stirrer, reflux condenser, thermometer, and with provision for adding solids, a mixture of 90 ml of 95% ethanol, 15 ml of 12N sodium hydroxide solution, and 7 gm (0.047 mole) of p-nitrostyrene is heated to reflux temperature with vigorous stirring while 25 gm (0.39 gm-atom) of zinc dust is added in small portions so that the reaction is kept under control. After the addition has been completed, reflux is continued for $\frac{1}{2}$ hr. Then the hot reaction mixture is filtered in such a manner that the filtrate drops directly into a large volume of cold water. The resultant gummy precipitate is filtered off, dried, and sublimed at 120°C (2 mm Hg). The sublimed solid is then recrystallized repeatedly from 95% ethanol to afford 0.95 gm (17%) of product, m.p. 138°–138.5°C.

3-2. Preparation of Benzo[c]cinnoline("2,2′-Azodiphenyl" or 3,4-Benzocinnoline) by Zinc–Sodium Hydroxide Reduction [89]

In an apparatus similar to the one described in Preparation 3-1, a mixture of 10 gm (0.041 mole) of 2,2-dinitrobiphenyl, 133.4 ml of absolute ethanol, and 13.4 ml of aqueous solution containing 10 gm of sodium hydroxide is warmed, with stirring, to 70°–80°C. At that temperature 30 gm (0.46 gm-atom) of 30 mesh granular zinc is added over a $\frac{1}{2}$ hr period. After the addition has been completed, warming between 70° and 80°C is continued for an additional $\frac{1}{2}$ hr. The hot reaction mixture is filtered through a Büchner funnel to separate the inorganic materials.

On cooling, a yellow precipitate forms in the filtrate. This precipitate is separated and washed with absolute ethanol. The wash solution, combined with the filtrate, is used to extract additional product from the inorganic materials previously separated. The combined yellow crystals are recrystallized from absolute ethanol; yield 4.1 gm (55.4%), m.p. 156°–158°C.

Occasionally, commercial zinc dust needs to be activated for this reduction. This may be accomplished by adding a small quantity of dilute hydrochloric acid to the zinc, filtering rapidly, and washing the zinc dust with water immediately. The metal thus activated must be used promptly [90].

While aromatic nitro and azoxy compounds have been reduced to azo compounds with lithium aluminum hydride, aliphatic nitro compounds produced only the corresponding aliphatic amines [91]. The usual technique involves dropwise addition of 1 mole of nitro compound in ether to 1.05–1.15 moles of lithium aluminum hydride in ether solution at Dry Ice temperatures followed by warming to room temperature. If the resulting product is only slightly soluble in ether, hydrolysis should be carried out with dilute sulfuric acid. Then the azo compound simply needs to be filtered off, washed with water, and dried. If the product is ether-soluble, the ether layer is separated, evaporated, and the residue is recrystallized [91, 92].

A variation of this procedure is given in Preparation 3-3, in which a mixed solvent is used in the preparation of benzo[c]cinnoline by a lithium aluminium hydride reduction [93].

3-3. *Preparation of Benzo[c]cinnoline by Lithium Aluminum Hydride Reduction* [93]

$$\text{(31)}$$

In an apparatus suitably protected against atmospheric moisture and carbon dioxide, to a solution of 4.0 gm (0.0164 mole) of 2,2′-dinitrobiphenyl in 35 ml of anhydrous benzene is cautiously added, with stirring, a solution of 3.0 gm (0.08 mole) of lithium aluminum hydride in 15 ml of anhydrous ether. The mixture is heated to reflux for 4 hr, during which time a brown precipitate forms. After the reaction mixture has been cooled, 5 ml of water is cautiously added and the inorganic materials are removed by filtration. The filtrate is washed in turn with a 5% sodium carbonate solution and with water. After drying with anhydrous sodium sulfate and filtration, the solvent is evaporated off. The residue is recrystallized from benzene: yield 2.7 gm (92%), pale-yellow blades, m.p. 156°C.

The melting points of some symmetrical aromatic azo compounds, prepared from the corresponding aromatic nitro compounds, are given in Table VIII.

B. Reduction of Azoxy Compounds

As mentioned previously, azoxy compounds may also be reduced to azo compounds. Lithium aluminum hydride appears to be the reagent. The primary importance of this reaction lies in structure determinations of the azoxy compound. In this connection it must be pointed out that both *cis*- and *trans*-azoxybenzenes are converted only to *trans*-azobenzene [100]. It has been postulated that the reduction of *cis*-azoxybenzene proceeds by way of an intermediate having a single bond between the nitrogen atoms. This intermediate, having free rotation about this bond, would naturally tend to produce the more stable *trans* product. Although this appears reasonable to us, the evidence presented consists of an experiment in which 0.22 gm of *cis*-

TABLE VIII

MELTING POINTS OF SYMMETRICALLY SUBSTITUTED AROMATIC AZO COMPOUNDS

Compound	M.p. (°C)	Ref.
Azobenzene	67–68	75, 85, 94
	66–67.5, 68	
o-Azotoluene	51, 53–54	75
m-Azotoluene	51	75
p-Azotoluene	142	67, 95
	144–145	96, 97
Azomesitylene	77	67
2,2'-Diethylazobenzene	44	98
2,2'-Dibromoazobenzene	125	75
4,4'-Dibromoazobenzene	204	99
2,2'-Dichloroazobenzene	136	75
3,3'-Dichloroazobenzene	101	75
	99	99
4,4'-Dichloroazobenzene	184–185	75
	185	99
4,4'-Dicarbethoxyazobenzene	143	99
2,2'-Difluoroazobenzene	98	99
4,4'-Difluoroazobenzene	100	99
2,2'-Diiodoazobenzene	151	99
4,4'-Diiodoazobenzene	235	99
4,4'-Divinylazobenzene	138–138.5	88
4,4'-Diphenylazobenzene	248	99
2,2'-Azonaphthalene	208	84
2,2'-Dimethoxyazobenzene	143	99
4,4'-Dimethoxyazobenzene	160	99
3,3'-Dinitroazobenzene	146–148	95
4,4'-Dinitroazobenzene	212–214	95
3,3'5,5'-Tetrachloroazobenzene	189	99
2,2',6,6'-Tetramethoxyazobenzene	110	99
3,3'-Dichloro-4,4',6,6'-tetramethoxyazobenzene	243	99
4,4'-Dichloro-2,2',5,5'-tetramethoxyazobenzene	184	99

azoxybenzene was reduced and only 0.06 gm of *trans*-azobenzene was isolated. Although no *cis*-azobenzene was found in the product fraction, we are curious about the fate of the unaccounted-for remainder. This may be another example of product loss by complexing with the inorganic materials produced when a lithium aluminum hydride reaction mixture is decomposed only with water, a matter we have observed in connection with other reductions using such a procedure.

An extension of the Wallach rearrangement [98] (conversion of azoxybenzene into *p*-hydroxyazobenzene induced by chlorosulfonic acid) has shown that azoxybenzene is deoxygenated (reduced) to azobenzene by several acid halides, which in turn undergo oxidation [101]. For example, phosphorus trichloride is oxidized to phosphorus oxychloride as azoxybenzene is reduced practically quantitatively to azobenzene. Other acid halides such as S_2Cl_2, $SOCl_2$, $AlCl_3$, and $FeCl_3$ are also effective. Acetyl chloride can be used as a solvent for several other solid acidic halides (AlI_3, $AlBr_3$, $ZnBr_2$, ZnI_2, and SbF_3). The use of aluminum chloride tends also to give some aromatic chlorination [102].

Upon refluxing azoxy compounds in 1,2-dimethoxyethane in the presence of a fresh preparation of molybdenum hexacarbonyl on alumina, azo compounds are formed. The benzene dispersion of azoxy compounds and a fresh preparation of triiron dodecacarbonyl has also been used to reduce the azoxy starting compound to the corresponding azo compounds. The reactions were run in the millimolar range. Yields up to 60% were reported in some cases. The hazards of handling metal carbonyls and, presumably, carbon monoxide should be carefully considered before general application of this reduction—particularly on a large scale—is attempted [103].

C. Reduction of Azines

The azines represent another class of organic compounds which, in principle, should be reducible to azo compounds. The method is attractive since, with the availability of anhydrous hydrazine, azines are readily prepared from a wide variety of ketones and aldehydes. Evidently, introduction of 1 gm-molecule of hydrogen into an azine molecule has been accomplished (see Section 6, Procedure 8) [104]. Two preparations involving the 1,4-addition of chlorine to an azine system have been carried out. These reactions led to chlorinated azo compounds.

3-4. Preparation of 1,1'-Dichloroazocyclohexane [105]

$$\text{(32)}$$

In a hood, through a rapidly stirred suspension of 76 gm (0.396 mole) of cyclohexanoneazine in 300 ml of petroleum ether (b.p. 60°–90°C) cooled to −60°C is passed a slow stream of gaseous chlorine until a slight excess of the gas is noted. The excess of chlorine is removed by ventilation at the water aspirator. Then the solution is concentrated to half-volume by gentle evaporation at reduced pressure. The reaction system is filtered free from tarry impurities and the filtrate is allowed to stand for 24 hr at room temperature. The product gradually separates out and is isolated by filtration. Evaporation of the mother liquor may afford another crop of product. The total yield is 81.5 gm (78%). The product, after recrystallization from petroleum ether, has a melting point of 66°C. (NOTE: Since aliphatic azo compounds are inherently unstable and may serve as free radical sources, the stability of the product should always be checked with due precautions, and excessive exposure to heat should always be avoided.)

This basic reaction has been extended to the preparation of α,α'-diacyloxy-azoalkanes in a single-step reaction [see Eq. (33)].

This reaction requires somewhat more careful control of the quantity of chlorine used, but otherwise appears to be quite straightforward. The method is outlined here in generalized terms only.

3-5. Preparation of α,α'-Diacyloxyazoalkanes [106]

$$
\begin{array}{c}
R^1 \\

\end{array}
C{=}N{-}N{=}C
\begin{array}{c}
R^1 \\
R^2
\end{array}
+ 2NaOCR^3 \xrightarrow[R^3CO_2H]{Cl_2}
R^3C{-}O{-}\underset{R^2}{\overset{R^1}{C}}{-}N{=}N{-}\underset{R^2}{\overset{R^1}{C}}{-}O{-}CR^3 \quad (33)
$$

In a hood, 4.1 moles of the sodium salt of a carboxylic acid is dissolved in just enough of the anhydrous free acid to make a mobile solution. Then 1 mole of the ketazine is added and, with stirring at 10°–20°C, one equivalent of chlorine is slowly passed through the solution. Sodium chloride precipitates rapidly during this addition. Stirring is continued for $\frac{1}{2}$ hr at 20°C after the addition has been completed. Then the reaction mixture is poured into 5–8 volumes of

TABLE IX

PROPERTIES OF α,α'-DIACYLOXYAZOALKANES [106]

$$
R^3{-}\overset{O}{\overset{\|}{C}}{-}O{-}\underset{R^2}{\overset{R^1}{C}}{-}N{=}N{-}\underset{R^2}{\overset{R^1}{C}}{-}O{-}\overset{O}{\overset{\|}{C}}{-}R^3
$$

Substituents			B.p., °C (mm Hg)	M.p. (°C)	n_D^{20}	Yield (%)
R^1	R^2	R^3				
CH_3	CH_3	H	40 (0.1)	—	1.4320	75
CH_3	C_2H_5	H	79 (0.4)	—	1.4445	45
CH_3	$i\text{-}C_4H_9$	H	93 (0.4)	—	1.4486	54
C_2H_5	C_2H_5	H	89 (0.25)	—	1.4530	57
$-(CH_2)_5$	—	H	—	100	—	27
CH_3	CH_3	CH_3	—	103	—	79
$-(CH_2)_7$	—	C_2H_5	—	58	—	32

an ice-water mixture and the product is extracted with ether. The product is separated from the solvent either by distillation or by crystallization.

Table IX gives properties and yields of a series of products prepared by this method.

The direct hydrogenation of azines in general to azo structures would be of considerable interest. Obviously, control of the reaction is critical, otherwise complete reduction to hydrazines will take place, as in the hydrogenation of desoxybenzoinazine over palladium-on-charcoal with an atmospheric pressure of hydrogen [107]. As noted in Section 6, Preparation 8 [104], such a catalytic hydrogenation over copper chromite has been accomplished recently with a macrocyclic azine.

4. OXIDATION REACTIONS

A. Oxidation of Amines

As is well known, the oxidation of primary aromatic amines is a highly complex process, depending on exact reaction conditions to produce well-defined products—usually complex dyes. A few synthetic procedures have been reported in which amines have been converted into azo compounds.

For example, the autoxidation of aromatic amines in the presence of equivalent amounts of potassium *t*-butoxide afforded azo compounds such as azomesitylene, 4,4'-azotoluene, 4-methyl-4'-methoxyazobenzene (by using a mixture of *p*-toluidine and *p*-anisidine), and 4-chloro-4'methoxyazobenzene, m.p. 121°–122°C (by using a mixture of *p*-chlorotoluidine and *p*-anisidine) in highly variable yields. To be noted is that formation of azo compounds take place in the presence of *ortho*- and *para*-directing substituents and in the presence of *meta*-directing substituents if the substituent is in a position *meta* to the amino group. *Meta*-directing substituents in the position *ortho* or *para* to the amino group interfere with the autoxidation. The transitory green colors observed during the reaction have been attributed to the intermediate formation of monomeric *C*-nitroso compounds [67].

As indicated, in the presence of mixtures of aromatic amines, some of the unsymmetrically substituted azo compounds are formed. However, both α- and β-naphthylamines are converted into phenazines. From the preparative standpoint, we consider this reaction as having only limited application [67].

The effect of substituents is dramatically indicated in the manganese dioxide oxidation of aromatic amines. Again aniline, *p*-chloroaniline, and *p*-toluidine are readily oxidized to the corresponding azo compounds (yields reported to be approximately 87%), whereas at least 95% of unreacted amines were recovered when the reaction was attempted with α- or β-naphthylamine or with *o*-, *m*-, or *p*-nitroaniline [51, 94].

Further study of the reaction has indicated that nitroanilines, as well as aminobenzoic acids, are not substantially converted into azo compounds, whereas polycyclic aromatic amines give polymeric products.

In general, it appears that para-substituted anilines react more rapidly than other position isomers and that the order of reactivity is $F > Cl > Br > I$.

Using activated manganese dioxide oxidations, even 2-aminopyridine has been converted into 2-azopyridine (m.p. 85°C). When mixtures of two anilines are oxidized together, the products isolated are only the symmetrically substituted azobenzenes [94].

4-1. General Procedure for Oxidation of Aromatic Amines with Active Manganese Dioxide [94]

$$2Ar-NH_2 \xrightarrow{\ MnO_2\ } Ar-N=N-Ar + 2H_2O \qquad (34)$$

(a) Preparation of active manganese dioxide [99, 102]. Commercial precipitated manganese dioxide has been used [94], but a freshly prepared form is to be preferred.

A solution of 84.0 gm of manganous sulfate monohydrate (analytical grade) in 150 ml of water and 117 ml of a 40% aqueous sodium hydroxide solution are added simultaneously over a 1 hr period to a stirred solution of 96 gm of potassium permanganate in 600 ml of water maintained at 80°–90°C. Stirring is continued for another hour at 90°C and the hot suspension is filtered and washed with a large amount of hot tap water until the precipitate is free from potassium permanganate. Washing is continued until the pH of the washings is neutral to litmus. The solid is dried at 125°C and ground to a powder, reheated to 125°C for 24 hr, and reground to pass a 60 mesh screen. Before use, the powder is heated for 24 hr at 100°–120°C and stored in a desiccator.

(b) Oxidation. In a reflux apparatus fitted with a Dean-Stark water separator and a mechanical stirrer, a suspension of 0.01 mole of the amine, 0.05–0.06 mole of dried manganese dioxide, and 50–60 ml of benzene is refluxed for approximately 6 hr as the formed water is removed. The hot solution is filtered and the manganese dioxide is washed free from colored product with hot benzene. The filtrate is evaporated to a small volume and hexane is added to crystallize the product from solution. The product is ultimately recrystallized either from a benzene–hexane mixture or from ethanol. Yields are normally about 90%.

Recently, 4′-amino-5′-tert-butylbenzo-18-crown-6 derivatives as indicated in structure X have been oxidatively coupled to form the corresponding trans-azo compounds. When $n = 2$ m.p. 143.7°–145°C, yield 40%; where $n = 3$, m.p. 173°–174°C, yield 33% [108].

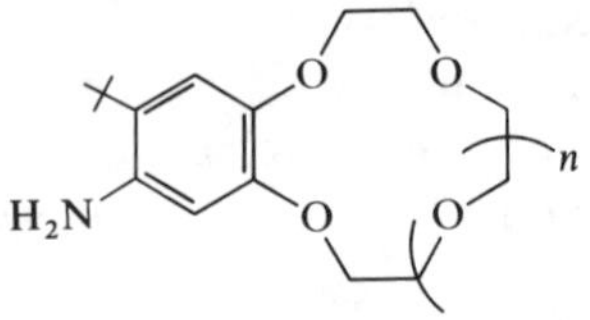

(X) $n = 2$ and 3

As indicated above [54], primary amines converted into the alkylamides of sulfuric acid can be oxidized with sodium hypochlorite to azo compounds. The reaction appears to proceed by way of an intermediate hydrazine, which is ultimately oxidized [54]. The reaction is suitable for the formation of symmetrically substituted azoalkanes. Highly branched primary aliphatic amines have been oxidized with sodium hypochlorite in an aqueous dioxane medium [109].

Aromatic amines, such as aniline, have been oxidized with sodium hypochlorite solution to produce a very complex mixture including azobenzene, aminophenols, and tars [110]. Polyfluoroaromatic amines, on the other hand, yield azo compounds in a straightforward manner [111]. To be sure, the conversion appears to be only of the order of 25%, but we suspect that, given the reaction of an aqueous oxidizing medium with a reactant of high density and the unique wetting and solubility characteristics of fluorinated compounds, we cannot expect to obtain better yields unless specialized surfactants are present to improve the contact between the oxidizing agent and the surface of the amine droplets.

4-2. *Preparation of Decafluoroazobenzene* [111]

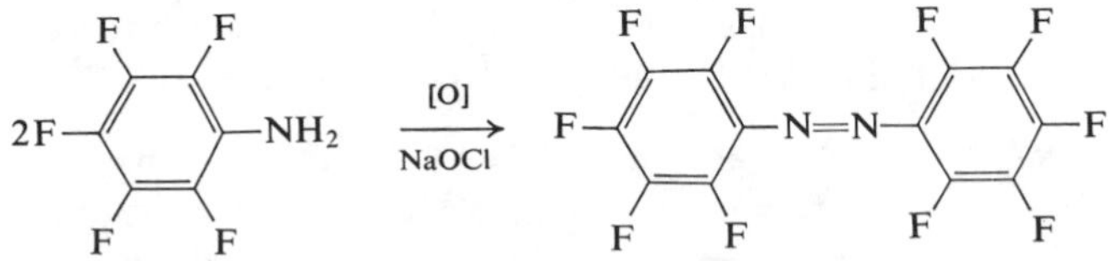

To 10.0 gm (0.0546 mole) of pentafluoroaniline is added 360 ml of a sodium hypochlorite solution containing 10–14% w/v available chlorine. The mixture is stirred vigorously at room temperature for 4 hr. Then the crude product is isolated by repeated extraction with ether. The ether extract is washed with deionized water until the water contains no significant levels of chloride ions. The ether solution is dried with sodium sulfate, filtered, and the filtrate is evaporated cautiously to dryness. The residue is recrystallized from ethanol: yield 2.5 gm (25.3%), m.p. 142°–143°C.

By similar procedures, 4,4'-*H*-octafluoroazobenzene (m.p. 118°C) and 4,4'-dimethyl-2,2',3,3',5,5',6,6'-octafluoroazobenzene (m.p. 159°C) were produced.

The oxidation of aromatic amines with peracids was the subject of some dispute. It has now been demonstrated that simple oxidation of aromatic amines with peracids produces azoxy compounds without the intermediate formation of azo compounds [96]. To be sure, small amounts of azo compounds are isolated from the reaction mixture, but this is considered a side reaction.

On the other hand, in the presence of small amounts of metallic salts such as cupric chloride, the course of the reaction changes to produce good yields of azo compounds [95]. The reaction, unfortunately, is not quite general. Thus the toluidines and aminophenols, for example, produce copious amounts of tarry products, while the nitroanilines, evidently, are smoothly converted into azo compounds.

4-3. Preparation of 3,3'-Dinitroazobenzene [95]

$$ \tag{36} $$

A solution of 2 gm (0.014 mole) of 3-nitroaniline in 40 ml of glacial acetic acid and 1.85 gm (0.028 mole) of peracetic acid is treated with 1 ml of an aqueous solution containing 5 mg of cupric chloride dihydrate. The reaction mixture is stirred for 16 hr at 17°C. After this time, the precipitated product is filtered off (0.8 gm). The mother liquor is added to 200 ml of a 3N sodium hydroxide solution, whereupon another 0.61 gm of product is precipitated. The crude products are combined: yield 1.42 gm (72%), m.p. 140°–144°C. On recrystallization the melting point is raised to 144°–147°C.

Aliphatic amines have been oxidized to aliphatic azo compounds with iodine pentafluoride [112–115]. In the aromatic series, this reagent evidently gives rise to tarry products when used to oxidize amines, although it has been used to oxidize hydrazobenzene to azobenzene.

The literature shows only highly hindered aliphatic amines as having been subjected to the reaction (t-butylamine [112] and cumylamine [113]). The applicability to other amines requires further study.

For safe handling of iodine pentafluoride, we recommend consultation with the manufacturer of the compound before it is used.

4-4. Preparation of Azocumene (2,2'-Diphenyl-2,2'-azopropane) [113]

To a stirred solution of 120 ml of methylene chloride, 18 ml of dry pyridine, and 5 ml of iodine pentafluoride maintained at −10°C to −20°C in a Dry Ice–carbon tetrachloride slurry is added a solution of 13.5 gm (0.1 mole) of cumyl-amine in 10 ml of methylene chloride over a 1 hr period. The reaction mixture

$$2 \; \bigcirc \!\!\! - \!\! \underset{\underset{CH_3}{|}}{\overset{\overset{CH_3}{|}}{C}} \!\! - \!\! NH_2 \; \xrightarrow{\text{IF}_5} \; \bigcirc \!\!\! - \!\! \underset{\underset{CH_3}{|}}{\overset{\overset{CH_3}{|}}{C}} \!\! - \!\! N \!\! = \!\! N \!\! - \!\! \underset{\underset{CH_3}{|}}{\overset{\overset{CH_3}{|}}{C}} \!\! - \!\! \bigcirc \qquad (37)$$

is stirred for another hour at $-10°C$, and then for 1 hr at $0°$. After this time, water is added to the reaction mixture and stirring is continued until the yellow solid which had formed is dissolved. The lower organic layer is separated and washed in turn with water, 1 N hydrochloric acid, a saturated sodium thiosulfate solution, and again with water. After drying with anhydrous magnesium sulfate and filtration, the product solution is partially evaporated by means of a rotary evaporator at a temperature below $30°C$. The brown solid obtained on cooling is separated and recrystallized twice from methylene chloride: yield 4.75 gm (17.9%), m.p. $86.9°–88.7°C$.

By a similar technique, t-butylamine was converted into 2,2'-dimethyl-2,2'-azopropane (b.p. $53°C$, 70 mm Hg, n_D^{20} 1.4133) [112].

Aromatic azo polymers have been prepared from aromatic diamines using cupric ions as catalyst for an oxygen-oxidation in pyridine or dimethylacetamide-pyridine solution [116].

B. Oxidation of Hydrazine Derivatives

Since N,N'-disubstituted hydrazines are readily available from a variety of sources, their dehydrogenation constitutes a widely applicable route to both aliphatic and aromatic azo compounds. Such oxidative procedures are of particular value in the aliphatic series because so many of the procedures applicable to aromatic compounds, such as the coupling with diazonium salts, have no counterpart. The oxidation reactions permit the formation not only of azoalkanes, but also of a host of azo compounds containing other functional groups, e.g., α-carbonyl azo compounds [117], α-nitrile azo compounds [118], azo derivatives of phosphoric acid [119], and phenylphosphoric acid derivatives [120].

One matter to which the recent literature has not paid adequate attention is the obvious problem that azo compounds can be converted into azoxy compounds by oxidation. For example, at a relatively low temperature, substituted hydrazines have been oxidized to azo compounds with hydrogen peroxide [121], while, under somewhat more drastic conditions, azo compounds have been converted into azoxy compounds also with hydrogen peroxide [93]. It seems to us that, in lieu of more detailed experimental observations, we must assume the formation of both azo and azoxy compounds when an oxygenating (as opposed to a hydrogen-abstracting) oxidizing agent is used to treat a hydrazine derivative. Consequently, control of reaction conditions and of the quantity of oxidizing agent used and chromatographic analysis of the product should all be employed in developing these syntheses further.

It must be reiterated that, whereas aromatic azo compounds are relatively stable thermally and can be subjected to typical reactions of aromatic compounds [93, 100, 122], the aliphatic compounds may be significantly less stable thermally. Aliphatic azo compounds, such as α,α'-azobis(isobutyronitrile), do decompose on heating and are used as free radical sources. Hence adequate safety precautions must be taken in handling them. This, by the way, does not mean that aliphatic azo compounds have not been subjected to distillation and to vapor phase chromatogaphy. Many have been distilled and, as will be pointed out in a subsequent section, their preparation by isomerization of hydrazone depends on a distillation technique.

4-5. Preparation of Azobis(diphenylmethane) by Hydrogen Peroxide Oxidation [121]

$$(38)$$

In a cold bath or in an explosion-proof refrigerator, a solution of 1.3 gm (approximately 0.0035 mole) of crude N,N'-benzhydrylhydrazine (m.p. 114°–130°C) in a mixture of 40 ml of acetone and 50 ml of ethanol and 1.25 gm (0.0096 mole) of 30% hydrogen peroxide is maintained at 10°–15°C for 8 hr. The solvent is rapidly removed by evaporation under reduced pressure with only moderate warming. The residue is rapidly recrystallized from ethanol. Yield 1 gm (76%), m.p. 115°C dec. (NOTE: On melting, azobis(diphenylmethane) decomposes with loss of nitrogen to form a solid with m.p. 205°–210°C, which has been identified as the expected 1,1,2,2-tetraphenylethane.)

The traditional method of oxidizing hydrazine derivatives makes use of halogens or hypohalites as oxidizing agents. The techniques range from the preparation of 1,1'-azobis(1-cyclohexanenitrile) by the addition of bromine to an alcoholic hydrochloric acid solution of the corresponding hydrazine [123], through the use of bromine water [124, 125], to oxidations with sodium, hypobromite [90], or sodium hypochlorite [126].

4-6. Preparation of Diethyl Azodicarboxylate [125]

$$\underset{\substack{\| \\ C_2H_5OC-NH-NH-COC_2H_5}}{\overset{O \qquad\qquad O}{}} \xrightarrow{Br_2/H_2O} \underset{\substack{\| \\ C_2H_5OC-N=N-COC_2H_5}}{\overset{O \qquad\qquad O}{}} \quad (39)$$

With suitable safety precautions, in a hood, to a well-stirred suspension of 3.52 gm (0.20 mole) of diethylhydrazine-N,N'-dicarboxylate in 250 ml of water which is maintained at ice temperature with an adequate bath is added, over a 15 min period, 12.3 ml of bromine. After the addition has been completed, stirring is continued for an additional 15 min, then the product is extracted with 400 ml of ether. The ether extract is cooled to 0°C and extracted, in turn, with the following reagents, all being maintained at 0°C: water, aqueous sodium carbonate, very dilute sodium thiosulfate, and water again. (NOTE: the product is unstable to thiosulfates.) The ether solution is then dried with calcium chloride. After filtration and evaporation of the solvent, the orange residue is distilled through a short fractionating column using an oil bath to heat the flask. The fraction boiling between 60°C and 70°C (0.1 mm Hg), m.p. 6°C, is taken as product: yield 22 gm (63%).

In preparation 4-6 above, the hydrazine was partially protected by conversion to the dicarboxylate, or perhaps more properly to the dicarbamate. By use of Diels–Alder reactions involving dimethyl azodicarboxylate, fairly complex bicyclic dicarbamates with structures such as that indicated in XI have been prepared:

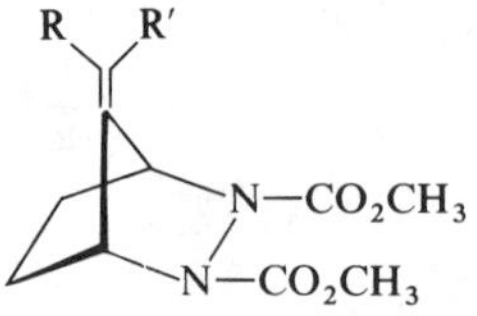

R	R'
CH$_3$	CH$_3$
—(CH$_2$)$_5$—	
H	Cl
Ph	Ph
C$_2$H$_5$	C$_2$H$_5$

(XI)

In a one-pot, two-step reaction, the corresponding bicyclic azo compounds have been prepared at or below room temperature in 60°–80°C. The reaction involves treatment with the lithium salt of n-propyl mercaptan or of methyl mercaptan followed by oxidation with aqueous potassium ferricyanate at 0°C. The reaction is carried out in hexamethylphosphoramide (HMPA) [127]. Since the latter solvent is highly toxic and suspected of being a carcinogenic agent, this reaction is mentioned here for reference only. This process should be reinvestigated with a view of finding a safe reaction medium.

4-7. *Preparation of Azobenzene* [90]

In a separatory funnel fitted with a stopcock which cannot slip out is placed a solution of 10 gm (0.054 mole) of hydrazobenzene in 60 ml of ether. To this solution is added, in small portions, an ice-cold solution of sodium hypo-

$$\text{Ph–NH–NH–Ph} \xrightarrow{\text{NaOBr}} \text{Ph–N=N–Ph} \quad (40)$$

bromite prepared by the dropwise addition of 10 gm of bromine to an ice-cold 8% aqueous sodium hydroxide solution. On completion of the addition of all the hypobromite solution, the reaction mixture is shaken for an additional 10 min. Then the aqueous layer is drawn off and the ether solution is evaporated cautiously. The residue is dissolved in 30–40 ml of boiling ethanol.

Water is added to the boiling solution dropwise until a small amount of product begins to separate. The mixture is then made homogeneous again by adding a few milliliters of ethanol, and the solution is cooled until the product crystallizes out. The filtered azobenzene is washed with an ice-cold 50% ethanol–water mixture: yield 7–9 gm (approximately 80%), m.p. 67°–68°C.

Azoalkanes have also been produced by oxidation with mercuric oxide. This reagent is considered the reagent of choice for azoalkanes [107, 121, 128], as well as for the preparation of cyclic azo compounds [3, 129, 130] and 1-pyrazolines [131]. When used in the preparation of α-carbonyl azo compounds, mercury complexes are evidently formed with the diacylhydrazine starting materials. These complexes are then treated with halogen (iodine or chlorine) to produce the azo compounds [126].

Although the literature usually does not specify the form of mercuric oxide used, we suggest that both red and yellow mercuric oxide be tested in this synthesis.

4-8. Preparation of Ethylazoisobutane [128]

$$C_2H_5NHNHCH_2-\underset{\underset{CH_3}{|}}{CH}-CH_3 \xrightarrow{\text{HgO}} C_2H_5N=NCH_2-\underset{\underset{CH_3}{|}}{CH}-CH_3 + H_2O \quad (41)$$

To a vigorously stirred suspension of 40 gm (0.138 mole) of mercuric oxide in 100 ml of water is added slowly 15 gm (0.134 mole) of N-ethyl-N'-isobutyl-hydrazine. After the addition has been completed, stirring is continued for an additional $\frac{1}{2}$ hr at room temperature. The mercury and mercury oxides are then removed by filtration. The precipitate is washed with ether and the aqueous phase is extracted with ether. The ether solutions are combined and dried with anhydrous magnesium sulfate. Then the ether is distilled off. The residue is distilled at 105°–107°C (741 mm Hg) yield 5.8 gm (40%). On redistillation, an appreciable quantity of a high-boiling residue is left behind, possibly because of thermal decomposition.

Table X lists some mixed azoalkanes and their physical properties.

By a similar technique, azoethane (b.p. 58°C) has been prepared in 97% yield [132]. In ether suspension, under nitrogen, yellow mercuric oxide has

TABLE X

PHYSICAL PROPERTIES OF MIXED AZOALKANES [123]

$$R^1N{=}NR^2$$

R^1	R^2	B.p., °C (741 mm Hg)	n_D^{25}	d_D^{25}	Yield (%)
C_2H_5	$n\text{-}C_3H_7$	86–87	1.3946	0.7603	23
C_2H_5	$i\text{-}C_3H_7$	74–75	1.3856	0.7410	59
C_2H_5	$n\text{-}C_4H_9$	119–120	1.4048	0.7755	36
C_2H_5	$i\text{-}C_4H_9$	105–107	1.3996	0.7643	40
$n\text{-}C_3H_7$	$i\text{-}C_3H_7$	101–102	1.3958	0.7549	46

been used to oxidize a series of 1-alkyl-2-phenylhydrazines to the correspond-ing phenylazoalkanes [133].

N-Bromosuccinimide has been used as an oxidizing agent, particularly for the oxidation of acylazo and more complex azo compounds such as azobis(diphenylphosphine oxide). Reaction conditions are somewhat vari-able. Some workers carry the reaction out at room temperature [119, 120, 134], others at ice or lower temperatures [135].

4-9. *Preparation of t-Butyl p-Bromophenylazoformate* [134]

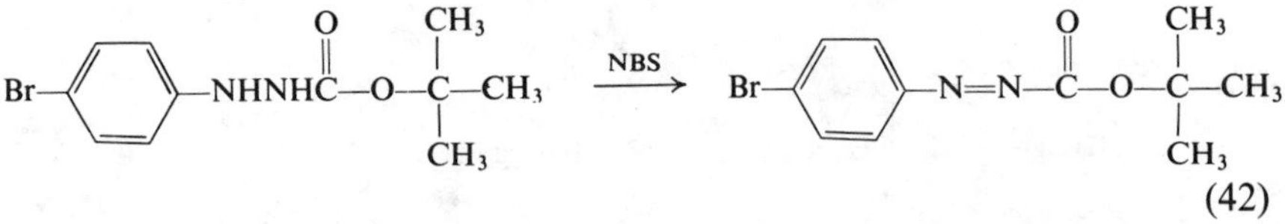

(42)

To a solution of 17.95 gm (0.0634 mole) of *t*-butyl-2-(*p*-bromophenyl) carbazate and 4.94 gm (0.0626 mole) of dry pyridine in 300 ml of methylene chloride is added, in small portions, over a 20 min period, 11.13 gm (0.0551 mole) of N-bromosuccinimide. The red solution is allowed to stand at room temperature for 3 hr. Then the reaction mixture is washed in turn with two portions of 100 ml of water, 125 ml of 10% aqueous sodium hydroxide, and another two portions of water. The product solution is then dried with anhyd-rous potassium carbonate. The solvent is removed by distillation under re-duced pressure, using a water bath at 50°C as the source of heat. On standing, the red liquid crystallizes to a yellow-orange solid which is dissolved in meth-anol, treated with charcoal, filtered, and the filtrate treated with just sufficient water to cause product precipitation: yield 15.34 gm (86%), yellow-orange crystals, m.p. 66°–67°C.

4-10. *Preparation of Phenylazodiphenylphosphine Oxide* [135]

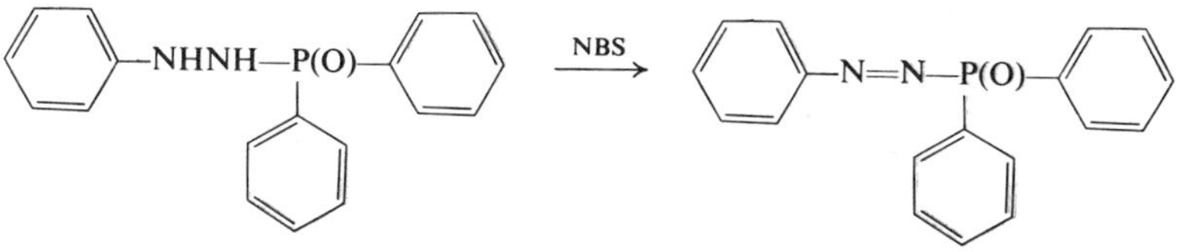

$$\text{Ph—NHNH—P(O)Ph}_2 \xrightarrow{\text{NBS}} \text{Ph—N=N—P(O)Ph}_2 \qquad (43)$$

To a stirred methylene chloride solution of 4.62 gm (15 mmoles) of 1-phenyl-2-(diphenylphosphine oxide)hydrazine maintained at $-20°C$ is added over a 10 min period 2.67 gm (15 mmoles) of N-bromosuccinimide. The solution is allowed to warm to room temperature, and stirring is continued for 10 min. The solids formed are removed by filtration and discarded. The solution is washed in turn with two portions of 5% aqueous sodium thiosulfate solution, 0.1 N hydrochloric acid, water, dilute aqueous potassium bicarbonate, and again water. The methylene chloride solution is dried over anhydrous sodium sulfate and filtered. The filtrate is evaporated to incipient crystallization at room temperature at reduced pressure: yield 4.1 gm (90%), m.p. 105°–106°C.

By a similar technique a large variety of azodicarboxylic acid derivatives have been prepared [136].

C. Hydrogen Abstraction with Diethyl Azodicarboxylate

Hückel's molecular orbital theory is said to indicate that azo compounds with strong electron-attracting substituents (e.g., alkoxycarbonyl, acyl, nitrile) should act as strong electron acceptors [137]. On this basis, it was found that diethyl azodicarboxylate behaves as an oxidizing agent which can only act by hydrogen abstraction in a nonphotochemical process. In its reaction this oxidizing agent is converted into diethyl hydrazodicarboxylate, whereas aromatic primary amines, and particularly hydrazobenzenes, are converted into azo compounds.

Unfortunately, the literature has only very general reaction details, yet, like the original authors, we believe that this reaction has considerable synthetic value and should be explored further. For this reason, we prefer not to relegate this method simply to our Miscellaneous Procedures section.

Three procedures are mentioned:

(1) The reaction of 1 equivalent of the azocarboxylate with the amine or hydrazo compound, in the dark, at room temperature, in anhydrous benzene, for up to 3 days. By this method, hydrazobenzene is converted into azobenzene in a 98% yield within $\frac{1}{2}$ hr.

(2) The reaction of a similar charge may be accelerated by refluxing for 0.5–10 hr. By this procedure, *p*-anisidine was converted into 4,4′-azoanisole in 28% yield in 5 hr.

(3) The reaction of a similar charge of reagents, but using chloroform as the solvent.

The reaction may be represented by Eq. (44).

(44)

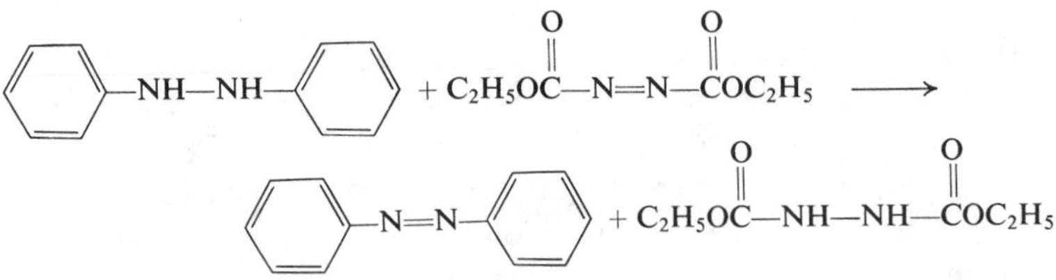

D. Oxidation with Nitric Acid and Oxides of Nitrogen

A variety of azo compounds have been produced by oxidation of the corresponding hydrazo compounds with nitric acid. The combination of thermally unstable azo compounds and strong nitric acid media is considered by us to be too hazardous to recommend it as a general procedure. However, by this means, azomethane (b.p. 1°–2°C), α,α′-azobis(isobutyronitrile) (m.p. 105°C), diethyl azodicarboxylate (b.p. 108°–110°C, 15 mm Hg; 93°–95°C, 5 mm Hg), m.p. 6°C), azodicarbonamide (m.p. 180°C), and azodibenzoyl (m.p. 118°–119°C) have been prepared [138, 139].

Similar azo compounds were also prepared by the oxidation of aliphatic hydrazo compounds with nitrogen dioxide [118].

E. Oxidation of Hydrazones

Hydrazones are readily prepared from a variety of ketones with hydrazine or one of a number of substituted hydrazines. Some hydrazones may be converted into azo compounds by an isomerization procedure (see below).

However, with lead tetraacetate, keto hydrazones may be oxidized to azoacetates according to Eq. (45) [140].

$$\begin{array}{c}R^1\\ \diagdown\\ \diagup\\ R^2\end{array}\!\!C{=}N{-}\underset{\displaystyle H}{\underset{|}{N}}{-}R^3 \xrightarrow{Pb(OCOCH_3)_4} R^2{-}\overset{\displaystyle R^1}{\overset{|}{\underset{\underset{\displaystyle \underset{\displaystyle O}{\overset{\|}{OCCH_3}}}{|}}{C}}}{-}N{=}N{-}R^3 \qquad (45)$$

The reaction of aldehyde hydrazones is more complex and requires further study [141]. It is usually carried out with a 90% solution of lead tetraacetate in acetic acid with sufficient additional solvent to bring about homogeneous reaction conditions initially.

4-11. *Preparation of 2-Acetoxyisopropylazobenzene* [140]

$$CH_3 \diagdown \atop CH_3 \diagup C=N-NH-\!\!\!\bigcirc \quad \xrightarrow{\text{Pb(OCOCH}_3)_4} \quad CH_3-\underset{\underset{O}{\overset{\displaystyle CH_3}{\overset{|}{\underset{|}{OCCH_3}}}}{\overset{|}{C}}-N=N-\!\!\!\bigcirc \qquad (46)$$

To a stirred solution containing 49 gm (0.11 mole) of lead tetraacetate (supplied as a 90% solution in acetic acid) in 200 ml of methylene chloride is added a solution of 14.8 gm (0.10 mole) of acetone phenylhydrazone in 25 ml of methylene chloride over a 15 min period while maintaining the reaction temperature between 0° and 10°C with an ice bath. After the addition has been completed, the reaction temperature is raised to 20°–25°C and stirring is continued for an additional 15 min. Then to the reaction mixture is added 200 ml of water, the inorganic solids are filtered off, and the methylene chloride layer is separated. This product layer is washed in turn with water and with dilute aqueous sodium bicarbonate until all the acetic acid has been removed. After drying the methylene chloride solution with anhydrous sodium sulfate, the solvent is evaporated off at reduced pressure. The residue is distilled under reduced pressure. The product has b.p. 89°C (1 mm Hg); yield 17.0 gm (83%).

5. ISOMERIZATION REACTIONS

A. Isomerization of Hydrazones

The problem of the tautomerism between hydrazones and azo compounds has been the subject of considerable controversy. The details of the problem have been reviewed by Buckingham [142]. Among the causes of the dispute are the fact that phenylhydrazones are readily oxidized in air to give phenylazohydroperoxides which resemble azo compounds in their spectra (both visible and ultraviolet) and the fact that the hydroperoxides may decompose to azo compounds [143].

Other papers pertinent to this controversy are those of Bellamy and Guthrie [133], Buckingham [142], and others [144–150]. The fact that all the compounds involved in this problem are also capable of geometric isomerism probably complicates the picture further.

It may well be, as Buckingham points out [142], that at room temperature, under neutral conditions, and for a limited number of structural types, such as phenylhydrazones, no tautomeric shift to azo compounds takes place. The experimental facts are, however, that, under forcing conditions, aliphatic hydrazones have indeed been converted into azo compounds although often only in modest yield.

Evidence has been presented that a genuine equilibrium exists between azo compounds and hydrazones, at least in the case of aliphatic cyclic compounds [151] (Eq. 47). The proposed mechanism was assumed to involved anion

$$C_nH_{2n}\text{—}C\text{=}N\text{—}NH\text{—}CH_3 \rightleftharpoons C_nH_{2n}\text{—}CH\text{—}N\text{=}N\text{—}CH_3 \qquad (47)$$

$$n = 4, 5, 6, 7$$

formation, as shown in Eq. (48). The mechanism does not, however, explain why the specific reagent required for the protropy is potassium hydroxide in molar amounts. Both sodium methylate and diethanolamine were ineffective.

$$-\overset{..}{\underset{H}{N}}\text{—}\overset{..}{N}\text{=}C\diagdown \rightleftharpoons :\overset{..}{\underset{H}{N}}\text{—}\overset{..}{N}\text{=}C\diagdown \rightleftharpoons :\overset{..}{N}\text{=}\overset{..}{N}\text{—}\overset{\ominus}{C}\text{—}H \qquad (48)$$

While the isomerization did not take place above 120°C, the usual boiling range of azoalkanes, it could be carried out at reduced pressure. It appears critical that a minimum amount of heat be applied to the system to cause refluxing by judicious use of low pressures and careful insulation of the apparatus. The method fails in an attempt to isomerize cyclooctanone methylhydrazone. The bond angle strain is said to effect the ease of rearrangement of the hydrazone molecules to the azo isomer. To be noted also is that the azo compounds have somewhat lower boiling points than the isomeric hydrazones. This contributes to the formation of the azo product by a careful distillation technique.

5-1. *General Procedure for the Isomerization of Hydrazones to Azoalkanes* [151]

$$C_nH_{2n}\text{—}C\text{=}N\text{—}NH\text{—}CH_3 \xrightarrow{\text{KOH}} C_nH_{2n}\text{—}CH\text{—}N\text{=}NCH_3 \qquad (49)$$

$$n = 4\text{–}7$$

(a) *Preparation of anhydrous potassium hydroxide.* Reagent grade potassium hydroxide pellets (6 gm) and a few small boiling chips are placed in a 50 ml round-bottomed flask connected through a condenser to a vacuum pump. The pressure in the flask is reduced and the content is slowly heated with an infrared lamp until the pellets melt. As the moisture is removed, the potassium hydroxide forms a white porous solid. Heating is continued for a minimum of 48 hr, after which the potassium hydroxide is used immediately.

(*b*) *Isomerization.*   In each synthesis approximately 15 gm (approximately 0.1 mole) of the hydrazone is added to the flask containing 6 gm (approximately 0.1 mole) of dried potassium hydroxide. The flask is fitted with a vacuum fractionation column, 15 cm long and 2.5 cm in diameter, glass-helix-packed to a height of approximately 3 cm. The top of the flask and the distillation column are carefully insulated with glass wool and aluminum foil, and heating is carried out with a Glas-Col heating mantle. The distillation head is connected to an oil pump through a vacuum gauge and a manostat. Heat is slowly applied at reduced pressure until refluxing takes place at a temperature below 120°C. Then, while maintaining a constant pressure, product is slowly removed through the distillation head. Collection of product is continued until a small increase in the head temperature is observed. Then the head is set for total reflux until the temperature in the head again stabilizes. Product is again collected and the process is continued. The use of an automatic distillation head whose takeoff valve is controlled by the head temperature rather than by the more usual time-cycling technique should be particularly useful in this preparation.

Isomerization times range from 11 hr (in the case of cyclohexylazomethane) to 72 hr (in the case of cycloheptylazomethane). Yields are in the range 84–87% except in the case of cyclooctylazomethane, which failed to isomerize. The azo compounds produced by this procedure were identified by elemental analysis, boiling point, infrared spectra, and ultraviolet spectra.

Table XI gives physical data for a series of alkyl azoalkanes produced by this procedure.

B. Preparation of Cis Azo Compounds

The normally prepared azo compounds are, as expected, predominantly in the trans form. However, on chromatographing typical azo compounds on an alumina column, a small, extraneous band had been observed which was subsequently identified as a small amount of the cis isomer.

On irradiation of *trans*-azobenzene (m.p. 68°C) with ultraviolet light, the concentration of *cis*-azobenzene was increased. It had a melting point of 71°C. A mixture of *cis*- and *trans*-azobenzenes had a melting range from 35°C to 45°C [72].

It was also observed that, for example, in the case of *p,p'*-azotoluene, on melting, the cis form was converted into the more stable trans form.

To be noted in the preparation of *cis-p,p'*-azotoluene is the chromatographic technique used in the separation. This is a method which has been used widely in work with azo compounds as a general means of purification.

TABLE XI

PHYSICAL PROPERTIES OF ALKYL AZOALKANES

$$R'-N=N-R''$$

R'	R"	B.p., °C (mm Hg)	d_{20} (gm/ml)	n_D^{20}	Ref.
C_5H_{10}	CH_3	104 (743)	$0.8647_{25°C}$	1.4414	151
C_6H_{12}	CH_3	76 (66)	—	1.4490	151
C_7H_{14}	CH_3	92 (20)	—	—	151
$(CH_3)_2CH$	CH_3	49.4–49.9[a]	0.7377	1.3797	149
$(CH_3)_2CHCH_2$	CH_3	79.5–80[a]	0.7583	1.3941	149
$(C_2H_5)(CH_3)CH$	CH_3	78.5–78.7[a]	0.7594	1.3943	149
$n\text{-}C_3H_7$	C_2H_5	87.5–87.7[a] (78)	0.7631	1.3969	149
$(CH_3)_2CH$	C_2H_5	75.5[a]	0.7468	1.3887	149
$(CH_3)_2CHCH_2$	C_2H_5	105.4–105.6[a]	0.7646	1.4015	149
$(CH_3)_2CH$	$n\text{-}C_3H_7$	101.8–102.3[a]	0.7596	1.3987	149
$(CH_3)_2CHCH_2$	$n\text{-}C_3H_7$	129.4–129.6[a]	0.7743	1.4094	149
$n\text{-}C_3H_7$	$n\text{-}C_3H_7$	112.4–112.7[a]	0.7730	1.4065	149
$(CH_3)_2CH$	$(CH_3)_2CH$	89.6–89.7	0.7419	1.3901	149
$C_6H_5CH_2$	CH_3	67.9–68.5 (10)	0.9635	1.5166	149

[a] Pressure 1 atm.

5-2. Preparation of cis-p,p'-Azotoluene [97]

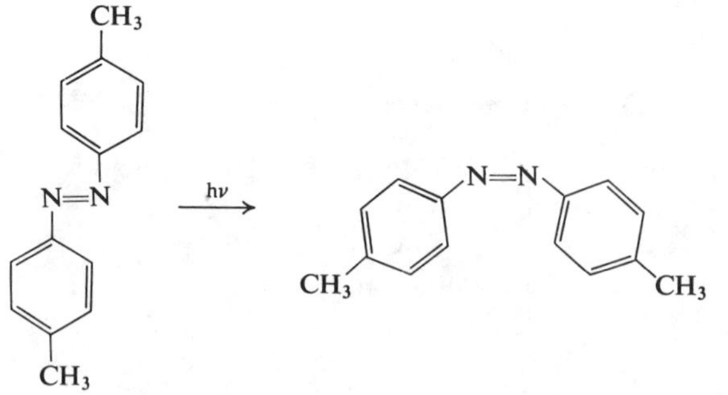

$$(50)$$

With adequate protection of personnel against stray radiation, a solution of 400 mg of *trans-p,p'*-azotoluene in 100 ml of petroleum ether is subject to the radiation from an unshielded quartz mercury lamp for 30 min at a distance of 30 cm.

The solution is then chromatographed by passage through a 20 cm × 2 cm column packed with aluminum oxide (Merck). The trans isomer is completely eluted from the column with 120–150 ml of petroleum ether. The *cis-p,p'*-azotoluene remains as a layer 1.5–2.0 cm long beginning about 1 cm from the top of the column. It is isolated by elution with 150 ml of petroleum ether containing 1 % of methanol. The resulting eluent is washed with water and dried with anhydrous sodium sulfate. The solvent is removed partially by evaporation under reduced pressure. From a low volume of solution the product is crystallized by cooling in an ice-water bath: yield 30 mg (7.5 %), m.p. 105°C, mixed m.p. with trans isomer 95°–100°C.

By a similar procedure, 1.0 gm of *p*-bromoazobenzene was isomerized in 10 ml of chloroform by irradiation with a Hanovia type 16200 ultraviolet lamp for 2 hr [152]. The same authors, using the same equipment, prepared 1.0 gm of *cis-p,p'*-azotoluene from 6.0 gm of the trans isomer by irradiation for 4 hr. They eluted the product from an aluminum oxide column with ether [152].

The synthesis of bis(crown ethers) with azo linkages normally produces the more stable trans isomers [87, 108, 20a]. Upon irradiation with a 500-W high-pressure mercury lamp, the trans form was converted to the cis and the kinetics of the process was studied. The cis isomer could be isomerized thermally to the trans.

A recent study attempted to differentiate between an inversion and a rotational mechanism for the trans → cis (anti → syn) isomerism. This work indicates that the isomerization proceeds by an inversional mechanism [152a]. Table XII lists the melting points of many of the azo compounds used in this study. Some of the symmetrically substituted azobenzenes used had melting points similar to those recorded in Table VIII.

Rau and Lüddeke [152b] also proposed a rotational mechanism on the basis of work with azobenzenophanes for which inversion is the only isomerization pathway.

TABLE XII

FURTHER PHYSICAL PROPERTIES OF AROMATIC AZO COMPOUNDS
[152a]

Compound	M.p. (°C)
4-Methoxyazobenzene	55.5–56
4-Methylazobenzene	70–72
4-(Trifluoromethyl)azobenzene	98.5–99
4,4'-Bis(trifluoromethyl)azobenzene	103–104
2,4',6-Trimethylazobenzene	Red oil
2,2',6,6-Tetramethylazobenzene	49–49.5
2,6-Dimethyl-4'-methoxyazobenzene	37.2–40.3
2,6-Dimethyl-4'-(trifluoromethyl)azobenzene	Red oil

6. MISCELLANEOUS PROCEDURES

(1) Oxidation of dialkylhydrazines with cupric chloride [153] (Eq. 51).

$$R—NH—NHR \xrightarrow{\text{CuCl}_2} R—N{=}N—R \cdot Cu_2Cl_2 \xrightarrow{\Delta} R—N{=}N—R + Cu_2Cl_2$$

$$(51)$$

(2) Oxidation of acylhydrazines with *t*-butyl hypochlorite in acetone at $-50°$ to $-70°C$ [154] (Eq. 52). This procedure is apparently suitable for azo

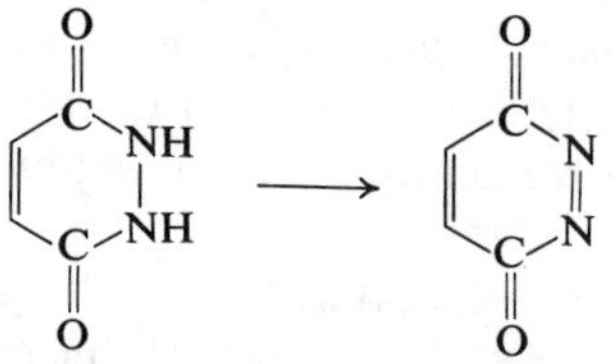

$$(52)$$

compounds which are sensitive to acids and bases. It has been applied to simpler reaction systems.

(3) Oxidation of hydrazines by complex formation with HgO followed by treatment with halogens [126].

(4) Oxidation of acylhydrazines with lead tetraacetate [155] (Eq. 53).

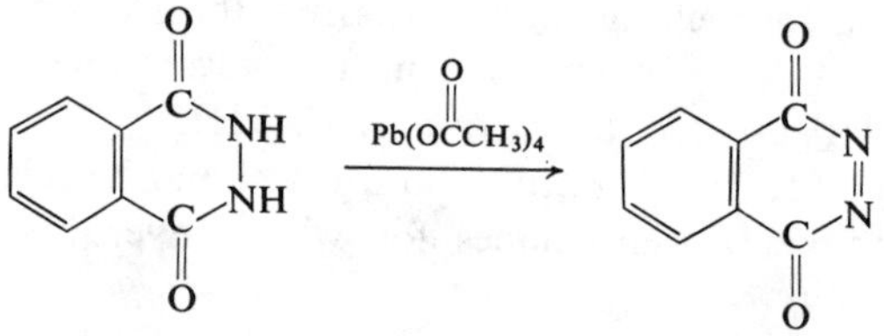

$$(53)$$

(5) Oxidation of hydrazines with cuprous chloride and air [156] (Eq. 54).

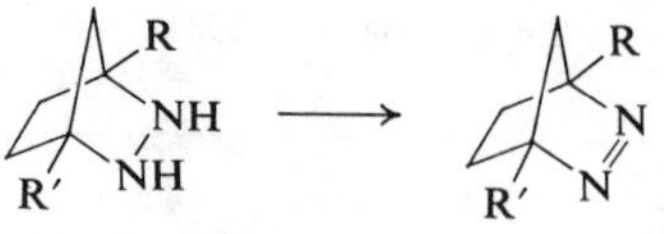

$$(54)$$

(6) Oxidation of hydrazines with iodine pentafluoride [112].

(7) Oxidation of isocyanates with 100% hydrogen peroxide [157] (Eq. 55).

$$RNCO + HOOH + RNCO \longrightarrow \left[RNH\overset{\text{O}}{\overset{\|}{C}}—O—O—\overset{\text{O}}{\overset{\|}{C}}—NHR \right] \longrightarrow$$

$$[2RNH—CO_2\cdot] \longrightarrow RNHNHR \xrightarrow{\text{H}_2\text{O}_2} RN{=}NR \quad (55)$$

(8) Catalytic hydrogenation of macrocyclic azines [104] (Eq. 56).

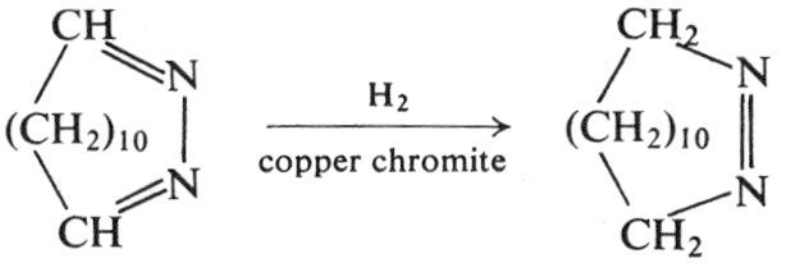

$$(56)$$

(9) Reduction of azoxy compounds with triethylphosphine [158] (Eq. 57).

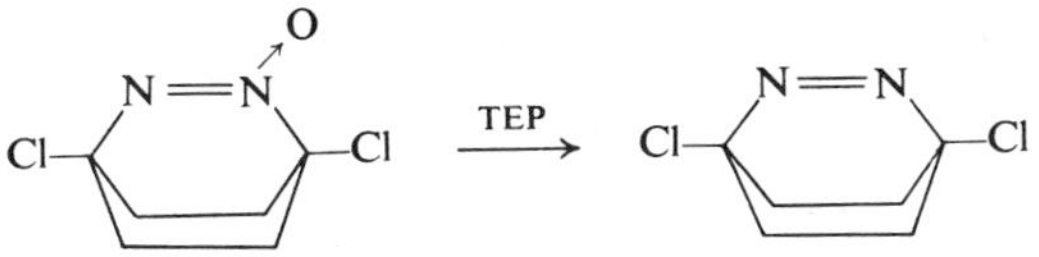

$$(57)$$

(10) Coupling of amidrazones with phenols, active methylene compounds, or aromatic amines [159] (Eq. 58).

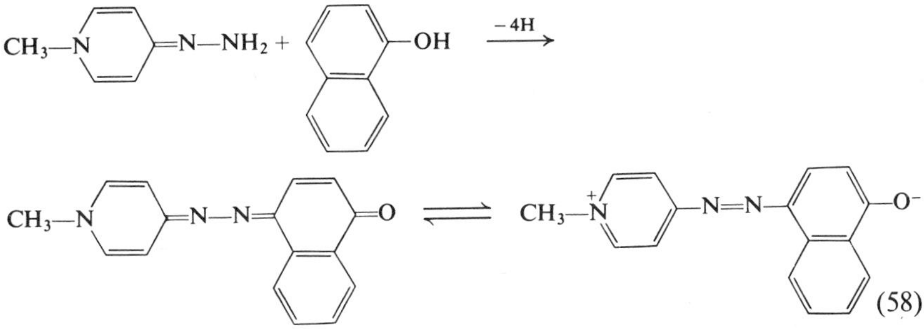

$$(58)$$

(11) Azo compounds from ketones with hydroxylamine sulfuric acid [160] (Eq. 59).

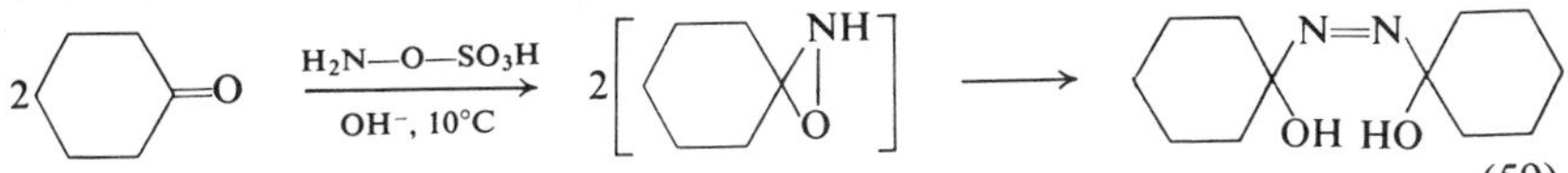

$$(59)$$

(12) From phenols with phenylhydrazines and perchloryl fluoride [161] (Eq. 60).

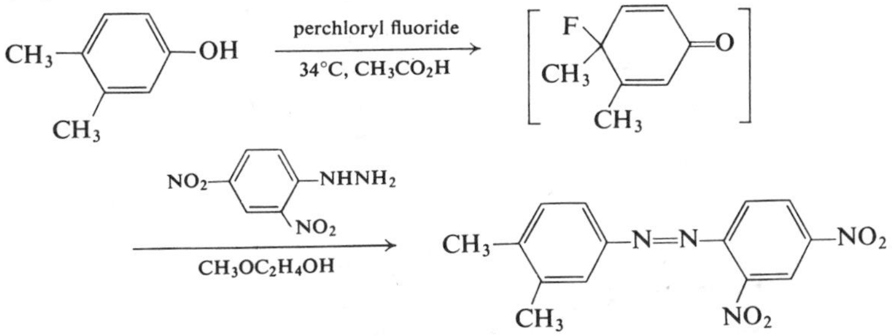

$$(60)$$

(13) Reduction of aromatic nitro compounds with carbon monoxide [162] (Eq. 61).

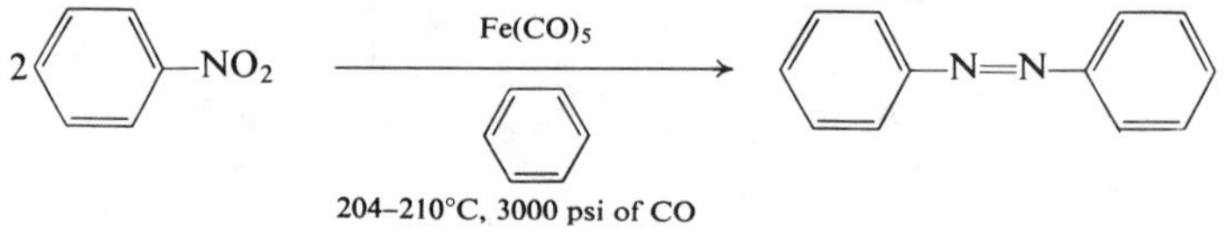

$$2 \underset{}{\bigcirc}-NO_2 \quad \xrightarrow[\underset{204-210°C, \ 3000 \ psi \ of \ CO}{}]{Fe(CO)_5} \quad \bigcirc-N{=}N-\bigcirc \qquad (61)$$

(14) Reduction of aromatic nitro compounds with stannous chloride in basic medium [104].

(15) The addition of 2-diazo-1,3-diketones to active methylene compounds [163] (Eq. 62).

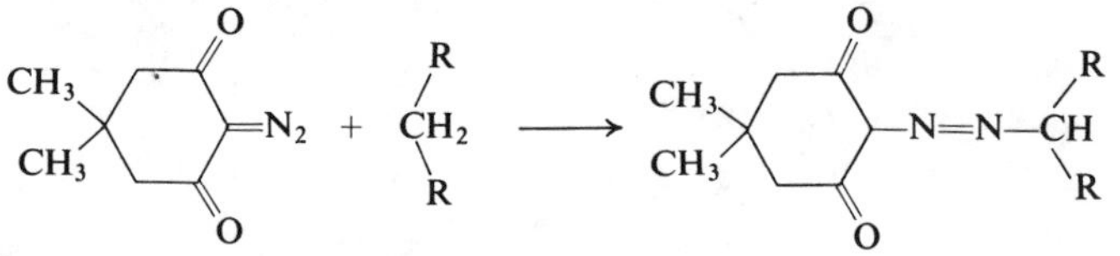

$$\text{(62)}$$

(16) Coupling of fluorinated nitriles [164] (Eqs. 63, 64) (see Table XIII).

$$CF_3CF_2CN \quad \xrightarrow[\underset{and \ pressure}{Br_2, \ heat}]{AgF} \quad CF_3CF_2-N{=}N-CF_2CF_3 \qquad (63)$$

$$R_fCN \quad \xrightarrow[\underset{[165]}{\underset{at \ -50°C \ to \ -40°C}{or \ BrF_3 \ and \ Br_2}}]{BrF_3} \quad R_fCF_2N{=}NCF_2R_f \qquad (64)$$

(17) Preparation of 1,1′-azobisformamides and biureas from dialkyl azodiformates [166] (Eq. 65).

$$C_2H_5OC{-}N{=}N{-}COC_2H_5 + RNH_2 \quad \longrightarrow \quad RNHC{-}N{=}N{-}CNHR \qquad (65)$$

TABLE XIII

PROPERTIES OF FLUORINATED AZO COMPOUNDS
[164] PREPARED BY COUPLING OF NITRILES

Compound	B.p., °C (mm Hg)
$(CF_3CF_2N)_2$	68–71[a]
$(CF_3CF_2CF_2CF_2N)_2$	113[a]
$[CH_3(CF_2)_6CF_2N]_2$	67–65 (0.2)
$(HCF_2CF_2CF_2N)_2$	106–108[a]
$[HCF_2(CF_2)_3CF_2N]_2$	112–115 (28)

[a] Pressure 1 atm.

(18) Preparation of polyazobenzenes [167] (Eq. 66).

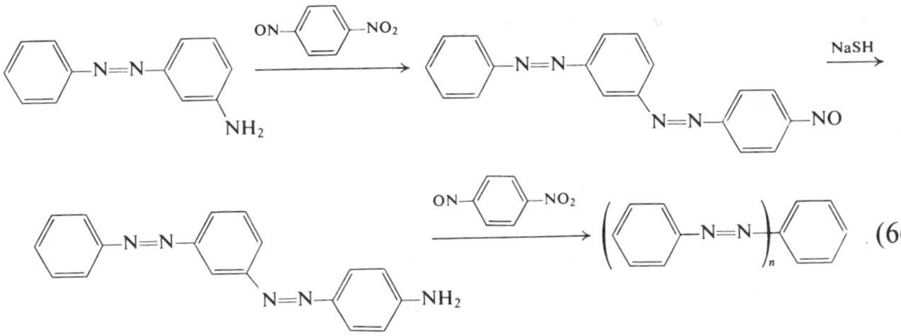

(19) Periodic acid oxidation of cyclic phenylhydrazine derivatives [168] (Eq. 67).

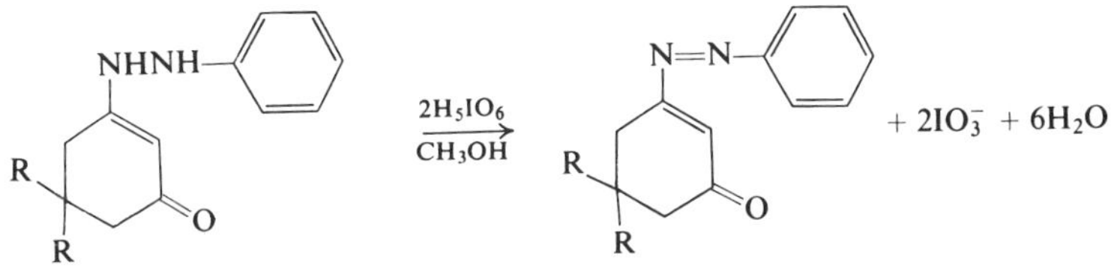

$$+ 2IO_3^- + 6H_2O \qquad (67)$$

(20) Preparation of α-azohydroperoxides (CAUTION: Explosive) by oxidation of phenylhydrazones with molecular oxygen [169–171].

(21) Preparation of 2,19-dithia[3.3](4,4′)-*trans*-diphenyldiazeno(2)phene and related azobenzenophanes [152b].

(22) Preparation of polycyclic azoalkanes by reaction of a triazoline dione (urazoles) by base-catalyzed hydrolysis followed by oxidation [172, 173] (Eq. 68).

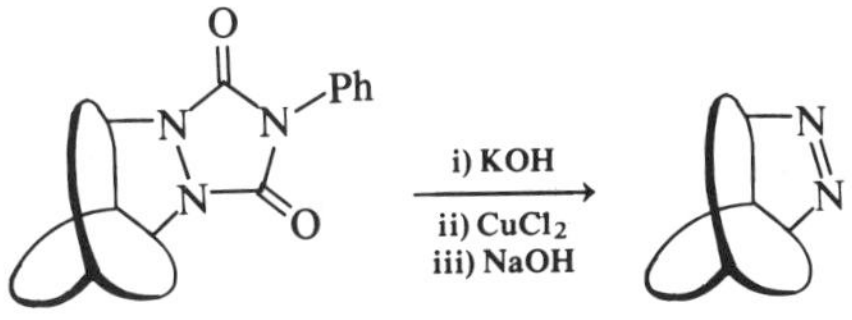

$$(68)$$

REFERENCES

1. "International Union of Pure and Applied Chemistry, Nomenclature of Organic Chemistry," Sect. C, Butterworth, London and Washington, D.C., 1965, p. 204 ff. IUPAC: Rule C-911; Chem. Abstr. Meth.: Rule C-912.
2. *Chemical Abstracts*, Index Guide, **86–95**, 118g (1977–1981).
3. C. J. Overberger and T. F. Merkel, *J. Org. Chem.* **46**, 442 (1981).
4. S. Patai, ed., "The Chemistry of the Hydrazo, Azo, and Azoxy Groups," Wiley, New York, 1975.
5. Society of Dyers and Colourists, "Colour Index," 3rd ed., American Association of Textile Chemists and Colourists, Research Triangle Park, N.C., 1971.
6. H. Zollinger, "Azo and Diazo Chemistry—Aliphatic and Aromatic Compounds," Wiley (Interscience), New York, 1961.

7. K. H. Saunders, "The Aromatic Diazo-Compounds and their Technical Applications," Longmans, Green, New York, 1949.
8. H. Bock, *Angew. Chem.* **77**, 469 (1965).
9. E. N. Abrahart, "Dyes and their Intermediates," 2nd ed., Edward Arnold, London, 1977.
10. P. Rys and H. Zollinger, "Fundamentals of the Chemistry and Application of Dyes," Wiley, London, 1972.
11. K. Venkatarama, "The Chemistry of Synthetic Dyes," Academic Press, New York, 1952.
12. J. A. Miller and E. C. Miller, *Advan. Cancer Res.* **1**, 339 (1953).
13. J. A. Miller and G. C. Finger, *Cancer Res.* **17**, 387 (1957).
14. R. Wistar and P. D. Bartlett, *J. Amer. Chem. Soc.* **63**, 413 (1941).
15. C. R. Hauser and D. S. Breslow, *J. Amer. Chem. Soc.* **63**, 418 (1941).
16. W. Bradley and J. D. Thompson, *Chimia* **15**, 147 (1961) (see *Nature* **178**, 1069 (1956) for preparation of organic soluble diazonium salts).
17. W. W. Hartman and J. B. Dickey, *Org. Syn. Coll. Vol.* **2**, 163 (1943).
18. E. F. Elslager, and D. F. Worth, *J. Med. Chem.* **13**, 370 (1970).
19. J. J. Randall, C. E. Lewis, and P. M. Slagan, *J. Org. Chem.* **27**, 4098 (1962).
20. J. F. Bunnett and G. B. Hoey, *J. Amer. Chem. Soc.* **80**, 3142 (1958).
20a. S. Shinkai, T. Minami, Y. Kusano, and O. Manabe, *J. Amer. Chem. Soc.* **104**, 1907 (1982).
21. E. F. Eislager, D. F. Worth, D. B. Capps, L. M. Werbel, and F. W. Short, U.S. Pat. 3,218,309 (Nov. 16, 1965).
22. K. H. Meyer and H. Tochtermann, *Chem. Ber.* **54**, 2283 (1921).
23. L. F. Fieser and W. P. Campbell, *J. Amer. Chem. Soc.* **60**, 1142 (1938).
24. A. G. Anderson and B. M. Heckler, *J. Amer. Chem. Soc.* **81**, 4941 (1959).
25. D. Vorländer, *Ann. Chem.* **320**, 122 (1902).
26. K. Clusius and F. Endtinger, *Helv. Chim. Acta* **43**, 566 (1960).
27. B. M. Bogoslovskii, *J. Gen. Chem. USSR* **16**, 193 (1946).
28. D. C. Freeman, Jr., and C. E. White, *J. Org. Chem.* **21**, 379 (1956).
29. M. Lange, Ger. Pat. 78,255 (March 28, 1894).
30. F. Suckfüll and H. Dittmer, *Chimia* **15**, 137 (1961); M. Christen, L. Funderbunk, E. A. Halevi, G. E. Lewis, and H. Zollinger, *Helv. Chim. Acta* **49**, 1376 (1966).
31. O. A. Stamm and H. Zollinger, *Chimia* **15**, 535 (1961).
32. S. M. Parmerter, The coupling of diazonium salts with aliphatic carbon atoms. *Org. React.* **10**, 1 (1959).
33. R. R. Phillips, The Japp-Klingemann reaction. *Org. React.* **10**, 143 (1959).
34. E. Gudriniece and G. Vanags, *Zh. Obshch. Khim.* **28**, 58 (1958).
35. L. Sakhar and E. Gudriniece, *Zh. Org. Khim.* **2**, 674 (1966).
36. R. Resnick and H. C. Yao, U.S. Pat. 3,249,598 (May 3, 1966).
37. W. Borsche and A. Herbert, *Ann Chem.* **546**, 293 (1941).
38. V. von Richter, *Chem. Ber.* **16**, 677 (1883).
39. A. Albert and A. Hampton, *J. Chem. Soc.* 4985 (1952).
40. R. Stoermer and O. Gans, *Chem. Ber.* **45**, 3104 (1912).
41. P. Ruggli and W. Wüst., *Helv. Chim. Acta* **28**, 781 (1945).
42. W. H. Nutting, R. A. Jewell, and H. Rapoport, *J. Org. Chem.* **35**, 505 (1970).
43. R. W. Faessinger and E. V. Brown, *J. Amer. Chem. Soc.* **73**, 4606 (1951).
44. D. W. Lamson, R. Sciarro, D. Hryb, and R. O. Hutchins, *J. Org. Chem.* **38**, 1952 (1973).
45. E. C. Taylor, C.-P. Tseng, and J. B. Rampal, *J. Org. Chem.* **47**, 552 (1982).
46. Y. M. Wu, L. Y. Ho, and C. H. Cheng, *J. Org. Chem.* **50**, 392 (1985).
47. A. Risaliti, S. Bozzini, and A. Steuer, *Tetrahedron* **25**, 143 (1969).
48. N. Ishikawa, M. J. Namkung, and T. L. Fletcher, *J. Org. Chem.* **30**, 3878 (1965).
49. M. J. Namkung, N. K. Naimy, C. A. Cole, N. Ishikawa, and T. L. Fletcher, *J. Org. Chem.* **35**, 728 (1970).

50. A. H. Dinwoodie and R. N. Hazeldine, *J. Chem. Soc.* 2266 (1965).
51. G. A. Russell, R. Konaka, E. T. Strom, W. C. Danen, K.-Y. Chang, and G. Kaupp, J. Amer. Chem. Soc. **90**, 4646 (1968).
52. H. E. Fierz-David, L. Blangey, and E. Merian, *Helv. Chim. Acta* **34**, 846 (1951).
53. A. Michaelis and K. Petou, *Chem. Ber.* **31**, 984 (1898).
54. R. Ohme and E. Schmitz, *Angew. Chem.* **77**, 429 (1965) *Angew. Chem. Int. Ed. Engl.* **4**, 433 (1965).
55. F. D. Greene, M. A. Berwick, and J. C. Stowell, *J. Amer. Chem. Soc.* **92**, 867 (1970); J. C. Stowell *J. Org. Chem.* **32**, 2360 (1967).
56. R. Ramart-Lucas, T. Builmart, and M. Martynoff, *Bull. Soc. Chim. Fr.* 415 (1947).
57. G. M. Badger and G. E. Lewis, *J. Chem. Soc.* 2151 (1953).
58. J. van Alphen, *Rec. Trav. Chim. Pays Bas* 109 (1945).
59. B. T. Gillis and J. D. Hagerty, J. Amer. Chem. Soc. **87**, 4576 (1965).
60. H. H. Hodgson and E. Marsden, *J. Chem. Soc.* 274 (1945).
61. D. Y. Curtin and J. A. Ursprung, *J. Org. Chem.* **21**, 1221 (1956).
62. R. Tarao and Y. Nomura, *Sci. Papers Coll. Gen. Educ. Univ. Tokyo* **11**, (2), 201 (1961); *Chem. Abstr.* **58**, 3337e (1963).
63. Y. Nomura, *Bull. Chem. Soc. Japan* **34**, 1648 (1960).
64. Y. Nomura and H. Anzai, *Bull. Chem. Soc. Japan* **35**, 111 (1962).
65. Y. Nomura, H. Anzai, R. Tarao, and K. Shoimi, *Bull. Chem. Soc. Japan* **37**, 967 (1964).
66. Y. Nomura and H. Anzai, *Bull. Chem. Soc. Japan* **37**, 970 (1964).
67. L. Horner and J. Dehnet, *Chem. Ber.* **96**, 786 (1963).
68. C. G. Overberger and J.-P. Anselme, *J. Amer. Chem. Soc.* **86**, 658 (1964).
69. C. G. Overberger, N. Weinshenker, and J.-P. Anselme, *J. Amer. Chem. Soc.* **86**, 5364 (1964).
70. D. E. McGreer, P. Morris, and G. Carmichael, *Canad. J. Chem.* **41**, 726 (1963).
71. D. E. McGreer, N. W. K. Chiu, M. G. Vinje, and K. C. Wong, *Canad. J. Chem.* **43**, 1407 (1965).
72. E. Cullen and P. L'Écuyer, *Canad. J. Chem.* **39**, 862 (1961).
73. D. A. Blackadder and C. Hinshelwood, *J. Chem. Soc.* 2898 (1957).
74. K. Tabai and K. Natou, *Bull. Chem. Soc. Japan* **39**, 2300 (1966); **40**, 1538 (1967).
75. W. Tadros, M. S. Ishak, and E. Bassili, *J. Chem. Soc.* 629 (1959).
76. E. Tauber, *Chem. Ber.* **24**, 3085 (1891).
77. B. T. Newbold, *J. Chem. Soc.* 4260 (1961).
78. H. C. Waterman and D. L. Vivian, *J. Org. Chem.* **14**, 289 (1949).
79. P. Z. Slack and R. Slack, *Nature* **160**, 437 (1947).
80. S. D. Ross, G. J. Kahan, and W. A. Leach, *J. Amer. Chem. Soc.* **74**, 4122 (1952).
81. J. L. Everett and W. C. J. Ross, *J. Chem. Soc.* 1972 (1949).
82. M. Busch and K. Schulz, *Chem. Ber.* **62B**, 1458 (1929).
83. T. Wohlfart, *J. Prakt. Chem.* [2] **65**, 295 (1902).
84. J. Meisenheimer and K. Witte, *Chem. Ber.* **36**, 4153 (1903).
85. H. E. Bigelow and D. B. Robinson, *Org. Syn. Coll. Vol.* **3**, 103 (1955).
86. G. M. Badger and G. E. Lewis, *J. Chem. Soc.* 2147 (1953).
87. S. Shinkai, T. Nakaji, T. Ogawa, K. Shigematsu, and O. Manabe, *J. Amer. Chem. Soc.* **103**, 111 (1951).
88. H. J. Shine and J. T. Chamness, *J. Org. Chem.* **28**, 1232 (1963).
89. J. Radell, L. Spialter, and J. Hollander, *J. Org. Chem.* **21**, 1051 (1956).
90. R. Adams and J. R. Johnson, "Laboratory Experiments in Organic Chemistry," 4th ed., p. 359, Macmillan, New York, 1949.
91. R. F. Nystrom and W. G. Brown, *J. Amer. Chem. Soc.* **70**, 3738 (1948).
92. M. J. S. Dewar and R. S. Goldberg, *Tetrahedron Lett.* 2717 (1966).
93. G. M. Badger, J. H. Seidler, and B. Thomson, *J. Chem. Soc.* 3207 (1951).

94. M. Z. Barakat, M. F. Abdel-Wahab, and M. M. El-Sadr, *J. Chem. Soc.* 4685 (1956).
95. E. Pfeil and K. H. Schmidt, *Ann. Chem.* **675**, 36 (1964).
96. D. Lefort, C. Four, and A. Pourchez, *Bull. Soc. Chim.* 2378 (1961).
97. A. H. Cook, *J. Chem. Soc.* 876 (1938).
98. O. Wallach and L. Belli, *Chem. Ber.* **13**, 525 (1880).
99. O. H. Wheeler and D. Gonzalez, *Tetrahedron* **20**, 189 (1964).
100. M. Badger, R. G. Buttery, and G. E. Lewis, *J. Chem. Soc.* 2143 (1953).
101. J. F. Vozza, *J. Org. Chem.* **34**, 3219 (1969).
102. E. F. Pratt and J. F. Ven de Castle, *J. Org. Chem.* **26**, 2973 (1961).
103. H. Alper and M. Gopal, *J. Org. Chem.* **46**, 2593 (1981).
104. J. Kossanyi, *C. R. Acad. Sci. Paris* **157**, 929 (1963).
105. S. Goldschmidt and B. Acksteiner, *Chem. Ber.* **91**, 502 (1958); *Ann. Chem.* **618**, 173 (1958).
106. E. Benzig, *Angew. Chem.* **72**, 709 (1960); see also E. Benzig, *Chimia* **13**, 89 (1959); *Ann. Chem.* **631**, 1 (1960).
107. S. E. Scheppele and S. Seltzer, *J. Amer. Chem. Soc.* **90**, 358 (1968).
108. S. Shinkai, T. Ogawa, Y. Kusano, O. Manabe, C. Kikukawa, T. Goto, and T. Matsuda, *J. Amer. Chem. Soc.* **104**, 1960 (1982).
109. E. I. du Pont de Nemours & Co., Brit. Pat. 1,168,406 (October 22, 1969).
110. E. Bamberger and F. Tschirner, *Chem. Ber.* **31**, 1523 (1898).
111. J. Burdon, C. J. Morton, and D. F. Thomas, *J. Chem. Soc.* 2621 (1965).
112. T. E. Stevens, *J. Org. Chem.* **26**, 2531 (1961).
113. S. F. Nelsen and P. D. Bartlett, *J. Amer. Chem. Soc.* **88**, 137 (1966).
114. J. W. Timberlake and J. C. Martin, *J. Org. Chem.* **33**, 4054 (1968).
115. J. R. Shelton, J. F. Gormish, C. K. Liang, P. L. Samuel, P. Kovacic, and L. W. Hayes, *Canad. J. Chem.* **46**, 1149 (1968).
116. H. C. Bach, *Amer. Chem. Soc. Div. Polymer Chem. Preprints* **8**, (1), 610 (1967).
117. E. Fahr and H. Lind, *Angew. Chem.* **78**, 376 (1966); *Angew. Chem. Int. Ed. Engl.* **5**, 372 (1966).
118. F. D. Vidal and V. D. Sarli, U.S. Pat. 3,192,196 (June 29, 1965).
119. H. Bock and G. Rudolph, *Chem. Ber.* **98**, 2273 (1965).
120. H. Bock and E. Baltin, *Chem. Ber.* **98**, 2844 (1965).
121. S. G. Cohen and C. H. Wang, *J. Amer. Chem. Soc.* **77**, 2457 (1955).
122. G. M. Badger and G. E. Lewis, *J. Chem. Soc.* 2147 (1953).
123. C. G. Overberger, P. T. Huang, and M. B. Berenbaum, *Org. Syn. Coll. Vol.* **4**, 66 (1963).
124. M. C. Chaco and N. Rabjohn, *J. Org. Chem.* **27**, 2765 (1962).
125. G. W. Kenner and R. J. Stedman, *J. Chem. Soc.* 2089 (1952).
126. D. Y. Curtin and T. C. Miller, *J. Org. Chem.* **25**, 885 (1960).
127. R. D. Little and M. G. Venegas, *J. Org. Chem.* **43**, 2921 (1978).
128. L. Spialter, D. H. O'Brien, G. L. Untereiner, and W. A. Rush, *J. Org. Chem.* **30**, 3278 (1965).
129. C. G. Overberger, J. W. Stoddard, C. Yaroslavsky, H. Katz and J.-P. Anselme, *J. Amer. Chem. Soc.* **91**, 322 (1969).
130. C. G. Overberger and M.-S. Chi, *J. Org. Chem.* **46**, 303 (1981).
131. R. J. Crawford, A. Mishra, and R. J. Dummel,, *J. Amer. Chem. Soc.* **88**, 3959 (1966).
132. R. Renaud and L. C. Leitch, *Canad. J. Chem.* **32**, 549 (1954).
133. A. J. Bellamy and R. D. Guthrie, *J. Chem. Soc.* 2788 (1965).
134. L. A. Carpino, P. H. Terry, and P. J. Crowley, *J. Org. Chem.* **26**, 4336 (1961).
135. H. Bock, G. Rudolph, and E. Baltin, *Chem. Ber.* **98**, 2054 (1965).
136. H. Bock and J. Kroner, *Chem. Ber.* **99**, 2039 (1966); E. Baltin and J. Kroner, *ibid.* **99**, 337 (1966).
137. F. Yoneda, K. Suzuki, Y. Nitta *J. Amer. Chem. Soc.* **88**, 2328 (1966).
138. J. P. Picard and J. C. Boivin, *Canad. J. Chem.* **29**, 223 (1951).

139. J. C. Kauer, *Org. Syn. Coll. Vol.* **4**, 411 (1963).
140. D. C. Iffland, L. Salisbury, and W. R. Schafer, *J. Amer. Chem. Soc.* **83**, 747 (1961).
141. W. A. F. Gladstone, *Chem. Commun.* 179 (1969).
142. J. Buckingham, *Quart. Rev.* **23**, 37 (1969).
143. A. V. Chernova, R. R. Shagidullin, and Yu. P. Kitaev, *Zh. Org. Khim.* **3**, 916 (1967); *Chem. Abstr.* **67**, 53349k.
144. R. O'Connor, *J. Org. Chem.* **26**, 4375, 5208 (1961).
145. R. O'Connor and G. Henderson, *Chem. Ind. (London)* **850** (1965).
146. A. J. Bellamy and R. D. Guthrie, *Chem. Ind. (London)* 1575 (1964).
147. A. J. Bellamy and R. D. Guthrie, *J. Chem. Soc.* 2788, 3528 (1965).
148. B. V. Ioffe, Z. I. Sergeeva, and V. S. Stopski, U.S.S.R. Pat. 174,188 (Aug. 28, 1965).
149. B. V. Ioffe, Z. I. Sergeeva, and V. S. Stopski, *Dokl. Akad. Nauk SSSR* **167**, 831 (1966); *Chem. Abstr.* **65**, 3733c (1966).
150. B. V. Ioffe and V. S. Stopskii, *Tetrahedron Lett.* 1333 (1968).
151. J. J. Scheloske, Wright Air Development Center, AD 603684, 109 pp. (1964).
152. D. L. Webb and H. H. Jaffe, *J. Amer. Chem. Soc.* **86**, 2419 (1964).
152a. J. P. Otruba III and R. G. Weiss, *J. Org. Chem.* **48**, 3448 (1983).
152b. H. Rau and E. Lüddeke, *J. Amer. Chem. Soc.* **104**, 1616 (1982).
153. F. P. Jahn, *J. Amer. Chem. Soc.* **59**, 1761 (1937).
154. T. J. Kealy, *J. Amer. Chem. Soc.* **84**, 966 (1962); R. C. Cookson, S. S. H. Gilani, and I. D. R. Stevens, *Tetrahedron Lett.* 615 (1962); J. Sauer and B. Schröder, *Angew. Chem.* **77**, 736 (1965); *Angew. Chem. Int. Ed. Engl.* **4**, 711 (1965).
155. R. A. Clement, *J. Org. Chem.* **25**, 1724 (1962); O. C. Chapman and S. J. Dominianni, *ibid.* **31**, 3862 (1966).
156. S. G. Cohen, R. Zand, and C. Steel, *J. Amer. Chem. Soc.* **83**, 2895 (1961).
157. H. Esser, K. Rastädter, and G. Reuter, *Chem. Ber.* **89**, 685 (1956). For preparation of 100% H_2O_2 see W. Eggersglüss, "Organische Peroxyde," p. 77. Verlag Chemie, G.m.b.H., Weinheim/Bergstr., 1951.
158. W. Luttke and V. Schabacker, *Ann. Chem.* **687**, 236 (1965).
159. S. Hunig, *Chimia* **15**, 133 (1961); *Angew. Chem.* **70**, 215 (1958).
160. E. Schmitz, R. Ohme, and S. Schramm, *Angew. Chem.* **75**, 208 (1963).
161. E. Hecker and M. Hopp, *Ann. Chem.* **692**, 174 (1966).
162. J. E. Kmiecik, *J. Org. Chem.* **30**, 2014 (1965).
163. M. Regitz and D. Stadler, *Angew. Chem.* **76**, 920 (1964).
164. W. J. Chambers, C. W. Tullock, and D. D. Coffman, *J. Amer. Chem. Soc.* **84**, 2337 (1962); W. J. Chambers, U.S. Pat. 3,117,996 (Jan. 14, 1964).
165. S. V. Sokolov and S. A. Mazalov, U.S.S.R. Pat. 172,769 (July 7, 1965); *Chem. Abstr.* **65**, PC 605e (1966).
166. C. M. Kraebel and S. M. Davis, *J. Chem. Eng. Data.* **14**, 133 (1969).
167. K. Ueno, *J. Amer. Chem. Soc.* **74**, 4508 (1952).
168. A. J. Fatiadi, *J. Org. Chem.* **35**, 831 (1970).
169. H. C. Yao and P. Resnick, *J. Org. Chem.* **30**, 2832 (1965).
170. R. Criegee and J. Lohaus, *Ber. Dtsch. Chem. Ges.* **84**, 219 (1951).
171. A. L. Baumstark and P. O. Vasquez, *J. Org. Chem.* **48**, 65 (1983).
172. W. Adam, O. De Lucchi, and I. Erden, *J. Amer. Chem. Soc.* **102**, 4806 (1980).
173. W. Adam and O. De Lucchi, *J. Org. Chem.* **46**, 4133 (1981).

CHAPTER 15 / **AZOXY COMPOUNDS**

1. INTRODUCTION

The chemistry of azoxy compounds was reviewed some time ago by Bigelow [1]. While his paper deals primarily with aromatic azoxy compounds, it does give examples of most of the general preparative procedures available. It is particularly valuable in its review of the pioneering work by Angeli on the structure of azoxy compounds which was published in the somewhat inaccessible journal, *Atti della Accademia Nazionale dei Lincei.*

Other references reviewing various aspects of the chemistry properties, and applications of azoxy compounds are [2-7]. References to toxic, mutagenic, and carcinogenic properties of some azoxy compounds are included in [3].

Early interest included, *inter alia*, the Wallach rearrangement of aromatic azoxy compounds under acid conditions to *p*-hydroxyazoaromatic products [8-10]. A more recent study of 4,4′-disubstituted azoxybenzenes indicates that if the substituents are electron-withdrawing groups, such a methoxy groups, the rearrangement products are mixtures of 4,4′-dimethoxy-3-hydroxyazobenzene and 4,4′-dimethoxy-2-hydroxyazobenzene [11].

The current interest in azoxy compounds is in connection with their physiological activities (see [3] and other papers by Snyder and co-workers) and their ubiquitous utilization in liquid crystals. Despite this, few, if any, novel methods for the synthesis of azoxy compounds have been proposed in recent years.

One of the reactions which has been used to prepare azoxy compounds is the condensation of *C*-nitroso compounds with hydroxylamines. In the aliphatic series this reaction is quite general and permits the preparation of unsymmetrical azoxy compounds. In the case of aromatic compounds, however, only symmetrical azoxy compounds can be synthesized reliably. In the reaction of dissimilar aromatic nitroso compounds and aromatic hydroxylamines, a complex mixture of azoxy products is obtained.

Aliphatic azoxy compounds have also been prepared by the condensation of *C*-nitroso compounds with amines, oximes, and even diazomethane derivatives.

Grignard reagents have been reacted with diimide dioxides prepared from nitrosohydroxylamines and with derivatives of nitrosohydroxylamines to prepare unsymmetrical azoxy compounds, including aliphatic–aromatic types.

The oxidation of both aliphatic and aromatic azo compounds to the

414

corresponding azoxy derivative may be carried out with a variety of reagents. While older techniques favored chromic or nitric acid as the oxidizing agent, newer methods make use of various organic peracids or hydrogen peroxide. In the oxidation of aliphatic azo compounds, relatively weak peracids are favored to reduce the possibility of acid-catalyzed isomerization of azo compounds to hydrazones. Under controlled conditions cis azo compounds may be converted into cis azoxy compounds.

The problem of the position which the entering oxygen will occupy on oxidation of an azo compound has not been fully resolved. There is evidence that, in the case of some aliphatic azo compounds in which one side of the azo bridge is a methyl radical and the other side is a more complex aliphatic radical, the final azoxy compound bears the oxygen on the nitrogen atom farthest from the methyl radical. The effect of substituents on the oxidation of aromatic azo compounds has not been studied extensively.

The oxidation of aromatic phenylhydrazones leads to cis azoxy compounds. At one time, these products were designated "hydrazone oxides," but more recent work has established them as azoxy compounds. Aliphatic hydrazones give more complex oxidation products. At low temperatures, azoxy compounds form; at reflux temperatures, hydrazides are isolated.

Some aromatic amines may be oxidized to azoxy compounds with peracids provided even traces of metallic ions are carefully excluded. In the presence of metallic ions, the oxidation product is the corresponding azo compound.

Aromatic hydroxylamines, oxidized in the presence of metallic ions, are converted into nitroso compounds. In the absence of such ions, azoxy compounds form.

The oxidation of indazole oxides constitutes a method for the preparation of unsymmetrical aromatic azoxy compounds with the position of the oxygen atom unequivocally established.

Among the reductive methods of preparing azoxy compounds is the reduction of aliphatic nitroso compounds with stannous chloride. Triethyl phosphite has been used for the bimolecular reduction of fully fluorinated aromatic nitroso compounds.

The bimolecular reduction of aromatic nitro compounds, depending on reaction conditions, may produce azoxy compounds, azo compounds, hydrazo compounds (1,2-diarylhydrazines), benzidines, or amines. Whereas the reduction with zinc and sodium hydroxide leads to azo compounds, zinc and acetic acid/acetic anhydride produces azoxy compounds. Other reducing agents suggested are stannous chloride, magnesium with anhydrous methanol, a sodium–lead alloy in ethanol, thallium in ethanol, and sodium arsenite.

Alcoholic potassium hydroxide and sodium alcoholate in the presence of alcohols such as benzyl alcohol have been used as reducing agents for aromatic nitro compounds.

A particularly interesting and general reducing system for the preparation of azoxy compounds uses a strongly alkaline glucose solution as a reducing agent.

Substituted aromatic nitro compounds having a substituent with a positive Hammett sigma constant can be reduced with potassium borohydride to the azoxy stage.

The cis isomers of the normal trans azoxy compounds may be produced by cautious oxidation of cis azo compounds, oxidation of aromatic phenylhydrazones, or by treatment of trans azoxy compounds with a base.

NOMENCLATURE

The naming of azoxy compounds in which the position of the azoxy oxygen atom is immaterial because the parent groups attached to the azoxy bridge are identical, or in which the position of the oxygen is unknown, follows the principles used in naming azo compounds except that "azoxy" is substituted for "azo."

In an unsymmetrically substituted compound, the position of the azoxy oxygen must be specified. The IUPAC [12] rules are as follows:

(1) When both groups attached to the azoxy radical are cited, the prefix *NNO* or *ONN* is used before the term "azoxy": the prefix *NNO* specifies that the second of the two groups is attached to the *NO* group of the azoxy bridge (structure I), while *ONN* specifies that the first group is attached to the *NO*

$$\underset{\text{(I) Phenyl-}NNO\text{-azoxy-}p\text{-toluene}}{C_6H_5\text{—}N\overset{\overset{\displaystyle O}{\uparrow}}{=}N\text{—}C_6H_4CH_3(p)}$$

group; for example, structure I may also be named *p*-tolyl-*ONN*-azoxybenzene.

(2) When only one parent compound is cited, *ONN* specifies that the primed substituent is directly attached to *NO*, while *NNO* specifies that the unprimed substituent is directly attached to *NO* (structures II, III).

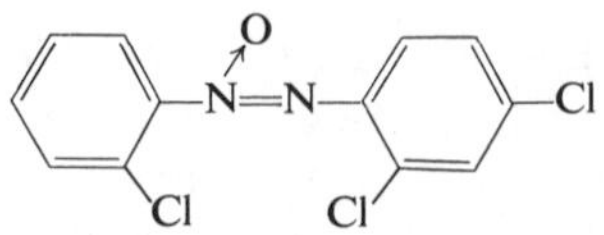

(II) 2,2′,4-Trichloro-*ONN*-azoxybenzene

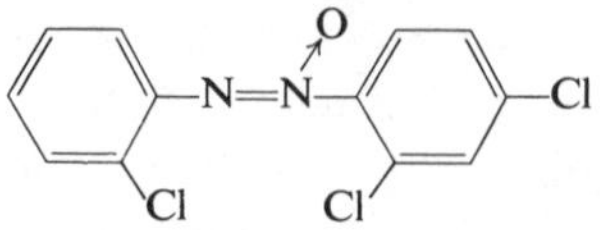

(III) 2,2′,4-Trichloro-*NNO*-azoxybenzene

(3) To distinguish it from substances in which the position of the oxygen is

well established, a compound in which the exact position of the azoxy oxygen is uncertain or unknown may be designated by use of the prefix *NON*. (Unfortunately, Gillis and Hagarty [13] and Gillis and Schimmel [14] compound confusion by using the symbol *NON* to refer to *ONN* as defined above.)

While this system of nomenclature is reasonably straightforward for many azoxy compounds, it is not adequate for more complex systems. It has therefore been suggested [15] that the parent azo compounds be designated *N,N'*-diimides (for R—N=N—R') and that the azoxy compounds derived therefrom be termed oxides (or monoxides).

This system of nomenclature then also simplifies the naming of unsymmetrical compounds, such as (IV), i.e., the dimers of the nitroso compounds, as

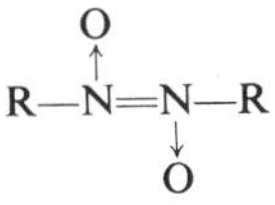

$$R-\underset{\underset{O}{\downarrow}}{\overset{\overset{O}{\uparrow}}{N}}=N-R$$

(IV)

diimide dioxides. In this system of nomenclature, structure (V) becomes

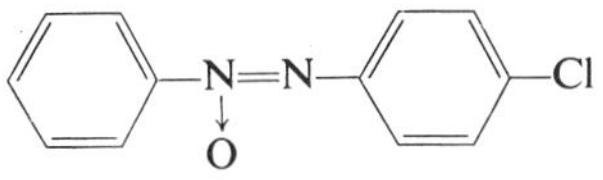

(V)

N-phenyl-*N'*-[4-chlorophenyl]diimide *N*-oxide, and structure (VI) becomes

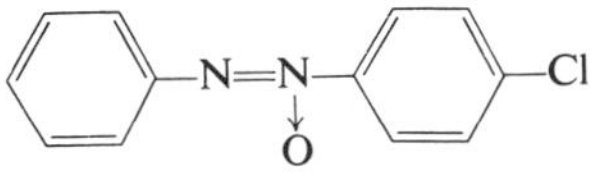

(VI)

N-phenyl-*N'*-[4-chlorophenyl]diimide *N'*-oxide. The term "diazine" has also been used instead of "diimide" [16].

Another system of nomenclature, developed with the cooperation of the editors of the *Journal of the Chemical Society* (London), will be found in Brough *et al.* [17]. It is reasonably simple and straightforward. Another system is used by Spence *et al.* [18]. The current *Chemical Abstracts* system uses designations such as "Diazene, diphenyl-1-oxide" [19].

Particularly in the older literature, α and β designations are found. In many cases, this designation may simply indicate that the authors isolated two isomeric azoxy compounds of unspecified structure. Where the identity of the isomers has been established the α term implies that the NO group is directly attached to the parent radical in the name, i.e., α-4-chlorophenylazoxybenzene is equivalent to *N*-phenyl-*N'*-[4-chlorophenyl]diazene *N*-oxide, while β implies that the NO group is not attached to the parent radical, i.e., β-4-chlorophenylazoxybenzene is equivalent to N-phenyl-N'-[4-chlorophenyl]-diazene N'-oxide.

Cis–trans isomerism is usually indicated by conventional nomenclature rules, although it must be pointed out that the older literature mentions such geometric isomerism rarely.

In this chapter our nomenclature usually follows the preference of the original authors, although we tend to favor the "azoxy" nomenclature over the diimide monoxide or diazene-1-oxide convention for the simpler structures.

2. CONDENSATION REACTIONS

A. Reaction of *C*-Nitroso Compounds with Hydroxylamines and Related Compounds

As has already been indicated in the discussion of the nomenclature of azoxy compounds, the syntheses of these compounds concern themselves with such problems as the position of the azoxy oxygen in unsymmetrically substituted products,* cis–trans isomerism, and whether one or both of the parent hydrocarbons are aromatic or aliphatic.

---

* In the present discussion the terms "symmetrically substituted" and "unsymmetrically substituted" products refer to the nature of the parent hydrocarbons attached in a strictly linear fashion to the azo compound from which the azoxy compounds may be derived. For example, in this context the following structures are considered "symmetrically substituted" azoxy compounds:

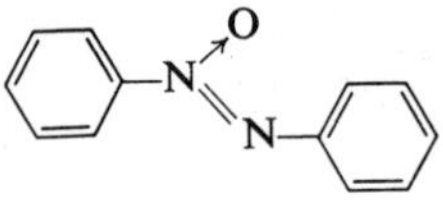

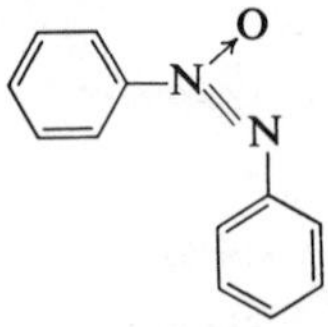

Examples of "unsymmetrically" substituted azoxy compounds would be

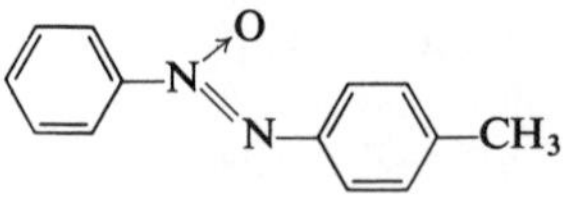

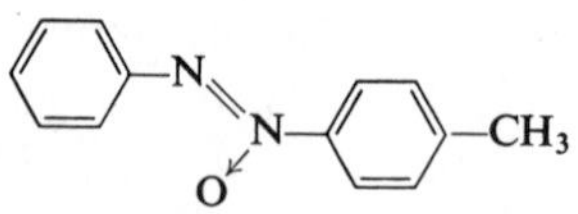

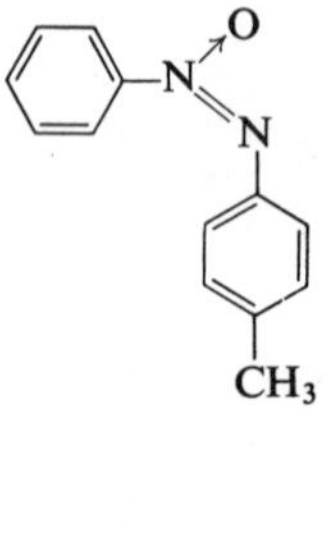

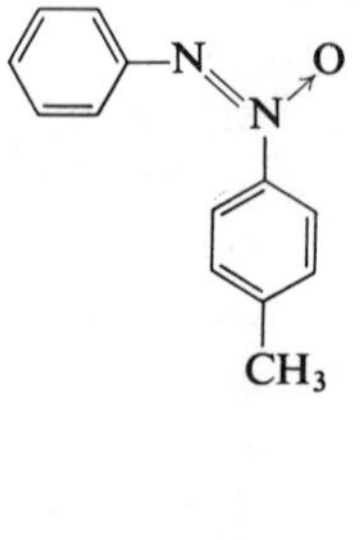

It is self-evident that one of the simpler methods of preparing unsymmetrically substituted azoxy compounds must involve the condensation of two distinctly different starting materials. In principle, the reaction of *C*-nitroso compounds with hydroxylamines meets this requirement (Eq. 1).

$$R\text{—}NO + R'NHOH \longrightarrow R\text{—}NON\text{—}R' + H_2O \tag{1}$$

Historically this reaction developed from the assumption that the formation of azoxy compounds by the reduction of aromatic nitro compounds probably involved the intermediate formation of *C*-nitroso compounds and hydroxylamines. In the all-aliphatic series, this reaction appears to be quite general. Symmetrically and unsymmetrically substituted azoxy compounds have been prepared by it, the only major problems being the usual ones of developing procedures that afford good yields and of determining the exact position of the azoxy oxygen in unsymmetrically substituted products.

In this reaction the source of the azoxy oxygen appears to be the nitroso group [16]. The preparation of *t*-butyl-*ONN*-azoxymethane (*N*-methyl-*N'*-*t*-butyldiazene-*N'*-oxide) is an example of a preparation of an unsymmetrical azoxy compound which is quite generally applicable. The structure assignment is based on NMR data.

2-1. *Preparation of t-Butyl-ONN-azoxymethane (N-Methyl-N'-t-butyldiazene-N'-oxide)* [16, 2]

$$\underset{\underset{CH_3}{|}}{\overset{\overset{CH_3}{|}}{CH_3\text{—}C\text{—}NO}} + CH_3NHOH\cdot HCl \xrightarrow{\ KOH\ } \underset{\underset{CH_3}{|}}{\overset{\overset{CH_3\ \ O}{| \ \nearrow}}{CH_3C\text{—}N\text{=}N\text{—}CH_3}} + HCl + H_2O \tag{2}$$

(SAFETY NOTE: *C*-Nitroso compounds and hydroxylamine derivatives must be handled with due caution. The final product may also have adverse physiological properties [3].)

To a stirred suspension of 1.9 gm (0.033 mole) of powdered potassium hydroxide in 25 ml of anhydrous ether in a reflux setup is added gradually 2.6 gm (0.03 mole) of *N*-methylhydroxylamine hydrochloride. The ether is evaporated at reduced pressure, leaving a slightly yellow, curdy solid. To this residue is rapidly added 2.4 gm (0.03 mole) of *t*-nitrosobutane. The reaction mixture is cautiously warmed. The exothermic reaction which may develop is moderated by cooling the flask as required. After the reaction has been brought under control, the reaction system is heated first for 1 hr at 85°C, followed by heating at 110°C for 2 hr. After the reaction mixture has been cooled, it is diluted with approximately 10 ml of water. The product is separated from the aqueous layer by repeated extraction with ether. The ether extracts are combined and dried over anhydrous sodium sulfate. After filtration and removal of the ether by evaporation, the residue is distilled under reduced pressure. The colorless product is isolated at 60°–62°C (110 mm Hg): yield 2.2 gm (63%), n_D^{20} 1.4265.

By a similar procedure the symmetrical 2-azoxy-2,3-dimethylhexane was also produced [20]. The properties of aliphatic azoxy compounds are given in Table II (see pp. 450–451).

The preparation of symmetrical aromatic azoxy compounds by the condensation of a C-nitroso compound with a hydroxylamine using a procedure similar to the one given above is quite straightforward when suitable modifications are made to handle the solid products [21]. The problem is more complex when an attempt is made to prepare an unsymmetrical aromatic compound by the condensation of starting materials having different parent hydrocarbon radicals. A typical product mixture obtained from such a reaction is given in Eq. (3) [22]. The percentages given represent percentages of material found in the isolated and separated product mixture.

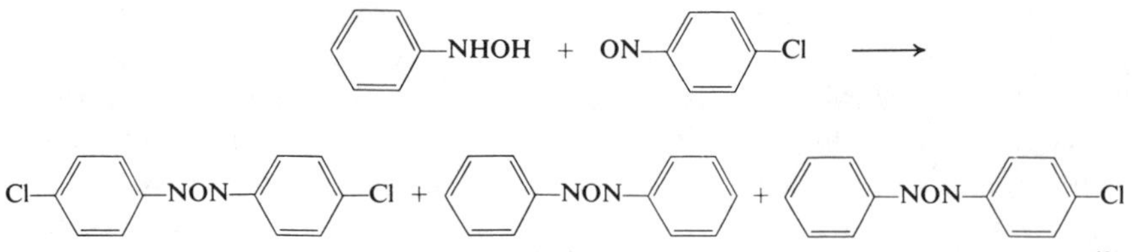

$$ \text{(50\%)} \qquad \text{(32\%)} \qquad \text{(18\%)} \qquad (3) $$

To be noted here is that the expected product (p-chlorophenylazoxybenzene) constituted only 18% of the isolated product. We may assume that the exact composition of such product mixtures will vary considerably with the nature and position of the substituents on the aromatic rings of either the nitroso or the hydroxylamine compound. The position of the azoxy oxygen may also be influenced by these substituents.

The mechanism of the formation of this complex product mixture is subject to some discussion in the literature.

One mechanism postulates a rapid preliminary equilibrium between the reactants which, in effect, would be a mutual oxidation–reduction equilibrium

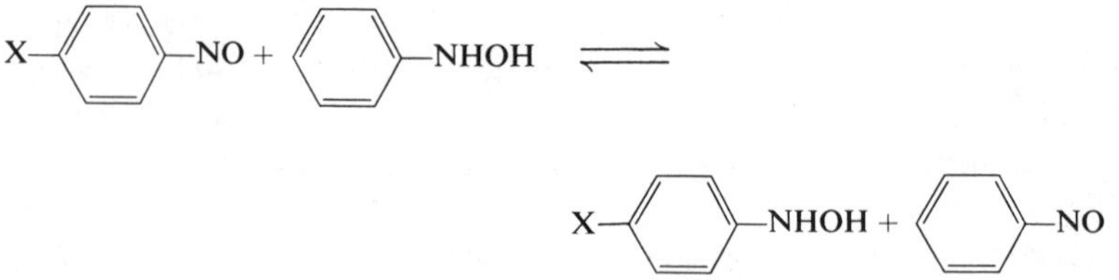

$$ (4) $$

(Eq. 4), which is thought to involve the intermediate formation of an ionic species [23, 25] (Eq. 5). This postulate is based on results obtained with isotopically labeled starting materials. When ^{15}N was used in one of the starting materials, a statistically scrambled distribution of ^{15}N was found in

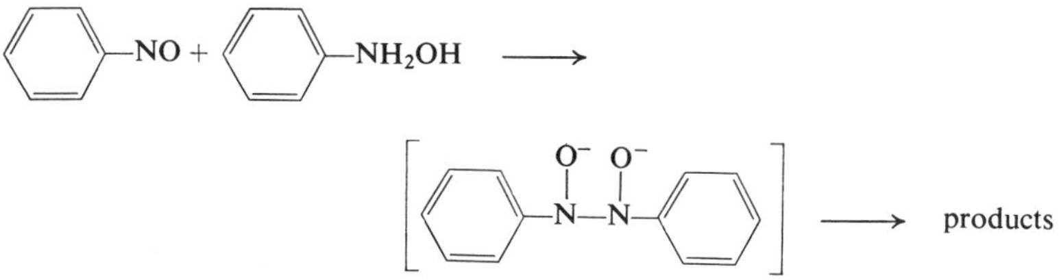

$$\text{C}_6\text{H}_5\text{-NO} + \text{C}_6\text{H}_5\text{-NH}_2\text{OH} \longrightarrow$$

$$\left[\text{C}_6\text{H}_5-\underset{O^-}{N}-\underset{O^-}{N}-\text{C}_6\text{H}_5 \right] \longrightarrow \text{products} \quad (5)$$

the product; when ^{18}O in phenylhydroxylamine was used, the resulting azoxybenzene showed only a 50% enrichment under a variety of conditions, implying that the source of the azoxy oxygen in aromatic compounds is the hydroxylamine only to the extent of 50% [25].

The formation of the ionic species in Eq. (5) may be the result of the intermediate existence of radical anions (Eqs. 6–8).

$$\text{C}_6\text{H}_5\text{-NO} + \text{C}_6\text{H}_5\text{-NHOH} \xrightarrow{\text{B}^-} 2\,\text{C}_6\text{H}_5\text{-NO}^{\cdot-} \quad (6)$$

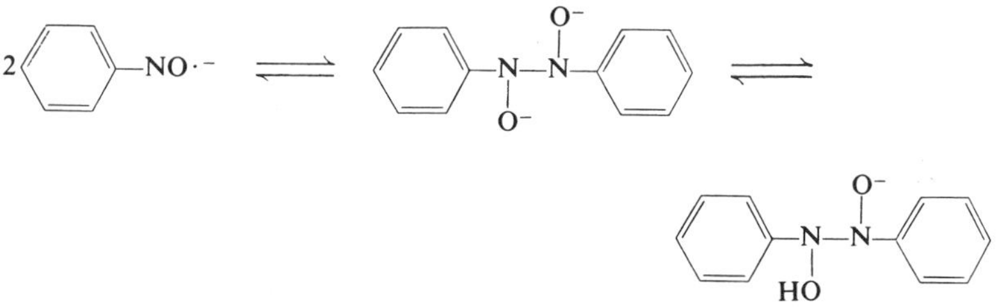

$$2\,\text{C}_6\text{H}_5\text{-NO}^{\cdot-} \rightleftharpoons \text{C}_6\text{H}_5-\underset{O^-}{\overset{O^-}{N}}-N-\text{C}_6\text{H}_5 \rightleftharpoons$$

$$\text{C}_6\text{H}_5-N-\underset{HO}{\overset{O^-}{N}}-\text{C}_6\text{H}_5 \quad (7)$$

$$\text{C}_6\text{H}_5-\underset{HO}{\overset{O^-}{N}}-N-\text{C}_6\text{H}_5 \longrightarrow \text{C}_6\text{H}_5-N=\overset{O}{N}-\text{C}_6\text{H}_5 + OH^- \quad (8)$$

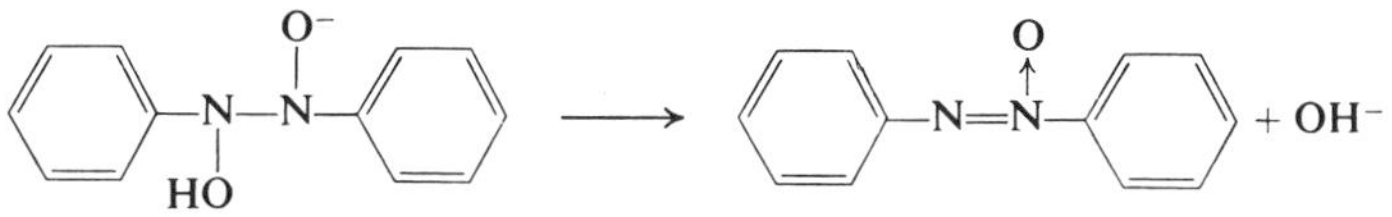

The last step of this mechanism (Eq. 8) is said to be favored by the presence of alcohols and prevented or slowed in dimethyl sulfoxide [26]. In this reference the role of solvents, strong bases, etc., is discussed. However, more facts are required to develop this condensation to the point where the formation of unsymmetrically aromatic azoxy compounds can be controlled.

Aliphatic C-nitroso compounds have also been condensed with aliphatic amines and oximes. Among the products isolated have been fair yields of alipathic azoxy compounds [27]. The detailed analysis to account for the stoichiometry of the reactions is still lacking. These reactions may require further development, and only one example is given here for reference.

2-2. Preparation of Azoxycyclododecane [27]

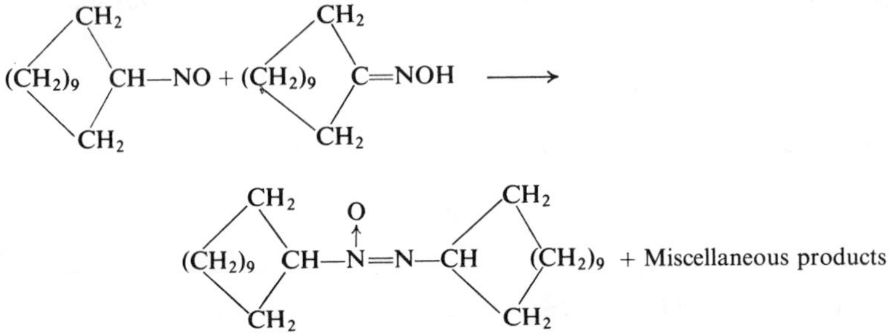

$$
\begin{array}{c}
\text{CH}_2 \\
(\text{CH}_2)_9 \quad \text{CH--NO} + (\text{CH}_2)_9 \quad \text{C=NOH} \longrightarrow \\
\text{CH}_2 \qquad\qquad\qquad \text{CH}_2
\end{array}
$$

$$
\begin{array}{c}
\text{CH}_2 \qquad\quad \text{O} \qquad\qquad \text{CH}_2 \\
\quad\uparrow \\
(\text{CH}_2)_9 \quad \text{CH=N--N--CH} \quad (\text{CH}_2)_9 \; + \text{Miscellaneous products} \qquad (9) \\
\text{CH}_2 \qquad\qquad\qquad\qquad \text{CH}_2
\end{array}
$$

With due precautions for the handling of C-nitroso compounds, oximes, and azoxy compounds, a stirred mixture of 10 gm (0.05 mole) of nitrosocyclododecane and 10.1 gm (0.051 mole) of cyclododecanone oxime in 100 ml of propionic acid is heated to reflux for 3 hr. Then the reaction mixture is distilled under reduced pressure to remove approximately 70 ml of the propionic acid. The distillation residue is cooled to room temperature, diluted with 80 ml of methanol, and cooled in an ice bath to $0°–5°C$ to hasten the crystallization of the product. The product is separated by filtration: yield 4.1 gm (21.3 %), m.p. $75°–78°C$.

Some 2-nitrosoacylbenzene derivatives have been reacted with phenyldiazine in toluene to produce mixtures of benz[c]isoxazoles and azoxy compounds (Eq. 10) [28].

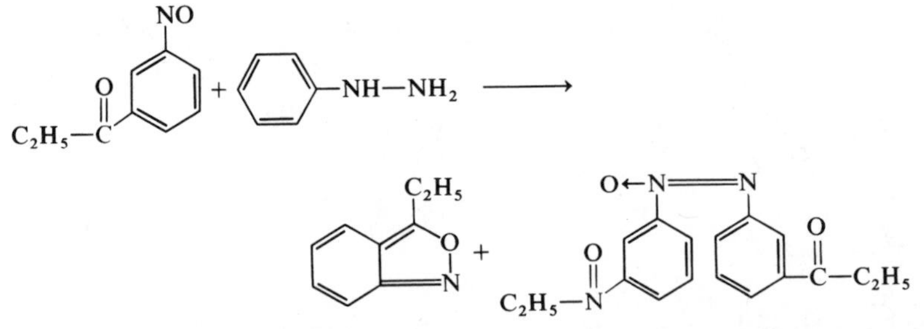

$$ (10) $$

Nitrosobenzene has been reacted with complex amines in the presence of an acid chloride (presumably to protect the amino group) and iodine to form a phenyldiazene oxide [29].

Haszeldine and co-workers [30] have treated fluorinated nitroso compounds with diazomethane derivatives to produce azoxy compounds derived from the nitroso starting material (e.g., Eq. 11).

$$
\text{2CF}_3\text{NO} \xrightarrow{\text{Ph}_2\text{CN}_2} \overset{\overset{\text{O}}{\uparrow}}{\text{CF}_3\text{--N=N--CF}_3} + \text{Ph}_2\text{CO} \qquad (11)
$$

B. Preparations Involving Organo-metallic Reagents

A method for the preparation of unsymmetrical azoxy compounds involves the reaction of certain diimide dioxides with Grignard reagents [15]. This reaction has somewhat limited applicability because the diimide dioxides which were used were prepared by alkylation of organonitrosohydroxylamines, a class of compounds of which "cupferron" is perhaps the best-known example. The reaction is, in effect, a reduction of a diimide dioxide to an azoxy compound by use of a Grignard reagent. The overall process is represented by Procedure 2-3. Since the starting materials are, in effect, unsymmetrically substituted "nitroso dimers," extension of the reaction to nitroso dimers would be interesting.

2-3. Preparation of Methyl-NNO-azoxybenzene [15]

| Cupferron | | N-Methyl-N'-phenyl-
diimide dioxide |

$$\text{(12)}$$

NOTE: In Eq. (12) we indicate the fate of the methyl group from the Grignard reagent by use of the symbol CH_3^*.

(SAFETY NOTE: Dimethyl sulfate is considered carcinogenic. Other materials used also may be hazardous.)

To a solution of 4.2 gm (0.0276 mole) of N-methyl-N'-phenyldiimide dioxide in 50 ml of anhydrous ether is added 185 ml of a 0.21 N solution of methyl magnesium chloride in anhydrous ether (0.0389 mole) over a 3 hr period while maintaining the reaction mixture at 20°C. Then the reaction mixture is cautiously heated with 100 ml of a saturated aqueous solution of ammonium sulfate. At this point 37 mg of an unidentified deep-red solid (m.p. 126°–127°C dec. after recrystallization, m.p. 140.5°C dec. separates. This material is removed by filtration and discarded.

The ether layer is cautiously steam-distilled. The cooled distillate is extracted with ether and the ether extract is discarded. The ether-insoluble portion is then crystallized from petroleum–ether (b.p. 40°–60°C) to afford 0.54 gm (14%) of product, m.p. −3° to 0°C. On repeated crystallization, the melting point may be raised to 6.5°–8.5°C, n_D^{20} 1.556.

A more convenient variation of this reaction involves preparing the toluene sulfonyl derivatives of the nitrosohydroxylamines and treating these "tosylates" with Grignard reagents. This procedure permits the preparation of a variety of unsymmetrical azoxy compounds [31]. The cited reference shows only the preparation of aromatic azoxy compounds and arylazoxyalkanes by the reaction of Grignard reagents with tosylates derived from aromatic nitrosohydroxylamines. Aliphatic nitrosohydroxylamine tosylates did not undergo this reaction, thus precluding the possibility of using this approach for the preparation of totally aliphatic azoxy compounds.

The tosylates may be prepared by treating the nitrosohydroxylamine with p-toluenesulfonyl chloride in aqueous sodium bicarbonate solution, in acetone–aqueous sodium hydroxide mixtures, or in benzene solution with the preformed salts of the nitrosohydroxylamines.

The Grignard reagents used in the reaction may be either those derived from aryl halides or those formed from alkyl halides. Phenyllithium reacted with the tosylate to give sulfones rather than azoxy compounds.

Traces of azo compounds were detected in the reaction mixture. They probably were formed by the reduction of the azoxy compound by the Grignard reagent.

2-4. Preparation of 4'-Methylazoxybenzene (p-Tolyl-NNO-azoxybenzene) [31]

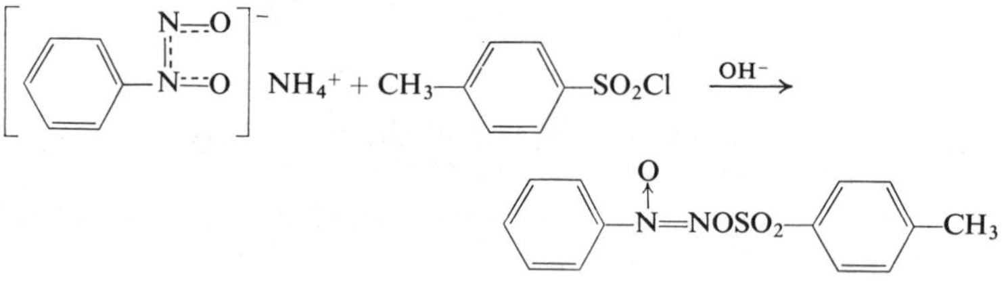

$$\text{(13)}$$

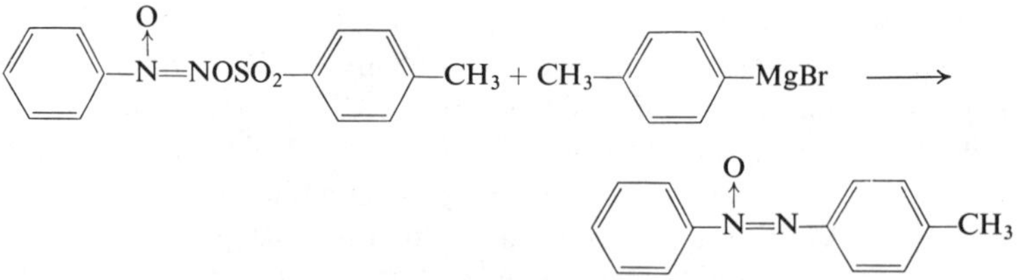

$$\text{(14)}$$

(a) Preparation of phenylnitrosohydroxylamine tosylate. To a stirred solution of 16 gm (0.10 mole) of Cupferron (ammonium salt of N-phenyl-N-nitrosohydroxylamine) in 200 ml of a 10% sodium bicarbonate solution in water at room temperature is added 22 gm (0.11 mole) of p-toluenesulfonyl chloride. The reaction mixture is vigorously stirred overnight and extracted with methylene chloride.

After removal of the methylene chloride from the extract by evaporation, the dark residue is stirred with 30 ml of methanol. The solid which forms is separated by filtration. Recrystallization of the solid affords 12 gm of N-phenyl-N′-tosyloxydiimide N-oxide (65%), m.p. 130°–137°C dec.

(b) Preparation of 4′-methylazoxybenzene.   To a stirred solution of 2.20 gm (7.5 mmoles) of N-phenyl-N′-tosyldiimide N-oxide in 40 ml of tetrahydrofuran at room temperature is added dropwise 9 ml of a 1.2 M solution of p-tolyl-magnesium bromide in tetrahydrofuran. After the addition has been completed, the reaction mixture is stirred at 50°–60°C for 2 hr, cooled, and then poured into a mixture of ice and dilute hydrochloric acid. The crude product is extracted with methylene chloride. After concentration of the organic layer by evaporation at the aspirator, the residue is percolated through a silica-gel column. Elution of the column with pentane–methylene chloride (3:1) affords first 0.06 gm of 4-methylazobenzene (m.p. 69°–70°C). Continued elution first with pentane–methylene chloride (2:1) and then at a ratio of 1:1 finally gives 1.16 gm of 4′-methylazoxybenzene: yield 73%. After recrystallization from hexane, 1.05 gm of product is isolated, m.p. 50°–51°C.

Alkyllithium and Grignard reagents have been reacted with N-nitroso-O,N-dialkylhydroxylamines to form regiospecific azoxy alkanes. Usually the (Z)-stereoisomers were isolated. However, in one case the (E)-form was the major product [32].

Treatment of N-chloro-N-alkoxy-N-tert-alkylamines with Grignard reagents afforded N-alkyl-N-alkoxy-N-tert-alkylamines and a "dimeric" azoxy compound (Eq. 15) [33].

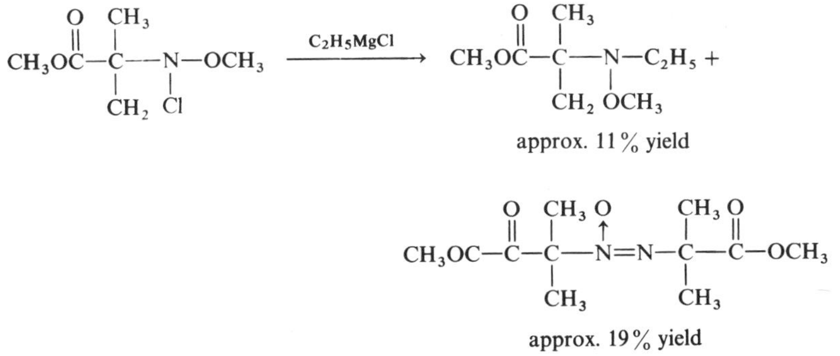

approx. 11% yield

approx. 19% yield

Aryliminodimagnesium reagents reacts with aromatic nitro compounds to produce a mixture of six symmetrical and eight unsymmetrical azoxyarenes (Eq. 16) [34]. According to a related Japanese patent, such reagents may form mixtures of azoxy and azo compounds, probably by deoxygenation of the azoxy compounds by an excess of aryliminodimagnesium.

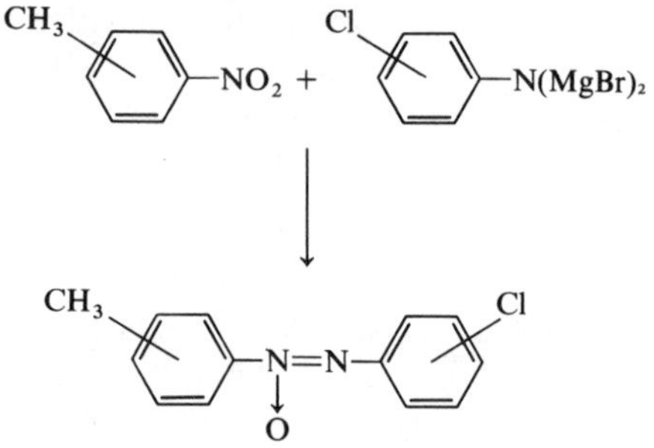

$$(16)$$

3. OXIDATION REACTIONS

A. Oxidation of Azo Compounds

Since both symmetrical and unsymmetrical azo compounds may be prepared by a variety of procedures, it is self-evident that synthetic methods for the introduction of an oxygen atom on the azo bridge would be a useful approach to azoxy compound preparation. Although several such methods exist, surprisingly little attention has been paid to the following problems.

(1) Since azo compounds may be prepared by the oxidation of hydrazo compounds *via* procedures similar to those used in oxidizing azo compounds to azoxy compounds, better definition of reaction conditions is required to control the formation of either type of compound. The existing literature rarely, if ever, indicates that the oxidation of a 1,2-disubstituted hydrazine could conceivably produce a mixture of azo and azoxy compounds.

(2) In the case of the oxidation of unsymmetrically substituted azo compounds, we would expect, *a priori*, that invariably a mixture of -*ONN*- and -*NNO*- products would be produced. This has rarely been observed, possibly because of the difficulties of structural studies or because of inadequate separation techniques [36, 37].

(3) Although one laboratory has studied which of the two nitrogen atoms preferentially accepts the entering oxygen in the aliphatic series and another has made a start on the aromatic series, much more systematic work in this area is required.

It would seem to us that, through careful attention to accounting for the fate of the starting material by quantitatively separating all the products of the reaction, through the use of modern chromatographic and kinetic techniques (see, for example, Badger and Lewis [38, 39]), and through the application of modern instrumentation, answers to many of these related subsidiary questions should be uncovered.

While such oxidizing agents as chromic acid and nitric acid have been used to convert azo compounds into azoxy compounds, the more recent techniques have involved perbenzoic acid [17, 38–42], peracetic acid, or hydrogen peroxide in glacial acetic acid [13, 43–47], and hydrogen peroxide in trifluoroacetic acid [48]. Use of the various organic peroxides and hydroperoxides does not appear to have been studied.

The possibility exists that strong acidic reaction systems such as hydrogen peroxide in glacial acetic acid may cause isomerization to hydrazones, particularly in the oxidation of aliphatic or aliphatic–aromatic azo compounds. Therefore, the much milder perbenzoic acid in an inert solvent has been suggested as an oxidizing agent [17]. Peracetic acid (40 % solution) has also been used in conjunction with indifferent solvent [13, 16].

3-1. Preparation of trans-ω-Azoxytoluene [16]

$$ \xrightarrow[\text{C}_6\text{H}_5\text{CO}_3\text{H}]{\text{[O]}} \tag{17} $$

To a solution of 2.67 gm (0.02 mole) of perbenzoic acid in 50 ml of chloroform which is being stirred and maintained at 0°C is added over a 30 min period a solution of 4.2 gm (0.02 mole) of ω-azotoluene in 50 ml of chloroform. The mixture is stirred at 0°–5°C overnight. Then the vigorously stirred reaction mixture is treated with a slight excess of a 10 % solution of potassium iodide in water followed immediately by an excess of a 0.1 N sodium thiosulfate solution. The organic layer is separated and washed in turn with ice water, cold 1 N sodium hydroxide solution, and again with water.

The product solution is dried with anhydrous magnesium sulfate, filtered, and then freed from solvent by evaporation under reduced pressure. The oily residue is crystallized from ethanol: yield 3 gm (67 %), m.p. 42°–43°C.

Under similar conditions, cis-azobenzene could be oxidized to cis-azoxybenzene. Evidently the only major precaution to be taken in this preparation is the exclusion of ultraviolet light (by carrying the reaction out in a dark room) [41]. Whether this precaution is truly required is open to some doubt since cis-azoxybenzenes were prepared more recently by oxidation while warming with a heating lamp [46]. The isomerization by ultraviolet light is probably an equilibrium process in which equilibrium constants have a pronounced dependence on the chemical constitution of the materials involved. Therefore, variations in the observations of the stability of the products are not entirely surprising.

3-2. Preparation of cis-Azoxybenzene [41]

$$
\begin{array}{c}
\text{azobenzene} \xrightarrow[\text{C}_6\text{H}_5\text{CO}_3\text{H}]{[\text{O}]} \text{azoxybenzene}
\end{array} \tag{18}
$$

In a dark room, to 50 ml of a 0.78 N solution of perbenzoic acid in chloroform is added with stirring, at room temperature, 1.5 gm (0.008 mole) of *cis*-azobenzene. The mixture is stirred for $2\frac{1}{2}$ hr. Then it is extracted in turn with three 40 ml portions of 5% aqueous sodium hydroxide solution and one 50 ml portion of water. The chloroform solution is rapidly dried with anhydrous sodium sulfate, filtered, and evaporated at room temperature under reduced pressure. The trans isomer is separated from the desired cis isomer by slurring and filtering the resulting solid six times with 20 ml portions of petroleum ether (b.p. 40°–60°C). The residue is the desired product: yield approx. 0.75 gm (50%), m.p. 87°C.

The oxidation of azo compounds with hydrogen peroxide in an acetic acid medium or with peracetic acid has been carried out by many investigators. For example, in a study of the 4,4'-dialkoxyazoxybenzenes, which are of interest in the field of liquid crystals, the corresponding alkoxynitrobenzenes were reduced with lithium aluminum hydride to the corresponding azo stage and, after decomposition of the reducing agent and removal of the solvent, the product residue was taken up in acetic acid and oxidized at 65°C for 36 hr with 30% hydrogen peroxide. By this method a 75% yield of 4,4'-dihexyloxyazoxybenzene was obtained, m.p. 81°C, nematic-liquid transition point 128°C. To be noted is that the crude reaction product contained approximately 30% of the unoxidized azo compound. Separation was accomplished by chromatography on aluminum oxide [43].

While the reported yields from the hydrogen peroxide oxidation are highly variable, the accessibility of starting materials and the simplicity of the procedure are significant factors in considering the reaction.

In one of the few studies to determine the position of the entering oxygen atom in the case of unsymmetrically substituted aromatic azo compounds, it was found that, at least for compounds with one hydroxyl group ortho to the azo bridge, the entering oxygen attaches itself to the nitrogen nearest the hydroxyl group, as in Eq. (19). Only when this hydroxyl group is acylated or tosylated does the oxygen also attack the nitrogen atom removed from the acylated substituent as in Eq. (20) [45].

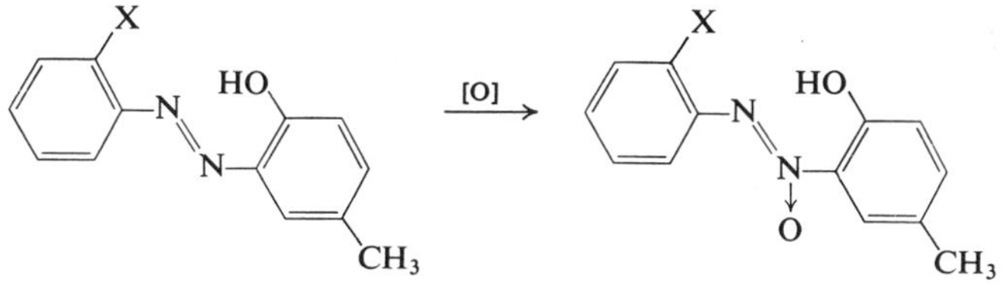

$$X = H, OCH_3, CH_3, Br, NO_2$$

(19)

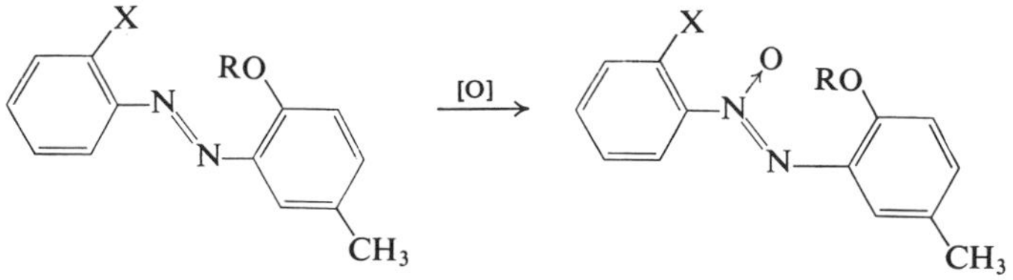

(20)

and

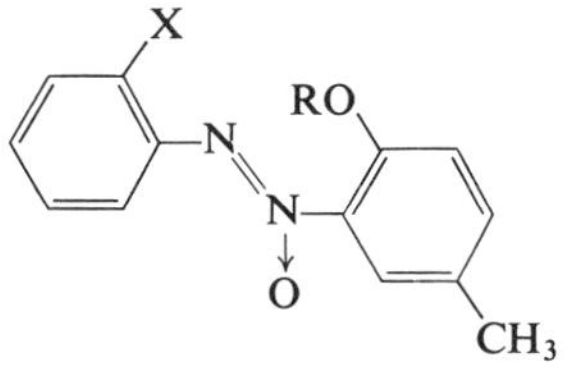

(21)

$$X = OCH_3, CH_3, Br, NO_2$$
$$R = COCH_3, COC_6H_5, \text{ or } SO_2C_6H_4CH_3(p).$$

In view of the poor yield reported and the lack of details on structure deter-minations in this particular paper, the structure assignment needs amplifica-tion.

Preparation 3-3, that of (2-bromophenyl)-*NNO*-azoxy(2-hydroxy-5-methyl-benzene), illustrates the general procedure followed.

3-3. Preparation of (2-Bromophenyl)-NNO-azoxy(2-hydroxy-5-methylbenzene) [45]

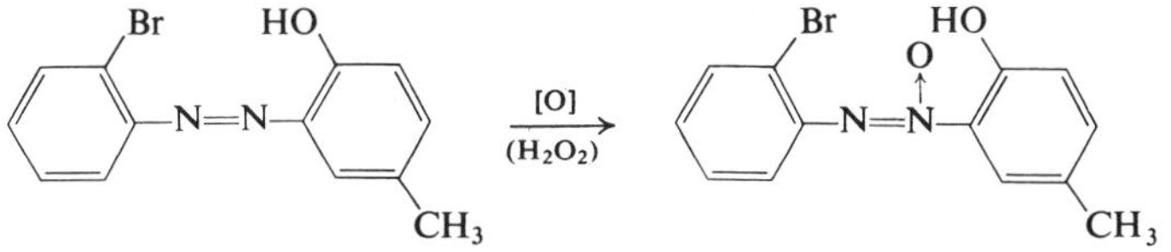

(22)

CAUTION: 30% hydrogen peroxide must be handled with safety precau-tions supplied by the manufacturer.

To a solution of 1.4 gm (4.8 mmoles) of (2-bromophenyl)azo(2-hydroxyl-5-methylbenzene) in 100 ml of glacial acetic is added 16.7 ml of 30% hydrogen peroxide. The mixture is warmed between 75°C and 80°C for 18 hr. During this time the color of the solution changes slowly from dark red to orange. After cooling and filtration to remove unidentified insoluble materials, the solution is poured over 200 gm of ice. The precipitated crude product is collected on a filter and washed with distilled water until no longer acid to Congo red. The filter cake is then pressed dry and air-dried in the dark. Then the crude product is recrystallized from 40 ml of a solution consisting of 30 ml of acetone and 10 ml of water: yield 0.2 gm (15%), m.p. 96°C. No other products were isolated from the oxidation.

Cis azobenzenes have been oxidized with more concentrated hydrogen peroxide solutions, as illustrated in Procedure 3–4.

3-4. Preparation of cis-p-Azoxytoluene [46]

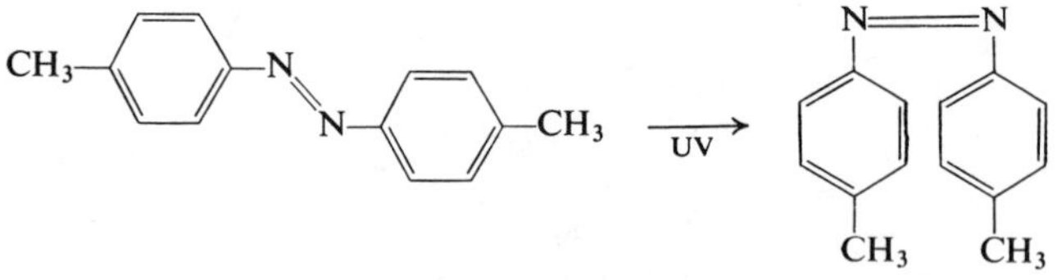

(23)

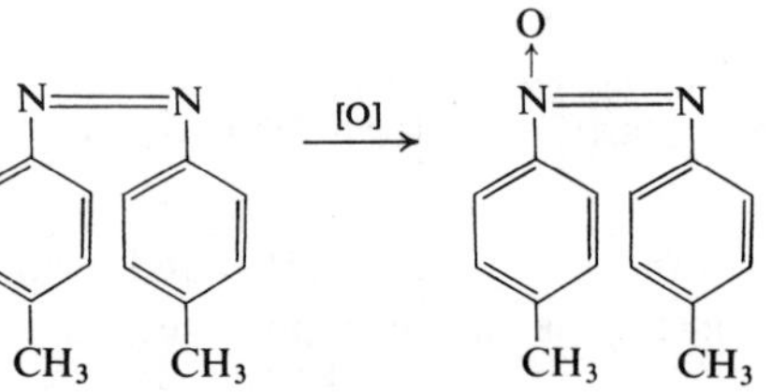

(24)

CAUTION: Proper precautions must be taken in handling 98% hydrogen peroxide.

(a) *Preparation of oxidizing mixture.* A solution of 140 ml of glacial acetic acid and 6.0 ml (248 mmole) of 98% hydrogen peroxide is allowed to stand overnight. The mixture is then extracted with three 20 ml portions of chloroform. The chloroform layer is drawn off, dried with calcium sulfate, filtered through a glass-wool plug, and used promptly.

(b) *Isomerization of p-azotoluene.* A solution of 6.0 gm of p-azotoluene in ligroin is irradiated with a Hanovia ultraviolet lamp (type 16200) for 4 hr. The product mixture is then separated by chromatography on an aluminum oxide column. In this procedure the trans isomer is eluted with ligroin, the cis isomer with diethyl ether. The recovered trans isomer is concentrated and reprocessed until sufficient cis isomer has been accumulated.

A solution of 1.00 gm (5.27 moles) of *cis-p*-azotoluene taken up in 5 ml of chloroform and 50 ml of the oxidizing solution is placed in a flask which had been covered with black enamel. The reaction flask is placed in a Dewar flask containing an ice bath maintained at 0°C for 6 hr. Then the product solution is poured into 100 ml of ice-cold water and the organic phase is separated. From this stage on, all solutions are precooled to −5°C and the chromatographic column is maintained at 0°–3°C by use of a cooling jacket.

The chloroform solution is extracted in turn with two 50 ml portions of water, one 50 ml portion of a 10% aqueous sodium carbonate solution, and another 50 ml portion of water. The chloroform solution is then dried over calcium sulfate, filtered, and evaporated to dryness in an air stream.

The residue is taken up in approximately 15 ml of redistilled 40°–60°C ligroin and placed on a cooled chromatographic column (12 × 150 mm) filled with 200 mesh silicic acid. The trans azo and trans azoxy compounds are eluted first with ligroin and a solution of 2% (by volume) of ether in ligroin. The cis azoxy compound is finally eluted with a 10% (by volume) solution of ether in ligroin. The product solution is evaporated at 0°C in an air stream.

The product residue may be washed with cold solvents: yield 541 mg (48%).

In the oxidation of unsymmetrical α,β-unsaturated aliphatic azo compounds with 40% peracetic acid it was found that the long-held notion that the oxygen will always attack the nitrogen attached to a methyl group in an azomethane derivative is, in fact, not correct [13]. At least in some of the examples given, the preferred oxidation is at what would seem to be the more sterically hindered side of the azo bridge. Preparation 3-5 illustrates this point. Similarly, 1-(methylazo)cyclohexene was oxidized to 1-(methyl-*NNO*-azoxy)cyclohexene. However, 2-(methylazo)isobutene was converted into 2-(methyl-*ONN*-azoxy)isobutene.

3-5. *Preparation of 2-(Methyl-NNO-azoxy)propene* [13]

$$CH_3-N=N-\underset{\underset{CH_3}{|}}{C}=CH_2 \quad \xrightarrow{[O]} \quad CH_3-N\overset{\overset{\displaystyle O}{\uparrow}}{=}N-\underset{\underset{CH_3}{|}}{C}=CH_2 \qquad (25)$$

CAUTION: Due safety precautions must be taken when handling 40% peracetic acid.

A stirred solution of 7.2 gm (86 mmole) of 2-(methylazo)propene in 100 ml of anhydrous diethyl ether is gradually treated with 22 gm (116 mmole) of 40% peracetic acid. The addition rate is controlled so that gentle refluxing takes place until the color of the azo compound has faded. After completion of the addition, the reaction mixture is allowed to stand for several hours. The reaction mixture is neutralized by the cautious addition of powdered sodium carbonate, and then filtered. Ultimately traces of residual acid are removed by treatment of the filtered reaction mixture with 10% sodium carbonate solution. The ether layer is separated and dried with anhydrous magnesium sulfate.

After filtration, the ether is removed by evaporation in an air stream. The final product is recovered by distillation under reduced pressure (behind a shield because possible hydroperoxides of ether have not been adequately removed): yield 3.8 gm (44%), b.p. 84°–86°C (150 mm Hg), n_D^{25} 1.4660.

The oxidation of a fluorinated aromatic azo compound has been carried out with a mixture of 89% hydrogen peroxide and trifluoroacetic anhydride; it may be considered a trifluoroperacetic acid oxidation [48].

3-6. *Preparation of Octafluoro-4,4'-dimethoxyazoxybenzene* [48]

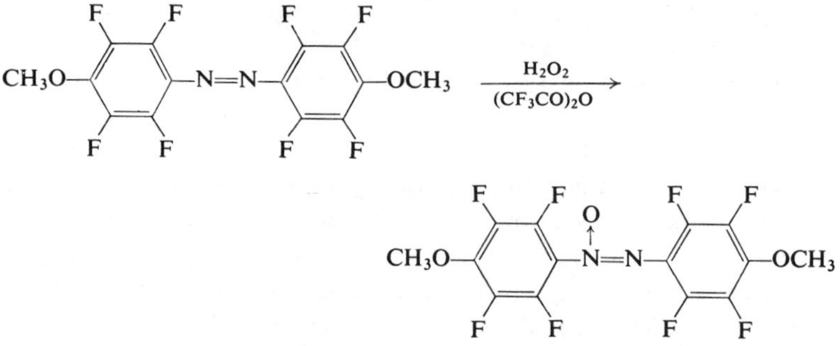

CAUTION: Due precautions must be taken when handling 89% hydrogen peroxide and when carrying out this trifluoroperacetic acid oxidation.

To a solution of 0.7 gm (1.96 mmoles) of octafluoro-4,4'-dimethoxyazo-benzene in 30 ml of methylene chloride is added a cooled solution prepared from 8 ml of 89% hydrogen peroxide and 5 ml of trifluoroacetic anhydride. After refluxing the red solution for 15 min, the expected change to a yellow color has taken place. The reaction mixture is added to a large volume of cold water, the organic layer is separated and dried over anhydrous magnesium sulfate. After filtration and evaporation of the solvent, 0.64 gm (82%) of crude product remains. After recrystallization from methanol, the m.p. is 130°C.

B. Oxidation of Hydrazones

A number of years ago it was observed that the perbenzoic acid oxidation of benzaldehyde phenylhydrazone afforded a high-melting product of low solubility which was termed "benzaldehyde phenylhydrazone oxide." More recent work has shown that this, as well as related oxidation products of other phenylhydrazones, is, in fact, the cis isomer of the corresponding azoxy compounds [14, 17].

In this connection it is interesting to note that the rapid oxidation of phenylhydrazones with peracids generally leads to cis azoxy compounds, while the slower air oxidation (particularly when exposed to light) evidently affords hydroperoxides of some thermal stability which decompose to azo compounds [49].

Also to be noted is that the position of the entering oxygen in the azoxy product is reasonably certain. A possible mechanism for the oxidation has been proposed [14] (Eqs. 27–29).

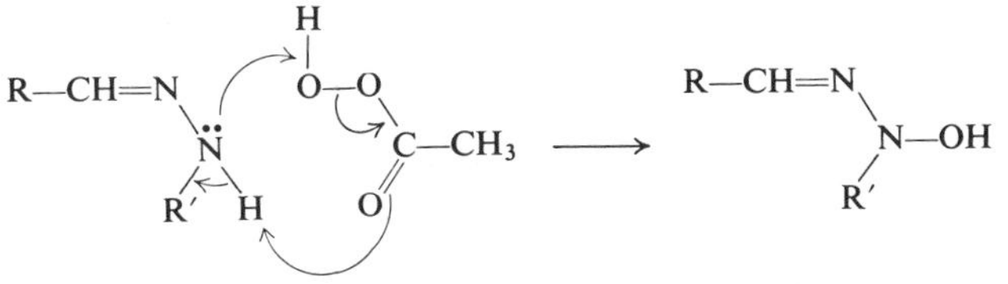

$$\text{(27)}$$

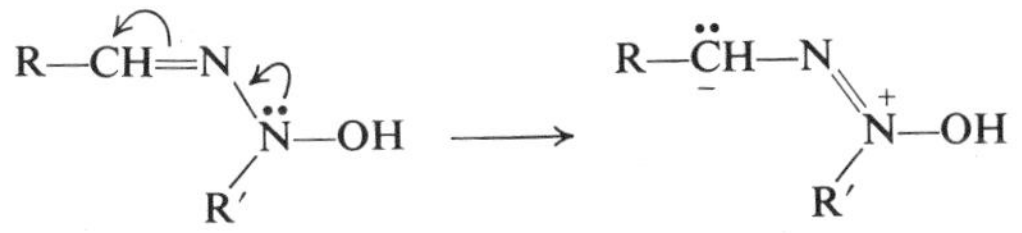

$$\text{(28)}$$

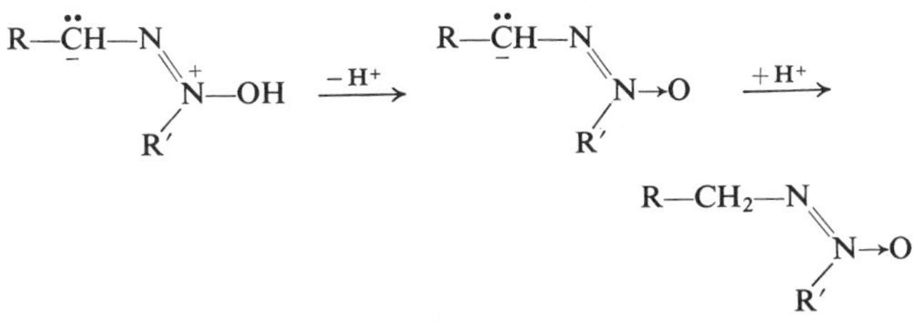

$$\text{(29)}$$

In effect, this mechanism indicates that the hydrogen of the NH group is replaced by the entering oxygen. The fact that cis isomers appear to be formed invariably is not explained by this mechanism. The unusually high melting point of cis aliphatic azoxy compounds may be attributed to the fact that either these cis compounds have a higher dielectric constant than the corresponding trans isomers, or partial association to a dimeric structure takes place to give a possible structure such as (VIII) [14].

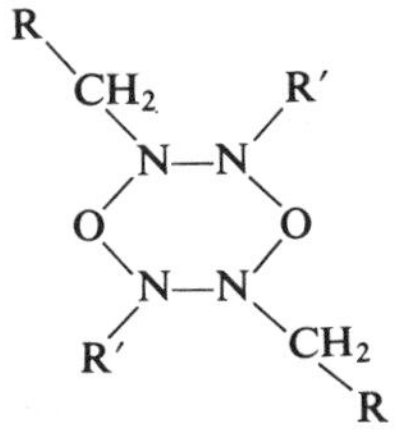

(VIII)

Evidently phenylhydrazones of aromatic aldehydes, on peracetic acid oxidation, are converted |to azoxy compounds in good| yield. The situation is more complex with aliphatic hydrazones of aromatic aldehydes (e.g., methylhydrazones and benzylhydrazones). In the case of methylhydrazones, oxidation at ice temperatures afford the expected cis azoxy compounds. At reflux temperatures the oxidation product is the isomeric hydrazide. Thus, for example, benzaldehyde methylhydrazone is converted into 1-benzoyl-2-methylhydrazine. Evidently the hydrazide is formed by rearrangement of the azoxy compound in the acidic reaction medium. Similarly, *p*-tolualdehyde benzylhydrazone, on oxidation, affords a mixture of α-(benzyl-*ONN*-azoxy)-*p*-xylene and 1-benzoyl-2-(α-*p*-xylyl)-hydrazine. Again, it was found that acids will bring about the rearrangement of the aliphatic azoxy compound to the hydrazide.

3-7. *Preparation of cis-α-(Phenyl-ONN-azoxy)toluene* [14]

(30)

CAUTION: Due safety precautions must be taken when handling 40% peracetic acid.

To a stirred solution of 19.6 gm (0.10 mole) of benzaldehyde phenylhydrazone in ether is slowly added a solution of 25.0 gm (0.13 mole) of 40% peracetic acid in ether. Stirring at room temperature is continued for 3 hr after completion of the addition. Since the product is quite insoluble, it is filtered off directly from the reaction mixture and washed in turn with chloroform and ether. After drying, the melting point is determined by inserting a melting-point tube into a bath preheated to 170°C, m.p. 187° dec.; yield 11.5 gm (54%).

(CAUTION: Disposal of the peroxide-containing ether and chloroform must conform to recommended safety procedures.)

C. Oxidation of Aromatic Amines

The oxidation of aromatic amines generally produces highly colored, tarry product mixtures. However, some aromatic amines have been successfully converted into azoxy compounds by peracetic acid oxidations. In regard to the factors influencing the reaction, several observations may be pertinent.

(1) In the oxidation of pentafluoroaniline with performic acid, along with the expected pentafluoronitrosobenzene, a 17% yield of decafluoroazoxybenzene was isolated. Separate experiments showed that the condensation of

the nitrosobenzene with the residual amine did not lead to the clean-cut preparation of the azoxy compound, whereas the thermal degradation of the nitroso compound did afford the azoxy compound. The implications of these observations are that either the azoxy product was formed, at least in part, by direct oxidation of the amine or the thermal history of the reaction permitted its formation from the intermediate nitroso compound [48].

(2) The oxidation of nonfluorinated aromatic amines with free peracids (not acyl peroxides) proceeded best with the basic amines. The reaction did not pass through the azo stage of oxidation since azo compounds could not be converted under the reaction conditions and hydrazo compounds were oxidized to azo compounds. Under the reaction conditions, *p*-toluidine was converted only into *p*-nitrotoluene [50].

(3) Metal ions in catalytic amounts exercise a profound influence on the course of the oxidation. In the absence of metal ions, the peracetic acid oxidation of 3-nitroaniline produces 3,3′-dinitroazoxybenzene. In the presence of traces of cupric ions and, to a lesser extent, in the presence of small quantities of iron, nickel, and rhodium salts, only 3,3′-dinitroazobenzene is formed. The oxidation of toluidines and aminophenols usually leads to tarry products [51].

In the light of the last set of observations, oxidations of amines to the azoxy stage should be carried out with the rigorous exclusion of metallic ions. Metal stirring rods must be avoided. In fact, since the aromatic amines may have been manufactured by a metal–acid reduction, a preliminary analysis for trace metals is indicated.

3-8. *Preparation of p-Azoxyanisole* [50]

$$CH_3O-\langle\ \rangle-NH_2 \xrightarrow[RCO_3H]{[O]} CH_3O-\langle\ \rangle-\overset{\overset{O}{\uparrow}}{N}=N-\langle\ \rangle-OCH_3 \quad (31)$$

A stirred mixture of 0.615 gm (5 mmoles) of *p*-anisidine (freshly recrystallized from carefully deionized water) in 50 ml of petroleum ether is cooled in an ice bath and maintained between 5°C and 7°C throughout the reaction. To the amine solution is added rapidly a solution of 1.08 gm (5 mmoles) of perlauric acid in 50 ml of petroleum ether, and stirring is continued until the reaction solution takes on a brownish color.

The product solution is then extracted in turn with a 10 % aqueous potassium hydroxide solution, a 5 *N* hydrochloric acid solution, and water. The organic layer is separated, dried with anhydrous magnesium sulfate, and filtered.

On evaporation of the organic solution to dryness, approximately 0.39 gm of the crude oxidation product is isolated (62 % yield).

By vapor-phase chromatography this residue was found to consist of 3% *p*-nitroanisole, 10% azoanisole, and 87% azoxyanisole.

To separate the components of this reaction mixture, the crude product is dissolved in a minimum quantity of petroleum ether. The solution is then passed through a 2 × 20 cm chromatography column packed with aluminum oxide. The nitro and azo compounds are eluted from the column first with sufficient petroleum ether. The azoxy compound is eluted with petroleum ether containing 1% of methanol. The eluting solvent is evaporated and the residual product is recrystallized from ethanol, m.p. 117°–118°C, nematic-liquid transition point 134°C.

Recently, *p,p'*-bis(perdeuterioalkoxy)azoxybenzenes were prepared by the hydrogen peroxide oxidation of the corresponding *p*-perdeuterioalkoxyaniline [52].

D. Oxidation of Hydroxylamines

The oxidation of aromatic hydroxylamines with peracids in the presence of cupric ions produces nitroso compounds. In the rigorous absence of metallic ions azoxy compounds are formed [51]. On the other hand, the air oxidation is strongly accelerated by metals, the approximate order of activity based on a kinetic study being: cupric ≫ ferric > manganous > nickel ≅ chromic > cobaltous ions. Silver and stannous ions appear to have no effect [53].

In the example cited, lead acetate was used as a catalyst [27].

3-9. *Preparation of Azoxycyclohexane* [27]

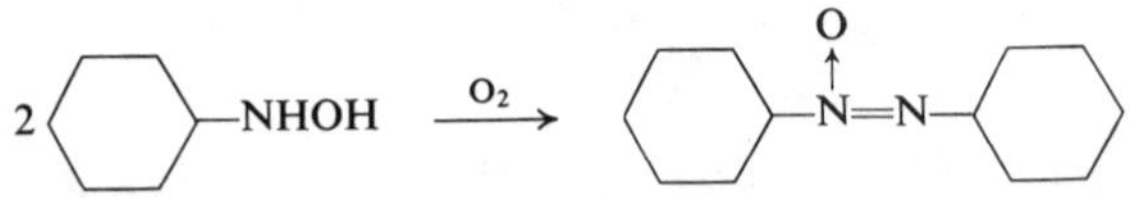

$$2 \left\langle \text{—NHOH} \right\rangle \xrightarrow{O_2} \left\langle \text{—N=N—} \right\rangle \overset{O}{\uparrow} \tag{32}$$

Through a stirred solution of 25 gm (0.22 mole) of cyclohexylhydroxylamine, 100 ml of methanol, and 2 gm of lead acetate maintained between 0°C and 10°C with an ice bath for 45 hr is bubbled a steady stream of air. During this period the evaporated solvent is replaced from time to time.

On vacuum fractional distillation of the product mixture, 1 gm of cyclohexanone oxime and 19.5 gm (85%) of crude azoxycyclohexane are isolated. After two fractional distillations of the product fraction, the following physical properties are observed for azoxycyclohexane: b.p. 160°–161°C (14 mm Hg), m.p. 22°–23°C, n_D^{20} 1.497, d_D^{20} 1.007.

E. Oxidation of Indazole Oxides

The assignment of the structures of unsymmetrical azoxy compounds was traditionally based on the results obtained from substitution reactions. This required assumptions about directing influences which were difficult to substantiate. The technique of oxidizing indazole oxides followed by decarboxylation represents an unequivocal synthetic procedure for the establishment of the position of the azoxy oxygen in the trans azoxy isomers. The reaction sequence used is given in Eq. (33) [54].

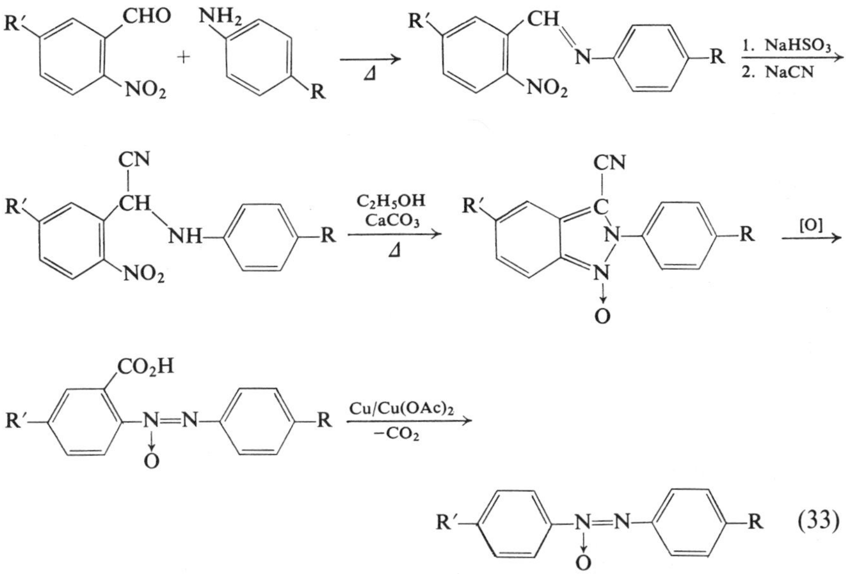

In this reaction sequence, barring the remote possibility of rearrangements, the source of the oxygen on the azoxy bridge is the nitro group of the original nitroaldehyde. Further experimental work had shown that migration of the nitro group did not take place. Consequently, this procedure represents an unequivocal synthesis of azoxy compounds of known structure with the restriction that the geometric isomerism is not rigorously defined by this method, but is assumed to produce only the trans isomer.

A more recent modification of the procedure considerably simplifies the reaction. In this procedure the condensation of a nitroaldehyde with an aniline derivative in acetic acid solution and in the presence of potassium cyanide formed the 3-cyano-2-phenylindazole 1-oxide directly. Although the reaction was used to prepare the two *m*-mononitroazoxybenzenes, presumably the reaction is of more general applicability [37].

3-10. Preparation of m-Nitrophenyl-NNO-azoxybenzene [37]

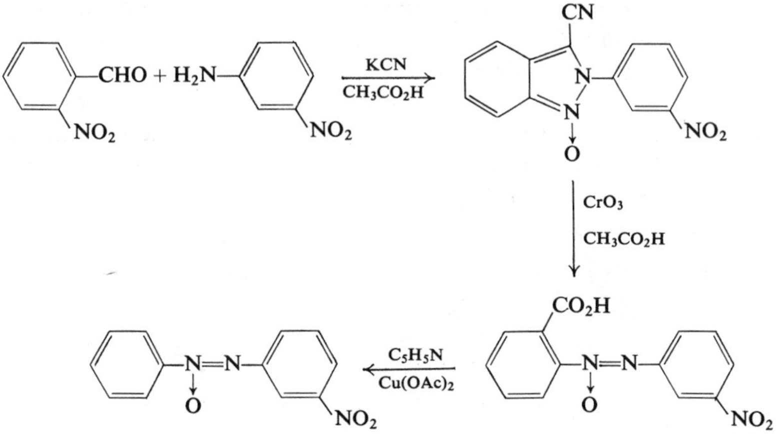

(34)

CAUTION: Due precautions must be taken in handling potassium cyanide. Particularly in acid media, hydrogen cyanide may also evolve.

(a) *Preparation of* 3-cyano-2-(m-nitrophenyl)indazole 1-oxide. In a hood, with suitable safety precautions, a solution of 5.0 gm (0.033 mole) of *o*-nitrobenzaldehyde and 4.6 gm (0.033 mole) of *m*-nitroaniline in 300 ml of glacial acetic acid is prepared by moderate warming of the mixture. The reaction mixture is then cooled to 25°C and 4.0 gm (0.062 mole) of potassium cyanide is added. The mixture is allowed to stand for 12 hr. Water (200 ml) is added to the dark-brown solution, and the resulting solution is placed in a refrigerator for 24 hr. The yellow precipitate is collected on a filter, washed with cold water, and dried: yield 5.0 gm (54%), m.p. 205°–208°C. After three recrystallizations from ethanol the m.p. is raised to 211°–213°C.

(b) *Oxidation to* m-nitrophenyl-NNO-azoxy-2-benzoic acid. A solution of 2.0 gm (0.0071 mole) of crude 3-cyano-2-(*m*-nitrophenyl)-indazole 1-oxide (m.p. 205°–208°C) in 75 ml of glacial acetic acid is prepared by warming, if necessary. To this solution is added, in small portions, 1.5 gm of chromium trioxide. The solution is warmed gently until evidence of oxidation is noted. Then it is allowed to stand at room temperature for 3 hr. The reaction mixture is diluted with 225 ml of cold water and the product, which partially precipitates, is extracted with several portions of ether. The combined ether extracts are washed repeatedly with cold water. Then the product is separated by extraction with 5% aqueous sodium hydroxide solution, followed by acidification with a 15% solution of phosphoric acid. The yellow acidic product is collected on a filter and washed free from residual phosphoric acid with water: yield 1.0 gm (47%), m.p. (recrystallized from benzene) 183°–184.5°C.

(c) *Decarboxylation to m-nitrophenyl-NNO-azobenzene* ("*3-nitroazoxy-benzene*"). To a solution of 0.5 gm (0.0017 mole) of crude *m*-nitrophenyl-*NNO*-azoxy-2-benzoic acid in pyridine, a few crystals of copper acetate are added and the mixture is heated for 4.5 hr at reflux temperature. After cooling to room temperature, the product solution is diluted with 100 ml of ether and then washed, in turn, with four portions of a 10% hydrochloric acid solution until the aqueous layer is slightly acid, with one portion of water, and finally with four portions of a 5% aqueous solution of sodium hydroxide. The ether layer is separated and dried with anhydrous magnesium sulfate, filtered, and then freed from solvent by evaporation. The residue is recrystallized from ethanol: yield 0.21 gm (50%), m.p. 121.5°–122°C.

The decarboxylation has also been carried out by substituting copper powder for copper acetate.

F. Oxidative Hydrolysis of N-Substituted Urazole Derivatives

In Chapter 14, on azo compounds, reference is made to the preparation of certain polycyclic azo compounds by the oxidative hydrolysis of addition products of bicycloalkadienes with 4-methyl-1,2,4-triazoline-3,5-dione (i.e., urazoles) in dimethyl sulfoxide with potassium *tert*-butoxide [55]. With more vigorous oxidizing conditions, such as heating with aqueous ethylene glycol with an excess of potassium hydroxide and 30% hydrogen peroxide in the case of *N*-phenylurazole derivatives, the corresponding *cis*-azoxy compounds are prepared in approximately 80% yield. (NOTE: The hydrolytic oxidation of *N*-methylurazoles is said to be exothermic when large amounts of material are processed, so caution is recommended [3, 56].) Equations (35) and (36) briefly outline the process.

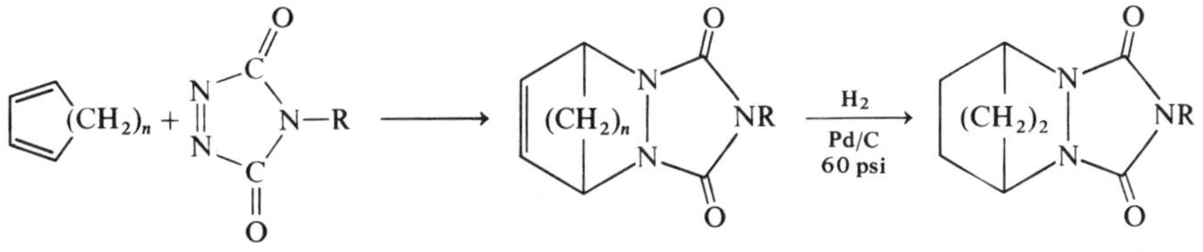

$$n = 1\text{-}4 \qquad R = CH_3, C_6H_5 \qquad\qquad (35)$$

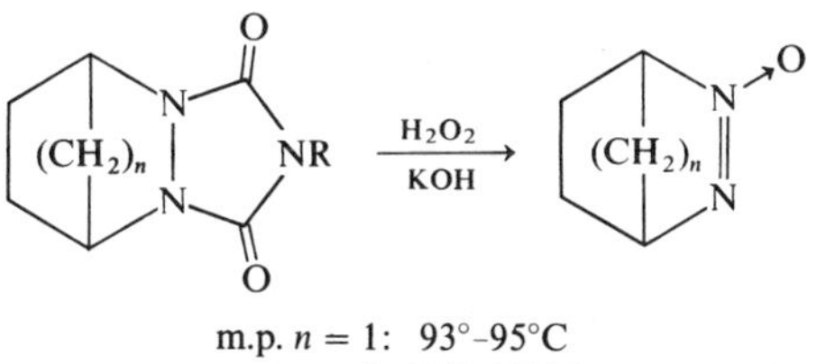

$$\qquad\qquad\qquad\qquad\qquad (36)$$

m.p. $n = 1$: 93°–95°C
2: 156°–159°C
3: 176°–177°C
4: 133°–134°C

4. REDUCTION REACTIONS

A. Reduction of Nitroso Compounds

The bimolecular reduction of aliphatic nitroso compounds is complex and somewhat unreliable. With careful control of reaction conditions, α-nitroso ketones (in dimeric form) may be reduced with stannous chloride in an acidic medium at room temperature to the azoxy compounds, while dimeric α-nitroso acid derivatives may be reduced at about 50°C [20, 57, 58]. Nitrosoalkanes, on the other hand, are decomposed at room temperature to alcohols and nitrogen, and are reduced to amines at 50°–60°C. It has been postulated that only the dimeric nitroso compounds can be reduced to azoxy compounds and, in fact, that the dimer has covalent nitrogen–nitrogen bond. Equations (37)–(40) summarize these data [20].

$$(\alpha\text{-nitroso ketones})_2 \xrightarrow[25°C]{\text{SnCl}_2/\text{HCl}} R-\overset{\overset{O}{\uparrow}}{N}=N-R \xrightarrow[\text{in excess}]{\text{SnCl}_2/\text{HCl}} 2RH + H_2NNH_2 \quad (37)$$

$$(\alpha\text{-nitroso esters})_2 \xrightarrow[55°C]{\text{SnCl}_2/\text{HCl}} R-\overset{\overset{O}{\uparrow}}{N}=N-R \quad (38)$$

$$(\text{nitrosoalkanes})_2 \xrightarrow[25°C]{\text{hydrolysis}} N_2 + ROH \quad (39)$$

$$(\text{nitrosoalkanes})_2 \underset{\text{above } 50°C}{\rightleftharpoons} \text{nitrosoalkanes} \xrightarrow[50°C]{\text{SnCl}_2/\text{HCl}} 2RNH_2 \quad (40)$$

The behavior of α-nitrosonitriles is similar to that of the α-nitroso esters. In fact, the resultant azoxynitriles could be converted into azoxy esters by conventional alcoholysis procedures.

The preparation of ethyl α-azoxyisopropyl ketone is a typical example of the bimolecular reduction with stannous chloride [59].

4-1. Preparation of Ethyl α-Azoxyisopropyl Ketone [59]

$$\left[\begin{array}{c} O \quad\quad CH_3 \\ \| \quad\quad | \\ C_2H_5-C-C-NO \\ | \\ CH_3 \end{array} \right]_2 \xrightarrow{\text{SnCl}_2/\text{HCl}} C_2H_5-\overset{O}{\overset{\|}{C}}-\overset{CH_3}{\underset{CH_3}{\overset{|}{C}}}-\overset{O}{\overset{\uparrow}{N}}=N-\overset{CH_3}{\underset{CH_3}{\overset{|}{C}}}-\overset{O}{\overset{\|}{C}}-C_2H_5 \quad (41)$$

To a rapidly stirred solution of 10.8 gm (0.048 mole) of crystalline stannous chloride (SnCl$_2\cdot$2H$_2$O) in 15.8 ml of concentrated hydrochloric acid, while the reaction temperature is maintained between 30°C and 36°C, 10 gm (0.039 mole on a bimolecular basis) of dimeric ethyl α-nitrosoisopropyl ketone is added over a 20 min period. The reaction mixture is then nearly neutralized by the cautious addition of sodium carbonate.

The liquid product is separated by extraction with ether. The ether solution is dried over anhydrous potassium carbonate. After filtration, the ether is removed by evaporation and the residue is distilled, a fraction being collected between 126°C and 126.5°C (6 mm Hg), n_D^{20} 1.4587, d_D^{20} 1.0151: yield 5.18 gm (55%).

An interesting method of reducing nitroso compounds involves the use of triethyl phosphite as reducing agent [48, 60]. The generality of this procedure requires further exploration.

4-2. Preparation of Decafluoroazoxybenzene by Triethyl Phosphite Reduction [48]

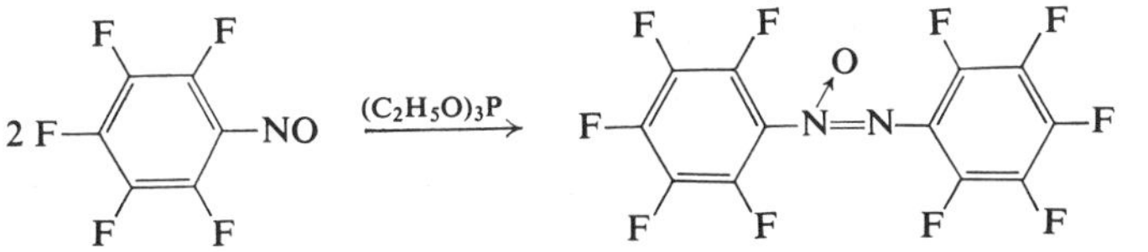

$$\tag{42}$$

A solution of 0.55 gm (2.8 mmoles) of pentafluoronitrosobenzene and 1 ml of triethyl phosphite in benzene is warmed for 30 min at 60°C. The reaction mixture is cooled to room temperature and extracted several times with water. The organic layer is then evaporated to dryness and the residue is crystallized from methanol: yield 0.42 gm (81%), m.p. 48°C.

B. Reduction of Nitro Compounds

The bimolecular reduction of aromatic nitro compounds, depending on reaction conditions, may proceed by way of azoxy and azo compounds to 1,2-diarylhydrazines (also referred to as hydrazo compounds). This may be illustrated by the schematic process in Eq. (43). Under strongly acidic conditions, the 1,2-diarylhydrazine may undergo rearrangement to produce benzidine.

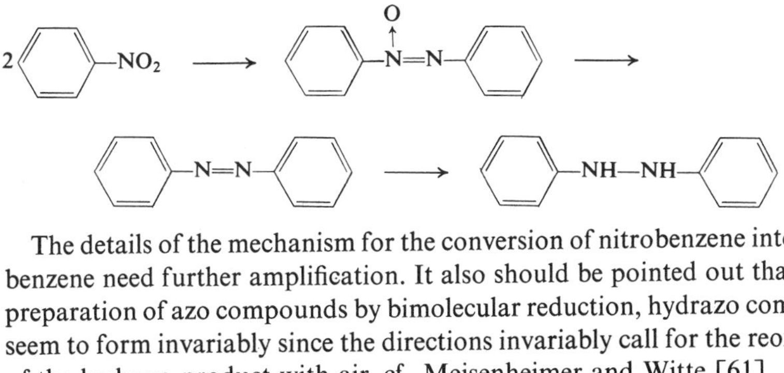

$$\tag{43}$$

The details of the mechanism for the conversion of nitrobenzene into azoxybenzene need further amplification. It also should be pointed out that, in the preparation of azo compounds by bimolecular reduction, hydrazo compounds seem to form invariably since the directions invariably call for the reoxidation of the hydrazo product with air, cf., Meisenheimer and Witte [61].

As indicated in the preceding chapter, the reduction of aromatic nitro compounds with zinc and sodium hydroxide solution leads to the azo product [61, 62]. On the other hand, in an acetic acid–acetic anhydride medium, reduction with zinc produces a symmetrical azoxy compound [63].

4-3. Preparation of o,o'-Dicyanoazoxybenzene [63]

$$2 \quad \text{(CN,NO}_2\text{ benzene)} \xrightarrow[\substack{(HOCOCH_3) \\ (CH_3CO)_2O}]{Zn} \text{(CN, O↑N=N, CN benzene)} \quad (44)$$

To a vigorously stirred solution of 9.0 gm (0.062 mole) of o-nitrobenzonitrile in 75 ml of acetic acid and 10 ml of acetic anhydride, while maintaining the reaction temperature between 30°C and 35°C by adjusting the addition rate and warming or cooling as needed, is gradually added 12.0 gm (0.19 gm-atom) of zinc dust over a $1\frac{1}{2}$ hr period. Then the pasty reaction mixture is diluted with 1 liter of ice water and the solid is separated by centrifugation.

The solid product is next treated for 15 min with 300 ml of a 10% aqueous sodium carbonate solution at 40°C, filtered, washed with water, refiltered, and pressed dry. The solid is then dissolved in warm ethanol. The solution is treated with decolorizing carbon and filtered. On cooling, the product crystallizes out: yield 3.9 gm (53%), m.p. 192°–193°C.

The reduction of aromatic nitro compounds to azoxy products with stannous chloride in a basic medium has also been reported; however, the final purification of the product appears to be tedious [61].

Among the metallic reducing systems [64, 65] which have been suggested are magnesium with anhydrous methanol [64] and an alloy of sodium and lead (sold as 'Drynap') [66, 67]. In the case of the magnesium methanol reduction, the product mixture usually contains azoxy-, azo-, hydrazo-, amino-, and hydroxylamine products, although approximately 75% of the product consists of a mixture of azoxy and azo compounds. The 'Drynap' reduction also seems to lead frequently to mixtures of azoxy and azo compounds.

McKillop et al. [64] reported that refluxing a mixture of an aromatic nitro compound with thallium metal and ethanol for $1\frac{1}{2}$ to 12 hr gave 64–93% yields of the corresponding azoxy compounds. The procedure seems simple and should find widespread use in the laboratory. However, it has the important limitation that groups such as CHO, COR, COOH, COOR, CN, phenolic OH, and amino groups totally inhibit the reaction. Nevertheless, high yields of azoxy compounds are obtained from nitro aromatics with ether or alkyl substituents in any position. Fluoro- and chloro-substituted aromatic nitro compounds afford fluoro- and chloroazoxy compounds. However, bromo- and iodo-substituted aromatic nitro compounds afford low yields of haloazoxy derivatives isolatable from complex reaction mixtures (Eq. 45).

$$2ArNO_2 + 6Tl + 6C_2H_5OH \longrightarrow Ar\overset{O^-}{\underset{+}{-N}}=NAr + 6TlOC_2H_5 + 3H_2O \quad (45)$$

A single pure product is obtained in this reaction, and thallium ethoxide is not effective in giving further reduction to the azo compound. However, heating for periods in excess of 12 hr or those shown in Table I could lead to the formation of azo compounds. Polyhalonitro compounds and 2,4-dichloronitrobenzene give the corresponding azo compound in high yields on refluxing the ethanol solution for 12 hr.

The reaction is generally carried out by refluxing a stirred mixture of 0.014 mole of the nitro compound and 0.042 mole (8.5 gm) of thallium in 75 ml of ethanol for the periods shown in Table I. The cooled solution is decanted to remove unreacted thallium. Then 8 gm of potassium iodide is added and the mixture is stirred for 1 hr at room temperature. The precipitated thallium iodide is filtered off, and the filtrate is concentrated under reduced pressure. The residue is dissolved in chloroform and then filtered through an alumina column (4 × 1 in.) to remove traces of inorganic salts. Chloroform is used as the eluent. The resulting solution is concentrated under reduced pressure to afford the solid azoxy compounds.

Alcoholic potassium hydroxide has been used to reduce iodonitrobenzene. While, from o-iodonitrobenzene only approximately a 1% yield of unsubstituted azoxybenzene was isolated, the m- and p-isomers produced substantial quantities of m-iodo- and p-iodoazoxybenzene, respectively. In view of the well-known equilibrium $KOH + C_2H_5OH \rightleftharpoons K^+ + OC_2H_5^- + H_2O$ and the fact that sodium alcoholates, particularly those of the higher alcohols, have also been used as reducing agents with isolation of oxidation products of the alcohol [68], it is reasonable to assume that either the free alcohol or the alcoholate ion acts as the reducing agent in these cases.

4-4. Preparation of m,m'-Diiodoazoxybenzene [62]

$$\quad (46)$$

To a stirred solution of 10 gm (0.040 mole) of m-iodonitrobenzene in 100 ml of ethanol is added 16 gm (0.30 mole) of potassium hydroxide. The mixture is

TABLE I[k]

CONVERSION OF SUBSTITUTED NITROBENZENES INTO AZOXY COMPOUNDS

Nitrobenzene derivative	Azoxy compound	Registry No.	Time[a] (hr)	Yield[b] (%)	M.p.[c] (°C)	Lit. m.p. (°C)
Nitrobenzene	Azoxybenzene	495-48-7	6	76	34.5-35.5	35[c]
2-Nitrotoluene	2,2'-Dimethylazoxybenzene	956-31-0	6	73	57-58	60[c]
3-Nitrotoluene	3,3'-Dimethylazoxybenzene	19618-06-5	6	77	33-35	38-39[d]
4-Nitrotoluene	4,4'-Dimethylazoxybenzene	955-98-6	5	77	66-68	68[c]
4-Ethylnitrobenzene	4,4'-Diethylazoxybenzene	23595-86-0	8½	80	B.p. 180-185 (0.7 mm)	B.p. 244 (16 mm)[e]
2,5-Dimethylnitrobenzene	2,2',5,5'-Tetramethylazoxybenzene	14381-98-7	7½	64	110-112	111.5-112.5[f]
2-Nitrobiphenyl	2,2'-Diphenylazoxybenzene	7334-10-3	4½	84	158-160	160-163[g]
2-Nitroanisole	2,2'-Dimethoxyazoxybenzene	13620-57-0	5½	80	79-80	81-82[c]
4-Nitroanisole	4,4'-Dimethoxyazoxybenzene	1562-94-3	5½	76	116.5-118.5, 134.5-135.5	118.5, 135[h]
4-n-Butyloxynitrobenzene	4,4'-Di-n-butyloxyazoxybenzene	17051-01-3	12	80	102-104, 136.5-137	107, 134[i]
4-n-Hexyloxynitrobenzene	4,4'-Di-n-hexyloxyazoxybenzene	2587-42-0	7	71	80-81.5, 128.5	81, 127[j]
4-Fluoronitrobenzene	4,4'-Difluoroazoxybenzene	326-04-5	12	89	84-86	86-87[c]
2-Chloronitrobenzene	2,2'-Dichloroazoxybenzene	13556-84-8	1½	86	53.5-55	55-56[c]
3-Chloronitrobenzene	3,3'-Dichloroazoxybenzene	139-24-2	4½	84	95.5-97	96[c]
4-Chloronitrobenzene	4,4'-Dichloroazoxybenzene	614-26-6	5	93	154-156	155-156[c]

[a] In most cases about 5-10% of the thallium was *not* consumed during the reaction. Increasing the time of reaction had no significant effect on the yield, and resulted in minor amounts of decomposition.   [b] No attempt was made to optimize yields.   [c] P. H. Gore and O. H. Wheeler, *J. Amer. Chem. Soc.* **78**, 2160 (1956).   [d] L. Zechmeister and P. Rom, *Ann. Chem.* **468**, 117 (1929).   [e] B. T. Newbold and D. Tong, *Canad. J. Chem.* **42**, 836 (1964).   [f] E. Bamberger, *Chem. Ber.* **59**, 418 (1926).   [g] E. Wenkert and B. F. Barnett, *J. Amer. Chem. Soc.* **82**, 4671 (1960).   [h] R. S. Porter and J. F. Johnson, *J. Phys. Chem.* **66**, 1826 (1962).   [i] C. Weygand and R. Gabler, *Chem. Ber.* **71B**, 2399 (1938).   [j] C. Weygand and R. Gabler, *J. Prakt. Chem.* **155**, 332 (1940).   [k] Reprinted from A. McKillop, R. A. Raphael, and E. C. Taylor, *J. Org. Chem.* **35**, 1671 (1970). (Copyright 1970 by the American Chemical Society. Reprinted with permission of the copyright owner.)

then heated to reflux for 1 hr. After this, the reaction mixture is subjected to steam distillation. The distillation residue is then extracted with boiling benzene. After cooling, the benzene layer is separated and evaporated to dryness. Yield 5.9 gm (60%), m.p. (after recrystallization from ethanol–benzene) 120°–121°C.

When sodium methylate or ethylate was prepared by direct reaction of sodium with an excess of alcohols and the resulting mixture was used as a dispersion in benzene to reduce aromatic nitro compounds, yields of azoxy compounds were quite low. With the higher alcohols, substantial production of azoxy compounds was observed. However, the reduction product mixture usually contained a 40% yield of amino compounds. In a few examples, where benzyl alcohol was used to prepare sodium benzylate, only azoxy products and no amino by-products were formed. The scope of this preparation requires further study.

A detailed procedure for the reduction of nitrobenzene with sodium arsenate according to Eq. (47) has been published [69].

$$4\,C_6H_5{-}NO_2 + 3As_2O_3 + 18NaOH \longrightarrow$$

$$2\,C_6H_5{-}\overset{\overset{\displaystyle O}{\uparrow}}{N}{=}N{-}C_6H_5 + 6Na_3AsO_4 + 9H_2O \quad (47)$$

The reduction of aromatic nitro compounds with glucose and other carbohydrates in an alkaline medium has been a much neglected reaction procedure (see, for example, Bigelow and Palmer [69, Note 7]). The reaction has much to recommend it: yields are usually high, the reaction is rapid, and the oxidation products derived from glucose are all water-soluble and readily separated from the product. The oxidation coproducts have not been completely identified. Among these products are acetic acid, lactic acid, and traces of oxalic acid; absent are formic, gluconic, glucuronic, saccharic, 5-ketogluconic, and tartaric acids [70].

It was found that between 70°C and 100°C the reaction temperature was not critical, that a reaction time of approximately 35 min was optimum, that the optimum concentration of sodium hydroxide was in the range 13–20%, that the optimum ratio of glucose to aromatic nitro compound was $1\frac{1}{2}$ moles to 1 mole; that the concentration of the reaction solution could be varied over a wide range, and that a high rate of stirring as well as the presence of an emulsifying agent increased the yield [70]. Since the reaction conditions have a resemblance to certain redox-emulsion polymerization recipes, it is the present authors' opinion that the effect of low concentration of polyvalent ions (e.g., Fe^{2+} or Cu^+, appropriately chelated, if necessary) should be evaluated. As a matter of fact, a stainless steel stirrer may furnish enough ions to catalyze the reaction.

The reduction to produce azoxy compounds appears to be successful for a wide range of aromatic nitro compounds. The following compounds, however, could not be reduced: 1-nitronaphthalene, *m*-dinitrobenzene, 3,5-dinitrobenzoic acid (although *o*-, *m*-, and *p*-nitrobenzoic acids were reduced smoothly), compounds containing an amino group *o*-, or *p*- to a nitro group (except sodium 3-nitro-6-aminobenzenesulfonate), nitroazo compounds, *p*-nitrosodimethylaniline, *p*-nitrophenylhydrazine, and 2,5-dinitrophenylhydrazine. In the light of more recent data, the reduction of *p*-nitrobenzoic acid is also doubtful [71]. Among the materials which could be reduced were *m*-nitrobenzenesulfonic acid, 3-chloro-5-nitrobenzenesulfonic acid, and 3-nitro-6-aminobenzenesulfonic acid. These three compounds are mentioned here because they were isolated as the appropriate salts which were subjected to nitrogen analysis but were not characterized further.

4-5. Preparation of 4,4'-Dichloro-2,2'-azoxytoluene [70]

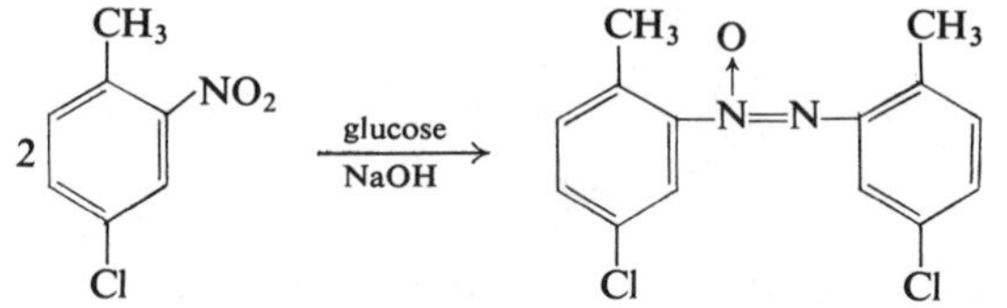

(48)

A rapidly stirred (1000 rpm) mixture of 17.2 gm (0.1 mole) of 2-nitro-4-chlorotoluene, 0.5 ml of Tergitol-08 (a surfactant), and 100 ml of a 30% aqueous solution of sodium hydroxide is warmed to 60°C. External heating is discontinued and a solution of 27 gm (0.15 mole) of glucose in 150 ml of water is added at a rate to maintain the reaction at 60°C (approximately 30 min). The reaction mixture is then chilled in an ice bath, and the crude product is removed by filtration.

After repeated washing with ice water, the crude product is air-dried and finally recrystallized from ethanol: yield 6.9 gm (71%), m.p. 126°–127°C.

It was found that certain substituents render nitrobenzene inert to reduction by potassium borohydride, whereas other substituents activate the reduction to azoxy compounds. The substituents which favored this reduction all had positive Hammett sigma constants (e.g., *p*-Cl, *p*-Br, *p*-I, *p*-COOH, *m*-Cl, *m*-Br, *m*-I, *m*-CHO reduced to *m*-azoxybenzyl alcohol and *m*-OC$_2$H$_5$). Among the by-products of the reaction of a *p*-halogen compound, when carried out in ethanol solution, was *p*-nitrophenetole. The reduction of *p*-fluoronitrobenzene afforded only *p*-nitrophenetole [71].

4-6. Preparation of m-Azoxybenzyl Alcohol [71]

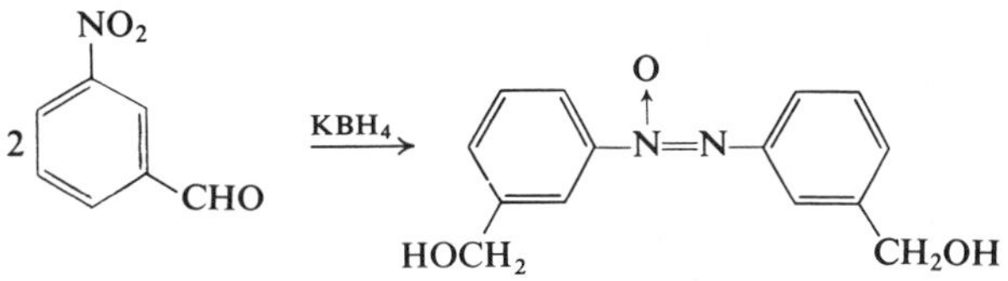

A rapidly stirred mixture of 10 gm (0.185 mole) of potassium borohydride and 250 ml of ethanol containing a small amount of potassium hydroxide is heated to the reflux temperature. Then a solution of 5.0 gm (0.033 mole) of *m*-nitrobenzaldehyde in a small amount of ethanol is added and heating is continued for 24 hr. After this period, half of the solvent is removed by distillation under reduced pressure. The residue is cooled and added, with stirring, to a mixture of ice and concentrated hydrochloric acid. The crude product is collected on a filter, washed free of excess acid with water, and dried. The product is ultimately recrystallized. Yield 4.3 gm (71%), m.p. 82°–83°C.

The reduction of many aromatic nitro compounds with *trans*-dihalobis-(pyridine)palladium (II) is a homogeneous process normally leading to the corresponding aromatic amine. However, in the case of *p*-chloronitrobenzene, bis(*p*-chlorophenyl)diazene-1-oxide is said to form [72].

5. ISOMERIZATION REACTIONS

It is generally assumed that under ordinary laboratory conditions the trans isomers of the azoxy compounds are formed. Although the cis isomers are generally much higher-melting and are much more insoluble than the corresponding trans isomers, their stabilities at room temperature appear to be highly variable. Consequently, only a few cis compounds have been isolated.

As indicated before, cis azo compounds may be oxidized to cis azoxy compounds, and certain hydrazones also may be oxidized to cis azoxy compounds (hence the old but incorrect designation "hydrazone oxides") [17].

In the base-catalyzed condensation of nitroso compounds with hydroxylamines at low temperatures, if the starting materials are present in equimolar quantities, modest yields of cis isomers may be isolated [73]. The present authors are not in a position to judge whether this procedure affords the cis isomers directly, or whether the normal trans isomer is formed initially and is subsequently converted into the cis form in the presence of the base. The latter sequence seems quite likely in view of other observations in which trans azoxy compounds were isomerized to the cis azoxy form in the presence of sodium methoxide [16, 17].

5-1. *Preparation of cis-ω-Azoxytoluene* [73]

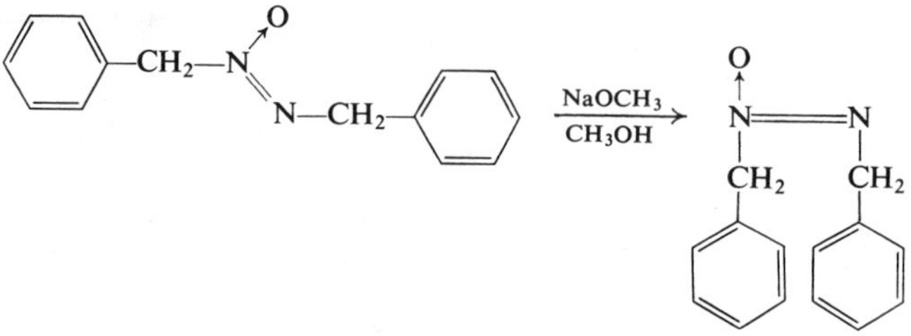

$$(50)$$

To a solution of 1 gm of sodium methoxide in 9 ml of methanol maintained at 42°–43°C is added 1 gm of *trans-ω-azoxytoluene*. After a short stirring period, a white precipitate begins to separate. The solid is collected on a filter and recrystallized from toluene–ethanol, m.p. 190°–192°C: yield not reported.

6. MISCELLANEOUS PREPARATIONS

(1) Condensation of aromatic nitro compounds with aromatic hydroxyla-mines [37] (Eq. 51).

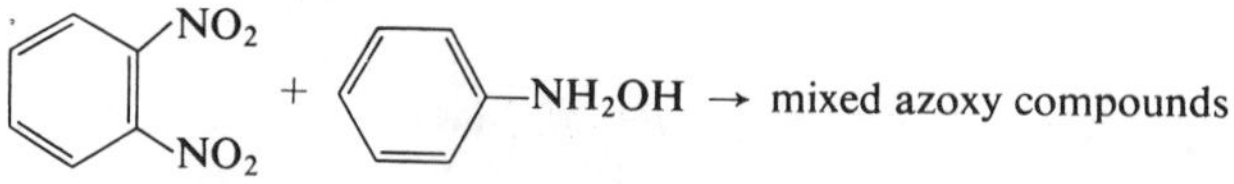

$$(51)$$

(2) Hydrogenation of dimeric nitroso compounds with a palladium-on-barium sulfate catalyst [27] (Eq. 52).

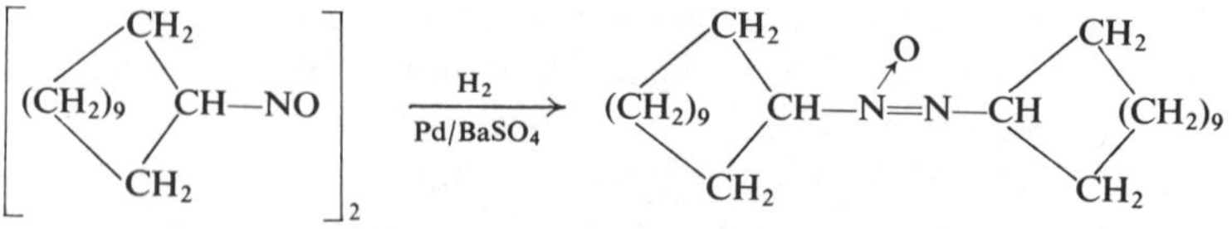

$$(52)$$

(3) Reduction of a mixture of amines and nitroalkanes with sodium [20] (Eq. 53).

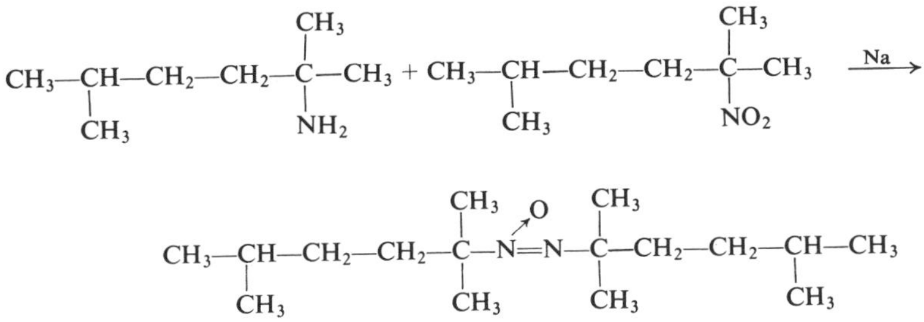

$$CH_3-CH-CH_2-CH_2-\overset{\overset{\displaystyle CH_3}{|}}{\underset{\underset{\displaystyle NH_2}{|}}{C}}-CH_3 + CH_3-\overset{\overset{\displaystyle CH_3}{|}}{\underset{\underset{\displaystyle CH_3}{|}}{CH}}-CH_2-CH_2-\overset{\overset{\displaystyle CH_3}{|}}{\underset{\underset{\displaystyle NO_2}{|}}{C}}-CH_3 \xrightarrow{Na}$$

$$CH_3-\overset{\overset{\displaystyle CH_3}{|}}{\underset{\underset{\displaystyle CH_3}{|}}{CH}}-CH_2-CH_2-\overset{\overset{\displaystyle CH_3}{|}}{\underset{\underset{\displaystyle CH_3}{|}}{C}}-\overset{O}{N}{=}N-\overset{\overset{\displaystyle CH_3}{|}}{\underset{\underset{\displaystyle CH_3}{|}}{C}}-CH_2-CH_2-\overset{\overset{\displaystyle CH_3}{|}}{\underset{\underset{\displaystyle CH_3}{|}}{CH}}-CH_3 \quad (53)$$

(4) Pyrolysis of nitroso compounds [48] (Eq. 54).

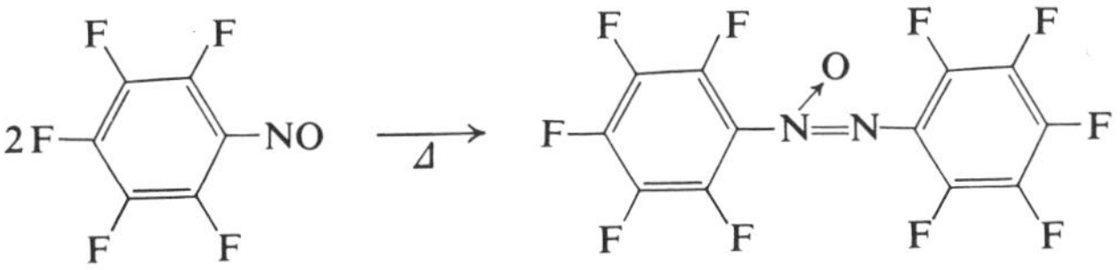

$$(54)$$

(5) Treatment of aromatic nitro compounds with potassium cyanide [63, 74] (Eqs. 55, 56).

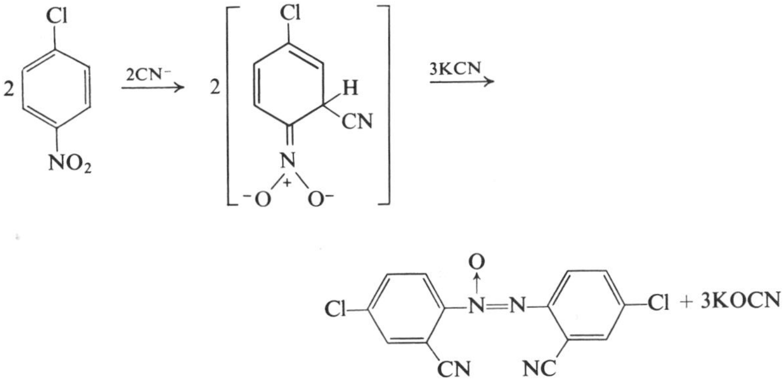

$$(55)$$

$$3KOCN \xrightarrow{6H_2O} 3NH_3 + 3KHCO_3 \qquad (56)$$

(6) Preparation of N-substituted N'-fluorodiimide N-oxides [75, 75] (Eq. 57).

$$R-NO \xrightarrow[\substack{or \\ HNF_2 \text{ in pyridine}}]{N_2F_2} R-\overset{O}{N}{=}NF \qquad (57)$$

(7) Reaction of nitroso compounds with chloramine T [76] (Eq. 58).

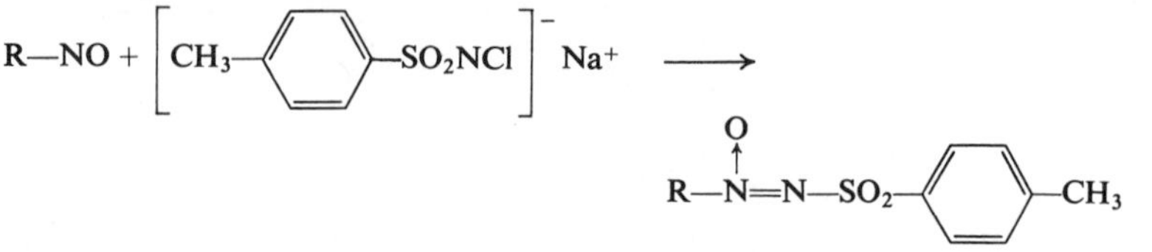

$$R\text{—}NO + \left[CH_3\text{—}\underset{}{\bigcirc}\text{—}SO_2NCl \right]^- Na^+ \longrightarrow$$

$$R\text{—}\overset{\overset{O}{\uparrow}}{N}\text{=}N\text{—}SO_2\text{—}\underset{}{\bigcirc}\text{—}CH_3 \quad (58)$$

(8) Electrolytic reduction of nitro compounds [77, 78].

(9) Reduction of nitro compounds with yellow phosphorus [79] (Eq. 59).

$$R\text{—}C_6H_4\text{—}NO_2 \xrightarrow[\text{aq. alkali}]{P \text{ (yellow)}} \text{azoxy compounds and amines} \quad (59)$$

(10) Reduction of nitro compounds with phosphine [80, 81].

(11) Reduction of N,N'-dioxides (bis nitrosyl deivatives or intramolecular nitro dimers) with hydrogen with a palladium-on-carbon catalyst [82] (Eq. 60).

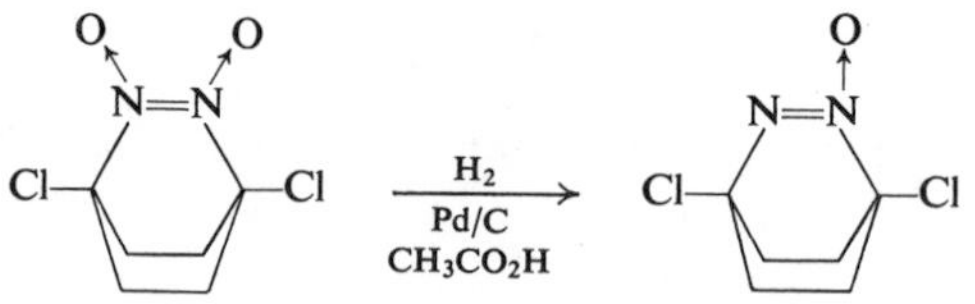

$$Cl\text{—}\underset{}{\bigcirc}\text{—}Cl \xrightarrow[\substack{Pd/C \\ CH_3CO_2H}]{H_2} Cl\text{—}\underset{}{\bigcirc}\text{—}Cl \quad (60)$$

(12) Preparation of polyesters derived, *inter alia*, from azoxybenzoic acid and azoxyphenols [83].

Tables II and III give the physical properties of a selected list of azoxy compounds.

TABLE II

PROPERTIES OF ALIPHATIC AND MIXED ALIPHATIC–AROMATIC
AZOXY COMPOUNDS

Compound	B.p., °C (mm Hg)	M.p. (°C)	n_D^t	Ref.
Azoxymethane	98	—	1.4300[17]	40
1-Azoxypropane	67 (20)	—	1.4365[20]	40
2-Azoxypropane	38 (14)	—	—	40
Azoxyisobutane	50 (20)	—	1.4208[20]	16
t-Butyl-*ONN*-azoxymethane	60–62 (110)	—	1.4265[20]	16
2-(Methyl-*NNO*-azoxy)-propene	84–86 (150)	—	1.4660[25]	13
2-(Methyl-*ONN*-azoxy)-isobutene	82–84 (90)	—	1.5215[25]	13
2-Azoxy-2,5-dimethylhexane	111 (5)	—	—	20
Methyl α-azoxyisopropyl ketone	—	60–61	—	57
Ethyl α-Azoxyisopropyl ketone	126–126.5 (6)	—	1.4587[20]	59
Isobutyl α-azoxyisopropyl ketone	—	30–31	—	59

TABLE II (*continued*)

Compound	B.p., °C (mm Hg)	M.p. (°C)	n_D^t	Ref.
2-Methyl-2-azoxypropionitrile	—	37	—	58
2-Methyl-2-azoxypropionic acid	—	128–129	—	58
Ethyl-2-methyl-2-azoxypropionate	142–144 (12)	—	1.4404^{20}	58
Azoxycyclohexane	160–161 (14)	22–23	1.497^{20}	27
	165–170 (21)	27	—	27
Azoxycyclododecane	—	76–78	—	27
Methyl-*NNO*-azoxybenzene	—	6.5–8.5	1.556^{20}	15
Ethyl-*NNO*-azoxybenzene	—	—	1.5434^{20}	31
Isopropyl-*NNO*-azoxybenzene	—	—	1.5296^{20}	31
n-Butyl-*NNO*-azoxybenzene	—	—	1.5280^{20}	31
Methyl-*NNO*-azoxy(*p*-chlorobenzene)	—	40	—	31
Isopropyl-*NNO*-azoxy(*p*-chlorobenzene)	—	—	1.5438^{20}	31
Methyl-*NNO*-2-azoxy-2,5-dimethyl-hexane	184 (738.5)	—	1.4330^{20}	2
trans-ω-Azoxytoluene	—	42–43	—	16
cis-ω-Azoxytoluene	—	190–192	—	16
Phenyl-*ONN*-azoxymethane	78 (3.5)	—	—	16
Benzyl-*NNO*-azoxybenzene	—	186–190	—	16
ω-Azoxy-*p*-chlorotoluene	—	103	—	40
cis-(4-Chlorobenzyl)-*NNO*-azoxybenzene	—	180	—	17
trans-(4-Chlorobenzyl)-*NNO*-azoxybenzene	—	37	—	17
trans-(4-Chlorobenzyl)-*ONN*-azoxybenzene	—	63	—	17
cis-(3-Bromo-4-chlorobenzyl)-*NNO*-azoxybenzene	—	176	—	17
trans-(3-Bromo-4-chlorobenzyl)-*NNO*-azoxybenzene	—	112	—	17
trans-(*p*-Chlorobenzyl)-*ONN*-azoxy-(*p*-bromobenzene)	—	107	—	7
1-(Methyl-*NNO*-azoxy)cyclohexene	132–135 (2)	—	1.5045^{30}	13
cis-α-(Phenyl-*ONN*-azoxy)toluene	—	187	—	14
cis-α-(Phenyl-*ONN*-azoxy)-*p*-nitrotoluene	—	172	—	14
cis-α-(Phenyl-*ONN*-azoxy)-*p*-phenyltoluene	—	188	—	14
cis-α-(Phenyl-*ONN*-azoxy)-*p*-methoxytoluene	—	162	—	14
cis-α-(Methyl-*ONN*-azoxy)-toluene	—	153	—	14
cis-(Methyl-*ONN*-azoxy)-*p*-toluene	—	116	—	14
cis-(Methyl-*ONN*-azoxy)-*p*-nitrobenzene	—	205	—	14
cis-(Methyl-*ONN*-azoxy)-*p*-bromo-benzene	—	185	—	14

TABLE III

PROPERTIES OF AROMATIC AZOXY COMPOUNDS

Compound	M.p. (°C)	Ref.
cis-Azoxybenzene	87, 86	41, 73
trans-Azoxybenzene	35, 36	22, 31
cis-o-Azoxytoluene	79–81	73
o-Azoxytoluene	60, 59	44, 73
cis-m-Azoxytoluene	87, 88–89	41, 73
trans-m-Azoxytoluene	37, 34–35.5	41, 73
cis-p-Azoxytoluene	84, 83–85	41, 73
trans-p-Azoxytoluene	74,[a] 69.5–70, 70–71	41, 50, 73
1,1′-Azoxynaphthalene	127, 122–123	39, 50
2,2′-Azoxynaphthalene	166, 160–161, 158–160	39, 50, 61
p,p′-Azoxybiphenyl	203–204	44
cis-o-Azoxyanisole	116	73
o-Azoxyanisole	81–82, 89–90	44, 67, 73
m-Azoxyanisole	52	75
p-Azoxyanisole	119,[b] 118[b]	43, 67
m-Azoxyphenetole	49–50	71
p-Azoxyphenetole	137[c]	43, 67
p,p′-Dihexyloxyazoxybenzene	81[d]	43
p,p′-Dichloroazoxybenzene	158, 155–156, 154–155	22, 44, 50
p-Chlorophenyl-*NON*-azoxybenzene	62[e]	22
p-Chlorophenyl-*ONN*-azoxybenzene	82	31
p-Bromophenyl-*ONN*-azoxybenzene	95, 92–93	31, 54
p-Tolyl-*NNO*-azoxybenzene	50	31
p-Tolyl-*NNO*-azoxy(*p*-chlorobenzene)	109.5	31
p-Tolyl-*NNO*-azoxy(*p*-bromobenzene)	125	31
p-Chlorophenyl-*NNO*-azoxybenzene	68	31
p-Chlorophenyl-*NNO*-azoxy(*p*-toluene)	107	31
p-Chlorophenyl-*NNO*-azoxy(*p*-bromobenzene)	160	31
p-Anisyl-*NNO*-azoxybenzene	72	31
p-Anisyl-*NNO*-azoxy(*p*-chlorobenzene)	145	31
1-Naphthyl-*NNO*-2-azoxynaphthalene	137	29
1-Naphthyl-*NNO*-azoxybenzene	84	29
2-Naphthyl-*NNO*-azoxybenzene	125	29
2-Naphthyl-*ONN*-azoxybenzene	117	29
Phenyl-*ONN*-azoxy-(4-dimethylaminobenzene oxide)	136–137	44
p,p′-Dinitroazoxybenzene	192	44, 50
p,p′-Difluoroazoxybenzene	86–87	44
p,p′-Dibromoazoxybenzene	172	44
p,p′-Diiodoazoxybenzene	198, 199–200	44, 62
m,m′-Dinitroazoxybenzene	143	44, 57
3,5,3′5′-Tetranitroazoxybenzene	185	44
m,m′-Difluoroazoxybenzene	51.5	44
m,m′-Dichloroazoxybenzene	96	44
m,m′-Dibromoazoxybenzene	111, 109–110	44, 71
m,m′-Diiodoazoxybenzene	120.5–121.5, 120–121, 118–119	44, 62, 71
o,o′-Dichloroazoxybenzene	55–56, 57	44, 73

TABLE III (*continued*)

Compound	M.p. (°C)	Ref.
o,o'-Dibromoazoxybenzene	115	44
o,o'-Diiodoazoxybenzene	148	44
(2-Bromophenyl)-*NNO*-azoxy(2-hydroxy-5-methylbenzene)	96	45
(2-Methoxyphenyl)-*NNO*-azoxy(2-hydroxy-5-methylbenzene)	117–117.5	45
(2-Methylphenyl)-*NNO*-azoxy(2-hydroxy-5-methylbenzene)	86–86.5	45
(2-Nitrophenyl)-*NNO*-azoxy(2-hydroxy-5-methylbenzene)	102–103	45
(2-Methoxyphenyl)-*ONN*-azoxy(2-hydroxy-5-methylbenzene)	69	45
Octafluoro-4,4'-dimethoxyazoxybenzene	130	48
4-*H*,4'-*H*-Octafluoroazoxybenzene	52	48
p,p'-Dinitroazoxybenzene	190–191	51
m-Nitrophenyl-*NNO*-azoxybenzene	91–91.5	37
m-Nitrophenyl-*ONN*-azoxybenzene	121.5–122	37
p-Bromophenyl-*NNO*-azoxybenzene	73–73.5	54
p-Ethoxyphenyl-*NNO*-azoxybenzene	73–74	54
p-Ethoxyphenyl-*ONN*-azoxybenzene	76–76.5	54
Decafluoroazoxybenzene	48, 49–51	48
o,o'-Dicyanoazoxybenzene	192–193	63
3,3'-Dichloro-2,2'-azoxytoluene	126–127	70
p,p'-Azoxybenzoic acid	240 dec.,*f* 350–355, 114–115	70, 71, 79
m,m'-Azoxybenzoic acid	320 dec., 76–77	70, 79
o,o'-Azoxybenzoic acid	248 dec., 79–84	70, 79
2,3,2',3'-Azoxyphthalic acid	360	70
6,6'-Dibromo-3,3'-azoxybenzenesulfonic acid	360	70
m,m'-Azoxyaniline	146–148	70
m,m'-Diamino-*p,p'*-dimethylazoxybenzene	155–155.5	70
m,m'-Diacetylazoxybenzene	131–131.5	70
6,6'-Azoxyquinoline	260–262	70
o,o'-Azoxyphenol	153–154	70
m-Azoxybenzyl Alcohol	82–83	71
cis-*o,o'*-Dichloroazoxybenzene	92–93	73
4,4'-Dimethoxy-2,2'-dicarbamylazoxybenzene	230–234	74
4,4'-Dimethoxyazoxybenzene-2,2'-dicarboxylic acid	222–229	74

[a] Gore and Wheeler [44] report the melting point of *p*-azoxytoluene as 68°C, citing Bamberger and Renauld [21] as reporting 69°C. This is in agreement with Lefort *et al.* [50].

[b] Nematic-liquid transition: 135°C.

[c] Nematic-liquid transition: 165°C.

[d] Nematic-liquid transition: 128°C.

[e] This material proved to be a 1:1 mixture of 4- and 4'-chloroazoxybenzene (Stevens [31]).

[f] The reported m.p. of 240°C dec. is that of *p*-nitrobenzoic acid. By two independent syntheses, Shine and Mallory [71] obtained a product of m.p. 350°–355°C dec.

REFERENCES

1. H. E. Bigelow, *Chem. Res.* **9**, 117 (1931).
2. J. G. Aston and D. M. Jenkins, *Nature (London)* **167**, 863 (1951).
3. J. P. Snyder, V. T. Bandurco, F. Darack, and H. Olsen, *J. Amer. Chem. Soc.* **96**, 5158 (1974).
4. S. Patai, ed. "The Chemistry of Hydrazo, Azo, and Azoxy Groups," Wiley, New York, 1975.
5. V. N. Yandovskii, B. V. Gidaspov, and I. V. Tselinskii, *Usp. Khim.* **49**, 449 (1980).
6. V. N. Yandovskii, B. V. Gidaspov, and I. V. Tselinskii, *Usp. Khim.* **50**, 296 (1981).
7. A. A. Nemodruk, I. K. Kleimenova, and I. M. Gibalo, *Zh. Anal. Khim.* **36**, 552 (1981).
8. O. Wallach and L. Belli, *Ber. Dtsch. Chem. Ges.* **13**, 525 (1880).
9. K. H. Schündehütte, in Houben-Weyl, "Methoden der Organischen Chemie," **10/3**, 771 (1965).
10a. R. A. Cox and E. Buncel, *Can J. Chem.* **51**, 3143 (1973).
10b. E. Buncel, S. R. Keum, M. Cygler, K. E. Varughese, and G. E. Birnbaum, *Can. J. Chem.* **62**, 1628 (1984).
11. I. Shimao and S. Oae, *Bull. Chem. Soc. Jpn.* **56**, 643 (1983).
12. "International Union of Pure and Applied Chemistry, Nomenclature of Organic Chemistry," Sect. C., P. 204 ff. Butterworth, London and Washington, D.C., 1965.
13. B. T. Gillis and J. D. Hagarty, *J. Org. Chem.* **32**, 95 (1967).
14. B. T. Gillis and K. F. Schimmel, *J. Org. Chem.* **27**, 413 (1962).
15. M. V. George, R. W. Kierstead, and G. F. Wright, *Cand. J. Chem.* **37**, 679 (1959).
16. J. P. Freeman, *J. Org. Chem.* **28**, 2508 (1963).
17. J. N. Brough, B. Lythgoe, and P. Waterhouse, *J. Chem. Soc.* 4069 (1954).
18. G. G. Spence, E. C. Taylor, and O. Buchardt, *Chem. Rev.* **70**, 231 (1970).
19. *Chemical Abstracts*, Index Guide, **86–95**, 119G (1977–1981).
20. J. G. Aston and D. E. Ailman, *J. Amer. Chem. Soc.* **60**, 1930 (1938).
21. E. Bamberger and E. Renauld, *Chem. Ber.* **30**, 2278 (1897).
22. Y. Ogata, M. Tsuchida, and Y. Takagi, *J. Amer. Chem. Soc.* **79**, 3397 (1957).
23. G. A. Russell, E. J. Geels, F. Smentowski, K.-Y. Chang, J. Reynolds, and G. Kaupp, *J. Amer. Chem. Soc.* **89**, 3821 (1967).
24. T. Kauffmann and S. M. Hage, *Angew. Chem. Int. Ed. Engl.* **2**, 156 (1963); *Angew. Chem.* **75**, 295 (1963).
25. L. A. Neiman, V. I. Maimid, and M. M. Shemyakin, *Tetrahedron Lett.* 3157 (1965).
26. G. A. Russell and E. J. Geels, *J. Amer. Chem. Soc.* **87**, 122 (1965).
27. H. Meister, *Ann. Chem.* **679**, 83 (1964).
28. S. S. Mochalov, A. M. Fedotov, and Yu. S. Shabarov, *Zh. Org. Khim.* **15**, 947 (1979); *Chem. Abstr.* **91**, 91274k (1979).
29. D. H. R. Barton, G. Lamotte, W. B. Motherwell, and S. C. Narang, *J. Chem. Soc. Perkin Trans.* **1**(8), 2030 (1979).
30. R. E. Banks, W. T. Flowers, and R. N. Haszeldine, *J. Chem. Soc. Perkin Trans.* **1**(11), 2765 (1979).
31. T. E. Stevens, *J. Org. Chem.* **29**, 311 (1964).
32. A. C. M. Meesters, H. Rueger, K. Rajeswari, and M. H. Benn, *Can. J. Chem.* **59**, 264 (1981).
33. V. G. Shtamburg, V. F. Rudchenko, Sh. S. Nasibov, I. I. Chervin, A. P. Pleshkova, and R. G. Kostyanovskii, *Izv. Akad. Nauk SSSR Ser. Khim.* **1981** (10), 2370 (1981); *Chem Abstr.* **96**, 51733f (1982).
34. M. Okubo and K. Koga, *Bull. Chem. Soc. Jap.* **58**, 203 (1983).
35. Chisso Corp. Jpn. Kokai Tokkyo Koho JP 57,197,251 (82,197,261) (1982).
36. C.-S. Hahn and H. H. Jaffé, *J. Amer. Chem. Soc.* **84**, 949 (1962).
37. L. C. Behr, E. G. Alley, and O. Levand, *J. Org. Chem.* **27**, 65 (1962).
38. G. M. Badger and G. E. Lewis, *J. Chem. Soc.* 2147 (1953).
39. G. M. Badger and G. E. Lewis, *J. Chem. Soc.* 2151 (1953).

40. B. W. Langley, B. Lythgoe, and L. S. Rayner, *J. Chem. Soc.* 4191 (1952).
41. G. M. Badger, R. G. Buttery, and G. E. Lewis, *J. Chem. Soc.* 2143 (1953).
42. G. M. Badger and R. G. Buttery, *J. Chem. Soc.* 2156 (1953).
43. M. J. S. Dewar and R. S. Goldberg, *Tetrahedron Lett.* (24), 2717 (1966).
44. P. H. Gore and O. H. Wheeler, *J. Amer. Chem. Soc.* **78**, 2160 (1956).
45. V. M. Dzimoko and K. A. Dunaevskaya, *J. Gen. Chem. USSR* **31**, 3155 (1961); *Zh. Obshch. Khim.* **31**, 3385 (1961).
46. D. L. Webb and H. H. Jaffé, *J. Amer. Chem. Soc.* **86**, 2419 (1964).
47. S. V. Blokhina, G. G. Maidachenko, L. I. Mineev, and V. I. Klopov, *Zh. Org. Khim.* **17**, 151 (1981); *Chem. Abstr.* **95**, 24412y (1981).
48. J. Burdon, C. J. Morton, and D. F. Thomas, *J. Chem. Soc.* 2621 (1965).
49. A. V. Chernova, R. R. Shagidullin, and Yu. P. Kitaev, *Chem. Abstr.* **67**, 53349k (1967); *Zh. Org. Khim.* **31**, 916 (1967).
50. D. Lefort, C. Four, and A. Pourchez, *Bul. Soc. Chim. Fr.* 2378 (1961).
51. E. Pfeil and K. H. Schmidt, *Ann. Chem.* **675**, 36 (1964).
52. A. Suszko-Purzycka, M. Ossowska-Chrusciel, and J. Chruaciel, Pol. PL 121,096 (1983); *Chem. Abstr.* **101**, 23109k (1984).
53. Y. Ogata and T. Morimoto, *J. Org. Chem.* **30**, 597 (1965).
54. L. C. Behr, *J. Amer. Chem. Soc.* **76**, 3672 (1954).
55. W. Adam, O. DeLucchi, and I. Erden, *J. Amer. Chem. Soc.* **102**, 4806 (1980). W. Adam and O. DeLucchi, *J. Org. Chem.* **46**, 4133 (1981).
56. H. Olsen and J. P. Snyder, *J. Amer. Chem. Soc.* **99**, 1524 (1977).
57. J. G. Aston, D. F. Menard, and M.G. Mayberry, *J. Amer. Chem. Soc.* **54**, 1535 (1932).
58. J. G. Aston and G. T. Parker, *J. Amer. Chem. Soc.* **56**, 1387 (1934).
59. D. E. Ailman, *J. Amer. Chem. Soc.* **60**, 1933 (1938).
60. P. J. Bunyan and J. I. G. Cadogan, *Proc. Chem. Soc.* 78 (1962).
61. J. Meisenheimer and K. Witte, *Chem. Ber.* **36**, 4153 (1903).
62. B. T. Newbold, *J. Chem. Soc.* 4260 (1961).
63. E. Cullen and Ph. L'Écuyer, *Can. J. Chem.* **39**, 862 (1961).
64. L. Zechmeister and P. Ron, *Ann. Chem.* **468**, 117 (1929).
65. A. McKillop, R. A. Raphael, and E. C. Taylor, *J. Org. Chem.* **35**, 1670 (1970).
66. K. Tabei and K. Natou, *Chem. Soc. Japan* **39**, 2300 (1966).
67. K. Tabei and M. Yamaguchi, *Bull. Chem. Soc. Japan* **40**, 1538 (1967).
68. C. M. Suter and F. B. Dains, *J. Amer. Chem. Soc.* **50**, 2733 (1928).
69. H. E. Bigelow and A. Palmer, *Org. Syn. Coll. Vol.* **2**, 57 (1943).
70. H. W. Galbraith, E. F. Degering, and E. F. Hitch, *J. Amer. Chem. Soc.* **73**, 1323 (1951).
71. H. J. Shine and H. E. Mallory, *J. Org. Chem.* **27**, 2390 (1962).
72. S. Bhattacharya, P. K. Santra, and C. R. Saha, *Ind. J. Chem. Sect. A* **23A**, 724 (1984).
73. E. Müller and W. Kreutzmann, *Ann. Chem.* **495**, 132 (1932).
74. M. M. Rauhut and J. F. Bunnett, *J. Org. Chem.* **21**, 939 (1956).
75. T. E. Stevens and J. P. Freeman, *J. Org. Chem.* **29**, 2279 (1964).
76. T. E. Stevens, *J. Org. Chem.* **33**, 855 (1968).
77. M. V. King, *J. Org. Chem.* **26**, 3323 (1961).
78. S. Swann, Jr., *in* "Techniques of Organic Chemistry," (A. Weissberger, ed.), Vol. II, pp. 478–481. Wiley (Interscience), New York, 1956.
79. N. S. Kozlov and V. A. Soshin, *Uch. Zap. Permsk. Gos. Ped. Inst.* No. 32, 84 (1965); *Chem. Abstr.* **66**, 2290k (1967).
81. S. A. Buckler, L. Doll, F. K. Lind, and M. Epstein. *J. Org. Chem.* **27**, 794 (1962).
82. W. Lüttke and V. Schabacker, *Ann. Chem.* **687**, 236 (1965).
83. K. Iimura, N. Koida, and R. Ohta, *Proc. IUPAC, IUPAC Macromol. Symp., 28th* **1982**, 825 (1982); *Chem. Abstr.* **99**, 159128p (1983).

1. INTRODUCTION

Substances in which a nitroso group is directly attached to a carbon atom are termed *C*-nitroso compounds in this chapter. The basic procedures for the preparation of aromatic nitroso compounds were developed during the last quarter of the nineteenth century and the first few years of the present century. Since that time, with a few notable exceptions, only minor improvements on these procedures have been developed. We may assume that the lack of variety in synthetic procedures has gone hand in hand with a lack of interest in the application of this class of compounds.

The situation with regard to aliphatic nitroso compounds was confused by the observation that oximes were frequently isolated from nitrosation reactions. In fact, many oximes were called "isonitroso" compounds. Without adequate experimental evidence, it had generally been believed that only tertiary aliphatic nitroso compounds (which obviously would not rearrange to oximes) had reasonable stability. More recent work has shown that a wide variety of aliphatic nitroso compounds can be produced and that they exhibit adequate stability. It was also discovered that many nitroso compounds are dimeric in nature under ordinary conditions.

The equilibrium between the monomeric and dimeric nitroso compounds may be represented as in Eq. (1). Early evidence for the existence of such an

$$2R\text{—}NO \;\rightleftharpoons\; R\text{—}N{=}N\text{—}R \overset{O}{\underset{O}{}} \tag{1}$$

equilibrium was obtained from cryoscopic measurements of molecular weights in benzene and in naphthalene. Additional indications are the color changes which are frequently observed. Aliphatic monomeric nitroso compounds are usually blue in crystalline form, in solution, and often even in the vapor phase. Aromatic monomeric nitroso compounds are usually green or blue-green. The dimers are white or cream-colored.

Many para- and meta-substituted nitrosobenzenes are monomeric. On the other hand, ortho-substituted nitrosobenzenes (i.e., compounds with a substituent greater than H in the ortho-position) are dimeric.

In this chapter an attempt is made to indicate whether the products discussed are monomers or dimers; however, since the literature is not always clear on this point, an element of uncertainty persists.

A number of reviews, papers, and one book of theoretical and synthetic interest have appeared in recent years on various aspects of nitroso chemistry. The work of Turney and Wright [1], Touster [2], Gowenlock and Lüttke [3], Hamer *et al.* [4], Ogata [5], Dietrich and Crowfoot Hodgkin [6], Ridd [7], Griffis and Henry [8], Theilacker [9], Cook [10], Boyer [11], Feuer [12a], Patai [12b], Hoffman and Woodward [13], and Gilchrist [14] represents a selection of pertinent material. The reader is directed particularly to Gowenlock and Lüttke [3].

One of the best-known methods for the preparation of aromatic nitroso compounds is the Fischer–Hepp rearrangement of N-nitrosoamines (Eq. 2). In effect, this reaction is the C-nitrosation of secondary aromatic amines.

$$Ar—NHR \cdot HCl + HNO_2 \longrightarrow \underset{\underset{NO}{|}}{Ar—N—R} + H_2O + HCl$$

$$\underset{\underset{NO}{|}}{Ar—N—R} \xrightarrow{HCl} p\text{-}NO—Ar—NHR \qquad (2)$$

Fischer–Hepp rearrangement

Tertiary aromatic amines may be nitrosated directly with nitrous acid, whereas primary aromatic amines normally are diazotized under these conditions. Some phenols may also be nitrosated under conditions similar to those used for the nitrosation of tertiary amines.

Typical nitrosating agents are sodium nitrite with mineral acids or pentyl nitrite with mineral acids. A relatively new reagent, nitrosylsulfuric acid, has been used to nitrosate primary aromatic amines. On the other hand, trifluoroacetic acid and sodium nitrite could not be used as a nitrosating system. With benzene or toluene only 2 to 3 % of the corresponding nitro (not nitroso) compounds were isolated [15].

Treatment of aliphatic active methylene compounds with nitrosating reagents normally leads to oxime formation. An exception is the nitrosation of compounds with active tertiary carbon atoms such as ethyl isopropyl ketone. These are convertible into C-nitroso compounds.

Some aromatic hydroxycarboxylic acids, on treatment with sodium nitrite, lose carbon dioxide with the introduction of a nitroso group to replace the carboxylate group.

Certain olefins have been converted into 1-chloro-2-nitrosoalkane dimers on treatment with nitrosyl chloride (Eq. 3). Terminal olefins appear to be active only if they are not allylic in nature. The addition of nitrosyl chloride to norbornene is of particular interest, and details are presented below. Nitrosyl formate also adds to certain olefins in a manner resembling the addition of nitrosyl chloride (Eq. 3).

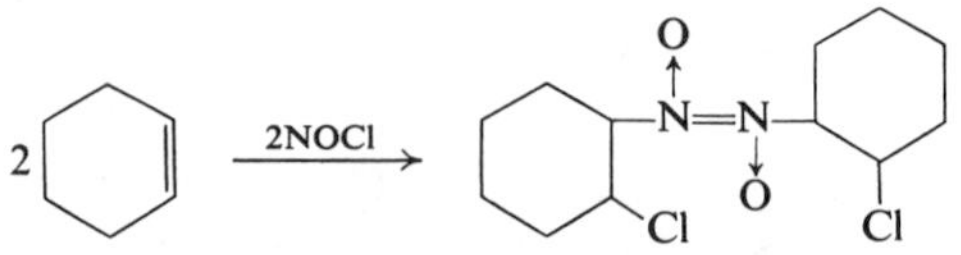

(3)

A variety of oxidative procedures exist for the preparation of both aromatic and aliphatic nitroso compounds.

Primary aromatic amines may be oxidized with Caro's acid or a variety of peracids (Eq. 4).

$$ArNH_2 \xrightarrow{[O]} ArNO + H_2O \tag{4}$$

Aliphatic amines have also been oxidized with peracids.

A method of preparation of wide applicability is the oxidation of hydroxylamines with such diverse reagents as bromine water, hypobromide ions, dichromate ions, lead tetraacetate, peracids, and ferric chloride (Eq. 5).

$$2RNHOH \xrightarrow{[O]} (R{-}NO)_2 \tag{5}$$

Particularly for the preparation of aliphatic nitroso compounds, the oxidation of oxaziranes and imines with peracetic acid (Eqs. 6) probably represents the most original and most widely applicable method of preparation.

$$\begin{array}{rcl} (C_2H_5)_2CO + RNH_2 & \longrightarrow & RN{=}C(C_2H_5)_2 \\ RN{=}C(C_2H_5)_2 & \xrightarrow{CH_3CO_3H} & R{-}\overset{O}{N{-}C}(C_2H_5)_2 \\ R{-}\overset{O}{N{-}C}(C_2H_5)_2 & \xrightarrow{CH_3CO_3H} & [RNO] \\ 2[RNO] & \longrightarrow & (RNO)_2 \end{array} \tag{6}$$

An interesting reaction, the Baudisch reaction, involves the formation of nitrosophenols by the action of hydroxylamine hydrochloride and hydrogen peroxide, in the presence of metallic ions or certain Werner complexes, on aromatic compounds. The products are primarily *o*-nitrosophenol complexes of the metallic ion. Unfortunately, this reaction requires further development before it can be considered a reliable preparative procedure.

The field of reductive preparations for the formation of nitroso compounds has not yet been adequately explored. For example, only indirect evidence exists that the electrolytic reduction of *t*-nitroalkanes to tertiary alkylhydroxylamines proceeds by way of nitroso compounds.

In this chapter a number of other preparative procedures are also given. They either are not of general applicability or have not yet been developed fully.

a. NOMENCLATURE

The naming of monomeric nitroso compounds is generally quite straightforward. In this discussion a dimer is simply indicated as "nitrosoalkane dimer" or as "nitrosoarene dimer" if reasonable evidence for the isolation of the dimer is at hand. In the works of Müller and Metzger [16] terms such as "bis(nitrosoalkane)" are used to indicate the dimers.

In another system, the trans isomer of dimeric 1-chloro-2-nitrosocyclohexane would be called "*trans*-2,2'-dichloro-*trans*-azodioxycyclohexane" [17]. In line with a system used in naming azoxy compounds, the dimers may also be called diimide dioxides.

The structures of nitroso dimers are generally regarded as resonance hybrids of structures such as (I), (II), and (III).

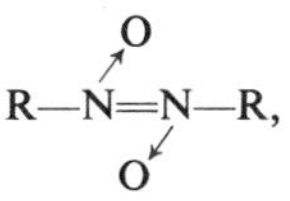

(I) 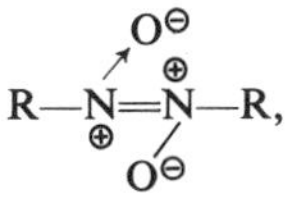   (II) 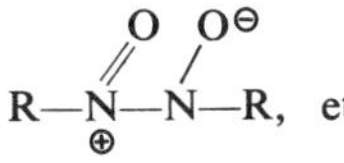   (III)

Two classes of compounds related to the dimers of nitroso compounds are the furoxanes (IV) and the pseudo-nitroso compounds (V). These compounds

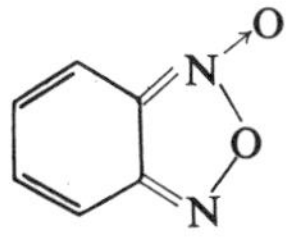

(IV) Furoxane 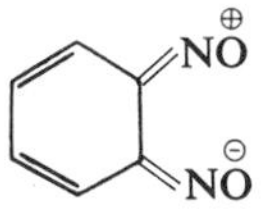   (V) Pseudonitroso compound

are isomers of the normal *o*-dinitrosobenzene (VI) [19].

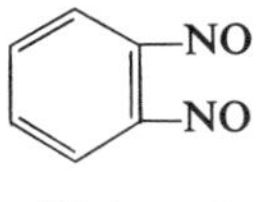

(VI) *o*-Dinitrosobenzene

b. NOTE ON SAFETY CONSIDERATIONS

Many nitroso compounds are reported to melt with decomposition. For example, *p*-nitrosophenol is unstable and may ignite at elevated temperatures [20]. Some nitroso compounds are reported to have disagreeable odors, some are lachrymators, and some nitrosophenols may cause rashes [20].

Whereas the *N*-nitroso compounds are quite definitely considered carcinogenic, no such general statement can yet be made about the *C*-nitroso class. Yet all nitroso compounds must be handled with considerable care because of their potential toxicity. For example, one report indicates that 2-nitrosonaphthalene may be an active urinary carcinogen [21].

2. CONDENSATION REACTIONS

A. Nitrosation of Aromatic Amines

Although many nitrosation reactions are represented as reactions involving free nitrous acid, the actual existence of the free acid has only been established in dilute solutions [21]. The acid is a weak one ($k = 4.5 \times 10^{-4}$), about halfway in acid strength between acetic acid and chloroacetic acid. The liberation of "free" nitrous acid from sodium nitrite in acetic acid, a well-known nitrosation system, is therefore questionable. Since, however, our purpose is to present synthetic procedures rather than the intimate and transitory details of reaction mechanisms, we frequently follow the literature practice of indicating the course of the reaction as if nitrous acid were indeed involved.

The treatment of aromatic amines with a nitrosation mixture has been known for nearly a century and has been summarized as follows.

Treatment of Primary Aromatic Amines with Nitrous Acids.

$$ArNH_2 + HNO_2 \xrightarrow{\text{HX}} [ArN_2]^+ \; X^- + H_2O \tag{7}$$

Treatment of Secondary Aromatic Amines with Nitrous Acids.

$$ArNHR \cdot HCl + HNO_2 \longrightarrow \underset{\underset{R}{|}}{ArN}-NO + H_2O + HCl \tag{8}$$

Followed by the Fischer–Hepp rearrangement (Eq. 9).

$$\underset{\underset{R}{|}}{ArN}-NO \xrightarrow{\text{HCl}} p\text{-NO}-Ar-NHR \tag{9}$$

Treatment of Tertiary Aromatic Amines with Nitrous Acids.

$$Ar-NR'R'' + HNO_2 \longrightarrow p\text{-NO}-Ar-NR'R'' + H_2O \tag{10}$$

In the aliphatic series, primary amines are deaminated by nitrous acid, secondary aliphatic amines are *N*-nitrosated, while tertiary aliphatic amines form salts in the acidic medium followed by secondary reactions such as dealkylation [23, 24].

Although Eqs. (7)–(10) may be valid for a wide range of aromatic amines treated in the usual manner in a mineral acid medium with sodium nitrite, direct nuclear nitrosation of certain primary and secondary aromatic amines was found possible when nitrosylsulfuric acid in concentrated sulfuric acid was used as the nitrosation agent [25].

The primary aromatic amines which will be preferentially nitrosated rather than diazotized are those which do not bear strongly electron-withdrawing substituents and are capable of coupling with diazonium salts to give azo dyes.

Typical primary amines which undergo such nitrosation are: m-toluidine, p-xylidine, m-anisidine, 2-amino-4-methoxytoluene, 3-amino-4-methoxy-toluene, m-aminophenol, α-naphthylamine, 1-naphthylamine-2-, -6-, -7-, and -8-monosulfonic acids, and 1-naphthylamine-4-monosulfonic acid (which reacts with displacement of the sulfonic acid group). The secondary amines derived from these primary amines also can be nitrosated directly (i.e., without the intermediate formation of an N-nitroso compound which needs to be sub-jected to the Fischer-Hepp rearrangement). The entering nitroso group appears to substitute exclusively in the para position.

The preparation of 1-amino-4-nitrosonaphthalene is given here by way of an illustration of the technique used in this reaction. 1-Aminonaphthalene has been withdrawn from the market because of its suspected tumor-producing characteristics and therefore this example is not to be considered an illustration of a safe reaction. It hardly needs to be emphasized that, quite generally, due precautions must be taken when working with the indicated concentrated sul-furic acid system. In carrying out this reaction, a large excess of nitrosation reagent must be avoided to reduce the possibility of diazotization of the pri-mary amine.

Under similar conditions, secondary aromatic amines, such as N-ethyl-1-naphthylamine, may also be nitrosated directly.

2-1. Preparation of 1-Amino-4-nitrosonaphthalene [25]

$$\text{[naphthalene, } NH_2 \cdot H_2SO_4] + NOSO_4H \xrightarrow{H_2SO_4} \text{[naphthalene, } NH_2 \cdot H_2SO_4, NO] \quad (11)$$

$$\text{[naphthalene, } NH_2 \cdot H_2SO_4, NO] + NaHCO_3 \longrightarrow \text{[naphthalene, } NH_2, NO] + NaHSO_4 + H_2O + CO_2 \quad (12)$$

With due precautions for handling concentrated sulfuric acid and a carcinogenic amine, in a five-necked flask fitted with a sealed stirrer, a thermometer, a gas inlet tube, a calcium chloride drying tube, and an addition port is placed 200 ml of concentrated sulfuric acid. The acid is stirred rapidly in an ice bath while 17.5 gm (0.25 mole) of dried, finely ground sodium nitrite is added at such a rate that the internal temperature never exceeds 10°C. After the addition has been completed, the flask content is warmed slowly to 60°C and stirred at this temperature until a clear solution has been obtained. Then it is rapidly cooled again with an ice bath and maintained between 0° and 5°C while carbon dioxide is passed into the flask, and 50 gm (approximately 0.2 mole) of dry, powdered α-naphthylamine sulfate (CAUTION: said to be tumor-producing) is added through the addition port (care must be taken to prevent sulfuric acid from splashing out of the opening and to prevent the powdered amine salt from being blown out of the apparatus by the carbon dioxide pressure). The addition port is then closed off and stirring is continued until a test in a Lunge nitrometer no longer shows the presence of nitrous acid.

With vigorous stirring, 1 liter of cold water is cautiously added to the reaction flask. The resultant solid is filtered off, washed in turn with cold water, alcohol, and then with ether. The yield (after drying at a modest temperature) is 51 gm (91%) of the sulfate of the product.

The free base is prepared by grinding the product with 25 gm of sodium bicarbonate with a small amount of water until a brown paste forms. This mixture is repeatedly extracted with ether. The ether extract is evaporated under reduced pressure. The crude base, which remains behind, is recrystallized from carbon tetrachloride, m.p. 144°–145°C dec. This product is presumed not to be quite pure because of the possibility of α-naphthaquinonimide oxime formation.

The nitrosation of secondary aromatic amines by more conventional procedures first affords the *N*-nitroso derivatives. On treatment with acids, particularly with hydrochloric acid, these derivatives undergo the Fischer-Hepp rearrangement to the aromatic *p*-nitroso secondary amines [26–30] (see Eqs. 8, 9). The mechanism of this rearrangement was assumed to be an intermolecular reaction involving either nitrosyl chloride (NOCl) or $NO \cdot OH_2^+$ ion. However, when nitrosation was carried out in the presence of $Na^{15}NO_2$, it was found that the *N*-nitroso group does not become detached from the ring either as NOCl or as an ionic species such as $NO \cdot OH_2^+$ [31]. This experiment suggests that an intramolecular mechanism, or possibly a trans nitrosation of one molecule by another, prevails. An earlier investigation [32] reported that *N*-nitroso-*N*-octadecylaniline did not undergo the Fischer-Hepp rearrangement but was denitrosated. In fact, even *N*-nitroso-*N*-hexylaniline afforded only a poor yield of *p*-nitro-*N*-hexylaniline. This was believed to show that long alkyl chains blocked the available open positions on the ring and thus interfered with the migration of the free reactive species. This steric argument may also apply, it seems to us, to the intramolecular hypothesis. Here the long alkyl chain may serve to screen the N-nitroso group from whatever reagent may be necessary to bring about the reaction. As a matter of fact, the argument that the accepting positions are blocked is weakened by experiments of the same author in which he demonstrated that tertiary aromatic amines with one long N-alkyl substituent similar to the one used in the secondary aromatic amine can can be directly nitrosated in the para position in the normal manner.

Within the limitations on chain length indicated above, the nature of the N-alkyl group which may be present on the secondary nitrogen in a Fischer-Hepp rearrangement may be varied widely. For example, 3-anilinopropionitrile and β-anilinopropionic acid and its ethyl ester have been subjected to the reaction. While the esters undergo alcoholysis or trans esterification under the reaction conditions, the nitriles are not hydrolyzed [33]. Substitution in the meta position of the aromatic ring evidently does not interfere with the rearrangement.

On the other hand, N-aryl-N-nitrosoaminonitriles undergo interesting secondary reactions. Under acidic conditions, sydnone imine salts form (Eq. 13); under basic conditions, a rearrangement to an isomeric oxime takes place, which, on treatment with phosgene in pyridine leads to 4-substituted-3-cyano-1,2,4-oxadiazolinones (Eq. 14) [34].

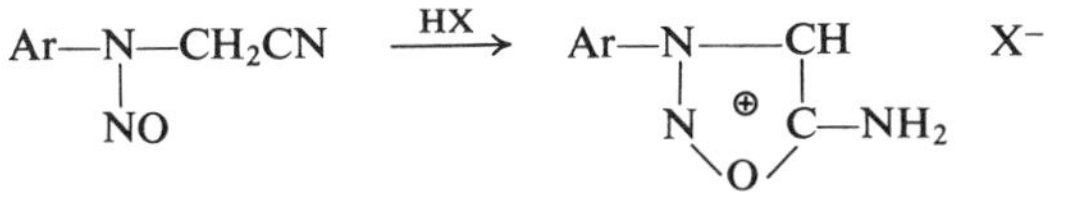

$$\text{Ar—N—CH}_2\text{CN} \xrightarrow{\text{HX}} \text{Ar—N——CH} \quad \text{X}^- \tag{13}$$

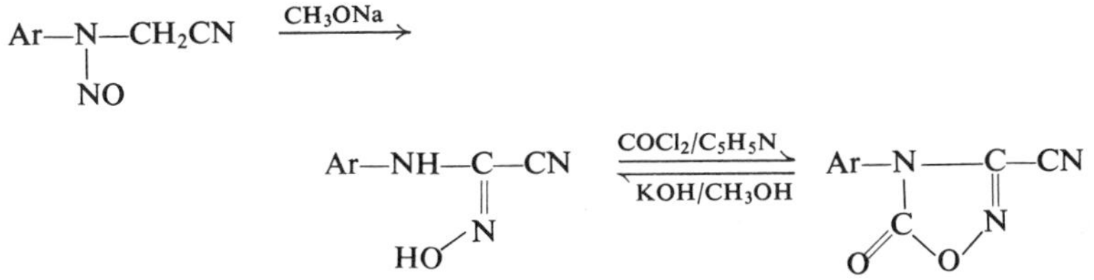

$$\text{Ar—N—CH}_2\text{CN} \xrightarrow{\text{CH}_3\text{ONa}}$$

$$\text{Ar—NH—C—CN} \xrightleftharpoons[\text{KOH/CH}_3\text{OH}]{\text{COCl}_2/\text{C}_5\text{H}_5\text{N}} \text{Ar—N——C—CN} \tag{14}$$

2-2. *Preparation of N-Methyl-m-chloro-p-nitrosoaniline* [34]

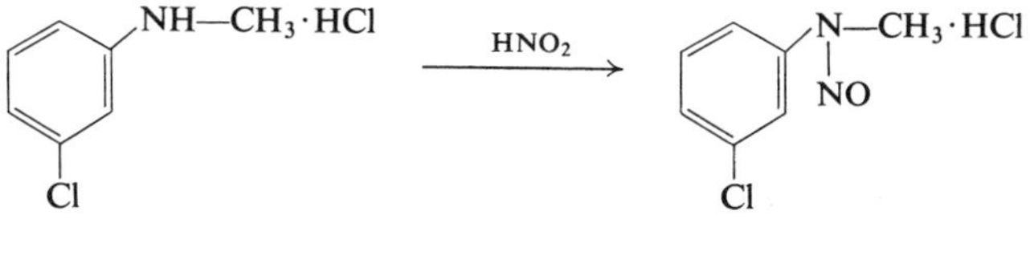

$$\xrightarrow{\text{HNO}_2} \tag{15}$$

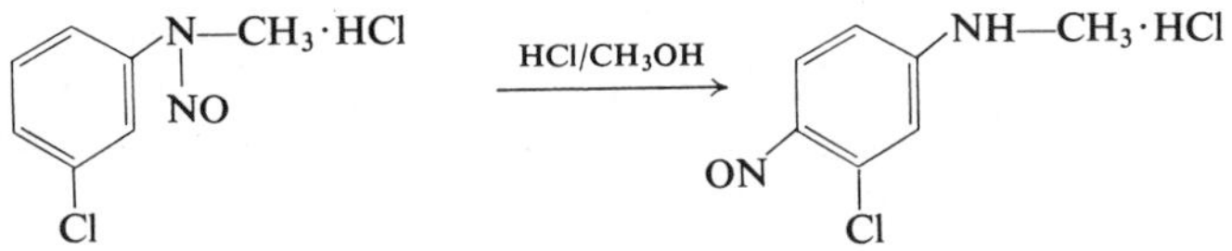

$$(16)$$

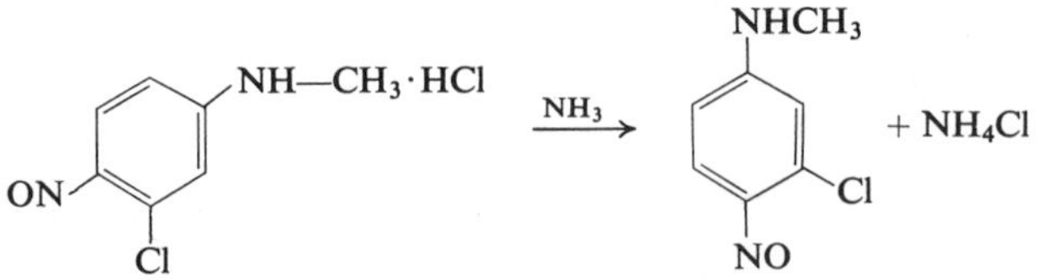

$$+ NH_4Cl \qquad (17)$$

To 100 gm of a stirred 40% solution of hydrogen chloride in methanol maintained at 0°C with an ice bath is added, 14.2 gm (0.1 mole) of *N*-methyl-*m*-chloroaniline over a 30 min period at such a rate that the temperature never exceeds 15°C. The reaction system is then cooled to 5°C, and 8 gm (0.11 mole) of 97% sodium nitrite is added all at once. Stirring is continued for 4 hr while the temperature is allowed to rise to 30°C. The reaction mixture is then filtered and the solid hydrochloride salt is washed well with 50 ml of ether. This is followed by air-drying.

The hydrochloride salt is stirred with 200 gm of ice water, and concentrated aqueous ammonia is added dropwise while maintaining a temperature between 0° and 20°C, until the pH of the aqueous phase reaches 8. The mixture is stirred for an additional hour at 10°–20°C. The solid product is filtered off, washed with 100 ml of cold water, and air-dried at 25°–30°C: yield 14.5 gm (85%), m.p. 128°–130°C dec. In an attempt to recrystallize a sample of this product, it decomposed.

As indicated above, tertiary aromatic amines are directly *C*-nitrosated. The usual reagents are sodium nitrite and dilute hydrochloric acid, sodium nitrite and glacial acetic acid containing concentrate hydrochloric acid, and nitrite esters with hydrochloric acid [23, 35]. While tertiary amines with such complex alkyl groups as found in *N,N*-di(3,5,5-trimethylhexyl)aniline are readily nitrosated [32], of the four *N*-butyl-*N*-methylaniline isomers, *N-t*-butyl-*N*-methylaniline does not undergo the reaction, and even the nitroso compounds which did form were only unstable oils [35].

In reactions of *N,N*-dimethylaniline, care must be taken in the final isolation of the free *p*-nitrosoamine from its salt, since prolonged treatment with excessive alkali leads to the formation of *p*-nitrosophenol [23]. By the way, the initial proof that the product of nitrosation is the para isomer was simply based on the observation that, after prolonged acid hydrolysis of *p*-nitrosophenol, a product formed which had an odor that resembled that of 1,4-benzoquinone. It is to be hoped that more substantial structure proofs are used today.

By nitrosation of *m*-dialkylaminophenols with hydrochloric acid and sodium nitrite at 0°–5°C, good yields of 5-dialkylamino-2-nitrosophenols have been prepared [24].

Whereas many tertiary aromatic amines are nitrosated quite simply at low temperatures with sodium nitrite and hydrochloric acid, with no particular care being taken as to preparative details, some amines have to be nitrosated under conditions which are quite critical. Often the final workup of the product must be rapid. The example of the following preparation emphasizes some of the necessary details.

2-3. *Preparation of N-Methyl-N-(3,5,5-trimethylhexyl)-p-nitrosoaniline* [32]

$$\text{C}_6\text{H}_5\text{—N(CH}_3\text{)—CH}_2\text{CH}_2\text{—CH(CH}_3\text{)—CH}_2\text{—C(CH}_3\text{)}_2\text{—CH}_3 \cdot \text{HCl} \xrightarrow{\text{HNO}_2}$$

$$\text{ON—C}_6\text{H}_4\text{—N(CH}_3\text{)—CH}_2\text{CH}_2\text{—CH(CH}_3\text{)—CH}_2\text{—C(CH}_3\text{)}_2\text{—CH}_3 \cdot \text{HCl} \qquad (18)$$

$$\text{ON—C}_6\text{H}_4\text{—N(CH}_3\text{)—CH}_2\text{CH}_2\text{—CH(CH}_3\text{)—CH}_2\text{—C(CH}_3\text{)}_2\text{—CH}_3 \cdot \text{HCl} \xrightarrow{\text{K}_2\text{CO}_3}$$

$$\text{ON—C}_6\text{H}_4\text{—N(CH}_3\text{)—CH}_2\text{CH}_2\text{—CH(CH}_3\text{)—CH}_2\text{—C(CH}_3\text{)}_2\text{—CH}_3 + \tfrac{1}{2}\text{KCl} + \text{CO}_2 + \tfrac{1}{2}\text{H}_2\text{O} \qquad (19)$$

To a stirred mixture of 8.7 gm (0.037 mole) of *N*-methyl-*N*-(3,5,5-trimethylhexyl)aniline in 10 ml of concentrated hydrochloric acid, 10 ml of water, and 20 ml of glacial acetic acid maintained at 0°C ± 2°C is added, under the liquid surface, over a 20 min period, a solution of 3.5 gm (0.5 mole) of sodium nitrite in 10 ml of water. After the addition at 0°C ± 2°C has been completed, stirring is continued for another 20 min. The dark-red solution is then made slightly alkaline by adding powdered potassium carbonate. The green viscous oil which separates is immediately taken up in ether. The ether solution is quickly washed with cold water, dried with anhydrous sodium sulfate, filtered, and evaporated. The residue, on strong cooling and scratching, finally crystallizes to yield 7.5 gm (77%), m.p. 41°–44°C. After several recrystallizations from petroleum ether (b.p. 40°–60°C), green crystals are obtained, m.p. 48°C.

Tertiary aromatic amines have also been nitrosated with pentyl nitrite in hydrochloric acid [23].

Table I gives the melting points of several *p*-nitrosoanilines.

B. Nitrosation of Phenols

The nitrosation of phenols proceeds in a manner similar to that of tertiary amines. For example, 1-nitroso-2-naphthol has been prepared from the sodium salt of *β*-naphthol by treatment with sodium nitrite and sulfuric acid near 0°C [37]. This general procedure, suitably modified, has been used to prepare other nitrosophenols such as *p*-nitrosophenol (m.p. 135°–136°C) [38].

TABLE I[a]

PROPERTIES OF *p*-NITROSOANILINES

R	R′	M.p. (°C dec.)	Ref.
H	*n*-Propyl	59	28
H	Isopropyl	82.5, 86, 84–85	32
H	Isobutyl	93–94	28
H	*n*-Hexyl	47–48	32
H	3,5,5-Trimethylhexyl	93–94	32
H	Phenyl	98 dec., 144–145	27, 36
H	2-Methylphenyl	148	36
H	3-Methylphenyl	124	36
H	4-Methylphenyl	172–173	36
H	2-Methoxyphenyl	164–166	36
H	4-Ethoxyphenyl	158–160	36
H	4-Chlorophenyl	157–160	36
H	4-Aminophenyl	168	36
H	4-Dimethylaminophenyl	175	36
H	4-Carboxyphenyl	360	36
H	4-Nitrophenyl	185	36
H	4-Nitrilophenyl	196	36
Methyl	Methyl	85	4
Methyl	3,5,5-Trimethylhexyl	48	32
Ethyl	3,5,5-Trimethylhexyl	Green oil at room temp.	32
Hexyl	Methyl	33.5–34	32
Hexyl	Hexyl	Viscous oil at room temp.	32
3,5,5-trimethylhexyl	3,5,5-Trimethylhexyl	48–48.5	32a

[a] Properties of miscellaneous aromatic nitroso compounds are given in Table IV.

A variation of this procedure is the nitrosation of a phenol such as thymol in an alcohol solution with hydrochloric acid and sodium nitrite. This procedure is said to avoid the evolution of oxides of nitrogen, since it may involve the intermediate formation of ethyl nitrite as the nitrosating agent [39]. However, from the safety standpoint, both the oxides of nitrogen and nitrite esters must be considered hazardous.

At low acid concentrations, nitric oxide tends to form. This evidently may attack nitrosophenol to form diazonium compounds directly. The diazonium salts, in turn, may couple with unreacted phenol to give colored products. Nitrous acid may also produce nitrophenols from phenols. The mechanism of this reaction may involve oxidation of initially formed nitrosophenols, homolytic attack by nitrogen dioxide, or nucleophilic attack by nitrite ions [1].

Particularly in the older literature, the structure of the nitrosophenols was not clearly established. Thus, for example, the dinitroso derivative of resorcinol is presumed to be the tautomeric dioxime shown in Eq. (20) [44].

(20)

Furthermore, the possibility of the existence of dimeric nitroso compounds was not recognized until relatively recently. The characteristic light absorption attributed to the nitroso group is at about 700–750 mμ. The possibility of dimerization or oxime formation or both reduces the molar extinction to a small value [38].

The structures assigned to some of the nitroso phenols may require re-examination by more modern techniques.

C. Nitrosation of Active Methylene Compounds

The nitrosation of aliphatic carbon atoms, particularly of carbon atoms activated by adjacent carbonyl, carboxyl, nitrile, or nitro groups, has been reviewed in great detail [2]. Judging from this review, with few exceptions, nitrosation of active methylene compounds leads to the formation of oximes (unfortunately termed "isonitroso" compounds in the older literature). The few exceptional cases cited in which true nitroso compounds (or their dimers) were formed involved tertiary carbon atoms in which no hydrogen atoms were available to permit tautomerism to the oxime or involved a reaction which was carried out under neither acidic nor basic conditions.

Nitrosomethane, nitroscyclohexane, and nitrosodibenzoylmethane have been prepared and appear to exist as reasonably stable dimers.

A kinetic study has indicated that under acidic conditions α-ketonitrosoalkanes may rearrange rapidly to the oxime [41]. Amines also bring about the isomerization of nitrosocyclohexane [42]. These observations along with the older observations mentioned in Touster's review [2] imply that the whole question of the nitrosation of aliphatic carbon atoms should be reexamined with modern analytical techniques to establish the reaction conditions under which the true aliphatic nitroso compounds (or their dimers) can be isolated.

In the nitrosation of ethyl isopropyl ketone (Preparation 2-4), acetyl chloride is used as a catalyst. While aqueous hydrochloric acid has also been used to catalyze this reaction, anhydrous hydrogen chloride or acetyl chloride evidently is more effective as far as the yield is concerned. To be noted here is that the greater yield is that of the true nitroso compound. The secondary carbon atom of the ethyl group is converted into an oxime on nitrosation.

2-4. *Preparation of 2-Methyl-2-nitroso-3-pentanone Dimer and 2-Methyl-4-oximino-3-pentanone* [43]

$$
2CH_3CH_2\overset{\overset{\displaystyle O}{\|}}{C}-\underset{\underset{\displaystyle CH_3}{|}}{CH}-CH_3 \quad \xrightarrow[\text{CH}_3\text{COCl}]{\text{C}_2\text{H}_5\text{ONO}}
$$

$$
\left[CH_3CH_2\overset{\overset{\displaystyle O}{\|}}{C}-\underset{\underset{\displaystyle CH_3}{|}}{\overset{\overset{\displaystyle NO}{|}}{C}}-CH_3 \right]_2 \quad \text{and} \quad CH_3-\overset{\overset{\displaystyle O}{\|}}{C}-\underset{\underset{\displaystyle NOH}{\|}}{C}-\underset{\underset{\displaystyle CH_3}{|}}{CH}-CH_3 \quad (21)
$$

In an apparatus suitably protected against atmospheric moisture and fitted with a gas-inlet tube, mechanical stirrer, and an efficient reflux condenser, to a solution of 45 gm (0.45 mole) of ethyl isopropyl ketone and 5 gm of freshly distilled acetyl chloride is added, through the gas-inlet tube, 18 gm (0.24 mole) of ethyl nitrite at 45°–55°C over a $2\frac{1}{2}$ hr period. The reaction mixture is stored overnight in a refrigerator, whereupon 15.2 gm (48.7% based on ethyl nitrite used) of 2-methyl-2-nitroso-3-pentanone dimer (bimolecular ethyl α-nitrosoisopropyl ketone) deposits. The product is isolated by filtration, m.p. 122°–123°C.

The mother liquor is cooled in an ice bath and shaken with 40 ml of a cold 10% aqueous solution of sodium hydroxide. Thereupon 9 gm of a nonaqueous upper layer is separated and discarded. The aqueous layer is extracted with ether and the ether extract is discarded. The aqueous layer is then cooled in an ice bath and acidified to a pH of 7 with dilute sulfuric acid. The precipitate which forms is filtered and washed with water. This precipitate represents 5.7 gm (18 % based on ethyl nitrite used) of 2-methyl-4-oximino-3-pentanone, m.p. 92°–93°C.

D. Free Radical Nitrosation

Much of the pioneering research on nitrosoalkane dimers is based on reactions involving the formation of free radicals. Most of the reactions are of little value from the preparative standpoint, either because a highly specialized apparatus (e.g., photolysis equipment, high-vacuum trains, even a Van de Graaff generator) is used or because complex mixtures of products are produced. However, this work is of such importance in the historical development of aliphatic nitroso chemistry that it merits a brief review here rather than relegation to Section 5.

The first preparation of nitrosomethane dimer (m.p. 122°–122.2°C corr.) was carried out by the photolytic decomposition of t-butyl nitrite at 25°C and at a pressure of 50 mm using a 300 watt mercury vapor lamp (General Electric "Uviarc") as a source of radiation [44].

Pyrolytic decomposition of t-butyl nitrite in a flow system gave a dimeric nitrosomethane which was reasonably stable at low temperatures and had a melting point of 97.5°C. At room temperature this material converted to another dimeric nitrosomethane with m.p. 122°C. It was shown that the lower-melting form was the cis isomer, while the higher-melting form was the trans isomer [47]. On extending this work to the pyrolysis of a variety of alkyl nitrites, every possible C_1 to C_4 and one C_5 (probably 3-nitrosopentane) nitrosoalkane dimer was produced—usually as the cis isomer, except for the dimer of 2-methyl-2-nitrosopropane, which could only exist as the trans dimer on steric grounds [46]. Table II gives the melting points of these products and of related nitrosoalkane dimers.

A method for the isomerization of trans dimeric compounds to cis dimeric isomers is indicated in Section 5, Miscellaneous Preparation 11 [18].

The results of a three-dimensional X-ray study have shown that 2-nitrosoethane dimer exists mainly in the trans configuration [47].

A procedure which may be of value from the preparative standpoint involves the preparation of trans-nitrosomethane dimer by adding a solution of diacetyl peroxide in sec-butyl nitrite to warm sec-butyl nitrite [59]. From the product of the reaction it has been assumed that this preparation involves the generation of free methyl radicals which react with the nitrite to give nitrosomethane and alkoxy radicals. The latter disproportionate to ketones and alcohols, while the nitroso compound dimerizes.

Pyrolysis of the di-n-butylmercury in the presence of nitric oxide in a flow system produced mixtures of oximes and nitroso compounds [60].

TABLE II

PROPERTIES OF NITROSOALKANE DIMERS

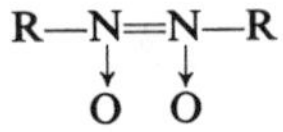

$$R—N{=}N—R$$
$$\downarrow \quad \downarrow$$
$$O \quad O$$

Dimer of	Cis dimer, m.p. (°C)	Trans dimer, m.p. (°C)	Ref.
Nitrosomethane	97.5	122	46
Nitrosoethane	84	—	46
1-Nitrosopropane	76	—	46
2-Nitrosopropane	60	52, 53–56	46, 48
1-Nitrosobutane	72	—	46
1-Nitroso-1-methylpropane	61	—	46
1-Nitroso-2-methylpropane	80	40, 45–6	46, 49
2-Methyl-2-nitrosopropane	—	70, 83–84, 76–76.2	46, 48, 50
$C_5H_{11}NO$ (probably 3-nitroso-pentane)	42	—	46
2-Methyl-2-nitrosobutane	—	50–50.5, 50.3–51.2	50, 51
4-Nitrosoheptane	—	49–50	52
2-Nitrosooctane	—	n_D^{20} 1.4565	48
2,3,3-Trimethyl-2-nitrosopentane	—	63–64	48
n-Nitrosododecane	—	76–78, 75–77	48, 53
n-Nitrosooctadecane	—	87–89	48
Nitrosocyclohexane	—	116–117, 115.5–116.5, 86–88	48, 52, 54
Nitrosocycloheptane	—	96–97	55
Phenylnitrosomethane (ω-nitroso-toluene)	—	116–118, 120–120.5	48, 52
1-Phenyl-2-nitrosoethane	—	98–100	48
trans-α-Nitrosodecalin	—	148–149	56
cis-β-Nitrosodecalin	132–133.5	—	56
trans-β-Nitrosodecalin	—	113.5–134.5	50
2-Methyl-2-nitroso-1-acetoxypropane	—	67–69	51
Trichloronitrosomethane monomer	—	(b.p. 55–5 dec.)	57
1-Methoxy-2-nitroso-1,1,2,3,3,3-hexafluoropropane[a]	—	(b.p. 60)[a]	58
1-Methoxy-1-chloro-2-nitroso-1,2,3,3,3-pentafluoropropane[a]	—	(b.p. 43 (160 mm Hg))	58
1-Methoxy-1-chloro-2-nitroso-1,2,2-trifluoroethane[a]	—	(b.p. 74)	58
Ethyl 2-nitroso-2,3,3,3-tetrafluoro-propionate[a]	—	(b.p. 38 (100 mm Hg))	58

[a] Literature does not specify whether a monomeric or dimeric product is described.

In ultraviolet light, a variety of alkanes have been subjected to the action of nitrogen oxide in the presence of chlorine. The preferred ratio of hydrocarbon:molecular chlorine:nitric oxide was 1:1.5:1. The products were *gem*-chloronitroso compounds. Thus, for example, cyclohexane was converted into 1-chloro-1-nitrosocyclohexane, *n*-heptane into 4-chloro-4-nitrosoheptane, and toluene into ω-chloronitrosotoluene. The later product ultimately was converted into a diphenylfuroxane (Eq. 22).

$$2 \langle C_6H_5 \rangle\!\!-\!CH_3 + 2NO + 3Cl_2 \xrightarrow{\text{uv}}$$

$$2 \langle C_6H_5 \rangle\!\!-\!\underset{\underset{Cl}{|}}{\overset{\overset{NO}{|}}{CH}} \longrightarrow \langle C_6H_5 \rangle\!\!-\!\underset{\underset{O}{\diagdown N}}{C}\!\!-\!\underset{\underset{O}{N\diagup}}{C}\!\!-\!\langle C_6H_5 \rangle \quad (22)$$

On reduction of the chloronitrosoalkanes with hydrogen, lithium aluminum hydride, or sodium borohydride in red light, the corresponding dehalogenated ketoximes are formed [16, 61, 62].

Table III gives physical properties of some of the *gem*-chloronitrosoalkanes produced by this method. It is to be noted that all were blue liquids and monomeric in nature, except as noted. Other monohalonitrosoalkanes are also given in this table.

At low concentrations of chlorine, dimeric nitrosoalkanes free from chlorine are produced when alkanes are treated also with nitric oxide. Under these circumstances, molecular chlorine is first converted into atomic chlorine, which attacks the alkane to form alkyl radicals and hydrogen chloride. The alkyl radicals, in turn, form nitrosoalkanes with nitric oxide. This reaction is most effectively carried out when the ultraviolet radiation is between 380 and 430 nm [52, 55].

The question of the necessary conditions for isomerization of nitrosocyclohexane to cyclohexanone oxime is of considerable commercial importance. The Beckmann rearrangement of cyclohexanone oxime leads to lactams which are Nylon intermediates. The starting materials for such a synthesis, cyclohexane and oxides of nitrogen, are very inexpensive raw materials. Hence the interest in this aspect of nitroso chemistry.

It has been shown that the tautomerization is favored by the presence of gaseous hydrogen chloride, particularly at a wavelength of 300 nm [67]. When cyclohexane is saturated with hydrogen chloride, treated with nitric oxide, and exposed to a source of ultraviolet radiation, the oxime forms along with a trace of 1-chloro-1-nitrosocyclohexane [68]. Cyclooctane seems to form the corresponding oxime and the chloronitroso compounds, but under no circumstances nitrosocyclooctane [68].

TABLE III

PROPERTIES OF MONOHALONITROSOALKANES

Compound	B.p., °C (mm Hg)	n_D^{20}	Ref.
2-Bromo-2-nitrosobutane	28 (19)	—	63
2-Chloro-2-nitrosobutane	34 (60)	1.4108	64
2-Chloro-2-nitrosopentane	32 (23)	1.4179	64
3-Chloro-3-nitrosopentane	32 (20)	1.4180	64
2-Chloro-2-nitrosohexane	49 (20)	1.4192	64
2-Chloro-2-nitroso-4-methylpentane	42 (18)	1.4141	64
2-Chloro-2-nitroso-3,3-dimethylbutane	(m.p. 112)	—	64
2-Chloro-2-nitrosoheptane	37 (4)	1.4270	64
4-Chloro-4-nitrosoheptane	57 (16)	1.4321, 1.4270	61, 64
2-Chloro-2-nitrosooctane	75–76 (10)	1.4290	64
4-Chloro-4-nitroso-2,6-dimethylheptane	66–67 (10)	1.4325	64
6-Chloro-6-nitrosoundecane	94 (9)	1.4412	64
1-Chloro-1-nitrosocyclopentane	33 (12)	1.4561	64
1-Chloro-1-nitrosocyclohexane	51 (12)	1.4629	61
1-Chloro-1-nitroso-2-methylcyclohexane	63 (15)	1.4649	61
1-Chloro-2-nitrosocyclopentane dimer	(m.p. 86–87)	—	65
1-Chloro-4-nitroso-2-cyclopentene dimer	(m.p. 130–133)	—	65
1-Chloro-2-nitrosocyclohexane dimer	(m.p. 154–155)	—	65

When nitrosyl chloride and nitric oxide in the ratio 2:6 react with cyclohexane in the presence of ultraviolet radiation, only nitrosocyclohexane dimer (m.p. 115.5°–116.5°C) is formed, while nitrosyl chloride alone and cyclohexane, under similar conditions, afford only cyclohexanone oxime [54]. Using a 2-MeV Van de Graaff generator, cyclohexane and nitric oxide react to give nitrosocyclohexane dimer. If hydrogen chloride is present, the hydrochlorine salt of the oxime is formed. Similarly, nitrosocyclohexane dimer, in the presence of hydrogen chloride, is converted into the same oxime salt under these conditions of high-energy radiation [69]. This equipment may lend itself to a continuous commercial operation.

Trifluoronitrosomethane (b.p. 86.0°C, 767 mm Hg) is of considerable interest as a component of high-temperature-resistant elastomers. This compound has been prepared by treatment of trifluoroiodomethane, in the presence of mercury, with nitric oxide in a photochemical reactor whose mercury lamp emitted radiation at 253.7 nm. The preparation is particularly sensitive to the initial pressure of the gases, reactant ratio, irradiation time, intensity of the ultraviolet radiation, reaction temperature, and method of removal of nitric oxide from the product [70].

E. Nitroso Decarboxylation (Elimination-Type Reaction)

It has been a relatively recent observation that certain aromatic hydroxy-carboxylic acids form nitroso compounds (presumably in dimeric form) on treatment with sodium nitrite [71] (Eq. 23).

The scope of the reaction has not yet been developed. It is known that salicylic acid, 3-, 4-, and 5-methylsalicylic acids, and p-hydroxybenzoic acid undergo the reaction, but 3,5-dibromosalicylic acid, 3-nitrosalicylic acid, 5-nitrosaliclic acid, 3,5-dinitrosalicylic acid, and 3,5-dinitro-4-hydroxysalicylic acid are not nitroso-decarboxylated [72].

2-5. *Preparation of 2,6-Dibromo-4-nitrosophenol* [71]

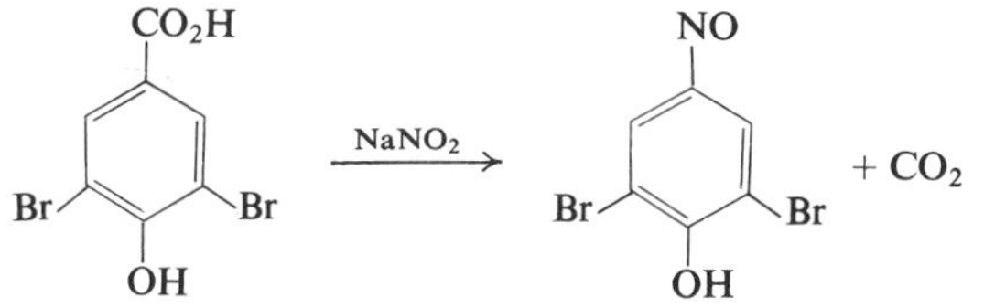

$$ \text{(23)} $$

To a solution of 2.96 gm (0.1 mole) of 3,5-dibromo-4-hydroxybenzoic acid in 15 ml of water and 25 ml of ethanol is added 0.7 gm (0.1 mole) of sodium nitrite. An immediate evolution of carbon dioxide takes place while the reaction solution becomes yellow-green. After the solution has been allowed to stand at room temperature for 16 hr, it is warmed on a steam bath for 10 min, cooled to 5°C, and acidified with 1 ml of concentrated hydrochloric acid.

The tan precipitate which forms is removed by filtration and washed with cold water. By diluting the mother liquor with water, a second crop of product may be obtained. Total yield 2.29 gm (81%). The product, after recrystallization from aqueous methanol, has a decomposition temperature of 168°–169°C.

CAUTION: Previous literature indicates that the product darkens about 160°C and detonates between 168° and 175°C.

A reaction which has some resemblance to this nitroso decarboxylation is the pyrolytic or photolytic decarboxylation of fluorinated nitrites (Eqs. 24 and 25) [73]. The nitroso compounds derived from these nitrite esters appear to be monomeric in nature (since they are described as blue liquids). The overall yields of the preparation are low.

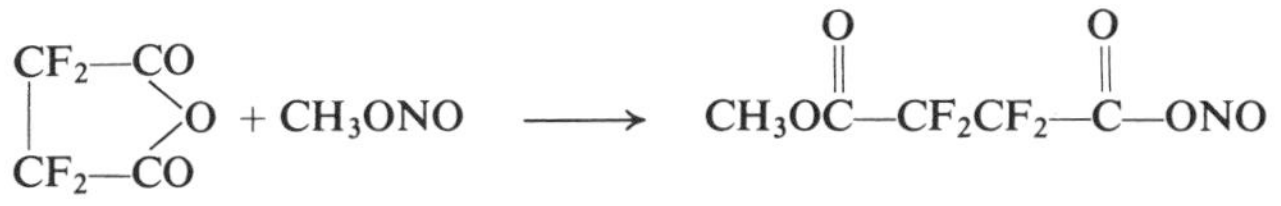

$$ \text{(24)} $$

$$ \text{(25)} $$

F. Nitrosation of Organometallic Compounds

The reaction of organometallic compounds with nitrosyl chloride or nitric oxide has received relatively little attention.

Treatment of diphenylmercury with nitrosyl chloride produced some nitrosobenzene [74]. The reaction of phenylmagnesium chloride with nitrosyl chloride also produced this compound [75]. However, cyclohexylmagnesium chloride and either nitrosyl chloride or nitric oxide produce the corresponding *N*-nitrosohydroxylamine (a cupferron free acid) and cyclohexylhydroxylamine. The *N*-nitrosohydroxylamine, under acidic conditions, decomposes to afford nitrosocyclohexane dimer [76] (see Preparation 2-6).

Some mercuric acetylides have been treated with nitrosyl chloride to an inert solvent to give solutions of highly reactive nitrosoacetylenes (Eq. 26) [77]. The solution of this compound was moderately stable at $-78°C$, but it decomposed at room temperature.

$$[CH_3(CH_2)_3—C\equiv C]_2Hg \xrightarrow[\substack{CHCl_3 \text{ or} \\ THF}]{2NOCl} 2CH_3(CH_2)_3—C\equiv C—NO + HgCl_2 \qquad (26)$$

G. Decomposition of *N*-Nitrosohydroxylamines (Free Acid) (Elimination-Type Reaction)

An example of the decomposition of an *N*-nitrosohydroxylamine is given in Preparation 2-6.

2-6. *Preparation of Nitrosocyclohexane Dimer* [76]

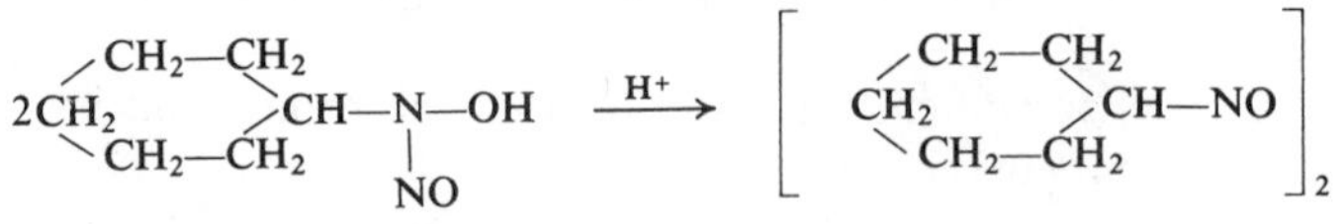

$$2CH_2\underset{\substack{CH_2—CH_2}}{\overset{\substack{CH_2—CH_2}}{\diagdown\diagup}}CH—N—OH \xrightarrow{H^+} \left[CH_2\underset{\substack{CH_2—CH_2}}{\overset{\substack{CH_2—CH_2}}{\diagdown\diagup}}CH—NO \right]_2 \qquad (27)$$

In a hood, to a cooled solution of 0.344 gm (0.0024 mole) of *N*-nitroso-*N*-cyclohexylhydroxylamine in 3 ml of glacial acetic acid is added 1 drop of concentrated hydrochloric acid. While there is a strong evolution of gas (68% nitrogen, 27% nitric oxide, and 5% oxygen), the solution, which initially was blue, finally becomes yellow.

After the reaction mixture is diluted with water, the product is separated by repeated extraction with ether. The ether solution is evaporated and the residual product is dried on an unglazed clay plate. The yield of nitrosocyclohexane dimer is 0.07 gm (26%), m.p. 116°–117°C after recrystallization from acetone.

H. Reactions of Nitrosyl Chloride with Olefins

The nitrosochlorination of olefins has been known for a century and has materially contributed to the early development of the chemistry of terpenes [78]. Studies of the scope of the reaction were not undertaken until relatively recently and results appear to be somewhat fragmentary. While 1-olefins such as 1-hexene, 1-heptene, and 1-octene do not react with nitrosyl chloride, the corresponding 2-isomers add the reagent in conformity with Markovnikov's rule as if NO^+ and Cl^- moieties were involved [79].

Terminal olefins appear to be reactive only if they are not allylic in nature (e.g., styrene and 2,4,4-trimethyl-1-pentene). Allylbenzene (3-phenyl-1-propene) is inert toward nitrosyl chloride, whereas propenylbenzene (1-phenyl-1-propene) reacts. The preparations are usually carried out at low temperatures. When molecular weights of the products are determined at 5°C, they correspond to dimeric structures. At the melting point of naphthalene, the products are predominantly monomeric. This observation is reasonably general for nitroso compounds [79].

The reaction may be carried out either by the addition of nitrosyl chloride or by the generation of this reagent *in situ* from pentyl nitrite and hydrochloric acid.

In a study of the addition of nitrosyl chloride or nitrosyl bromide to norbornene and norbornadiene, it was observed that (a) there was no structural rearrangement during the reaction, (b) a cis addition had taken place, (c) nucleophilic solvents such as ethanol or acetic acid were not incorporated in the products. These facts seem to speak against an ionic addition mechanism, while a free radical initiated by NO·radicals was considered unlikely since nitric oxide is inactive toward norbornadiene. Therefore a four-center mechanism has been suggested [80]. However, when a relatively simple, unstrained olefin such as Δ^9-octalin was subjected to the reaction, only blue, crystalline, monomeric 9-nitroso-10-chlorodecalin was produced (m.p. 90°-91°C). This product had a trans configuration. Thus it is evident that the structure of the olefin has a significant bearing on the steric course of the addition [81].

The following two preparations illustrate the addition of nitrosyl halides to norbornene. In the first one, nitrosyl chloride itself is used; in the second, nitrosyl bromide is generated *in situ*.

2-7. *Preparation of cis-exo-2-Chloro-3-nitrosonorbornane Dimer* [81]

$$(28)$$

In a well-ventilated hood, through a stirred solution of 12 gm (0.127 mole) of norbornene in 150 ml of chloroform cooled to −60°C with a Dry Ice–acetone bath is bubbled nitrosyl chloride until the blue color is replaced by the yellow-brown color associated with excessive nitrosyl chloride. Then 360 ml of hexane, cooled to −70°C, is added and stirring is continued for $\frac{1}{2}$ hr. The precipitating product is removed by filtration and washed with hexane: yield 13.5 gm (65%), m.p. 145°–150°C. On recrystallization from a chloroform–hexane mixture, the melting point is raised to 155.5°–156.5°C.

It is to be noted that the infrared spectrum of the crude product is different from that of the recrystallized product; however, the properties of the latter are in agreement with that reported in the earlier literature.

2-8. *Preparation of cis-exo-2-Bromo-3-nitrosonorbornane Dimer* [81]

$$2 \quad \underset{\text{HBr}}{\overset{C_5H_{11}ONO}{\longrightarrow}} \quad \left[\begin{array}{c} \text{NO} \\ \text{H} \\ \text{Br} \\ \text{H} \end{array} \right]_2 \tag{29}$$

In a well-ventilated hood, to a stirred solution of 4.7 gm (0.08 mole) of norbornene and 7.02 gm (0.06 mole) of isopentyl nitrite in 10 ml of ethanol, cooled to 0°C in an ice–salt bath, is added dropwise, over a $\frac{1}{2}$ hr period, a solution of 10 gm (0.59 mole) of 48% hydrobromic acid in 12 ml of ethanol. After the addition has been completed, stirring in the ice bath is continued for another 20 min. The product is isolated by filtration and washed with ethanol. Yield 8.1 gm (80%), m.p. 134°–136°C. On recrystallization from chloroform, the melting point is raised to 138°–139°C.

It is of interest to note that the addition of nitrosyl chloride to a molecule such as *cis,trans,trans*-1,5,9-cyclododecatriene takes place only at one of the trans bonds [82]. It would appear from this that, in a competitive reaction between cis and trans double bonds, the reaction at the trans bond is favored. However, further work is required to substantiate this generalization, particularly in view of the fact that in an experiment involving both *cis* and *trans*-stilbene the nature of the nitrosyl chloride adduct was not fully determined [65].

The reaction of nitrosyl chloride with cyclopentene, cyclohexene, and cycloheptene in carbon tetrachloride solution at 5°C afforded the corresponding adducts in good yield, although during the cyclohexene reaction considerable hydrolysis to 2-chlorocyclohexanone took place. In the case of cycloheptene, the crude product was not identified but rather was immediately subjected to levulinic acid hydrolysis to 2-chlorocycloheptene. From the products isolated, it may be inferred that cyclopentadiene adds nitrosyl chloride in a 1,4- manner in moderate yields [55].

The addition or nitrosyl chloride to cyclohexene has been examined in some detail. The results indicate that careful product analysis is required in every new preparation because the nature of the products is strongly dependent on both the solvent and the reaction temperature [17].

Thus, in this particular system, reaction in liquid sulfur dioxide at $-30°C$ led to *trans*-1-chloro-2-nitrosocylohexane dimer exclusively.

In carbon terachloride at 20°C only a small amount of the expected 1-chloro-2-nitrosocyclohexane dimer was formed. The major product fraction consisted of 1,2-dichloro-1-nitrosocyclohexane, 1-nitro-2-chlorocyclohexane, and 2-chlorocyclohexanone, with some 1,2-dichlorocyclohexane and two other minor, unidentified components.

In carbon tetrachloride at $-30°C$ the product mixture consisted of 95% of *trans*-1-chloro-2-nitrosocyclohexane dimer and 5% of *trans*-1-nitro-2-nitroso-cyclohexane dimer. In trichloroethylene at $-30°C$, a 50–50 mixture of these two products was formed [17].

It is interesting to note that nitrosyl chloride addition to steroid-5-enes resulted in 5-α-chloro-6-β-nitrosteroids exclusively [83].

2-9. Preparation of trans-1-Chloro-2-nitrosocyclohexane Dimer (trans-2,2'-Dichloro-trans-azodioxycyclohexane) in Sulfur Dioxide [17]

(30)

(*a*) *Procedure for nitrosyl chloride addition.* In a hood, gaseous nitrosyl chloride from a commercial cylinder is passed, in turn, through a U-tube filled with a 50–50 mixture of powdered sodium nitrite and potassium chloride and another U-tube filled with anhydrous calcium chloride, into a trap cooled to $-12°C$ which has been calibrated to known liquid volumes at this temperature (d_4^{-12} of nitrosyl chloride is 1.417). When the requisite volume of liquid has been collected, the gas flow is stopped and the liquid is transferred under nitrogen pressure to the solvent–olefin mixture which has been chilled to $-40°C$ and which is contained in a suitable apparatus fitted with a mechanical stirrer, a condenser protected against moisture, and a low-temperature thermometer. The reaction flask is then allowed to come to the reaction temperature by use of a large Dry Ice–isopropanol bath. Since the reaction is exothermic, careful control of the reaction temperature is necessary.

(b) *Addition reaction.*   To a cold, stirred mixture of 10.15 ml (0.1 mole) of cyclohexene and 50 ml of liquid sulfur dioxide is added 4.62 ml (0.1 mole) of nitrosyl chloride. The reaction mixture is stirred for 3 hr at −30°C. The precipitated, slightly green solid is filtered off and washed immediately with cold methanol to yield 12.55 gm (85%), m.p. 139°–144°C. On recrystallization from absolute ethanol, a white product is isolated, m.p. 152°–153°C.

I. Reaction of Nitrosyl Formate with Olefins

Nitrosyl formate may be generated *in situ* by treating isopentyl nitrite with anhydrous formic acid. This reagent evidently adds to olefins such as cyclohexene, styrene, norbornene, *trans*-3-hexene, and 2,3-dimethylbutene to give nitrosoformates. While the first three olefins were converted into dimeric products, 2,3-dimethylbutene produced a 50–50 mixture of the monomer and the dimer (as a blue oil). The product of the reaction with *trans*-3-hexene contained some of the corresponding oximino formate.

The saponification of these dimers with aqueous alcoholic potassium hydroxide produced the corresponding hydroxynitrosoalkane dimers [84].

2-10. Preparation of 2-Formoxynitrosocyclohexane Dimer [84]

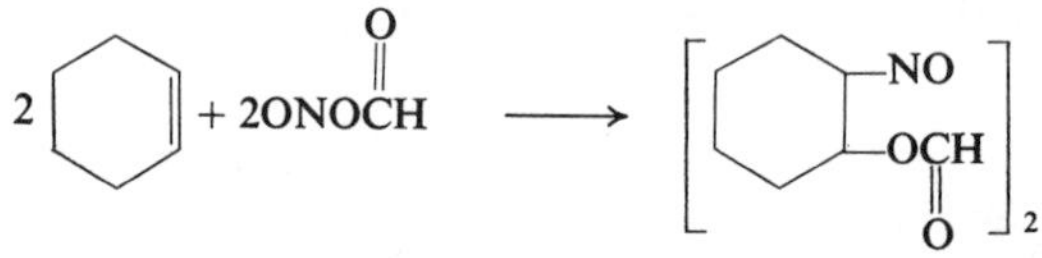

$$(31)$$

To 150 ml of cold anhydrous formic acid is added dropwise, with stirring, a solution of 16.4 gm (0.2 mole) of cyclohexene and 46.8 gm (0.4 mole) of isopentyl nitrite. The white solid which forms is filtered off and recrystallized from ethanol to yield 18.3 gm (59%), m.p. 149°–150°C.

On saponification of this product with aqueous alcoholic potassium hydroxide, a 92% yield of 2-hydroxynitrosocyclohexane dimer is obtained, m.p. 162.5°C.

3. OXIDATION REACTIONS

Both aromatic and aliphatic nitroso compounds have been prepared by oxidative procedures. While few of the methods can be considered generally applicable, a sufficient variety of reagents have been proposed that it would appear reasonable to state that virtually any nitroso compound may be prepared by one of these procedures. The organic substrates which have been used are: oxaziranes and imines, amines, hydroxylamines, and oximes. A by-product of the oxidation of 4-methylcinnoline (an azo compound) has also been identified as a dimeric nitroso compound.

The Baudisch reaction permits conversion of aromatic hydrocarbons to nitrosophenols.

A. Oxidation of Oxaziranes and Imines

In a study of the chemistry of oxaziranes [85], it was discovered that this class of novel heterocyclic compounds is readily oxidized with peracetic acid to give dimeric nitroso compounds. Since the oxaziranes are, in turn, prepared by a peracetic acid oxidation of imines, the crude imines can, by proper adjustment of the level of the oxidizing agent, be converted directly into dimeric nitroso compounds. Equations (32)–(35) show the general synthetic scheme [48].

$$(C_2H_5)_2CO + RNH_2 \longrightarrow RN{=}C(C_2H_5)_2 \tag{32}$$

$$RN{=}C(C_2H_5)_2 \xrightarrow{[O]} R{-}\overset{\displaystyle O}{\overset{\diagup\ \diagdown}{N{-}C}}(C_2H_5)_2 \tag{33}$$

$$R{-}\overset{\displaystyle O}{\overset{\diagup\ \diagdown}{N{-}C}}(C_2H_5)_2 \xrightarrow{[O]} [R{-}NO] \tag{34}$$

$$2[R{-}NO] \longrightarrow (R{-}NO)_2 \tag{35}$$

As this reaction scheme shows, in effect, a primary amine is oxidized to the corresponding nitroso compound. The amino group may be attached to primary and secondary aliphatic carbon atoms. Long-chained, short-chained, and alicyclic compounds have been used. This procedure appears to be the most generally applicable method for the preparation of aliphatic nitroso compounds. Yields generally range between 30 and 90%. Among the by-products of the reaction are the isomeric oximes. t-Nitrosoalkanes were not successfully prepared by the imime oxidation. The reaction appears to be related to the relatively high volatility of the products.

3-1. Preparation of 2-Nitrosopropane Dimer [48]

$$2\ CH_3{-}\underset{\underset{CH_3}{|}}{CH}{-}\overset{\displaystyle O}{\overset{\diagup\ \diagdown}{N{-}C}}\overset{\displaystyle CH_3}{\underset{\displaystyle CH_3}{\diagdown}} \xrightarrow{CH_3CO_3H} \left(\underset{CH_3}{\overset{CH_3}{\diagdown}}CH{-}NO\right)_2 \tag{36}$$

Behind a shield, and with the extreme precautions necessary when working with peracetic acid, peracetic acid is prepared at 0°C by adding dropwise to 20 ml of rapidly stirred methylene chloride, 3 ml (0.11 mole) of 90% hydrogen peroxide, and 1 drop of sulfuric acid followed by 13.5 gm (0.132 mole) of acetic anhydride over a 10 min period while maintaining the mixture at 0°C. Then the mixture is stirred at room temperature for 15 min, and to the clear solution is added 0.5 gm of sodium acetate (to neutralize the acid catalyst).

This peracetic acid dispersion is added dropwise to a solution of 11.5 gm (0.1 mole) of 2-isopropyl-3,3-dimethyloxazirane (prepared by the reaction of peracetic acid with *N*-isopropylideneisopropylamine [85] in 10 ml of methylene chloride which is cooled in an ice bath. The reaction mixture is allowed to warm to room temperature and is preserved for 16 hr. Then the mixture is treated with 100 ml of cold 20% aqueous ammonia. The organic layer is separated and dried with anhydrous magnesium sulfate.

The oil which separates on evaporation of the solvent is dissolved in 40 ml of ether. The resulting solution is cooled in a Dry Ice–acetone bath for 2 hr. The precipitated product is filtered to yield 2.4 gm (33%), m.p. 53°–55°C.

3-2. *Preparation of α-Nitrosotoluene Dimer* [48]

$$2\langle\!\!\!\bigcirc\!\!\!\rangle\text{—CH}_2\text{NH}_2 + 2C_2H_5COC_2H_5 \longrightarrow 2\langle\!\!\!\bigcirc\!\!\!\rangle\text{—CH}_2\text{N}\!\!=\!\!\text{C}(C_2H_5)_2 + 2H_2O \qquad (37)$$

$$2\langle\!\!\!\bigcirc\!\!\!\rangle\text{—CH}_2\text{N}\!\!=\!\!\text{C}(C_2H_5)_2 \xrightarrow{\text{CH}_3\text{CO}_3\text{H}} \left(\langle\!\!\!\bigcirc\!\!\!\rangle\text{—CH}_2\text{NO}\right)_2 \qquad (38)$$

(*a*) *Preparation of crude 3-butylidinebenzylamine.* A mixture of 60.4 gm (0.7 mole) of diethyl ketone, 53.5 gm (0.5 mole) of benzylamine, and 200 ml of benzene is distilled azeotropically until no further water is formed. The remainder of the volatile solvents is then removed under reduced pressure to leave nearly 0.5 mole of the crude imine.

(*b*) *Preparation of α-nitrosotoluene dimer.* With the precautions and by the technique described in Preparation 3-1, a solution of peracetic acid is prepared from 30 ml of methylene chloride, 6 ml (0.22 mole) of 90% hydrogen peroxide, 1 drop of sulfuric acid, and 27 gm (0.264 mole) of acetic anhydride, neutralized with 0.5 gm of sodium acetate.

This peracetic acid solution is added dropwise to an ice-cooled solution of 17.7 gm (0.1 mole) of the imine prepared above in 50 ml of methylene chloride. The reaction mixture is then stored at 0°C for 16 hr, washed in turn with 200 ml of ice water and 200 ml of a cold 20% aqueous sodium bicarbonate solution. The organic layer is dried and the solvent is evaporated under reduced pressure. The residue is covered with 25 ml of ethanol and stored in a refrigerator overnight.

After filtration, the product yield is 4.5 gm (37%), m.p. 114°–115°C. After recrystallization from acetone, the melting point is raised to 116°–118°C.

From the ethanolic filtrate, there may be isolated 7.4 gm (60%) of an oil which is substantially benzaldoxime.

B. Oxidation of Amines with Caro's Acid

As is well known, the oxidation of aniline with various oxidizing agents leads to a variety of products, usually highly colored polymeric materials. However, the reaction of Caro's acid (permonosulfuric acid, H_2SO_5) with aniline produces nitrosoaniline rapidly [80]. With this reagent, many aromatic amines have been oxidized to the corresponding nitroso compounds, e.g., the three nitronitrosobenzenes were prepared from the corresponding nitroanilines [87, 88]. The reaction is normally carried out in an aqueous medium. In fact, if the Caro's acid oxidation were carried out at low temperatures in an ethereal solution of aniline, phenylhydroxylamine would be formed. It is assumed that this product is protected from further oxidation because the solvent reduces the possibility of contact with the aqueous oxidizing agent [89]. Primary aliphatic amines, in which the amino group is directly attached to a tertiary carbon atom are also converted into the corresponding nitroso compounds [50].

It is to be noted that some nitroso compounds are quite volatile. For example, t-nitrosobutane codistills with diethyl ether although its melting point is reported to be 76°–76.2°C [50].

The procedure for using aqueous Caro's acid as an oxidizing agent usually involves the preparation of the dilute acid from either potassium persulfate or ammonium persulfate. The following preparation illustrates the method [90].

Ammonium persulfate is usually the reagent of choice in these and other reactions involving persulfates since this salt is much more soluble in aqueous systems than potassium persulfate. Unfortunately, ammonium persulfate is not as stable to long-term storage as potassium persulfate. Consequently, prior to use, ammonium persulfate must be assayed for efficacy.

3-3. Preparation of 2-Nitro-4-chloronitrosobenzene Dimer [90]

$$ \tag{39} $$

(a) *Preparation of Caro's acid.* To 54 ml of cold, concentrated sulfuric acid is added slowly 145 gm (0.64 mole) of ammonium persulfate. The mixture is allowed to stand for 1 hr and is poured onto 355 gm of crushed ice. The ice is allowed to melt. Then the solution is ready for use.

(b) *Oxidation of 2-nitro-4-chloroaniline.* A cold suspension of 40 gm (0.23 mole) of 2-nitro-4-chloroaniline in 60 ml of ice-cold concentrated sulfuric acid and 10 ml of water is stirred mechanically with a glass stirrer in an ice bath for 1 hr. Then Caro's acid solution is added and stirring is continued without further cooling for 17 hr. The yellow precipitate which forms is collected on a filter, washed with water, dried in a vacuum desiccator, and recrystallized from dry acetone to yield 26.2 gm (61 %) m.p. 125.5°C dec. After four recrystallizations under nitrogen with dry acetone, the melting point is raised to 125.8°C dec. (evacuated capillary).

By a similar technique, *m*-trifluoromethylnitrosobenzene (m.p. 73°–74°C) [91] and nitrosobenzene-2,4,4-*d* (m.p. 63°–64°C) have been prepared [92].

In an example in which potassium persulfate was used to generate Caro's acid, to 558 ml of concentrated sulfuric acid was added, in one portion, 500 gm of potassium persulfate. After stirring for 1 hr, the mixture was cautiously added to 9 kg of ice. After adjusting the pH of the solution to pH 5.0 to 5.5 with approximately 1.5 kg of potassium carbonate, the solution was filtered. The cold filtrate (ca. 9.6 liters) was considered to contain between 90 and 93 gm of persulfuric acid (Caro's acid).

The addition of a solution of 50 gm of 4-aminobenzonitrite in 200 ml of dioxan at 20°C was followed by stirring at room temperature for 20 hr. The light brown solid which had formed was filtered off and then steam distilled. From the distillate 25 gm of 4-nitrosobenzonitrite (44.5%) was isolated (recrystallized from ethanol to afford yellow needles, m.p. 136°–137°C). The nonvolatile solid residue was identified as 4,4'-dicyanoazoxybenzene (19 gm, 38%, m.p. 229°–230°C). In general, it seems that aromatic nitroso compounds are considered more volatile than related azoxy compounds [93].

By a similar procedure, ethyl *p*-nitrosobenzoate has been prepared [94].

C. Oxidation of Amines with Peracids

Certain aromatic amines, when treated with equivalent amounts of peracetic acid, form azo compounds in the presence of cupric ions and azoxy compounds in the absence of such metallic ions (cf. Chapters 14 and 15). Other amines give rise only to tarry products [95]. Some nitroso compounds, however, also may be prepared with peracetic acid. Obviously, the conditions under which these related functional groups are prepared need further elucidation.

A number of dihalogenated aromatic amines have been conveniently converted into the corresponding nitroso compounds by room-temperature oxidation with peracetic acid with and without the presence of catalytic amounts of sulfuric acid. Care must be taken to maintain mild reaction conditions to prevent the conversion of the nitroso product into a nitro compound [96]. The example cited here for the preparation of 2,6-dichloronitrosobenzene dimer does afford an excellent yield.

3-4. Preparation of 2,6-Dichloronitrosobenzene Dimer [96]

$$2\,Cl\!-\!\!\underset{}{\overset{NH_2}{\bigcirc}}\!\!-\!Cl \xrightarrow{CH_3CO_3H} \left(Cl\!-\!\!\underset{}{\overset{NO}{\bigcirc}}\!\!-\!Cl\right)_2 \tag{40}$$

Behind a shield and with the precautions mentioned in Preparation 3-1, to a mixture of 400 ml of glacial acetic acid and 80 ml (0.70 mole) of 30% aqueous hydrogen peroxide is added 16.2 gm (0.10 mole) of 2,6-dichloroaniline. The mixture is maintained at room temperature for 48 hr. The straw-yellow product which forms is removed by filtration and recrystallized from hot glacial acetic acid to yield 16.1 gm (91.3%) of an almost white product, m.p. 173°–175°C (to a pale-green melt). On recrystallization in turn from ethanol, twice from benzene, and following this by vacuum sublimation, the melting point may be raised to 175.5°–176°C.

Instead of the peracetic acid oxidation just described, performic acid (prepared from 98% formic acid and 90% hydrogen peroxide) may also be used as an oxidizing agent for the preparation of nitroso compounds. With this reagent, pentafluoronitrosobenzene and 4-nitrosotetrafluorobromobenzene have been prepared from the respective amines [97–99].

The use of perbenzoic acids, specifically m-chloroperbenzoic acid or its anion, will be outlined in Section 5, Miscellaneous Preparations.

Properties of a group of miscellaneous aromatic nitroso compounds are given in Table IV.

In the aliphatic series, the peracid oxidation appears to be limited to t-alkyl primary amines [98].

D. Oxidation of Hydroxylamines

A very common method of preparing nitroso compounds involves the oxidation of hydroxylamines with one of a large variety of oxidizing agents. The starting hydroxylamines are frequently prepared by reduction of readily accessible aromatic nitro compounds (for example, see Preparation 3-7). Aliphatic hydroxylamines have also been oxidized to nitroso compounds.

a. OXIDATION WITH BROMINE WATER AND HYPOBROMIDE IONS

Secondary aliphatic nitroso compounds such as 4-nitrosoheptane dimer [52], nitrosocycloalkane dimer [55] (e.g., nitrosocycloheptane dimer), and

TABLE IV

<small>PROPERTIES OF MISCELLANEOUS AROMATIC NITROSO COMPOUNDS</small>

Compound[a]	M.p. (°C)	Ref.
o-Nitronitrosobenzene dimer	126–126.5, 125.8–126.2, 125	80, 86a, 87, 90, 101
m-Nitronitrosobenzene	89.5–90.5	87
p-Nitronitrosobenzene	118.5–119	90
2-Nitro-4-chloronitrosobenzene dimer	125.8 dec.	90
2-Nitro-5-chloronitrosobenzene dimer	136.4	90
m-Trifluoromethylnitrosobenzene	73–74	91
Nitrosobenzene	65–66, 66–68, 66.5–67	92, 102, 94
p-Nitrosodiphenyl ether	31–32	92
p-Nitrosoacetophenone	108–109	92
Nitrosobenzene-2,4,6-d_3	63–64	92
2,6-Dichloronitrosobenzene dimer	175.5–176	96
2,4,6-Trichloronitrosobenzene dimer	145–146, 152	84, 86a, 96, 101
2,4,6-Tribromonitrosobenzene dimer	122–123	96
2,6-Dibromonitrosobenzene dimer	135–136	96
2,6-Dibromo-4-chloronitrosobenzene dimer	110–111	96
3,5-Dichloro-4-nitrosotoluene dimer	163–164.5	96
2,3,5-Trichloro-4-nitrosotoluene dimer	189–190	96
3,5-Dibromo-4-nitrosotoluene dimer	136.5–138	96
Ethyl 3,5-dibromo-4-nitrosobenzoate dimer	133–134	96
Ethyl 3,5-dichloro-4-nitrosobenzoate dimer	154–155	96
3,5-Dichloro-4-nitrosobenzonitrile dimer	220–221	96
3,5-Dibromo-4-nitrosobenzonitrile dimer	190–191	96
Pentafluoronitrosobenzene dimer	44.5–45	97
4-Bromo-2,3,5,6-tetrafluoronitrosobenzene dimer	39–40	97
m-Dinitrosobenzene	144–145	19
p-Nitrosotoluene	44–48, 48, 46	101, 102, 94
p-Chloronitrosobenzene	86–88, 89–91, 89.5	102, 94, 100
2,6-Dimethylnitrosobenzene	137	101
2,4,6-Trimethylnitrosobenzene	120	101
4-Iodonitrosobenzene	104	101
2,6-Dimethyl-4-iodonitrosobenzene	153	101
5-Iodo-2-nitrosotoluene	113	101
3,5-Dichloro-4-iodonitrosobenzene	87	101
4-Methoxynitrosobenzene (4-nitrosoanisole)	23, 25	20, 101
2,6-Dimethyl-4-methoxynitrosobenzene	122	101
4-Bromonitrosobenzene	94	4, 101
2,6-Dimethyl-4-bromonitrosobenzene	155	101
5-Bromo-2-nitrosotoluene	101	101
4-Nitrosobiphenyl	74	101
2-Nitrosobiphenyl	101	101
Benzo[c]cinnoline dioxide	240 dec., 233–236	103, 104
2,2′-Dinitroso-4,4′-bis(trifluoromethyl)biphenyl	267 dec.	104

TABLE IV (*continued*)

Compound[a]	M.p. (°C)	Ref.
p-Nitrosophenol	120–135 dec.	20
p-Ethoxynitrosobenzene (p-nitrosophenetole)	34–35	20
p-Butoxynitrosobenzene	–4 to –6 (b.p. 55°–56°C, 40 mm Hg)	20
p-Fluoronitrosobenzene	39–40	94
m-Nitrosotoluene	50–51.5	94
m-Methoxynitrosobenzene	48	94
m-Chloronitrosobenzene	71.5–72.5	94
Ethyl p-nitrosobenzoate	82	94
o-Nitrosotoluene	72.5	100
3,5-Dimethylnitrosobenzene	59	100
o-Ethylnitrosobenzene	61	100
m-Ethylnitrosobenzene	22	100
Isopropyl p-nitrosobenzoate	61–62	100
o-Chloronitrosobenzene	56	100
m-Chloronitrosobenzene	72	100
o-Bromonitrosobenzene	97	100
o-Iodonitrosobenzene	117	100
m-Iodonitrosobenzene	77	100
o-Methoxynitrosobenzene	101.5	100

[a] The dimeric character of compounds listed here is indicated only where literature evidence exists for it. In other cases, the compounds may be either monomeric or dimeric.

several nitrosodecalin dimers [56] have been prepared by oxidation with bromine water. The preparation of 4-nitrosoheptane dimer is typical of the procedure.

3-5. *Preparation of 4-Nitrosoheptane Dimer (Bromine Water Oxidation)* [52]

$$2\,CH_3(CH_2)_2\!-\!\underset{\underset{NHOH}{|}}{CH}\!-\!(CH_2)_2CH_3 \xrightarrow[HCl]{Br_2/H_2O} \left(CH_3(CH_2)_2\!-\!\underset{\underset{NO}{|}}{CH}\!-\!(CH_2)_2CH_3\right)_2 \quad (41)$$

A solution of 2.45 gm (0.019 mole) of N-heptyl-4-hydroxylamine is prepared in 2 N hydrochloric acid. To this solution is added, with vigorous stirring, bromine water until the bromine color just persists. During this addition a colorless precipitate of the product begins to form. After the addition has been completed, stirring is continued for a short time. Then the product is filtered off, washed with a small quantity of ice-cold water, and dried on an unglazed clay plate: yield 1.93 gm (80%), m.p. 46°–48°C. The product may be recrystallized from acetone or distilled under reduced pressure (b.p. 90°–93°C, 0.4 mm Hg). The final melting point is reported as 49°–50°C.

By a similar procedure, nitrosocyclohexane dimer (m.p. 116°–117°C) [52] and nitrosocycloheptane dimer (m.p. 96°–97°C) [55] have been prepared.

The nitrosodecaline dimers were also prepared by this method. They were: *trans-α*-nitrosodecalin dimer (m.p. 148°–149°C), *trans-β*-nitrosodecalin dimer (m.p. 133.5°–134.5°C), and *cis-β*-nitrosodecalin dimer (m.p. 132°–133.5°C). In these dimers, the prefixes *cis* and *trans* refer to the conformation about the bridgehead carbons. The nitrogen–nitrogen bond of the dimer is presumed to be the usual trans bond [56].

An example of the oxidation of a hydroxylamine with hypobromide ions in basic solution is the preparation of 2-methyl-2-nitrosopropane dimer, which follows.

3-6. Preparation of 2-Methyl-2-nitrosopropane Dimer (Hypobromide Oxidation) [48]

$$2\,CH_3\!-\!\underset{\underset{CH_3}{|}}{\overset{\overset{CH_3}{|}}{C}}\!-\!NHOH \xrightarrow{\;OBr^-\;} \left(CH_3\!-\!\underset{\underset{CH_3}{|}}{\overset{\overset{CH_3}{|}}{C}}\!-\!NO\right)_2 \qquad (42)$$

To a well-stirred, ice-cooled solution of 19.2 gm (0.12 mole) of bromine in a solution of 12 gm (0.3 mole) of sodium hydroxide and 75 ml of water is added dropwise a solution of 8.9 gm (0.1 mole) of *t*-butylhydroxylamine in 25 ml of water over a 5 min period. Stirring of the mixture (initially blue) is continued for 2 hr. Then the crystalline dimer is filtered off, washed with ice-cold water, and dried to yield 7.5 gm (86%), m.p. 83°–84°C.

Calder *et al.* [105] gave details for the preparation of the *t*-butylhydroxylamine from *t*-butylamine by oxidation to the corresponding nitro compound, followed by aluminum-amalgam reduction to the hydroxylamine. Their description of the hydrobromide oxidation to the nitroso derivative calls for reaction at $-20°$ to $0°C$. Their yields range between 75 and 85%. One important point made in this preparation, which should be applied generally to the handling of nitroso compounds, is that the crude product should be carefully washed with water to free it of alkali. Otherwise, the dimeric nitroso compound is said to decompose to volatile materials on standing. In general, the product should be stored at $0°C$ in the dark.

b. Oxidation with Dichromate Ions in Sulfuric Acid

In some cases in which the Caro's acid oxidation of amines was not satisfactory, the corresponding hydroxylamines have been oxidized with acidified dichromate solutions [51]. Both aliphatic and aromatic nitroso compounds have been prepared by this method [19, 51, 92, 106]. Frequently, the reaction mixture from the reduction of a nitro compound is treated directly with the oxidizing medium without isolation of the intermediate hydroxylamine. The method has been called the "nitro reduction oxidation technique" [92], a terminology we cannot condone.

3-7. Preparation of m-Dinitrosobenzene (Dichromate Oxidation) [19]

$$\text{(NO}_2\text{, NO}_2\text{-benzene)} \xrightarrow[\text{NH}_4\text{Cl}]{\text{Zn}} \text{(NHOH, NHOH-benzene)} \qquad (43)$$

$$\text{(NHOH, NHOH-benzene)} \xrightarrow[\text{H}_2\text{SO}_4]{\text{Na}_2\text{Cr}_2\text{O}_7} \text{(NO, NO-benzene)} \qquad (44)$$

To a vigorously stirred mixture of 20 gm (0.12 mole) of m-dinitrobenzene and a solution of 15 gm (0.28 mole) of ammonium chloride in 500 ml of water is added 37 gm (0.52 gm-atom) of zinc dust over a 5 min period. After 5 min the temperature of the mixture begins to rise, and after about 20 min it reaches a peak of 45°C. Then the product solution is removed by filtration. The zinc oxide residue is washed with 1 liter of hot water. The filtrate and the water washes are combined and immediately cooled to 0°C by adding enough ice that a minimum of 500 gm of unmelted ice remains in the mixture.

Then, with due precautions, 175 ml of ice-cold sulfuric acid is added with rapid stirring, followed by the rapid addition of a solution of 17 gm of sodium dichromate (technical grade) in 50 ml of water. The mixture is vigorously stirred for a few minutes and is then filtered. The brown precipitate is washed with 1 liter of cold water and transferred to a steam-distillation apparatus. The product, being volatile with steam, is purified by steam distillation. The purified m-dinitrosobenzene solidifies in the distillate and is finally isolated by filtration: yield 8.3 gm (51 %), m.p. 144°-145°C (this product is said to be monomeric in nature).

c. Oxidation with Lead Tetraacetate

The oxidation of N-acyl-N-arylhydroxylamines with lead tetraacetate is very rapid even at very low temperatures. The product obtained is the corresponding aromatic nitroso compound. The most favorable reaction conditions involve propionic acid or ethanol–acetic acid as a solvent and reaction times of less than 10 sec at temperatures of $-20°C$ or lower [102]. The use of ethanol–acetic acid is particularly recommended for several reasons. First, since the product is best isolated by steam distillation, the solvent assists in steam distilling rapidly. The ethanol in the distillate helps minimize clogging of the condenser and also solubilizes small quantities of impurities that may be entrained.

In Preparation 3-8, use is made of a reagent containing 94% lead tetraacetate and 6% acetic acid. If the dry, powdered oxidizing agent is used, the reaction solution must be cooled to $-40°C$ or lower before its addition if comparable results are to be obtained. Yields are definitely reduced if the reaction time is prolonged much beyond 10 sec, and if the steam distillation is not completed within a very short time. For this reason this preparation is probably suitable only if small quantities of product are desired. Yields vary from modest to good for the preparation of nitrosobenzene, *p*-nitrosotoluene, and *p*-nitrosochlorobenzene. An attempt to prepare nitrosocyclohexane by this method did not lead to identifiable products.

3-8. *Preparation of Nitrosobenzene (Lead Tetraacetate Oxidation)* [102]

$$\text{(45)}$$

In an apparatus set up for steam distillation and containing a rapidly stirred solution of 5.0 gm (0.023 mole) of *N*-benzoylphenylhydroxylamine (*N*-phenylbenzohydroxamic acid) in 100 ml of a 1:1 solution of ethanol–acetic acid, cooled to $-20°C$, is added 11.5 gm (0.024 mole based on 94% purity) of lead tetraacetate in one portion. After approximately 10 sec (when the initial green color just begins to darken), 100 ml of water is added and the brown or black mixture is rapidly subjected to steam distillation. The distillate is collected in a receiver filled with chopped ice. The product is isolated by filtration of the distillate. Finally, it is dried by pressing dry between filter papers to yield 1.4–2.0 gm (56–80%), m.p. 66°–68°C.

d. REACTIONS WITH OTHER OXIDIZING AGENTS

In the presence of cupric ions, *m*-nitrophenylhydroxylamine is oxidized by peracetic acid to the corresponding *m*-nitronitrosobenzene. In the absence of the metallic ion, the corresponding azoxy compound forms [95].

The filtered reaction mixtures from the zinc-ammonium chloride reductions of aromatic nitro compounds have been added to aqueous solutions of ferric chloride. Within 10–15 min the oxidation to nitroso compounds was completed. In the oxidation of nine different hydroxylamines, yields ranged from 30 to 60% [101, 107].

Silver carbonate has been used to oxidize propylhydroxylamine to *trans*-nitrosopropane dimer, m.p. 52°C [108].

Ferric ion is another oxidizing agent that has been used. Yost [109] treated

0.029 mole of 2-fluorenylhydroxylamine in 600 ml of DMF with 100 ml of a cold aqueous solution of 0.043 mole of ferric ammonium sulfate under a nitrogen atmosphere for 15 min. The orange precipitate (2,2′-azoxyfluorene, m.p. 262°C) was filtered off. To the filtrate, 1300 ml of ice-cold water was added. The green precipitate that formed was collected, washed with cold water, and dried under reduced pressure. Yield: 2.60 gm (47%), m.p. 72°–78°C. Upon recrystallization from hexane, the melting range was raised to 78°–79°C.

E. Oxidation of Oximes with Halogens

In his pioneering work on nitroso compounds, Piloty [63, 110, 111] described their formation by treatment of oximes with bromine according to Eq. (46).

$$\underset{R'}{\overset{R}{\diagdown}}C=NOH + Br_2 \xrightarrow{C_5H_5N} R-\underset{\underset{R'}{|}}{\overset{\overset{Br}{|}}{C}}-NO + HBr \tag{46}$$

A more recent procedure recommends the careful exclusion of light during the oxidation of oximes with dry chlorine [61, 62].

3-9. *Preparation of 4-Chloro-4-nitrosoheptane (Oxidation with Chlorine)* [61]

$$CH_3CH_2CH_2-\underset{\underset{NOH}{||}}{C}-CH_2CH_2CH_3 \xrightarrow{Cl_2} CH_3CH_2CH_2-\underset{\underset{NO}{|}}{\overset{\overset{Cl}{|}}{C}}-CH_2CH_2CH_3 \tag{47}$$

In a hood, with careful exclusion of daylight, dry gaseous chlorine is bubbled through a solution of 12.0 gm (0.093 mole) of 4-heptanone oxime in 150 ml of ether with vigorous stirring. The addition of chlorine is continued until the colorless oil which separates initially is redissolved and a blue mixture forms. The reaction mixture (lachrymatory) is cooled and extracted with 2 N sodium hydroxide solution.

The ethereal product solution is dried with anhydrous sodium sulfate and filtered. Then the ether is evaporated from the filtrate. The residue is purified by fractional distillation under reduced pressure. The main fraction has a boiling range of 56°–58°C (15 mm Hg). The product is a blue liquid of unpleasant odor which is also lachrymatory: yield 12.6 gm (82%), n_D^{20} 1.4321.

a. Oxidation of Oximes with *N*-Bromosuccinimide

In a pyridine medium, the Piloty oxidation mentioned above often leads to difficulties in the isolation of liquid products of low molecular weight. Handling problems are much simpler when *N*-bromosuccinimide in an aqueous medium is used. The following preparation (of 1-bromo-1-nitrosocyclohexane) illustrates the techniques used. Unfortunately, the final isolation of the product was not described, because the product solution was oxidized directly to the corresponding bromonitro compound. Since the latter was formed in a 63% overall yield, it may be assumed that the intermediate bromonitroso product was formed in good yield [112].

3-10. *Preparation of 1-Bromo-1-nitrosocyclohexane (Oxidation with NBS)* [112]

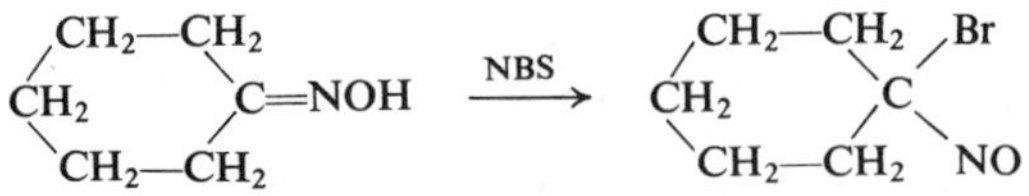

$$
\begin{array}{c}
\underset{\displaystyle CH_2-CH_2}{\underset{\displaystyle \overset{\displaystyle CH_2-CH_2}{}}{}} C=NOH \xrightarrow{\text{NBS}} \underset{\displaystyle CH_2-CH_2}{\underset{\displaystyle \overset{\displaystyle CH_2-CH_2}{}}{}} C\overset{Br}{\underset{NO}{\diagup}}
\end{array} \qquad (48)
$$

To a vigorously stirred suspension of 54.5 gm (0.3 mole) of *N*-bromosuccinimide in 150 ml of water, cooled to 10°C with an ice bath, is rapidly added a solution consisting of 11.3 gm (0.1 mole) of cyclohexanone oxime, 25.3 gm (0.3 mole) of sodium bicarbonate, and 150 ml of water over a 15 min period, while maintaining the reaction temperature at 10°C.

After stirring has been continued for an additional 15 min, the product is extracted with four 50 ml portions of petroleum ether (b.p. 35°–37°C). The combined extracts (CAUTION: lachrymatory) are concentrated by evaporation under reduced pressure. When approximately 30–50 ml of the blue solution remains, 1–2 gm of the white dimer begins to separate. As indicated above, the product from this reaction has not been isolated. It is presumed that procedures of isolation similar to those given for other preparations in this chapter will be satisfactory.

F. Oxidation of Azo Compounds

Azo compounds may be considered to have the carbon-nitrogen framework of dimeric nitroso compounds. The oxidation of one of the nitrogen atoms of an azo compound to an azoxy compound has been discussed in Chapter 15. The oxidation of the second nitrogen would give rise to one of the resonance forms of a nitroso dimer. Indeed, on oxidation of 4-methylcinnoline with hydrogen peroxide in acetic acid, a 4% yield of 4-methylcinnoline-1,2-dioxide, m.p. 168°–169°C, was isolated (Eq. 49) [113]. This product may be considered an intramolecular dimeric nitroso compound.

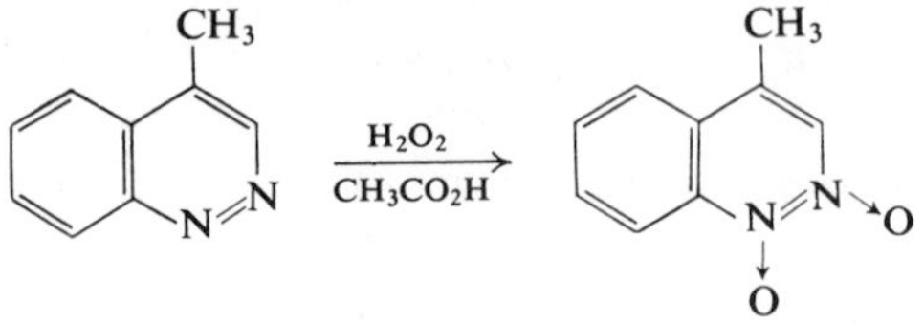

$$ (49) $$

Further studies of the oxidation of azo, azoxy, and hydrazine derivatives to the nitroso dimer stage would be of considerable interest. In such research the problems of unsymmetrically substituted nitroso dimers might also be considered.

G. The Baudisch Reaction

An intriguing method for preparing phenolic nitroso compounds was discovered by Baudisch [114]. Interestingly enough, the product mixture from the reaction appears to be primarily the ortho-substituted phenol, a class of compounds of which very few examples seem to have been described.

The reaction appears to be applicable to a wide range of aromatic starting materials, exceptions being aromatic aldehydes and aromatic primary amines [115]. In effect, the aromatic compound to be converted is stirred with an aqueous solution of hydroxylamine hydrochloride in the presence of metallic ions such as copper ions [114–116], or complex ions such as the pentacyanoammine ferrate (II) ion [87]. The mixture is then treated with 30% hydrogen peroxide and a highly colored complex of the nitrosophenol forms. Presumably, the free nitrosophenol may be isolated by treatment of the complex with an acid (Eq. 50).

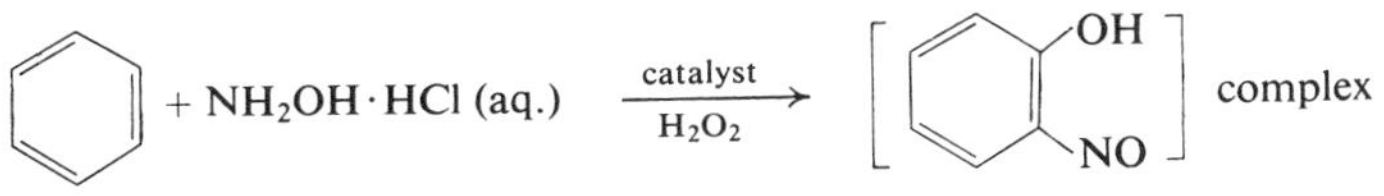

$$\text{(benzene)} + \text{NH}_2\text{OH} \cdot \text{HCl (aq.)} \xrightarrow[\text{H}_2\text{O}_2]{\text{catalyst}} \left[\text{(o-nitrosophenol: OH, NO)} \right] \text{complex} \qquad (50)$$

Although the references cited mention the preparation of a large number of new o-nitrosophenols by this method, this work had been done primarily to study the reaction mechanism or the properties of the complexes. Detailed directions for typical preparations, yield data, and properties of the final products are lacking.

4. REDUCTION REACTIONS

It is natural to presuppose that the reduction of nitro compounds should lead to the nitroso compounds, at least as an intermediate stage. Until quite recently, no reductive processes for the formation of nitrosoalkanes were known [3]. More recently, some indirect evidence is said to show that, on electrolytic reduction of tertiary aliphatic nitro compounds, the final t-alkyl-hydroxylamines are produced by the intermediate formation of nitroso compounds which were not isolated [99].

In the aromatic series, 2,2′-dinitrobiphenyl has been reduced with zinc and potassium hydroxide [103] or with sodium sulfide to give low yields of benzo[c]cinnoline dioxide, an intramolecular, dimeric, nitroso compound [104, 119] (Eq. 51).

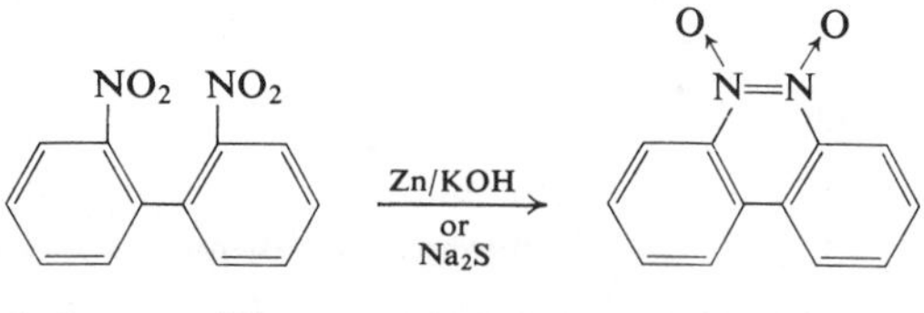

$$\text{(51)}$$

It was believed that *m*-trifluoromethylnitrobenzene could be reduced either with ethylmercaptan or with zinc in aqueous ammonium chloride to the corresponding nitroso compound [120]. Subsequent work showed however, that the product isolated was *m*-trifluromethylazoxybenzene [91].

It would appear, therefore, that the field of reductive preparations of nitroso compounds has not yet been adequately explored.

5. MISCELLANEOUS PREPARATIONS

(1) The Barton reaction [121]. A photochemical exchange process between a nitrite group and a proximate nonactivated hydrogen has been called the Barton reaction. The process appears to be limited to aliphatic compounds (Eq. 52).

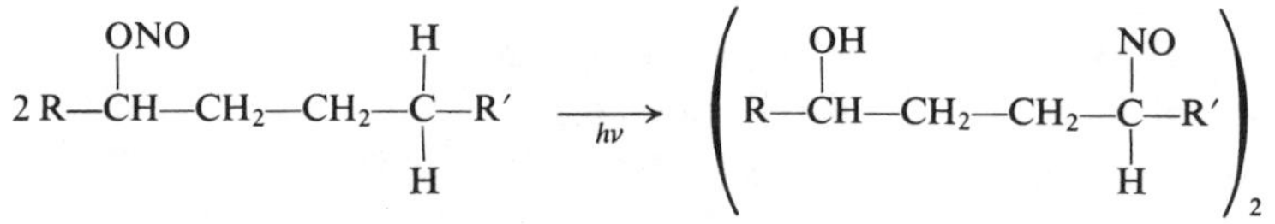

$$\text{(52)}$$

(2) The Kabsalkalian–Townley reaction [122]. A photochemical exchange process between a nitrite group and a hydrogen atom in alicyclic compounds which usually leads to linear α-nitroso-ω-aldehyde dimers (Eq. 53).

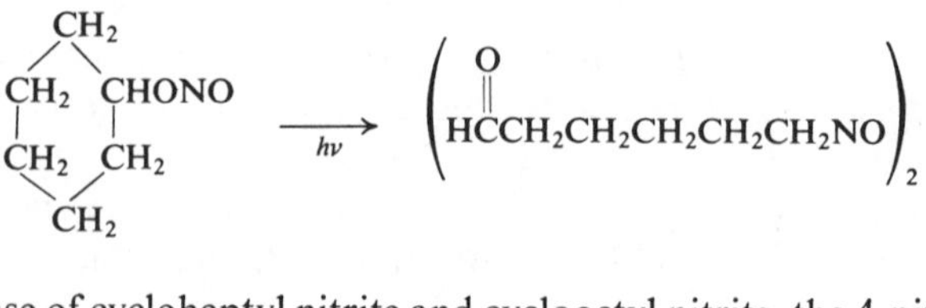

$$\text{(53)}$$

In the case of cycloheptyl nitrite and cyclooctyl nitrite, the 4-nitroso-1-cyclic alcohol dimers expected from the Barton reaction are isolated.

(3) Direct *C*-nitrosation of *m*-halophenols in acetic acid [123] (Eq. 54).

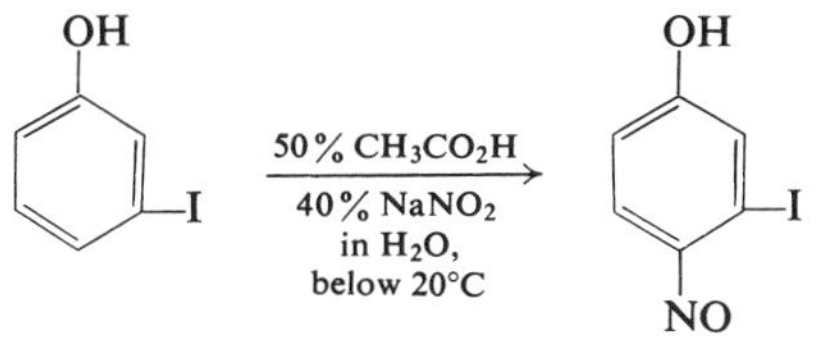

$$(54)$$

(4) Nitrosation of pyrimidines and aminouracil [124–126] (Eqs. 55, 56).

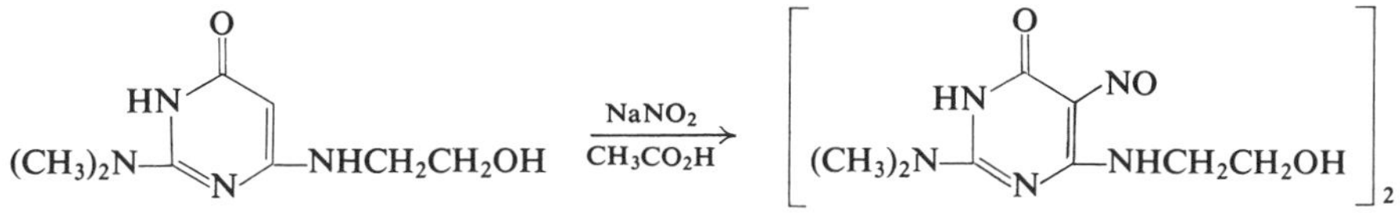

m.p. 246°–247°C
[124, 125]

$$(55)$$

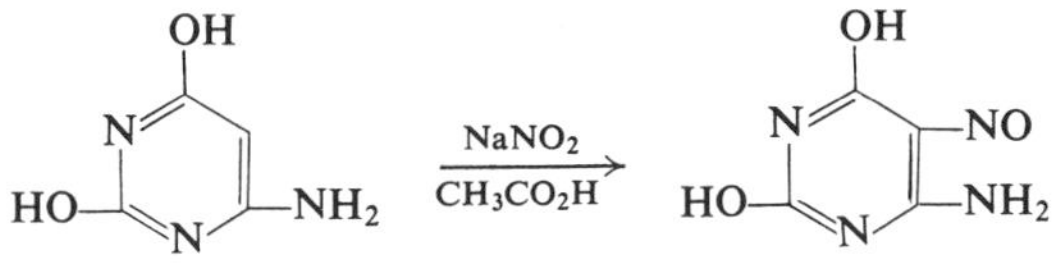

[126]

$$(56)$$

(5) Etherification of *p*-nitrosophenol [20] (Eq. 57).

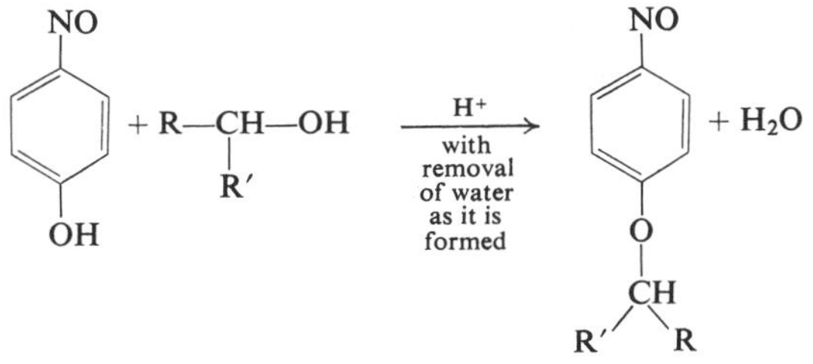

$$(57)$$

R' = H or alkyl

(6) Amination of *p*-nitrosophenol ethers with primary aromatic amines [36] (Eq. 58).

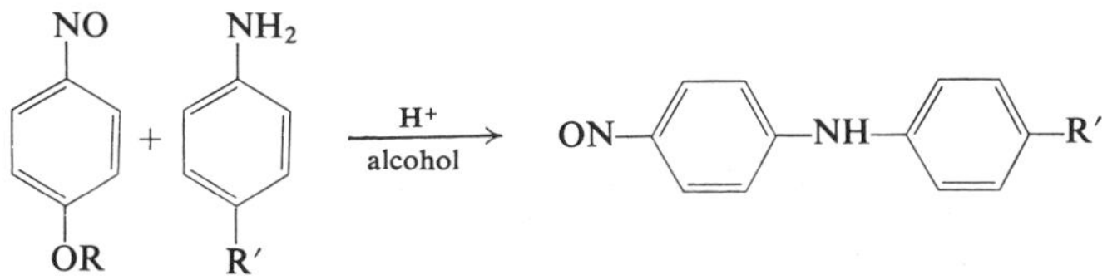

$$(58)$$

(7) Reaction of sulfinates with nitrosyl chloride [57] (Eq. 59).

$$CCl_3SO_2Na + NOCl \xrightarrow{-70° \text{ to } 0°C} CCl_3NO + SO_2 + NaCl \qquad (59)$$

(8) Addition of dinitrogen trioxide to olefins [127] (Eqs. 60, 61).

$$
\begin{array}{c}
\overset{\displaystyle ONO}{|} \\
CH_2{=}C{-}CH_2Cl \xrightarrow{N_2O_3} ON{-}CH_2{-}C{-}CH_2Cl \\
\underset{\displaystyle CH_3}{|} \underset{\displaystyle CH_3}{|}
\end{array}
\qquad (60)
$$

$$
CH_2{=}CHCH_2Cl \xrightarrow{N_2O_3} ONO{-}CH_2{-}\underset{\displaystyle NO}{\underset{\displaystyle |}{C}H}{-}CH_2Cl
\qquad (61)
$$

(monomer and dimer)

(9) Addition of nitric oxide to fluoroolefins [128–131]. The addition of nitric oxide to tetrafluoroolefins produces mixtures of nitroso, nitro, and other compounds. In the presence of ferric chloride, substantial quantities of certain nitroso products are formed (Eqs. 62, 63).

$$CF_2{=}CF_2 + NO \xrightarrow{FeCl_3} ClCF_2{-}CF_2NO \qquad (62)$$

(71%)

$$CF_2{=}CFCl + NO \xrightarrow{FeCl_3} ClCF_2{-}CFClNO \qquad (63)$$

(76%)

(10) Addition of nitrosyl chloride to fluoroolefins [131–133] (Eqs. 64, 65).

$$CF_2{=}CF_2 + NOCl \xrightarrow{FeCl_3} ClCF_2{-}CF_2NO \qquad (64)$$

(79.6%)   [132]

$$CF_2{=}CFCl \xrightarrow{NOCl} ClCF_2CFClNO, \ ClCF_2CFClNO_2, \text{ and } ClCF_2CFCl_2 \quad (65)$$

[133]

(11) Isomerization of *trans* nitroso isomers to cis nitroso isomers [18] (Eq. 66).

$$
\begin{array}{c}
O_{\diagdown} R \\
 N{=}N \\
R^{\diagup} {}_{\diagdown}O
\end{array}
\xrightarrow[\substack{\text{3. Irradiated with} \\ \text{uv} \rightarrow \text{monomer} \\ \text{4. Warm to } -78°C}]{\substack{\text{1. Dissolved} \\ \text{in 2-methyl} \\ \text{tetrahydrofuran} \\ \text{2. Cooled in liquid} \\ \text{nitrogen}}}
\begin{array}{c}
O_{\diagdown} {}_{\diagup}O \\
 N{=}N \\
R^{\diagup} {}_{\diagdown}R
\end{array}
\qquad (66)
$$

(12) Treatment of 4-chloroaniline with chloroperoxidase as a catalyst and hydrogen peroxide in a phosphate buffer at pH 4.4 led to the formation of 4-chloronitrosobenzene [134, 135].

(13) Hindered amines such as 2-aminobiphenyl, 3-methoxyfluoren-2-amine, and 3-bromofluoren-2-amine have been oxidized with m-chloroperbenzoic acid in chloroform to the corresponding nitroso compounds [136].

(14) Heterocyclic amino compounds as well as some aromatic amines have been reacted with dimethyl sulfide and N-chlorosuccinimide, and deprotonated with sodium methoxide to the S,S,-dimethylsulfilimines, which may be oxidized with m-chloroperbenzoic acid to produce nitroso compounds (Eq. 67) [137, 138].

$$R{-}NH_2 \xrightarrow[\text{NCS}]{(CH_3)_2S} \xrightarrow{NaOCH_3} R{-}\overset{\ominus}{N}{-}\overset{\oplus}{S}(CH_3)_2$$

$$\Big\downarrow \text{\scriptsize m-chloroperbenzoic acid} \qquad (67)$$

$$R^-NO$$

Among products prepared by this method are 4-methyl-2-nitrosopyridine, m.p. $137°-138°C$ (50% yield), and the unstable 1-nitrosoisoquinoline, 2-nitrosopyrimidine, and 2-nitrosopyrazine. These were characterized by conversion to appropriate derivatives.

(15) Nitrosoalkenes have been prepared from α-chloroketoximes treated with triethylamine and less readily from α-chlorosilyloximes with tetrabutylammonium fluoride. Those nitrosoalkenes which cannot tautomerize are stable and readily formed even from silyloxime [139, 140].

REFERENCES

1. T. A. Turney and G. A. Wright, *Chem. Rev.* **59**, 497 (1959).
2. O. Touster, *Org. React.* **7**, 327 (1953).
3. B. G. Gowenlock and W. Lüttke, *Quart. Rev.* **12**, 321 (1958).
4. J. Hamer, M. Ahmad, and R. E. Holliday, *J. Org. Chem.* **28**, 3034 (1963).
5. Y. Ogata, *Yuki Gosei Kogaku Kyokai Shi* **19**, 438 (1961).
6. H. Dietrich and D. Crowfoot Hodgkin, *J. Chem. Soc.* 3686 (1961).
7. J. H. Ridd, *Quart. Rev.* **15**, 418 (1961).
8. C. B. Griffis and M. C. Henry, Mater. Sym. Nat. SAMPE Symp., 7th Los Angeles (12), 15 pp. (1964); *Chem. Abstr.* **62**, 5418f (1965).
9. W. Theilacker, *Angew. Chem. Int. Ed. Engl.* **4**, 688 (1965); *Angew. Chem.* **77**, 717 (1965).
10. E. W. Cook, Jr., *Diss Abstr. B* **28**(1), 95 (1967); University Microfilms (Ann Abror, Mich.), Order No. 66-9413.
11. J. H. Boyer, NASA Access, N-67-11725 Avail. CFSTI (1966).
12a. H. Feuer, ed., "The Chemistry of the Nitro and Nitroso Groups," Part I. Wiley (Interscience), Wiley, New York, 1969.
12b. S. Patai, ed., "The Chemistry of the Nitro and Nitroso Groups," Wiley, New York, 1969.
13. R. Hoffmann and R. B. Woodward, *Science* **167**, 825 (1970).
14. T. G. Gilchrist, *Chem. Soc. Rev.* **12**, 53 (1983).
15. U. A. Spilzer and R. Stewart, *J. Org. Chem.* **39**, 3936 (1974).
16. E. Müller and H. Metzger, *Chem. Ber.* **87**, 1282 (1954).

17. B. W. Ponder, T. E. Walton, and W. J. Pollock, *J. Org. Chem.* **33**, 3957 (1968).

18. A. Mackor, Th. A. J. W. Wajer, and Th. J. de Boer, *Tetrahedron Lett.* **29**, 2757 (1967).

19. J. H. Boyer, U. Toggweiler, and G. A. Stoner, *J. Amer. Chem. Soc.* **79**, 1748 (1957).

20. J. T. Hays, E. H. de Butts, and H. L. Young, *J. Org. Chem.* **32**, 153 (1967).

21. S. L. Radomski and E. Brill, *Science* **167**, 992 (1970).

22. N. V. Sidgewick, "The Chemical Elements and Their Compounds," Vol. 1, P. 694. Oxford Univ. Press, London and New York, 1950; A. F. Wells, "Structural Inorganic Chemistry," 3rd ed., p. 624. Oxford Univ. Press, London and New York, 1962.

23. A. Baeyer and H. Caro, *Chem. Ber.* **7**, 809, 963 (1874).

24. E. F. Elslager and D. F. Worth, *J. Med. Chem.* **13**, 370 (1970).

25. L. Blangey, *Helv. Chim. Acta.* **21**, 1579 (1938).

26. O. Fischer and E. Hepp, *Chem. Ber.* **19**, 2991 (1886).

27. M. Ikuta, *Ann. Chem.* **243**, 272 (1888).

28. L. Wacker, *Ann. Chem.* **243**, 290 (1888).

29. E. Kock, *Ann. Chem.* **243**, 310 (1888).

30. B. T. Baliga, *J. Org. Chem.* **35**, 2031 (1970).

31. G. Steel and D. L. H. Williams, *Chem. Commun.* 975 (1969).

32. J. Willenz, *J. Chem. Soc.* 1677 (1955).

33. J. J. D'Amico, C. C. Tung, and L. A. Walker, *J. Amer. Chem. Soc.* **81**, 5957 (1959).

34. H. U. Daeniker, *Helv. Chim. Acta* **47**, 33 (1964); *Angew. Chem.* **76**, 760 (1964).

35. T. C. VanHoek, P. E. Verkade, and B. M. Wepster, *Rec. Trav. Chim.* **77**, 559 (1958).

36. J. T. Hays, H. C. Young, and H. H. Espy, *J. Org. Chem.* **32**, 158 (1967).

37. C. S. Marvel and P. K. Porter, *Org. Syn. Coll. Vol.* **1**, 411 (1932).

38. Y. Ogata and H. Tezuka, *J. Org. Chem.* **33**, 3179 (1968).

39. E. Kremers, N. Wakeman, and R. M. Hixon, *Org. Syn. Coll. Vol.* **1**, 511 (1932).

40. W. R. Ornsdorff and M. L. Nichols, *J. Amer. Chem. Soc.* **45**, 1536 (1923).

41. K. Singer and P. A. Vamplew, *J. Chem. Soc.* 3052 (1957).

42. A. Di Giacomo, *J. Org. Chem.* **30**, 2614 (1965).

43. J. G. Aston and M. G. Mayberry, *J. Amer. Chem. Soc.* **57**, 1888 (1935).

44. C. S. Coe and T. F. Doumani, *J. Amer. Chem. Soc.* **70**, 1516 (1948).

45. B. G. Gowenlock and J. Trotman, *Chem. Ind.* (*London*) 538 (1955); *J. Chem. Soc.* 4190 (1955).

46. B. G. Gowenlock and J. Trotman, *J. Chem. Soc.* 1670 (1956).

47. F. P. Boer and J. W. Turley, *J. Amer. Chem. Soc.* **89**, 1034 (1967).

48. W. D. Emmons, *J. Amer. Chem. Soc.* **79**, 6522 (1957).

49. H. Krimm and K. Hamann, Ger. Pat. 953,069 (Nov. 29, 1956).

50. E. Bamberger and R. Seligman, *Chem. Ber.* **36**, 685 (1903).

51. J. R. Schwartz, *J. Amer. Chem. Soc.* **79**, 4353 (1957).

52. E. Müller and H. Metzger, *Chem. Ber.* **88**, 165 (1955).

53. P. W. K. Flanagan, U.S. Pat. 3,205,273 (Sept. 7, 1965).

54. E. Müller and G. Schmid, *Chem. Ber.* **92**, 514 (1959).

55. E. Müller, D. Fries, and H. Metzger, *Chem. Ber.* **88**, 1891 (1955).

56. E. Müller and U. Heuschkel, *Chem. Ber.* **92**, 63 (1959).

57. H. Sutcliffe, *J. Org. Chem.* **30**, 3221 (1965).

58. B. L. Dyatkins, R. A. Bekker, and I. L. Knunyants, *Izv. Akad. Nauk SSSR Ser Khim.* 1121 (1956) (6); *Chem. Abstr.* **63**, 8181b (1965).

59. M. S. Kharasch, T. H. Meltzer, and W. Nudenberg, *J. Org. Chem.* **22**, 37 (1957).

60. H. T. J. Chilton and B. G. Gowenlock, *Nature* **172**, 73 (1953); *J. Chem. Soc.* 3174 (1954).

61. E. Müller and H. Metzger, *Chem. Ber.* **87**, 1449 (1954).

62. M. Kosinski, *Lodz. Tow. Nauk. Pr. Wydz, 3, Acta Chim.* **9**, 93 (1964).

63. O. Piloty and A. Stock, *Chem. Ber.* **35**, 3093 (1902).

64. M. Kosinski, *Lodz. Tow. Pr. Nauk. Wydz. 3*; *Acta Chim.* **9**, 93 (1964); *Chem. Abstr.* **62**, 11674a (1965).

65. B. W. Ponder and D. R. Walker, *J. Org. Chem.* **32**, 4136 (1967).

66. E. Müller and H. Metzger, *Chem. Ber.* **89**, 396 (1956); H. Metzger and E. Müller, *ibid.* **90**, 1179 (1957).

67. H. Metzger and E. Müller, *Chem. Ber.* **90**, 1185 (1957).

68. E. Müller, D. Fries and H. Metzger, *Chem. Ber.* **90**, 1188 (1957).

69. E. Müller and G. Schmid, *Chem. Ber.* **94**, 1364 (1961).

70. A. H. Dimwoodie and R. N. Haszeldine, *J. Chem. Soc.* 1675 (1965).

71. R. A. Henry, *J. Org. Chem.* **23**, 648 (1958).

72. K. M. Ibne-Rasa, *J. Amer. Chem. Soc.* **84**, 4962 (1962).

73. E. C. Stump, W. H. Oliver, and C. D. Padgeti, *J. Org. Chem.* **33**, 2102 (1968).

74. A. Beyer, *Chem. Ber.* **7**, 1638 (1874).

75. B. Oddo, *Gazz. Chim. Ital.* **39**, 659 (1909).

76. E. Müller and H. Metzger, *Chem. Ber.* **89**, 396 (1956).

77. E. Robson and J. M. Tedder, *Proc. Chem. Soc.* **13** (1963).

78. W. A. Tilden, *J. Chem. Soc.* **28**, 514 (1875); O. Wallace, "Terpene und Campher," 2nd ed. Viet und Co., Leipzig, 1914.

79. N. Thorne, *J. Chem. Soc.* 4271 (1956).

80. J. Meinwald, Y. C. Meinwald, and T. N. Baker, III, *J. Amer. Chem. Soc.* **85**, 2513 (1963).

81. J. Meinwald, Y. C. Meinwald, and T. N. Baker, III, *J. Amer. Chem. Soc.* **86**, 4074 (1964).

82. M. Ohno, M. Okamoto, and N. Naruse, *Tetrahedron Lett.* 1971 (1965).

83. A. Hassner and C. Heathcock, *J. Org. Chem.* **29**, 1350 (1964).

84. H. C. Harmann and D. Swern, *Tetrahedron Lett.* 3303 (1966).

85. W. D. Emmons, *J. Am. Chem. Soc.* **79**, 5739 (1957).

86. H. Caro, *Z. Angew. Chem.* **11**, 845 (1898).

87. E. Bamberger and R. Hübner, *Chem. Ber.* **36**, 3803 (1903).

88. R. Kuhn and W. v. Klaveren, *Chem. Ber.* **71**, 779 (1938).

89. E. Bamberger and F. Tschirner, *Chem. Ber.* **32**, 1675 (1899).

90. F. B. Mallory, K. E. Schueller, and C. S. Wood, *J. Org. Chem.* **26**, 3312 (1961).

91. R. R. Holmes, R. P. Bayer, L. A. Errede, H. R. Davis, A. W. Wiesenfeld, P. M. Bergman, and D. L. Nicholas, *J. Org. Chem.* **30**, 3837 (1965).

92. A. B. Sullivan, *J. Org. Chem.* **31**, 2811 (1966).

93. J. N. Ashley and S. S. Berg, *J. Chem. Soc.* 3089 (1957).

94. B. A. Della Coletta, J. G. Frye, T. L. Youngless, J. P. Zeigler, and R. G. Landolt, *J. Org. Chem.* **42**, 305 (1977).

95. E. Pfeil and K. H. Schmidt, *Ann. Chem.* **675**, 36 (1964).

96. R. R. Holmes and R. P. Bayer, *J. Amer. Chem. Soc.* **82**, 3454 (1960); R. R. Holmes, *J. Org. Chem.* **29**, 3076 (1964); R. R. Holmes, R. P. Bayer, and D. L. Nicholas, *J. Org. Chem.* **32**, 2912 (1967).

97. J. A. Castellano, J. Green, and J. M. Kauffman, *J. Org. Chem.* **31**, 821 (1966).

98. G. M. Brooke, J. C. Burdon, and J. C. Tatlow, *Chem. Ind.* (*London*) 832 (1961).

99. R. A. Abramovitch, S. R. Challand, and Y. Yamada, *J. Org. Chem.* **40**, 1541 (1975).

100. R. E. Lutz and M. R. Lytton, *J. Org. Chem.* **2**, 68 (1937–1938).

101. W. J. Mijs, S. E. Hoekstra, R. M. Ulmann, and E. Havinga, *Rec. Trav. Chim.* **77**, 746 (1958).

102. H. E. Baumgarten, A. Staklis, and E. M. Miller, *J. Org. Chem.* **30**, 1203 (1965).

103. E. Tauber, *Chem. Ber.* **24**, 3081 (1891).

104. S. D. Ross, G. J. Kahan, and W. A. Leah, *J. Amer. Chem. Soc.* **74**, 4122 (1952).

105. A. Calder, A. R. Forrester, and S. P. Hepburn, *Org. Syn.* **52**, 77 (1972).

106. T. Parsons, Jr., and J. C. Bailar, Jr., *J. Amer. Chem. Soc.* **58**, 268 (1936).

107. W. H. Nutting, R. A. Jewell, and H. Rapoport, *J. Org. Chem.* **35**, 505 (1970).

108. J. A. Maasen and T. J. DoBoer, *Trav. Chim. Pays-Bas* **90**, 373 (1971).
109. Y. Yost, *J. Med. Chem.* **12**, 961 (1969).
110. O. Piloty, *Chem. Ber.* **31**, 1878 (1898).
111. O. Piloty and B. Graf Schwerin, *Chem. Ber.* **34**, 1863 (1901).
112. D. C. Iffland and G. X. Criner, *J. Amer. Chem. Soc.* **75**, 4047 (1953).
113. M. H. Palmer and E. R. R. Russell, *Chem. Ind. (London)* 157 (1966).
114. O. Baudisch, *Science* **92**, 336 (1940); **108**, 443 (1948); *J. Amer. Chem. Soc.* **63**, 622 (1941).
115. G. Cronheim, *J. Org. Chem.* **12**, 1, 7, 20 (1947).
116. J. O. Konecny, *J. Amer. Chem. Soc.* **77**, 5748 (1955).
117. K. Maruyama, I. Tanimoto, and R. Goto, *J. Org. Chem.* **32**, 2516 (1967).
118. P. E. Iversen and H. Lund, *Tetrahedron Lett.* 4027 (1967).
119. S. D. Ross and I. Kuntz, *J. Amer. Chem. Soc.* **74**, 1297 (1952).
120. L. A. Errede and H. R. Davis, *J. Org. Chem.* **28**, 1430 (1963).
121. D. H. R. Barton, J. M. Beaton, L. E. Geller, and M. M. Pechet, *J. Amer. Chem. Soc.* **82**, 2640 2641 (1960).
122. P. Kabasakalian and E. R. Townley, *J. Org. Chem.* **72**, 2918 (1962); *J. Amer. Chem. Soc.* **84**, 2711, 2716, 2719 2723 (1962).
123. H. H. Hodgson and D. E. Nicholson, *J. Chem. Soc.* 1808 (1939).
124. D. J. Brown, *Rev. Pure Appl. Chem.* **3**, 115 (1953).
125. W. Pfleider and H. Walter, *Ann. Chem.* **677**, 113 (1964).
126. W. R. Sherman and E. C. Taylor, Jr., *Org. Syn. Coll. Vol.* **4**, 247 (1963).
127. J. R. Park and D. L. H. Williams, *Chem. Commun.* 332 (1969).
128. J. D. Park, A. P. Stefani, G. H. Crawford, and J. R. Lacher, *J. Org. Chem.* **26**, 3316 (1961); J. D. Park, A. P. Stefani, and J. R. Lacher, *J. Org. Chem.* **26**, 3319 (1961).
129. J. M. Birchall, A. J. Bloom, R. N. Haszeldine, and C. J. Willis, *J. Chem. Soc.* 3021 (1962).
130. E. E. Griffin and R. N. Haszeldine, *J. Chem. Soc.* 1398 (1960).
131. C. A. Seymone and F. D. Greene, *J. Org. Chem.* **47**, 5226 (1982).
132. J. D. Park, A. P. Stefani, and J. R. Lacher, *J. Org. Chem.* **26**, 4017 (1961).
133. D. E. O'Connor and P. Tarrant, *J. Org. Chem.* **29**, 1793 (1964).
134. M. D. Corbett, B. R. Chipko, and D. G. Baden, *Biochem. J.* **175**, 353 (1978).
135. M. D. Corbett and B. R. Corbett, *J. Org. Chem.* **46**, 466 (1981).
136. Y. Yost and H. R. Gutmann, *J. Chem. Soc.* 2497 (1970).
137. S.-L. Huang and D. Swern, *J. Org. Chem.* **44**, 2510 (1979).
138. E. C. Taylor, C.-P. Tseng, and J. B. Rampal, *J. Org. Chem.* **47**, 552 (1982).
139. S. E. Denmark and M. S. Dappen, *J. Org. Chem.* **49**, 798 (1984).
140. A. W. Denmark, M. S. Dappen, and J. A. Sternberg, *J. Org. Chem.* **49**, 4743 (1984).

1. INTRODUCTION

The treatment of a large variety of organic nitrogen compounds with typical nitrosating agents may produce the corresponding *N*-nitroso compounds. Among the nitrogen compounds which have been nitrosated are aromatic and aliphatic amines, aminoketones, *N*-substituted amides, urethanes, sulfonamides, lactams, diketopiperazines, ureas, thioureas, oxazolidones, ketimines, alkylhydrazines, 3-nitroguanidines, guanylhydrazines, carbamylhydrazines, and hydroxylamines.

From the standpoint of theoretical research, extensive studies have been reported. In the first edition of this work, we stated that from the standpoint of applications, *N*-nitroso compounds were of limited interest. For example, the exhaust gases of the early lunar module (LM) evidently contained nitroso derivatives of unsymmetrical dimethylhydrazine and hydrazine, as well as methylamine [1]. Since that time, the interest in this class of compounds has grown with the discovery of evidence that *N*-nitrosoamines may be extremely carcinogenic and that some *N*-nitrosoureas may be of value as anticancer agents.

In the course of a number of well-known organic syntheses, such as the formation of diazonium salts from aromatic amines by reaction with nitrous acid, undoubtedly the intermediate formation of *N*-nitroso compounds is involved. The Demjanov and Tiffeneau–Demjanov ring expansions also involve *N*-nitroso compounds [2]. Some *N*-nitroso compounds have been used as sources of free radicals and as blowing agents.

In certain cases, *N*-nitroso derivatives of secondary amines have been used as a means of purifying amines [3]. In this procedure, *N*-nitroso derivatives of amines are prepared, isolated, and then reduced with stannous chloride and hydrochloric acid. Isolation of the amines from the reducing medium evidently produces compounds purer than those purified by more conventional means.

N-Nitrosation reactions should be of interest also because they are methods of establishing covalent nitrogen-to-nitrogen bonds directly. One method of preparing substituted hydrazines depends on this procedure (see Volume I, Chapter 14).

The thermal stability of *N*-nitroso derivatives is highly variable; consequently it is suggested that any effort to prepare such derivatives be carried out with extreme care. In particular, care must be exercised in avoiding contact of *N*-nitroso compounds with strong alkaline media because the unstable diazo-alkanes are generally prepared by this method (see Volume 1, Chapter 15).

According to one patent [4], some unstable nitroso compounds may be stabilized by addition of such inert materials as ammonium bicarbonate, ammonium carbonate, magnesium acid phosphate, ammonium or sodium acid phosphate, calcium lactate, ammonium alum, and calcium silicate.

Ridd [5] and Kalatzis and Ridd [6] have written an interesting series of studies and reviews on the interrelationship of nitrosoation, diazotization, and deamination. Other reviews covering various aspects of *N*-nitroso chemistry are given in [7–22]. Chapter 16 should also be reviewed for background material on nitrosations.

Although there are exceptions, which will be mentioned in the appropriate section of this chapter, *N*-nitrosation is usually carried out on the nitrogen compounds in which the nitrogen bears only one hydrogen.

The *N*-nitroso derivatives frequently are thermally or photolytically un-stable or both. Many of them have a corrosive action on the skin or on the mucous membranes, and some are thought to be highly carcinogenic (e.g., see [19]). Since this matter has only recently come to light, much of the older literature makes mention neither of safety precautions nor of health hazards in handling nitroso compounds. A report by V. M. Craddock entitled "Nitrosamines and human cancer: proof of an association?" [23] outlines the complexity of evaluating the many factors that have to be considered. According to this article, many participants at a meeting on *N*-nitroso compounds agree with K. Preussmann of the General Cancer Center that there is proof of an association of *N*-nitrosamines and human cancer.

In the literature, safety precautions are only briefly described, although several authors have offered to assist other workers in setting up appropriate safety procedures. One of the clearer statements was made by Kupper and Mechejda [24]. They stated that experiments "were carried out in an efficient fume hood". Nitrosamine wastes were burned in a high-temperature inciner-ator. Whether this protocol is adequate needs further study. In our own experience with relatively harmless but odoriferous compounds we have found that frequently fume hoods of several laboratories are interconnected. Consequently, vapors from one hood may find their way into another. In another case, atmospheric conditions on the roof caused vapors from one exhaust duct to reenter the building through another duct.

Furthermore, since simply to exhaust toxic vapors into the environment from a fume hood exhaust no longer can be considered an acceptable procedure, clearly the exhausts from the hoods must be decontaminated prior to release to the environment or even to interconnecting ducts. Higher temperature incineration would appear to be a useful technique provided there is certainty that the hazardous compounds do not volatilize out of the combustion zone prior to total destruction. We generally suggest extreme caution in handling *N*-nitroso compounds, their reaction solvents, wastes, filter papers, laboratory equipment, etc.

The nitrosations are usually carried out with conventional reagents. The most common procedures involve: (a) sodium nitrite with acids in an aqueous system, (b) sodium nitrite in acetic anhydride, (c) nitrogen trioxide, (d) nitrosyl chloride in the presence of a base, and (e) dinitrogen tetroxide in the presence of a base.

Mixtures of nitrogen dioxide and nitric oxide, nitrosonium tetrafluoroborate and alkyl nitrites have also been used occasionally in specific situations.

NOTE: In the nitroso compound literature, frequent reference is made to the *Liebermann test*. This test may very well have been one for identifying phenols by the dyes which are formed in the presence of nitrous acid.

The test consists of warming a sample with phenol in concentrated sulfuric acid to give an intensely colored solution (usually dark red). The test solution is then diluted with water and treated with an excess of potassium hydroxide. A blue color is formed when a nitroso compound was present in the sample [25]. This is based on the presumption that many, if not all, *C*-nitroso and *N*-nitroso compounds are decomposed by concentrated sulfuric acid to give the nitrous acid required for the reaction.

2. *N*-NITROSAMINES

The nitrosation of amines leads to a variety of reaction products depending on the degree of substitution of the amino group. The nitrosation of aromatic primary amines normally leads to the formation of diazonium salts (see Volume I, Chapter 15), while the nitrosation of aliphatic primary amines may produce a variety of deamination products [2, 5, 6, 25–27]. Evidently, 1,2-ethanediamine coordinated in a platinum (IV) complex may be nitrosated in boiling water by adding a boiling solution of potassium nitrate [28].

The reaction of secondary amines generally stops at the *N*-nitrosamine stage, although derivatives of aromatic amines may undergo the Fischer–Hepp rearrangement (see Chapter 16). In some cases, tertiary aromatic and aliphatic amines have been known to dealkylate and form *N*-nitrosamines. Aromatic tertiary amines may also be nitrosated in the para position, a reaction discussed in Chapter 16.

In the reaction of amines with nitrous acid, one would expect the primary reaction to be the formation of an alkylammonium nitrite salt. Despite a few observations on the formation of such salts in the late nineteenth and early twentieth centuries, the formation of such salts was largely ignored until recently [29]. To decrease the possibility of formation of nitrosamines, the preparation of nitrite salts is carried out in methanol at a low temperature and in the presence of carbon dioxide, which seems to form compounds with formulas corresponding to $(RNH_2)_2CO_2$ as intermediates. Once formed, however, many nitrite salts appear to be quite stable. Some exhibit relatively high melting points. Other nitrites are said to be unstable.

Since the neutralization reaction of an amine with nitrous acid is presumably instantaneous while the nitrosation of an amine proceeds at a lower rate, further investigations would be of interest to elucidate whether nitrite salt formation is a necessary preliminary step to the formation of *N*-nitrosamines or whether the covalent product forms independently. In the latter case, there would be competitive reactions of significantly different reaction rates and mechanisms. The yield of *N*-nitrosamines may be influenced by the extent to which the more water-soluble nitrite salts may be present in the course of a preparation.

A. *N*-Nitrosation of Primary Amines

The best known example of the treatment of a primary aliphatic amine with nitrous acid involves the reaction of esters of glycine hydrochloride with sodium nitrite to form esters of diazoacetic acid. This reaction is carried out at low temperatures and under such reaction conditions that any *N*-nitroso primary amine which might have been formed is immediately converted to the diazoacetate [30, 31]. Treatment of 1-methyl-2,2,2-trifluoroethylamine hydrochloride, another primary amine, with sodium nitrite in an aqueous system also evidently leads to the corresponding diazoalkane [32].

The ease with which diazo compounds are formed in these reactions should constitute reasonable warning that the nitrosation of all amines is fraught with considerable hazard. Not only can the unstable diazo compounds be formed, but also the intermediate nitrosamines themselves may be explosive in nature. Therefore, each particular case must be carefully examined for safety hazards. As a matter of fact, even though a reaction may have run successfully at one time or another, this does not necessarily imply that it is indeed safe (see safety note in Redemann *et al.* [33]).

In the case of β-aminoalkanols as well as 3-aminoalkanols, recent studies have shown that treatment of such primary amines with nitrous acid leads to deamination to generate aldehydes capable of reacting with more of the aminoalkanol starting material to form oxazolidines, which, in turn, are then *N*-nitrosated (Eq. 2) [26].

$$HOCH_2CH_2NH_2 \xrightarrow{HONO} [HOCH_2CH_2\overset{\oplus}{N}_2] \xrightarrow[-H]{-N_2} \overset{O}{\overset{\|}{H}}CCH_3 \qquad (1)$$

$$
\underset{\text{O}}{\overset{\text{O}}{\text{H}\overset{\|}{\text{C}}\text{CH}_3}} + \text{HOCH}_2\text{CH}_2\text{NH}_2 \xrightarrow{\text{HONO}} \begin{array}{c} \text{CH}_2\text{—CH}_2 \\ | \qquad\qquad | \\ \text{O} \qquad \text{N—NO} \\ \diagdown \;\diagup \\ \text{CH} \\ | \\ \text{CH}_3 \end{array} \qquad (2)
$$

Primary amines may also be converted to α-nitrosoaminoalkyl ethers by treating a primary amine with aldehydes in the presence of acetic acid and sodium nitrite about 5°C [27].

B. *N*-Nitrosation of Secondary Amines

The nitrosation of secondary aliphatic amines with sodium nitrite and hydrochloric acid has been well described [34].

The preparation of *N*-nitroso-β-alkylaminoisobutyl ketones is of particular interest as a method for preparing the intermediates for the diazoalkane synthesis. The preparation of these compounds is based on the addition of a primary amine to mesityl oxide to give a secondary amino ketone which is then nitrosated (see Volume I, Chapter 15, Section 2A,g). This preparation uses an acetic acid/sodium nitrite reagent instead of the better-known hydrochloric acid/sodium nitrite reagent [35, 36].

2-1. *Preparation of Methyl N-Nitroso-β-methylaminoisobutyl Ketone* [36]

$$
\text{CH}_3\text{NH}_2\cdot\text{HCl} + \underset{\underset{\text{CH}_3}{|}}{\overset{\overset{\text{O}}{\|}}{\text{CH}_3\text{C}\text{CH}=\text{C}\text{—CH}_3}} \xrightarrow{\text{NaOH}} \underset{\underset{\text{CH}_3}{|}}{\overset{\overset{\text{CH}_3}{|}}{\text{CH}_3\text{NH}\text{—C}\text{—CH}_2\overset{\overset{\text{O}}{\|}}{\text{C}}\text{CH}_3}} \qquad (3)
$$

$$
\underset{\underset{\text{CH}_3}{|}}{\overset{\overset{\text{CH}_3}{|}}{\text{CH}_3\text{NH}\text{—C}\text{—CH}_2\overset{\overset{\text{O}}{\|}}{\text{C}}\text{CH}_3}} \xrightarrow[\text{CH}_3\text{CO}_2\text{H}]{\text{NaNO}_2} \underset{\underset{\text{NO}\;\;\text{CH}_3}{|\quad\;|}}{\overset{\overset{\text{CH}_3}{|}}{\text{CH}_3\text{—N}\text{—C}\text{—CH}_2\overset{\overset{\text{O}}{\|}}{\text{C}}\text{CH}_3}} \qquad (4)
$$

With suitable safety precautions, an ice-cold solution of 135 gm (2.0 moles) of methylamine hydrochloride in 200 ml of water is treated with a solution of 80 gm (2.0 moles) of sodium hydroxide in 200 ml of water with vigorous stirring while maintaining the temperature below 20°C. Over a period of 1 hr, to this solution is added, dropwise, 195 gm (1.8 moles) of mesityl oxide. Stirring is continued for an additional hour. Then the reaction is neutralized cautiously at a temperature below 7°C with approximately 120 gm of glacial acetic acid. Thereafter a solution of 360 gm (3.9 moles) of 90% pure sodium nitrite in 450 ml of water followed by 180 gm of glacial acetic acid is added without special cooling. Stirring is continued for 4 hr and the temperature is allowed to rise, but not to exceed 35°C.

The product is extracted from the reaction mixture by two extractions with ether. The aqueous reaction mixture is then saturated with sodium chloride and again extracted with ether. The combined ether extracts are washed in turn with two portions of dilute acetic acid and one portion of a 10% calcium chloride solution. Finally the solution is dried over calcium chloride. Low-boiling solvents and impurities are separated by evaporation followed by cautious distillation up to 100° (30 mm Hg). The residual product weighs 213–228 gm (75–80%), m.p. 21°–22°C.

A variation of the nitrosation with acetic acid/sodium nitrite is illustrated in the preparation of *N*-nitrosodicyclohexylamine. In this case, dicyclohexylamine was converted into its acetate salt (m.p. 115°–116°C) and the salt was warmed with an aqueous solution of acetic acid and sodium nitrite. The insoluble precipitate was filtered and recrystallized from acetone, yielding the nitrosoamine in the form of colorless crystals, m.p. 104°–105°C [29].

Not only may acetic acid/sodium nitrite be used to nitrosate secondary amines but also mineral acids and sodium nitrite [37–40]. There are also trans-nitrosation procedures, which will be discussed below.

The relatively complex cyclic system camphidine has been converted into *N*-nitrosocamphidine with the acetic acid/sodium nitrite reagent [41]. This reagent has also been used to prepare [^{14}C]dimethylnitrosamine from the corresponding [^{14}C]dimethylamine [42].

On the other hand, cimetidine has been nitrosated with hydrochloric acid/sodium nitrite. In this connection it is important to point out that the exact reaction conditions must be carefully controlled, as will be indicated below. Under one set of conditions the simple *N*-nitrosocimetidine hydrate was formed. With an excess of sodium nitrite, and in the presence of atmospheric oxygen, *N*-nitrosocimetidine nitrate was isolated [43].

2-2. *Preparation of N-Nitrosocimetidine Hydrate and Nitrate* [43]

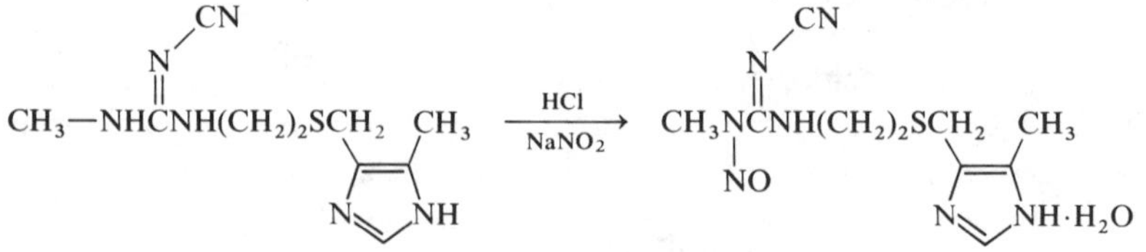

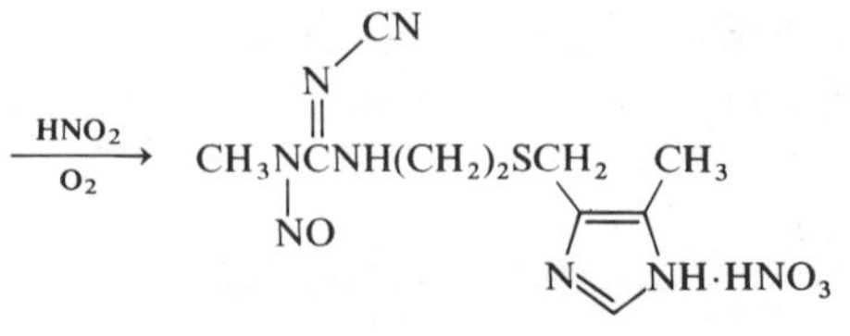

With suitable safety precautions, in a 100-ml Erlenmeyer flask fitted with a magnetic stirrer is placed a solution of 2.25 gm (9 mmole) of cimetidine in 45 ml of 2 *M* hydrochloric acid. The flask is closed with a serum cap; the air in the flask is displaced with nitrogen. The flask is cooled to 0°C. Then a solution of 2.1 gm (30 mmole) of sodium nitrite in 15 ml of water is injected through the serum cap. Stirring at 0°C is continued for 40 min. The flask is then opened and made basic to pH 10 with potassium carbonate.

The product is extracted from the reaction mixture with three 25-ml portions of ethyl acetate. The combined extracts are washed in turn with a saturated sodium chloride solution and water. After drying the ethyl acetate solution over anhydrous sodium sulfate, the solvent is removed on a rotary evaporator. The pale yellow oily residue is stored at 0°C overnight to assist in its solidification. The product had a m.p. of 23°–28°C; ir (neat) 1440 (NO), 1630 (C=N), 2150 (C=N), 3300 cm^{-1} (NH). Yield: 85%.

After nitrosation for a period of 2 hr at 0°C with five equivalents of sodium nitrite in an open system with atmospheric oxygen freely circulating, the nitrosocimetidine nitrate was isolated as a yellow solid in 75% yield, m.p. 143°–144°C (from ethanol).

The tetranitrosation of 1,4,5,8-tetraazadecalin is of interest not only as a synthesis but also from the standpoint of stereochemistry. The compound may exist in four isomeric forms. The reaction evidently produces a mixture which, in DMSO, consists of 88% in form I, 12% in form II, and 0.5% in form III (based on the ^{1}H NMR spectrum). With ^{15}N and ^{13}C NMR spectroscopy evidence of possibly 0.5% of form IV was found [44]:

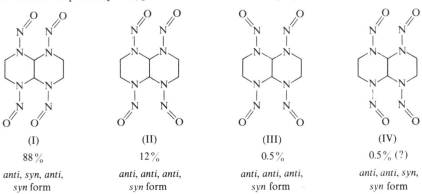

(I)	(II)	(III)	(IV)
88%	12%	0.5%	0.5% (?)
anti, syn, anti, syn form	*anti, anti, anti, syn* form	*anti, anti, anti, syn* form	*anti, anti, syn, syn* form

2-3. Preparation of 1,4,5,8-Tetranitroso-1,4,5,8-tetraazodecalin [44]

$$\text{(7)}$$

With suitable safety precautions, in a 125-ml Erlenmeyer flask fitted with a magnetic stirrer, a solution of 3.45 gm (50 mmole) of sodium nitrite in 20 ml of water* and 1.42 gm (10 mmole) of 1,4,5,8-tetraazodecalin. The solution is cooled to $-2°C$ and then maintained below $5°$ while 50 ml of $1N$ hydrochloric acid is added over a period of 1 min. A white precipitate forms almost immediately. The mixture is stirred for 30 min at $0°C$ and then at room temperature for 1 hr. The product is collected on a Büchner funnel, washed thoroughly with water, and dried overnight in a vacuum oven. Yield: 2.35 gm (91%); decomposition point $211°-212°C$. The product is recrystallized from DMF/water to yield light yellow needles; ir (KBr) 2900 (w), 1475 (m), 1450 (sh), 1410 (m), 1370 (n), 1310 (m), 1300 (m), 1275 (m), 1260 (m), 1210 (m), 1190 (m), 1110 (m), 1050 (m), 975 (w), 935 (m), 895 (w), 830 (w), 735 (m).

By similar procedures, trans-10-methyl-1,8-dinitroso-1,8-diazadecalin (m.p. $68°-70°C$) and trans-10-methyl-1,8-dinitroso-1,8-diazadecalin (m.p. $82°-84°C$) were prepared [45].

Secondary amines containing at least one aromatic group are readily nitrosated in an aqueous medium using sodium nitrite and hydrochloric acid. The preparation of N-nitrosophenylglycine, an intermediate in the preparation of 3-phenylsydnone [46, 47], is a case in point.

2-4. Preparation of N-Nitrosophenylglycine [46]

$$\tag{8}$$

A suspension of 50 gm (0.3 mole) of phenylglycine in 600 ml of water is cooled in an ice bath to $0°C$. When the suspension has reached $0°C$, a solution of 25 gm (0.36 mole) of sodium nitrite in 150 ml of water is added dropwise at such a rate that the temperature remains at $0°C$. Stirring is continued until a clear red solution is obtained. Then the solution is rapidly filtered. To the solution is added sufficient concentrated hydrochloric acid with rapid stirring to render it acidic (about 100 ml of concentrated hydrochloric acid). Stirring is continued while the fluffy product precipitates. The resultant product is filtered with suction, washed repeatedly with ice-cold water, and air-dried: yield 54.5 gm (91%), m.p. $102°-103°C$.

Since many N-nitrosoaniline derivatives are liquids, suitable modifications have to be made for the isolation of the product. In many cases the nitrosation is carried out with a dispersion of N-substituted anilines in hydrochloric acid and crushed ice to maintain the low temperatures of the reaction. Concentrated

* The original preparation given in [44] mentions only the formation of a solution of sodium nitrite in tetraazodecalin. We assume that water was also used, by analogy to [45].

hydrochloric acid and sodium nitrite solutions are then added in turn while maintaining temperatures between 0° and 10°C. The product may be isolated by extraction with suitable solvents, such as benzene [48] or ether [49, 50], followed by evaporation of the solvent and distillation of the product under reduced pressure.

The nitrosation of 2,5-diphenylpyrrolidine in an ice-cold solution of ethanol and hydrochloric acid with sodium nitrite produced the expected *N*-nitroso-2,5-diphenylpyrrolidine. The product could be separated into trans and cis isomers by fractional crystallization using acetone–water mixtures. The ratio of trans to cis isomer was found to be 2.5:1. When the nitrosation was carried out in an acetic acid solution, the ratio of trans to cis isomer remained the same although the yield of the identified products was somewhat lower (71 vs 56%) [51]. The ratio of the isomers of the starting pyrrolidine was not reported.

Purified *trans-N*-nitroso-2,5-diphenylpyrrolidine, when treated with 30.5% aqueous potassium hydroxide solution in a nitrogen atmosphere, was partially isomerized to *cis-N*-nitroso-2,5-diphenylpyrrolidine (21% yield of cis isomer) [51]. The identity of the product was confirmed by infrared spectra and elemental analysis. The configurations of the two isomers were assigned on the basis of NMR studies.

2-5. *Preparation of trans- and cis-N-Nitroso-2,5-diphenylpyrrolidine* [51]

$$\text{(9)}$$

(cis and trans isomers)

In a 1 liter, three-necked flask equipped with a mechanical stirrer, addition funnel, and a gas outlet, a solution of 138.4 gm (0.62 mole) of 2,5-diphenyl-pyrrolidine in 210 ml of ethanol and 100 ml of concentrated hydrochloric acid is prepared. While vigorous stirring is continued, the solution is cooled in an ice bath and, over a 2 hr period, a solution of 61.5 gm (0.891 mole) of sodium nit-rite in 100 ml of water is added dropwise. Stirring is continued overnight while the reaction mixture comes to room temperature. To isolate the product, the reaction mixture is poured into 1 liter of water and the resultant precipitate is collected on a filter. The crude product is thoroughly washed with water. Yield of crude product is 160 gm.

By fractional recrystallization from acetone, 79.9 gm (51%) of a high-melting isomer, m.p. 139.5°–142.5°C and 31.7 gm (20%) of a low-melting isomer, m.p. 96.0°–98.0°C are isolated. On further crystallization, the higher-melting trans isomer was found to have a melting point of 140.0°–140.9°C, while the lower-melting cis isomer was found to have a melting point of 97.0°–98.0°C.

Syntheses of α,β-(vinyl nitrosamines or N-nitrosoenamines) and β,γ-(allylic)-unsaturated N-nitrosoamines have recently been studied. The dehydrohalogenation of alkyl(β-chloroethyl)nitrosamines with potassium hydroxide in methanol tends to form nitrosoenamines. However, with longer β-chloroalkyl chains, allylic nitrosamines form. It turns out that there is an equilibrium between the allylic and vinylic isomers in which the allylic isomer is converted to the N-nitrosoenamine on heating. It was also found that solid potassium hydroxide in 18-crown-6-ether is an effective catalyst for these elimination reactions. The chloride is a relatively poor leaving group compared to a tosyloxy leaving group. A third method for preparing vinylic nitrosamines makes use of the acidity of the α-hydrogens of a nitrosamine itself. With lithium diisopropylamide–HMPA (4 equivalents) in THF, the α-carbanion forms. Reaction of this composition with phenylselenyl chloride at $-80°C$ leads to the formation of an α-phenylselenyl nitrosamine, which can be isolated after extraction of side-products with $1\,M$ hydrochloric acid, drying the organic solution, and concentrating under reduced pressure. The selenium derivative is then oxidized with m-chloroperbenzoic acid to form phenyl selenoxides, which decompose spontaneously to generate α,β-unsaturated nitrosamines in good to excellent yields [52]. Reference [52] gives somewhat general directions for the use of these three procedures for generating N-nitrosoenamines. These compounds are reactive toward reagents such as dialkylcopper lithium and enolate ions (nucleophilic reagents). They also participate in electrophilic reactions such as acid-catalyzed additions. Since the nitroso group may be readily cleaved, it may be considered a protective group for secondary amines [24]. The preparation of 1-nitroso-1,2,3,4-tetrahydropyridine from 1-nitroso-1,2,3,6-tetrahydropyridine by base-catalyzed isomerization is a recent example of this process [53] (Eq. 10).

A number of N-nitrosamines can act as nitrosating agents of other secondary amines. Since, in the process, the original N-nitrosamine loses its nitroso group while the other reagent gains it, the process may be considered a trans-nitrosation. The compound N-nitroso-3-nitrocarbazole may be used to convert N-methylaniline to N-nitrosomethylaniline in benzene solution. The reagent is particularly useful when an acid-free reaction system is required (e.g., when aziridine derivatives, which polymerize rapidly in acids, are to be nitrosated) [54].

2-6. Preparation of N-Nitroso-N-methylaniline by trans-Nitrosation [54]

(11)

With due safety precautions, in a 250-ml flask fitted with a reflux condenser, to a solution of 9.5 gm (39.4 mmole) of N-nitroso-3-nitrocarbazole in 160 ml of dry benzene is added 2.16 gm (20.2 mmole) of N-methylaniline. The solution is heated under reflux for 20 min. The reaction mixture is cooled to room temperature and filtered. The solid formed is 3-nitrocarbazole; m.p. 211°C, weight 4.0 gm. The filtrate is washed with 10% hydrochloric acid to afford another 1.9 gm of 3-nitrocarbazole. The filtrate is again washed with 10% hydrochloric acid followed by a wash with 10% aqueous sodium hydroxide and repeated water washes.

The washed benzene solution was concentrated on a rotary evaporator to leave a dark yellow oil and another 0.6 gm of 3-nitrocarbazole. The oil was distilled under reduced pressure through a semimicro distillation column, yielding 2.9 gm (55%) of N-nitroso-N-methylaniline, b.p. 83°–86°C/2 mm.

Singer et al. [55, 56] have studied the transnitrosations extensively. They observed that many alicyclic nitrosamines, nitrosamino acids, nitrosamides, monoalkyl nitrosoureas, nitrosourethanes, N-alkylnitrosoguanidines, and trialkylnitrosoureas may act as transnitrosating agents at pH values of 1 to 3 in the presence of nucleophilic catalysts such as thiocyanate ions. Aliphatic acyclic nitrosamines tend to be less reactive or nonreactive. In general it should be noted that whereas a given nitrosamine may be noncarcinogenic in animal tests, it may still transnitrosate other amines in vivo to produce harmful products [55].

A few miscellaneous methods for the preparation of N-nitrosoamines have been reported. For example, various oxides of nitrogen have been used to nitrosate secondary amines. Dinitrogen tetroxide (N_2O_4), prepared by passing nitrogen dioxide (NO_2) and oxygen over phosphorus pentoxide into a trap at −80°C, has nitrosated secondary amines at low temperatures using methylene chloride dried with phosphorus pentoxide as the solvent. Dinitrogen tetroxide is also available from suppliers of reagent gases. It has been recommended that this reaction be carried out with the careful exclusion of oxygen, water, and carbon dioxide. Yields are reported to be generally quite good [30].

In view of the general applicability of dinitrogen tetroxide as a nitrosating agent for amides, the application of this reagent for reaction with other nitrogen compounds is strongly indicated.

Several patents exist on the use of mixtures of nitrogen dioxide (NO_2) and nitric oxide (NO) as the reactant for the preparation of *N*-nitroso compounds of secondary amines, both in the liquid and in the vapor state, usually in a flow reactor [58, 59]. These reactions are not limited to dialkyl amines, as the earlier literature seems to indicate. Even aromatic compounds like diphenylamine evidently have been nitrosated with the mixed oxides of nitrogen.

Nitrosonium tetrafluoroborate has been prepared from nitrosyl fluoride (NOF) and boron trifluoride in Freon-113. With this reagent, in a nitromethane solution, pyridine was converted into *N*-nitrosopyridinium tetrafluoroborate, m.p. 155°–158°C [60].

In the vapor phase, perfluoro-2-azopropene has been oxidized at high temperatures with oxygen using rubidium fluoride as a catalyst to produce *N*-nitrosobis(trifluoromethyl)amine, b.p. -3°C to -4°C [60a] (Eq. 12).

$$CF_3N{=}CF_2 \xrightarrow[\substack{RbF \\ 400°-600°C}]{O_2} (CF_3)_2N{-}NO \quad \substack{(40-60\% \\ yields)} \quad (12)$$

Physical constants for some of the *N*-nitroso derivatives prepared by this method are given in Table I.

C. *N*-Nitrosation of Tertiary Amines

Contrary to common belief, tertiary aliphatic amines react with aqueous nitrous acid to undergo dealkylation to form a carbonyl compound, a secondary nitrosamine, and nitrous oxide [66]. Base-weakening groups markedly reduce nitrosative cleavage, and quaternization may prevent it completely. Several examples of this reaction are shown in Table II.

Treatment of certain other tertiary aromatic amines with nitrous acid may lead to either *C*-nitroso compounds, nuclear nitro compounds, or *N*-nitrosamines with loss of an alkyl group. In the case of the nitrodimethylanilines, the latter two types of reaction may occur. The formation of nitro-*N*-nitrosomethylanilines predominates at room temperature, whereas the formation of polynitro compounds predominates at more elevated temperatures. The formation of nitrosamines from *N*,*N*-dimethylanilines appears to be particularly favored when both ortho positions are occupied by nitro groups, although *N*-nitroso compounds were also obtainable from other nitrodimethylanilines. The product of the reaction, of course, is an *N*-nitroso secondary amine.

The example in Preparation 2-7 is given here as one which affords a very good yield of the nitroso derivative without the formation of any other nitro compounds [61].

TABLE I

N-Nitroso Derivative of	B.p., °C (mm Hg)	M.p. (°C)	Ref.
Piperidine	109 (24)	—	37
Pyrolidine	99–100 (15)	—	37
Di-n-butylamine	112–114 (13)	—	37
Morpholine	110 (14)	—	37
Dicyclohexylamine	—	102–104	29,37
Diisopropylamine	—	42–44	37
Methylaniline	135–137 (13)	—	48
N-Hexylaniline	127–130 (1.5)	—	50
N-Octylaniline	—	50	50
N-Methyl-3-nitroaniline	—	77	61
N-Methyl-4-nitroaniline	—	104	61
N-Methyl-3,6-dinitroaniline	—	128	61
N-Methyl-4-bromo-3-nitroaniline	—	78	61
N-methyl-4-bromo-2-nitroaniline	—	73	61
Camphidine	—	166.5–167	61
N-Cyclohexyl-N′-phenyl-p-phenylenediamine (di-NO deriv.)	—	114–116	62
N-β-Naphthyl-N′-cyclohexyl-p-phenylene-diamine (di-NO deriv.)	—	87	62
N-Phenyl-N′-cyclopentyl-p-phenylenediamine (di-NO deriv.)	—	64–65	62
N-Phenyl-N′-isopropyl-p-phenylenediamine (di-NO deriv.)	—	74–75	62
N-Phenyl-N′-propyl-p-phenylenediamine (di-NO deriv.)	—	105–106	62
N-Phenyl-N′-isobutyl-p-phenylenediamine (di-NO deriv.)	—	92–93	62
N-Phenyl-N′-sec-butyl-p-phenylenediamine (di-NO deriv.)	—	57–58	62
N-Phenyl-N′-butyl-p-phenylenediamine (di-NO deriv.)	—	64–65	62
N-Phenyl-N′-ethyl-p-phenylenediamine (di-NO deriv.)	—	94–95	62
N-Phenyl-N′-methyl-p-phenylenediamine (di-NO deriv.)	—	111–112	62
N-Phenyl-N′-n-heptyl-p-phenylenediamine (di-NO deriv.)	—	69–70	62
N-Benzylglycine (mono-NO deriv.)	—	142.5–143	63
N-Benzyl-α-phenylglycine)(mono-NO-deriv.)	—	120–123	63
N-(p-Chlorobenzyl')-α-phenylglycine (mono-NO deriv.)	—	118–119.5	63
1,3,5-Hexahydrotriazine(tri-NO deriv.)	—	105–106	64
3-Nitro-2,3-dehydropiperidine (mono-NO deriv.)	—	56–57	(65)

TABLE II

Nitrosation of Tertiary Amines in Aqueous Acetic Acid at pH 4–5 [66]

Tertiary amine[a]	Yield (%)	Nitrosoamine	Yield (%)	Carbonyl compound
$(C_6H_5CH_2)_3N$	38	$(C_6H_5CH_2)_2N-NO$	—	C_6H_5CHO
$(C_6H_5CH_2)_2N-C_6H_4-Cl\,(p)$	60	$(C_6H_5CH_2)(p\text{-}Cl-C_6H_4)N-NO$	40	C_6H_5CHO
$(C_6H_5CH_2)_2N-C_6H_4-NO_2\,(p)$	—	—	—	—
$(C_6H_5CH_2)_3\overset{+}{N}CH_3NO_3^-$	—	—	—	—
$C_6H_5CH_2N(C_2H_5)_2$	57.5	$C_6H_5CH_2N(C_2H_5)NO$	16	C_6H_5CHO
$C_6H_5CH(CH_3)N(C_2H_5)_2$	38	$C_6H_5CH(CH_3)N(C_2H_5)NO$	3	$C_6H_5\underset{\displaystyle \parallel}{C}CH_3$ ($=O$)
$C_6H_5CH(COOC_2H_5)N(C_2H_5)_2$	57	$C_6H_5CH(COOC_2H_5)N(C_2H_5)NO$	16.4	$C_6H_5C-COOC_2H_5$ ($=O$)
Quinuclidine	—	—	—	—

[a] The amine (0.1 mole) is dissolved in a buffered (pH 4–5) solution of 500 ml of 60% aqueous acetic acid and 68 gm of sodium acetate. The reaction mixture is warmed to 90°C. Then 69 gm (1.0 mole) of sodium nitrite dissolved in 100 ml of water is added dropwise over a 45 min period while heating at 90°C is continued. After the addition, the reaction mixture is heated for 2 hr, cooled, poured into 200 ml of cold water, and extracted three times with 200 ml portions of ether. The ether was washed with 10% potassium carbonate solution until basic, then with saturated sodium chloride solution, dried, stripped, and distilled to obtain the products shown in the table.

2-7. Preparation of N-Methyl-N-nitroso-3,6-dinitroaniline [61]

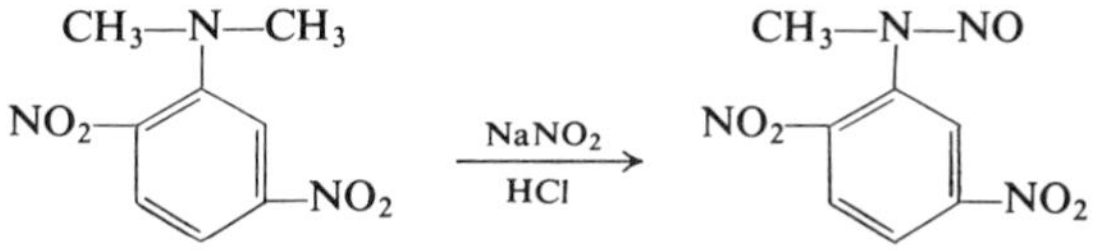

$$\text{(13)}$$

To a solution of 10 gm (0.0474 mole) of *N,N*-dimethyl-3,6-dinitroaniline in 120 ml of hydrochloric acid diluted with 40 ml of water, with rapid stirring at room temperature, 20 ml of a 50% aqueous solution of sodium nitrite is added. The resulting reaction mixture is allowed to stand at room temperature overnight. After the solution becomes deep yellow, the nitrosamine separates, often with the evolution of oxides of nitrogen. The product is collected on a filter, washed with cold water, and recrystallized from ethanol: yield 7.8 gm (73%), m.p. 128°C, orange needles or plates.

A study of the stereoelectronic effects in the nitrosation of tertiary amines may be summarized by Eqs. (14), (15), and (16) [67].

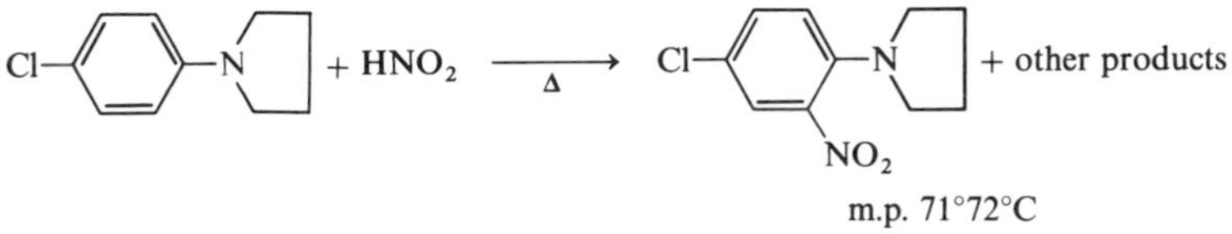

$$\text{(14)}$$

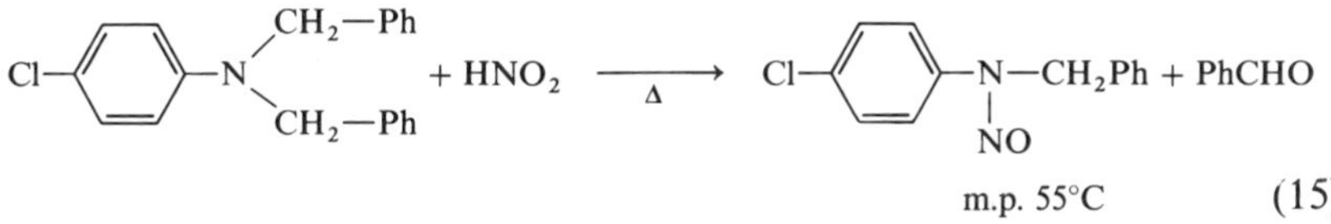

m.p. 71°72°C

m.p. 55°C $$\text{(15)}$$

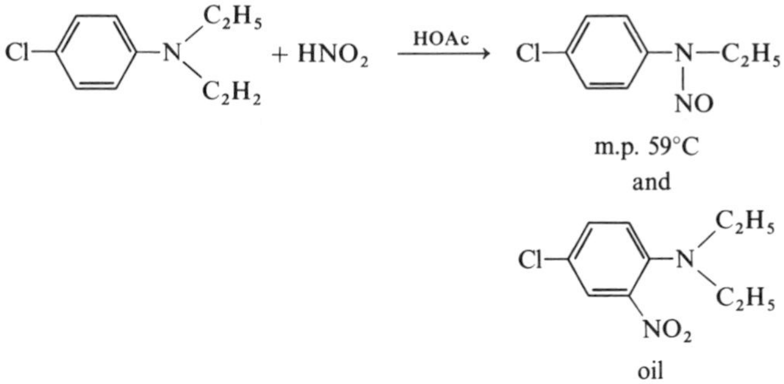

m.p. 59°C
and $$\text{(16)}$$

oil

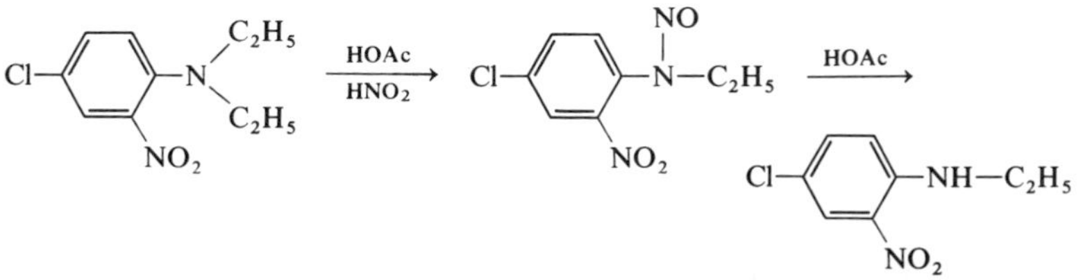

$$\text{(17)}$$

3. N-NITROSOAMIDES

The nitrosation products of amides have been studied extensively.

Aside from the explosive hazard of N-nitrosoamides, it has been observed that many N-nitroso compounds have a serious physiological effect on mucous membranes and on the skin. Apparently, this corrosive action is not observed in the case of N,N'-dimethyloxalamide [68, 69]. Even so, considering that some N-nitroso compounds are reputed to be carcinogenic, due care should be exercised in the handling of all nitroso compounds.

In general, it must be pointed out that, while many of the N-nitrosoamides have been distilled, this must be done with great caution and only in very small quantities because some of the products are subject to detonation on heating. The solid derivatives may also decompose explosively on melting [70, 71]. The thermal stability appears to be dependent on the structure of the parent amine. In the case of amides derived from aliphatic amines, the order of increasing stability appears to be: amides of tertiary carbinamines $(R^1R^2R^3C—NH_2) >$ secondary carbinamines $(R^1R^2CHNH_2) >$ primary carbinamines $(R—CH_2NH_2)$ [72].

The stability of N-nitroso derivatives of aromatic amines varies over a wide range. This variation appears to be related to the substituents on the aromatic nucleus. Nitrosoamides also may decompose photolytically, particularly in an acid medium [39].

Five methods for the nitrosation of amides will be mentioned here. Of these, the methods using dinitrogen tetroxide appears to have the widest applicability in terms of the classes of amides which can be nitrosated and the purity of the products obtained. However, the other methods also have special utility.

Table III lists some of the physical properties of a representative series of N-nitroso-N-substituted amides.

A. Nitrosation in Aqueous Media

In an aqueous solution, water-soluble amides have been nitrosated in the presence of sodium nitrite on acidification with mineral acids. This procedure appears to be restricted to amides which are water-soluble. Amides derived from primary carbinamines may be prepared by this method. The reaction is somewhat slow, requires a large excess of sodium nitrite, evolves large quantities of nitric oxide and nitrogen dioxide, and generally is said to lead to somewhat impure products. It has the advantage, however, that the procedure can be carried out on a relatively large scale. The more recent literature mentions use of this procedure [71, 72], but no experimental details have

TABLE III

Properties of *N*-Nitroso-*N*-Substituted Amides[a]

Parent amide	B.p., °C (mm Hg)	M.p. (°C)	Ref.
n-Butylpropionamide	63–65 (6)	—	74
n-Butylbutyramide	72–76 (7)	—	74
Isopentylpropionamide	63–65 (6)	—	74
Fumardianilide (dinitroso deriv.)	—	121 dec., explosively	76
Acetanilide	—	50–51 dec.	77
N-(*p*-Tolyl)acetamide	—	77–78 dec.	77
N-(*p*-Bromophenyl)acetamide	—	84–85 dec.	77
N-(*p*-Chlorophenyl)acetamide	—	79–80 dec.	77
N-(*p*-Nitrophenyl)acetamide	—	68–70 dec.	77
Formanilide	—	45–46 dec.	77
Propionanilide	—	53 dec.	77
Butyranilide	—	39 dec.	77
Isobutyranilide	—	35 dec.	77
5-Acetamido-2-methoxypyridine	—	51.5–52.5 dec.	78
1-Methoxy-2-acetamidonaphthalene	—	73 dec.	79
2-Methoxy-1-acetamidonaphthalene	—	86–88 dec.	79
4-Methoxy-3-acetamidonaphthalene	—	Viscous oil	79
N-*n*-Butylacetamide	125 (13)	—	70
N-Isobutylacetamide	125 (20)	—	70
N-*sec*-Butylacetamide	119 (18)	—	70
N-*t*-Butylacetamide	—	99–100	70
N-*n*-Butyl-3,5-dinitrobenzamide	—	109–110	70
N-Isobutyl-3,5-dinitrobenzamide	—	161–162	70
N-*sec*-Butyl-3,5-dinitrobenzamide	—	173–174	70
N-*sec*-Butylbenzamide	—	95–96	70
N-Cyclohexylacetamide	—	106–107	70
Ethyl *N*-cyclohexylcarbamate	—	57–58	70
N-α-Phenylethylacetamide	—	78–91	70
N-α-Phenylethylbenzamide	—	57–58 dec.	80
N,*N'*-Dimethyloxalamide	—	68 dec.	68
2,5-Diketopiperazine (dinitroso derivative)	—	183 dec.	73

[a] See text concerning potential hazards.

been given for the treatment of linear amides. Procedure 4-1 gives details of this method as applied to the nitrosation of lactams.

B. Nitrosation with Sodium Nitrite in Acetic Anhydride

Since acetic acid and acetic anhydride mixtures are excellent solvents for amides derived from primary carbinamines and cyclohexylamines, *N*-nitroso derivatives can be prepared in this solvent medium with sodium nitrite. The method appears to be somewhat more rapid than that carried in an aqueous medium, but fails when amides derived from secondary carbinamines are to be nitrosated. As the reaction medium is one which will also acetylate primary amines, appropriate amines may be dissolved in the reaction medium and, in turn, acetylated and nitrosated without the isolation of the intermediate amide.

Instead of acetic acid as a component in the reaction medium, phosphoric acid has been substituted. This appears to have the advantage that smaller quantities of the oxides of nitrogen are evolved during the reaction. However, on occasion, only the acetylation reaction was observed rather than the nitrosation [70].

3-1. General Preparation of N-Alkyl-N-nitrosoamides with Sodium Nitrite in Acetic Anhydride [70]

$$\underset{\substack{\text{O}\\\|}}{\text{R}'\text{—C—NHCH}_2\text{R}} \xrightarrow[\text{CH}_3\text{CO}_2\text{COCH}_3]{\text{NaNO}_2} \underset{\substack{\text{O}\\\|\\\text{NO}}}{\text{R}'\text{—C—NCH}_2\text{R}} \qquad (18)$$

In a well-ventilated hood with other safety precautions, to a solution of 0.01 mole of an amide derived from a primary carbinamine or a cyclohexyl-amine in 10 ml of glacial acetic acid and 50 ml of acetic anhydride cooled to 0°C, over a 5-hr period is added 15 gm (0.22 mole) of sodium nitrite.

After maintaining the reaction mixture for 10 hr at 0°C, the reaction mixture is allowed to warm over a 30 min period to 10°–15°C. Then the product mixture is poured over a mixture of ice and water. After the ice has melted, the product is extracted with ether. The ether extract is washed in turn with water, a 5% aqueous solution of sodium carbonate, and again with water. Then the solution is dried over anhydrous sodium sulfate. After removal of the solvent by evaporation under reduced pressure, a final purification of the nitrosoamide is accomplished either by very cautious distillation under reduced pressure at temperatures below 40°C, if a liquid, or by recrystallization from mixtures of ether and pentane. The yields are usually quite good.

C. Nitrosation with Nitrogen Trioxide (N_2O_3)

Nitrogen trioxide may be obtained by the acidification of an aqueous solution of sodium nitrite or by the reduction of nitric acid.

Since large amounts of nitric oxide and nitrogen dioxide are generated in the course of the reaction, usually a large excess of nitrogen trioxide has to be employed. This preparation has been carried out either in an ether solution or suspension of the appropriate amide or in a solution of acetic acid containing a small amount of acetic anhydrides to dissolve the amide completely [73–75].

3-2. *Preparation of N-Nitroso-N-n-propylacetamide* [73]

$$\underset{\substack{\| \\ O}}{CH_3-C}-NH-CH_2CH_2CH_3 \quad \xrightarrow[CH_3CO_2H]{N_2O_3} \quad \underset{\substack{\| \\ O}}{CH_3-C}-\underset{\substack{| \\ NO}}{N}-CH_2CH_2CH_3 \quad (19)$$

In a well-ventilated hood, with other safety precautions, to 20 gm (0.20 mole) of *N-n*-propylacetamide in 30 gm of glacial acetic acid, to which has been added a small quantity of acetic anhydride, cooled in an ice-salt bath, is added nitrogen trioxide in a generous stream in order to saturate the solution.

The green solution is then cooled in a refrigerator for at least 12 hr. During this time the color of the solution changes to a shade of orange.

To the reaction mixture is then added an equal volume of cold water. The resultant red oil is separated by decantation, washed repeatedly with water, and finally dried with sodium sulfate. After filtration, 23 gm of product is obtained (yield 90%). No physical constants for the product have been reported, presumably because of decomposition near the boiling point.

While the *O*-amides are readily converted into the corresponding *N*-nitroso derivatives with nitrogen trioxide, the analogous thioamides undergo a complex series of reaction, which, depending on the exact reaction conditions, may lead to products with the elimination of either sulfur or the NO or N_2. For example, *n*-butylthioacetamide, when treated with well-dried nitrogen trioxide at 0°C in ether solution, is converted in fairly good yields into *N*-nitroso-*N-n*-butylacetamide [75].

D. Nitrosation with Nitrosyl Chloride

Nitrosyl chloride is a sufficiently powerful reagent to permit the formation of *N*-nitroso derivatives of amides from secondary carbinamines. In this regard the reagent differs markedly from those used in the first three methods mentioned above. The use of the reagent is quite general and has even been employed to prepare dinitroso derivatives of such compounds as fumardianilide.

On the other hand, nitrosyl bromide is ineffective. This may be attributed to the fact that the reagent is extensively dissociated to afford nitric oxide and molecular bromine. Furthermore, in the case of some amides, reaction with nitrosyl bromide results in the formation of complex crystalline compounds containing reactive bromine.

The reaction is usually carried out in the presence of glacial acetic acid, acetic anhydride, and acetate salt as well as a small amount of phosphorus pentoxide [76–79]. The preparation of *N,N'*-dinitrosofumardianilide is an example of the procedures used.

3-3. *Preparation of N,N′-Dinitrosofumardianilide* [46]

$$
\begin{array}{c}
\underset{\displaystyle \overset{\textstyle O}{\|}}{C_6H_5-NH-C-C-H} \\
H-C-C-NH-C_6H_5 \\
\underset{O}{\|}
\end{array}
\quad\xrightarrow{\text{NOCl}}\quad
\begin{array}{c}
\underset{\displaystyle}{\overset{\textstyle NO\ O}{C_6H_5-N-C-CH}} \\
HC-C-N-C_6H_5 \\
\underset{O\ \ NO}{\|}
\end{array}
\qquad (20)
$$

In a well-ventilated hood, with other safety precautions, behind an explosion shield, to a finely divided suspension of 2.6 gm (0.01 mole) of fumardianilide in 90 ml of glacial acetic acid, 30 ml of acetic anhydride, and also containing 8 gm of freshly fused potassium acetate and 2 gm of phosphorus pentoxide, cooled to $+7°C$ is added, slowly, with vigorous stirring, 25 ml of a 20% solution of nitrosyl chloride (5 gm of nitrosyl chloride, 0.08 mole) in acetic anhydride. After the addition has been completed, stirring is continued for 2 hr while the reaction temperature is maintained between 7° and 12°C. During this period, much of the suspended material dissolves.

In the hood, behind the shield, the reaction mixture is then poured into 500 ml of vigorously stirred ice water. The nitroso compound begins to separate as a fine, yellow precipitate. The product is removed by filtration and washed thoroughly with about 1 liter of cold water. After pressing the crystals dry, they are air-dried for 3 hr. Finally they are dried over calcium chloride in a desiccator: yield 2.4 gm (88%), m.p. 121°C (decomposes explosively).

In the nitrosation of benzylformamide-d_1 in acetic acid/acetic anhydride, with sodium nitrite, the resulting product contained significant quantities of *N*-nitroso-*N*-benzylacetamide along with the *N*-benzylformamide. When the reaction was carried out at $-55°$ to $-60°C$ by the dropwise addition of a nitrosyl chloride solution in dichloromethane to a dispersion of benzylformamide-d_1 in pyridine and dichloromethane, a 91% yield of the desired *N*-nitroso-*N*-benzylformamide-d_1 was isolated (ir:C—D stretch at 2190 cm^{-1}, NO band at 1530 cm^{-1}) [81].

The *N*-nitroso bile acid conjugates *N*-nitrosotaurocholic acid and *N*-nitrosoglycocholic acid were prepared by treating the sodium salt of the former and the free acid of the latter in a solution of anhydrous sodium acetate in glacial acetic acid with a slow stream of nitrosyl chloride at 10°–15°C for approximately 31 min. The reaction mixture was treated with methanol to form methyl nitrite with the excess nitrosyl chloride and from any steroidal nitrite esters which may have been present in the crude product. Methyl nitrite was removed by sparging with nitrogen. The final workup of these particular compounds is rather specific for these bile acid derivatives and beyond the general scope of this chapter. For details, the reader should consult the original literature [82].

E. Nitrosation with Dinitrogen Tetroxide (N_2O_4)

According to White [71], nitrosation involving the use of dinitrogen tetroxide ("nitrogen tetroxide") represents the most general method of nitrosating amides. The method is quite rapid, yields are high, and purity of products is excellent. Since nitric acid is a coproduct of the reaction, the normal procedure involves the use of an excess of anhydrous sodium acetate as a base. If the base is not added, dinitrosation takes place.

Dinitrogen tetroxide is available from suppliers of reagent gases. It is a poisonous gas and should only be handled in a well-ventilated hood. Since its boiling point is 21°C, the cylinder in which the material is supplied may be cooled to 0°C and the cold liquefied dinitrogen tetroxide may be poured into the desired cold solvent. Usually the solvent is kept at −20°C. The density of dinitrogen tetroxide at 0°C is 1.5 gm/ml [80].

3-4. *General Procedure for Nitrosation of Amides with Dinitrogen Tetroxide* [70, 71]

$$R-NH-\overset{\overset{\displaystyle O}{\|}}{C}-R' \xrightarrow[\text{NaOCOCH}_3]{N_2O_4} R-\underset{\underset{\displaystyle NO}{|}}{N}-\overset{\overset{\displaystyle O}{\|}}{C}-R' \tag{21}$$

In a well-ventilated hood, behind an explosion shield, to a solution of 0.015 mole of dinitrogen tetroxide (nitrogen tetroxide) in either carbon tetrachloride or acetic acid at −60°C is added 0.03 mole of anhydrous sodium acetate. The mixture is warmed to 0°C and, with vigorous stirring, 0.01 mole of the amide is added. After stirring at 0°C for 10–20 min, the mixture is poured into an excess of a slurry of water and ice.

The workup of the final product is similar to that outlined in Procedure 3-1, i.e., extraction of the product with ether followed by a wash with cold water, aqueous 5% sodium carbonate solution, and water followed by drying with anhydrous sodium sulfate. The solvent then is removed under reduced pressure and, depending on the properties of the final product, purification is accomplished either by distillation under reduced pressure at temperatures below 40°C or by recrystallization from an ether–pentane mixture. In the case of unstable nitrosoamides, these operations have to be carried out at 0°C.

N,N'-Dimethyloxalamide has been nitrosated with dinitrogen tetroxide in 75% yield by a procedure similar to the one just outlined. The nitrosation was also carried out with nitrogen trioxide with similar results when carbon tetrachloride was the solvent. However, when acetic anhydride was the solvent, the yield dropped to 32% [68].

When the higher N,N'-dialkyloxalamides were nitrosated with dinitrogen tetroxide, the resultant nitroso compounds were found to be red oils which tended to decompose when attempts were made to distill them. These compounds were, apparently, reasonably stable when stored in the cold in dark bottles.

N,N'-trimethylenebisbenzamide was nitrosated with dinitrogen tetroxide–acetic acid–acetic anhydride to give a 95% yield of N,N'-trimethylenebis-(N-nitrosobenzamide) [69].

N-(2-Chloroethyl)-N-nitrosoacetamide has been prepared by nitrosating N-(2-chloroethyl)acetamide in ether in the presence of sodium bicarbonate by the dropwise addition of an ethereal solution of dinitrogen tetroxide. The product was a dark yellow oil; yield: 66%. ir: $(-O)$ 1510 cm^{-1} [83].

The method of White [70, 71] (Preparations 3–4) has been used recently to prepare a large series of N-nitrosoamides. In these papers, NO$_2$ was used instead of N$_2$O$_4$. This gas was generated by the dropwise addition of concentrated nitric acid to copper wire. The products were only characterized by NMR and ir data [84].

4. *N*-NITROSOLACTAMS

A few lactams have been nitrosated by procedures similar to those used for the nitrosation of linear amides. For example, ϵ-caprolactam has been nitrosated with aqueous sodium nitrite and mineral acid [73]. 2-Pyrrolidone has been nitrosated with dinitrogen tetroxide [85].

4-1. *Preparation of N-Nitroso-ε-caprolactam* [73]

$$
\begin{array}{ccc}
\underset{\underset{\underset{\underset{N}{|}}{\underset{H}{|}}}{\underset{CH_2}{|}}{\overset{CH_2}{\overset{/\backslash}{CH_2\ \ CH_2}}} \underset{C=O}{|} & \xrightarrow[\text{HCl}]{\text{NaNO}_2} & \underset{\underset{\underset{\underset{N}{|}}{\underset{NO}{}}}{\underset{CH_2}{|}}{\overset{CH_2}{\overset{/\backslash}{CH_2\ \ CH_2}}} \underset{C=O}{|}
\end{array} \tag{22}
$$

In a well-ventilated hood, and with other safety precautions, to a vigorously stirred solution of 11.3 gm (0.1 mole) of ϵ-caprolactam and 23 gm (0.3 mole) of technical grade sodium nitrite in 50 ml of water, cooled in an ice-salt bath, is added dropwise 25 ml of concentrated hydrochloric acid at such a rate that the temperature never exceeds $-10°C$. The reaction mixture is then cooled in a refrigerator for 2 hr. Impurities are rapidly removed by filtration and the product is extracted with ether.

The ether extract is washed in turn with an ice-cold sodium bicarbonate solution and with ice water until the aqueous washes are neutral. The ether solution is dried over calcium chloride for 1 hr in a refrigerator.

After filtration, the ether is evaporated under reduced pressure without application of heat. The residue is cooled to −70°C, taken up in cold ether, and precipitated with cold petroleum ether. Yield 5.5 gm (35%), m.p. 11°C.

By a similar procedure, *N,N′*-dinitroso-2,5-diketopiperazine was also prepared [73].

The nitrosation of 2-pyrrolidone with dinitrogen tetroxide is said to be capricious, particularly when attempted for the first time. The problems appear to occur during the carbonate washing, but solutions for the problem have not been worked out. Furthermore, the nitroso product may detonate when distillation is attempted. Therefore the product is used without purification [85].

4-2. Preparations of N-Nitroso-2-pyrrolidone [85]

$$
\begin{array}{c}
\underset{\substack{| \quad\quad |\\ \underset{\displaystyle H}{\overset{\displaystyle CH_2}{\underset{\displaystyle N}{\diagdown}}\overset{\displaystyle }{\diagup}C=O}}}{CH_2-CH_2}
\quad \xrightarrow{\ N_2O_4\ } \quad
\underset{\substack{| \quad\quad |\\ \underset{\displaystyle NO}{\overset{\displaystyle CH_2}{\underset{\displaystyle N}{\diagdown}}\overset{\displaystyle }{\diagup}C=O}}}{CH_2-CH_2}
\end{array}
\tag{23}
$$

CAUTION: Capricious reaction, product may detonate.

In a well-ventilated hood, and with extreme precautions, to a rapidly stirred mixture of 50.2 gm (0.546 mole) of dinitrogen tetroxide and 123 gm (1.5 mole) of anhydrous sodium acetate in 150 ml of carbon tetrachloride prepared at Dry Ice temperature is added, over a 1 hr period, a solution of 46.5 gm (0.545 mole) of 2-pyrrolidone in 50 ml of methylene chloride while maintaining a temperature of −60°C. The reaction mixture is allowed to stand at −60°C for 1 hr, followed by $\frac{1}{2}$ hr at −20°C (color change from blue to green to yellow observed). Then the mixture is cooled to −30°C. With precautions (see text above) the mixture is treated with a solution of 138 gm (1 mole) of potassium carbonate in 150 ml of water while the temperature is maintained at −20°C.

The water layer is separated from the organic layer. The aqueous layer is extracted several times with methylene chloride. The extracts are combined with the organic layer.

After drying of the organic layer over anhydrous sodium sulfate and filtering, the solvent is cautiously evaporated. The product is a deep-red oil; yield 31.6–34.2 gm (85–93%); the boiling point has been reported to be 86°C (0.3 mm) (susceptible to detonation).

5. *N*-NITROSOSULFONAMIDES

In the preparation of diazomethane, *p*-tolylsulfonylmethylnitrosoamide has been suggested as an intermediate. The preparation of this compound has been well described [86, 87]. The general method of preparation resembles that of other amides.

Another example, using an acetic acid–acetic anhydride solution with sodium nitrite, is the preparation of *N*-nitroso-*N*-benzyl-*p*-toluenesulfonamide [88]. This nitrosoamide (Preparation 7-1) is said to be quite stable at room temperature.

5-1. *Preparation of N-Nitroso-N-benzyl-p-toluenesulfonamide* [88]

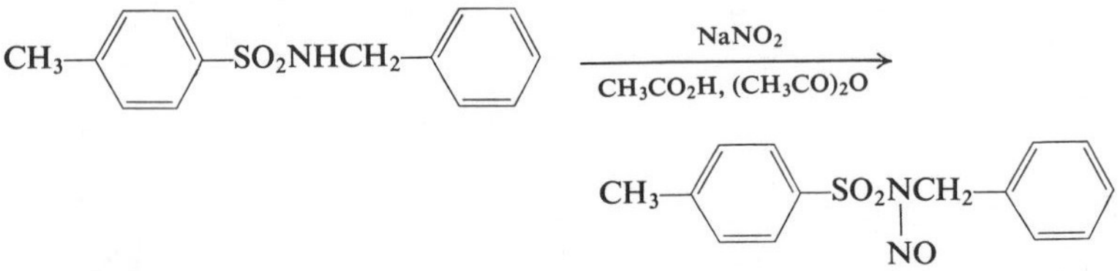

$$CH_3 - \langle\rangle - SO_2NHCH_2 - \langle\rangle \xrightarrow[CH_3CO_2H,\ (CH_3CO)_2O]{NaNO_2}$$

$$CH_3 - \langle\rangle - SO_2NCH_2 - \langle\rangle \quad (24)$$
$$\underset{NO}{|}$$

In a well-ventilated hood, behind a safety shield, in a three-necked flask equipped with a mechanical stirrer, thermometer, and provisions for adding solids, to a solution of 10.5 gm (0.04 mole) of *N*-benzyl-*p*-toluenesulfonamide in 50 ml of glacial acetic acid and 200 ml of acetic anhydride, cooled to 5°C, is added, over 6 hr, in small portions, 60 gm (0.85 mole) of finely powdered sodium nitrite. During this addition period the temperature is kept below 10°C at all times. After the addition has been completed, the green reaction mixture is stirred overnight. Then the mixture is poured, with vigorous stirring, into a large volume of ice water. The resultant mixture is then cooled for 1 hr in an ice bath. The crude product is filtered, washed several times with cold water, and dried under reduced pressure. After recrystallization from ethanol, the yield is 9.4 gm (81%), m.p. 90°–92°C.

N-Nitrosotoluenesulfonamides may also be prepared by using the dinitrogen tetroxide method, such as Procedure 3-4. *N*-Nitroso-*p*-toluenesulfonamide has been prepared by this method from the corresponding toluenesulfonamide [83].

6. *N*-NITROSOUREAS AND *N*-NITROSOTHIOUREAS

Substituted ureas have been nitrosated in aqueous systems to obtain intermediates for the preparation of diazoalkanes [89, 90]. The base-induced decomposition of *N*-nitroso-*N*-(2,2-diphenylcyclopropyl) urea has been studied [91], and a series of 1,3-bis(2-chloroethyl)-1-nitrosoureas and 1,5-bis(2-chloroethyl)-1-nitrosobiurets have been studied as potential anticancer agents [92].

The nitrosation in the presence of sodium nitrite and mineral acid resembles the analogous procedure for the nitrosation of amides and has been well detailed [89, 90].

While both *N*-nitrosomethylurea (m.p. 123°–124°C dec.) and *N*-nitroso-*N*-*n*-propylurea (m.p. 76.0°–76.5°C dec.) are readily prepared, nitrosation of *N*-isopropylurea even at −15°C resulted in the rapid evolution of gases [90].

To prepare nitrosoureas in which the oxygen of the NO group is labeled with an ^{18}O, sodium (0.46 gm-atoms) and 3 ml of 99% $H_2^{18}O$ were reacted. By addition of 10 mmole of nitrosonium tetrafluoroborate at 0°C, $NaN^{18}O_2$ was prepared in solution. The addition of a solution of hydrogen chloride in ether formed a mixture from which [^{18}O]nitrosyl chloride could be isolated for reaction with substituted ureas [93].

The preparation of *N*-(*n*-butyl)-*N'*,*N'*-dimethyl-*N*-nitrosourea is an example of the nitrosation of a trialkylurea which is generally applicable [94].

6-1. Preparation of N-(n-Butyl)-N',N'-dimethyl-N-nitrosourea [94]

$$
\underset{\underset{CH_3}{\overset{\overset{O}{\parallel}}{C_4H_9\!-\!NH\!-\!C\!-\!N\!-\!CH_3}}}{}
+ N_2O_4 \xrightarrow[CH_2O_2]{NaOAc}
\underset{\underset{NO\quad CH_3}{\overset{\overset{O}{\parallel}}{C_4H_9\!-\!N\!-\!C\!-\!N\!-\!CH_3}}}{}
\quad (25)
$$

In a hood, with suitable safety precautions, to a stirred slurry of 2.5 gm (30 mmole) of powdered anhydrous sodium acetate and 2 ml (65 mmole) of dinitrogen tetroxide in 20 ml of methylene chloride maintained at −40°C is added a solution of 774 mg (5.4 mmole) of *N*-(*n*-butyl)-*N'*,*N'*-dimethylurea in 10 ml of methylene chloride. The solution is warmed to 0°C and stirred for 35 min. The mixture is then poured into approximately 25 gm of crushed ice. The organic layer is separated and preserved; the aqueous layer is extracted twice with 25-ml portions of methylene chloride. The organic layers are combined, washed in turn with 25 ml of cold water, two 25-ml portions of cold 5% aqueous sodium bicarbonate, and two 25-ml portions of cold water. After drying with sodium sulfate, the solvent is removed on a rotary evaporator. The residue is a yellow oil weighing 943 mg. Upon distillation in a micro still with a pot temperature of 80°C at 0.1 mm, 156 mg of a pale yellow oil is isolated. Yield: 17%. Infrared (thin film): 1700, 1400 cm^{-1}.

By a similar procedure, *N'*,*N'*-dimethyl-*N*-(1-norbornyl)urea was nitrosated. The product, *N*-(1-norbornyl-*N'*,*N'*-dimethyl-*N*-nitrosourea was reported to have a melting point of 35.5°–36.5°C(dec.). Infrared (in CHCl$_3$) 1705, 1406 cm^{-1} [94].

A urea which was nitrosated in 98% formic acid at 0°C with sodium nitrite was 1-[2-(2-chloroethoxy)ethyl]-3,3-cyclohexylurea. The yield of the nitrosourea was 54%; ir (CHCl$_3$): 3400 (NH), 1700 (C=O), 1520 (N=O) cm^{-1} [83].

The nitrosation of thioureas is difficult because the sulfur of the thioureas may be more readily attacked by a nucleophile such as NO$^+$ than the nitrogen. Evidently in hydrochloric acid the formation of unprotonated nitrous acid and N$_2$O$_3$ is favored. These species favor electrophilic attack on the nitrogen of thioureas. The products are sensitive to heat, light, and atmospheric oxygen [95–97].

6-2.   *General Procedure for the Preparation of N-Nitrosothioureas* [95]

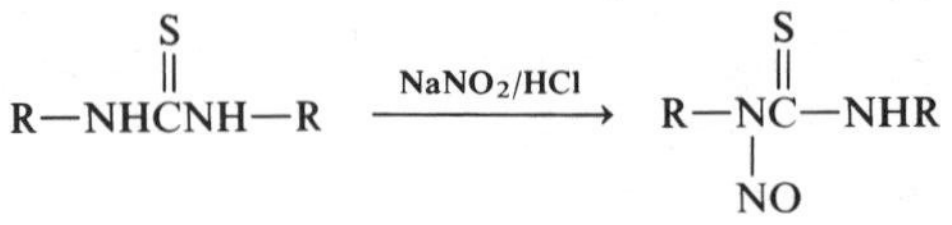

$$
\underset{\substack{\|\\ \text{R}-\text{NHCNH}-\text{R}}}{\overset{\text{S}}{}} \xrightarrow{\text{NaNO}_2/\text{HCl}} \underset{\substack{|\\ \text{NO}}}{\overset{\substack{\text{S}\\\|}}{\text{R}-\text{NC}-\text{NHR}}} \qquad (26)
$$

TABLE IV

PROPERTIES OF *N*-NITROSOTHIOUREAS

Parent thiourea	Yield (%)	M.p. (°C)	Ref.
1,3-Dimethyl	90	46	95
1,3-Diethyl	68	An oil	95
1,3-Dipropyl	77	An oil	95
1,3-Bis(2-methoxyethyl)	67	An oil	95
1,3-Bis(2-fluoroethyl)	70	23	95
3-Cyclohexyl-1-methyl	90	35—36	95
3-Cyclohexyl-1-ethyl	93	42	95
3-Cyclohexyl-1-propyl	68	22—23	95
3-Cyclohexyl-1-(2-methoxyethyl)	65	An oil	95
3-Cyclohexyl-1-(2-fluoroethyl)	60	51—52	95
3-Cyclohexyl-1-(2-hydroxyethyl)	65	53—57	95
More complex nitroso components			
1,3-Dimethyl-1-nitroso-[^{15}N]thiourea-^{13}C	68		95
3-Nitrosoimidazoline-2-thione			96
1-(2'-Hydroxyethyl)-3-nitrosoimidazoline-2-thione	85	135	96
^{15}-N-labeled thioureas are described in			96
3-(2-Hydroxyethyl)-3-methyl-1-propyl-1-nitrosothiourea	61	Heavy oil	97
3-Cyclohexyl-1-(2,2-dideuterio-2-hydroxyethyl-1-nitrosothiourea	70	55	97

With suitable safety precautions, under a nitrogen atmosphere, to a suspension of 20–50 mmole of the thiourea and 20–50 mmole of sodium nitrite in 200–400 ml of methylene chloride, with cooling to between $-10°$ and $-5°C$, with mechanical stirring is added dropwise over a 1 hr period 200–500 ml of 0.07 to 0.1 N aqueous hydrochloric acid. Upon completion of the addition, the reaction mixture is warmed to 5°C with stirring. The organic layer is separated, washed with water, and dried with sodium sulfate. After filtration, the solvent is removed on a rotary evaporator. The residue is crystallized from petroleum or by chromatography over Florisil.

Table IV lists the properties of a number of *N*-nitrosothioureas.

7. *N*-NITROSO-*N*-ALKYLURETHANES (ALKYL *N*-NITROSO-*N*-ALKYLCARBAMATES)

A warning has been published for insertion in *Organic Syntheses* in connection with the preparation of *N*-nitrosomethylurethane (b.p. 59°–61°C, 10 mm Hg) to the effect that the compound has been found to be a potent carcinogen [98]. Consequently, we also must caution the reader about this specific compound and suggest that other *N*-nitrosourethanes also be handled with extreme caution. These compounds also act as skin irritants [98, 99].

The method of preparation given by Hartman and Phillips [98] involves nitrosation with sodium nitrite and ice-cooled nitric acid as the acid. The method has been adopted by other workers for related derivatives [99, 100].

Carbamates have also been nitrosated with "nitrous fumes" in ether solution [101] and with aqueous sodium nitrite in acetic acid [102].

Suspensions of sodium bicarbonate in ether containing carbamates have been nitrosated with dinitrogen tetroxide at $-30°C$ [83, 103].

8. *N*-NITROSO-2-OXAZOLIDONES

A series of 2-oxazolidones of the general structure (V) has been nitrosated.

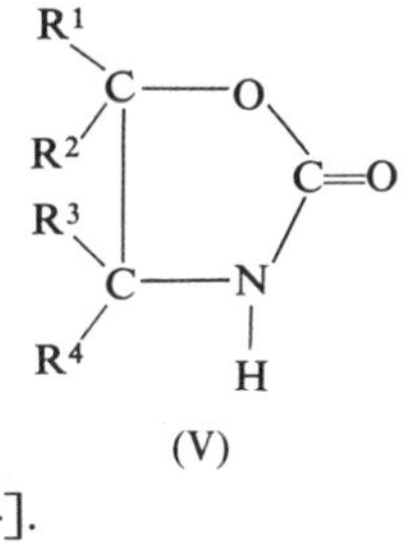

(V)

by one of two methods [104].

If the 2-oxazolidones were sufficiently soluble in cold dilute hydrochloric acid, such solutions were treated directly with sodium nitrite solutions. Slightly soluble 2-oxazolidones were nitrosated in pyridine solution with nitrosyl chloride.

The nitroso derivatives are generally reasonably stable, except the parent compound of the series, 3-nitroso-2-oxazolidone (m.p. 50°–53°C dec.) and 5-phenyl-3-nitroso-2-oxazolidone (m.p. 76.5°–77.5°C dec.), both of which are considered dangerous materials. Probably other unstable members of this class of compounds exist.

8-1. *Preparation of 3-Nitroso-5,5-diphenyl-2-oxazolidone* [104]

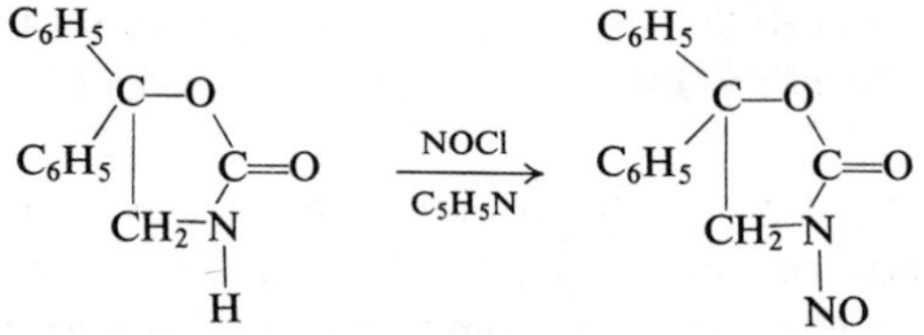

$$\begin{array}{c}C_6H_5\\[-2pt]C_6H_5\end{array}\!\!\!\underset{CH_2-N}{\overset{C-O}{\big|}}\!\!C=O \quad\xrightarrow[C_5H_5N]{NOCl}\quad \begin{array}{c}C_6H_5\\[-2pt]C_6H_5\end{array}\!\!\!\underset{CH_2-N}{\overset{C-O}{\big|}}\!\!C=O \tag{27}$$

In a well-ventilated hood, to a vigorously stirred dispersion of 29 gm (0.12 mole) of 5,5-diphenyl-2-òxazolidone in 200 ml of dried pyridine maintained

TABLE V

PROPERTIES OF 3-NITROSO-2-OXAZOLIDONES [56]

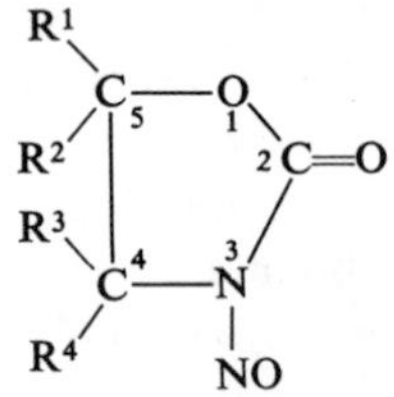

$$\underset{R^4}{\overset{R^1}{\underset{R^3}{\overset{R^2}{\Big|}}}}\ \ [104]$$

Substituents	M.p. (°C)
None[a]	50–53[a]
5-Cyclopentyl	82.8–83.2
5-Cyclohexyl	82.6–83.5
5,5-Dimethyl	87.7–89.8
5,5-Diethyl	3 (b.p. 106°–107°C, 1 mm Hg)
5-Phenyl[b]	76.5–77.5(d)[b]
5-Phenyl-5-methyl	116.5–117.4
5,5-Diphenyl	107.5–108.5
4,5,5-Triphenyl	115.8–117.5
4,5-Diphenyl erythro isomer	115.2–117.5
4,5-Diphenyl threo isomer	106.5–108.0

[a] Hazardous chemical; ignites spontaneously.
[b] Hazardous chemical; decomposes quite rapidly at room temperature.

between 10° and 15°C is added, over a 20-min period, 28 ml of a solution of 4.8 N nitrosyl chloride (0.13 mole) in acetic anhydride. Stirring is continued for an additional 5 min. Then the deep-red mixture is poured into a stirred slurry of 250 gm of ice and water.

The yellow solid product is separated by filtration, washed with ice water, and dried under reduced pressure over concentrated sulfuric acid. The product is recrystallized from a benzene–petroleum ether (b.p. range 95°–100°C) mixture: yield 29 gm (92%), m.p. 107.5°–108.5°C.

Table V gives the properties of a representative series of 3-nitroso-2-oxazolidones.

9. N-NITROSO-1,3-OXAZOLIDINES

The oxazolidines are numbered as shown in structure VI:

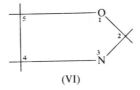

(VI)

The N-nitrosooxazolidines have been formed by heating ethanolamine with sodium nitrite in acid solutions. If no further aldehyde is added, some of the free ethanolamine is deaminated to form acetaldehyde, which reacts in the system to produce N-nitroso-2-methyl-1,3-oxazolidine in 10–27% yield, b.p. 81°C/0.5 mm Hg. From 2-amino-2-methyl-1-propanol, N-nitroso-2-isopropyl-4,4-dimethyl-1,3-oxazolidine if formed in a 12% yield, b.p. 45°C/0.2 mm Hg.

Treatment of 1-amino-2-propanol with propionaldehyde in acetic acid with sodium nitrite led to a 56% yield of N-nitroso-2-ethyl-5-methyl-1,3-oxazolidine, b.p. 44°–45°C/0.25 mm Hg, as a mixture of cis and trans isomers (with respect to the 5-methyl and 2-ethyl groups) which are exclusively the E rotamer with respect to the 2-hydrogen and the nitroso oxygen. Treatment of 1-amino-2-propanol and acetone with acetic acid and sodium nitrite led to N-nitroso-2,2,5-trimethyl-1,3-oxazolidine in 11% yield, b.p. 64°C (0.2 mm Hg) [26].

10. 1-NITROSO-3-NITROGUANIDINES

The 1-alkyl-1-nitroso-3-nitroguanidines (structure VII) have been suggested as intermediates for the preparation of diazoalkanes [105]. While a variety of 1-alkyl-3-nitroguanidines have been prepared successfully, of these, the 1-isopropyl and the 1-cyclohexyl derivatives could not be nitrosated successfully. Evidently any nitroguanidine with a secondary carbon attached to the amino group could not be nitrosated under the usual reaction conditions.

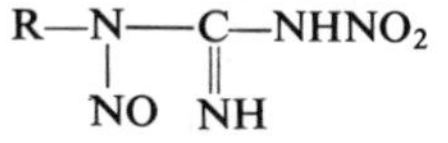

$$R—N—C—NHNO_2$$
$$\begin{array}{cc} | & \| \\ NO & NH \end{array}$$

(VII)

It was observed that the nitroso-nitroguanidines may cause skin irritation. The higher alkylnitrosonitroguanidines are particularly noxious. The preparation of 1-benzyl-1-nitroso-3-nitroguanidine is typical of the techniques used.

10-1. *Preparation of 1-Benzyl-1-nitroso-3-nitroguanidine* [105]

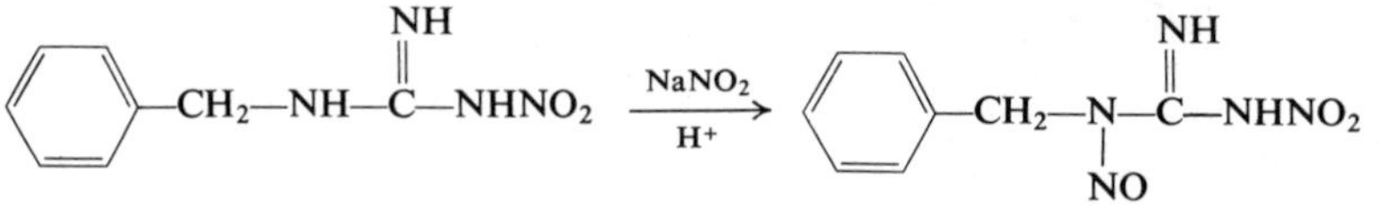

(28)

In a well-ventilated hood, behind a safety shield, with equipment protective against skin irritation, a solution of 10 gm (0.0516 mole) of benzylnitroguanidine in 165 ml of concentrated nitric acid is diluted with 100 ml of water. To the solution, cooled to 14°C, is added slowly a solution of 7.12 gm (0.103 mole) of sodium nitrite in 20 ml of water.

The yellow precipitate is separated by filtration and washed with cold water until the aqueous washes are nearly neutral: crude yield 7.9 gm (69%), m.p. 114.5°–115°C. On recrystallization from methanol, the melting point is raised to 117.5°–118°C.

The melting point of 1-*n*-propyl-1-nitroso-3-nitroguanidine, prepared by a similar technique, is reported to be 118°C.

N-(2-Chloroethyl-*N'*-*nitro*-*N*-nitrosoguanidine was prepared in 61% yield, m.p. 94°–96°C [84].

11. *N*-NITROSOHYDRAZINE DERIVATIVES

There appears to have been no systematic study of the nitrosation of substituted hydrazines. Work at the beginning of the century indicated that the nitrosation of primary hydrazines may result in the intermediate formation of nitroso compounds, but that these compounds subsequently react further, in the presence of nitrous acid, to give a variety of products. 1-Benzyl-1-nitrosohydrazine evidently has been prepared and subsequently benzoylated. The resultant *N*-nitrosohydrazide apparently is reasonably stable [100]. More recently, *N*-nitroso-*N*-alkylhydrazides have been prepared by the ring cleavage of compounds such as *N*-benzamidopiperidine and *N*-benzamidopyrrolidine. This method, however, does not appear to have very general applicability. Using the sodium nitrite-hydrochloric acid method of nitrosation, a series of 1-nitroso-1-alkyl-2-guanylhydrazines and 1-nitroso-1-alkyl-2-carbamyl-hydrazines has been prepared and characterized [108].

12. *N*-NITROSOKETIMINES

In cold carbon tetrachloride, a number of ketimines have been nitrosated with nitrosyl chloride. The *N*-nitrosoketimines exhibit varying degrees of stability. This stability is enhanced in the case of *N*-nitrosobenzophenonimines by electron-withdrawing substituents on the benzene ring and by the presence of bulky substituents on the ketimine carbon atom. Unhindered *N*-nitrosoketimines may decompose at room temperature or even below. This observation is considered to be consistent with a low activation energy for the decomposition and a substantial contribution of the 1,4-dipolar structure VIII, to the ground state. Structures VIII, IX and X represent possible resonance structures of the *N*-nitrosoketimines [109].

$$\begin{matrix} R \\ \diagdown \\ C{=}N{-}N{=}O \\ \diagup \\ R' \end{matrix} \longleftrightarrow \begin{matrix} R \\ \diagdown \\ C{=}\overset{+}{N}{=}N{-}\overset{-}{O} \\ \diagup \\ R' \end{matrix} \longleftrightarrow \begin{matrix} R \\ \diagdown \\ \overset{+}{C}{-}N{=}N{-}\overset{-}{O} \\ \diagup \\ R' \end{matrix}$$

(VIII)   (IX)   (X)

Preparation 12-1 gives a general procedure for the preparation of *N*-nitrosoketimines. This method was also used in [110] for the preparation of *N*-nitrosobenzophenoneimine and *N*-nitroso-2-camphanimine.

12-1. *General Preparation of N-Nitrosoketimines* [109]

$$\begin{matrix} R \\ \diagdown \\ C{=}NH \\ \diagup \\ R' \end{matrix} \xrightarrow[\text{NaOCOCH}_3,\ \text{CCl}_4]{\text{NOCl}} \begin{matrix} R \\ \diagdown \\ C{=}NNO \\ \diagup \\ R' \end{matrix} \qquad (29)$$

In a well-ventilated hood, behind a safety shield, a suspension of 14.8 gm (0.145 mole) of anhydrous sodium acetate and 40 ml of dry carbon tetrachloride is cooled to below $-10°C$ in a Dry Ice–acetone bath. With vigorous stirring, 4.00 gm (60 mmoles) of nitrosyl chloride is added. When the mixture has reached $-10°C$ or lower, with vigorous stirring a solution of 40 mmole of the

TABLE VI

Properties of *N*-Nitrosoketimines [113]

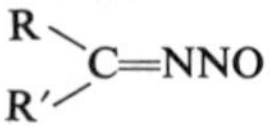

$$\begin{matrix} R \\ \diagdown \\ \diagup \\ R' \end{matrix} C{=}NNO \qquad [111]$$

R	R′	M.p. (°C) dec.	Color
o-Tolyl	Isopropyl	Oil	Red
t-Butyl	*o*-Tolyl	Oil	Red
Phenyl	Phenyl	50–53	Blue
p-Chlorophenyl	*p*-Chlorophenyl	61–63	Blue
p-Tolyl	*o*-Tolyl	43	Purple
p-Chlorophenyl	*o*-Chlorophenyl	69–71	Purple

ketimine in 10 ml (or more if necessary) of dry carbon tetrachloride, which has previously been cooled to −10°C, is added at such a rate that the internal temperature in the apparatus is maintained at −7°C or below. Stirring of the dark reaction mixture is continued for 20 min at −10°C.

Then 100 ml of an ice-cold 10% aqueous potassium bicarbonate solution is added quickly and the aqueous layer is drawn off. The product solution layer is washed with 200 ml of ice water. The carbon tetrachloride solution is dried with anhydrous magnesium sulfate and filtered. The solvent is evaporated under reduced pressure, and the residual dark-colored oil is dissolved in 10 ml of anhydrous ether and cooled to −68°C in a Dry Ice Bath. If a solid forms, it is recrystallized from petroleum ether, which may be warmed to 0°C to effect dissolution followed by cooling to Dry Ice temperatures for precipitation of the purified product.

The *N*-nitrosoketimines which were formed as red oils in the cited report decomposed when attempts were made to purify them either by vacuum distillation or by chromatography on silica gel.

Table VI lists properties of some *N*-nitrosoketimines.

13. *N*-NITROSOHYDROXYLAMINES

The nitrosation of *N*-substituted hydroxylamines affords an interesting series of compounds, the *N*-alkylnitrosohydroxylamines (structure XI).

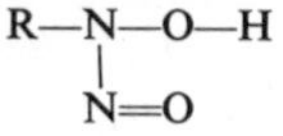

$$R{-}N{-}O{-}H$$
$$| $$
$$N{=}O$$

(XI)

As we may anticipate, a molecule with two nitrogens attached to each other, and each also part of a nitrogen-oxygen bond system, leads to interesting theoretical and structural problems. The hydrogen of the hydroxyl group is acidic in nature. On treatment with bases, *N*-nitrosohydroxylamines are converted into salts whose electronic configuration does not conform to classical structures. The resonance structures of these salts is frequently given as shown in structure (XII). Because they are chelating agents, these salts have been used as precipitants of metals in inorganic analysis. The best known reagent of this class is *N*-nitroso-β-phenylhydroxylamine, ammonium salt, cupferron.

$$\left[\begin{array}{c} R-N-O \\ | \\ N-O \end{array} \right]^{-} M^{+}$$

(XII)

The mechanism and kinetics of nitrosation of hydroxylamines have been discussed [112-114].

Recently, *O*-alkyl-substituted hydroxylamine derivatives have also been *N*-nitrosated to give unstable products (see Preparation 13-1).

The nitrosation of β-phenylhydroxylamine is typical of the procedures used in the nitrosation of hydroxylamines. The reaction is carried out in ether in the presence of a large exess of gaseous ammonia at 0°C. The nitrosation itself is carried out with either pentyl nitrite [115] on butyl nitrite [67]. The reaction must be carried out at low temperatures. Benzene may be substituted for ether as the solvent. While it is possible to add the nitrosating agent all at once when relatively small amounts of hydroxylamine are to be nitrosated, when a large-scale preparation is to be carried out it is preferable that the nitrosating agent be added gradually and that good control of the reaction temperature be maintained. The method of preparation using butyl nitrite is well described by Marvel [116].

More recently, the chemistry of acetylated *O*-alkylated *N*-nitrosohydroxylamine derivatives has been studied. For example, *N*-acetyl-*N*-nitroso-*O*-*t*-butylhydroxylamine has been prepared by the low-temperature nitrosation of the corresponding *N*-acetyl-*O*-*t*-butylhydroxylamine with nitrosyl chloride [117, 118] (Preparation 13-1).

13-1. *Preparation of N-Acetyl-N-nitroso-O-t-butylhydroxylamine* [118]

$$\underset{\text{O}}{\overset{\parallel}{CH_3-C}}-NH-O-C(CH_3)_3 \xrightarrow{\text{NOCl}} \underset{\substack{\text{O} \\ | \\ \text{NO}}}{\overset{\parallel}{CH_3-C}}-N-O-C(CH_3)_3 \qquad (30)$$

In a well-ventilated hood, behind a safety shield, to a stirred solution·of 523.6 mg (4.00 mmoles) of *N*-acetyl-*O*-*t*-butylhydroxylamine and 0.48 ml of dry pyridine in 6 ml of carbon tetrachloride, cooled to −10°C, is added dropwise a saturated solution of nitrosyl chloride in 12 ml of carbon tetrachloride. After addition of the nitrosating agent has been completed, the reaction mixture is stirred for an additional hour at −10°C. Then, while maintaining the system at 0°C, the reaction mixture is washed in turn with 25 ml portions of water, a 10% aqueous solution of hydrochloric acid, a 10% solution of sodium bicarbonate solution, and finally with water again. The organic layer is separated and dried over anhydrous magnesium sulfate.

After filtration of the reaction mixture, the concentration of the nitrosated product in the solvent is estimated by NMR, using dioxane as an internal standard. Because of the extreme instability of the nitrosated product, it is never isolated and the solution is always kept as cold as possible. At −20°C, dilute solutions evidently are reasonably stable. Concentrated solutions, on the other hand, explode on warming to room temperature.

N-Nitroso-*N'*-benzylhydroxylamine has been prepared by treating *N*-benzylhydroxylamine hydrochloride in ether with aqueous sodium nitrite at 0°–3°C. Yield 76%; m.p. 75°–77°C. This compound may be converted to its sodium salt by treating its ethanolic solution with sodium ethoxide, m.p. 238°–239°C [63].

14. MISCELLANEOUS PREPARATIONS

(1) Preparation of 3,3′-bis-*N*-(2-chloroethyl)-*N*-nitrosocarbamoyl propyl disulfide [83].

(2) Preparation of 2-chloroethyl-2-(3-cyclohexyl-1-nitrosoureido)ethyl sulfoxide [83].

REFERENCES

1. B. R. Simoneit, A. L. Burlingame, D. A. Flory, I. D. Smith, *Science* **166**, 733 (1969).
2. P. A. S. Smith and D. R. Baer, *Org. React.* **11**, 157 (1960).
3. J. S. Buck and C. W. Ferry, *Org. Syn. Coll. Vol.* **2**, 290 (1943).
4. Monsanto Co., Brit. Pat. 981,631 (Jan. 27, 1965); *Chem. Abstr.* **64**, PC 3103b (1966).
5. J. H. Ridd, *Quart. Rev.* **15**, 418 (1961).
6. E. Kalatzis and J. H. Ridd, *J. Chem. Soc.* **13**, 529 (1966).
7. H. Leotte, *Rev. Port. Quim.* **6**, (3), 108 (1964); *Chem. Abstr.* **64**, 1910g (1966).
8. C. G. Overberger, J.-P. Anselme, and J. G. Lombardino, "Organic Compounds with Nitrogen–Nitrogen Bonds." Ronald Press, New York, 1966.
9. J. H. Dusenberg and R. E. Powell, *J. Amer. Chem. Soc.* **73**, 3266, 3269 (1951).
10. M. A. Lashkarev and R. M. Vasyunas, *Izv. Vyssh. Ucheb. Zaved. Khim. Khim. Tekhnol.* **6** (2), 236 (1963); *Chem. Abstr.* **59**, 7071e (1963).
11. A. Mannschreck, H. Münsch, and A. Matthews, *Angew. Chem. Int. Ed. Engl.* **5**, 728 (1966); *Angew. Chem.* **78**, 751 (1966).

12. P. Rademacher, R. Stølevik, and W. Lüttke, *Angew. Chem.* **80**. 842 (1968).
13. P. A. S. Smith, "Open Chain Nitrogen Compounds," vol. II, pp. 455–464, Benjamin, New York, 1966.
14. A. L. Fridman, F. M. Mukhametshin, and S. S. Novikov, *Russ. Chem. Rev.* (Engl. transl.) **40**, 34 (1971).
15. H. Fener, ed., "Chemistry of the Nitro and Nitroso Groups," Interscience, New York, 1969.
16. J. P. Anselme, ed., "*N*-Nitrosamines," ACS Symp. Ser., American Chemical Society, Washington, D.C., 1978.
17. R. A. Scanlan and S. R. Tannenbaum, eds., "*N*-Nitroso Compounds," American Chemical Society, Washington, D.C., 1981.
18. F. H. C. Stewart, The Chemistry of Sydnones, *Chem. Rev.* **164**, 129 (1964).
19. C. E. Searle, ed., "Chemical Carcinogens," American Chemical Society, Washington, D.C., 1976 (cf. P. N. Magee, R. Montesano, and R. Preussmann, *ibid.*, pp. 461–625, on *N*-nitrosamines as potent animal carcinogens).
20. Proceedings of the 7th New Drug Symposium on Nitrosoureas, *Cancer Treat. Rep.* **60**, 651 (1976).
21. E. A. Walker, L. Griciute, M. Castegnaro, and M. Borzsonyi, eds., "*N*-Nitroso Compounds: Analysis, Formation, and Occurrence," IARC Sci. Publ. 31, International Agency for Research on Cancer, Lyon, France, 1980 (cf. G. M. Singer, *ibid.*, p. 139).
22. J. S. Snyder and L. M. Stock, *J. Org. Chem.* **45**, 1990 (1980).
23. V. M. Craddock, *Nature* **306**, 638 (1983).
24. R. Kupper and C. J. Mechejda, *J. Org. Chem.* **45**, 2921 (1980).
25. Test No. 2494, "The Merck Index," 5th ed., p. 813, Merck & Co., Rahway, N.J., 1940.
26. J. E. Saavedra, *J. Org. Chem.* **46**, 2610 (1981).
27. J. E. Saavedra, *J. Org. Chem.* **48**, 2388 (1983).
28. W. A. Freeman, *J. Amer. Chem. Soc.* **105**, 2725 (1983).
29. J. K. Wolfe and K. L. Temple, *J. Amer. Chem. Soc.* **70**, 1414 (1948).
30. E. B. Womack and A. B. Nelson, *Org. Syn. Coll. Vol.* **3**, 392 (1955).
31. N. E. Searle, *Org. Syn. Coll. Vol.* **4**, 424 (1963).
32. R. A. Shepard and P. L. Sciaraffa, *J. Org. Chem.* **31**, 964 (1966).
33. C. E. Redemann, F. O. Rice, R. Roberts, and H. P. Ward, *Org. Syn. Coll. Vol.* **3**, 244, (1955); J. Kenner and Co-workers, *J. Chem. Soc.* 363 (1933); 286 (1936); 181 (1939).
34. H. H. Hatt, *Org. Syn. Coll. Vol.* **2**, 211 (1943).
35. M. Berenbom and W. S. Fones, *J. Amer. Chem. Soc.* **71**, 1629 (1949).
36. D. W. Adamson and J. Kenner, *J. Chem. Soc.* 1551 (1937).
37. Y. L. Chow, *Canad. J. Chem.* **43**, 2711 (1965).
38. Y. L. Chow, C. Colon, and S. C. Chen, *J. Org. Chem.* **32**, 2109 (1967).
39. M. V. George and G. F. Wright, *J. Amer. Chem. Soc.* **80**, 1200 (1958).
40. E. J. Corey, H. S. Sachdev, J. Z. Gougoutas, and W. Saenger, *J. Amer. Chem. Soc.* **92**, 2488 (1967).
41. S. Marburg, *J. Org. Chem.* **30**, 2843 (1965).
42. A. H. Dutton and D. F. Heath, *J. Chem. Soc.* 1892 (1956).
43. M. A. Abou-Gharbia, H. Pylypiw, G. W. Harrington, and D. Swern, *J. Org. Chem.* **46**, 2193 (1981).
44. R. L. Willer, D. M. Moore, and L. F. Johnson, *J. Amer. Chem. Soc.* **104**, 3951 (1982).
45. R. L. Willer, C. K. Lowe-Ma, and D. W. Moore, *J. Org. Chem.* **49**, 1481 (1984).
46. J. C. Earl and A. W. Mackney, *J. Chem. Soc.* 899 (1935).
47. C. J. Thomas and D. J. Voaden, *Org. Syn. Coll. Vol.* **45**, 96 (1965).
48. W. W. Hartman and L. J. Roll, *Org. Syn. Coll. Vol.* **2**, 460 (1943).
49. J. J. D'Amico, C. C. Tung, and L. A. Walker, *J. Amer. Chem. Soc.* **81**, 5957 (1959).
50. J. Willenz, *J. Chem. Soc.* 1677 (1955).

51. C. G. Overberger, M. Valentine, and J.-P. Anselme, *J. Amer. Chem. Soc.* **91**, 687 (1969).
52. R. Kupper and C. J. Michejda, *J. Org. Chem.* **44**, 2326 (1979).
53. R. E. Lyle, W. E. Krueger, and V. E. Gunn, *J. Org. Chem.* **48**, 3574 (1983).
54. C. L. Bumgardner, K. S. McCallum, and J. P. Freeman, *J. Amer. Chem. Soc.* **83**, 4417 (1961).
55. S. S. Singer, G. M. Singer, and B. C. Cole, *J. Org. Chem.* **45**, 4931 (1980).
56. S. S. Singer and B. C. Cole, *J. Org. Chem.* **46**, 3461 (1981).
57. E. White and W. R. Feldman, *J. Amer. Chem. Soc.* **79**, 5832 (1957).
58. I. M. Roberts, U.S. Pat. 3,340,303 (Sept. 5, 1967); *Chem. Abstr.* **68**, P21516s (1968).
59. G. V. Mock, Brit. Pat. 772,081 (April 10, 1957); *Chem. Abstr.* **51**, P14799e (1957).
60. G. A. Olah, J. A. Olah, and N. A. Overchuk, *J. Org. Chem.* **30**, 3373 (1965).
60a. J. A. Young, S. N. Tsoukalas, and R. D. Dresdner, *J. Amer. Chem. Soc.* **82**, 396 (1960).
61. W. G. Macmillan and T. H. Reade, *J. Chem. Soc.* 2863 (1929).
62. C. C. Tung, U.S. Pat. 2,938,922 (May 31, 1960); *Chem. Abstr.* **55**, P455f (1961).
63. M. Nakajima and J.-P. Anselme, *J. Org. Chem.* **48**, 1444 (1983).
64. A. T. Nielsen, D. W. Moore, M. D. Ogan, and R. L. Atkins, *J. Org. Chem.* **44**, 1678 (1979).
65. R. H. Smith, Jr., M. B. Kroeger-Koepke, and C. J. Michejda, *J. Org. Chem.* **47**, 2910 (1982).
66. P. A. S. Smith and R. N. Loeppky, *J. Amer. Chem. Soc.* **89**, 1147 (1967).
67. R. N. Leoppky and W. Tomasik, *J. Org. Chem.* **48**, 2751 (1983).
68. H. Reimlinger, *Chem. Ber.* **94**, 2547 (1961).
69. H. Hart and J. L. Brewbaker, *J. Amer. Chem. Soc.* **91**, 706 (1967).
70. E. White, *J. Amer. Chem. Soc.* **77**, 6008 (1955).
71. E. White, *J. Amer. Chem. Soc.* **77**, 6011 (1955).
72. Y. L. Chow and A. C. H. Lee, *Canad. J. Chem.* **45**, 311 (1967).
73. K. Heyns and O. F. Woyrsch, *Chem. Ber.* **86**, 76 (1953).
74. K. Heyns and W. von Bebenburg, *Ann. Chem.* **595**, 55 (1955).
75. K. Heyns and W. von Bebenburg, *Chem. Ber.* **89**, 1303 (1956).
76. A. T. Blomquist, J. R. Johnson, H. J. Sykes, *J. Amer. Chem. Soc.* **65**, 2446 (1943).
77. D. H. Hey, J. Stuart-Webb, G. H. Williams, *J. Chem. Soc.* 4657 (1952).
78. Y. Ahmad and D. H. Hey, *J. Chem. Soc.* 4516 (1954).
79. Y. Ahmad and D. H. Hey, *J. Chem. Soc.* 3819 (1959).
80. E. White, *Org. Syn.* **47**, 44 (1967).
81. N. Nakajima and J.-P. Anselme, *J. Org. Chem.* **48**, 2492 (1983).
82. D. E. G. Shuker, S. R. Tannenbaum, and J. S. Wishnok, *J. Org. Chem.* **46**, 2092 (1981).
83. J. W. Lown, A. V. Joshua, and L. W. McLaughlin, *J. Med. Chem.* **23**, 798 (1980).
84. J. Garcia, J. Gonzáler, R. Segura, F. Urpi, and J. Vilarrasa, *J. Org. Chem.* **49**, 3322 (1984).
85. C. D. Gutsche and I. Y. C. Tao, *J. Org. Chem.* **32**, 1778 (1967).
86. Th. J. DeBoer and H. J. Baker, *Rec. Trav. Chim.* **73**, 229 (1954).
87. Th. J. DeBoer and H. J. Baker, *Org. Syn. Coll. Vol.* **3**, 943 (1963).
88. C. G. Overberger and J.-P. Anselme, *J. Org. Chem.* **28**, 592 (1963).
89. F. Arndt, *Org. Syn. Coll. Vol.* **2**, 461, 462 (1943).
90. J. R. Dyer, R. B. Randall, Jr., and H. M. Deutsch, *J. Org. Chem.* **29**, 3423 (1964).
91. T. K. Tandy, Jr., and W. M. Jones, *J. Org. Chem.* **30**, 4257 (1965).
92. T. P. Johnston *et al.*, *J. Med. Chem.* **9**, 892 (1966); *ibid.* **10**, 668, 675 (1967).
93. J. W. Lown and S. M. S. Chauhan, *J. Org. Chem.* **47**, 851 (1982).
94. E. H. White, T. J. Ryan, B. S. Hahn, and R. H. Erickson, *J. Org. Chem.* **49**, 4860 (1984).
95. J. W. Lown and S. M. S. Chauhan, *J. Org. Chem.* **48**, 507 (1983).
96. J. W. Lown and S. M. S. Chauhan, *J. Org. Chem.* **48**, 513 (1983).
97. J. W. Lown and S. M. S. Chauhan, *J. Org. Chem.* **48**, 3901 (1983).
98. W. W. Hartmann and R. Phillips, *Org. Syn. Coll. Vol.* **2**, 464 (1943).
99. C. D. Gutsche and H. E. Johnson, *Org. Syn. Coll. Vol.* **4**, 780 (1963).
100. A. L. Wilds and A. L. Meader, Jr., *J. Org. Chem.* **13**, 763 (1948).

101. L. Hellerman and R. L. Garner, *J. Amer. Chem. Soc.* **57**, 139 (1935).
102. W. R. Benson and R. J. Gajan, *J. Org. Chem.* **31**, 2498 (1966).
103. C. S. Cooper, A. L. Peyton, and R. J. Weinkam, *J. Org. Chem.* **48**, 4116 (1983).
104. M. S. Newman and A. Kutner, *J. Amer. Chem. Soc.* **73**, 4199 (1951).
105. A. F. McKay, W. L. Ott, G. W. Taylor, M. N. Buchanan, and J. F. Crooker, *Canad. J. Res.* **28**[B], 683 (1950).
106. J. Thiele, *Ann. Chem.* **376**, 239, 250 (1910).
107. P. A. S. Smith and H. G. Pars, *J. Org. Chem.* **24**, 1325 (1959).
108. W. G. Finnegan and R. A. Henry, *J. Org. Chem.* **30**, 567 (1965).
109. C. J. Thoman and I. M. Hunsberger, *J. Org. Chem.* **33**, 2852 (1968).
110. E. H. White, A. A. Wilson, J. P. Anhalt, R. J. Baumgarten, and J. I. Choca, *J. Org. Chem.* **47**, 2892 (1982).
111. C. J. Thoman and I. M. Hunsberger, *Angew. Chem.* **80**, 972 (1968).
112. M. N. Hughes and G. Stedman, *J. Chem. Soc.* 2824 (1963).
113. M. H. Hughes, T. D. B. Morgan, and G. Stedman, *Chem. Commun.* (8), 241 (1966).
114. T. D. B. Morgan, G. Stedman, and M. N. Hughes, *J. Chem. Soc. B* 344 (1968).
115. C. S. Marvel and O. Kamm, *J. Amer. Chem. Soc.* **41**, 280 (1919).
116. C. S. Marvel, *Org. Syn. Coll. Vol.* **1**, 177 (1932).
117. T. Koenig and M. Deinzer, *J. Amer. Chem. Soc.* **88**, 4518 (1966).
118. T. Koenig and M. Deinzer, *J. Amer. Chem. Soc.* **90**, 7014 (1968).

Index

A

Acetophenone
  N-Arylketimine formation, 305
  α-Halogenated ketimine formation, 308
  imine formation with aniline, 305
Acetophenone oxime, reduction to imine, 314
2-Acetoxyisopropylazobenzene, 401
Acetyl azide *see* Acyl azides
1-Acetyl-1-methyl-4-phenylsemicarbazide, 222
Acid azides, *see* Acyl azides
Acyl azides, 330–333
  aliphatic, 330
  biochemical applications, 341
  preparation of aromatic, 331
Acylhydrazines,
  oxidation with *t*-butyl hypochlorite, 406
  oxidation with lead tetraacetate, 406
N-Acetyl-N-nitroso-O-*t*-butylhydroxylamine,
    531–532
Active methylene compounds, nitrosation,
    457, 467–469
Adrenocorticotropic hormone, coupling to
    bovine serum albumine, 308
Aldehyde ammonias, 308
Aldimine, *see* Imines
Aldol condensation, of azidoacetophenones,
    327
Aliphatic amines,
  oxidation with peracids, 458
  reactions with nitrous acid, 460
Aliphatic azoxy compounds, properties, 450–
    451
Aliphatic-aromatic azoxy compounds, proper-
    ties, 450–451
Aliphatic secondary nitroso compounds, 483
Alkoxynitrobenzenes, reduction of, 428
Alkylammonium nitrite salts, 501–502
O-Alkylated-N-nitrosohydroxylamines, 531
Alkyllithium, reaction with N-nitroso-O,N-di-
    alkyl-hydroxylamines, 425
N-Alkyl-N-nitrosamides, synthesis with sodium
    nitrite in acetic anhydride, 516

N-Alkylnitroso hydroxylamine, 530–532
t-Alkyl primary amines, peracid oxidation of,
    483
Allene dipropargyl phosphonate, 32
Allenes, 1–47, *see also* names of specific
    allenes, 1,2-Dienes
  cyclic allenes, 4–9
  from *gem*-dihalocyclopropanes, 2–7
  uses, 1
  optical isomers, 1
  by rearrangement reactions, 13–39
Allyl mesylate, 40
N-Allyl-2-ethylcyclohexylideneamine, 202
β,α-(Allylic)-N-nitrosamines, 508
Almadori rearrangement, 312
Amidrazones, 407
Amines,
  aliphatic, oxidation to imines, 314
  aromatic, bimolecular oxidation with man-
    ganese dioxide, 391
  nitrosation, 457, 460–466
  oxidation of, 355, 390–394, 479, 481–483
  oxidation to azoxy compounds, 415, 428,
    434–436
  oxidation to C-nitroso compounds, 479,
    481–483
  oxidation with Caro's acid, 458, 481–482
  oxidation with Peracids, 482–483
  primary, N-nitrosation, 502–503
  secondary,
    N-nitrosation, 503–510
    purification via N-nitroso derivatives, 499
4'-Aminobenzo-18-crown-6, diazotized, cou-
    pled to N-*n*-butylaniline, 362
4'-Amino-5'-*tert*-butylbenzo-18-crown-6, ox-
    idation, 391–392
β-Aminoethylsulfuric acid, 315, 316
Aminonaphthalenes, *see* Naphthylamines
1-Amino-4-nitrosonaphthalene, 461–462
Aminouracil,
  C-nitrosation, 493
Anil, *see* Imines
Anti-Bredt imines, 319

N